Lecture Notes in Computer Science 13847

Founding Editors

Gerhard Goos
Juris Hartmanis

Editorial Board Members

The series Lecture Notes in Computer Science (LNCS), including its subseries Lecture Notes in Artificial Intelligence (LNAI) and Lecture Notes in Bioinformatics (LNBI), has established itself as a medium for the publication of new developments in computer science and information technology research, teaching, and education.

LNCS enjoys close cooperation with the computer science R & D community, the series counts many renowned academics among its volume editors and paper authors, and collaborates with prestigious societies. Its mission is to serve this international community by providing an invaluable service, mainly focused on the publication of conference and workshop proceedings and postproceedings. LNCS commenced publication in 1973.

Lei Wang · Juergen Gall · Tat-Jun Chin ·
Imari Sato · Rama Chellappa
Editors

Computer Vision – ACCV 2022

16th Asian Conference on Computer Vision
Macao, China, December 4–8, 2022
Proceedings, Part VII

 Springer

Editors
Lei Wang 🆔
University of Wollongong
Wollongong, NSW, Australia

Juergen Gall 🆔
University of Bonn
Bonn, Germany

Tat-Jun Chin 🆔
University of Adelaide
Adelaide, SA, Australia

Imari Sato
National Institute of Informatics
Tokyo, Japan

Rama Chellappa 🆔
Johns Hopkins University
Baltimore, MD, USA

ISSN 0302-9743   ISSN 1611-3349 (electronic)
Lecture Notes in Computer Science
ISBN 978-3-031-26292-0   ISBN 978-3-031-26293-7 (eBook)
https://doi.org/10.1007/978-3-031-26293-7

Preface

The 16th Asian Conference on Computer Vision (ACCV) 2022 was held in a hybrid mode in Macau SAR, China during December 4–8, 2022. The conference featured novel research contributions from almost all sub-areas of computer vision.

For the main conference, 836 valid submissions entered the review stage after desk rejection. Sixty-three area chairs and 959 reviewers made great efforts to ensure that every submission received thorough and high-quality reviews. As in previous editions of ACCV, this conference adopted a double-blind review process. The identities of authors were not visible to the reviewers or area chairs; nor were the identities of the assigned reviewers and area chairs known to the authors. The program chairs did not submit papers to the conference.

After receiving the reviews, the authors had the option of submitting a rebuttal. Following that, the area chairs led the discussions and final recommendations were then made by the reviewers. Taking conflicts of interest into account, the area chairs formed 21 AC triplets to finalize the paper recommendations. With the confirmation of three area chairs for each paper, 277 papers were accepted. ACCV 2022 also included eight workshops, eight tutorials, and one grand challenge, covering various cutting-edge research topics related to computer vision. The proceedings of ACCV 2022 are open access at the Computer Vision Foundation website, by courtesy of Springer. The quality of the papers presented at ACCV 2022 demonstrates the research excellence of the international computer vision communities.

This conference is fortunate to receive support from many organizations and individuals. We would like to express our gratitude for the continued support of the Asian Federation of Computer Vision and our sponsors, the University of Macau, Springer, the Artificial Intelligence Journal, and OPPO. ACCV 2022 used the Conference Management Toolkit sponsored by Microsoft Research and received much help from its support team.

All the organizers, area chairs, reviewers, and authors made great contributions to ensure a successful ACCV 2022. For this, we owe them deep gratitude. Last but not least, we would like to thank the online and in-person attendees of ACCV 2022. Their presence showed strong commitment and appreciation towards this conference.

December 2022

Lei Wang
Juergen Gall
Tat-Jun Chin
Imari Sato
Rama Chellappa

Organization

General Chairs

Gérard Medioni	University of Southern California, USA
Shiguang Shan	Chinese Academy of Sciences, China
Bohyung Han	Seoul National University, South Korea
Hongdong Li	Australian National University, Australia

Program Chairs

Rama Chellappa	Johns Hopkins University, USA
Juergen Gall	University of Bonn, Germany
Imari Sato	National Institute of Informatics, Japan
Tat-Jun Chin	University of Adelaide, Australia
Lei Wang	University of Wollongong, Australia

Publication Chairs

Wenbin Li	Nanjing University, China
Wanqi Yang	Nanjing Normal University, China

Local Arrangements Chairs

Liming Zhang	University of Macau, China
Jianjia Zhang	Sun Yat-sen University, China

Web Chairs

Zongyuan Ge	Monash University, Australia
Deval Mehta	Monash University, Australia
Zhongyan Zhang	University of Wollongong, Australia

AC Meeting Chair

Chee Seng Chan   University of Malaya, Malaysia

Area Chairs

Aljosa Osep	Technical University of Munich, Germany
Angela Yao	National University of Singapore, Singapore
Anh T. Tran	VinAI Research, Vietnam
Anurag Mittal	Indian Institute of Technology Madras, India
Binh-Son Hua	VinAI Research, Vietnam
C. V. Jawahar	International Institute of Information Technology, Hyderabad, India
Dan Xu	The Hong Kong University of Science and Technology, China
Du Tran	Meta AI, USA
Frederic Jurie	University of Caen and Safran, France
Guangcan Liu	Southeast University, China
Guorong Li	University of Chinese Academy of Sciences, China
Guosheng Lin	Nanyang Technological University, Singapore
Gustavo Carneiro	University of Surrey, UK
Hyun Soo Park	University of Minnesota, USA
Hyunjung Shim	Korea Advanced Institute of Science and Technology, South Korea
Jiaying Liu	Peking University, China
Jun Zhou	Griffith University, Australia
Junseok Kwon	Chung-Ang University, South Korea
Kota Yamaguchi	CyberAgent, Japan
Li Liu	National University of Defense Technology, China
Liang Zheng	Australian National University, Australia
Mathieu Aubry	Ecole des Ponts ParisTech, France
Mehrtash Harandi	Monash University, Australia
Miaomiao Liu	Australian National University, Australia
Ming-Hsuan Yang	University of California at Merced, USA
Palaiahnakote Shivakumara	University of Malaya, Malaysia
Pau-Choo Chung	National Cheng Kung University, Taiwan

Qianqian Xu	Key Laboratory of Intelligent Information Processing, Institute of Computing Technology, Chinese Academy of Sciences, China
Qiuhong Ke	Monash University, Australia
Radu Timofte	University of Würzburg, Germany and ETH Zurich, Switzerland
Rajagopalan N. Ambasamudram	Indian Institute of Technology Madras, India
Risheng Liu	Dalian University of Technology, China
Ruiping Wang	Institute of Computing Technology, Chinese Academy of Sciences, China
Sajid Javed	Khalifa University of Science and Technology, Abu Dhabi, UAE
Seunghoon Hong	Korea Advanced Institute of Science and Technology, South Korea
Shang-Hong Lai	National Tsing Hua University, Taiwan
Shanshan Zhang	Nanjing University of Science and Technology, China
Sharon Xiaolei Huang	Pennsylvania State University, USA
Shin'ichi Satoh	National Institute of Informatics, Japan
Si Liu	Beihang University, China
Suha Kwak	Pohang University of Science and Technology, South Korea
Tae Hyun Kim	Hanyang Univeristy, South Korea
Takayuki Okatani	Tohoku University, Japan/RIKEN Center for Advanced Intelligence Project, Japan
Tatsuya Harada	University of Tokyo/RIKEN, Japan
Vicky Kalogeiton	Ecole Polytechnique, France
Vincent Lepetit	Ecole des Ponts ParisTech, France
Vineeth N. Balasubramanian	Indian Institute of Technology, Hyderabad, India
Wei Shen	Shanghai Jiao Tong University, China
Wei-Shi Zheng	Sun Yat-sen University, China
Xiang Bai	Huazhong University of Science and Technology, China
Xiaowei Zhou	Zhejiang University, China
Xin Yu	University of Technology Sydney, Australia
Yasutaka Furukawa	Simon Fraser University, Canada
Yasuyuki Matsushita	Osaka University, Japan
Yedid Hoshen	Hebrew University of Jerusalem, Israel
Ying Fu	Beijing Institute of Technology, China
Yong Jae Lee	University of Wisconsin-Madison, USA
Yu-Chiang Frank Wang	National Taiwan University, Taiwan
Yumin Suh	NEC Laboratories America, USA

Yung-Yu Chuang   National Taiwan University, Taiwan
Zhaoxiang Zhang   Chinese Academy of Sciences, China
Ziad Al-Halah   University of Texas at Austin, USA
Zuzana Kukelova   Czech Technical University, Czech Republic

Additional Reviewers

Abanob E. N. Soliman	Atsushi Shimada	Chao Liu
Abdelbadie Belmouhcine	Attila Szabo	Chao Shi
Adrian Barbu	Aurelie Bugeau	Chaowei Tan
Agnibh Dasgupta	Avatharam Ganivada	Chaoyi Li
Akihiro Sugimoto	Ayan Kumar Bhunia	Chaoyu Dong
Akkarit Sangpetch	Azade Farshad	Chaoyu Zhao
Akrem Sellami	B. V. K. Vijaya Kumar	Chen He
Aleksandr Kim	Bach Tran	Chen Liu
Alexander Andreopoulos	Bailin Yang	Chen Yang
Alexander Fix	Baojiang Zhong	Chen Zhang
Alexander Kugele	Baoquan Zhang	Cheng Deng
Alexandre Morgand	Baoyao Yang	Cheng Guo
Alexis Lechervy	Basit O. Alawode	Cheng Yu
Alina E. Marcu	Beibei Lin	Cheng-Kun Yang
Alper Yilmaz	Benoit Guillard	Chenglong Li
Alvaro Parra	Beomgu Kang	Chengmei Yang
Amogh Subbakrishna	Bin He	Chengxin Liu
Adishesha	Bin Li	Chengyao Qian
Andrea Giachetti	Bin Liu	Chen-Kuo Chiang
Andrea Lagorio	Bin Ren	Chenxu Luo
Andreu Girbau Xalabarder	Bin Yang	Che-Rung Lee
Andrey Kuehlkamp	Bin-Cheng Yang	Che-Tsung Lin
Anh Nguyen	BingLiang Jiao	Chi Xu
Anh T. Tran	Bo Liu	Chi Nhan Duong
Ankush Gupta	Bohan Li	Chia-Ching Lin
Anoop Cherian	Boyao Zhou	Chien-Cheng Lee
Anton Mitrokhin	Boyu Wang	Chien-Yi Wang
Antonio Agudo	Caoyun Fan	Chih-Chung Hsu
Antonio Robles-Kelly	Carlo Tomasi	Chih-Wei Lin
Ara Abigail Ambita	Carlos Torres	Ching-Chun Huang
Ardhendu Behera	Carvalho Micael	Chiou-Ting Hsu
Arjan Kuijper	Cees Snoek	Chippy M. Manu
Arren Matthew C.	Chang Kong	Chong Wang
Antioquia	Changick Kim	Chongyang Wang
Arjun Ashok	Changkun Ye	Christian Siagian
Atsushi Hashimoto	Changsheng Lu	Christine Allen-Blanchette

Christoph Schorn
Christos Matsoukas
Chuan Guo
Chuang Yang
Chuanyi Zhang
Chunfeng Song
Chunhui Zhang
Chun-Rong Huang
Ci Lin
Ci-Siang Lin
Cong Fang
Cui Wang
Cui Yuan
Cyrill Stachniss
Dahai Yu
Daiki Ikami
Daisuke Miyazaki
Dandan Zhu
Daniel Barath
Daniel Lichy
Daniel Reich
Danyang Tu
David Picard
Davide Silvestri
Defang Chen
Dehuan Zhang
Deunsol Jung
Difei Gao
Dim P. Papadopoulos
Ding-Jie Chen
Dong Gong
Dong Hao
Dong Wook Shu
Dongdong Chen
Donghun Lee
Donghyeon Kwon
Donghyun Yoo
Dongkeun Kim
Dongliang Luo
Dongseob Kim
Dongsuk Kim
Dongwan Kim
Dongwon Kim
DongWook Yang
Dongze Lian

Dubing Chen
Edoardo Remelli
Emanuele Trucco
Erhan Gundogdu
Erh-Chung Chen
Rickson R. Nascimento
Erkang Chen
Eunbyung Park
Eunpil Park
Eun-Sol Kim
Fabio Cuzzolin
Fan Yang
Fan Zhang
Fangyu Zhou
Fani Deligianni
Fatemeh Karimi Nejadasl
Fei Liu
Feiyue Ni
Feng Su
Feng Xue
Fengchao Xiong
Fengji Ma
Fernando Díaz-del-Rio
Florian Bernard
Florian Kleber
Florin-Alexandru
  Vasluianu
Fok Hing Chi Tivive
Frank Neumann
Fu-En Yang
Fumio Okura
Gang Chen
Gang Liu
Gao Haoyuan
Gaoshuai Wang
Gaoyun An
Gen Li
Georgy Ponimatkin
Gianfranco Doretto
Gil Levi
Guang Yang
Guangfa Wang
Guangfeng Lin
Guillaume Jeanneret
Guisik Kim

Gunhee Kim
Guodong Wang
Ha Young Kim
Hadi Mohaghegh
  Dolatabadi
Haibo Ye
Haili Ye
Haithem Boussaid
Haixia Wang
Han Chen
Han Zou
Hang Cheng
Hang Du
Hang Guo
Hanlin Gu
Hannah H. Kim
Hao He
Hao Huang
Hao Quan
Hao Ren
Hao Tang
Hao Zeng
Hao Zhao
Haoji Hu
Haopeng Li
Haoqing Wang
Haoran Wen
Haoshuo Huang
Haotian Liu
Haozhao Ma
Hari Chandana K.
Haripriya Harikumar
Hehe Fan
Helder Araujo
Henok Ghebrechristos
Heunseung Lim
Hezhi Cao
Hideo Saito
Hieu Le
Hiroaki Santo
Hirokatsu Kataoka
Hiroshi Omori
Hitika Tiwari
Hojung Lee
Hong Cheng

Hong Liu
Hu Zhang
Huadong Tang
Huajie Jiang
Huang Ziqi
Huangying Zhan
Hui Kong
Hui Nie
Huiyu Duan
Huyen Thi Thanh Tran
Hyung-Jeong Yang
Hyunjin Park
Hyunsoo Kim
HyunWook Park
I-Chao Shen
Idil Esen Zulfikar
Ikuhisa Mitsugami
Inseop Chung
Ioannis Pavlidis
Isinsu Katircioglu
Jaeil Kim
Jaeyoon Park
Jae-Young Sim
James Clark
James Elder
James Pritts
Jan Zdenek
Janghoon Choi
Jeany Son
Jenny Seidenschwarz
Jesse Scott
Jia Wan
Jiadai Sun
JiaHuan Ji
Jiajiong Cao
Jian Zhang
Jianbo Jiao
Jianhui Wu
Jianjia Wang
Jianjia Zhang
Jianqiao Wangni
JiaQi Wang
Jiaqin Lin
Jiarui Liu
Jiawei Wang

Jiaxin Gu
Jiaxin Wei
Jiaxin Zhang
Jiaying Zhang
Jiayu Yang
Jidong Tian
Jie Hong
Jie Lin
Jie Liu
Jie Song
Jie Yang
Jiebo Luo
Jiejie Xu
Jin Fang
Jin Gao
Jin Tian
Jinbin Bai
Jing Bai
Jing Huo
Jing Tian
Jing Wu
Jing Zhang
Jingchen Xu
Jingchun Cheng
Jingjing Fu
Jingshuai Liu
JingWei Huang
Jingzhou Chen
JinHan Cui
Jinjie Song
Jinqiao Wang
Jinsun Park
Jinwoo Kim
Jinyu Chen
Jipeng Qiang
Jiri Sedlar
Jiseob Kim
Jiuxiang Gu
Jiwei Xiao
Jiyang Zheng
Jiyoung Lee
John Paisley
Joonki Paik
Joonseok Lee
Julien Mille

Julio C. Zamora
Jun Sato
Jun Tan
Jun Tang
Jun Xiao
Jun Xu
Junbao Zhuo
Jun-Cheng Chen
Junfen Chen
Jungeun Kim
Junhwa Hur
Junli Tao
Junlin Han
Junsik Kim
Junting Dong
Junwei Zhou
Junyu Gao
Kai Han
Kai Huang
Kai Katsumata
Kai Zhao
Kailun Yang
Kai-Po Chang
Kaixiang Wang
Kamal Nasrollahi
Kamil Kowol
Kan Chang
Kang-Jun Liu
Kanchana Vaishnavi
   Gandikota
Kanoksak Wattanachote
Karan Sikka
Kaushik Roy
Ke Xian
Keiji Yanai
Kha Gia Quach
Kibok Lee
Kira Maag
Kirill Gavrilyuk
Kohei Suenaga
Koichi Ito
Komei Sugiura
Kong Dehui
Konstantinos Batsos
Kotaro Kikuchi

Kouzou Ohara
Kuan-Wen Chen
Kun He
Kun Hu
Kun Zhan
Kunhee Kim
Kwan-Yee K. Wong
Kyong Hwan Jin
Kyuhong Shim
Kyung Ho Park
Kyungmin Kim
Kyungsu Lee
Lam Phan
Lanlan Liu
Le Hui
Lei Ke
Lei Qi
Lei Yang
Lei Yu
Lei Zhu
Leila Mahmoodi
Li Jiao
Li Su
Lianyu Hu
Licheng Jiao
Lichi Zhang
Lihong Zheng
Lijun Zhao
Like Xin
Lin Gu
Lin Xuhong
Lincheng Li
Linghua Tang
Lingzhi Kong
Linlin Yang
Linsen Li
Litao Yu
Liu Liu
Liujie Hua
Li-Yun Wang
Loren Schwiebert
Lujia Jin
Lujun Li
Luping Zhou
Luting Wang

Mansi Sharma
Mantini Pranav
Mahmoud Zidan
 Khairallah
Manuel Günther
Marcella Astrid
Marco Piccirilli
Martin Kampel
Marwan Torki
Masaaki Iiyama
Masanori Suganuma
Masayuki Tanaka
Matan Jacoby
Md Alimoor Reza
Md. Zasim Uddin
Meghshyam Prasad
Mei-Chen Yeh
Meng Tang
Mengde Xu
Mengyang Pu
Mevan B. Ekanayake
Michael Bi Mi
Michael Wray
Michaël Clément
Michel Antunes
Michele Sasdelli
Mikhail Sizintsev
Min Peng
Min Zhang
Minchul Shin
Minesh Mathew
Ming Li
Ming Meng
Ming Yin
Ming-Ching Chang
Mingfei Cheng
Minghui Wang
Mingjun Hu
MingKun Yang
Mingxing Tan
Mingzhi Yuan
Min-Hung Chen
Minhyun Lee
Minjung Kim
Min-Kook Suh

Minkyo Seo
Minyi Zhao
Mo Zhou
Mohammad Amin A.
 Shabani
Moein Sorkhei
Mohit Agarwal
Monish K. Keswani
Muhammad Sarmad
Muhammad Kashif Ali
Myung-Woo Woo
Naeemullah Khan
Naman Solanki
Namyup Kim
Nan Gao
Nan Xue
Naoki Chiba
Naoto Inoue
Naresh P. Cuntoor
Nati Daniel
Neelanjan Bhowmik
Niaz Ahmad
Nicholas I. Kuo
Nicholas E. Rosa
Nicola Fioraio
Nicolas Dufour
Nicolas Papadakis
Ning Liu
Nishan Khatri
Ole Johannsen
P. Real Jurado
Parikshit V. Sakurikar
Patrick Peursum
Pavan Turaga
Peijie Chen
Peizhi Yan
Peng Wang
Pengfei Fang
Penghui Du
Pengpeng Liu
Phi Le Nguyen
Philippe Chiberre
Pierre Gleize
Pinaki Nath Chowdhury
Ping Hu

Ping Li	Rui Wang	Shin-Jye Lee
Ping Zhao	Rui Zhu	Shishi Qiao
Pingping Zhang	Ruibing Hou	Shivam Chandhok
Pradyumna Narayana	Ruikui Wang	Shohei Nobuhara
Pritish Sahu	Ruiqi Zhao	Shreya Ghosh
Qi Li	Ruixing Wang	Shuai Yuan
Qi Wang	Ryo Furukawa	Shuang Yang
Qi Zhang	Ryusuke Sagawa	Shuangping Huang
Qian Li	Saimunur Rahman	Shuigeng Zhou
Qian Wang	Samet Akcay	Shuiwang Li
Qiang Fu	Samitha Herath	Shunli Zhang
Qiang Wu	Sanath Narayan	Shuo Gu
Qiangxi Zhu	Sandesh Kamath	Shuoxin Lin
Qianying Liu	Sanghoon Jeon	Shuzhi Yu
Qiaosi Yi	Sanghyun Son	Sida Peng
Qier Meng	Satoshi Suzuki	Siddhartha Chandra
Qin Liu	Saumik Bhattacharya	Simon S. Woo
Qing Liu	Sauradip Nag	Siwei Wang
Qing Wang	Scott Wehrwein	Sixiang Chen
Qingheng Zhang	Sebastien Lefevre	Siyu Xia
Qingjie Liu	Sehyun Hwang	Sohyun Lee
Qinglin Liu	Seiya Ito	Song Guo
Qingsen Yan	Selen Pehlivan	Soochahn Lee
Qingwei Tang	Sena Kiciroglu	Soumava Kumar Roy
Qingyao Wu	Seok Bong Yoo	Srinjay Soumitra Sarkar
Qingzheng Wang	Seokjun Park	Stanislav Pidhorskyi
Qizao Wang	Seongwoong Cho	Stefan Gumhold
Quang Hieu Pham	Seoungyoon Kang	Stefan Matcovici
Rabab Abdelfattah	Seth Nixon	Stefano Berretti
Rabab Ward	Seunghwan Lee	Stylianos Moschoglou
Radu Tudor Ionescu	Seung-Ik Lee	Sudhir Yarram
Rahul Mitra	Seungyong Lee	Sudong Cai
Raül Pérez i Gonzalo	Shaifali Parashar	Suho Yang
Raymond A. Yeh	Shan Cao	Sumitra S. Malagi
Ren Li	Shan Zhang	Sungeun Hong
Renán Rojas-Gómez	Shangfei Wang	Sunggu Lee
Renjie Wan	Shaojian Qiu	Sunghyun Cho
Renuka Sharma	Shaoru Wang	Sunghyun Myung
Reyer Zwiggelaar	Shao-Yuan Lo	Sungmin Cho
Robin Chan	Shengjin Wang	Sungyeon Kim
Robin Courant	Shengqi Huang	Suzhen Wang
Rohit Saluja	Shenjian Gong	Sven Sickert
Rongkai Ma	Shi Qiu	Syed Zulqarnain Gilani
Ronny Hänsch	Shiguang Liu	Tackgeun You
Rui Liu	Shih-Yao Lin	Taehun Kim

Takao Yamanaka
Takashi Shibata
Takayoshi Yamashita
Takeshi Endo
Takeshi Ikenaga
Tanvir Alam
Tao Hong
Tarun Kalluri
Tat-Jen Cham
Tatsuya Yatagawa
Teck Yian Lim
Tejas Indulal Dhamecha
Tengfei Shi
Thanh-Dat Truong
Thomas Probst
Thuan Hoang Nguyen
Tian Ye
Tianlei Jin
Tianwei Cao
Tianyi Shi
Tianyu Song
Tianyu Wang
Tien-Ju Yang
Tingting Fang
Tobias Baumgartner
Toby P. Breckon
Torsten Sattler
Trung Tuan Dao
Trung Le
Tsung-Hsuan Wu
Tuan-Anh Vu
Utkarsh Ojha
Utku Ozbulak
Vaasudev Narayanan
Venkata Siva Kumar
    Margapuri
Vandit J. Gajjar
Vi Thi Tuong Vo
Victor Fragoso
Vikas Desai
Vincent Lepetit
Vinh Tran
Viresh Ranjan
Wai-Kin Adams Kong
Wallace Michel Pinto Lira

Walter Liao
Wang Yan
Wang Yong
Wataru Shimoda
Wei Feng
Wei Mao
Wei Xu
Weibo Liu
Weichen Xu
Weide Liu
Weidong Chen
Weihong Deng
Wei-Jong Yang
Weikai Chen
Weishi Zhang
Weiwei Fang
Weixin Lu
Weixin Luo
Weiyao Wang
Wenbin Wang
Wenguan Wang
Wenhan Luo
Wenju Wang
Wenlei Liu
Wenqing Chen
Wenwen Yu
Wenxing Bao
Wenyu Liu
Wenzhao Zheng
Whie Jung
Williem Williem
Won Hwa Kim
Woohwan Jung
Wu Yirui
Wu Yufeng
Wu Yunjie
Wugen Zhou
Wujie Sun
Wuman Luo
Xi Wang
Xianfang Sun
Xiang Chen
Xiang Li
Xiangbo Shu
Xiangcheng Liu

Xiangyu Wang
Xiao Wang
Xiao Yan
Xiaobing Wang
Xiaodong Wang
Xiaofeng Wang
Xiaofeng Yang
Xiaogang Xu
Xiaogen Zhou
Xiaohan Yu
Xiaoheng Jiang
Xiaohua Huang
Xiaoke Shen
Xiaolong Liu
Xiaoqin Zhang
Xiaoqing Liu
Xiaosong Wang
Xiaowen Ma
Xiaoyi Zhang
Xiaoyu Wu
Xieyuanli Chen
Xin Chen
Xin Jin
Xin Wang
Xin Zhao
Xindong Zhang
Xingjian He
Xingqun Qi
Xinjie Li
Xinqi Fan
Xinwei He
Xinyan Liu
Xinyu He
Xinyue Zhang
Xiyuan Hu
Xu Cao
Xu Jia
Xu Yang
Xuan Luo
Xubo Yang
Xudong Lin
Xudong Xie
Xuefeng Liang
Xuehui Wang
Xuequan Lu

Xuesong Yang
Xueyan Zou
XuHu Lin
Xun Zhou
Xupeng Wang
Yali Zhang
Ya-Li Li
Yalin Zheng
Yan Di
Yan Luo
Yan Xu
Yang Cao
Yang Hu
Yang Song
Yang Zhang
Yang Zhao
Yangyang Shu
Yani A. Ioannou
Yaniv Nemcovsky
Yanjun Zhu
Yanling Hao
Yanling Tian
Yao Guo
Yao Lu
Yao Zhou
Yaping Zhao
Yasser Benigmim
Yasunori Ishii
Yasushi Yagi
Yawei Li
Ye Ding
Ye Zhu
Yeongnam Chae
Yeying Jin
Yi Cao
Yi Liu
Yi Rong
Yi Tang
Yi Wei
Yi Xu
Yichun Shi
Yifan Zhang
Yikai Wang
Yikang Ding
Yiming Liu

Yiming Qian
Yin Li
Yinghuan Shi
Yingjian Li
Yingkun Xu
Yingshu Chen
Yingwei Pan
Yiping Tang
Yiqing Shen
Yisheng Zhu
Yitian Li
Yizhou Yu
Yoichi Sato
Yong A.
Yongcai Wang
Yongheng Ren
Yonghuai Liu
Yongjun Zhang
Yongkang Luo
Yongkang Wong
Yongpei Zhu
Yongqiang Zhang
Yongrui Ma
Yoshimitsu Aoki
Yoshinori Konishi
Young Jun Heo
Young Min Shin
Youngmoon Lee
Youpeng Zhao
Yu Ding
Yu Feng
Yu Zhang
Yuanbin Wang
Yuang Wang
Yuanhong Chen
Yuanyuan Qiao
Yucong Shen
Yuda Song
Yue Huang
Yufan Liu
Yuguang Yan
Yuhan Xie
Yu-Hsuan Chen
Yu-Hui Wen
Yujiao Shi

Yujin Ren
Yuki Tatsunami
Yukuan Jia
Yukun Su
Yu-Lun Liu
Yun Liu
Yunan Liu
Yunce Zhao
Yun-Chun Chen
Yunhao Li
Yunlong Liu
Yunlong Meng
Yunlu Chen
Yunqian He
Yunzhong Hou
Yuqiu Kong
Yusuke Hosoya
Yusuke Matsui
Yusuke Morishita
Yusuke Sugano
Yuta Kudo
Yu-Ting Wu
Yutong Dai
Yuxi Hu
Yuxi Yang
Yuxuan Li
Yuxuan Zhang
Yuzhen Lin
Yuzhi Zhao
Yvain Queau
Zanwei Zhou
Zebin Guo
Ze-Feng Gao
Zejia Fan
Zekun Yang
Zelin Peng
Zelong Zeng
Zenglin Xu
Zewei Wu
Zhan Li
Zhan Shi
Zhe Li
Zhe Liu
Zhe Zhang
Zhedong Zheng

Zhenbo Xu
Zheng Gu
Zhenhua Tang
Zhenkun Wang
Zhenyu Weng
Zhi Zeng
Zhiguo Cao
Zhijie Rao
Zhijie Wang
Zhijun Zhang
Zhimin Gao
Zhipeng Yu
Zhiqiang Hu
Zhisong Liu
Zhiwei Hong
Zhiwei Xu

Zhiwu Lu
Zhixiang Wang
Zhixin Li
Zhiyong Dai
Zhiyong Huang
Zhiyuan Zhang
Zhonghua Wu
Zhongyan Zhang
Zhongzheng Yuan
Zhu Hu
Zhu Meng
Zhujun Li
Zhulun Yang
Zhuojun Zou
Ziang Cheng
Zichuan Liu

Zihan Ding
Zihao Zhang
Zijiang Song
Zijin Yin
Ziqiang Zheng
Zitian Wang
Ziwei Yao
Zixun Zhang
Ziyang Luo
Ziyi Bai
Ziyi Wang
Zongheng Tang
Zongsheng Cao
Zongwei Wu
Zoran Duric

Contents – Part VII

Segmentation and Grouping

Motion and Tracking

Generative Models for Computer Vision

DAC-GAN: Dual Auxiliary Consistency Generative Adversarial Network for Text-to-Image Generation

Zhiwei Wang, Jing Yang$^{(\boxtimes)}$, Jiajun Cui, Jiawei Liu, and Jiahao Wang

East China Normal University, Shanghai, China
{wlf,liujiawei,jhwang}@stu.ecnu.edu.cn,
jyang@cs.ecnu.edu.cn

Abstract. Synthesizing an image from a given text encounters two major challenges: the integrity of images and the consistency of text-image pairs. Although many decent performances have been achieved, two crucial problems are still not considered adequately. (i) The object frame is prone to deviate or collapse, making subsequent refinement unavailable. (ii) The non-target regions of the image are affected by text which is highly conveyed through phrases, instead of words. Current methods barely employ the word-level clue, leaving coherent implication in phrases broken. To tackle the issues, we propose DAC-GAN, a Dual Auxiliary Consistency Generative Adversarial Network (DAC-GAN). Specifically, we simplify the generation by a single-stage structure with dual auxiliary modules. (1) Class-Aware skeleton Consistency (CAC) module retains the integrity of image by exploring additional supervision from prior knowledge and (2) Multi-label-Aware Consistency (MAC) module strengthens the alignment of text-image pairs at phrase-level. Comprehensive experiments on two widely-used datasets show that DAC-GAN can maintain the integrity of the target and enhance the consistency of text-image pairs.

1 Introduction

Cross-modal tasks are rapidly evolving and text-to-image generation [1] is one of the significant branches with a broad range of applications, like image-editing, computer-aided inpainting, etc. Literally, the task is defined as to generate a text-consistent image with fidelity from a given caption. Existing methods have achieved great success, but obstacles stand in the way of two principal aspects. First, deflection or collapse of the target object is a frequent occurrence, leading to slow convergence and poor quality of synthesized images. Second, the semantics of the text is embodied in the non-target objects of the image and word-level correspondence undermines semantic coherence.

As shown in Fig. 1(a–ii), the main frame of the object deviates or even deforms within the generation phase, which leads to slow convergence and low quality of images. Similar to the process of human painting upon a given text,

L. Wang et al. (Eds.): ACCV 2022, LNCS 13847, pp. 3–19, 2023.
https://doi.org/10.1007/978-3-031-26293-7_1

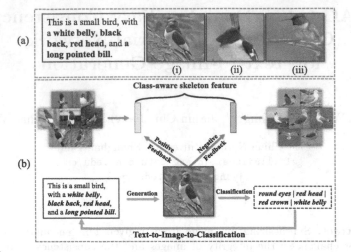

Fig. 1. Given the text description, existing methods yielded some unsatisfactory results. Compared with the ground-truth image in (a–i), the holistic frame of the target in (a–ii) has deviated during the training phase. And in (a–iii), some key attributes in the text are represented in non-target objects. To overcome the defects, the proposed DAC-GAN in (b) extracts the overall skeleton of each class to supervise the generation phase (the top half) and employs image multi-label classification to strengthen semantic consistency (the bottom half).

we associate the class-aware frame of target objects at first and then generate high-quality, category-accurate images under this specific framework. In order to make the model perceive the discriminative attributes related to the category, we propose Class-Aware skeleton Consistency (CAC) module. The CAC leverages an image prior knowledge extractor (IPKE) to obtain a class-aware skeleton feature as additional supervision so as to retain the structural logic integrity of the image skeleton in the training process. Note that our skeleton features are unnecessarily needed to be trained with GAN, which accelerates the convergence of the model, and it can be easily transplanted to other networks.

As shown in Fig. 1(a–iii), the text semantics not only changes the target but also affects non-target objects. What's more, the semantics is explicitly expressed by phrases, whereas the common practice leverage the word-level clue, i.e. splitting the sentence into discrete words. Word-level semantics breaks the coherency. For example, in the sentence "this small bird has a red crown and a white belly.", here, "small bird", "red crown" and "white belly" are the key elements that should not be split. To quantify the relevance between the generated image and the text, if such phrase-level attributes can be expressed from the image, it means that the latent manifold transformation is satisfactory. Then we introduce Multi-label-Aware Consistency (MAC), a module that embraces text prior knowledge extractor (TPKE) to tighten up the alignment of text and image while mitigating the impact of word-level semantics on context in an intuitive and concise manner.

In addition, the majority of methods adopted multi-stage structure, sacrificing computational complexity, to generate images from coarse to fine like [2]. Inspired by [3], we adjust the primitive Conditional Batch Normalization (CBN) module with sentence-level and phrase-level clues. In the paper, we introduce Dual Conditional Batch Normalization (DCBN) as the backbone of the generator to form an end-to-end paradigm.

In this article, motivated by the aforementioned observations, we propose a novel model in Fig. 1(b), called Dual Auxiliary Consistency Generative Adversarial Network (DAC-GAN).

Contributions in this article are expended as follows:

- The Class-Aware skeleton Consistency (CAC) leverages IPKE to distill the class-aware skeleton feature from prior knowledge to maintain the integrity of target object.
- The Multi-label-Aware Consistency (MAC) embraces TPKE to enhance the correspondence between text and image at the phrase-level.
- We propose an integral birdy structure to generate text-related images of high quality. The DCBN is the backbone to synthesize images. The CAC and the MAC arc wings of birds to coordinate in the generation procedure. The extensive experimental results demonstrate that our method can obtain more integral images and higher correspondence between the image and text.

2 Related Work

Due to the successful application of GANs [4,5] in the field of image generation [6–11], a great quantity of works have been devoted to more complex tasks, such as text to Image (T2I), image inpainting, etc. T2I is an interesting branch of image synthesis and one of its major difficulties lies in how to combine text semantics with image features.

Concatenating. Reed et al. [1] first attempted to synthesize photographic images from text descriptions by simply concatenating text vectors to visual features. To decompose the difficulty of generating high-resolution problems, Zhang et al. [2] stacked cascaded GANs to refine images from low resolution to high resolution, and introduced a common technique named Conditioning Augmentation. In order to stabilize the training phase and improve sample diversity, [12] arranged the multiple generators and discriminators in a tree-like structure. Different from StackGAN which required multi-stage training, HDGAN [13] introduced the hierarchically-nested discriminators to leverage mid-level representations of CNN in an end-to-end way.

Cross-Modal Attention. Xu et al. [14] took advantage of the attention mechanism to help the model obtain more fine-grained information. Observing poor correlation between text and generated image, Qiao et al. [15] constructed a symmetrical structure like a mirror, text-to-image-to-text, to maintain the consistency between image and text. In order to alleviate the dependence on the initial generated image, Zhu et al. [16] introduced an additional module Memory

Network [17] to dynamically rectify the quality of the image. Cheng et al. [18] made the most of the captions in the training dataset and enriched the given caption from prior knowledge to provide more visual details. [19] employed contrastive loss in image to sentence, image region to word, and image to image to enforce alignment between synthesized pictures and corresponding captions.

Conditional Batch Normalization (CBN). Yin et al. [20] used a Siamese scheme [21] to implicitly disentangle high-level semantics from distinct captions and first adopted CBN [3,22] for visual feature generation. For the efficiency of training, DF-GAN [23] decomposed the affine transformation from CBN and designed a Deep text-image Fusion Block (DFBlock) to enhance semantic fusion. As the backbone of DAC-GAN, DCBN leverages the whole caption and strong-feature phrases to strengthen the fusion between text and image.

3   Dual Auxiliary Consistency Generative Adversarial Network

As Fig. 2 shows, the architecture of our DAC-GAN is integrated like a bird structure. (1) CAC and (2) MAC are similar to the wings of birds which manipulate the progress from their respective perspectives. To maintain the integrity of the target object, the CAC obtains class-aware skeleton features as additional supervision by IPKE. The MAC leverages TPKE module to enhance semantic consistency at phrase-level. (3) As the bone of the generator, DCBN plays a principal role in image generation with sentence-level and phrase-level clues. Details of the model are introduced below (Fig. 3).

3.1   CAC: Class-Aware Skeleton Consistency

Within the training phase, the main framework of the target will be deviated or even distorted, resulting in poor quality of the visual image. Taking the bird dataset as an example, results such as multi-headed birds, swordfish birds, and multiple pairs of eyes will appear. In analogy to painters with different painting styles, they can draw according to their imagination or outline the overall framework of the object first and then refine under this frame. In order to keep the structure of the generated image logical and complete, we obtain a common feature from multiple images under each category and use the feature as a skeleton feature in the generation, thus constraining the image generation procedure to continuously have the underlying structure of such species. The images I of dataset is organised as: $I = \{I_c | c = 0, \cdots, C - 1\}$ where C represents the number of total species of dataset. $I_c = \{x_i | i = 0, \cdots, n - 1\}$ and n notes the sum of pictures under c-th species.

In detail, the class-aware feature is a low-dimensional vector. By combining contrastive learning and CNN, we not only integrate a feature that distinguishes deep semantics in inter-class images, but also contains the diversity of intra-class images. It is different from a trainable class-aware embedding obviously. The latter can not capture such distinctness for it treats intra-class images as

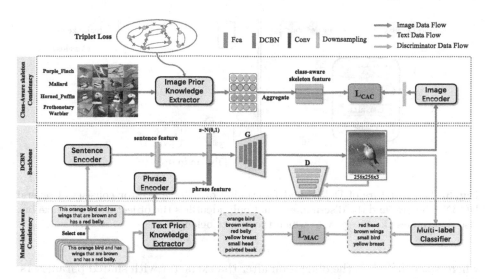

Fig. 2. The DAC-GAN architecture for text-to-image generation. We make the most of the caption by Dual Conditional Batch Normalization (DCBN). The Class-Aware skeleton Consistency (CAC) module is introduced to supervise which objects are currently being drawn and focus more on the main frame. We leverage the Multi-label-Aware-Consistency (MAC) module to strengthen the semantic consistency.

a single. It is also noted that, during the experiments we also apply clustering, and attention methods to integrate information. However, the improvement in effectiveness is limited, but it adds significant time complexity. We therefore go straight to the simplest averaging method. Taking categories as supervision, we customize the metric learning method [24–29] to increase the inter-class distance and decrease the intra-class distance by triplet loss [28, 30, 31]. At the same time, the triplet loss will implicitly act as a data augmentation. The triplet loss is defined as:

$$L\left(x_a, x_p, x_n\right) = \\ \max\left(0, m + \|f\left(x_a\right) - f\left(x_p\right)\| - \|f\left(x_a\right) - f\left(x_n\right)\|\right), \tag{1}$$

where x_a, x_p, x_n indicate anchor, positive, negative samples respectively. The m is a margin constant.

Based on metric learning, an Image Prior Knowledge Extractor ($IPKE$) transforms multiple pictures under each category into features and then aggregates a skeleton feature sk_c from the features. In order to improve efficiency, we have adopted the simplest method to obtain the average value of the features, which is used as additional skeleton-level supervision in the generation stage. The class-skeleton-aware feature f_{cls-sk} is plainly calculated as:

$$f_{cls-sk} = \{sk_c \mid c = 0, \ldots, C-1\}, \\ sk_c = \frac{1}{n} \cdot \sum_{i=1}^{n} IPKE(x_i), \tag{2}$$

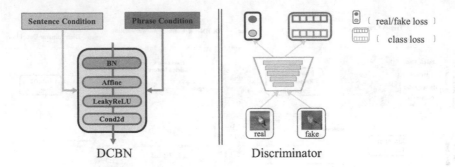

Fig. 3. The structure of DCBN and Discriminator. The generator consists of several DCBN modules, and the discriminator not only identifies the object as real or fake but also determines the category to which the object belongs.

where $sk_c \in R^D$, D is the dimension after the extractor with shared parameters. We train the extractor and then obtain the skeleton feature of each category in advance, and the parameters of the extractor are fixed during the training stage.

3.2 MAC: Multi-label-Aware Consistency

A very challenging aspect of the T2I task is to maintain the coherency between text and image. MirrorGAN [15] employed text-to-image-to-text structure to enhance semantic consistency. The uncertainty associated with the passing of the results of two generation tasks is enormous and it inevitably introduced massive noise which degenerates models. However, it is simple to give from the perspective of human thinking. Take Fig. 1 as an example, people intuitively observe salient features including red head, black wings, and white belly. If they can be matched with the real labels, it means that the generated image and text description are highly related.

By imitating the concise insight, we embrace a universal CNN-based model to classify images with multi-label. Note that, the labels are phrases instead of words. Taking a phrase as an example, "red throat" will be split into separate words "red" and "throat", altering the coherent semantics. We utilize a Text Prior Knowledge Extractor ($TPKE$) to extract phrases from all n captions as target labels (n indicates the number of captions per image in the dataset), and then a ResNet model [32] is employed to encode the synthesized image x_i into visual features f_i. The MAC module is expressed as:

$$P_i = TPKE(S_i^j), j \in \{1, \ldots, m-1\},$$
$$f_i = CNN(x_i), prob = \sigma(f_i), \tag{3}$$

where m denotes the number of captions per image in datasets. The σ denotes the sigmoid function, and $prob$ is the predicted probability distribution over phrases. The $TPKE$ consists of the dependency parsing module of the NLP library Spacy and a rule-based approach.

When we perform multi-label classification tasks using CE-based loss, the category imbalance issue is inevitably encountered, which disturbs the model's concentration on the labels with less frequency. Inspired by [33], we introduce Circle Loss to overcome the imbalance. Modified multi-label circle loss treats multi-label classification as a pairwise comparison between the target category Ω_{pos} and the non-target category Ω_{neg}. The final Multi-label-Aware Consistency (MAC) loss is promoted as:

$$L_{MAC} = \log \left(1 + \sum_{i \in \Omega_{ne.g.}} e^{s_i} \right) + \log \left(1 + \sum_{j \in \Omega_{pos}} e^{-s_j} \right), \qquad (4)$$

where s_i denotes the score of $i-th$ target category. Meanwhile, with the help of the good property of logsumexp, the weight of each item is automatically balanced.

3.3 DCBN: Dual Conditional Batch Normalization

The majority of previous methods enhanced the fusion of text and image by multi-stage structure or attention scheme, which complicated the training phase. Inspired by several works [20,20,23,34], we introduce the Dual Conditional Batch Normalization (DCBN) module to strengthen semantic fusion while simplifying the typical multi-stage structure. The global sentence contains coherent semantics, however, the noise in it affects implications while phrases contain explicit semantics and less noise. Through the analyses, we refine CBN with dual clues, sentence-level clue and phrase-level clue.

The naive CBN is a class-condition variant of Batch Normalization (BN) and the core of this change is that linguistic embeddings can be wielded to modulate the scaling up and down of the feature map. It inputs the linguistic vector $x \in R^{N \times C \times H \times W}$ into a multi-layer perceptron to obtain γ and β. Since the parameters depend on the input feature, cross-modal information interaction between text and image can be achieved. The CBN formula is defined as:

$$\begin{cases} y = \frac{x - E[x]}{\sqrt{Var[x] + \varepsilon}} \cdot \gamma_{new} + \beta_{new} \\ \gamma_{new} = \gamma + \gamma_c \\ \beta_{new} = \beta + \beta_c, \end{cases} \qquad (5)$$

where $E[x]$ and $Var[x]$ are the mean and variance for each channel, and γ_c, β_c are modulation parameters of condition c. Our refined DCBN function is formatted as:

$$\begin{cases} \gamma_{new} = \gamma + \lambda_1 \cdot \gamma_s + \lambda_2 \cdot \gamma_p \\ \beta_{new} = \beta + \lambda_1 \cdot \beta_s + \lambda_2 \cdot \beta_p, \end{cases} \qquad (6)$$

where λ_1 and λ_2 are weights of condition s and p.

As shown in Fig. 2, a sentence encoder *S–Encoder* [35] is used to extract global sentence embedding s from the given caption and a phrase encoder *P–Encoder* [36] to embed strong-feature phrases into a semantic vector p by

$$\begin{cases} s = S\text{–}Encoder(S) \\ p = P\text{–}Encoder(P), \end{cases} \tag{7}$$

where $P = \{P_l \mid l = 0, \cdots, L-1\}$ and L represents the number of phrases extracted from the whole given sentence S.

In previous studies [1,29], the linguistic embedding was high dimensional which caused discontinuity in latent semantics because of lacking data. To transmute the vector into a desirable manifold, we follow the conventional technique in [2]. The condition augmentation marked by F_{ca}, yields more pairs of image-text under the small data limitation, thereby promoting the robustness to a tiny disturbance on the condition manifold. The equation is defined as:

$$\begin{cases} s_{ca} = F_{ca}(s) \\ p_{ca} = F_{ca}(p). \end{cases} \tag{8}$$

3.4   Objective Functions

We note binary cross entropy and multi-label cross entropy with label smoothing as BE and CE_S respectively. The formula of CE_S is calculated by:

$$CE_S = -\sum_{i=1}^{K} p_i \log q_i,$$

$$p_i = \begin{cases} (1-\varepsilon), & \text{if } (i = y) \\ \frac{\varepsilon}{K-1}, & \text{if } (i \neq y) \end{cases}. \tag{9}$$

Here, ε is a small constant and K denotes the number of labels.

Following the discriminator loss in [37], the discriminator D not only distinguishes real data distribution from synthetic distribution but also classifies generated sample to a specific category. The training loss L_{D_S} related to the source of the input (real, fake, or mismatch) is expressed as

$$L_{D_S} = BE \underbrace{(D_S(I_i, l_j), 1)}_{i \sim real, j \sim real} + BE \underbrace{(D_S(I_i, l_j), 0)}_{i \sim fake, j \sim real} + BE \underbrace{(D_S(I_i, l_j), 0)}_{i \sim mis, j \sim real}, \tag{10}$$

where l_i denotes the text captions. Similarly, L_{D_C} relates to which the input is supposed to pertain. The formulation is defined by

$$L_{D_C} = CE_S \underbrace{(D_C(I_i, l_j), C_r)}_{i \sim real, j \sim real} + CE_S \underbrace{(D_C(I_i, l_j), C_r)}_{i \sim fake, j \sim real} + CE_S \underbrace{(D_C(I_i, l_j), C_w)}_{i \sim mis, j \sim real}, \tag{11}$$

the D is trained by minimizing the loss (L_D) as follows:

$$L_D = L_{D_S} + L_{D_C}. \tag{12}$$

Besides common condition loss which is denoted as:

$$L_{G_C} = BE \underbrace{\left(D_S\left(I_i, l_j\right), 1\right)}_{i \sim fake, j \sim real} + CE_S \underbrace{\left(D_C\left(I_i, l_j\right), C_r\right)}_{i \sim fake, j \sim real}, \tag{13}$$

we further introduce a cosine-based class-aware skeleton consistency loss (L_{CAC}) to maintain the frame of the objects during training as follows:

$$L_{CAC} = CS\left(sk_c, f_c^j\right), j \in \{1, \dots, n-1\}, \tag{14}$$

where CS represents cosine similarity and f_c^j indicates the feature of j-th image which belongs to c-th species. Meanwhile, we employ L_{MAC} to align the text-image semantics. Mathematically, the generation loss is expressed as:

$$L_G = L_{G_C} + \lambda_3 \cdot L_{CAC} + \lambda_4 \cdot L_{MAC}, \tag{15}$$

in which λ_3, λ_4 are the modulating weights of class-aware skeleton consistency loss and multi-label-aware consistency loss.

3.5 Implementation Details

Following [14,23], a pre-trained bi-directional LSTM is employed to yield a global sentence embedding s, and we use an embedding layer in the CNN-RNN [36] framework to embed phrases into a semantic vector p in a dimension of 256. Our generator consists of 7 DCBN blocks conditioned with the sentence and phrase embeddings, then we set $\lambda_1 = 0.4, \lambda_2 = 0.6$ to weight the dual conditions. As to the CAC module, we use a metric-learning based method to extract the features from images and average the multiple features into a 256-dimensional class-aware skeleton feature. In MAC, we utilize the powerful NLP library Spacy and rule-based methods to extract the phrases and leverage the phrases with more than 200 frequencies in CUB (50 in Oxford-102) as the multi-label of the image. The target labels are all phrases extracted from the corresponding ten captions, instead of a single one. Specifically, we set hyper-parameters $\lambda_3 = 1.6, \lambda_4 = 0.8$, and the learning rates of discriminator and generator are set to 0.0001, 0.0002 respectively.

4 Experiments

In practice, we evaluate our proposed model qualitatively and quantitatively. According to different datasets, we compare our model with various state-of-the-art methods and validate key components of the model through ablation studies.

Table 1. The Inception Score (higher is better) and Fréchet Inception Distance (lower is better) of the state-of-the-art on CUB and Oxford-102. Note that, the FID on CUB of MirrorGAN and the IS and FID on Oxford-102 of DF-GAN are calculated by reproducing from the open source code. And the FID of HDGAN is calculated from released weights.

Method	CUB		OX-ford102	
	IS↑	FID↓	IS↑	FID↓
GAN-INT-CLS [1]	2.88 ± .04	68.79	2.66 ± .03	79.55
StackGAN [2]	3.70 ± .04	51.89	3.20 ± .01	55.28
TAC-GAN [37]	(-)	(-)	2.88 ± .04	(-)
HDGAN [13]	4.15 ± .05	(-)	3.45 ± .07	37.19
AttnGAN [14]	4.36 ± .03	23.98	(-)	(-)
MirrorGAN [15]	4.56 ± .05	23.47	(-)	(-)
DAE-GAN [38]	4.42	15.19	(-)	(-)
DF-GAN [23]	**5.10**	14.81	3.32 ± .03	39.69
DAC-GAN	4.86 ± .06	**14.77**	**3.59 ± .06**	**35.31**

4.1 Datasets and Evaluation Metrics

Datasets. We evaluate the proposed model on fundamental datasets, CUB bird dataset [39] and Oxford-102 flower dataset [40]. Following previous works [2,14, 16,23], we process these datasets into class-disjoint training and testing sets. The CUB bird dataset contains 200 species with 11788 images in which 8855 images of birds from 150 categories are employed as training data while the left-over 2933 images from 50 categories are for testing. The Oxford-102 flower contains 7034 training images and 1155 testing data belonging to 82 and 20 categories separably. Each image in both datasets has ten text descriptions.

Evaluation Metrics. We leverage two evaluation metrics as same as previous works [41,42]. The Inception Score (IS) and Fréchet Inception Distance(FID) quantitatively measure the quality and diversity of the images to a certain extent.

In order to measure the correspondence of text description and visual image, Xu et al. [14] first introduced the evaluation of image retrieval, named R-precision, into the text-to-image generation task. Given an image, the R-precision is calculated by retrieving correlated text in a text description set. Then we calculate the cosine distance between the global image feature and 100 texts consisting of 99 random samples and 1 ground truth. For each retrieval, if r relevant items are in top R ranked results and the $R-precision = r/R$. Following previous works, we set $R = 1$.

4.2 Experiment Results

We compare our method with the state-of-the-art methods on CUB and Oxford-102 datasets from both quantitative and qualitative perspectives. For each dataset, we compare different methods. The detailed results are shown in Table 1

Table 2. The performances of R-precision on the CUB and Oxford-102 datasets. We compare different modules of our DAC-GAN with Mirror-GAN and DF-GAN. The Baseline denotes that we barely utilize the naive CBN for image generation.

Methods	CUB	Oxford-102
MirrorGAN	19.84	(-)
HDGAN	(-)	16.95
DF-GAN	19.47	18.94
Baseline+DCBN	20.89	**20.51**
Baseline+CAC	20.30	18.85
Baseline+MAC	<u>21.17</u>	19.46
DAC-GAN	**21.62**	<u>19.64</u>
Ground Truth	27.35	21.14

(a) CUB   (b) Oxford-102

Fig. 4. Visual comparison with distinct state-of-the-art methods on CUB and Oxford-102. We compare the DAC-GAN with MirrorGAN [15] and DF-GAN [23] on CUB (on the left). As to Oxford-102, we choose HDGAN [13] and DF-GAN [23] as a comparison (on the right).

and Table 2. Then we use a subjective visual comparison in Fig. 4 and an elaborated human evaluation in Fig. 5 to validate the integrity of images and text-related degree.

4.3 Quantitative Results

As shown in Table 1, higher inception score and lower fréchet inception distance mean better quality and diversity. Our DAC-GAN achieves 4.86 IS and 14.77 FID on CUB. Compared with DF-GAN, DAC-GAN decreases the FID from 14.81 to 14.77 which outperforms other methods by a large margin. As shown in Table 1, we conduct the performance on the Oxford-102 dataset. Compared with HDGAN, DAC-GAN improves IS from 3.45 to 3.59. We measure the IS and FID of DF-GAN by reproducing and DAC-GAN decreases FID from 39.69 to 35.31 (11.04% reduction). The results demonstrate that DAC-GAN achieves better quality and diversity of images.

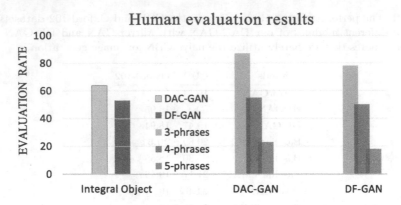

Fig. 5. The results of human evaluation on the integrity of objects and alignment of text-to-image. The higher score of the integral object means the better quality of synthesized images. n-phrases (n = 3, 4, 5) indicates that the number of matched phrase-labels is n.

As Table 2 shows, the DAC-GAN improves the R-precision by 8.97% compared with MirrorGAN and 11.04% compared with DF-GAN on CUB. For Oxford-102, the improvements are 21.00% and 8.29% compared to HDGAN and DF-GAN respectively. Higher R-precision indicates that semantic consistency between text description and synthesized images are better.

4.4 Qualitative Results

Visual Evaluation: We visualize the results to get a better sense of different models. We compare with DF-GAN on both datasets. In addition, we replenish MirrorGAN on CUB and HDGAN on Oxford-102. First of all, it can be seen that an interesting phenomenon in the first line of Fig. 4. The attributes of the text expression are not only reflected in the target object but also reflected in the background or non-target objects. It means the semantics has shifted after the sentence is divided into words. For example, attributes like color are confused about whether to decorate the background or the target. And then, let's note the second row in Fig. 4. Leaving aside the details, the overall structure of the target object deviates greatly from the real. This is not to mention the detail essence, and such a phenomenon will seriously affect the convergence and the quality of images. DAC-GAN can obtain images with more integral structure and more accurate text correspondence.

Human Test: For the CUB dataset, we designed a manual evaluation method. In the first stage, we selected ten random sets from the generated data and five real data sets, about 300 images in all. We hired 30 employees of different professions to perform a simple sensory evaluation of the fifteen sets (the employees do not know which are the real groups). In the second stage, we selected employees who successfully distinguished real from generated data in the first phase. These employees observe 10 groups of DF-GAN and 10 groups of DAC-GAN generated images while marking whether each image contains a complete object or

Table 3. The Inception Score and Fréchet Inception Distance on two benchmarks with the different modules of DAC-GAN, including DCBN, CAC, MAC. Different from DAC-GAN, DAC-GAN-word denotes that we split the sentence into words.

Metric	IS ↑		FID ↓	
	CUB	Oxford-102	CUB	Oxford-102
Baseline	$4.45 \pm .06$	$3.22 \pm .03$	19.93	40.58
Baseline+DCBN	$4.62 \pm .02$	$3.48 \pm .04$	18.61	39.23
Baseline+CAC	$4.68 \pm .03$	$3.46 \pm .05$	18.46	37.23
Baseline+MAC	$4.65 \pm .04$	$3.47 \pm .02$	17.10	36.73
DAC-GAN-word	$4.57 \pm .04$	$3.46 \pm .03$	19.12	37.42
DAC-GAN	$\mathbf{4.86 \pm .06}$	$\mathbf{3.59 \pm .06}$	**14.77**	**35.31**

not. As shown in Fig. 5, DAC-GAN improves the integrity ratio from 0.53 to 0.64 (20.75% improvement) compared with DF-GAN. In the third stage, to verify the correlation between text and image intuitively and concisely, we extracted the corresponding key phrases from the text as the labels of the image. We provided this prior knowledge to the employees who only need to observe whether the corresponding labels can be found in the image. We counted the number of matched labels for 3,4 and 5 respectively. As shown in the Fig. 5, DAC-GAN improves the n-phrases ($n = 3, 4, 5$) by a large margin (11.54%, 10.00%, 27.78%). The results verify that the DAC-GAN generates more pictures of the integral object and correlates them more closely with corresponding text.

4.5   Ablation Study

In this section, we conduct ablation studies on DAC-GAN and its variants. We define DAC-GAN with a naive CBN module as our baseline, and verify the performance of our DCBN, CAC, and MAC by including or excluding related modules.

As Table 3 shows, by including different components, IS and FID are consistently improved, which indicates that the distinct modules are effective. In addition, we compare word-level DAC-GAN to phrase-level DAC-GAN. The results of word-level DAC-GAN demonstrate that phrases have more explicit semantic information than words. As Table 2 shows, the R-precision of Baseline+DCBN and Baseline+MAC are higher than Baseline+MAC, for the former explicitly leverages phrase-level semantics.

We visualize samples generated by different modules in Fig. 6. Baseline+DCBN and Baseline+MAC focus more on textual correspondence while Baseline+CAC focuses more on the integrity of the target. It indicates that phrase-level knowledge can strengthen the cross-modal correspondence and the class-aware skeleton feature maintains integrity of the target image. By integrating all modules, DAC-GAN adjusts text relevance and target integrity to obtain better quality images with more details.

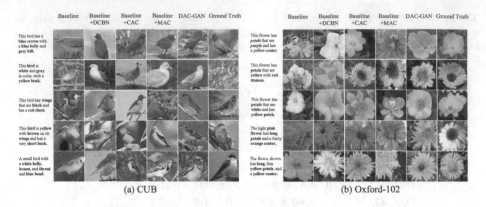

(a) CUB   (b) Oxford-102

Fig. 6. The images are synthesized by different modules of DAC-GAN including DCBN, CAC, and MAC on CUB and Oxford-102. The Baseline indicates that we barely employ naive CBN to generate images.

4.6   Limitation and Discussion

Although our model has achieved good results, there are still some shortcomings worth discussing. First, the process we extract phrases does not take the multi-hop phrases into consideration, which could be further improved. Moreover, two benchmark datasets we experiment on only have a single object and a simple scene which far from the real world. In view of the large gap between the target objects in the coco dataset [43], the CAC and MAC modules can have a greater effect on the results. The problem is how to obtain the class-aware skeleton features of different targets, which is also the direction of our future research.

5   Conclusion

In this paper, we design a novel Dual Auxiliary Consistency Generative Adversarial Network (DAC-GAN) for text-to-image generation task. To maintain the integrity of target object, the CAC module leverages an IPKE module to distill the class-aware skeleton features as additional supervision. The MAC employs a TPKE module to enhance the alignment of text-to-image. Compared with other methods, DAC-GAN can maintain the integrity of target objects and the correspondence between image and text. Moreover, The DCBN employs sentence-level and phrase-level to strengthen the fusion between language and visual.

Acknowledgement. This research is funded by the Science and Technology Commission of Shanghai Municipality 19511120200.

References

1. Reed, S., Akata, Z., Yan, X., Logeswaran, L., Schiele, B., Lee, H.: Generative adversarial text to image synthesis. In: International Conference on Machine Learning, pp. 1060–1069. PMLR (2016)

2. Zhang, H., et al.: StackGAN: text to photo-realistic image synthesis with stacked generative adversarial networks. In: Proceedings of the IEEE International Conference on Computer Vision, pp. 5907–5915 (2017)
3. De Vries, H., Strub, F., Mary, J., Larochelle, H., Pietquin, O., Courville, A.: Modulating early visual processing by language. arXiv preprint arXiv:1707.00683 (2017)
4. Goodfellow, I., et al.: Generative adversarial nets. In: Advances in Neural Information Processing Systems, vol. 27 (2014)
5. Salimans, T., et al.: Improved techniques for training GANs. In Lee, D., Sugiyama, M., Luxburg, U., Guyon, I., Garnett, R. (eds.) Advances in Neural Information Processing Systems, vol. 29. Curran Associates, Inc. (2016)
6. Brock, A., Donahue, J., Simonyan, K.: Large scale gan training for high fidelity natural image synthesis. arXiv preprint arXiv:1809.11096 (2018)
7. Tang, H., Xu, D., Sebe, N., Wang, Y., Corso, J.J., Yan, Y.: Multi-channel attention selection GAN with cascaded semantic guidance for cross-view image translation. In: Proceedings of the IEEE/CVF Conference on Computer Vision and Pattern Recognition, pp. 2417–2426 (2019)
8. Zhang, H., Goodfellow, I., Metaxas, D., Odena, A.: Self-attention generative adversarial networks. In: International Conference on Machine Learning, pp. 7354–7363. PMLR (2019)
9. Karras, T., Laine, S., Aila, T.: A style-based generator architecture for generative adversarial networks. In: Proceedings of the IEEE/CVF Conference on Computer Vision and Pattern Recognition, pp. 4401–4410 (2019)
10. Tang, H., Xu, D., Liu, G., Wang, W., Sebe, N., Yan, Y.: Cycle in cycle generative adversarial networks for keypoint-guided image generation. In: Proceedings of the 27th ACM International Conference on Multimedia, pp. 2052–2060 (2019)
11. Karras, T., Laine, S., Aittala, M., Hellsten, J., Lehtinen, J., Aila, T.: Analyzing and improving the image quality of stylegan. In: Proceedings of the IEEE/CVF Conference on Computer Vision and Pattern Recognition, pp. 8110–8119 (2020)
12. Zhang, H., et al.: StackGAN++: realistic image synthesis with stacked generative adversarial networks. IEEE Trans. Pattern Anal. Mach. Intell. 41, 1947–1962 (2018)
13. Zhang, Z., Xie, Y., Yang, L.: Photographic text-to-image synthesis with a hierarchically-nested adversarial network. In: Proceedings of the IEEE Conference on Computer Vision and Pattern Recognition, pp. 6199–6208 (2018)
14. Xu, T., et al.: AttnGAN: fine-grained text to image generation with attentional generative adversarial networks. In: Proceedings of the IEEE Conference on Computer Vision and Pattern Recognition, pp. 1316–1324 (2018)
15. Qiao, T., Zhang, J., Xu, D., Tao, D.: MirrorGAN: learning text-to-image generation by redescription. In: Proceedings of the IEEE/CVF Conference on Computer Vision and Pattern Recognition, pp. 1505–1514 (2019)
16. Zhu, M., Pan, P., Chen, W., Yang, Y.: DM-GAN: dynamic memory generative adversarial networks for text-to-image synthesis. In: Proceedings of the IEEE/CVF Conference on Computer Vision and Pattern Recognition, pp. 5802–5810 (2019)
17. Weston, J., Chopra, S., Bordes, A.: Memory networks. arXiv preprint arXiv:1410.3916 (2014)
18. Cheng, J., Wu, F., Tian, Y., Wang, L., Tao, D.: RiFeGAN: rich feature generation for text-to-image synthesis from prior knowledge. In: Proceedings of the IEEE/CVF Conference on Computer Vision and Pattern Recognition, pp. 10911–10920 (2020)

19. Zhang, H., Koh, J.Y., Baldridge, J., Lee, H., Yang, Y.: Cross-modal contrastive learning for text-to-image generation. In: Proceedings of the IEEE/CVF Conference on Computer Vision and Pattern Recognition, pp. 833–842 (2021)
20. Yin, G., Liu, B., Sheng, L., Yu, N., Wang, X., Shao, J.: Semantics disentangling for text-to-image generation. In: Proceedings of the IEEE/CVF Conference on Computer Vision and Pattern Recognition, pp. 2327–2336 (2019)
21. Chung, D., Tahboub, K., Delp, E.J.: A two stream siamese convolutional neural network for person re-identification. In: Proceedings of the IEEE International Conference on Computer Vision, pp. 1983–1991 (2017)
22. Dumoulin, V., Shlens, J., Kudlur, M.: A learned representation for artistic style. arXiv preprint arXiv:1610.07629 (2016)
23. Tao, M., Tang, H., Wu, F., Jing, X.Y., Bao, B.K., Xu, C.: DF-GAN: a simple and effective baseline for text-to-image synthesis. arXiv e-prints (2020)
24. Kulis, B., et al.: Metric learning: a survey. Found. Trends® Mach. Learn. 5, 287–364 (2013)
25. Li, J., Lin, X., Rui, X., Rui, Y., Tao, D.: A distributed approach toward discriminative distance metric learning. IEEE Trans. Neural Netw. Learn. Syst. 26, 2111–2122 (2014)
26. Hoffer, E., Ailon, N.: Deep metric learning using triplet network. In: Feragen, A., Pelillo, M., Loog, M. (eds.) SIMBAD 2015. LNCS, vol. 9370, pp. 84–92. Springer, Cham (2015). https://doi.org/10.1007/978-3-319-24261-3_7
27. Ding, Z., Fu, Y.: Robust transfer metric learning for image classification. IEEE Trans. Image Process. 26, 660–670 (2016)
28. Hermans, A., Beyer, L., Leibe, B.: In defense of the triplet loss for person re-identification. arXiv preprint arXiv:1703.07737 (2017)
29. Reed, S.E., Akata, Z., Mohan, S., Tenka, S., Schiele, B., Lee, H.: Learning what and where to draw. Adv. Neural. Inf. Process. Syst. 29, 217–225 (2016)
30. Dong, X., Shen, J.: Triplet loss in siamese network for object tracking. In: Proceedings of the European Conference on Computer Vision (ECCV), pp. 459–474 (2018)
31. Ge, W.: Deep metric learning with hierarchical triplet loss. In: Proceedings of the European Conference on Computer Vision (ECCV), pp. 269–285 (2018)
32. He, K., Zhang, X., Ren, S., Sun, J.: Deep residual learning for image recognition. In: Proceedings of the IEEE Conference on Computer Vision and Pattern Recognition, pp. 770–778 (2016)
33. Sun, Y., et al.: Circle loss: a unified perspective of pair similarity optimization. In: Proceedings of the IEEE/CVF Conference on Computer Vision and Pattern Recognition, pp. 6398–6407 (2020)
34. Faghri, F., Fleet, D.J., Kiros, J.R., Fidler, S.: VSE++: improving visual-semantic embeddings with hard negatives. arXiv preprint arXiv:1707.05612 (2017)
35. Hochreiter, S., Schmidhuber, J.: Long short-term memory. Neural Comput. 9, 1735–1780 (1997)
36. Wang, J., Yang, Y., Mao, J., Huang, Z., Huang, C., Xu, W.: CNN-RNN: a unified framework for multi-label image classification. In: Proceedings of the IEEE Conference on Computer Vision and Pattern Recognition, pp. 2285–2294 (2016)
37. Dash, A., Gamboa, J.C.B., Ahmed, S., Liwicki, M., Afzal, M.Z.: TAC-GAN-text conditioned auxiliary classifier generative adversarial network. arXiv preprint arXiv:1703.06412 (2017)
38. Ruan, S., Zhang, Y., Zhang, K., Fan, Y., Chen, E.: DAE-GAN: dynamic aspect-aware GAN for text-to-image synthesis (2021)

39. Wah, C., Branson, S., Welinder, P., Perona, P., Belongie, S.: The Caltech-UCSD birds-200-2011 dataset (2011)
40. Nilsback, M.E., Zisserman, A.: Automated flower classification over a large number of classes. In: 2008 Sixth Indian Conference on Computer Vision, Graphics & Image Processing, pp. 722–729. IEEE (2008)
41. Salimans, T., Goodfellow, I., Zaremba, W., Cheung, V., Radford, A., Chen, X.: Improved techniques for training GANs. Adv. Neural. Inf. Process. Syst. **29**, 2234–2242 (2016)
42. Heusel, M., Ramsauer, H., Unterthiner, T., Nessler, B., Hochreiter, S.: GANs trained by a two time-scale update rule converge to a local nash equilibrium. In: Advances in Neural Information Processing Systems, vol. 30 (2017)
43. Lin, T.-Y., et al.: Microsoft COCO: common objects in context. In: Fleet, D., Pajdla, T., Schiele, B., Tuytelaars, T. (eds.) ECCV 2014. LNCS, vol. 8693, pp. 740–755. Springer, Cham (2014). https://doi.org/10.1007/978-3-319-10602-1_48

Learning Internal Semantics with Expanded Categories for Generative Zero-Shot Learning

Xiaojie Zhao[1], Shidong Wang[2], and Haofeng Zhang[1](✉)

[1] School of Computer Science and Engineering, Nanjing University of Science and Technology, Nanjing 210094, China
{zhaoxj,zhanghf}@njust.edu.cn
[2] School of Engineering, Newcastle University, Newcastle upon Tyne NE1 7RU, UK
shidong.wang@newcastle.ac.uk

Abstract. In recent years, generative Zero-Shot Learning (ZSL) has attracted much attention due to its better performance than traditional embedding methods. Most generative ZSL methods exploit category semantic plus Gaussian noise to generate visual features. However, there is a contradiction between the unity of category semantic and the diversity of visual features. The semantic of a single category cannot accurately correspond to different individuals in the same category. This is due to the different visual expression of the same category. Therefore, to solve the above mentioned problem we propose a novel semantic augmentation method, which expands a single semantic to multiple internal sub-semantics by learning expanded categories, so that the generated visual features are more in line with the real visual feature distribution. At the same time, according to the theory of Convergent Evolution, the sub-semantics of unseen classes are obtained on the basis of the expanded semantic of their similar seen classes. Four benchmark datasets are employed to verify the effectiveness of the proposed method. In addition, the category expansion is also applied to three generative methods, and the results demonstrate that category expansion can improve the performance of other generative methods. Code is available at: https://github.com/njzxj/EC-GZSL.

Keywords: Generative zero-shot learning · Category expansion · Semantic augmentation · Convergent evolution

1 Introduction

Deep learning has driven the rapid development of classification, retrieval, positioning and other fields. However, this development depends on a large number of manually labeled datasets, which are often labor-intensive and time-consuming. In order to mitigate this problem, Zero-Shot Learning (ZSL) [20,29] has been proposed to recognize unseen classes. ZSL makes the training model suitable for unseen classes that do not exist in the training set through category semantics. With the popularity of generative network in the field of image, an increasing

L. Wang et al. (Eds.): ACCV 2022, LNCS 13847, pp. 20–36, 2023.
https://doi.org/10.1007/978-3-031-26293-7_2

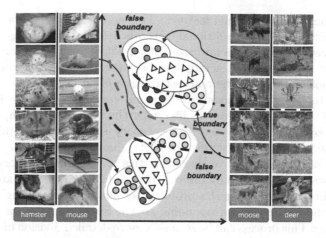

Fig. 1. The motivation of our method. Green and yellow represent moose and mouse respectively. (Color figure online)

number of generative ZSL methods emerge in recent years. Generative ZSL [40] uses semantics and seen classes to train a generator, and then generates samples of unseen classes with the learned generator to make up for missing samples of unseen classes in the classifier training process. However, the semantic of each class is unique, which leads to the lack of diversity of visual features generated by traditional methods using a single semantic.

The semantic of a category is the summary of the attributes of all individuals in the category, and it is equivalent to a mathematical description of the characteristics of a specific category. From the macro point of view, this description is reasonable for the whole category, but from the micro point of view, it is unreasonable for individuals that it is a biased description. For example, there are three kinds of mouse hair colors: black, white and yellow, so these three color dimensions are marked in the semantic of mouse, and the dimension value representing these attributes is not 0. If we use the semantic that is not 0 in the white attribute dimension to represent the black mouse, then it is wrong. Similarly, the male deer has antlers while the female deer has no antlers, but the semantic of deer is not 0 in the dimension of antlers, so that of deer cannot be used to describe the female deer.

Different individuals in the same category have relativity in the performance of characters, that is, Biological Relative Character. The semantic of a category includes all the characters of the category, but it cannot accurately describe a single individual. Generative ZSL generally generates visual features through category semantics. Because category semantics cannot accurately describe individuals with different relative characters, there must be differences in the distribution of generated visual features and real visual features. We know that the more the generated samples match the real samples, the more beneficial it is to the training of the final classifier. In Fig. 1, a category has three internal classes, and the visual feature distribution generated by a single semantic cannot fully fit the distribution of real visual features. This can lead to inaccurate classification boundaries. Therefore, traditional methods using a single category semantic to generate visual features are unreasonable. That is to say, category semantics cannot reflect the relative characteristics of organisms, and the generated visual features do not accord with the real distribution of visual features.

In order to make the semantic description correct for specific individuals, so as to generate visual features in line with the real distribution, an obvious idea is to recombine the attributes of category semantics to obtain multiple extended category semantics, so that the new semantics can correctly refer to the performance of different relative characteristics of the same category. However, this reorganization is technically difficult if no additional manual annotation is introduced. In terms of solving this problem, we aim to diversify single semantics and obtain the semantics corresponding to the specific expression of relative characteristics. This process can be seen as a more detailed division of a class, and then obtain the semantics of each internal class. A simplified example is to obtain the mouse semantics expressed by different color traits. For example, the other color dimensions of white mouse semantics are 0. Of course, the actual situation is more complicated, because relative characteristics cannot only appear in color.

The problem is transformed into obtaining the semantics of the internal class, that is, obtaining the semantics that can represent white mouse in the mouse category. An effective method is to divide a single class into internal categories, that is, to expand categories for each class. Specifically, firstly, clustering the visual features of each category separately, so that we can obtain the visual prototypes of the internal categories. Then we use a trained mapper to map the visual prototype of the internal classes to the semantic space. In this way, we get the semantics of the internal category.

The method of obtaining the internal class semantics of seen classes is not applicable to unseen classes. In biology, there is the concept of Convergent Evolution [36], that is, different species change their overall or partial morphological structures in the same direction due to similar lifestyles. Figure 1 shows this. Hamster and mouse have the same characteristic expression in hair color. Moose and deer also have similar character expression on antlers. We believe that the characteristics of internal classes of similar species are also similar. Based on this assumption, we propose a method based on semantic similarity, which transfer the semantic expanded results of seen classes to obtain the internal class semantics of unseen classes.

Replace the original category semantics with the expanded semantics to achieve the purpose of generating more real visual features, that is, the generated visual features are more in line with the distribution of real visual features. At the same time, the semantic attributes mapped by visual features are cleverly used in the process of final classification. The reconstructed visual features are obtained from the mapped semantics through the trained generator, and the visual features before mapping are concatenated with reconstructed visual features for the training of the final classifier. This method effectively improves the performance, because the reconstructed visual features eliminate irrelevant information for classification. In summary, our main contributions are as follows:

- Based on the principle of Relative Character, we propose a category expansion method to learning internal semantics of seen class. In addition, inspired by the thought of Convergent Evolution, the expanded semantics of unseen classes are also learned from those of seen classes.
- Concatenation of synthesized features and reconstructed features is employed to train the final classifier, which can effectively eliminate irrelevant information for classification, thereby improving the performance of the classifier.
- The proposed model is evaluated on four popular datasets and obtains competing results. Furthermore, for the category expansion part, we verified its effectiveness on three classical generative models.

2 Related Work

2.1 Zero-Shot Learning

The whole sample dataset is divided into seen and unseen parts in Zero-Shot Learning. The unseen classes are used in the training phase, and the unseen class is only used in the testing phase. According to whether the test sample contains seen classes, ZSL can be further categorized as conventional ZSL (CZSL) and generalized ZSL (GZSL) [3]. CZSL contains only unseen class samples in the test phase, while GZSL includes both unseen and seen class samples in the test phase. Since in more practical situations, the trained model needs to be applicable to both seen and unseen classes, GZSL has become the mainstream research point. Besides, according to whether to generate unseen visual features to convert ZSL into a fully supervised task, ZSL can also be divided into non-generative methods and generative methods.

Non-generative methods mainly exploits embedding strategies to associate visual features and semantics. Early methods train a projector to map visual features to semantic space [30,33]. However, later researchers consider that projecting visual features into semantic space will cause the serious hubness problem [2,45], so the way of mapping semantics to visual space is adopted to construct visual prototypes. For example, [4,18] project visual features and semantics to public space for classification. [1,10,25] uses a hybrid model based on mapping both semantic and visual features to hidden space. They mainly adopt the joint embedding method of multiple visual features and multiple text representations to connect the attributes with different areas of the image. In recent years, non-generative methods have begun to introduce attention mechanisms [17,42] to strengthen the semantics learning. Furthermore, knowledge graph has also been used to train classifier parameters [11,37], *e.g.*, MSDN [5] adopts the method of knowledge distillation to collaboratively learn attribute-based visual features and visual-based attribute features .

Generative ZSL trains a generator that can generate corresponding visual features according to its class semantics, and then uses the generator to generate samples of unseen classes to make up for the missing samples of unseen classes. f-CLSWGAN [40] exploits the Generative Adversarial Networks (GAN) [12] plus a classifier to train the generative network. [13,14,28] optimizes the model on this basis by adding more constraints for feature alignment. Cycle-CLSWGAN [9] introduces cyclic consistency loss for feature generator. F-VAEGAN-D2 [41] employs Variational Auto-Encoder (VAE) and uses the distribution obtained by the encoder to replace the Gaussian noise as the input for GAN. Some also directly utilizes VAE to generate visual features [23,26,34]. Besides, IZF [35] adopts the generative flow model instead of generative adversarial network to circumvent the fixed format of Gaussian noise.

2.2 Single Attribute and Relative Character

There is only one semantic for a single category, but the visual features of the same category are diverse. Semantics can be regarded as mathematical description of categories, and they include all the possible characteristics of categories. Even in the same category, the expression of a visual semantic, that is, the theory of relative character in biology, is different. Relative Character refers to that the same species often have different performances in characters. Therefore, a single semantic cannot well represent each individual in a category. Most non generative methods map visual features to semantic space. In this case, the mapping is biased. Some non generative method [22,24,37] has put forward this problem and alleviated this problem to a certain extent.

Most generative methods generate visual features by means of a single semantic and Gaussian noise, so as to enrich visual features for unseen classes. However, the semantic description of visual features generated in this way is still less diverse, because they are all generated under the same semantic description. This method cannot generate visual features that conform to the real distribution, because the expression of different individual characters in the same category is different, so the corresponding semantics should also be different. In order to solve the problem that a single semantic cannot correctly correspond to the representation of different visual features, we propose a category expansion based feature generation method.

2.3 Convergent Evolution

Convergent Evolution means that different species show the same or similar characteristics under the influence of specific conditions. For example, mouse and hamster are two different categories, and they have similarities in the expression of hair color. Moose and deer are different species, and they have similarities in the expression of antlers. If we can expand the semantics of seen classes, we can expand the semantics of similar unseen classes according to the theory of Convergent Evolution.

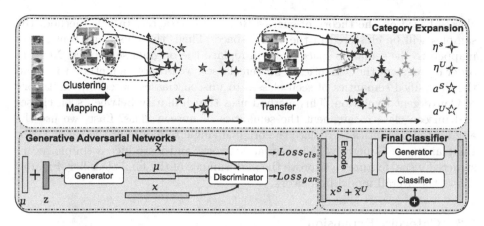

Fig. 2. The architecture of our proposed method. Category Expansion module shows how we can get the internal classes of seen and unseen classes.

3   Methodology

3.1   Problem Definition

Suppose there are S seen classes for training and U unseen classes for testing. We use C_s to represent seen classes and C_u to represent unseen classes, where $C_s \cap C_u = \varnothing$. $X^s = \left\{ x_1^s, x_2^s, x_3^s, \cdots, x_{N_s}^s \right\} \subset \mathbb{R}^{d_x \times N_s}$ are the visual features of seen classes, where d_x is the dimension of visual features and N_s is the number of instances of seen classes. And the corresponding sample class labels are $Y^s = \left\{ y_1^s, y_2^s, y_3^s, \cdots, y_{N_s}^s \right\}$. $X^u = \left\{ x_1^u, x_2^u, x_3^u, ..., x_{N_u}^u \right\} \subset \mathbb{R}^{d_x \times N_u}$ is the visual feature of the sample in unseen classes, where N_u is the number of instances with unseen classes. And the corresponding sample class label is $Y^u = \left\{ y_1^u, y_2^u, y_3^u, \cdots, y_{N_u}^u \right\}$. $A = \left\{ a^1, a^2, ...a^s, a^{S+1}, ..., a^{S+U} \right\} \subset \mathbb{R}^{d_a \times (S+U)}$ represents category semantics, where d_a is the dimension of semantics. The first S vectors are seen classes and the last U are unseen classes. Our goal is to learn a classifier $f : x \to C$ to classify the visual feature x in the category space. For CZSL x only belongs to unseen classes and $C = C_s \cup C_u$, while x belongs to both seen and unseen classes and $C = C_u$ for GZSL.

3.2   Overall Idea

In order to diversify the generated visual features, traditional generative GZSL methods mostly use a single semantic plus Gaussian noise to generate synthesized unseen samples. However, the visual features obtained in this way do not match the feature distribution of real samples. This is because the category semantic corresponding to the generated visual features are single, while the real semantics corresponding to the real sample features are diverse. In order to make the semantics corresponding to the generated visual features more diverse, we propose a method to augment the semantics by expanding categories of seen classes based on the theory of species Relative Character. In this method, the samples will be clustered in the visual space. Then, the clustering centers are mapped to the semantic space and will replace the original semantics. Next, in order to augment the semantics of unseen classes, we propose a method to extend the diversified semantics of seen classes to unseen classes based on the theory of Convergent Evolution. This method uses the similarity between seen classes and unseen class to augment the semantics of unseen class. Last, we use the augmented semantics to replace the original single semantic of the category for the training of Generative Adversarial Network, and introduce a simple reconstructed visual feature to improve the performance of the final classifier. Figure 2 shows the overall model framework.

3.3   Category Expansion

Category semantics are determined according to the character expression of the categories. It can be regarded as the attribute description of the category. For example, if a category has characters such as angle and claw, the corresponding semantic dimension will have numerical value. In order to uniformly describe a class, as long as any individual of the class has a specific character expression, its corresponding semantic dimension exists and is a fixed value. A typical example is that male deer have antlers and female deer have no antlers, but the dimension of antlers in deer semantics is fixed.

A single category semantic is the sum of the expression of all characters in the category, while different individuals in the same category have different performance of relative characters, which leads to the inability of a single semantic to generate true and accurate visual features. We cannot generate white mouse with the semantics of non-zero values in the dimensions corresponding to black, white and yellow. Of course, black mouse cannot be generated. Similarly, using the semantics of deer can generate male deer with antlers, but not female deer. Obviously, traditional generative methods that using this strategy are unreasonable. For a category, the visual features generated using a single semantic are inaccurate, such as mice of different colors, and incomplete, such as female deer without antlers. In Generative ZSL, the closer the generated visual features are to the real visual features, the better the performance of the generator, and it is more beneficial to downstream tasks. In order to solve this problem, we need to obtain the semantics of individuals expressing different characters in the same category. Therefore, we need to expand category, in other words, to obtain the internal semantics of different categories of mouse.

In order to obtain the semantics of different characters of the same category, that is, the semantics of internal categories, we use the method of clustering the samples of each category in visual space and then mapping them to the semantic space. We assume that the number of internal classes in a category is k, in other words, we assume that the number of clusters of each class in the visual space is k. Let $D_i = \left\{ x_1^i, x_2^i, x_3^i, \cdots, x_{N_i}^i \right\}$ represent the features of the samples belonging the ith class, and use the k-means algorithm to minimize the square error of the visual cluster division $C^i = \{C_1^i, C_2^i, C_3^i, \cdots, C_k^i\}$ of this class:

$$L = \sum_{j=1}^{k} \sum_{x^i \in C_j^i} \left\| x^i - \mu_j^i \right\|_2^2, \tag{1}$$

where $\mu_j^i = \frac{1}{|C_j^i|} \sum_{x^i \in C_j^i} x^i$ is the mean vector of cluster C_j^i. The clustering center of each internal class is regarded as the visual prototype of the internal class. We use μ^S to represent all μ_j^i for seen classes.

Then, we need a mapper to map visual prototypes to semantic space. The semantic information of visual features is extracted by a mapper E. We use $a' = E(x^s)$ to represent the semantic of extracted visual features. In order to make the mapped semantics correct, we use the category semantics to constrain the mapper. The loss function is as follows:

$$Loss_e = \frac{1}{N_S} \sum_{i=1}^{N_S} \left\| a_i' - a_i \right\|_F^2, \tag{2}$$

where a_i is the category semantics corresponding to the feature x_i^s. When the mapper E is obtained, we map the visual prototype μ^S of the seen category to the semantic space:

$$\eta^S = E(\mu^S), \tag{3}$$

where η^S represents the semantics of internal seen classes. In this way, we expand the category of each seen class and obtain the semantics expressed by different characters of the same class.

3.4 Augmentation Transfer

Because unseen class samples are not available in the training stage, we cannot obtain the semantics of their internal classes by using the method which used in seen classes. However, the hair color difference of mice is also reflected in hamsters, and the antler difference of deer is also reflected in moose. This phenomenon is called Convergent Evolution, that is, different species evolve into phenomena with similar morphological features or structures due to similar environment and other factors. Based on Convergent Evolution, we can transfer the semantics of seen classes through the similarity measurement between seen and unseen classes, so as to obtain the semantics of the expanded categories of unseen classes.

In order to migrate the differences between the internal classes of seen class to unseen class, we first need to obtain the differences between the internal classes of the seen class. We use $Z = \{z_1^1, \cdots, z_k^1, z_1^2, \cdots, z_k^2, \cdots, z_j^i, \cdots z_k^s\}$ to represent the distance between the jth internal class of the ith seen class and the original semantic a^i:

$$z_j^i = a^i - \eta_j^i. \tag{4}$$

Based on the theory of Convergent Evolution, similar categories have similar character expression. In order to transfer the expanded semantics, we need to obtain the seen class that is most similar to each unseen class. We calculate category similarity based on category semantics. For the semantic a^p of each unseen class, we calculate its semantic similarity with each seen class:

$$h_{pq} = \frac{a^p \cdot a^q}{\|a^p\| \|a^q\|} \ (S+1 \le p \le S+U, 1 \le q \le S), \tag{5}$$

where h_{pq} represents the semantic similarity between pth unseen class and qth seen class. We define the tags:

$$\tau_p = argmax_{q \in (1,2,...,S)} h_{pq}. \tag{6}$$

We think that seen class τ_p and unseen class p are the most similar in character expression. Obviously, their internal class distribution also have similarities according to class semantics. Therefore, we can get the internal class semantics of unseen classes:

$$\eta^p = \{\eta_1^p, \eta_2^p, \cdots, \eta_k^p\} = \{a^p - \alpha h_{p\tau_p} z_1^{\tau_p}, \cdots, a^p - \alpha h_{p\tau_p} z_k^{\tau_p}\}, \tag{7}$$

where α is a hyper-parameter to reduce the deviation caused by the semantic transfer process.

3.5   Feature Generation

When we have got the semantics of the internal classes, we will use them to replace the original semantic of each sample. We use the replaced semantics to train the Generative Adversarial Networks. The internal class semantics η and Gaussian noise z are employed to generate visual feature through generator G:

$$\tilde{x} = G(\eta, z), \tag{8}$$

where $z \sim N(0,1)$ is the Gaussian noise. The generated visual features and the real visual features are discriminated by discriminator D, which is optimized with WGAN [27]. The total loss is as follows:

$$Loss_{gan} = \mathbb{E}[D(x, \eta)] - \mathbb{E}[D(\tilde{x}, \eta)] - \lambda \mathbb{E}\left[(\|\nabla_{\hat{x}} D(\hat{x}, \eta)\|_2 - 1)^2\right], \tag{9}$$

where η is the internal class semantics of the generated sample $\tilde{x}$, and it is also the internal class semantic corresponding to the real visual feature x. The last one is the gradient penalty, where $\hat{x} = \alpha x + (1 - \alpha)\tilde{x}$ with $\alpha \sim U(0;1)$ and λ is the penalty coefficient.

At the same time, in order to make the generated samples more authentic, we follow f-CLSWGAN [40] to constrain the classification loss:

$$Loss_{cls} = -\mathbb{E}[log P(y|\tilde{x}; \theta)], \tag{10}$$

where y is the real class of $\tilde{x}$, not the internal class. θ is the parameter of the pre-trained classifier on the seen class. It is worthy noted that we do not use internal category tags for classification. Then the total loss is:

$$Loss_{total} = Loss_{gan} + \beta Loss_{cls}, \tag{11}$$

where β is the balancing coefficient.

3.6 Remove Irrelevant Features

The trained mapper can map the visual features to the semantic space. This mapping eliminates irrelevant information which is useless or even harmful to the final classification in a certain extent. We know that the generator can generate visual features for the corresponding semantics. Therefore, the mapped semantics and the reconstructed visual features obtained by the generator are free of irrelevant information. Based on this, we get reconstructed visual features:

$$\check{x} = G\left(E\left(x\right);0\right), \tag{12}$$

where x contains the features of seen class and the generated features of unseen class. It is noted that the Gaussian noise during generation is set to 0.

3.7 Classification

Now we have obtained the visual features of seen classes, the visual features of the generated unseen class and the reconstructed visual features of both. Considering the bias of the mapper to the seen class, in order to prevent the generated unseen reconstructed visual features from losing classification information in the process of reconstruction, we concatenate the reconstructed visual features with the original visual features for the training of the final classifier. Figure 2 shows the overall process. Let $\bar{x}$ represent the features after concatenating as $\bar{x} = x \oplus \check{x}$, where $\oplus$ represent the feature concatenation, the finally classifier loss is:

$$Loss_{final} = -\mathbb{E}\left[logP\left(y|\bar{x}, \theta_f\right)\right], \tag{13}$$

where θ_f is classifier parameters. Note that in the process of classification, we use the original category label. In other words, we classify each visual feature into corresponding category instead of internal category.

4 Experiments

4.1 Datasets and Setting

We evaluated our method on four datasets. **AWA2** [39] contains 37322 instances of 50 classes, and **aPY** [8] has 32 classes with a total of 15339 instances. The fine-grained dataset **CUB** [38] contains 11788 bird instances of 200 classes, and **SUN**

[31] contains 14340 instances of 717 classes. For all datasets, we use 2048 dimensional visual features extracted with ResNet-101 [15]. It should be noted that we use the newly extracted 1024 dimensional semantic attribute for CUB [32]. There are three hyper-parameters. We set $\alpha = 0.5$ for coarse-grained datasets, $\alpha = 0.2$ for fine-grained datasets, and $\beta = 0.01$ for all. It is worthy noted that all hyper-parameters are obtained with cross validation. To be specific, we separate a certain number of the seen classes as the validational unseen. For example, we randomly divide 40 seen classes in AWA2 into 30 seen classes and 10 validational unseen classes multiple times, and select the hyper-parameters that can achieve the best mean performance for final training. Although this operation is a bit different from k-fold cross-validation, it is an effective way for ZSL hyper-parameter selection. In addition, to increase the generalization ability, L_2 regularization is added to train the generator.

4.2 Comparison with Baselines

Table 1 shows the performance comparison of our proposed method with other methods. It can be seen that on the coarse-grained datasets, the results we have obtained are higher than the models proposed in recent years, achieving the state-of-the-art performance, and our result is 70.3% for AWA2 and 45.4% for aPY. For fine-grained datasets, we obtained quite good results on CUB, which is 65.4%. For SUN, the result is not the best, but it remains at the average level.

It should be noted that the internal categories of unseen classes are obtained based on the principle of Convergent Evolution, which is biological. SUN and aPY do not belong to the biological category datasets. This shows that our proposed category expansion method is still effective on a non-biological dataset, which means that the internal categories between similar categories also have the same characteristic on non-biological datasets.

4.3 Verification on Other Generative Models

In order to verify that the Category Expansion we proposed is generally applicable, we use the semantics of expanded categories to replace the original semantics, and then apply them to other generative models. We verify this strategy on three classical generative models, including F-VAEGAN-D2 [41], RFF-GZSL [14] and CE-GZSL [13].

In order to fairly compare the impact of using the new semantics and original semantics on the performance of different models, we reproduce the above three models. We follow the parameters provided in the three articles, but because some parameters are not provided and the experimental platform is different, the reproduced results are a bit different from the original reported results.

For f-VAEGAN-D2, we use 1024 dimensional semantic attributes in the CUB dataset, and other settings follow the parameters given in the paper. For RFF-GZSL, we set batch size to 512. For CE-GZSL, because its batch size has a great impact on performance and has high requirements for GPU card, we cannot set

Table 1. The results on four datasets. U represents the accuracy of the unseen class, S represents the accuracy of the seen class, and H represents the harmonic mean. The best value of each column is in bold, and '–' means not reported.

Method	AWA2			aPY			SUN			CUB		
	U	S	H	U	S	H	U	S	H	U	S	H
f-CLSWGAN [40]	–	–	–	32.9	61.7	42.9	42.6	36.6	39.4	43.7	57.7	49.7
RFF-GZSL [14]	–	–	–	–	–	–	45.7	38.6	41.9	52.6	56.6	54.6
cycle-CLSWGAN [9]	–	–	–	–	–	–	49.4	33.6	40.0	45.7	61.0	52.3
IZF [35]	60.6	77.5	68.0	–	–	–	**52.7**	57.0	**54.8**	52.7	**68.0**	59.4
GDAN [16]	32.1	67.5	43.5	30.4	**75.0**	43.4	38.1	89.9	53.4	39.3	66.7	49.5
CE-GZSL [13]	63.1	78.6	70.0	–	–	–	48.8	38.6	43.1	63.9	66.8	65.3
GCM-CF [44]	60.4	75.1	67.0	**37.1**	56.8	44.9	47.9	37.8	42.2	61.0	59.7	60.3
LisGAN [22]	–	–	–	–	–	–	42.9	37.8	40.2	46.5	57.9	51.6
FREE [6]	60.4	75.4	67.1	–	–	–	47.4	37.2	41.7	55.7	59.9	57.7
SE-GZSL [19]	**80.7**	59.9	68.8	–	–	–	40.7	45.8	43.1	60.3	53.1	56.4
HSVA [7]	59.3	76.6	66.8	–	–	–	48.6	39.0	43.3	52.7	58.3	55.3
E-PGN [43]	52.6	**86.5**	64.6	–	–	–	–	–	–	52.0	61.1	56.2
Disentangled-VAE [23]	56.9	80.2	66.6	–	–	–	36.6	47.6	41.4	54.1	58.2	54.4
Ours	67.0	73.9	**70.3**	33.5	70.3	**45.4**	48.5	36.6	41.7	**68.3**	62.8	**65.4**

Table 2. The results of expanded categories on three baseline models. The upper part is the result obtained by using the original semantics, and the lower part is the result obtained using the internal class semantics. 'EC' stands for Expanded Categories.

Method	AWA2			aPY			SUN			CUB		
	U	S	H	U	S	H	U	S	H	U	S	II
f-VAEGAN-D2	57.5	**68.7**	62.6	**32.9**	61.7	42.9	44.0	**39.8**	41.8	64.0	**67.0**	65.5
f-VAEGAN-D2+EC	**59.8**	68.5	**63.8**	31.7	**67.2**	**43.1**	**49.0**	37.3	**42.4**	**69.0**	64.9	**66.9**
CE-GZSL	57.0	74.9	64.7	9.85	**88.4**	17.3	40.9	**35.4**	37.9	66.3	**66.6**	66.4
CE-GZSL+EC	**62.5**	**75.1**	**68.2**	**35.0**	57.3	**43.5**	**49.7**	32.3	**39.5**	**67.6**	66.3	**66.9**
RFF-GZSL	54.1	**77.7**	63.8	21.0	**87.5**	33.8	42.6	**37.8**	40.0	66.3	63.1	64.7
RFF-GZSL+EC	**60.1**	72.2	**65.6**	**31.6**	71.8	**43.9**	**47.4**	36.4	**41.2**	**67.3**	**63.3**	**65.2**

the same batch size given in the paper, which makes the deviation of reproduction results too large. We set the batch size to 128.

However, the above parameter settings do not affect our verification because we follow the method of fixed variables. After replacing the original semantics, the parameters of the model are not changed, and the experimental results are compared only on the basis of modifying semantics.

Table 2 shows our experimental results. It can be seen that for coarse-grained datasets AWA2 [21] and aPY, the performance has been significantly improved after replacing semantics. For fine-grained datasets, although the performance is also improved, the effect is not as obvious as that of coarse-grained datasets. The classification on fine-grained datasets is meticulous, if each class is divided into new internal classes, the semantic difference of internal classes is also limited. For this reason, we think the result is reasonable. The experimental results fully

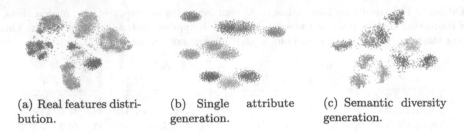

(a) Real features distribution.   (b) Single attribute generation.   (c) Semantic diversity generation.

Fig. 3. t-SNE illustration of visual features generated by traditional single attribute generation method and attribute diversity method.

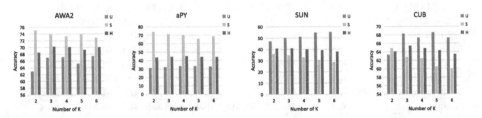

Fig. 4. Results under different number of internal classes.

prove that our proposed Category Expansion method can be applied to different generation models as a general method and improve the performance of the model.

4.4 Feature Generation

In order to verify that the generated visual features are more consistent with the real visual feature distribution, we visualize the generated visual features. Figure 3 (a) shows the distribution of real unseen visual features under t-SNE. Figure 3 (b) shows the unseen visual features generated by traditional single semantics, to be noted that here we use the classical f-CLSWGAN model. Figure 3 (c) shows the visual features generated by our proposed method. It can be seen that for the visual features generated by a single semantic, each category is close to a regular ellipse. The visual features generated by our category expansion method are more irregular, which is in line with the irregular distribution of real visual feature shown in Fig. 3 (a).

4.5 Number of Internal Classes

Different clustering centers also have different impact on final performance. Figure 4 shows the experimental results for different numbers of internal classes. We can see that the best results are $k = 3$ on AWA2, $k = 4$ on aPY, $k = 3$ on CUB and $k = 3$ on SUN. It can seen that after the accuracy H reaches the highest point, the influence of the value of k on the coarse-grained datasets

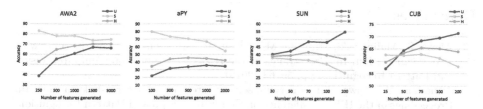

Fig. 5. Results of generating different numbers of unseen visual features.

begin to decrease and tend to be stable. But it has a negative impact on fine-grained datasets. Coarse-grained datasets have higher tolerance for the division of internal classes, because coarse-grained datasets have more space for the division of internal classes, while the fine-grained datasets have less space for class division, and the wrong internal class division will have a certain impact on the performance of the final classifier.

4.6   Number of Features Generated

The number of unseen visual features generated has an impact on the final experimental results. Figure 5 shows the comparison of results under different numbers of visual features generated. We can see that with the increase of the generated number, the accuracy of seen classes shows a downward trend, and the accuracy of unseen classes shows an upward trend. When the number of visual features generated is 3000 on AWA2, 500 on aPY, 70 on SUN and 75 on CUB, the highest accuracy is achieved due to the balance between the number of unseen and seen classes.

5   Conclusion

In this paper, we have discussed the problem of generating diverse feature with a single semantic in generative ZSL. The visual features generated by traditional semantics do not accord with the distribution of real visual features. Therefore, we have proposed a category expansion method based on Relative Character, and extend the results of semantic augmentation of seen classes to unseen classes based on Convergent Evolution. At the same time, we have employed a simple and efficient way to eliminate irrelevant information of visual features. Our method has achieved good performance on four benchmark datasets. On this basis, we have tested the semantic augmentation module as a general method on three generative ZSL methods, and verified that this semantic augmentation is generally applicable and can improve the performance of generative ZSL.

Acknowledgements. This work was supported in part by the National Natural Science Foundation of China (NSFC) under Grants No. 61872187, No. 62072246 and No. 62077023, in part by the Natural Science Foundation of Jiangsu Province under Grant No. BK20201306, and in part by the "111" Program under Grant No. B13022.

References

1. Akata, Z., Malinowski, M., Fritz, M., Schiele, B.: Multi-cue zero-shot learning with strong supervision. In: Proceedings of the IEEE Conference on Computer Vision and Pattern Recognition, pp. 59–68 (2016)
2. Annadani, Y., Biswas, S.: Preserving semantic relations for zero-shot learning. In: Proceedings of the IEEE Conference on Computer Vision and Pattern Recognition, pp. 7603–7612 (2018)
3. Chao, W.-L., Changpinyo, S., Gong, B., Sha, F.: An empirical study and analysis of generalized zero-shot learning for object recognition in the wild. In: Leibe, B., Matas, J., Sebe, N., Welling, M. (eds.) ECCV 2016. LNCS, vol. 9906, pp. 52–68. Springer, Cham (2016). https://doi.org/10.1007/978-3-319-46475-6_4
4. Chen, L., Zhang, H., Xiao, J., Liu, W., Chang, S.F.: Zero-shot visual recognition using semantics-preserving adversarial embedding networks. In: Proceedings of the IEEE Conference on Computer Vision and Pattern Recognition, pp. 1043–1052 (2018)
5. Chen, S., et al.: MSDN: mutually semantic distillation network for zero-shot learning. arXiv preprint arXiv:2203.03137 (2022)
6. Chen, S., et al.: FREE: feature refinement for generalized zero-shot learning. In: Proceedings of the IEEE/CVF International Conference on Computer Vision, pp. 122–131 (2021)
7. Chen, S., et al.: HSVA: hierarchical semantic-visual adaptation for zero-shot learning. In: Advances in Neural Information Processing Systems, vol. 34 (2021)
8. Farhadi, A., Endres, I., Hoiem, D., Forsyth, D.: Describing objects by their attributes. In: 2009 IEEE Conference on Computer Vision and Pattern Recognition, pp. 1778–1785. IEEE (2009)
9. Felix, R., Vijay Kumar, B.G., Reid, I., Carneiro, G.: Multi-modal cycle-consistent generalized zero-shot learning. In: Ferrari, V., Hebert, M., Sminchisescu, C., Weiss, Y. (eds.) ECCV 2018. LNCS, vol. 11210, pp. 21–37. Springer, Cham (2018). https://doi.org/10.1007/978-3-030-01231-1_2
10. Fu, Z., Xiang, T., Kodirov, E., Gong, S.: Zero-shot object recognition by semantic manifold distance. In: Proceedings of the IEEE Conference on Computer Vision and Pattern Recognition, pp. 2635–2644 (2015)
11. Geng, Y., Chen, J., Ye, Z., Yuan, Z., Zhang, W., Chen, H.: Explainable zero-shot learning via attentive graph convolutional network and knowledge graphs. Semantic Web (Preprint), 1–28 (2020)
12. Goodfellow, I., et al.: Generative adversarial nets. In: Advances in Neural Information Processing Systems, vol. 27 (2014)
13. Han, Z., Fu, Z., Chen, S., Yang, J.: Contrastive embedding for generalized zero-shot learning. In: Proceedings of the IEEE/CVF Conference on Computer Vision and Pattern Recognition, pp. 2371–2381 (2021)
14. Han, Z., Fu, Z., Yang, J.: Learning the redundancy-free features for generalized zero-shot object recognition. In: Proceedings of the IEEE/CVF Conference on Computer Vision and Pattern Recognition, pp. 12865–12874 (2020)
15. He, K., Zhang, X., Ren, S., Sun, J.: Deep residual learning for image recognition. In: Proceedings of the IEEE Conference on Computer Vision and Pattern Recognition, pp. 770–778 (2016)
16. Huang, H., Wang, C., Yu, P.S., Wang, C.D.: Generative dual adversarial network for generalized zero-shot learning. In: Proceedings of the IEEE/CVF Conference on Computer Vision and Pattern Recognition, pp. 801–810 (2019)

17. Huynh, D., Elhamifar, E.: Fine-grained generalized zero-shot learning via dense attribute-based attention. In: Proceedings of the IEEE/CVF Conference on Computer Vision and Pattern Recognition, pp. 4483–4493 (2020)
18. Jiang, H., Wang, R., Shan, S., Chen, X.: Learning class prototypes via structure alignment for zero-shot recognition. In: Ferrari, V., Hebert, M., Sminchisescu, C., Weiss, Y. (eds.) ECCV 2018. LNCS, vol. 11214, pp. 121–138. Springer, Cham (2018). https://doi.org/10.1007/978-3-030-01249-6_8
19. Kim, J., Shim, K., Shim, B.: Semantic feature extraction for generalized zero-shot learning. arXiv preprint arXiv:2112.14478 (2021)
20. Lampert, C.H., Nickisch, H., Harmeling, S.: Learning to detect unseen object classes by between-class attribute transfer. In: 2009 IEEE Conference on Computer Vision and Pattern Recognition, pp. 951–958. IEEE (2009)
21. Lampert, C.H., Nickisch, H., Harmeling, S.: Attribute-based classification for zero-shot visual object categorization. IEEE Trans. Pattern Anal. Mach. Intell. **36**(3), 453–465 (2013)
22. Li, J., Jing, M., Lu, K., Ding, Z., Zhu, L., Huang, Z.: Leveraging the invariant side of generative zero-shot learning. In: Proceedings of the IEEE/CVF Conference on Computer Vision and Pattern Recognition, pp. 7402–7411 (2019)
23. Li, X., Xu, Z., Wei, K., Deng, C.: Generalized zero-shot learning via disentangled representation. In: Proceedings of the AAAI Conference on Artificial Intelligence, vol. 35, pp. 1966–1974 (2021)
24. Liu, Y., Guo, J., Cai, D., He, X.: Attribute attention for semantic disambiguation in zero-shot learning. In: Proceedings of the IEEE/CVF International Conference on Computer Vision, pp. 6698–6707 (2019)
25. Long, Y., Liu, L., Shao, L., Shen, F., Ding, G., Han, J.: From zero-shot learning to conventional supervised classification: unseen visual data synthesis. In: Proceedings of the IEEE Conference on Computer Vision and Pattern Recognition, pp. 1627–1636 (2017)
26. Ma, P., Hu, X.: A variational autoencoder with deep embedding model for generalized zero-shot learning. In: Proceedings of the AAAI Conference on Artificial Intelligence, vol. 34, pp. 11733–11740 (2020)
27. Arjovsky, M., Chintala, S., Bottou, L.: Wasserstein GAN. In: Proceedings of ICML 2017 (2017)
28. Ni, J., Zhang, S., Xie, H.: Dual adversarial semantics-consistent network for generalized zero-shot learning. In: Advances in Neural Information Processing Systems, vol. 32 (2019)
29. Palatucci, M., Pomerleau, D., Hinton, G.E., Mitchell, T.M.: Zero-shot learning with semantic output codes. In: Advances in Neural Information Processing Systems, vol. 22 (2009)
30. Palatucci, M., Pomerleau, D., Hinton, G.E., Mitchell, T.M.: Zero-shot learning with semantic output codes. In: Advances in Neural Information Processing Systems, vol. 22 (2009)
31. Patterson, G., Xu, C., Su, H., Hays, J.: The sun attribute database: beyond categories for deeper scene understanding. Int. J. Comput. Vision **108**(1), 59–81 (2014)
32. Reed, S., Akata, Z., Lee, H., Schiele, B.: Learning deep representations of fine-grained visual descriptions. In: Proceedings of the IEEE Conference on Computer Vision and Pattern Recognition, pp. 49–58 (2016)
33. Romera-Paredes, B., Torr, P.: An embarrassingly simple approach to zero-shot learning. In: International Conference on Machine Learning, pp. 2152–2161. PMLR (2015)

34. Schonfeld, E., Ebrahimi, S., Sinha, S., Darrell, T., Akata, Z.: Generalized zero- and few-shot learning via aligned variational autoencoders. In: Proceedings of the IEEE/CVF Conference on Computer Vision and Pattern Recognition, pp. 8247–8255 (2019)

35. Shen, Y., Qin, J., Huang, L., Liu, L., Zhu, F., Shao, L.: Invertible zero-shot recognition flows. In: Vedaldi, A., Bischof, H., Brox, T., Frahm, J.-M. (eds.) ECCV 2020. LNCS, vol. 12361, pp. 614–631. Springer, Cham (2020). https://doi.org/10.1007/978-3-030-58517-4_36

36. Stern, D.L.: The genetic causes of convergent evolution. Nat. Rev. Genet. **14**(11), 751–764 (2013)

37. Wang, X., Ye, Y., Gupta, A.: Zero-shot recognition via semantic embeddings and knowledge graphs. In: Proceedings of the IEEE Conference on Computer Vision and Pattern Recognition, pp. 6857–6866 (2018)

38. Welinder, P., Branson, S., Mita, T., Wah, C., Schroff, F., Belongie, S., Perona, P.: Caltech-UCSD birds 200 (2010)

39. Xian, Y., Lampert, C.H., Schiele, B., Akata, Z.: Zero-shot learning-a comprehensive evaluation of the good, the bad and the ugly. IEEE Trans. Pattern Anal. Mach. Intell. **41**(9), 2251–2265 (2018)

40. Xian, Y., Lorenz, T., Schiele, B., Akata, Z.: Feature generating networks for zero-shot learning. In: Proceedings of the IEEE Conference on Computer Vision and Pattern Recognition, pp. 5542–5551 (2018)

41. Xian, Y., Sharma, S., Schiele, B., Akata, Z.: F-VAEGAN-D2: a feature generating framework for any-shot learning. In: Proceedings of the IEEE/CVF Conference on Computer Vision and Pattern Recognition, pp. 10275–10284 (2019)

42. Xie, G.S., et al.: Attentive region embedding network for zero-shot learning. In: Proceedings of the IEEE/CVF Conference on Computer Vision and Pattern Recognition, pp. 9384–9393 (2019)

43. Yu, Y., Ji, Z., Han, J., Zhang, Z.: Episode-based prototype generating network for zero-shot learning. In: Proceedings of the IEEE/CVF Conference on Computer Vision and Pattern Recognition, pp. 14035–14044 (2020)

44. Yue, Z., Wang, T., Sun, Q., Hua, X.S., Zhang, H.: Counterfactual zero-shot and open-set visual recognition. In: Proceedings of the IEEE/CVF Conference on Computer Vision and Pattern Recognition, pp. 15404–15414 (2021)

45. Zhang, L., Xiang, T., Gong, S.: Learning a deep embedding model for zero-shot learning. In: Proceedings of the IEEE Conference on Computer Vision and Pattern Recognition, pp. 2021–2030 (2017)

ADEL: Adaptive Distribution Effective-Matching Method for Guiding Generators of GANs

Jungeun Kim[1] (ID), Jeongeun Park[2] (ID), and Ha Young Kim[2](✉) (ID)

[1] Department of AI, Yonsei University, Seoul, South Korea
jekim5418@yonsei.ac.kr
[2] Graduate School of Information, Yonsei University, Seoul, South Korea
{park.je,hayoung.kim}@yonsei.ac.kr

Abstract. Research on creating high-quality, realistic fake images has engendered immense improvement in GANs. However, GAN training is still subject to mode collapse or vanishing gradient problems. To address these issues, we propose an adaptive distribution effective-matching method (ADEL) that sustains the stability of training and enables high performance by ensuring that the training abilities of the generator and discriminator are maintained in balance without bias in either direction. ADEL can help the generator's training by matching the difference between the distribution of real and fake images. As training is ideal when the discriminator and generator are in a balanced state, ADEL works when it is out of a certain optimal range based on the loss value. Through this, ADEL plays an important role in guiding the generator to create images similar to real images in the early stage when training is difficult. As training progresses, it naturally decays and gives model more freedom to generate a variety of images. ADEL can be applied to a variety of loss functions such as Kullback-Liebler divergence loss, Wasserstein loss, and Least-squares loss. Through extensive experiments, we show that ADEL improves the performance of diverse models such as DCGAN, WGAN, WGAN-GP, LSGAN, and StyleGANv2 upon five datasets, including low-resolution (CIFAR-10 and STL-10) as well as high-resolution (LSUN-Bedroom, Church, and ImageNet) datasets. Our proposed method is very simple and has a low computational burden, so it is expandable and can be used for diverse models.

Keywords: Deep learning · Generative adversarial networks · Training stabilization · Probability distribution matching

J. Kim and J. Park—These authors contributed equally.
This work was supported by the Korea Agency for Infrastructure Technology Advancement (KAIA) grant funded by the Ministry of Land, Infrastructure and Transport (22CTAP-C163908-02). This work was also supported by the Institute of Information & Communications Technology Planning & Evaluation (IITP) grant funded by the Korea government (MSIT) (No. 2020-0-01361, Artificial Intelligence Graduate School Program (Yonsei University)).

Supplementary Information The online version contains supplementary material available at https://doi.org/10.1007/978-3-031-26293-7_3.

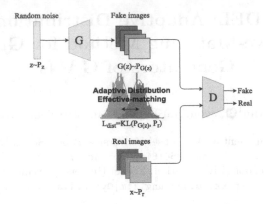

Fig. 1. Overall framework of ADEL. G denotes a generator and D is a discriminator. During training, when the discriminator loss is out of a certain optimal range, the distribution matching loss (L_{dist}) is applied, which encourages generator to match the actual data distribution. The generator trains based on this L_{dist} as well as L_G.

1   Introduction

Generative Adversarial Network (GAN), which is composed of a generator and discriminator, has been the focus of considerable research since it was first proposed by Ian Goodfellow [1]. GANs have demonstrated remarkable achievements in the field of computer vision [2–6] and continue to develop rapidly by being applied in the fields of natural language processing [7,8] and audio generation [9,10]. However, they still face many challenges. In particular, it is difficult to train a GAN stably, and mode collapse or vanishing gradient can often occur. To address these challenges, several methods have been devised to allow the model to learn the boundaries of the entire data distribution evenly and to keep it memorized [11–13].

Furthermore, helping the discriminator and the generator to learn evenly when training the GAN can considerably improve its performance. Therefore, various methods have tried to balance the training between the discriminator and generator [14–17]. For example, Wasserstein-GAN [18] applied Earth-Mover (EM) distance to a GAN and tried to stabilize it by clipping weights during back-propagation. BEGAN [19] used the auto-encoder instead of the discriminator to match the distribution of loss to allow the two networks to be in equilibrium. For EBGAN [19] and χ^2GAN [20], the distribution of data was matched to achieve Nash equilibrium during training. However, despite these efforts, maintaining training balance between a generator and discriminator is still challenging. In addition, most of these studies focus on adjusting the discriminator, or controlling the data distribution.

We propose a new training method, adaptive distribution effective-matching method (ADEL), to stabilize GAN training and improve its performance. Figure 1 shows our overall framework. Through the newly introduced distribution matching loss (L_{dist}), ADEL encourages the training of the generator by matching the distribution between the real images and the fake images created by the generator. By providing hints to the generator about the real image distribution so that the performance of the two networks is not skewed to one side, ADEL guides the GAN to create images that are more "realistic". However, this distribution matching method may interfere with the generator's degree of freedom. That is, the model to be trained may not cover the entire actual data distribution and may experience a loss of diversity, and vanishing gradient or a mode collapse may occur. Therefore, in this study, the level of distribution matching loss was adjusted as the training progressed, eventually allowing the generator to freely train without any constraints to cover the actual data distributions.

We demonstrate through comprehensive experiments that GAN training can be stabilized and better performance can be achieved by reducing the difference between the distribution of the fake images produced by the generator and the real images. In the Results section, we report the Inception Score (IS) [11], Fréchet Inception Score (FID) [21], and Kernel Inception Distance (KID) [22] scores to show quantitative results, while generated images demonstrating qualitative results are given in Experiments section.

Our main contributions are as follows:

- We propose a novel distribution matching method to guide generator's training of GANs. This method has a regularization effect, which helps the GAN to be trained stably.
- ADEL is a straightforward and effective method that can be easily applied to existing GAN models with a low computational burden.
- We show that ADEL can provide good performance in GANs with various losses and in small- as well as large-scale datasets.

2 Related Work

Various attempts have been made in many studies to address the challenges faced by GAN, such as that experiences mode collapse or the gradient of the generator is not updated because the loss of the discriminator rapidly converges to 0. Among them, some methods that change the model's structure or modify the loss function are stated below.

2.1 GAN Architectures

DCGAN [15] dramatically improved a GAN's image generation ability by applying the structure of convolutional neural networks (CNNs) to the GAN. Owing to the advantages of CNN, DCGAN produces images with stable yet high resolution. Progressive GAN (PGGAN) [2] creates a high-resolution image by starting from a low-resolution image and gradually increasing the image resolution by adding network layers. Through the method of gradually stacking layers, learning stabilization and time reduction were achieved. However, PGGAN has limited ability to control specific detailed features, such as the color of the hair or eyes of the generated image.

This limitation is overcome by StyleGAN [4]. StyleGAN adds a mapping network and Adaptive Instance Normalization to the generator to create more realistic fake images. After that, StyleGANv2 [23] was recently introduced, which resolved the artifacts problem of StyleGAN, and contributed astonishing improvement of the generated images quality. SAGAN [24] applied a self-attention module in a GAN, which improved its performance in natural language processing considerably. This proves that it is possible to effectively model long-range dependencies within the internal representations of images, and thus this approach can outperform existing models. Zhang et al. also showed that spectral normalization stabilizes the training of GANs, and that the two time-scale update rule improves the training speed of the discriminator. BigGAN [5] also improved SAGAN structurally; they showed that a higher performance could be achieved only by increasing the number of channels and batch size of SAGAN. Furthermore, shared embedding, hierarchical latent space, and orthogonal regularization methods were added to provide even better performance. However, there is still room for improvement in stabilizing training. In addition to these studies, bi-directional GANs have been proposed, which introduce a bidirectional mechanism to generate latent space z from images [25]. There have been studies in which the number of generators or discriminators used is different [26,27]. The adversarial auto-encoder takes GAN a step further by combining a variational auto-encoder with the GAN [28]. A recent study attempted to improve GAN's performance by applying a U-Net-based [29] architecture to the discriminator [30]. Another study achieved SOTA performance by applying the structure of a transformer to a GAN [31].

2.2   GAN Loss Functions

In the conventional GAN, the sigmoid cross entropy loss function was used for the discriminator. However, this loss function causes a vanishing gradient problem when updating the generator, which cannot be resolved by a structural change to the model. However, approaches that modified the loss function of the GAN appeared to alleviate this problem. Least squares GAN (LSGAN) [32] allows the generator to create images similar to real images by assigning a penalty to samples far from the decision boundary using least squares loss. By simply changing the loss function of GAN, LSGAN not only improves the quality of the generated image, but also enables stable training. While LSGAN uses the least squares loss, Banach Wasserstein GAN [33] simply replaces this least square norm with a dual norm, providing a new degree of freedom to select the appropriate Sobolev space, allowing WGAN to be placed in the Banach space. Relativistic GAN [34] modifies the loss as a relative evaluation by applying a probability indicating how realistic the real data is to the generated image. Another approach improves the performance of the model by applying the Hinge loss to the GAN [4,35,36]. As such, various studies have investigated the loss function of the GAN as a way to solve the vanishing gradient problem and eliminate mode collapse.

Table 1. Performance comparison of baseline models with and without ADEL method applied. Large and small values are preferred for ↑ and ↓ respectively

Model	Dataset	Method	IS (↑)	FID (↓)	KID (↓)	Memory (MB)	Training Time (sec/iteration)
DCGAN	CIFAR-10	Baseline	**5.47±0.09**	20.91	0.0100±0.0005	1604	0.07
		ADEL	5.32±0.23	**20.74**	**0.0085±0.0005**	2113	0.08
	STL-10	Baseline	4.70±0.24	63.86	0.0385±0.0009	2012	0.13
		ADEL	**5.03±0.19**	**60.37**	**0.0361±0.0009**	3945	0.16
WGAN	CIFAR-10	Baseline	5.22±0.07	20.23	**0.0076±0.0005**	2025	0.12
		ADEL	**5.23±0.17**	**19.26**	0.0081±0.0005	2031	0.13
	STL-10	Baseline	5.34±0.13	38.67	0.0194±0.0006	3795	0.23
		ADEL	**5.46±0.16**	**37.91**	**0.0190±0.0005**	3949	0.25
WGAN-GP	CIFAR-10	Baseline	5.24±0.08	23.75	0.0129±0.0006	2333	0.13
		ADEL	**5.46±0.09**	**23.08**	**0.0126±0.0006**	2462	0.13
	STL-10	Baseline	4.97±0.11	55.36	0.0360±0.0006	3807	0.23
		ADEL	**4.97±0.14**	**54.91**	**0.0336±0.0006**	3936	0.24
LSGAN	LSUN-B	Baseline	-	63.27	0.0663±0.0007	3864	0.74
		ADEL	-	**40.44**	**0.0359±0.0005**	4064	0.76
	ImageNet	Baseline	8.05±0.22	**46.30**	0.0354±0.0006	3864	1.01
		ADEL	**8.21±0.33**	46.32	**0.0343±0.0006**	4054	1.06

3   Methodology

The purpose of the generator and discriminator in GAN is to generate fake images that seem real and to distinguish between real and fake images, respectively. In general, while training a GAN, the discriminator converges faster than the generator, and this causes the vanishing gradient problem [37,38]. Based on this observation, we propose an adaptive distribution matching method. Our method is not only easy to apply by adding few lines of code but also simple to understand.

3.1   ADEL

When training a GAN, the generator creates fake images using the latent vector as input while the discriminator D distinguishes between the real images and

Algorithm 1. Overall ADEL algorithm

1: **Input:** max iteration T, batch size m, dimension of latent vector d, learning rate
   η, weight of generator loss α, weight of discriminator loss β, global optimal
   solution δ, optimal range ϵ
2: Initialize θ_G and θ_D, i.e., the parameters of the generator and discriminator,
   respectively
3: **for** $c = 1 \cdots T$ **do**
4:     Sample random latent vector $z \sim N(0,1)$, $z \in \mathbb{R}^d$
5:     Sample a batch of real images $r \sim P_r$, $r \in \mathbb{R}^m$
6:     Update $\theta_D \leftarrow \theta_D - \eta * \nabla_{\theta_D}(\beta * L_D)$
7:     $\gamma, L_{dist} \leftarrow 0$
8:     **if** $|L_D - \delta| > \epsilon$ **then**
9:         $P_{G(z)}, P_r \leftarrow Softmax(G(z)), Softmax(r)$
10:        $\gamma \leftarrow \exp^{-(c/T)}$
11:        $L_{dist} \leftarrow D_{KL}(P_{G(z)}, P_r)$
12:    **else**
13:        $pass$
14:    **end if**
15:    Update $\theta_G \leftarrow \theta_G - \eta * \nabla_{\theta_G}(\alpha * L_G + \gamma * L_{dist})$
16: **end for**

the fake images generated by the generator G, i.e., it identifies which images are
real and which are fake. The loss function of a general GAN is as Eq. 1.

$$L_G = \mathbb{E}_{z \sim P_z(z)}[log(1 - D(G(z)))]$$
$$L_D = \mathbb{E}_{r \sim P_r(r)}[logD(r)] + \mathbb{E}_{z \sim P_z(z)}[log(1 - D(G(z)))]$$
(1)

Here, r is a real data sample, P_r is a real data distribution, z is a random latent
vector, P_z is a standard normal distribution, L_G is a generator loss, and L_D is
a discriminator loss.

The discriminator only needs to determine whether a given image is real or fake
(1 or 0), but the generator needs to generate entire images (e.g. $28 \, times \, 28 \times 3$
[H × W × C]), so there exists a difference in the relative difficulties of the two tasks.
That is, the task of the generator that needs to generate a complex image is more
laborious than that of the discriminator in deciding whether an input image is
real. Furthermore, given that the generator's training depends only on the decision of the discriminator, the information that the generator can obtain is limited. In the view of the aforementioned two reasons, it becomes difficult for the
two networks to maintain a balance during training. Therefore, we provide additional information about the distribution of real images to the generators performing tasks that are more difficult. To this end, the proposed method introduces
a distribution matching loss (L_{dist}) described in Eq. 2, based on matching the

distribution between real images and fake images to balance the evenness of the training of the generator and discriminator.

$$L_{dist} = \begin{cases} D_{KL}(P_{G(z)}||P_r) = \mathbb{E}_{x \sim P_{G(z)}}[log\frac{P_{G(z)}(x)}{P_r(x)}] & \text{, if } |L_D - \delta| > \epsilon \\ 0 & \text{, otherwise} \end{cases} \quad (2)$$

L_{dist} is computed using the Kullback-Liebler (KL) divergence (D_{KL}) to match between real and generated data distribution, $P_{G(z)}$. ϵ is a predefined constant and δ is the global optimal solution of each GAN model, i.e. the discriminator's value at $P_{G(z)} = P_r$. Those will be described in detail below.

In this study, we define the objective functions of the generator ($J(\theta_G)$) and the discriminator ($J(\theta_D)$) as given below in Eq. 3.

$$J(\theta_G) = \alpha * L_G(\theta_G) + \gamma * L_{dist}(\theta_G)$$
$$J(\theta_D) = \beta * L_D(\theta_D) \quad (3)$$

α, β, and γ are weights to the generator, discriminator, and distribution matching loss. Also, θ_G and θ_D are the parameters of the generator and discriminator, respectively. During distribution matching, the ratio of L_G, L_D, and L_{dist} and whether to perform distribution matching is an essential issue. The value used as standard for distribution matching is also an important concern. To address these concerns, we propose a distribution matching method that considers not only the generator's information, using generated data distribution, but also the discriminator's information. Therefore, we introduce ϵ as the criterion for the difference between L_{dist} and δ. This serves as a threshold to measure how poorly the generator has trained. In Eq. 2, when the discriminator loss exceeds the range of $[\delta-\epsilon, \delta+\epsilon]$, L_{dist} exists. The general GAN loss consists of L_G and L_D, and the training of the discriminator dominates, so that L_D quickly converges to zero. However, in ADEL, when L_D is within the above-specified range, there exists an L_{dist} value. Therefore, in this scenario, the generator learns based on L_{dist} as well as L_G.

In Eq. 3, if the value of γ is increasing, then the generator will focus more on matching the distribution between real and fake data and interfere the degree of freedom available to the generator. Therefore, we design γ as an adaptive function based on the training iteration. The function γ is defined as follows.

$$\gamma := \gamma(c) = \exp^{-(c/T)} \quad (4)$$

where c is a current iteration and T is a max iteration. In Eq. 4, γ exponentially decreases as training progresses. As it is a large value at the beginning of the training, γ gives the generator a lot of information to guide it to generate similar images. When an iteration proceeds by decreasing the ratio of distribution matching loss, it gives the generator the degree of freedom to create various images, which is one of the goals of GAN models. To see the effect of L_{dist}, we fixed α and β as 1. In addition, the distribution of real and generated data (P_r and $P_{G(z)}$ in Eq. 2) are computed using a softmax function, as per Eq. 5. We applied the softmax function pixel by pixel to obtain the distribution of each image in every batch. That is, x_i is an image pixel value, N is the total number of pixels in an image and τ is set to 1.

$$P_d(x_i) = Softmax(x_i) = \frac{\exp^{\frac{x_i}{\tau}}}{\sum_{i=1}^{N} \exp^{\frac{x_i}{\tau}}} \text{ ,where } d \in \{r, G(z)\} \qquad (5)$$

Algorithm 1 is the procedure for our proposed method, ADEL.

4 Experiments

In this section, we introduce our experimental environments and evaluations using various datasets and models. All experiments were conducted in a single GeForce RTX 3090 GPU.

4.1 Environments

Dataset. We apply DCGAN, WGAN, and WGAN-GP on small-scale images (CIFAR-10 and STL-10) and LSGAN, and StyleGANv2 with large-scale images (LSUN-Bedroom, Church, and ImageNet). CIFAR-10 contains 32×32 images belonging to 10 classes, with a total of 60k images divided into 50k training data and 10k test data. STL-10 also has 10 classes, with 96×96 images, and consists of 500 training images and 800 test images for each class. ImageNet is a representative large-scale dataset. It has over 1M images belonging to 1000 classes. LSUN-Bedroom and LSUN-Church is also a large-scale dataset of 128×128 and 256×256 resolution, respectively.

Baselines. We applied our proposed method to the following GAN models:

- DCGAN: GAN architecture with CNN, which replaced the fully-connected neural network.
- WGAN: WGAN explains the limitation of KL divergence loss and introduces Wasserstein loss which is a new type of loss.
- WGAN-GP: WGAN does not work well with momentum-type optimizers such as Adam. WGAN-GP address this issue by applying the gradient penalty.
- LSGAN: LSGAN proposed least square loss to solve the vanishing gradients problem in the traditional GAN's discriminator. This helps to generate a higher quality of image and stability during training.
- StyleGANv2: StyleGAN is a variation of GAN model, which is possible to being able to adjust the style when generating images. StyleGAN2 is a version that removed unnatural parts when generating images in StyleGAN and create higher quality images.

Training Details. We trained WGAN using an RMSProp [41] and other models with an Adam optimizer ($\beta_1 = 0.5$, $\beta_2 = 0.999$) [40]. DCGAN and WGAN-GP are trained for 200k iterations. WGAN and LSGAN are trained for 1M iterations. The generator of DCGAN, WGAN and WGAN-GP produces 32×32 and 48×48 images for CIFAR-10 and STL-10, respectively. Like the LSGAN study, we generated 112×112 resolution images for LSNU-Bedroom and ImageNet. We set the learning rate as 0.0002 for DCGAN and LSGAN, 0.00005 for WGAN, and 0.0001 for WGAN-GP. All models are trained with a batch size of 64 using a single GPU. When applying our proposed method, we fixed all the hyper-parameters the same as baseline models, and we experimented by changing ϵ. The optimal ϵ value is obtained by grid search. For each model, the value showing the best performance was selected. In the case of StyleGANv2, the experimental environment is the same as all settings of StyleGANv2.

Evaluations. We used three metrics, IS, FID, and KID as indicators to evaluate the performance of the GAN at generating various images and how realistic they were. IS is applicable only to class-conditional generation, so we do not calculate the LSUN-Bedroom score. To compute each metric score, 10k real images and 10k fake images, produced by the generator, were used. Metrics were calculated at a constant interval. Reported scores, when two or more metric scores were superior, were selected and compared. Since, StyleGAN [4] reflects quantitative as well as qualitative quality in the high-resolution dataset of the GAN model by suggesting a new metric called Perceptual Path Length (PPL). Therefore, we reported the PPL metric in this study.

5   Results

5.1   Base Model Performance

To examine the effectiveness of the ADEL method, we observed how the performance changes by applying ADEL to each model. As presented in the results in Table 1, the performance of each model is further improved when ADEL is applied regardless of the dataset. Except in three cases, i.e., the IS for DCGAN's CIFAR-10, the KID for WGAN's CIFAR-10, and the FID for LSGAN's ImageNet, all three computational indicators consistently indicated improved performance in all models. In addition, DCGAN (ADEL) shows a noticeable performance improvement in the STL-10 dataset experiment. These results imply that when the generator is trained, ADEL guides the generator to produce images similar to real images by providing instructive information about the distribution of real images. In particular, in the LSUN-Bedroom dataset experiment of the LSGAN model, the FID and KID indicators showed performance improvements of 22.84 and 0.03, respectively. We also compared the difference between the computational costs and training time when ADEL was applied to baseline models and when it was not. The GPU usage memory increased from a minimum of 1.003 to a maximum of 1.961 times, and the training time took at least 0.003 s and a maximum of 0.05 s for one iteration. Thus, given that the additional workload or training time caused by applying ADEL is very small, ADEL can be applied to various models without any noticeable additional burden. Memory and training time is computed on a single RTX 3090 GPU.

In this study, only the FID metric showing the most change was presented. Figures 2 and 3 are graphs comparing the performance of each model and the

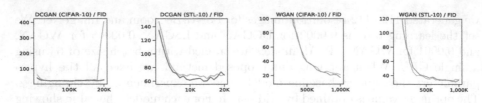

Fig. 2. Performance comparison between DCGAN and DCGAN (ADEL), WGAN and WGAN (ADEL) for all iterations.

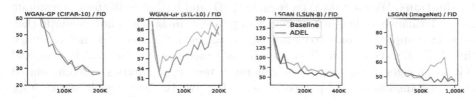

Fig. 3. Performance comparison between WGAN-GP and WGAN-GP (ADEL), LSGAN and LSGAN (ADEL) for all iterations.

model to which ADEL is applied at each evaluation iteration. In the graph, the colored line indicates the performance when ADEL is applied. It can be seen that each model has a section where FID is kept lower a specific point. It should be noted that mode collapse occurs after 400k iterations in the LSGAN (LSUN-Bedroom) experiment and 190k iterations in the DCGAN (CIFAR-10) experiment. However, the training progresses stably without mode collapse, when ADEL is applied during the same evaluation iteration. Furthermore, as can be observed in the LSGAN ImageNet graph, training is unstable at 600k in baseline models. By contrast, when ADEL is applied, it tends to be relatively stable.

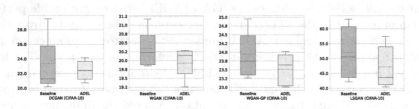

Fig. 4. Boxplots of FID scores of each model for five replicates. The red dotted line represents the average scores. (Color figure online)

To show that ADEL's performance is consistent, we replicated experiments five times for all models and datasets. In Fig. 4, we provide boxplots to compare the deviations of FID scores, which is a representative metric for GANs. It can be seen that ADEL demonstrates less deviation than the baselines. This was repeated for all metrics and models, but we could not include all the figures owing to page limitations. In the case of LSGAN ImageNet, where the difference between baseline and ADEL was the largest in all metrics, IS, FID, and KID values are 0.38, 3.82, and 0.0032, respectively. As shown in the figure, ADEL serves to make the model more robust because of reduced oscillation. Furthermore, although not presented in this paper, we find that ADEL has a low incidence of mode collapse. In particular, in the DCGAN CIFAR-10 experiment, mode collapse occurred with a 40% probability in the baseline model. By contrast, in ADE, training progressed stably without mode collapse. Thus, our proposed ADEL has the advantages of performance improvement and training stabilization. Besides these advantages, our proposed ADEL is a very simple approach with low computational burden. Hence, it is easily extendable to various GANs

Table 2. Performance comparison of StyleGANv2 with and without ADEL.

Model	Dataset	Method	PPL ($\downarrow$)	FID ($\downarrow$)
StyleGANv2	LSUN-Church 256 × 256	Baseline	342	3.86
		ADEL	**303**	**3.82**

5.2 Performance in the State-of-the-Art Model

We selected baseline models that serve as milestones with various loss functions. As these models are the basis of the latest GAN studies, we determined that if ADEL is applied in these models, even the latest derived studies would be covered. Additional experiments were conducted to prove that ADEL is also effectiveness in the SOTA model. Specifically, ADEL is applied to StyleGANv2 [23], a SOTA model in LSUN-Church, a single class and higher resolution dataset that is suitable for our experimental purpose. As presented in Table 2, StyleGANv2 improved FID and PPL. However, the KID score has not been officially reported from StyleGANv2, and it is difficult to make a fair comparison, so the score is not presented. According to Table 2, FID had a relatively marginal performance improvement. However, there was a very large performance improvement in PPL, the main metric of StyleGANv2 (Fig. 5).

5.3 Analysis of Adjusting Functions

ADEL adjusts the γ adaptively as per the discriminator's loss as in Eq. 4 when matching the distribution. Thus, it plays the role of guiding the generator by increasing the matching in the early stage of training and increasing the degree of freedom of the generator by ensuring a lower amount of matching towards the latter part of training to help generate various images. Before ADEL was chosen as the final matching method, we experimented with three different methods to compare the performance. ADEL's performance was the best of all. Table 3 compares the performance of the proposed method ADEL with three other

Fig. 5. (Top) examples of StyleGANv2 without ADEL applied, and (bottom) ADEL is applied.

distribution matching methods using a different adjusting function and a condition. Detailed explanations are as follows:

- The linearly decaying distribution matching method (LDDM) for every iteration without considering the condition given in Eq. 2 is as follows:

$$\gamma(c) = 1 - c/T \tag{6}$$

- The exponentially decaying distribution matching method (EDDM) for every iteration without considering the condition of Eq. 2 and γ is the same as Eq. 4.
- A method to adaptively adjust the degree of distribution matching in the form of a penalty based on the discriminator loss value (DLDM) is given as follows:

$$\gamma(L_D) = \exp^{|L_D - \delta|} \tag{7}$$

LDDM and EDDM are matching methods based on the current iteration versus max iteration, and DLDM is a method based on the current iteration's discriminator loss.

Table 3. Performance comparison between baseline, LDDM, EDDM, DLDM, and our proposed ADEL method.

Method	CIFAR-10			STL-10		
	IS	FID	KID	IS	FID	KID
Baseline	5.47	20.91	0.0100	4.70	63.86	0.0385
LDDM	5.20	21.19	0.0087	4.82	65.89	0.0374
EDDM	**5.54**	20.82	0.0097	5.02	61.73	**0.0341**
DLDM	5.36	23.72	0.0108	4.82	62.81	0.0368
ADEL (ours)	5.32	**20.74**	**0.0085**	**5.03**	**60.37**	0.0361

ADEL uses a function where γ decays exponentially. That is, in the early stage of training, the large γ value makes matching large to provide additional information to the generator. As iteration goes, ADEL has a small γ value and reduces the matching so that the generator can create more diverse images.

In order to check the effect of the exponential decaying function, we compared the LDDM method, which manually changes the max iteration linearly, and the EDDM method, which changes it exponentially without any restrictions on the condition. As observed from the results summarized in Table 3, the result of EDDM exceeds that of LDDM, and the result of LDDM does not even reach the baseline. This implies that adjusting the decaying function exponentially is more effective.

In addition, the performance of ADEL exhibits to EDDM in terms of applying conditions is compared. Applying a condition as in Eq. 2 has a more positive effect on performance improvement than when no such condition is applied. Finally, we compared ADEL and DLDM. DLDM is based on the discriminator loss; the discriminator loss tends to diverge toward the latter half of training, so the γ value is applied in the direction of increasing rather than continuously attenuating. Based on the results, ADEL's performance is much better than DLDM, which indicates that adjusting the decaying function based on the iteration is more effective than based on the loss value. These results indicates that it is helpful to give the model a degree of freedom to generate more diverse images at the end of training.

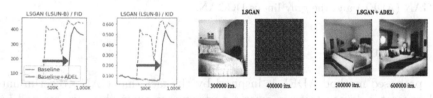

(a) Comparison between LSGAN and ADEL.

(b) Left of the dashed line: LSGAN with mode collapse. Right of the dashed line: ADEL is applied.

Fig. 6. Result when ADEL is applied to LSGAN where mode collapse has occurred.

5.4 Stabilizing and Robustness Effect of ADEL

In order to determine the stabilization effect of ADEL, we examined the kind of change that appeared by applying ADEL to LSGAN (LSUN-Bedroom) where mode collapse occurred. In Fig. 6 a), the gray dotted line represents the performance of the existing LSGAN baseline model, while the green line represents the changed performance after applying ADEL to the LSGAN baseline model before mode collapse occurs. Previously, mode collapse occurs at 400k evaluation iterations, but when ADEL is applied at the $350k^{th}$ iteration, it can be confirmed that training proceeds stably without mode collapse until about 750k evaluation iterations. Thus, ADEL has the effect of making the model more robust to mode collapse to find the optimal performance. In Fig. 6 b), the images on the left side of the dashed lines are example images generated by baseline LSGAN, while those on the right are example images generated when LSGAN is continuously trained by applying ADEL before mode collapse occurs.

Figure 7 shows the training loss graph of DCGAN and DCGAN (ADEL). In the graph of the DCGAN baseline model, L_G diverges and L_D is unstable, so that the training of the discriminator does not proceed smoothly. By contrast, the training of the discriminator in DCGAN (ADEL) dominates, resulting in L_D almost converging to 0. This can make it difficult to train the generator. However, when L_D converges to 0, $|L_D - \delta| > \epsilon$ is satisfied, because we set ϵ to a very small value. Therefore, L_{dist} still exists. We demonstrate this experimentally in Fig. 7. According to Algorithm 1, the generator is trained based on L_G and L_{dist} so that the gradient is continuously updated. Therefore, when ADEL is applied, it is robust to the vanishing gradient problem.

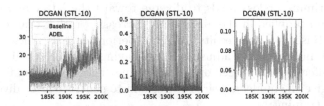

Fig. 7. Training loss graphs of L_G, L_D, and L_{dist} in order from left. The pink line is DCGAN (ADEL) and the gray line is DCGAN.

6   Conclusion

The proposed method ADEL is inspired by the fact that GAN's discriminator and generator demonstrate improved performance when they are trained in a balanced manner. ADEL works when the discriminator's loss exceeds a certain optimal range. At this time, the γ value, which controls the degree of ADEL, decreases exponentially. Therefore, it is applied strongly at the beginning of training to provide more guidance for the actual images to the generator, and it is applied weakly at the end of training to give the generator a degree of freedom to create more diverse images. To examine the effect of ADEL, five GAN models using different objective functions were selected as baselines, and the performances with and without ADEL applied to these models were compared. It was observed that the overall model performance improved when ADEL was applied. Furthermore, ADEL has the advantage that training is stable and robust to vanishing gradient problems. ADEL is not very burdensome because of its low computational cost and short training time.

References

1. Goodfellow, I.J., et al.: Generative adversarial networks. arXiv (2014). https://doi.org/10.48550/ARXIV.1406.2661
2. Karras, T., Aila, T., Laine, S., Lehtinen, J.: Progressive growing of GANs for improved quality, stability, and variation. arXiv (2018). https://doi.org/10.48550/ARXIV.1710.10196

3. Wang, T.C., Liu, M.Y., Zhu, J.Y., Tao, A., Kautz, J., Catanzaro, B.: High-resolution image synthesis and semantic manipulation with conditional GANs. In: Proceedings of the IEEE Conference on Computer Vision and Pattern Recognition, Salt Lake City, pp. 8798–8807. IEEE (2018)
4. Karras, T., Laine, S., Aila, T.: A style-based generator architecture for generative adversarial networks. In: Proceedings of the IEEE/CVF Conference on Computer Vision and Pattern Recognition, Long Beach, pp. 4401–4410. IEEE (2019)
5. Brock, A., Donahue, J., Simonyan, K.: Large scale GAN training for high fidelity natural image synthesis. arXiv (2019). https://doi.org/10.48550/ARXIV.1809.11096
6. Lee, K.S., Tran, N.T., Cheung, M.: Infomax-GAN: improved adversarial image generation via information maximization and contrastive learning. In: Proceedings of the IEEE/CVF Winter Conference on Applications of Computer Vision, Hawaii, pp. 3942–3952 (2021)
7. Guo, J., Lu, S., Cai, H., Zhang, W., Yu, Y., Wang, J.: Long text generation via adversarial training with leaked information. In: Proceedings of the AAAI Conference on Artificial Intelligence, New Orleans, vol. 32. AAAI Press (2018)
8. Dash, A., Gamboa, J.C.B., Ahmed, S., Liwicki, M., Afzal, Z.: TAC-GAN - text conditioned auxiliary classifier generative adversarial network. arXiv (2017). https://doi.org/10.48550/ARXIV.1703.06412
9. Gao, Y., Singh, R., Raj, B.: Voice impersonation using generative adversarial networks. In: 2018 IEEE International Conference on Acoustics, Speech and Signal Processing (ICASSP), Calgary, pp. 2506–2510. IEEE (2018)
10. Kumar, R., Kumar, K., Anand, V., Bengio, Y., Courville, A.: NU-GAN: high resolution neural upsampling with GAN. arXiv (2020). https://doi.org/10.48550/ARXIV.2010.11362
11. Salimans, T., Goodfellow, I., Zaremba, W., Cheung, V., Radford, A., Chen, X.: Improved techniques for training GANs. In: Advances in Neural Information Processing Systems, vol. 29. Morgan Kaufmann Publishers Inc., Barcelona (2016)
12. Metz, L., Poole, B., Pfau, D., Sohl-Dickstein, J.: Unrolled generative adversarial networks. arXiv (2017). https://doi.org/10.48550/ARXIV.1611.02163
13. Goodfellow, I.: NIPS 2016 tutorial: generative adversarial networks. arXiv (2017). https://doi.org/10.48550/ARXIV.1701.00160
14. Nowozin, S., Cseke, B., Tomioka, R.: F-GAN: training generative neural samplers using variational divergence minimization. In: Advances in Neural Information Processing Systems, vol. 29, pp. 271–279. Morgan Kaufmann Publishers Inc. (2016)
15. Radford, A., Metz, L., Chintala, S.: Unsupervised representation learning with deep convolutional generative adversarial networks. arXiv (2016). https://doi.org/10.48550/ARXIV.1511.06434
16. Gulrajani, I., Ahmed, F., Arjovsky, M., Dumoulin, V., Courville, A.C.: Improved training of Wasserstein GANs. In: Advances in Neural Information Processing Systems, vol. 30, Morgan Kaufmann Publishers Inc. (2017)
17. Wei, X., Gong, B., Liu, Z., Lu, W., Wang, L.: Improving the improved training of Wasserstein GANs: a consistency term and its dual effect. arXiv (2018). https://doi.org/10.48550/ARXIV.1803.01541
18. Arjovsky, M., Chintala, S., Bottou, L.: Wasserstein GAN. arXiv (2017). https://doi.org/10.48550/ARXIV.1701.07875
19. Berthelot, D., Schumm, T., Metz, L.: Began: boundary equilibrium generative adversarial networks. arXiv (2017). https://doi.org/10.48550/ARXIV.1703.10717

20. Tao, C., Chen, L., Henao, R., Feng, J., Duke, L.C.: Chi-square generative adversarial network. In: International Conference on Machine Learning, PMLR (2018), pp. 4887–4896. Curran Associates, Inc. (2018)
21. Heusel, M., Ramsauer, H., Unterthiner, T., Nessler, B., Hochreiter, S.: GANs trained by a two time-scale update rule converge to a local nash equilibrium. In: Proceedings of the Advances in Neural Information Processing Systems, Long Beach, vol. 30. Morgan Kaufmann Publishers Inc. (2017)
22. Binkowski, M., Sutherland, D.J., Arbel, M., Gretton, A.: Demystifying MMD GANs. arXiv (2018). https://doi.org/10.48550/ARXIV.1801.01401
23. Karras, T., Laine, S., Aittala, M., Hellsten, J., Lehtinen, J., Aila, T.: Analyzing and improving the image quality of styleGAN. In: Proceedings of the IEEE/CVF Conference on Computer Vision and Pattern Recognition, pp. 8110–8119. IEEE, Virtual (2020)
24. Zhang, H., Goodfellow, I., Metaxas, D., Odena, A.: Self-attention generative adversarial networks. In: Proceedings of the International Conference on Machine Learning (PMLR), pp. 7354–7363 (2019)
25. Donahue, J., Krähenbühl, P., Darrell, T.: Adversarial feature learning. arXiv (2017). https://doi.org/10.48550/ARXIV.1605.09782
26. Hoang, Q., Nguyen, T.D., Le, T., Phung, D.: Multi-generator generative adversarial nets. arXiv (2017). https://doi.org/10.48550/ARXIV.1708.02556
27. Nguyen, T.D., Le, T., Vu, H., Phung, D.: Dual discriminator generative adversarial nets. arXiv (2017). https://doi.org/10.48550/ARXIV.1709.03831
28. Yu, X., Zhang, X., Cao, Y., Xia, M.: VAEGAN: a collaborative filtering framework based on adversarial variational autoencoders. In: Proceedings of the International Joint Conference on Artificial Intelligence (IJCAI), Macau, pp. 4206–4212. Morgan Kaufmann Publishers Inc. (2019)
29. Ronneberger, O., Fischer, P., Brox, T.: U-Net: convolutional networks for biomedical image segmentation. In: Navab, N., Hornegger, J., Wells, W.M., Frangi, A.F. (eds.) MICCAI 2015. LNCS, vol. 9351, pp. 234–241. Springer, Cham (2015). https://doi.org/10.1007/978-3-319-24574-4_28
30. Schonfeld, E., Schiele, B., Khoreva, A.: A U-net based discriminator for generative adversarial networks. In: Proceedings of the IEEE/CVF Conference on Computer Vision and Pattern Recognition, pp 8207–8216. IEEE, Virtual (2020)
31. Jiang, Y., Chang, S., Wang, Z.: TransGAN: two pure transformers can make one strong GAN, and that can scale up. In: Advances in Neural Information Processing Systems vol. 34, pp. 14745–14758. Morgan Kaufmann Publishers Inc., Virtual (2021)
32. Mao, X., Li, Q., Xie, H., Lau, R.Y., Wang, Z., Paul Smolley, S.: Least squares generative adversarial networks. In: Proceedings of the IEEE International Conference on Computer Vision, Venice, pp. 2794–2802. IEEE (2017)
33. Zhang, H., Goodfellow, I., Metaxas, D., Odena, A.: Self-attention generative adversarial networks. In: International Conference on Machine Learning, PMLR, pp. 7354–7363. Curran Associates Inc. (2019)
34. Jolicoeur-Martineau, A.: The relativistic discriminator: a key element missing from standard GAN. arXiv (2018). https://doi.org/10.48550/ARXIV.1807.00734
35. Lim, J.H., Ye, J.C.: Geometric GAN. arXiv (2017). https://doi.org/10.48550/ARXIV.1705.02894
36. Kavalerov, I., Czaja, W., Chellappa, R.: A multi-class hinge loss for conditional GANs. In: Proceedings of the IEEE/CVF Winter Conference on Applications of Computer Vision, pp. 1290–1299, Virtual (2021)

37. Arjovsky, M., Bottou, L.: Towards principled methods for training generative adversarial networks. arXiv (2017). https://doi.org/10.48550/ARXIV.1701.04862
38. Wiatrak, M., Albrecht, S.V., Nystrom, A.: Stabilizing generative adversarial networks: a survey. arXiv (2019). https://doi.org/10.48550/ARXIV.1910.00927
39. Duchi, J., Hazan, E., Singer, Y.: Adaptive subgradient methods for online learning and stochastic optimization. J. Mach, Learn. Res. **12**, 2121–2159 (2011)
40. Kingma, D.P., Ba, J.: Adam: a method for stochastic optimization. arXiv (2017). https://doi.org/10.48550/ARXIV.1412.6980
41. Hinton, G., Srivastava, N., Swersky, K.: Neural networks for machine learning lecture 6a overview of mini-batch gradient descent. Cited On. **14**(8), 2 (2012)

A Diffusion-ReFinement Model
for Sketch-to-Point Modeling

Di Kong, Qiang Wang, and Yonggang Qi[✉]

Beijing University of Posts and Telecommunications, Beijing, China
{dikong,wanqqiang,qiyg}@bupt.edu.cn

Abstract. Diffusion probabilistic model has been proven effective in generative tasks. However, its variants have not yet delivered on its effectiveness in practice of cross-dimensional multimodal generation task. Generating 3D models from single free-hand sketches is a typically tricky cross-domain problem that grows even more important and urgent due to the widespread emergence of VR/AR technologies and usage of portable touch screens. In this paper, we introduce a novel Sketch-to-Point Diffusion-ReFinement model to tackle this problem. By injecting a new conditional reconstruction network and a refinement network, we overcome the barrier of multimodal generation between the two dimensions. By explicitly conditioning the generation process on a given sketch image, our method can generate plausible point clouds restoring the sharp details and topology of 3D shapes, also matching the input sketches. Extensive experiments on various datasets show that our model achieves highly competitive performance in sketch-to-point generation task. The code is available at https://github.com/Walterkd/diffusion-refine-sketch2point.

1 Introduction

The challenge of being able to obtain a 3D model from a single sketch has been studied for decades. Our goal is to provide precise and intuitive 3D modeling for users with limited drawing experience. And we try to propose a method that is tailored for point cloud reconstruction from a single free-hand sketch, complementing to existing single-view reconstruction methods. Taking into account the abstraction and distortion common in sketches by novice users, a sketch contains far less information than an image due to its simplicity and imprecision. This leads to a lack of important information such as color, shading and texture. To tackle this cross-domain problem, a number of learning-based methods [27,49] have been proposed that are trained by comparing rendered silhouettes or depth maps with ground truth ones, with no involvement of the ground truth in 3D shape. After obtaining depth and normal maps from the sketches, they use the maps to reconstruct the 3D shapes. The absence of the ground truth 3D shapes highly limits the capabilities of many sketch based 3D model generation networks to restore the topology and fine details in 3D shapes.

Supplementary Information The online version contains supplementary material available at https://doi.org/10.1007/978-3-031-26293-7_4.

© The Author(s), under exclusive license to Springer Nature Switzerland AG 2023
L. Wang et al. (Eds.): ACCV 2022, LNCS 13847, pp. 54–70, 2023.
https://doi.org/10.1007/978-3-031-26293-7_4

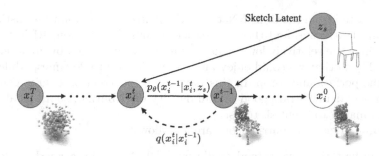

Fig. 1. The directed graphical model of the reconstruction process for point clouds.

We find that denoising diffusion probabilistic models (DDPM) [16] and their variants [23, 28, 45] have achieved great success in the generation tasks. Especially in the 3D point cloud completion and generation task [28, 29]. Combined with non-equilibrium thermodynamics, the reverse diffusion process can well simulate the generation process of 3D point cloud from disorder to order. While remarkable progress has been made in the point cloud generation tasks via GANs [1, 37, 41], the GANs based methods have some inherent limitations for modeling point clouds. Compared with GANs based generative work, the training procedure of diffusion probabilistic model is more stable and probabilistic generative model can achieve a better generation quality. However, the DDPM has rarely been studied in multimodal generation tasks. The conditional DDPM [4, 23] is also only applied in the homo-dimensional multimodal generation task and does not demonstrate its effectiveness in cross-dimensional problem. Sketch to 3D point cloud generation task can be treated as a conditional generation problem in the framework of DDPM [28, 50]. Indeed, we find the point clouds generated by a conditional DDPM often have a good overall distribution that uniformly covers the shape of the object. Nonetheless, due to the probabilistic nature of DDPM and the lack of a suitable network architecture to train the conditional DDPM for sketch to 3D point cloud generation in the previous works, we find DDPM reconstructed point clouds often lack flat surfaces and sharp details, which is also reflected by their high evaluation metric loss compared with state-of-the-art sketch to point cloud reconstruction methods in our experiments.

In this paper, we extend conditional DDPM to generate point cloud from a single free-hand sketch, adapting the score function in the reverse diffusion process to obtain the desired point distributions. Analogous to particles thermally diffusing in the 3D space, we use the diffusion process to simulate the transformation from the clean point distribution to a noise distribution. Likewise, we consider the reverse diffusion process to model the variation of point distribution in point cloud reconstruction, through which we recover the target point cloud from the noise. Our diffusion and sampling paradigm is shown in Fig. 2. Firstly, we use the Conditional Sketch-to-Point Reconstruction Network (CS2PRNet) to generate a point cloud by the DDPM conditioned on the input sketch image. It can iteratively move a set of Gaussian noise towards a desired and clean point cloud, corresponding to the input sketch. Following, the ReFinement Network (RFNet), a shape discriminator, further refines the desired point cloud generated from the CS2PRNet with the help of its capability to distinguish the ground truth ones and the reconstructed ones. In addition, RFNet can be used to refine the relatively low quality point clouds generated by an accelerated DDPM [22], so that we could enjoy an acceleration up to 20 times, while minimizing the performance drop. In this way, the reconstruction results generated by our model demonstrate good overall density distribution, sharp local details and good match to input sketches.

To summarize, our contributions are as follows:

- For the first time, we extend conditional DDPM to be a good model with an effective and efficient loss function to generate point cloud from a single free-hand sketch, which provides a high-quality and easy-to-use 3D content creation solution.
- We address the importance of sketch latent representation in the sketch-based reconstruction task, and design a CS2PRNet to condition the generation process explicitly on sketch shape latent. And by using RFNet to refine the reconstructed point clouds, we can generate point clouds with both sharp 3D characteristic details and good matching to input sketches.
- Extensive experiments on various datasets demonstrate that our model achieves competitive performance in sketch based 3D modeling task.

2   Related Work

2.1   Single-View 3D Reconstruction

Restoration of 3D geometry from a single image is an ill-posed problem. Early approaches utilize shadings [17] and textures [46] to obtain clues about surface orientations. With the emergence of large-scale 3D model datasets [2], data-driven approaches are developed to infer category-specific shapes directly from image features [3,5,9,10,19,30,33,44], in the formats of voxels [5,10], point clouds [9], mesh patches [8,19,33,44], and implicit representation [3,30]. Recently, neural rendering techniques [21,24,31,40,48] are proposed to alleviate the necessity for ground truth 3D models in training, which is achieved by comparing the estimated shape silhouette with an input image, thus enabling supervision with 2D images only. Our method is based on the conditional generative encoder-decoder network architecture of [9]. We extend the original approach by disentangling image features into a latent shape space and utilizing the diffusion probabilistic model [16,38] for reconstruction process.

2.2   Sketch-Based Modeling

Modeling based on sketches is a long established problem that has been investigated before deep learning methods become widespread. The earlier method [7], inspired by lofting technique, modeled shapes from a single image and user input strokes. Recent works [6,13,14,18,27,36,39,42,43,49] utilizing deep learning methods to guide the 3D modeling from user inputs. Only a handful of them, however, focused on reconstructions from free-hand sketches [42, 43, 49]. Wang *et al..* [43] presented a retrieval-based method to reconstruct 3D shapes. Wang *et al..* [42] adopted [9] for sketch-based 3D reconstruction by proposing an additional image translation network that aims at sketching style standardization to account for the variability of sketching styles. Sketch2model [49] solved the sketch ambiguities by introducing a view code. Different from their work, we regard the sketch latent representation, extracted from free-hand sketches, as a condition, and rely on it to guide the reverse diffusion process.

2.3 Diffusion Models

The diffusion models considered in our work can be interpreted as diffusion-based generative models, consisting of diffusion probabilistic models [38] and denoising diffusion probabilistic models [16]. Diffusion models are learned with two fixed Markov chains, controlling diffusion and reverse diffusion process. They can produce better samples than those of GANs. To tackle the generative learning trilemma, some denoising diffusion GANs have been proposed [25, 47], which reduce the sampling cost in diffusion models and perform better mode coverage and sample diversity compared to traditional GANs. In these works, a diffusion model whose reverse process is parameterized by condition GANs. Specifically to reduce the generation cost of diffusion-based generative models, Wang et al. [45] presented their latest Diffusion-GAN. The main distinction from denoising diffusion GANs is that Diffusion-GAN does not require a reverse diffusion chain during training and generation. Unlike their work, we introduced a shape discriminator in the final step of reverse diffusion process, which we used to help us better control the final 3D shape quality. Additionally, to address the slow sampling rate of the diffusion model, we apply [22], a more faster sampling strategy, through a defined bijection to construct the approximate diffusion process with less steps S. The length of the approximate reverse process S is relatively small.

Besides solving the problem of high sampling costs, denoising diffusion GANs can also be used in the study of multimodal generation task, such as text-to-speech (TTS) synthesis [25], multi-domain image translation [47]. They are capable to be applied in cross-domain tasks is because they model each denoising step using a multimodal conditional GAN. In parallel to the above GAN involved approaches, conditional DDPM [4, 23] has also been demonstrated to work for cross-modal generation tasks. However, they can currently only perform generative tasks in the same dimension, while our method can perform 2D-to-3D generation process by matching each shape latent variable with give 2D reference sketch image.

3   Methods

Given a single hand-drawn sketch in the form of line drawings, our method aims to reconstruct a 3D point cloud. We utilize the diffusion probabilistic model for the generation of 3D point cloud from a single free-hand sketch. Then we

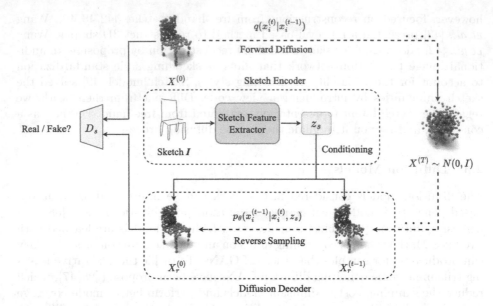

Fig. 2. The overview of our proposed model' s framework. It illustrates the training process and the reconstruction process. Our method utilizes the encoder-decoder architecture. The output of the decoder is refined by a shape discriminator D_s, obtaining a high quality point cloud by one step from a coarse point cloud.

extend the diffusion model architecture by decomposing sketch image features into a latent shape space, and condition the generation process on the sketch shape latent. To better accommodate the cross-modal generation task, a new conditional reconstruction network is also provided. And a refinement network is applied to preserve the reconstructed shape quality.

3.1   Background on Diffusion Probabilistic Models

We assume $q_{data} \sim q(x_i^0)$ to be the distribution of the groundtruth point cloud x_i in the dataset. And $q_{ultimate} = \mathcal{N}(\mathbf{0}_{3N}, \mathbf{I}_{3N \times 3N})$ to be the ultimate latent distribution, where $\mathcal{N}$ is the standard Gaussian normal distribution and N is the amount of points per point cloud. Then, the conditional diffusion probabilistic model of T steps consists of a diffusion process and a reverse sampling process.

The Diffusion Process. The diffusion process is implemented by a forward Markov chain. We use the superscript to denote the diffusion step t. For conciseness, we omit the subscription i in the following discussion. From clean data x^0, the diffusion process is to add Gaussian noise to get $x^{1:T}$:

$$q(x^{1:T}|x^0) = \prod_{t=1}^{T} q(x^t|x^{t-1}), \text{ where } q(x^t|x^{t-1}) = \mathcal{N}(x^t; \sqrt{1-\beta_t}x^{t-1}, \beta_t I) . \quad (1)$$

We define the Markov diffusion kernel as $q(x^t|x^{t-1})$. The role of the kernel is to add small Gaussian noise to disrupt the distribution of x^{t-1}. The whole process slowly injects Gaussian noise into the clean data distribution q_{data} until the output distribution is deformed to $q_{ultimate}$ according to a predefined variance schedule hyper-parameters $\beta_t, t = 1, ..., T$, which control the step sizes of the diffusion process.

The Reverse Sampling Process. The points are sampled out of a noise distribution $p(x^T)$ which is an approximation to $q(x^T)$. Let $p(x^T) \sim p_{start}$ be the input noise variable. The reverse process, conditioned on sketch shape latent z_s, converts x^T to x_r^0 through a backward Markov chain:

$$p_\theta(x_r^{0:T-1}|x^T, z_s) = \prod_{t=1}^{T} p_\theta(x^{t-1}|x^t, z_s) . \tag{2}$$

$$p_\theta(x^{t-1}|x^t, z_s) = \mathcal{N}(x^{t-1}; \mu_\theta(x^t, z_s, t), \sigma_t^2 \boldsymbol{I}) . \tag{3}$$

The mean $\mu_\theta(x^t, z_s, t)$ is a neural network parameterized by θ and the variance σ_t^2 is a time-step dependent constant closely connected to β_t. To generate a sample conditioned on z_s, we first sample from the starting distribution $p(x^T) \sim \mathcal{N}(\boldsymbol{0}_{3N}, \boldsymbol{I}_{3N \times 3N})$, then draw x^{t-1} via $p_\theta(x^{t-1}|x^t, z_s)$, where t decreases from T to 1. And x_r^0 is the sampled target shape.

Training. To make likelihood $p_\theta(x^0)$ tractable to calculate, we use the variational inference to optimize the negative log-likelihood $-logp_\theta(x^0)$. [16] introduced a certain parameterization for μ_θ that can largely simplify the training objective, known as variational lower bound (ELBO). We use the notation $\alpha_t = 1 - \beta_t$, and $\bar{\alpha}_t = \prod_{i=1}^{t} \alpha_i$. The parameterization is $\sigma_t^2 = \frac{1-\bar{\alpha}_{t-1}}{1-\bar{\alpha}_t}\beta_t$, and $\mu_\theta(x^t, z_s, t) = \frac{1}{\sqrt{\alpha_t}}(x^t - \frac{\beta_t}{\sqrt{1-\bar{\alpha}_t}}\epsilon_\theta(x^t, z_s, t))$, where ϵ_θ is a neural network taking noisy point cloud $x^t \sim q(x^t|x^0) = \mathcal{N}(x^t; \sqrt{\bar{\alpha}_t}x^0, (1 - \bar{\alpha}_t)I)$, diffusion step t, and conditioner z_s as inputs. The neural network ϵ_θ learns to predict the noise ϵ added to the clean point cloud x^0, which can be used to denoise the noisy point cloud $x^t = \sqrt{\bar{\alpha}_t}x^0 + \sqrt{1-\bar{\alpha}_t}\epsilon$. Then we minimizing the training objective L by adopting the variational bound:

$$L(\theta) = \mathbb{E}_q[\sum_{t=2}^{T}\sum_{i=1}^{N} D_{KL}(q(x_i^{t-1}|x_i^t, x_i^0)||p_\theta(x_i^{t-1}|x_i^t, z_s)) - \sum_{i=1}^{N} logp_\theta(x_i^0|x_i^1, z_s)] . \tag{4}$$

To be computable, we expand the ELBO into a sum of KL divergences, each of which compares two Gaussian distributions and therefore they can be computed in closed form. The detailed derivations, including the definition of $q(x_i^{t-1}|x_i^t, x_i^0)$, are provided in the supplementary material.

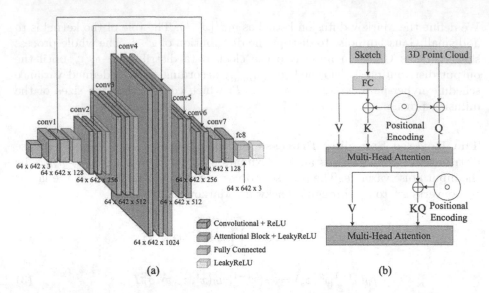

Fig. 3. The illustration of the Conditional Sketch-to-Point Reconstruction Network. Our reconstruction network uses long range convolution skip connections. We use two main types of blocks, convolutional block and attentional block.

3.2 Conditional Sketch-to-Point Reconstruction Network

In this section, we present the architecture of Conditional Sketch-to-Point Reconstruction Network ϵ_θ (see Fig. 3 for an illustration). With inputting the noisy point cloud x^t, the sketch shape latent z_s, the diffusion step t and the variance schedule β_t, the output of network ϵ_θ is per-point difference between x^t and x^{t-1}. In addition, ϵ_θ should also effectively incorporate information from z_s. The goal is to infer not only the overall shape but also the fine-grained details based on z_s. We design a neural network that achieves these features.

Before the introduction of the network, we need to resolve the role of the condition. Since the work in [28] has been shown to model the complex conditional distribution in 3D shape domain, we adopt them to approximate the true reverse sampling distribution $p_\theta(x^{t-1}|x^t, z_s)$, formulated similar to [16]. Unlike the [28], to achieve the goal of multimodal reconstruction, we use the latent shape representation extracted from sketch image. The main difference is that, in [28], x_0 is predicted as a deterministic mapping of x_t conditioned on a 3D object shape latent encoded in x_0 itself, while in our case x_0 is produced by the generator with latent variable z_s extracted from a sketch image corresponding to its 3D object. This is the key difference that allows our reverse sampling distribution $p_\theta(x^{t-1}|x^t, z_s)$ to become multimodal and complex in contrast to unimodal denoising diffusion probabilistic model in [28].

In [28], its MLP based generative network predicts the mean value of the next coordinate distribution of a point based on the latent shape representation z, with input of the coordinate x_i^{t+1} from the previous step. However, since the information extracted from a sketch image is not as much as that encoded in a 3D object, we think that the MLP is not suitable in our task, because it may lose some information about the accurate positions of the points. Thus we need a generative network that can better effectively form an associative mapping between the information extracted from the 2D sketch image and the coordinates of the 3D object. We designed a conditional generative network to accomplish our goal satisfactorily. The overall architecture is shown in Fig. 3. The reverse diffusion kernel is parameterized by $\epsilon_\theta(x_i^t, t, z_s)$. We put the detail of reverse diffusion kernel to the supplementary material. We use a fixed linear variance schedule β_1, ..., β_T to represent the timetable of the reconstruction process. Time embedding vector is comprised of $[\beta_t, sin(\beta_t), cos(\beta_t)]$ and is used to ensure conditioning on t. We implement the reconstruction network using a variant of PVCNN [26], which consists of a series of ConcatSquashConv1d layers [12]. The dimension of the ConcatSquash-CNN used in our model is 3-128-256-512-1024-512-256-128-3, and we use the LeakyReLU nonlinearity between the layers. The input to the first layer is the 3D position of points x_i^t. And we use the quaternion $c = [\beta_t, sin(\beta_t), cos(\beta_t), z_s]$ as the context embedding vector. Then the quaternion c is inserted to every level of the reconstruction network. The features from z_s are transformed and then aggregated to the point x_i^t through attention mechanism. The attentional block are applied four times and features are eventually propagated to the original input point cloud. More details about convolutional block and attentional block are provided in supplementary material.

3.3 ReFinement Network: A Shape Discriminator

After training the diffusion process, the model is able to reconstruct the point cloud from a single free-hand sketch. However, due to the limited information extracted from sketch images, point clouds reconstructed from our model trained only by the diffusion process will show local distortions and deformations. Therefore, we introduce a shape discriminator D_s to alleviate such distortions. By introducing an adversarial loss, the shape discriminator is trained in an adversarial manner together with the encoder and decoder. It functions as a trade-off between denoising and shape quality. Under the influence of the discriminator, our generative network may not produce a shape in an exact match to the input sketch at a certain angle, but is more capable of taking into account some of the characteristics that a 3D object has, such as topological structure, to the generation results.

Given x^T, our conditional reconstruction network first generates x_r^0. The discriminator is trained to distinguish between the real x^0 and fake x_r^0. Then we train the conditional 3D point cloud generator $p_\theta(x^{t-1}|x^t, z_s)$ to approximate the true reverse sampling distribution $q(x^{t-1}|x^t)$ with an adversarial loss that minimizes a divergence D_{adv} in the last reverse sampling step:

$$\min_\theta \mathbb{E}_q[\sum_{i=1}^N D_{adv}(q(x_i^0|x_i^1)||p_\theta(x_i^0|x_i^1, z_s))] , \qquad (5)$$

where fake samples from $p_\theta(x_r^0|x^1, z_s)$ are contrasted against real samples from $q(x^0|x^1)$. We denote the discriminator as $D_\phi(x_r^0, x^1)$, with parameters ϕ. It takes the 3-dimensional x_r^0 and x^1 as inputs, and decides whether x_r^0 is a plausible reverse sampled version of x^1. Given the discriminator, we train the generator by $max_\theta E_q E_{p_\theta}[log(D_\phi(x_r^0, x^1))]$, which updates the generator with the non-saturating GAN objective [11]. To summarize, the discriminator is designed to be diffusion-step-dependent and 3D topology-aware to aid the generator to achieve high-quality sketch-to-point reconstruction.

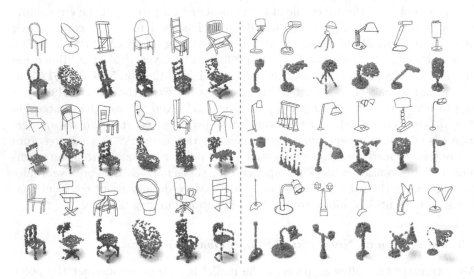

Fig. 4. Some representative examples of point clouds reconstructed by our model.

4 Experiments

In this section, we evaluate our proposed model's performance on sketch-to-point generation problem. We first perform case studies to show the effectiveness of our multimodal Diffusion-ReFinement model. Quantitative and qualitative evaluations on both synthetic and free hand-drawn sketches are presented in Sect. 4.3. We also provide some additional insight into our model by conducting an ablation study in Sect. 4.4 and proposing Feature Map module in Sect. 4.5.

4.1 Experimental Setup

Datasets. For sketch to 3D point cloud generation experiments, we employ the 3D shapes from ShapeNetCorev2 [2] to match our corresponding sketch datasets. Every point cloud has 642 points. And every category in the dataset is randomly split into training, testing and validation sets by the ratio 80%, 15%,

5% respectively. For the corresponding sketch datasets, at the pretrain stage, we use ShapeNet-Synthetic sketches to train our model. We use the same approach as [49] to create the ShapeNet-Synthetic dataset using rendered images provided by [20]. We use this dataset to pretrain the model, under the same train/test/validation split as said before. At the fine-tuning stage, to quantitatively evaluate our method on free-hand sketches and benefit further research, we use ShapeNet-Amateur sketches [34] to fine-tune our pretrained model. It contains a chair dataset with 902 sketch-3D shape quadruplets and a lamp dataset with 496 sketch-3D shape quadruplets.

Evaluation Metrics. Following prior works [1], we use two evaluation metrics to compare the quality of the reconstructed 3D point clouds to reference shapes: the Chamfer Distance (CD) and the Earth Mover's Distance (EMD) [35]. Chamfer distance measures the squared distance between each point in one point set to its nearest neighbor in the other set, while the EMD is the solution of the optimization problem that aims at transforming one set to the other.

4.2 Implementation Details

As the general training objective and algorithms in the previous section lay the foundation, we implement a model to reconstruct a point cloud from a single free-hand sketch based on the probabilistic model. We use a pretrained ResNet-18 [15] as our sketch image feature extractor, also can be called the sketch shape latent encoder and leverage the reverse sampling process presented in Sect. 3.1 for decoding. Expanding on Eq. (4) and (5), we train the model by minimizing the following adapted objective:

$$L(\theta) = \mathbb{E}_q[\sum_{t=2}^{T}\sum_{i=1}^{N} D_{KL}(q(x_i^{t-1}|x_i^t, x_i^0)||p_\theta(x_i^{t-1}|x_i^t, z_s)) - \sum_{i=1}^{N} logp_\theta(x_i^0|x_i^1, z_s)$$

$$+ \sum_{i=1}^{N} D_{adv}(q(x_i^0|x_i^1)||p_\theta(x_i^0|x_i^1, z_s))] .$$

$$(6)$$

To decode a point cloud conditioned on the latent code z_s, we sample some points x_i^T from the noise distribution $p(x_i^T)$ and pass the points through the reverse Markov chain $p_\theta(x_i^{0:T}|z_s)$ defined in Equation (2) to acquire the reconstructed point cloud $X^0 = \{x_i^0\}_{i=1}^{N}$. For diffusion model, we set the variance schedules to be $\beta_1 = 0.0001$ and $\beta_T = 0.05$, and β_t' s ($1 < t < T$) are linearly interpolated. For diffusion optimization, we use Adam optimizer with learning rate starting from e^{-3} and ended at e^{-4}. For discriminator, we also use Adam optimizer with learning rate e^{-4}. And we train a separate model for each category respectively.

4.3 Comparisons and Evaluations

We quantitatively and qualitatively compare our method with the following state-of-the-art single sketch image reconstruction models: TMNet [32],

Table 1. Comparison of single sketch image to point cloud reconstruction performance. CD is multiplied by 10^3 and EMD is multiplied by 10^2.

Category	CD						EMD					
	TMNet	Pixel2Mesh	PSGN	3D-R2D2	OccNet	Ours	TMNet	Pixel2Mesh	PSGN	3D-R2D2	OccNet	Ours
AmaChair	3.716	5.084	**2.977**	4.145	3.450	3.250	12.59	14.34	10.23	13.69	**10.08**	10.19
AmaLamp	6.856	9.339	6.453	7.947	6.293	**6.152**	17.75	18.81	17.38	17.60	16.73	**16.33**
AmaMean	5.286	7.212	4.715	6.046	4.872	**4.701**	15.17	16.58	13.81	15.65	13.41	**13.26**
SynAirplane	1.788	2.656	1.577	2.147	1.834	**1.448**	7.03	8.25	7.55	6.74	6.61	**6.39**
SynBench	2.871	3.477	2.755	3.153	**2.425**	2.623	9.27	14.01	10.22	9.81	8.97	**8.84**
SynCabinet	5.106	5.859	4.936	5.533	4.824	**4.760**	11.03	12.50	12.09	11.95	11.19	**10.64**
SynCar	2.840	3.312	**2.116**	3.129	2.417	2.291	7.31	8.29	7.76	7.66	**7.08**	7.26
SynChair	3.190	4.340	**2.692**	3.079	2.913	2.865	9.92	12.57	10.15	9.75	9.87	**9.67**
SynMonitor	3.957	4.481	3.833	4.059	3.974	**3.713**	9.84	11.96	10.48	10.03	9.84	**9.77**
SynLamp	6.023	7.706	5.865	6.975	5.778	**5.564**	16.88	17.63	16.14	17.04	16.25	**15.92**
SynSpeaker	5.725	6.479	5.654	5.942	5.514	**5.323**	12.71	13.87	13.12	12.82	12.35	**12.29**
SynRifle	1.425	1.874	1.392	1.454	**1.238**	1.374	7.58	7.12	7.07	**6.81**	6.90	
SynSofa	4.357	4.865	4.229	4.257	4.231	**4.152**	11.91	13.57	12.42	12.19	12.05	**11.62**
SynTable	4.581	5.827	4.428	5.077	**4.024**	4.164	**11.43**	12.90	12.54	12.05	11.89	11.70
SynTelephone	2.589	3.183	2.241	2.745	2.451	**2.172**	7.66	8.96	8.35	7.73	7.62	**7.51**
SynVessel	2.259	3.258	**2.041**	2.423	2.174	2.146	7.62	9.21	8.90	8.17	7.61	**7.23**
SynMean	3.593	4.409	3.366	3.844	3.369	**3.304**	9.97	11.64	10.53	10.23	9.86	**9.67**

Table 2. Model's performance on ShapeNet-Synthetic dataset and ablation study for Discriminator's effectiveness.

Training strategy III	Airplane		Bench		Cabinet		Car		Chair		Monitor		Lamp	
	CD	EMD	CD	EMD	CD	EMD	CD	EMD	CD	EMD	CD	EMD	CD	EMD
Trained on synthetic sketches	1.889	6.78	2.955	9.27	4.936	11.50	2.851	8.16	3.121	9.91	4.112	10.96	6.019	16.18
After Discriminator	**1.448**	**6.39**	**2.623**	**8.84**	**4.760**	**10.64**	**2.291**	**7.26**	**2.865**	**9.67**	**3.713**	**9.77**	**5.564**	**15.92**

Training strategy III	Speaker		Rifle		Sofa		Table		Telephone		Vessel		Mean	
	CD	EMD	CD	EMD	CD	EMD	CD	EMD	CD	EMD	CD	EMD	CD	EMD
Trained on synthetic sketches	5.542	12.78	1.893	7.38	4.722	12.07	4.477	11.90	2.524	7.86	2.532	8.03	3.659	10.21
After discriminator	**5.323**	**12.29**	**1.374**	**6.90**	**4.152**	**11.62**	**4.164**	**11.70**	**2.172**	**7.51**	**2.146**	**7.23**	**3.304**	**9.67**

Pixel2Mesh [44], PSGN [9], 3D-R2N2 [5], OccNet [30], using point clouds from thirteen categories in ShapeNet. The comparison is performed on ShapeNet-Synthetic and ShapeNet-Amateur datasets. For volume-based methods TMNet and Pixel2Mesh, both metrics are computed between the ground truth point cloud and 642 points uniformly sampled from the generated mesh. Since the outputs of Pixel2Mesh are non-canonical, we align their predictions to the canonical ground truth by using the pose metadata available in the dataset. Results are shown in Table 1. On chair of both datasets, our method outperforms other methods except PSGN when measured by CD. The EMD score of our method pushes closer towards the OccNet performance when tested on Amateur dataset and reaches the best performance when tested on Synthetic dataset. While on category of lamp, our approach outperforms other methods in both evaluation metrics for two datasets. Including these two categories, our approach outperforms other five baselines on most of the categories. Notably, from Tables 1 and 2, when both training and testing are conducted on the ShapeNet-Synthetic dataset, our model can have a better performance in both CD and EMD. Also, the visualization of reconstructed point clouds in Fig. 5 validates the effectiveness of our model compared with other baselines. While OccNet can reconstruct

Table 3. Comparison of 2D silhouette IoU score on ShapeNet-Synthetic and ShapeNet-Amateur datasets. Generated shapes are projected to the ground truth view and we calculate the IoU score between the projected silhouettes and the ground truth ones.

Category	Airplane	Bench	Cabinet	Car	Chair	Monitor	Lamp	Speaker	Rifle	Sofa	Table	Telephone	Vessel	Mean
TMNet	0.593	0.625	0.810	0.821	0.709(0.683)	0.784	0.606(0.585)	0.790	**0.672**	0.796	0.703	0.813	0.706	0.713
Pixel2Mesh	0.532	0.564	0.734	0.772	0.675(0.652)	0.729	0.548(0.530)	0.713	0.596	0.738	0.625	0.750	0.635	0.653
PSGN	0.652	0.633	0.832	0.866	**0.744**(0.710)	0.803	0.619(0.599)	0.808	0.633	0.813	0.712	0.848	0.724	0.733
3D-R2D2	0.565	0.573	0.786	0.796	0.718(0.707)	0.765	0.579(0.554)	0.766	0.618	0.805	0.678	0.797	0.688	0.693
OccNet	0.641	**0.684**	**0.873**	0.839	0.736(0.712)	0.788	0.643(0.622)	0.812	0.655	0.802	**0.745**	0.823	0.739	0.741
Ours	**0.679**	0.667	0.858	**0.868**	0.732(**0.715**)	**0.812**	**0.662(0.627)**	**0.827**	0.647	**0.820**	0.731	**0.854**	**0.756**	**0.750**

the rough shapes, it fails to capture the fine details of the geometry and is not able to model the topology of surface. PSGN performs generally better than OccNet in terms of the capability of modeling the fine structures. However, due to the limitations of the vanilla architecture, it struggles to reconstruct shapes with complex topology. In comparison, we believe that in the vast majority of cases, our approach has surpassed other approaches in terms of visual quality and is better at restoring detailed information of the corresponding sketches. We are able to generate point clouds with complex topology while maintaining high reconstruction accuracy.

We also compare projected silhouettes of generated 3D models with ground truth silhouettes, and show the results in Table 3. It shows our model is more powerful at matching input sketches.

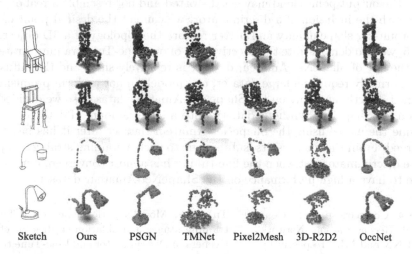

Sketch   Ours   PSGN   TMNet   Pixel2Mesh   3D-R2D2   OccNet

Fig. 5. Qualitative comparisons with other five baseline methods on free-hand sketches from the shapenet-amateur sketch dataset.

4.4 Ablation Study

In this section, we conduct controlled experiments to validate the importance of different components. Table 4 presents the comparison of the generator network using CNN and MLP respectively. Our approach has significantly outperformed

Input Sketch

Before SD

After SD

Fig. 6. Ablation study for shape discriminator. Without shape discriminator, the output shape may not resemble real objects(middle row).

the MLP based method across all datasets. In Table 5 and Table 2, we investigate the impact of the discriminator and validate the effectiveness of discriminator design. As is illustrated in Sect. 3.2 and Sect. 3.3, our generation result relies on both input sketch image and a discriminator refinement. With the adjustment of the discriminator, the network is expected to generate finer details to match the input sketch. Tables 5 and 2 show that with the discriminator being put into use, both metrics improved in all categories. Figure 6 shows the importance of the proposed shape discriminator to the generation results. Without shape discriminator, the output point cloud may get distorted and not resemble a real object. While with the inclusion of a discriminator, we can get the desired point cloud with promising shape quality and better restore the topology of a 3D object. In Table 5, we also demonstrate the effectiveness of our Fine-Tune training strategy. Since the size of ShapeNet-Amateur dataset is relatively small and the diffusion model normally requires a large dataset, our model cannot perform promisingly if trained directly and only on the ShapeNet-Amateur dataset, so we decided to train on the ShapeNet-Synthetic dataset with a larger amount of data, and then fine-tune the model using the ShapeNet-Amateur dataset after it has basically converged. From the final results, subsequent to such a training strategy, applying the discriminator last when the fine-tuning has almost converged, our model is able to have a high performance on the ShapeNet-Amateur dataset.

Table 4. Comparison of the Conditional Reconstruction Network using CNN and MLP respectively.

Tested dataset	Metric	MLP based	CNN based
AmaChair	CD	4.739	**3.25**
	EMD	10.84	**10.19**
SynChair	CD	3.983	**2.865**
	EMD	10.55	**9.67**
AmaLamp	CD	7.467	**6.152**
	EMD	17.42	**16.33**
SynLamp	CD	6.571	**5.564**
	EMD	16.75	**15.92**

Table 5. Model's performance on different training stages and ablation studies for effectiveness of Discriminator and Fine-Tune training strategy.

Training strategy I	AmaChair		AmaLamp	
	CD	EMD	CD	EMD
Trained on synthetic sketches	9.620	18.33	18.790	30.83
Fine tune on amateur sketches	3.766	10.35	6.473	17.18
After discriminator	**3.250**	**10.19**	**6.152**	**16.33**
Training strategy II	AmaChair		AmaLamp	
	CD	EMD	CD	EMD
Trained on Amateur sketches	6.376	12.78	8.078	19.45
After discriminator	5.575	12.08	7.867	16.65

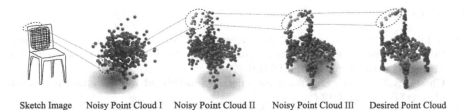

Sketch Image   Noisy Point Cloud I   Noisy Point Cloud II   Noisy Point Cloud III   Desired Point Cloud

Fig. 7. The Feature Map module maps features from the sketch image to the noisy point cloud and demonstrates how features guide the movement of points from the noisy point cloud to the desired point cloud.

4.5   Sketch-Point Feature Map (FM)

Further, to study how the sketch image influence the point cloud reconstruction process, we implement a Feature Map (FM) module. The FM module transmits information from the sketch image to the denoising process. The FM module maps the features from sketch image to noisy points in the reconstruction network, which are $\{x_i | 1 \leq i \leq N\}$. Figure 7 illustrates this process. In this way, the reconstruction network can utilize local features of the sketch image to manipulate the input noisy point cloud to form a clean and desired point cloud. The key step in this process is to map features from sketch image to $\{x_i | 1 \leq i \leq N\}$. We adopt a similar strategy from [29], except that we replace incomplete point clouds with sketches. Features of sketch image are transformed through a shared MLP, and then aggregated to the points x_i through the attention mechanism, which is a weighted sum of the features from part of the sketch. We set a large distance to define sketch parts in FM module. This makes FM module have large receptive fields, so that we can query a large part of the sketch image. And we leverage the spatial correspondence between the different parts of the sketch image and the point cloud through the proposed Feature Map module to infer high level 2D-to-3D structural relations.

5   Conclusions

In this paper, we propose the Sketch-to-Point Diffusion-ReFinement model for cross-domain point cloud reconstruction. A novel conditional reconstruction network is presented, to condition the generation process explicitly on sketch shape latent, which emphasizing the importance of sketch latent and brings controllability to the output point cloud. From observation and evaluation, a refinement network provides users to restore the point clouds with sharp 3D characteristic details and topology. Experimental results demonstrate that the proposed model achieves the state-of-the-art performance in sketch to point cloud generation task. We hope our method can inspire further researches in cross-domain 3D reconstruction and sketch-related areas.

References

1. Achlioptas, P., Diamanti, O., Mitliagkas, I., Guibas, L.: Learning representations and generative models for 3d point clouds. In: International Conference On Machine Learning, pp. 40–49. PMLR (2018)
2. Chang, A.X., et al.: Shapenet: An information-rich 3d model repository. CoRR abs/ arXiv: 1512.03012 (2015)
3. Chen, Z., Zhang, H.: Learning implicit fields for generative shape modeling. Proceedings of IEEE Conference on Computer Vision and Pattern Recognition (CVPR) (2019)
4. Choi, J., Kim, S., Jeong, Y., Gwon, Y., Yoon, S.: Ilvr: Conditioning method for denoising diffusion probabilistic models. arXiv preprint arXiv:2108.02938 (2021)
5. Choy, C.B., Xu, D., Gwak, J.Y., Chen, K., Savarese, S.: 3D-R2N2: a unified approach for single and multi-view 3d object reconstruction. In: Leibe, B., Matas, J., Sebe, N., Welling, M. (eds.) ECCV 2016. LNCS, vol. 9912, pp. 628–644. Springer, Cham (2016). https://doi.org/10.1007/978-3-319-46484-8_38
6. Delanoy, J., Aubry, M., Isola, P., Efros, A.A., Bousseau, A.: 3d sketching using multi-view deep volumetric prediction. Proc. ACM Comput. Graph. Interactive Techn. 1(1), 1–22 (2018)
7. Deng, C., Huang, J., Yang, Y.L.: Interactive modeling of lofted shapes from a single image. Comput. Vis. Media 6(3), 279–289 (2020)
8. Deng, Z., Liu, Y., Pan, H., Jabi, W., Zhang, J., Deng, B.: Sketch2pq: freeform planar quadrilateral mesh design via a single sketch. IEEE Trans. Visuali. Comput. Graph. (2022)
9. Fan, H., Su, H., Guibas, L.J.: A point set generation network for 3d object reconstruction from a single image. In: Proceedings of the IEEE Conference on Computer Vision and Pattern Recognition, pp. 605–613 (2017)
10. Girdhar, R., Fouhey, D.F., Rodriguez, M., Gupta, A.: Learning a predictable and generative vector representation for objects. In: Leibe, B., Matas, J., Sebe, N., Welling, M. (eds.) ECCV 2016. LNCS, vol. 9910, pp. 484–499. Springer, Cham (2016). https://doi.org/10.1007/978-3-319-46466-4_29
11. Goodfellow, I., et al.: Generative adversarial nets. In: Advances in Neural Information Processing Systems 27 (2014)
12. Grathwohl, W., Chen, R.T., Bettencourt, J., Sutskever, I., Duvenaud, D.: Ffjord: Free-form continuous dynamics for scalable reversible generative models. arXiv preprint arXiv:1810.01367 (2018)
13. Han, X., Gao, C., Yu, Y.: Deepsketch2face: a deep learning based sketching system for 3d face and caricature modeling. ACM Trans. Graph. (TOG) 36(4), 1–12 (2017)
14. Han, Z., Ma, B., Liu, Y.S., Zwicker, M.: Reconstructing 3d shapes from multiple sketches using direct shape optimization. IEEE Trans. Image Process. 29, 8721–8734 (2020)
15. He, K., Zhang, X., Ren, S., Sun, J.: Deep residual learning for image recognition. In: Proceedings of the IEEE Conference On Computer Vision And Pattern Recognition, pp. 770–778 (2016)
16. Ho, J., Jain, A., Abbeel, P.: Denoising diffusion probabilistic models. Adv. Neural. Inf. Process. Syst. 33, 6840–6851 (2020)
17. Horn, B.K.: Shape from shading: A method for obtaining the shape of a smooth opaque object from one view (1970)
18. Huang, H., Kalogerakis, E., Yumer, E., Mech, R.: Shape synthesis from sketches via procedural models and convolutional networks. IEEE Trans. Visual Comput. Graphics 23(8), 2003–2013 (2016)

19. Huang, Q., Wang, H., Koltun, V.: Single-view reconstruction via joint analysis of image and shape collections. ACM Trans. Graph. (TOG) **34**(4), 1–10 (2015)
20. Kar, A., Häne, C., Malik, J.: Learning a multi-view stereo machine. In: Advances in Neural Information Processing Systems 30 (2017)
21. Kato, H., Ushiku, Y., Harada, T.: Neural 3d mesh renderer. In: Proceedings of the IEEE Conference On Computer Vision And Pattern Recognition, pp. 3907–3916 (2018)
22. Kong, Z., Ping, W.: On fast sampling of diffusion probabilistic models. arXiv preprint arXiv:2106.00132 (2021)
23. Kong, Z., Ping, W., Huang, J., Zhao, K., Catanzaro, B.: Diffwave: A versatile diffusion model for audio synthesis. arXiv preprint arXiv:2009.09761 (2020)
24. Liu, S., Li, T., Chen, W., Li, H.: Soft rasterizer: A differentiable renderer for image-based 3d reasoning. In: Proceedings of the IEEE/CVF International Conference on Computer Vision, pp. 7708–7717 (2019)
25. Liu, S., Su, D., Yu, D.: Diffgan-tts: High-fidelity and efficient text-to-speech with denoising diffusion gans. arXiv preprint arXiv:2201.11972 (2022)
26. Liu, Z., Tang, H., Lin, Y., Han, S.: Point-voxel cnn for efficient 3d deep learning. In: Advances in Neural Information Processing Systems 32 (2019)
27. Lun, Z., Gadelha, M., Kalogerakis, E., Maji, S., Wang, R.: 3d shape reconstruction from sketches via multi-view convolutional networks. In: 2017 International Conference on 3D Vision (3DV), pp. 67–77. IEEE (2017)
28. Luo, S., Hu, W.: Diffusion probabilistic models for 3d point cloud generation. In: Proceedings of the IEEE/CVF Conference on Computer Vision and Pattern Recognition (CVPR) (June 2021)
29. Lyu, Z., Kong, Z., Xu, X., Pan, L., Lin, D.: A conditional point diffusion-refinement paradigm for 3d point cloud completion. arXiv preprint arXiv:2112.03530 (2021)
30. Mescheder, L., Oechsle, M., Niemeyer, M., Nowozin, S., Geiger, A.: Occupancy networks: Learning 3d reconstruction in function space. In: Proceedings of the IEEE/CVF Conference on Computer Vision and Pattern Recognition, pp. 4460–4470 (2019)
31. Niemeyer, M., Mescheder, L., Oechsle, M., Geiger, A.: Differentiable volumetric rendering: Learning implicit 3d representations without 3d supervision. In: Proceedings of the IEEE/CVF Conference on Computer Vision and Pattern Recognition, pp. 3504–3515 (2020)
32. Pan, J., Han, X., Chen, W., Tang, J., Jia, K.: Deep mesh reconstruction from single rgb images via topology modification networks. In: Proceedings of the IEEE/CVF International Conference on Computer Vision, pp. 9964–9973 (2019)
33. Pontes, J.K., Kong, C., Sridharan, S., Lucey, S., Eriksson, A., Fookes, C.: Image2Mesh: a learning framework for single image 3D reconstruction. In: Jawahar, C.V., Li, H., Mori, G., Schindler, K. (eds.) ACCV 2018. LNCS, vol. 11361, pp. 365–381. Springer, Cham (2019). https://doi.org/10.1007/978-3-030-20887-5_23
34. Qi, A., et al.: Toward fine-grained sketch-based 3d shape retrieval. IEEE Trans. Image Process. **30**, 8595–8606 (2021)
35. Rubner, Y., Tomasi, C., Guibas, L.J.: The earth mover's distance as a metric for image retrieval. Int. J. Comput. Vision **40**(2), 99–121 (2000)
36. Shen, Y., Zhang, C., Fu, H., Zhou, K., Zheng, Y.: Deepsketchhair: Deep sketch-based 3d hair modeling. IEEE Trans. Visual Comput. Graphics **27**(7), 3250–3263 (2020)
37. Shu, D.W., Park, S.W., Kwon, J.: 3d point cloud generative adversarial network based on tree structured graph convolutions. In: Proceedings of the IEEE/CVF International Conference On Computer Vision, pp. 3859–3868 (2019)

38. Sohl-Dickstein, J., Weiss, E., Maheswaranathan, N., Ganguli, S.: Deep unsupervised learning using nonequilibrium thermodynamics. In: International Conference on Machine Learning, pp. 2256–2265. PMLR (2015)
39. Su, W., Du, D., Yang, X., Zhou, S., Fu, H.: Interactive sketch-based normal map generation with deep neural networks. Proc. ACM Comput. Graph. Interactive Tech. **1**(1), 1–17 (2018)
40. Tulsiani, S., Zhou, T., Efros, A.A., Malik, J.: Multi-view supervision for single-view reconstruction via differentiable ray consistency. In: Proceedings of the IEEE Conference on Computer Vision and Pattern Recognition, pp. 2626–2634 (2017)
41. Valsesia, D., Fracastoro, G., Magli, E.: Learning localized generative models for 3d point clouds via graph convolution. In: International Conference on Learning Representations (2018)
42. Wang, J., Lin, J., Yu, Q., Liu, R., Chen, Y., Yu, S.X.: 3d shape reconstruction from free-hand sketches. arXiv preprint arXiv:2006.09694 (2020)
43. Wang, L., Qian, C., Wang, J., Fang, Y.: Unsupervised learning of 3d model reconstruction from hand-drawn sketches. In: Proceedings of the 26th ACM international conference on Multimedia, pp. 1820–1828 (2018)
44. Wang, N., Zhang, Y., Li, Z., Fu, Y., Liu, W., Jiang, Y.-G.: Pixel2Mesh: generating 3D mesh models from single RGB images. In: Ferrari, V., Hebert, M., Sminchisescu, C., Weiss, Y. (eds.) ECCV 2018. LNCS, vol. 11215, pp. 55–71. Springer, Cham (2018). https://doi.org/10.1007/978-3-030-01252-6_4
45. Wang, Z., Zheng, H., He, P., Chen, W., Zhou, M.: Diffusion-gan: Training gans with diffusion. arXiv preprint arXiv:2206.02262 (2022)
46. Witkin, A.P.: Recovering surface shape and orientation from texture. Artif. Intell. **17**(1–3), 17–45 (1981)
47. Xiao, Z., Kreis, K., Vahdat, A.: Tackling the generative learning trilemma with denoising diffusion GANs. In: International Conference on Learning Representations (ICLR) (2022)
48. Yan, X., Yang, J., Yumer, E., Guo, Y., Lee, H.: Perspective transformer nets: Learning single-view 3d object reconstruction without 3d supervision. In: Advances in Neural Information Processing Systems 29 (2016)
49. Zhang, S.H., Guo, Y.C., Gu, Q.W.: Sketch2model: View-aware 3d modeling from single free-hand sketches. In: Proceedings of the IEEE/CVF Conference on Computer Vision and Pattern Recognition, pp. 6012–6021 (2021)
50. Zhou, L., Du, Y., Wu, J.: 3d shape generation and completion through point-voxel diffusion. In: Proceedings of the IEEE/CVF International Conference on Computer Vision, pp. 5826–5835 (2021)

TeCM-CLIP: Text-Based Controllable Multi-attribute Face Image Manipulation

Xudong Lou[ID], Yiguang Liu, and Xuwei Li[(✉)]

College of Computer Science, Sichuan University, Chengdu, Sichuan, China
{liuyg,lixuwei}@scu.edu.cn

Abstract. In recent years, various studies have demonstrated that utilizing the prior information of StyleGAN can effectively manipulate and generate realistic images. However, the latent code of StyleGAN is designed to control global styles, and it is arduous to precisely manipulate the property to achieve fine-grained control over synthesized images. In this work, we leverage a recently proposed Contrastive Language Image Pretraining (CLIP) model to manipulate latent code with text to control image generation. We encode image and text prompts in shared embedding space, leveraging powerful image-text representation capabilities pretrained on contrastive language images to manipulate partial style codes in the latent code. For multiple fine-grained attribute manipulations, we propose multiple attribute manipulation frameworks. Compared with previous CLIP-driven methods, our method can perform high-quality attribute editing much faster with less coupling between attributes. Extensive experimental illustrate the effectiveness of our approach. Code is available at https://github.com/lxd941213/TeCM-CLIP.

1 Introduction

Image manipulation is an interesting but challenging task, which has attracted the interest of researchers. Recent works on Generative Adversarial Networks (GANs) [9] have made impressive progress in image synthesis, which can generate photorealistic images from random latent code [14–16]. These models provide powerful generative priors for downstream tasks by acting as neural renderers. However, this synthesis process is commonly random and cannot be controlled by users. Therefore, utilizing priors for image synthesis and processing remains an exceedingly challenging task.

StyleGAN [15, 16], one of the most commonly used generative network frameworks, introduces a novel style-based generator architecture that can generate high-resolution images with unparalleled realism. Recent works [5, 30, 36] have demonstrated that StyleGAN's latent space, $\mathcal{W}$, is introduced from a learned piecewise continuous map, resulting in less entangled representations and more feasible operations. The superior properties of $\mathcal{W}$ space have attracted a host of researchers to develop advanced GAN inversion techniques [1, 2, 10, 29, 33, 34, 39] to invert real images back into the latent space of StyleGAN and perform meaningful operations. The most popular approach is to train an additional encoder to

L. Wang et al. (Eds.): ACCV 2022, LNCS 13847, pp. 71–87, 2023.
https://doi.org/10.1007/978-3-031-26293-7_5

map real images to $\mathcal{W}$ space, which not only enables faithful reconstructions but also semantically meaningful edits. In addition, several methods [1–3,29,34,43] recently proposed $\mathcal{W}+$ space based on $\mathcal{W}$ space, which can achieve better manipulation and reconstruction.

Existing methods for semantic control discovery include large amounts of annotated data, manual examination [10,30,36], or pre-trained classifiers [30]. Furthermore, subsequent operations are usually performed by moving along with directions in the latent space, using a parametric model (e.g. 3DMM in StyleRig [33]) or a trained normalized flow in StyleFlow [3]. Therefore, existing controls can only perform image operations along with preset semantic directions, which severely limits the creativity and imagination of users. Whenever additional unmapped directions are required, further manual work or large amounts of annotated data is required. Text-guided image processing [8,19,21–23,28] has become feasible thanks to the development of language representations across a modality range. Recently StyleCLIP [26] achieved stunning image processing results by leveraging CLIP's powerful image-text representation. However, StyleCLIP has the following shortcomings: 1) For each text prompt, the mapper needs to be trained separately, which lacks activity in practical applications; 2) Mapping all style codes results in poor decoupling. HairCLIP [35] is designed for hair editing, it has designed a framework to quickly process hairstyles and colors for images and edit directly with textual descriptions or reference images, eliminating the need to train a separate mapper for each attribute. However, HairCLIP just edits hair attributes, it maps all the style codes like StyleCLIP, which slows down the speed and increases the coupling of attributes.

In this work, we design a model that requires fewer operations and can be disentangled more thoroughly. Overall, our model is similar to StyleCLIP [26] and HairCLIP [35]. The obvious difference is that the first two models map all style codes, while our model only needs to map partial style codes. Through extensive experiments, we found that the attributes controlled by each style code are different. If we desire to change a certain attribute of the image (such as hairstyle, emotion, etc.), we only need to perform a partially targeted style codes mapping manipulation. Our method not only exceedingly facilitates image mapping manipulation but provides less entanglement.

The contributions of this work can be summarized as follows:

1) Through extensive experiments, we analyze the properties controlled by each style code. When we need to change an attribute, we only need to manipulate the style codes that control the attribute, which can reduce the coupling between attributes while reducing the workload.
2) To further reduce the influence on other attributes, we design several decoupling structures and introduce background and face loss functions.
3) Extensive experiments and analyses reveal that our method has better manipulation quality and less training and inference time.

2   Related Work

2.1   GAN Inversion

With the rapid development of GANs, quite a few methods have been proposed to understand and control their latent space. The most normal method is GAN inversion, where the latent code most accurately reconstructs the input image from a pre-trained GAN. StyleGAN [15,16] is used in quite a few methods due to its state-of-the-art image quality and latent spatial semantic richness. In general, there are currently three ways to embed images from the image space to the latent space: 1) learn an encoder that maps a given image to the latent space [6,29,34]; 2) select a random initial latent code and optimize it with gradient descent [1,2,18]; 3) Mix the first two methods. Between them, methods that perform optimization are better than encoders in terms of reconstruction quality but take longer. In addition, encoder-based methods can be divided into $\mathcal{W}$ and $\mathcal{W}+$ spaces after encoding. Among them, the $\mathcal{W}$ space is the latent space obtained by performing a series of fully connected layer transformations on $\mathcal{Z}$ space, which is generally considered to reflect the learning properties of entanglement better than $\mathcal{Z}$ space. $\mathcal{W}+$ space and $\mathcal{W}$ space are constructed similarly, but the latent code $w \in \mathcal{W}+$ fed to each layer of generators is different, which is frequently used for style mixing and image inversion.

2.2   Latent Space Manipulation

A host of papers [3,29,30,43] propose various methods to learn semantic manipulation of latent codes and then utilize pretrained generators for image generation. Specifically, the latent space in StyleGAN [15,16] has been manifested to enable decoupled and meaningful image manipulations. Tewari et al. [33] utilize a pretrained 3DMM to learn semantic face edits in the latent space. Nitzan et al. [24] train an encoder to obtain a latent vector representing the identity of one image and the pose, expression, and illumination of another. Wu et al. [36] proposed to use the StyleSpace $\mathcal{S}$, and demonstrated that it is better disentangled than $\mathcal{W}$ and $\mathcal{W}+$. Collins et al. [5] perform local semantic editing by manipulating corresponding components of the latent code. These methods quintessentially follow an "invert first, manipulate later" process, first embedding the image into the latent space, and then manipulating its latent space in a semantically meaningful way. In this paper, we use a pre-trained e4e [34] model to embed images into the $\mathcal{W}+$ space, while encoding text prompt using CLIP's powerful image-text representation capabilities. Our approach is general and can be used across multiple domains without requiring domain or operation-specific data annotations.

2.3   Text-Guided Image Generation and Manipulation

Reed et al. [28] generated text-guided images by training a conditional GAN conditioned on text embeddings obtained by a pre-trained encoder. Zhang et al. [40,41] used multi-scale GANs to generate high-resolution images from text descriptions

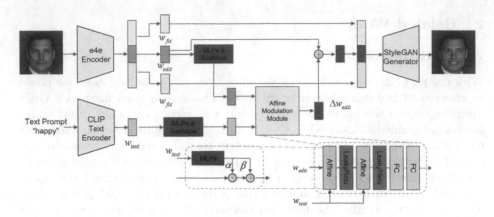

Fig. 1. We show the overall structure of our method by taking emotion ("happy") as an example. Our method supports the completion of corresponding sentiment editing according to the given text prompt, where the image is feature-mapped by pre-trained e4e [34], and the text is encoded into a 512-dimensional vector by CLIP's text encoder.

through a sketch-refinement process. AttnGAN [38] fused attention mechanisms between text and image features. Additional supervision is used in other works to further improve image quality. Several studies focus on text-guided image processing and kind of methods that use GAN-based encoder-decoder architectures to separate the semantics of input images and textual descriptions. ManiGAN [20] introduced a new text-image combination module that produces high-quality images. Different from the aforementioned works, StyleCLIP [26] proposed to combine the high-quality images generated by StyleGAN with the wealthy multi-domain semantics learned by CLIP [27] and use CLIP to guide the generation of images. TediGAN [37] encoded images and texts simultaneously into the latent space and then performed style mixing to generate corresponding images to realize text-guided image manipulation. Later improved versions of TediGAN [37] also used CLIP [27] for optimization operations to provide text-image similarity loss. Since StyleCLIP needs to train a separate mapper network for each text prompt, it lacks flexibility in practical applications. Therefore, HairCLIP [35] proposed a unified framework of text and image conditions to encode text or image conditions into the latent space to guide image generation.

In general, StyleCLIP [26], TediGAN [37], and HairCLIP [35] all work well for text-guided image generation, and HairCLIP outperforms the previous two models in speed. But the methods mentioned above do not decouple multiple properties well. Through extensive experiments, we discovered the properties controlled by each style code. Therefore, we merely perform feature mapping on partial style codes and achieve fewer entanglement and faster speed.

3 Proposed Method

As mentioned above, the latent space is divided into $\mathcal{Z}$ space, $\mathcal{W}$ space, and $\mathcal{W}+$ space, where $\mathcal{Z}$ space is normally distributed, and $\mathcal{W}$ space is obtained from the random noise vectors $z \in \mathcal{Z}$ via a sequence of fully connected layers. However, several studies [4,29,34,36,43] have demonstrated that inverting the images to a 512-dimensional vector $w \in \mathcal{W}$ cannot achieve accurate reconstruction. Subsequently, it has become more normal practice to encode images into an extended latent space $\mathcal{W}+$ consisting of concatenating 18 different 512-dimensional w vectors, one for each input layer of StyleGAN [15,16]. In this paper, we perform an inversion with a pre-trained e4e [34], which maps the images to latent space, $\mathcal{W}+$, so that all style codes generated by the encoder can be recovered at the pixel level and semantic level. Since each layer of the 18-layer $\mathcal{W}+$ space controls different attributes of the image, we manipulate the corresponding style codes from $w \in \mathcal{W}+$ according to the attributes to be manipulated. At the same time, the text embedding encoded by CLIP [27] is fused into the latent space of the image. Finally, the manipulated image can be generated from the StyleGAN generator and the specific network structure is shown in Fig. 1.

3.1 Image and Text Encoders

Our approach is based on the representative of a pre-trained StyleGAN [15,16] generator and $\mathcal{W}+$ latent space, which requires a powerful encoder to precisely encode the image into the latent domain. We select e4e [34] as our image encoder, which encodes the image into $\mathcal{W}+$ space. The latent code w is composed of 18 different style codes, which can represent the fine details of the original image more completely, accordingly, it is better than previous methods in reconstruction quality.

Unlike StyleCLIP [26], which only uses CLIP [27] as supervised loss, we refer to HairCLIP [35], which encoders the text prompts into 512-dimensional conditional embedding. Since CLIP is well trained on large-scale image-text pairs, it is easy to encode text prompts into conditional embedding and then fuse them with latent space effectively to achieve text-driven images.

3.2 Latent Mapper with Disentangled Latent Space

Quite a few works [15,16,31,37] have manifested that different StyleGAN layers correspond to different semantic-level information in the generated image. Style-CLIP and HairCLIP take the same strategy, with three mappers M_c, M_m, M_f with the same network structure, which are responsible for predicting manipulated Δw corresponding to different parts of the latent code $w = (w_c, w_m, w_f)$. But in fact, this mapping method cannot disentangle the associations between various attributes well. Through extensive experiments, we found that each style code $w \in \mathcal{W}+$ controls one or more attributes, and we only need to manipulate a few of the style codes (fusion text conditional embedding and feature mapping). As an example, the first four style codes can be processed for hair manipulation.

Specifically, we first perform a four-layer full connection mapping of the style codes to be processed and also perform a four-layer full connection mapping for the text conditional embedding, and the specific formula is as follows:

$$w'_{edit} = Reshape\left(MLP\left(w_{edit}\right)\right),$$
$$w'_{text} = Reshape\left(MLP\left(w_{text}\right)\right). \tag{1}$$

where MLP represents a four-layer fully connected layer, and $Reshape$ represents a deformation operation on the style code. $w_{edit} \in \mathbb{R}^{512}$ and $w_{text} \in \mathbb{R}^{512}$ are the style code that needs to be manipulated and the text conditional embedding, respectively. $w'_{edit} \in \mathbb{R}^{4\times512}$ and $w'_{text} \in \mathbb{R}^{4\times512}$ are the corresponding results after processing. Then processed by *Affine Modulation Module* (AMM) and then according to weights (learnable parameters) for additive fusion. We further add a two-layer fully-connected mapping for the fused style codes, which can do further information fusion. For other style codes that are not related to the attributes, we directly concatenate the original style codes and the modified style codes.

Affine Modulation Module. We design this module after referring to several methods [12,25,32] and making improvements. As shown in Fig. 1, after each mapper network goes through four fully connected layers, *Affine Modulation Module* (AMM) is used to fuse text-conditional embeddings to the style codes. AMM is a deep text-image fusion block that superimposes normalization layers, multiple affine transformation layers, and nonlinear activation layers (leaklyrelu) in the fusion block. AMM uses conditional embedding to modulate the style codes that were previously mapped and output in the middle. Its specific mathematical formula is as follows:

$$\alpha_1, \beta_1 = MLP(w'_{text}),$$
$$\alpha_2, \beta_2 = MLP(w'_{text}), \tag{2}$$

$$out = activation(\alpha_1 \times w'_{edit} + \beta_1),$$
$$out = activation(\alpha_2 \times out + \beta_2), \tag{3}$$

$$w_{out} = MLP(out). \tag{4}$$

where *out* is the intermediate variable.

3.3   Loss Function

Our goal is to manipulate an attribute (e.g., hairstyle, emotion, etc.) in a disentangled way based on text prompts, while other irrelevant attributes (e.g., background, identity, etc.) need to be preserved. We introduce CLIP loss, identity loss, and latent loss used in StyleCLIP. In addition, for some attribute manipulation tasks, such as hairstyle manipulation, we design a face loss; and for the task of facial emotion manipulation, we introduce a background loss.

CLIP Loss. Our mapper is trained to operate on image attributes indicated by desired text prompts while preserving other visual attributes of the input image.

We guide the mapper by minimizing the cosine distance of the generated image and text prompts in the CLIP latent space.

$$\mathcal{L}_{\text{clip}}(w) = D_{\text{clip}}\left(G\left(Concat(w_{fix}, w_{edit} + M(w_{edit}, w_{text}))\right), w_{text}\right). \quad (5)$$

where M is the AMM mentioned above, G is a pre-trained StyleGAN generator and D_{clip} is the cosine distance between the CLIP embeddings of its two arguments. In addition, w_{edit} and w_{fix} are the style codes that need to be manipulated and do not need to be manipulated, respectively.

Latent Loss. To preserve the visual attributes of the original input image, we minimize the L_2 norm of the manipulation step in the latent space.

$$\mathcal{L}_{\text{latent}}(w) = \|M(w_{edit}, w_{text})\|_2. \quad (6)$$

Identity Loss. Similarity to the input image is controlled by the L_2 distance in latent space, and by the identity loss

$$\mathcal{L}_{\text{id}}(w) = 1 - \langle R\left(G\left(Concat(w_{fix}, w_{edit} + M(w_{edit}, w_{text}))\right)\right), R(G(w)) \rangle. \quad (7)$$

where R is a pre-trained ArcFace [7] network for face recognition, and $\langle \cdot, \cdot \rangle$ computes the cosine similarity between it's arguments. G is the pre-trained and fixed StyleGAN generator.

Face Loss. Although only processing partial style codes could decrease attribute entanglement, the majority of style codes control multiple attributes, which leads to the issue of entanglement still not being well solved. Face loss is introduced which is mainly used to optimize and reduce the impact on the face when manipulating attributes that are not related to the face. Firstly, we use the pre-trained MTCNN [42] to detect faces and set all the values in the detected face position to 1, and all other positions to 0, and record it as the *mask*:

$$mask = MTCNN(G(w)), \quad (8)$$

$$w^{'} = Concat(w_{fix}, w_{edit} + M(w_{edit}, w_{text})),$$

$$\mathcal{L}_{\text{face}}(w) = \left\|G(w) \odot mask - G(w^{'}) \odot mask\right\|_2. \quad (9)$$

Background Loss. Background loss is the opposite of face loss. When we need to manipulate the face, the background loss can reduce the impact on attributes other than the face. The formula is as follows:

$$\mathcal{L}_{\text{background}}(w) = \|G(w) \odot (1 - mask) - G(w^{'}) \odot (1 - mask)\|_2. \quad (10)$$

Our total base loss function is a weighted combination of these losses:

$$\mathcal{L}_{\text{base}}(w) = \lambda_{\text{clip}}\mathcal{L}_{\text{clip}}(w) + \lambda_{\text{latent}}\mathcal{L}_{\text{latent}}(w) + \lambda_{\text{id}}\mathcal{L}_{\text{id}}(w). \quad (11)$$

where $\lambda_{clip}, \lambda_{latent}, \lambda_{id}$ are set to 0.6, 1.0, 0.2 respectively by default. For tasks related to face manipulation, such as emotional manipulation, age manipulation, gender manipulation, etc., we also introduce a background loss function, as follows:

$$\mathcal{L}_{\text{total}}(w) = \mathcal{L}_{\text{base}}(w) + \lambda_{\text{background}}\mathcal{L}_{\text{background}}(w). \quad (12)$$

And for the manipulation of hairstyle and hair color, we use face loss to reduce the impact on the face:

$$\mathcal{L}_{total}(w) = \mathcal{L}_{base}(w) + \lambda_{face}\mathcal{L}_{face}(w). \tag{13}$$

where both $\lambda_{background}$ and λ_{face} are set to 2.0.

Table 1. Experimental results of layer-by-layer analysis of the 18-layer StyleGAN generator. We list the attributes corresponding to all eighteen layers of style codes. Most style codes control multiple attributes. Among them, we use "micro features" to represent the style codes with low effects.

n-th	Attribute	n-th	Attribute
1	Eye glasses, head pose, hair length	10	Eye color, hair color
2	Eye glasses, head pose, hair length	11	Skin color
3	Eye glasses, head pose, hair length	12	Micro features
4	Head pose, hair length	13	Skin color, clothes color
5	Hairstyle, cheekbones, mouth, eyes, nose, clothes	14	Skin color, clothes color
6	Mouth	15	Micro features
7	Eyes, gender, mustache, forehead, mouth	16	Micro features
8	Eyes, hair color	17	Micro features
9	Hair color, clothes color, lip color, mustache color, eyebrow color, eye color, skin color	18	Clothes color

4   Experiments and Comparisons

To explore the efficiency and effectiveness of our method, we evaluate and compare our method with other methods on multiple image attribute manipulation tasks. The main comparison contents include hairstyle, hair color, emotion, age, gender, etc.

4.1   Implementation Details and Evaluation Metric

Implementation Details. We use e4e [34] as our inversion encoder and Style-GAN2 as our generator. We trained and evaluated our model on the CelebA-HQ dataset [13]. We collected text prompts from the internet describing attributes such as hairstyle, hair color, facial emotion, age, gender, etc. The number of text prompts for each attribute varies, ranging from 5 to 10. For each attribute mapper, it merely maps the style codes related to the attribute that needs to be manipulated and fuses the text conditional embeddings, which are randomly selected from the text set. For the training strategy, the base learning rate is 0.0001, the batch size is 1, and the number of training iterations for each attribute mapper is 20,000. We use the Adam optimizer [17], with β_1 and β_2 set to 0.9 and 0.999, respectively.

Fig. 2. Visual comparison with other methods. The left side of each row is the text prompt, the first column is the input image, and each subsequent column is the generated result of each method.

Evaluation Metric. Since our task is to edit the image, the pixel level of the image will change greatly, and we can only analyze the results according to the semantic level. Therefore, some indicators, such as PSNR, SSIM, FID [11], etc. that are commonly used in the field of image generation are not suitable for our task. We only evaluate whether the output satisfies the text prompt and whether the identity attribute is consistent with the input. For the metric of accuracy (ACC), we use CLIP [27] to calculate the similarity between text prompts and outputs, and introduce identity similarity (IDS) to evaluate the identity consistency between the outputs and the inputs.

4.2 Comparisons with State-of-the-Art Methods

We mainly conduct comprehensive comparisons on the manipulation of character hairstyles, hair color, facial emotion, and other attributes. Our method merely does mapping training once for the attributes to be manipulated. Compared with mainstream methods such as StyleCLIP [26] and HairCLIP, the speed is

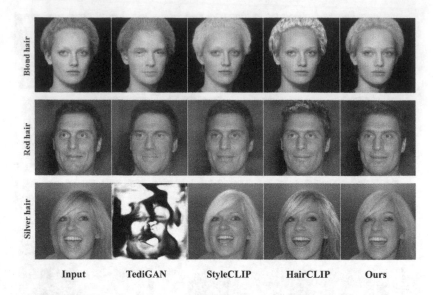

Fig. 3. Visual comparison with other methods on hair color editing.

faster and the coupling between properties is lower. As shown in Table 3, our method is less than StyleCLIP in both training and inference time. The properties controlled by each style code in the $W+$ space are given in Table 1. Through extensive experimentation, and with the help of more than 20 people, which we analyzed and discussed together, we finally determined the properties controlled by each style code in the $W+$ space.

Hairstyle Manipulation. We comprehensively compare our method with current state-of-the-art text-driven image processing methods TediGAN [37], Style-CLIP, and HairCLIP in hairstyle editing. According to Table 1, the *5-th* style code can control the hairstyle, while the first 4 style codes can control the length of the hair. Through extensive experiments, we found that jointly editing the first 5 layers of style codes can achieve better hairstyle editing results. Therefore, we only do feature mapping for the first 5 layers of style codes and embed text features, and do nothing for other style codes. According to the experimental results shown in Fig. 2, TediGAN fails to edit hairstyles. Both the qualitative and quantitative results of StyleCLIP are close to our method, but the issue with StyleCLIP is slow and requires training a mapper for each text prompt. And StyleCLIP does not match the text "Mohawk hairstyle" well. In general, HairCLIP has the best effect on hairstyle editing, but there is a clear difference in hair color, indicating that there is a coupling between attributes. We can find from the quantitative results in Table 2 that the results generated by Hair-CLIP are the closest to the text prompts but lose certain identity information compared to our method.

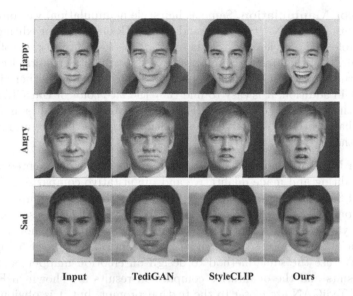

Fig. 4. Visual comparison with other methods on emotion editing.

Fig. 5. Visual comparisons with other methods on gender and age editing. Gender on the left, age on the right.

Table 2. Quantitative comparison of all attribute edits, where red font indicates the best result for each column.

Method	Hairstyle		Hair color		Emotion		Gender		Age	
	ACC	IDS	ACC	IDS	ACC	IDS	ACC	IDS	ACC	IDS
TediGAN	0.1842	0.5276	0.2032	0.3647	0.2135	0.5716	0.1719	0.4816	0.1926	0.6673
StyleCLIP	0.2119	0.8574	0.2436	0.8217	0.2169	0.7808	0.2346	0.7211	0.2153	0.8037
HairCLIP	0.2367	0.8731	0.2303	0.9237						
Ours	0.2325	0.8934	0.2377	0.8732	0.2478	0.8081	0.2153	0.7331	0.2410	0.7219

Hair Color Manipulation. Same as hairstyle manipulation, we compared our method with TediGAN, StyleCLIP, and HairCLIP on hair color editing. According to Table 1, the style codes of the *8-th*, *9-th*, *10-th* layers control the color of the hair. The qualitative comparisons of experimental results are shown in Fig. 3. Intuitively, TediGAN failed on silver hair editing, and other effects were mediocre. In addition, it is easy to find that the face color of HairCLIP is different from the face color of the input. From the quantitative results in Table 2, we can see that HairCLIP preserves the identity attribute better, and StyleCLIP matches the text prompt better.

Emotion Manipulation. In addition to hair, we also experimented with facial emotion editing and made qualitative and quantitative comparisons with TediGAN and StyleCLIP. Since HairCLIP cannot operate on attributes other than hair, we omit it here. Emotions are reflected by facial features, consequently, we edit style codes that control characters' facial features. As shown in Table 1, human facial features (nose, eyes, mouth) are controlled by layer 5, layer 6, and layer 7. We use the same method to design an emotion mapper, and the generated results and the qualitative comparison results are shown in Fig. 4. The results of TediGAN are closer to the textual prompt, but it is obvious that the identity attributes vary greatly. Both StyleCLIP and our method retain perfect identity attributes, but StyleCLIP does not work well for "Sad" emotion editing. It can also be found from the quantitative analysis in Table 2 that the IDS of StyleCLIP and our method are relatively high, reaching 0.7808 and 0.8081 respectively, indicating that the identity attributes are well preserved. And the ACC of our method is also the highest among the three, reaching 0.2478, which illustrates that our method is more matched with text prompts.

Other Manipulations. In addition to the above-mentioned manipulations, we also conducted experiments and comparisons on several attributes such as age and gender. For the edits of age, our method cannot accurately generate the specified age results, and can only be edited with approximate descriptions, such as old, middle-aged, young, etc. Gender editing, it's editing women as men and men as women. These two attributes are related to human facial features, consequently, we trained these two mappers in the same way as emotion editing. The specific experimental and comparison results are shown in Fig. 5 and Table 2. Both in terms of qualitative and quantitative results, StyleCLIP and our method are close, and both are better than TediGAN. StyleCLIP achieved the best results on both the ACC indicator of gender and the IDS indicator of age, reaching 0.2346 and 0.8037, respectively. Our method achieves the best results on the IDS indicator of gender and the ACC indicator of age, reaching 0.7331 and 0.2410, respectively.

4.3   Ablation Study

To demonstrate the effectiveness of each component in our overall approach, we perform ablation studies by evaluating the following subset models and loss functions:

Table 3. Quantitative time comparison on a single 1080Ti GPU, where red fonts indicate the best results.

Method	Train time	infer.time
StyleCLIP	10–12h	75 ms
Ours	8 h	43 ms

Table 4. Quantitative comparison of face loss and background loss, where red fonts indicate the best results.

Hairstyle			Emotion		
λ_{face}	ACC	IDS	λ_{bg}	ACC	IDS
0	0.2308	0.8636	0	0.2387	0.7988
1	0.2347	0.8742	1	0.2412	0.8010
2	0.2325	0.8934	2	0.2478	0.8081
3	0.2136	0.8898	3	0.2296	0.8103

Table 5. For the number of edited style codes, quantitative comparisons are made on emotion editing and model scale, respectively, where red fonts indicate the best results.

Method	Emotion		Mapping module	
	ACC	IDS	Params(M)	MFLOPs
Edit all style codes	0.2394	0.6083	170.2	170.1
Ours	0.2478	0.8081	28.37	28.35

Loss Function. We verify the effectiveness of face loss and background loss by controlling variables, the results are shown in Fig. 6. To verify the importance of the background loss, we did a comparative experiment on emotion editing, and the experimental results are shown in the left half of Fig. 6. It can be seen that when the background loss is not used, the generated results are significantly different from the input on the hair, and after using the background loss, the background similarity can be well maintained. Likewise, we conduct comparative experiments on hairstyle editing to verify the effectiveness of face loss. The eyes and mouth have changed to a large extent without using face loss, and after using face loss, these changes are gone. The above two sets of comparative experiments fully verify the effectiveness of our background loss and face loss. It should be pointed out that these two losses limit the editing ability of images. As shown in Table 4, face loss can reduce the loss of identity information, but it limits the editing ability of images to a certain extent. Compared with face loss, the effect of background loss is limited. When we do not use these two losses, our ACC and IDS metrics both exceed StyleCLIP, which can also reflect the effectiveness of our $\mathcal{W}+$ feature selection.

Mapping All Style Vectors. To prove the effectiveness and efficiency of using partial style codes, we map all style codes according to StyleCLIP. As can be seen from Fig. 7, when mapping all style codes, the generated results will look unnatural and cannot match the text prompts well. We also give the quantitative comparison results in Table 5. Mapping all style codes not only increased the number of parameters and FLOPs several times, but also worse the ACC and IDS indicators, even the IDS decreased by nearly 0.2, which also proves that our method not only reduces the number of parameters speeding up the training and inference time but also generate better results.

Input   Ours   w/o $\mathcal{L}_{background}$ Input   Ours   w/o $\mathcal{L}_{face}$

Fig. 6. Visual comparison of our generated results with methods and variants of our model. The top left and bottom left are comparisons of sentiment, two text prompts are "happy" and "angry", where the left is the original input, the middle is our result, and the right is the result without background loss. The top right and bottom right are comparisons of hairstyles, the two text prompts are "afro hairstyle" and "bowl-cut hairstyle", where the left is the original input, the middle is our result, and the right is the result without face loss.

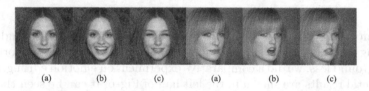

(a)   (b)   (c)   (a)   (b)   (c)

Fig. 7. Visual comparison of our generated results with methods and variants of our model. The text prompts are "happy" and "angry", respectively. (a) The input images. (b) Results were obtained with our method (map part of the style codes). (c) Results were obtained with "map all style codes".

5   Conclusions

We propose a new image attribute manipulation method, which combines the powerful generative ability of StyleGAN with the extraordinary visual concept encoding ability of CLIP, which can easily embed text prompts into images and guide image generation. Our model support high-quality manipulation of multiple attributes, including emotion editing, hairstyle editing, age editing, etc., in a decoupled manner. Extensive experiments and comparisons demonstrate that our method outperforms previous methods in terms of operational capability, irrelevant attribute preservation, and image realism.

It should be pointed out that our method sometimes fails to edit certain colors (blue, yellow, etc.) in hair color editing. We suspect that the editor's style codes are not complete enough, or that CLIP's text encoder does not work well with colors. Another point is that although our method is faster than StyleCLIP, one mapper cannot edit all attributes. In the future, we will continue to refine and improve these issues.

References

1. Abdal, R., Qin, Y., Wonka, P.: Image2StyleGAN: how to embed images into the StyleGAN latent space? In: Proceedings of the IEEE/CVF International Conference on Computer Vision, pp. 4432–4441 (2019)
2. Abdal, R., Qin, Y., Wonka, P.: Image2StyleGAN++: how to edit the embedded images? In: Proceedings of the IEEE/CVF Conference on Computer Vision and Pattern Recognition, pp. 8296–8305 (2020)
3. Abdal, R., Zhu, P., Mitra, N.J., Wonka, P.: StyleFlow: attribute-conditioned exploration of StyleGAN-generated images using conditional continuous normalizing flows. ACM Trans. Graph. (ToG) **40**(3), 1–21 (2021)
4. Alaluf, Y., Patashnik, O., Cohen-Or, D.: ReStyle: a residual-based StyleGAN encoder via iterative refinement. In: Proceedings of the IEEE/CVF International Conference on Computer Vision, pp. 6711–6720 (2021)
5. Collins, E., Bala, R., Price, B., Susstrunk, S.: Editing in style: uncovering the local semantics of GANs. In: Proceedings of the IEEE/CVF Conference on Computer Vision and Pattern Recognition, pp. 5771–5780 (2020)
6. Creswell, A., Bharath, A.A.: Inverting the generator of a generative adversarial network. IEEE Trans. Neural Netw. Learn. Syst. **30**(7), 1967–1974 (2018)
7. Deng, J., Guo, J., Xue, N., Zafeiriou, S.: ArcFace: additive angular margin loss for deep face recognition. In: Proceedings of the IEEE/CVF Conference on Computer Vision and Pattern Recognition, pp. 4690–4699 (2019)
8. Dong, H., Yu, S., Wu, C., Guo, Y.: Semantic image synthesis via adversarial learning. In: Proceedings of the IEEE International Conference on Computer Vision, pp. 5706–5714 (2017)
9. Goodfellow, I., et al.: Generative adversarial nets. In: Advances in Neural Information Processing Systems, vol. 27 (2014)
10. Härkönen, E., Hertzmann, A., Lehtinen, J., Paris, S.: GANSpace: discovering interpretable GAN controls. In: Advances in Neural Information Processing Systems, vol. 33, pp. 9841–9850 (2020)
11. Heusel, M., Ramsauer, H., Unterthiner, T., Nessler, B., Hochreiter, S.: GANs trained by a two time-scale update rule converge to a local Nash equilibrium. In: Advances in Neural Information Processing Systems, vol. 30 (2017)
12. Huang, X., Belongie, S.: Arbitrary style transfer in real-time with adaptive instance normalization. In: Proceedings of the IEEE International Conference on Computer Vision, pp. 1501–1510 (2017)
13. Karras, T., Aila, T., Laine, S., Lehtinen, J.: Progressive growing of GANs for improved quality, stability, and variation. arXiv preprint arXiv:1710.10196 (2017)
14. Karras, T., et al.: Alias-free generative adversarial networks. In: Advances in Neural Information Processing Systems, vol. 34, pp. 852–863 (2021)
15. Karras, T., Laine, S., Aila, T.: A style-based generator architecture for generative adversarial networks. In: Proceedings of the IEEE/CVF Conference on Computer Vision and Pattern Recognition, pp. 4401–4410 (2019)
16. Karras, T., Laine, S., Aittala, M., Hellsten, J., Lehtinen, J., Aila, T.: Analyzing and improving the image quality of StyleGAN. In: Proceedings of the IEEE/CVF Conference on Computer Vision and Pattern Recognition, pp. 8110–8119 (2020)
17. Kingma, D.P., Ba, J.: Adam: a method for stochastic optimization. arXiv preprint arXiv:1412.6980 (2014)
18. Kingma, D.P., Welling, M.: Auto-encoding variational bayes. arXiv preprint arXiv:1312.6114 (2013)

19. Koh, J.Y., Baldridge, J., Lee, H., Yang, Y.: Text-to-image generation grounded by fine-grained user attention. In: Proceedings of the IEEE/CVF Winter Conference on Applications of Computer Vision, pp. 237–246 (2021)

20. Li, B., Qi, X., Lukasiewicz, T., Torr, P.H.: ManiGAN: text-guided image manipulation. In: Proceedings of the IEEE/CVF Conference on Computer Vision and Pattern Recognition, pp. 7880–7889 (2020)

21. Li, W., et al.: Object-driven text-to-image synthesis via adversarial training. In: Proceedings of the IEEE/CVF Conference on Computer Vision and Pattern Recognition, pp. 12174–12182 (2019)

22. Liu, Y., et al.: Describe what to change: a text-guided unsupervised image-to-image translation approach. In: Proceedings of the 28th ACM International Conference on Multimedia, pp. 1357–1365 (2020)

23. Nam, S., Kim, Y., Kim, S.J.: Text-adaptive generative adversarial networks: manipulating images with natural language. In: Advances in Neural Information Processing Systems, vol. 31 (2018)

24. Nitzan, Y., Bermano, A., Li, Y., Cohen-Or, D.: Face identity disentanglement via latent space mapping. arXiv preprint arXiv:2005.07728 (2020)

25. Park, T., Liu, M.Y., Wang, T.C., Zhu, J.Y.: Semantic image synthesis with spatially-adaptive normalization. In: Proceedings of the IEEE Conference on Computer Vision and Pattern Recognition (2019)

26. Patashnik, O., Wu, Z., Shechtman, E., Cohen-Or, D., Lischinski, D.: StyleCLIP: text-driven manipulation of StyleGAN imagery. In: Proceedings of the IEEE/CVF International Conference on Computer Vision, pp. 2085–2094 (2021)

27. Radford, A., et al.: Learning transferable visual models from natural language supervision. In: International Conference on Machine Learning, pp. 8748–8763. PMLR (2021)

28. Reed, S., Akata, Z., Yan, X., Logeswaran, L., Schiele, B., Lee, H.: Generative adversarial text to image synthesis. In: International Conference on Machine Learning, pp. 1060–1069. PMLR (2016)

29. Richardson, E., et al.: Encoding in style: a StyleGAN encoder for image-to-image translation. In: Proceedings of the IEEE/CVF Conference on Computer Vision and Pattern Recognition, pp. 2287–2296 (2021)

30. Shen, Y., Gu, J., Tang, X., Zhou, B.: Interpreting the latent space of GANs for semantic face editing. In: Proceedings of the IEEE/CVF Conference on Computer Vision and Pattern Recognition, pp. 9243–9252 (2020)

31. Shi, Y., Yang, X., Wan, Y., Shen, X.: SemanticStyleGAN: learning compositional generative priors for controllable image synthesis and editing. In: Proceedings of the IEEE/CVF Conference on Computer Vision and Pattern Recognition, pp. 11254–11264 (2022)

32. Tao, M., Tang, H., Wu, F., Jing, X.Y., Bao, B.K., Xu, C.: DF-GAN: a simple and effective baseline for text-to-image synthesis. arXiv e-prints (2020)

33. Tewari, A., et al.: StyleRig: rigging StyleGAN for 3D control over portrait images. In: Proceedings of the IEEE/CVF Conference on Computer Vision and Pattern Recognition, pp. 6142–6151 (2020)

34. Tov, O., Alaluf, Y., Nitzan, Y., Patashnik, O., Cohen-Or, D.: Designing an encoder for StyleGAN image manipulation. ACM Trans. Graph. (TOG) **40**(4), 1–14 (2021)

35. Wei, T., et al.: HairCLIP: design your hair by text and reference image. In: Proceedings of the IEEE/CVF Conference on Computer Vision and Pattern Recognition, pp. 18072–18081 (2022)

36. Wu, Z., Lischinski, D., Shechtman, E.: StyleSpace analysis: disentangled controls for StyleGAN image generation. In: Proceedings of the IEEE/CVF Conference on Computer Vision and Pattern Recognition, pp. 12863–12872 (2021)
37. Xia, W., Yang, Y., Xue, J.H., Wu, B.: TediGAN: text-guided diverse face image generation and manipulation. In: Proceedings of the IEEE/CVF Conference on Computer Vision and Pattern Recognition, pp. 2256–2265 (2021)
38. Xu, T., et al.: AttnGAN: fine-grained text to image generation with attentional generative adversarial networks. In: Proceedings of the IEEE Conference on Computer Vision and Pattern Recognition, pp. 1316–1324 (2018)
39. Yang, C., Shen, Y., Zhou, B.: Semantic hierarchy emerges in deep generative representations for scene synthesis. Int. J. Comput. Vis. **129**(5), 1451–1466 (2021). https://doi.org/10.1007/s11263-020-01429-5
40. Zhang, H., et al.: StackGAN: text to photo-realistic image synthesis with stacked generative adversarial networks. In: Proceedings of the IEEE International Conference on Computer Vision, pp. 5907–5915 (2017)
41. Zhang, H., et al.: StackGAN++: realistic image synthesis with stacked generative adversarial networks. IEEE Trans. Pattern Anal. Mach. Intell. **41**(8), 1947–1962 (2018)
42. Zhang, K., Zhang, Z., Li, Z., Qiao, Y.: Joint face detection and alignment using multitask cascaded convolutional networks. IEEE Sig. Process. Lett. **23**(10), 1499–1503 (2016)
43. Zhu, J., Shen, Y., Zhao, D., Zhou, B.: In-domain GAN inversion for real image editing. In: Vedaldi, A., Bischof, H., Brox, T., Frahm, J.-M. (eds.) ECCV 2020. LNCS, vol. 12362, pp. 592–608. Springer, Cham (2020). https://doi.org/10.1007/978-3-030-58520-4_35

Gated Cross Word-Visual Attention-Driven Generative Adversarial Networks for Text-to-Image Synthesis

Borun Lai[1], Lihong Ma[1], and Jing Tian[2](✉)

[1] School of Electronics Information Engineering,
South China University of Technology, Guangzhou, China
eebrlai@mail.scut.edu.cn, eelhma@scut.edu.cn
[2] Institute of Systems Science, National University of Singapore,
Singapore 119615, Singapore
tianjing@nus.edu.sg

Abstract. The main objective of text-to-image (Txt2Img) synthesis is to generate realistic images from text descriptions. We propose to insert a gated cross word-visual attention unit (GCAU) into the conventional multiple-stage generative adversarial network Txt2Img framework. Our GCAU consists of two key components. First, a cross word-visual attention mechanism is proposed to draw fine-grained details at different subregions of the image by focusing on the relevant words (via the visual-to-word attention), and select important words by paying attention to the relevant synthesized subregions of the image (via the word-to-visual attention). Second, a gated refinement mechanism is proposed to dynamically select important word information for refining the generated image. Extensive experiments are conducted to demonstrate the superior image generation performance of the proposed approach on CUB and MS-COCO benchmark datasets.

1 Introduction

The objective of text-to-image (Txt2Img) is to generate a realistic image from a given text description that is consistent with the text semantics. Deep learning techniques, particularly, Generative Adversarial Networks (GANs), have become an effective generative approach in Txt2Img synthesis [1,2,11]. It has many significant applications, such as image enhancement [10], text-image matching [8].

The GAN-based approaches encode the text description as a global sentence vector and then apply it as a conditional constraint to generate an image that matches the text description. They can be classified into two categories: (i) one-stage methods, and (ii) multiple-stage methods. The one-stage methods generate images by adding up-sampling layers in a single generator. However, this may cause the generated image to be inconsistent with the input text description [15]. Thus, a matching-aware zero-centered gradient penalty method is proposed in [19] to make the generated image better match the text description. The multiple-stage methods generate an initial low-resolution image by the first generator and then refine it by subsequent generators to create high-resolution progressively, where the global sentence feature is used as a conditional constraint to the discriminator at each stage to ensure that the generated image matches the text description [24, 25].

Considering that fine-grained details are critical in the generated image, the attention mechanism has been exploited for Txt2Img generation. AttnGAN [22] drew image details by computing the attention distribution of all word feature vectors on each visual feature vector. However, the unchanged text representation is used at each stage of image refinement. Moreover, if the attention weights are wrongly estimated at the beginning, some important word information may be ignored. An attention regularization loss was proposed in SEGAN [17] to highlight important words. A threshold is set so that the attention weight of important words (above the threshold) can be gradually increased and the attention weight of irrelevant words (below the threshold) can be gradually decreased. The limit is that it is not easy to determine the appropriate value range of the threshold. A dynamic memory mechanism was proposed in DM-GAN [27] to refine image details dynamically. The memory writing gate would select important word information according to the global image information and word information, and save them in memory slots. The memory is addressed and read according to the correlation between each subregion of the image and the memory, thus gradually completing the refinement of the image. Its limitation is that it considers the contribution of all subregions of the image to each word as equal. KT-GAN [18] focused on adjusting attention distribution by using hyperparameters to extract important word information. However, KT-GAN is a time-consuming method, and it requires accurately-estimated attention weight of each word.

The fundamental challenge in Txt2Img synthesis is how to exploit the information from the input sentence, which guides details generation in the image. Our approach yields the following two contributions.

- Firstly, each word in the input sentence provides different information depicting the image content. The image information should be taken into account to determine the importance of every word, and the word information should also be considered to determine the importance of every subregion of the image. For that, we propose a cross word-visual attention mechanism. It draws details at different image subregions by focusing on the relevant words via visual-to-word (V2W) attention, and select important words by focusing on the relevant image subregions via word-to-visual (W2V) attention.
- Secondly, if the same word representation is utilized at multiple phases of image refinement, the procedure may become ineffective. For that, we propose a gated refinement mechanism to dynamically select the important word information from the updated word representation based on the updated image representation at multiple image refinement stages.

We propose to include these two contributions into a multiple-stage GAN-based Txt2Img synthesis framework by combining them to construct a gated cross word-visual attention unit.

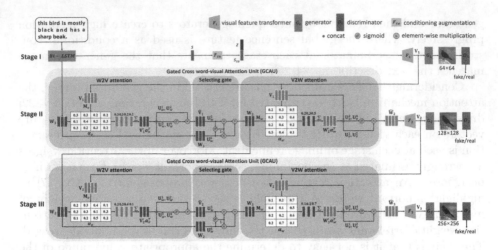

Fig. 1. The framework of our proposed approach. The proposed gated cross-word visual attention unit, which contains a W2V attention, a V2W attention, and a selecting gate, is used for the Stage II and Stage III.

The remainder of this paper is organized as follows. The proposed Txt2Img synthesis approach is presented in Sect. 2 by developing a gated cross word-visual attention method. It is evaluated in extensive experiments in Sect. 3. Finally, Sect. 4 concludes this paper.

2   Proposed Txt2Img Synthesis Approach

We leverage a conventional multiple-stage GAN-based Txt2Img framework, where a low-resolution initial image is firstly generated and then refined via several stages to obtain the final high-resolution synthesized image. Let $\mathbf{V}_i$ and $\mathbf{W}_i$ be visual features and word features, respectively. F_{ca} represents the Conditioning Augmentation [24] that converts the sentence vector to the conditioning vector, $z \sim N(0,1)$ is a random noise vector, F_i represents the visual feature transformer at the i-th stage, G_i represents the generator at the i-th stage, D_i represents the discriminator at the i-th stage.

As shown in Fig. 1, we propose to insert a gated cross word-visual attention unit (GCAU) at each stage (except the first stage) of this Txt2Img framework. Our GCAU contains a W2V attention, a V2W attention, and a selecting gate. These three components are described in detail as follows.

2.1   Cross Word-Visual Attention

Denote the word feature matrix $\mathbf{W}_i \in \mathbb{R}^{D_w \times N_w}$, the visual feature matrix $\mathbf{V}_i \in \mathbb{R}^{D_v \times N_v}$, where D_w and D_v are dimensions of a word feature vector and a visual feature vector, N_w and N_v are numbers of word feature vectors and visual feature vectors.

W2V Attention. Firstly, it transforms visual features from a visual semantic space to a word semantic space by a 1×1 convolution operator $\mathbf{M}_v(\cdot)$ to obtain a mapped visual feature matrix $\mathbf{V}'_i \in \mathbb{R}^{D_w \times N_v}$ as

$$\mathbf{V}'_i = \mathbf{M}_v\left(\mathbf{V}_i\right). \tag{1}$$

Then, it calculates a similarity matrix $\mathbf{W}_i^T \mathbf{V}'_i$ between the mapped visual feature matrix $\mathbf{V}'_i$ and the word feature matrix $\mathbf{W}_i$. By calculating the attention distribution $\boldsymbol{\alpha}_v \in \mathbb{R}^{N_w \times N_v}$ on each mapped visual feature vector, the normalized attention distribution is obtained as

$$\boldsymbol{\alpha}_v = \text{softmax}\left(\mathbf{W}_i^T \mathbf{V}'_i\right). \tag{2}$$

Next, according to the attention distribution $\boldsymbol{\alpha}_v$, each mapped visual feature vector is weighted and summed up to obtain the visual-context feature matrix as $\mathbf{V}'_i \boldsymbol{\alpha}_v^T$.

V2W Attention. It follows a similar procedure as W2V as follows. Firstly, it applies a 1×1 convolution operator $\mathbf{M}_w(\cdot)$ to obtain a mapped word feature matrix $\mathbf{W}'_i \in \mathbb{R}^{D_v \times N_w}$ as

$$\mathbf{W}'_i = \mathbf{M}_w\left(\mathbf{W}_i\right) \tag{3}$$

Then, it calculates the attention distribution $\boldsymbol{\alpha}_w \in \mathbb{R}^{N_v \times N_w}$ on each mapped word feature vector to obtain the normalized attention distribution as

$$\boldsymbol{\alpha}_w = \text{softmax}\left(\mathbf{V}_i^T \mathbf{W}'_i\right). \tag{4}$$

Next, each mapped word feature vector is weighted and summed up to obtain the word-context feature matrix as $\mathbf{W}'_i \boldsymbol{\alpha}_w^T$.

Finally, following the idea of the Attention on Attention (AoA) method [4], we further concatenate the visual-context feature matrix and word feature matrix, then apply two separate linear transformations conditioned on the concatenated result. Then we add another attention using element-wise multiplication to eventually obtain the W2V attentional information $\hat{\mathbf{V}}_i$ as

$$\hat{\mathbf{V}}_i = \left(\mathbf{U}_w^1 \mathbf{V}'_i \boldsymbol{\alpha}_v^T + \mathbf{U}_w^2 \mathbf{W}_i + b_w^1\right) \otimes \sigma\left(\mathbf{U}_w^3 \mathbf{V}'_i \boldsymbol{\alpha}_v^T + \mathbf{U}_w^4 \mathbf{W}_i + b_w^2\right), \tag{5}$$

where $\sigma(\cdot)$ is the sigmoid activation function, $\otimes$ denotes the element-wise multiplication, $\mathbf{U}_w^1, \mathbf{U}_w^2, \mathbf{U}_w^3, \mathbf{U}_w^4 \in \mathbb{R}^{D_w \times D_w}$, $b_w^1, b_w^2 \in \mathbb{R}^{D_w}$. It highlights visual subregions that each word should pay attention to, it measures the word importance that will be sent to the selecting gate in the gated refinement mechanism for important word selection. The AoA method [4] is also applied to obtain the V2W attentional information.

2.2   Gated Refinement

We propose a selecting gate to dynamically select the important word feature at different image refinement stages. It adopts a structure of a memory writing gate [27], but we modify it in two ways. Firstly, we use the word information refined from the previous stage as the input, instead of the fixed initial word information in [27]. Secondly, we adaptively combine features from different visual subregions according to the W2V attentional information, instead of treating them equally in [27].

Our selecting gate is defined as follows. It inputs the previous word information $\mathbf{W}_{i-1}$ and the W2V attentional information $\hat{\mathbf{V}}_{i-1}$, which are firstly transformed by linear transformations $\mathbf{U}_w^5, \mathbf{U}_w^6 \in \mathbb{R}^{1 \times D_w}$, and normalized by the sigmoid activation function $\sigma(\cdot)$ as

$$g(\mathbf{W}_{i-1}, \hat{\mathbf{V}}_{i-1}) = \sigma(\mathbf{U}_w^5 \mathbf{W}_{i-1} + \mathbf{U}_w^6 \hat{\mathbf{V}}_{i-1}). \tag{6}$$

Then, it removes the past word information to be forgotten and obtains the attentional information to be memorized as

$$\mathbf{W}_i = g(\mathbf{W}_{i-1}, \hat{\mathbf{V}}_{i-1})\hat{\mathbf{V}}_{i-1} + (1 - g(\mathbf{W}_{i-1}, \hat{\mathbf{V}}_{i-1}))\mathbf{W}_{i-1}. \tag{7}$$

2.3 Objective Function

The objective function is defined by combining all stages of image refinement as

$$L = \sum_i L_{G_i} + \lambda_1 L_{CA} + \lambda_2 \sum_i L_{DAMSM}(\mathbf{W}_i), \tag{8}$$

where L_{G_i} is an adversarial loss [22], λ_1 and λ_2 are the corresponding weights of a conditioning augmentation loss L_{CA} [27] and a loss $L_{DAMSM}(\mathbf{W}_i)$, which is modified from the DAMSM loss [22]. L_{G_i} encourages the generated image to be realistic and match the given text description, L_{CA} avoids overfitting in model training, and $L_{DAMSM}(\mathbf{W}_i)$ encourages each subregion of the image to match each word in the given text description as much as possible.

It is important to note that we use the refined word information $\mathbf{W}_i$ in the objective function. We compute the DAMSM loss between the generated image and the refined word information at each stage of image refinement. This is different from the initial word information $\mathbf{W}$ that is used in [22]. This modified loss enables the generated image to match the text description by ensuring the semantic consistency between the visual subregions and the selected word information at each stage. It also enables our W2V attention to accurately highlight visual subregions more relevant to the word.

2.4 Summary of Our Method

At the i-th stage of image refinement, our GCAU takes the previous word feature $\mathbf{W}_{i-1}$ and visual feature $\mathbf{V}_{i-1}$ as the inputs, and output the current word feature $\mathbf{W}_i$ and visual feature $\mathbf{V}_i$ as follows.

Step 1. Apply the W2V attention to calculate the current W2V attentional information $\hat{\mathbf{V}}_{i-1}$ from $\mathbf{W}_{i-1}$ and $\mathbf{V}_{i-1}$.

Step 2. Apply the selecting gate to select current important word information $\mathbf{W}_i$ from $\hat{\mathbf{V}}_{i-1}$ and $\mathbf{W}_{i-1}$.

Step 3. Apply the V2W attention to calculate the V2W attentional information $\hat{\mathbf{W}}_i$ from $\mathbf{W}_i$ and $\mathbf{V}_{i-1}$. Then we concatenate $\hat{\mathbf{W}}_i$ with itself and input to the visual feature transformer to obtain the updated visual feature $\mathbf{V}_i$.

Step 4. Generate the refined image $\mathbf{I}_i$ from $\mathbf{V}_i$ via $\mathbf{G}_i$.

3 Experimental Results

Extensive experiments in this section are carried out to compare our proposed approach with other previous state-of-the-art Txt2Img synthesis approaches to verify the performance of our approach. In addition, we conduct an ablation study to verify the performance of each component.

3.1 Datasets

We use two public benchmark datasets, including CUB [20] and MS-COCO [9] datasets. The CUB [20] dataset is a single-object dataset with 200 categories, in which the training set contains 8,855 images and the test set contains 2,933 images. There are ten text descriptions for each image. The MS-COCO [9] dataset is a multi-object dataset, in which the training set contains 82,783 images and the test set contains 40,470 images. There are five text descriptions for each image.

3.2 Implementation Details

We use the bidirectional LSTM as text encoder to encode the input text description to obtain the word features and the sentence feature. An image with 64×64 resolution is generated at the initial stage, and then refined to generate images with 128×128 and 256×256 resolution. D_v and D_w are set to 64 and 256, respectively. N_w is set to 64 and N_v is the resolution of the generated image at each stage, which is set to 64×64, 128×128, 256×256. The model is trained on a Nvidia GeForce RTX 2080 Ti GPU. The batch size is set to 16 on the CUB dataset and 12 on the MS-COCO dataset. All models are optimized with the ADAM optimizer [5], β_1 and β_2 are set to 0.5 and 0.999. The learning rate of generators and discriminators are set to 0.0002. The model is trained for 800 epochs on the CUB dataset and 200 epochs on the MS-COCO dataset. For λ_1 and λ_2 in Eq. (8), λ_1 is set to 1 on the CUB dataset and MS-COCO dataset, λ_2 is set to 5 on the CUB dataset and 50 on the MS-COCO dataset.

3.3 Evaluation Metrics

We evaluate the model generation performance by generating 30,000 images based on the text descriptions from unseen test set. There are three metrics used for evaluation: IS [16], FID [3], and R-precision [22]. IS [16] is used to evaluate the diversity of the generated images, FID [3] is used to evaluate the reality of the generated images, and R-precision [22] is used to evaluate how well the generated images match the text descriptions.

- The IS [16] metric calculates the Kullback-Leibler divergence between the conditional class distribution and the marginal class distribution. A higher IS means the generated images have more diversity.
- The FID [3] metric calculates the Fréchet distance between synthetic and real-world images. It first uses the Inception v3 network to extract features, then uses a Gaussian model to model the feature space, and finally calculates the distance between the two features. A lower FID means that the generated images are closer to real-world images.
- The R-precision [22] metric measures the cosine similarity between global image features and candidate sentence features. A higher R-precision means that the generated images match the text descriptions better.

3.4   Experimental Results

The comparison results of our approach with other previous state-of-the-art approaches on the test set are shown in Table 1, Table 2 and Table 3. The following is the analysis report of the comparison results.

Firstly, as seen in the IS performance in Table 1, our method performs only worse than TVBi-GAN [21] and DF-GAN [19] on the CUB dataset, and only worse than SD-GAN [23] on the MS-COCO dataset. Although SD-GAN [23] trains the model with multiple text descriptions, our method only uses a single text description, which may lead to a possible limitation of our method in the diversity of generated images. Moreover, SD-GAN [23] will fail to train if each image in the dataset contains only a single text description. In addition, SD-GAN [23] uses the siamese structure to extract text semantic information, which is more complex than our network and more powerful hardware equipment is required for training.

Secondly, as seen in the FID performance in Table 2, our method performs only worse than TVBi-GAN [21] on the CUB dataset and achieves the best performance on the MS-COCO dataset. Our method performs worse than TVBi-GAN [21] on the CUB dataset, but the CUB dataset is a single-object dataset, while the MS-COCO dataset is a multi-object dataset, and our method achieves the best performance on the MS-COCO dataset which proves that our method performs better in generating multi-object images.

Thirdly, as seen in the R-precision performance in Table 3, our method achieves the state-of-the-art performance on the CUB dataset and performs only worse than Obj-GAN [7] on the MS-COCO dataset. Obj-GAN [7] uses a discriminator based on the Fast R-CNN model to provide rich object-wise discrimination signals, which helps semantic alignment of text descriptions and images. This also complicates the network. Our method does not need to add additional networks, and the performance is very close.

Table 1. The performance of IS for our proposed method comparing with other methods on CUB and MS-COCO test sets. Higher IS means better performance.

Methods	CUB [20]	MS-COCO [9]
GAN-INT-CLS [15]	$2.88 \pm .04$	$7.88 \pm .07$
GAWWN [14]	$3.62 \pm .07$	–
StackGAN [24]	$3.70 \pm .04$	$8.45 \pm .03$
StackGAN++ [25]	$4.04 \pm .05$	–
HD-GAN [26]	$4.15 \pm .05$	$11.86 \pm .18$
AttnGAN [22]	$4.36 \pm .03$	$25.89 \pm .47$
MirrorGAN [13]	$4.56 \pm .05$	$26.47 \pm .41$
ControlGAN [6]	$4.58 \pm .09$	$24.06 \pm .60$
LeicaGAN [12]	$4.62 \pm .06$	–
SEGAN [17]	$4.67 \pm .04$	$27.86 \pm .31$
ObjGAN [7]	–	$30.29 \pm .33$
DM-GAN [27]	$4.75 \pm .07$	$30.49 \pm .57$
SD-GAN [23]	$4.67 \pm .09$	$35.69 \pm .50$
DF-GAN [19]	$4.86 \pm .04$	–
TVBi-GAN [21]	$5.03 \pm .03$	$31.01 \pm .34$
Ours	$\mathbf{4.79 \pm .05}$	$\mathbf{31.22 \pm .58}$

Table 2. The performance of FID for our proposed method comparing with other methods on CUB and MS-COCO test sets. Lower FID means better performance.

Methods	CUB [20]	MS-COCO [9]
StackGAN [24]	51.89	74.05
AttnGAN [22]	23.98	35.49
SEGAN [17]	18.17	32.28
DM-GAN [27]	16.09	32.64
TVBi-GAN [21]	11.83	31.97
Obj-GAN [7]	–	25.64
KT-GAN [18]	17.32	30.73
Ours	**15.16**	**25.49**

Table 3. The performance of R-precision for our proposed method comparing with other methods on CUB and MS-COCO test sets. Higher R-precision means better performance.

Methods	CUB [20]	MS-COCO [9]
AttnGAN [22]	67.82	85.47
MirrorGAN [13]	57.67	74.52
ControlGAN [6]	69.33	82.43
DM-GAN [27]	72.31	88.56
Obj-GAN [7]	–	91.05
Ours	**78.07**	**90.97**

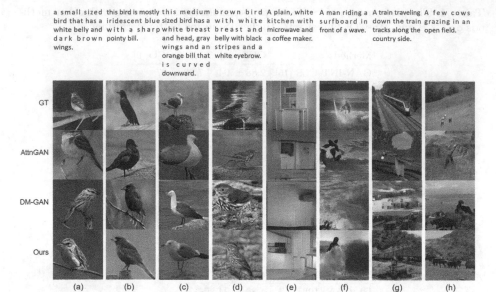

Fig. 2. The performance comparison of ground truth images and images generated by AttnGAN [22], DM-GAN [27] and our method. The four columns on the left are from the CUB [20] dataset, and the four columns on the right are from the MS-COCO [9] dataset.

Lastly, we compare our method with SEGAN [17] and DM-GAN [27]. Firstly, as seen in the IS performance in Table 1, our method achieves 2.57% higher than SEGAN [17], 0.84% higher than DM-GAN [27] on the CUB dataset, and 12.06% higher than SEGAN [17], 2.39% higher than DM-GAN [27] on the MS-COCO dataset. Secondly, as seen in the FID performance in Table 2, our method achieves 16.56% lower than SEGAN [17], 5.77% lower than DM-GAN [27] on the CUB dataset, and 21.03% lower than SEGAN [17], 21.90% lower than DM-GAN [27] on the MS-COCO dataset. Thirdly, as seen in the R-precision performance in Table 3, our method also achieves better performance than DM-GAN [27].

As can be seen from Fig. 2, for single-object generation and multi-object generation, the shapes of the generated images are more realistic and the generated images also have more details, such as the black stripes and white eyebrow in Fig. 2(d) and the microwave in Fig. 2(e). This verifies that our method can generate more realistic images with more details.

3.5   Ablation Study

Our work is to improve the V2W attention by integrating W2V attention and gated refinement, which enables V2W attention to pay more attention to important words. We conduct an ablation study to gradually integrate various components and evaluate the model performance using IS and FID based on the CUB

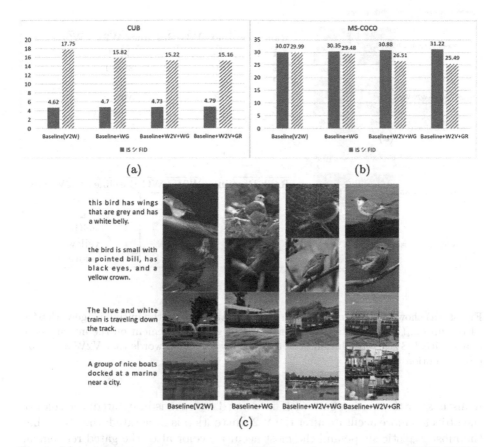

Fig. 3. (a) and (b) are the results of the ablation study on the CUB and MS-COCO datasets. (c) is the visualization of the ablation study. Baseline denotes that only V2W attention is integrated, WG denotes writing gate [27], W2V denotes our proposed W2V attention, and GR denotes our proposed gated refinement.

and MS-COCO dataset. As shown in Fig. 3(a) and Fig. 3(b), the performance of our model on IS and FID improved progressively with each component being integrated, which demonstrates the effective contribution of each component. We also show the images by gradually integrating various components. As shown in Fig. 3(c), for the first text description, the generated objects obviously do not yet have the correct shape, and the important words "bird", "wings", and "belly" have not been accurately positioned and highlighted when only the V2W attention is integrated; the shape of bird is highlighted after the writing gate with fixed word features is integrated; the shape of bird is formed and the important words "wings" and "belly" have been accurately positioned and highlighted after the W2V attention is integrated; the image details corresponding to "bird", "wings", and "belly" have been significantly enhanced after the gated refinement with refined word features is integrated. For the third text description, the shape of the object is more

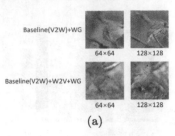

the bird has a small bill and a black eyering that is small.

Rank	Baseline+WG	Baseline+W2V+WG
1	the	bird
2	bird	small
3	small	the
4	is	eyering
5	that	has

(a)

a bird with predominantly black color has splashes of yellow and white here and there.

Rank	Baseline+W2V+WG	Baseline+W2V+GR
1	bird	bird
2	a	a
3	with	with
4	yellow	yellow
5	color	white

(b)

Fig. 4. (a) shows the refinement effect of W2V attention on the initial image with 64 × 64 resolution. (b) shows the refinement effect of gated refinement on the intermediate image with 128 × 128 resolution. The tables show the top-5 words that V2W attention pays attention to.

realistic after the writing gate with fixed word features is integrated; the color of the object is more accurate after the W2V attention is integrated; the object has the most realistic shape and the most accurate color after the gated refinement with refined word features is integrated.

To further verify how our proposed W2V attention and gated refinement improve V2W attention, we visualize the top-5 words that V2W attention pays attention to. As shown in Fig. 4(a), the attention weights of important words "bird", "small" and "eyering" are improved after the W2V attention is integrated, which means that V2W attention pays more attention to these important words. We can also see that the shape of the object in the initial image can be effectively improved after W2V attention is integrated, which is due to the fact that W2V attention can focus on the relevant subregions of the image to select important words, instead of treating each subregion of the image equally in the writing gate [27]. As shown in Fig. 4(b), V2W attention can still pays more attention to important words "bird" and "yellow" after gated refinement is integrated. In addition, the attention weight of the important word "white" is improved, and we can also see that the details on the wings of the object in the final image are richer, which is due to the fact that gated refinement can retain important word information selected at the previous stage.

4   Conclusions

A new Txt2Img synthesis approach has been proposed in this paper by incorporating a gated cross word-visual attention unit into the multiple-stage GAN-based image generation framework. Our approach reconstructs images with better quality and visually realistic images, as verified in our qualitative and quantitative results using two benchmark datasets.

Acknowledgement. The work described in this paper is supported by China GDSF No. 2019A1515011949.

References

1. Goodfellow, I., et al.: Generative adversarial nets. In: Advances in Neural Information Processing Systems, Montreal, Canada, pp. 2672–2680, December 2014
2. Gregor, K., Danihelka, I., Graves, A.: DRAW: a recurrent neural network for image generation. In: International Conference on Machine Learning, Lille, France, pp. 1462–1471, July 2015
3. Heusel, M., Ramsauer, H., Unterthiner, T., Nessler, B., Hochreiter, S.: GANs trained by a two time-scale update rule converge to a local Nash equilibrium. In: Advances in Neural Information Processing Systems, Long Beach, CA, USA, pp. 6626–6637, December 2017
4. Huang, L., Wang, W., Chen, J., Wei, X.: Attention on attention for image captioning. In: IEEE/CVF International Conference on Computer Vision, Seoul, Korea, pp. 4633–4642, October 2019
5. Kingma, D., Ba, J.: Adam: a method for stochastic optimization. In: International Conference on Learning Representations, vol. 5, San Diego, CA, May 2015
6. Li, B., Qi, X., Lukasiewicz, T., Torr, P.: Controllable text-to-image generation. In: Advances in Neural Information Processing Systems, Vancouver, BC, Canada, pp. 2065–2075, December 2019
7. Li, W.: Object-driven text-to-image synthesis via adversarial training. In: IEEE International Conference on Computer Vision and Pattern Recognition, Long Beach, CA, USA, pp. 12166–12174, June 2019
8. Li, W., Zhu, H., Yang, S., Zhang, H.: DADAN: dual-path attention with distribution analysis network for text-image matching. SIViP **16**(3), 797–805 (2022). https://doi.org/10.1007/s11760-021-02020-2
9. Lin, T.-Y., et al.: Microsoft COCO: common objects in context. In: Fleet, D., Pajdla, T., Schiele, B., Tuytelaars, T. (eds.) ECCV 2014. LNCS, vol. 8693, pp. 740–755. Springer, Cham (2014). https://doi.org/10.1007/978-3-319-10602-1_48
10. Lu, Z., Chen, Y.: Single image super-resolution based on a modified U-net with mixed gradient loss. SIViP **16**(5), 1143–1151 (2022). https://doi.org/10.1007/s11760-021-02063-5
11. Pathak, D., Krahenbuhl, P., Donahue, J.: Context encoders: feature learning by inpainting. In: IEEE International Conference on Computer Vision and Pattern Recognition, Las Vegas, NV, USA, pp. 2536–2544, June 2016
12. Qiao, T., Zhang, J., Xu, D., Tao, D.: Learn, imagine and create: text-to-image generation from prior knowledge. In: Neural Information Processing Systems (2019)

13. Qiao, T., Zhang, J., Xu, D., Tao, D.: MirrorGAN: learning text-to-image generation by redescription. In: IEEE International Conference on Computer Vision and Pattern Recognition, Long Beach, CA, USA, pp. 1505–1514, June 2019
14. Reed, S., Akata, Z., Mohan, S., Tenka, S., Schiele, B., Lee, H.: Learning what and where to draw, New Republic (2016)
15. Reed, S.E., Akata, Z., Yan, X.: Generative adversarial text to image synthesis. In: International Conference on Machine Learning, New York City, NY, USA, pp. 1060–1069, June 2016
16. Salimans, T., Goodfellow, I., Zaremba, W., Cheung, V., Radford, A., Xi, C.: Improved techniques for training GANs. In: Advances in Neural Information Processing Systems, Barcelona, Spain, pp. 2234–2242, December 2016
17. Tan, H., Liu, X., Li, X., Zhang, Y., Yin, B.: Semantics-enhanced adversarial nets for text-to-image synthesis. In: IEEE/CVF International Conference on Computer Vision, Seoul, Korea, pp. 10500–10509, October 2019
18. Tan, X., Liu, M., Yin, B., Li, X.: KT-GAN: knowledge-transfer generative adversarial network for text-to-image synthesis. IEEE Trans. Image Process. **30**, 1275–1290 (2021)
19. Tao, M., et al.: DF-GAN: deep fusion generative adversarial networks for text-to-image synthesis. arXiv preprint arXiv:2008.05865 (2020)
20. Wah, C., Branson, S., Welinder, P., Perona, P., Belongie, S.: The caltech-UCSD birds-200-2011 dataset. Technical report CNS-TR-2011-001, California Institute of Technology (2011)
21. Wang, Z., Quan, Z., Wang, Z.J., Hu, X., Chen, Y.: Text to image synthesis with bidirectional generative adversarial network. In: IEEE International Conference on Multimedia and Expo, London, UK, pp. 1–6, July 2020
22. Xu, T.: AttnGAN: fine-grained text to image generation with attentional generative adversarial networks. In: IEEE International Conference on Computer Vision and Pattern Recognition, Salt Lake City, UT, USA, pp. 1316–1324, June 2018
23. Yin, G., Liu, B., Sheng, L., Yu, N., Wang, X., Shao, J.: Semantics disentangling for text-to-image generation. In: IEEE International Conference on Computer Vision and Pattern Recognition, Long Beach, CA, USA, pp. 2322–2331, June 2019
24. Zhang, H., et al.: StackGAN: text to photo-realistic image synthesis with stacked generative adversarial networks. In: IEEE International Conference on Computer Vision, Venice, Italy, pp. 5907–5915, October 2017
25. Zhang, H., et al.: StackGAN++: realistic image synthesis with stacked generative adversarial networks. IEEE Trans. Pattern Anal. Mach. Intell. **41**(8), 1947–1962 (2019)
26. Zhang, Z., Xie, Y., Yang, L.: Photographic text-to-image synthesis with a hierarchically-nested adversarial network. In: 2018 IEEE/CVF Conference on Computer Vision and Pattern Recognition (2018)
27. Zhu, M., Pan, P., Chen, W., Yang, Y.: DM-GAN: dynamic memory generative adversarial networks for text-to-image synthesis. In: IEEE International Conference on Computer Vision and Pattern Recognition, Long Beach, CA, USA, pp. 5795–5803, June 2019

End-to-End Surface Reconstruction for Touching Trajectories

Jiarui Liu[1,2], Yuanpei Zhang[1], Zhuojun Zou[1], and Jie Hao[1,2]([✉])

[1] Institute of Automation, Chinese Academy of Sciences, Beijing, China
jie.hao@ia.ac.cn
[2] Guangdong Institute of Artificial Intelligence and Advanced Computing,
Guangzhou, China

Abstract. Whereas vision based 3D reconstruction strategies have progressed substantially with the abundance of visual data and emerging machine-learning tools, there are as yet no equivalent work or datasets with which to probe the use of the touching information. Unlike vision data organized in regularly arranged pixels or point clouds evenly distributed in space, touching trajectories are composed of continuous basic lines, which brings more sparsity and ambiguity. In this paper we address this problem by proposing the first end-to-end haptic reconstruction network, which takes any arbitrary touching trajectory as input, learns an implicit representation of the underling shape and outputs a watertight triangle surface. It is composed of three modules, namely trajectory feature extraction, 3D feature interpolation, as well as implicit surface validation. Our key insight is that formulating the haptic reconstruction process into an implicit surface learning problem not only brings the ability to reconstruct shapes, but also improves the fitting ability of the network in small datasets. To tackle the sparsity of the trajectories, we use a spatial gridding operator to assign features of touching trajectories into grids. A surface validation module is used to tackle the dilemma of computing resources and calculation accuracy. We also build the first touching trajectory dataset, formulating touching process under the guide of Gaussian Process. We demonstrate that our method performs favorably against other methods both in qualitive and quantitative way. Insights from the tactile signatures of the touching will aid the future design of virtual-reality and human-robot interactions.

1 Introduction

Studying the mechanics of rebuilding the surface of objects through touching will complement vision-based scene understanding. Humans have the ability to imagine the shape of an unknown object by sliding along the surface, some of them are even capable of identifying among different persons. This is usually accomplished by first exploring the surfaces of target objects, then integrating the information from different locations and imagining the whole shape according to prior knowledge. Although 3D reconstruction has been a classical research topic in vision and robotics applications, reconstruction from touching is still an

© The Author(s), under exclusive license to Springer Nature Switzerland AG 2023
L. Wang et al. (Eds.): ACCV 2022, LNCS 13847, pp. 101–116, 2023.
https://doi.org/10.1007/978-3-031-26293-7_7

uncharted territory to explore. Here in this paper, we design a data-driven algo-rithm to learn an uniform 3D representation space with touching information, and generate detailed 3D surface mesh as output.

Touching trajectories are usually treated as continuous curves, which can be discretized as a list of 3D coordinates. As the exploration can be started at any position and the local situation of the exploration is different each time, one target object may result in countless touching coordinate sequences. If our reconstruction framework is carried out in the order of exploration, it will be difficult to learn features that truly represent the global shape of objects. Our key insight is that the information contained in each trajectory is independent to the exploring orders–we can treat a touching trajectory as a special kind of point cloud that is dense in some local direction but extremely sparse in a global way.

In the recent few years, many neural network-based methods have been pio-neered to mine the information contained in disorganized data such as point clouds, namely Multi-Layer Perceptron based methods [24, 25], voxelization based meth-ods [3, 10, 15], and graph based methods[33, 34]. [39] takes partial point clouds as input, extract features with gridding and 3D convolution function, and obtain the final completed point cloud by predicting the coordinates after performing reverse gridding and feature sampling. However, none of these methods is designed to deal with touching trajectories. Besides, most of these methods still take more opera-tions for the final presentation of a shape, namely, a triangular mesh. Reconstruct-ing or generating from low-information domain brings extra difficulties, espe-cially when there is no bidirectional mapping between 3D shape domain and this domain. [41] and [18] learns geometric transformations between a low-information domain and the 3D surfaces domain without taking point-to-point correspon-dences, namely transforming between meso-skeletons and surfaces, partial and complete scans, etc. However, data from those two domains are required to follow a global one-to-one mapping relationship. In our situation, touching trajectories are generated in a more random way. There isn't a specific mapping function from a complete shape to a touching trajectory, on the contrary, each target shape cor-responds to a continuous touching space.

To address the issues mentioned above, we introduce an end-to-end haptic reconstruction network[1], which takes a sparse point cloud as input and generates triangular meshes as output. The framework of our method is illustrated in Fig. 1. Instead of learning the coordinates of new generated point clouds, we learn an implicit manifold for each target shape and use a random spatial sampling mechanism to supervise the generated implicit surface. During inference stage, we reconstruct a watertight triangular surface under the same framework by organizing the sampled points as the vertices of high-resolution grids. We also build a customized touching trajectory dataset to support the training of our network. We formulate an ordinary touching logic into a Gaussian process(GP) learning problem, and use the variance as a guide to explore the surface to somewhere that has not been touched. We assume the mapping function from the implicit surface to the touched surface points as a Gaussian Process, the goal of exploring the surface as much as possible is thus simplified to exploring

---

[1] Dataset and source code can be found: https://github.com/LiuLiuJerry/TouchNet.

towards high variances, which indicates a high uncertainty during touching. We demonstrate the effectiveness of our method in both quantitative and qualitive way. It has been proved that our method not only succeeds in accomplishing touching reconstruction task, but also outperforms other potential methods by a large margin.

Our contributions are listed as follows:

- To the best of our knowledge, we are the first to pioneer the reconstruction from touching information. Our touching-reconstruction method can be taken as the first step of haptic shape learning and multimodal understanding.
- We propose the first touching reconstruction network, which embeds irregular exploration paths into an uniform feature space and generates watertight triangle meshes as output.
- We present the first touching trajectory dataset to support more data-driven methods requiring touching trajectory data.
- This method will serve as a knowledge base for learning and rebuilding from very irregular point clouds.

As the first step of haptic-reconstruction, this framework can be used in haptic identification, 3D object creation and many other practical applications.

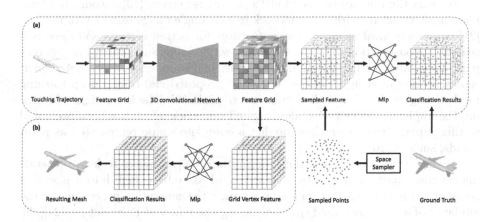

Fig. 1. Overview of our method. (a) Training process. (b) Reconstruction process.

2   Related Work

The existing relative research works involve both 3D reconstruction and haptic perception fields, which are detailed as follows.

3D Reconstruction. Existing 3D reconstruction algorithms are mostly designed to generate shapes from images or videos, while few of them generate from other domains such as contour lines, meso-skeletons or latent codes derived from a learned implicit manifold space. The internal mechanism can be further categorized into deformation-based methods and generation-based methods.

Deformation based methods assume the target shape is topological homeomorphism with a given shape, learning the deformation between the origin shape and the target shape. [17] optimizes object meshes for multi-view photometric consistency by posing it as a piecewise image alignment problem. [12] incorporates texture inference as prediction of an image, learning the deformable model with one single image. These kind of methods are widely used in human face and body reconstruction tasks, basing on the widely used 3D morphable face model 3DMM[6] and 3D morphable human shape model SMPL[19]. [8] first proposes to harness the power of Generative Adversarial Networks (GANs) to reconstruct the facial texture. [26] proposes a learning based method to learn complete 3D models with face identity geometry, albedo and expression parameters from images and videos. To avoid the statistical dependency contained in training dataset, [35] estimates 3D facial parameters using several independent networks, which results in a much bigger description space. In human shape reconstruction field, the problem tends to be more complicated with the disturbing information introduced by clothes. [1] learns to reconstruct the 3D human shape from few frames. It first encodes the images of the person into pose-invariant latent codes, then predicts the shape using both bottom-up and top-down streams.

To relax the reliance on the model's parameter space, [13] encodes the template mesh structure within the network from a single image using a GraphCNN. [21] constructs local geometry-aware features for octree vertices and designs a scalable reconstruction pipeline, which allows dividing and processing different parts in parallel for large-scale point clouds. [28] proposes to learn an implicit representation of each 3D shape. With the proposed end-to-end deep learning method, highly detailed clothed humans are digitized. The key insight of learning the implicit representation also inspired our work which has been presented in this paper. Other researchers predict a complete shape represented as point clouds, such as [20].

Besides generating from vision information, [16] proposes to learn generating semantic parts in a step by step way, and assembles them into a plausible structure. [18] augments 3D shapes by learning the generation process under the guidance of a latent manifold space. Utilizing the power of PointNet++ [24][25], [41] generates point surfaces from meso-skeletons by learning the displacement between two domains, without relying on point-to-point correspondences. However, this method still requires the shape of same domain organized in a uniform way, and the mapping is assumed to be bidirectional, which is not suitable in the touching procedural, while one shape will have countless touching results. [39] addresses the irregularity problems by proposing three differentiable layers: Gridding, Gridding Reverse, and Cubic Feature Sampling. The Gridding and Reverse Gridding operation has been proved to be an efficient way to learn both local and global features, but the generation framework is not capable of predicting smooth surface when the given point clouds are extremely sparse.

Haptic Perception. Tactile perception as well as tactile information based learning are commonly used in robot manipulation. For the purpose of object perception, plenty of work has been proposed to learn the feature of the local surface, such as [2,4,7,9,14,22,27,30,36]. Those work usually collects multimodal information using specially designed sensing components, using operation including pressing, sliding and tapping, and finally classifies the material or the shape of the objects using different mathematical methods, such as wavelet transform, dictionary learning, convolutional neural networks, etc. To interpretate tactile data in a global way, [29] proposes to treat tactile arrays obtained from different grasps using tactile glove as low-resolution single channel images, and utilizes the strong power of deep convolutional neural network to identify objects using filtered tactile frames. However, the grasping gestures as well as trajectories indicating spatial location information are not taken into consideration, which limits the perception of objects' 3D shape. [40] and [5] propose active learning frameworks based on optimal query paths to efficiently address the problem of tactile object shape exploration. Limited by the normal estimation algorithm as well as the flexibility of manipulators, those methods only work for objects with very primitive shapes, and the reconstruction results only indicate a rough outline of the actual objects. [32] and [31] incorporate depth and tactile information to create rich and accurate 3D models. However, the tactile information is only acquired as augmentation, while the majority of a target shape is still provided by depth camera. Here in our work, we demonstrate that using tactile information only has been sufficient for detailed high-quality shape reconstruction tasks.

3   Haptic Reconstruction Network

We treat haptic trajectory reconstruction as a particular shape completion problem, where the input data is extremely sparse in a global way, but is continuous along the exploration direction. Overall, it contains very limited information comparing to partial point clouds taken by a depth camara, and contains more ambiguity compared to contour lines or meso-skeletons. The size of the dataset is also limited by the big difficulty of collecting touching data, which challenges the ability of reconstruction network to learn the key feature of a "good shape" as well. Here we adapt the idea of gridding input touching trajectory points into a grid for feature extraction, and propose our own designment for the touching reconstruction task, including trajectory feature extraction, 3D feature interpolation and implicit surface validation.

3.1   Trajectory Feature Exaction

In recent years, many neural network-based methods have been pioneered to mine the information contained in unorganized data such as point clouds. One type of them is to use the Multi-Layer Perceptions and pooling functions to aggregate information across points. These kind of methods do not fully consider

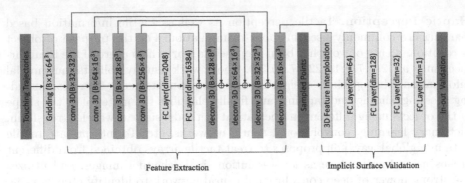

Fig. 2. Architecture of our touching reconstruction network. ⊕ denotes feature addition. The number of sampled points is set to 4000.

the connectivity across points and the context of neighboring points. Another type of approaches are to voxelize the point cloud into binary voxels, then apply 3D convolutional neural networks to those voxels. Limited by the resolution of the voxels, theses methods have to drop the high-precision location information, resulting in a waste of the precision of the original data. Inspired by GRNet[39], here we apply gridding operation to explicitly preserve the structural and context information.

Taken a touching trajectory as inputs, gridding operator assigns the trajectory points into regular grids. For each point of the point cloud, the corresponding grid cell the point lies in is found according to their coordinates. Then eight vertices of the 3D grid cell are weighed using the function that explicitly measures the geometric relations of the point cloud. By applying this operation, the point clouds are mapped into 3D grids without loss of data accuracy. Then we adopt a 3D convolutional neural network with skip connections to extract the global features indicated by the sparse touching trajectories. As is detailed in Fig. 2, the architecture of our network follows the idea of U-net [38]. It has four 3D convolutional layers, two fully connected layers and four transposed convolutional layers. Each convolutional layer has a bank of 4^3 filters with padding of 2, while each transposed convolutional layer has a bank of 4^3 filters with padding of 1 and stride of 2. The numbers of output channels of two fully connected layers are 2048 and 16384. Features are flattened into one-dimensional data before being taken into fully connected layers. Batch normalization and leaky ReLU activation are used to prevent the network from gradient vanishing problem. After that, a sampler as well as Multilayer Perceptron Layers are used to transform the learned feature manifold into *in-out* labels of each sampled points, which is detailed in Sect. 3.3.

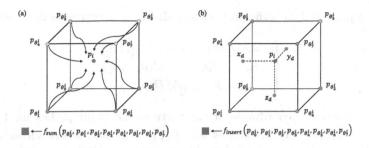

Fig. 3. Different feature aggregation methods. (a)Sum operation. (b)Trilinear interpolation.

3.2   3D Feature Interpolation

Feature aggregation, namely feature fusion and feature sampling, has been an important subject in neural network related researching. Common feature fusion operations are meanly accomplished by global concatenation and addition. Addition and concatenation of eight vertices of the 3D grid cell where each point lies in has been proved to aggregate the local context information of each point and build a connection between 3D gridding features and sampled points. However, this function does not take into account the relative distances between each point and the corresponding vertices of grid cell, which play important roles in traditional geometry descriptor computation (Fig. 3).

We introduce a new feature aggregation operator, called 3D Feature Interpolation, to integrate features of surrounding grid vertices to sampled points. This operation assumes that the gridding features are a regular discretization of a continuous feature manifold. A manifold is a topological space which has the property of Euclidean space locally, every local area of which is linear and homeomorphic to the Euclidean space. Built on this assumption, we perform interpolation based on the distance to the grid cell edges it lies in. Trilinear interpolation, widely known as a intuitive and explicable interpolation method in graphics, is chosen as our feature interpolation operator.

Let $\mathcal{S} = \{f_1^v, f_2^v, ..., f_{t^3}^v\}$ be the feature map of 3D CNN, where t^3 is the size of the feature map, and $f_i^v \in \mathbb{R}^c$. Given a sampled point p_i, its corresponding feature is signed as f_i^c, which is computed as

$$
\begin{aligned}
f_i^c = &\ f_{\theta_0^i}^c (1 - x_d)(1 - y_d)(1 - z_d) + \\
&\ f_{\theta_1^i}^c x_d (1 - y_d)(1 - z_d) + \\
&\ f_{\theta_2^i}^c (1 - x_d) y_d (1 - z_d) + \\
&\ f_{\theta_3^i}^c (1 - x_d)(1 - y_d) z_d + \\
&\ f_{\theta_4^i}^c x_d (1 - y_d) z_d + \\
&\ f_{\theta_5^i}^c (1 - x_d) y_d z_d + \\
&\ f_{\theta_6^i}^c x_d y_d (1 - z_d) + \\
&\ f_{\theta_7^i}^c x_d y_d z_d,
\end{aligned}
\tag{1}
$$

where x_d, y_d, and z_d is the proportional distance to corresponding cell edges:

$$\begin{cases} x_d = D(p_i, p_{\theta_0^i})/D(p_{\theta_1^i}, p_{\theta_0^i}) \\ y_d = D(p_i, p_{\theta_1^i})/D(p_{\theta_2^i}, p_{\theta_1^i}) \\ z_d = D(p_i, p_{\theta_2^i})/D(p_{\theta_3^i}, p_{\theta_2^i}). \end{cases} \tag{2}$$

$\{p_{\theta_j^i}\}_{j=1}^8$ denotes the coordinate of the vertices the point p_i lies in, $\{f_{\theta_j^i}\}_{j=1}^8$ denotes the features of eight vertices, and $D(,)$ computes the Euclidean distance between two coordinates. Note that this operator is simple and differentiable, which makes it easy to be extended to other tasks.

3.3 Implicit Surface Validation

Representing target objects as implicit surfaces, the formulation of the ground truth shape is written as:

$$S = \{x \in \mathbb{R}^3 | \mathcal{F}(x) = 1\}, \tag{3}$$

where $S \in \mathbb{R}^3$ is the one level set of function $\mathcal{F} : \Omega \in \mathbb{R}^3 \to \mathbb{R}$. The touching trajectories are denoted as set of point clouds $\mathcal{P} = \{P_i\}_{i=1}^w$.

To validate the implicit surface learned by our network, we use a rejection-sampling strategy to quantify the error between implicit surfaces and ground truth shapes. To strengthen the ability of the network to learn the details, we adapt an Gaussian distribution based adaptive sampler. It first samples points on the surface of ground truth objects, then adds random displacements to the sampled points under the guide of a Gaussian distribution. After that, we sample uniformly inside the whole space and randomly choose 4000 points as the sampling results.

The validation process is mathematically modeled as a Multi-Layer Perception function. The numbers of output channels of fully-connected layer are designed as 64, 128, 32, 1. Finally, an *in-out* label is computed using a sigmoid function. Compared to generating point cloud positions, our method can learn a continuous implicit surface which can be transformed into triangular meshes for further usage. Besides, this rejection sampling method turns the complex position regression problem into a simpler two-way classification problem, which is more suitable for a network to excel.

During reference, the validation module is integrated into an octree based marching cubes algorithm. During reconstruction, the generation space is discretized and the vertices are organized in an octree structure for different resolutions. The inside-outside values are first calculated in a low resolution, and then the cubes whose vertex values vary from each other are evaluated in a higher resolution. We find the isosurface by calculating the position coordinates of each intersection point.

3.4   Loss Function

For the convenience of calculation, we denote the ground truth label as:

$$S_{gt} = \begin{cases} 1, & p \in mesh \\ 0, & p \notin mesh \end{cases} \tag{4}$$

The loss function is defined as the L2 distance between the predicted label and the ground truth:

$$L = \sum_{i \in P} \left\| S_{pred}^i - S_{gt}^i \right\|_2^2 \tag{5}$$

Here P is the point set sampled around the target shapes, S_{pred}^i is the ith label predicted by our network, and S_{gt}^i is the corresponding label indicating the *in-out* relationship with the ground truth shape.

4   Dataset Acquisition

Different from traditional shape reconstruction task, one challenge we handled is, there's no such a dataset for us to validate our algorithm directly. Here we propose our own haptic dataset, and introduce our method for the acquisition. Acquisition of the dataset containing the trajectory generated by real hands requires specially designed hardware and is a waste of time. Here we formulate the touching procedural as a process of minimizing the uncertainty of the target object under the guide of Gaussian Process, and build the touching dataset in computer simulation.

4.1   Gaussian Process

Gaussian process has been proposed as a regression method for implicit surfaces [40]. It not only approximates the underlying surface but also provides an uncertainty measure for the mesh. A GP models the probability $P(F(x)|\mathcal{P})$ as a Gaussian distribution, where $P(F(x)|\mathcal{P})$ is the probability of the implicit surface function F conditioned on the tactile data:

$$\mu_F(X) = m + \kappa(x)^T b \tag{6}$$

the variance is computed as:

$$\mathbb{V}_F(\mathrm{x}) = k(\mathrm{x}, \mathrm{x}) - \kappa(\mathrm{x})^T G^{-1} \kappa(\mathrm{x}) \tag{7}$$

where κ, K and G represent matrixes of kernel functon k.

$\kappa(x) = (k(x; x_1), ..., k(x; x_w))^T$, $G = K + \sigma_n^2 I_w \in \mathbb{R}^{w \times w}$, $K = (k(x_i; x_j))_{i,j=1}^w$, $b = G^{-1}(Y - m) \in \mathbb{R}^w$, $Y = (c_1, ..., c_w)^T$. We set $m = 1$ as the prior mean, representing that most part of the space is empty.

For each time, we compute the contact points as well as the local normal at contact position, which will be used during exploration stage.

4.2   Touching Simulation

Aside from obtaining the surface from GP, at each time, our final goal is to compute a direction in the local Euclidean space and simulate the exploration result for the next time step. As what has been mentioned, we estimate the GP variance function as an evaluation for the uncertainty. Our assumption is that moving towards a most uncertain position will maximize the information obtained during the sliding and thus minimize the exploration time. We compute the gradient of the GP variance function as:

$$g_{\mathbb{V}_F}(\mathrm{x}) = \frac{\partial k}{\partial \mathrm{x}}(\mathrm{x}, \mathrm{x}) - 2\kappa(\mathrm{x})^T G^{-1} \frac{\partial \kappa}{\partial \mathrm{x}}(\mathrm{x}) \tag{8}$$

We donate the exploration direction as $g = \frac{g_{\mathbb{V}_F}(\mathrm{x})}{|g_{\mathbb{V}_F}(\mathrm{x})|}$. As shown in 4(a), the exploration direction g could point at any direction in 3D space. To make this direction meaningful, we project g on the tangent space of target shape at point P_i, and normalize the projected V_i' as V_i. The next point P_{i+1} is obtained by moving point P_i towards direction V_i and projecting the moved point P_{i+1}' on the surface, noted as P_{i+1}.

Exploring objects of complex shape in this method directly might cause local optimal results–the exploring point moves inside a low variance region back and forth, ignoring other unexplored regions. This is because the exploring point walks into a region surrounded by places which have been explored before, thus the point might find the variance low enough to quit the exploration. In this situation the generated path does not contain a global information of target object, discriminative information might be missed. To avoid this, we stop the local marching and choose a new start position where the variance is estimated to be the highest. We stop exploring once the exploration exceeds certain steps and use the explored paths represented as points as input (Fig. 4).

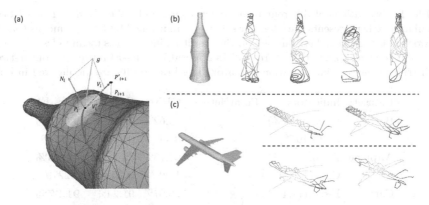

Fig. 4. Local trajectory planning and examples of touching simulation results. (a)Schematic diagram of local exploration direction calculation. (b)Touching simulation results of a bottle. (c)Touching simulation results of an airplane model.

5 Experiments

5.1 Implementation Details

We build the touching dataset from a subset of ShapeNet [37]. Before conducting touching simulation, we convert the triangle soups into watertight manifold surfaces using projection-based optimization method [11]. We organize three subjects of shapes, Bottle, Airplane, Car, which contains 106, 1972, 1989 target shapes respectively. For each shape, we simulate 4 touching trajectories with randomly chosen start points.

We implement our network using PyTorch[23] with Adam optimizer, and train the network with a batch size of 4 for 200 epochs on NVIDIA V100 GPUs. We use a multi-step learning rate and the initial learning rate is set to $1e-4$.

5.2 Quantitative Results

We conduct our evaluation on sampled points labeled as *in* for comparison. Let $T = (x_i, y_i, z_i)_{i=1}^{n_T}$ be the ground truth and $S = (x_i, y_i, z_i)_{i=1}^{n_S}$ be a reconstructed shape being evaluated, where n_T and n_S are the numbers of points of T and S, we use Chamfer Distance and F-Score as quantitative evaluation metrics.

Chamfer's distance. Following some existing generation methods [39,41], we take this as a metric to evaluate the global distance between two point clouds. Given two point clouds written as S and T, this distance is defined as:

$$L_{CD} = \frac{1}{n_S} \sum_{s \in S} min_{t \in T} \|s - t\|_2^2 + \frac{1}{n_T} \sum_{t \in T} min_{s \in S} \|s - t\|_2^2 \qquad (9)$$

Table 1. Evaluations of different reconstruction methods. We use $d = 1$ in F1-Score calculation. CD represents the Chamfer's distance multiplied by 10^{-3}. − means that the network fails to learn a distribution and the resulting indicator is meaningless. Ours-FA represents our network with feature addition operator, Ours-FI represents our network with trilinear feature interpolation operator. The best results are highlighted in bold.

Dataset	Indicators	PointNet++	GRNet	Ours-FA	Ours-FI
Bottle	F-score(@1%)	−	43.62%	90.08%	**92.74%**
	CD($\times 10^{-3}$)	−	0.44	0.2781	**0.2718**
Airplane	F-score(@1%)	−	62.28%	86.11%	**87.95%**
	CD($\times 10^{-3}$)	−	1.29	**0.2185**	0.2279
Car	F-score(@1%)	−	26.07%	91.24%	**91.37%**
	CD($\times 10^{-3}$)	−	0.7221	0.1234	**0.1203**

F1-score. To quantify the local distance between two shapes, we use F-score metric as:

$$L_{F-Score}(d) = 2\frac{PR}{P+R}, \tag{10}$$

where P and R are the precision and recall defined as the percentage of the points whose distance to the closest point in another shape(written as $min(,)$) is under threshold d:

$$P = \frac{1}{n_{\mathcal{S}}} \sum_{s \in \mathcal{R}} |min(s, \mathcal{T}) < d| \tag{11}$$

$$R = \frac{1}{n_{\mathcal{T}}} \sum_{t \in \mathcal{T}} |min(t, \mathcal{S}) < d| \tag{12}$$

Each metric mentioned above is evaluated on all these three dataset. We take the farthest point sampling method to sample the points into 2048-long points for alignment. Table 1 shows that our framework achieves good indicators even with an extremely small training set. In both cases, our method outperforms others by a large margin, indicating the excellence and effectiveness of our network.

5.3   Qualitative Results

The details of the ground truth shapes, simulated touching trajectories, results completed by GRNet, and the results of our methods are listed in Fig. 5. We demonstrate that our method can easily tackle the problem of tactile trajectory reconstruction. This method is compared with GRNet[39] and PointNet++-based displacement learning function[18]. The PointNet++-based algorithm can hardly learn the generation distribution of our irregular touching dataset. GRnet learn a rough shape of the target object, but the generated point clouds have blurred boundaries and lack detail. While sharing the gridding operation with GRNet, our implicit surface learning strategy outperforms coordinate generation strategy with more even points and smoother surfaces.

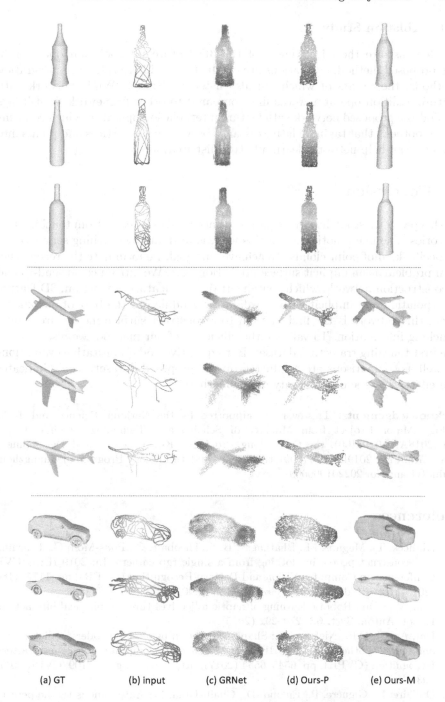

  (a) GT   (b) input   (c) GRNet   (d) Ours-P   (e) Ours-M

Fig. 5. Reconstruction results of Bottle and Airplane dataset. Column from left to right is: (a) Ground truth shape of target objects, (b) Touching trajectories generated from target objects, (c) Results from GRnet, (d) Sampled point clouds labeled as **in** by our network, (e) Surface mesh reconstructed using our method.

5.4 Ablation Study

To demonstrate the effectiveness of the 3D feature interpolation operator in the proposed method, evaluations are conducted on different feature integration methods, the results of which are also listed in Table 1. While network with feature addition operator works fine compared to other framework, it still lags behind the proposed network with feature interpolation operator, which confirms the hypothesis that taking relative distance between cube vertices and points into account can help networks learn a better distributions.

6 Conclusion

In this paper we study how to generate surface meshes directly from touching trajectories. The main motivation of this work is to treat the touching sequences as a specific kind of point clouds. To achieve this goal, we formulate the reconstruction problem as an implicit surface prediction task. We introduce our end-to-end reconstruction network, which contains trajectory feature extraction, 3D feature interpolation and implicit surface validation module. To the best of our knowledge, this network is the first method to reconstruct surface meshes from pure touching information. To validate the efficiency of our method, we also propose the first touching trajectory dataset. Both qualitive and quantitative evaluations as well as comparisons are conducted on the proposed dataset, which indicates the effectiveness and superiority of our method.

Acknowledgements. This work is supported by the National Science and Technology Major Project from Minister of Science and Technology, China (Grant No. 2018AAA0103100), the Guangdong Provincial Key Research and Development Plan(Grant No. 2019B090917009), the Science and Technology Program of Guangzhou, China(Grant No. 202201000009).

References

1. Alldieck, T., Magnor, M., Bhatnagar, B.L., Theobalt, C., Pons-Moll, G.: Learning to reconstruct people in clothing from a single rgb camera. In: 2019 IEEE/CVF Conference on Computer Vision and Pattern Recognition (CVPR), pp. 1175–1186 (2019). https://doi.org/10.1109/CVPR.2019.00127
2. Chu, V., et al.: Robotic learning of haptic adjectives through physical interaction. Robot. Auton. Syst. **63**, 279–292 (2015)
3. Dai, A., Qi, C.R., Nießner, M.: Shape completion using 3d-encoder-predictor cnns and shape synthesis. In: 2017 IEEE Conference on Computer Vision and Pattern Recognition (CVPR), pp. 6545–6554 (2017). https://doi.org/10.1109/CVPR.2017.693
4. Dallaire, P., Giguère, P., Émond, D., Chaib-Draa, B.: Autonomous tactile perception: A combined improved sensing and bayesian nonparametric approach. Robot. Auton. Syst. **62**(4), 422–435 (2014)

5. Driess, D., Englert, P., Toussaint, M.: Active learning with query paths for tactile object shape exploration. In: 2017 IEEE/RSJ International Conference on Intelligent Robots and Systems (IROS), pp. 65–72 (2017). https://doi.org/10.1109/IROS.2017.8202139
6. Egger, B., et al.: 3d morphable face models - past, present and future. arXiv: 1909.01815 (2019)
7. Erickson, Z., Chernova, S., Kemp, C.C.: Semi-supervised haptic material recognition for robots using generative adversarial networks. In: Conference on Robot Learning, pp. 157–166. PMLR (2017)
8. Gecer, B., Ploumpis, S., Kotsia, I., Zafeiriou, S.: GANFIT: generative adversarial network fitting for high fidelity 3d face reconstruction. arXiv: 1902.05978 (2019)
9. Giguere, P., Dudek, G.: A simple tactile probe for surface identification by mobile robots. IEEE Trans. Rob. **27**(3), 534–544 (2011)
10. Han, X., Li, Z., Huang, H., Kalogerakis, E., Yu, Y.: High-resolution shape completion using deep neural networks for global structure and local geometry inference. In: 2017 IEEE International Conference on Computer Vision (ICCV), pp. 85–93 (2017). https://doi.org/10.1109/ICCV.2017.19
11. Huang, J., Zhou, Y., Guibas, L.: Manifoldplus: A robust and scalable watertight manifold surface generation method for triangle soups. arXiv preprint arXiv:2005.11621 (2020)
12. Kanazawa, A., Tulsiani, S., Efros, A.A., Malik, J.: Learning category-specific mesh reconstruction from image collections. arXiv: 1803.07549 (2018)
13. Kolotouros, N., Pavlakos, G., Daniilidis, K.: Convolutional mesh regression for single-image human shape reconstruction. In: CVPR (2019)
14. Kursun, O., Patooghy, A.: An embedded system for collection and real-time classification of a tactile dataset. IEEE Access **8**, 97462–97473 (2020)
15. Li, D., Shao, T., Wu, H., Zhou, K.: Shape completion from a single rgbd image. IEEE Trans. Visual Comput. Graphics **23**(7), 1809–1822 (2017). https://doi.org/10.1109/TVCG.2016.2553102
16. Li, J., Niu, C., Xu, K.: Learning part generation and assembly for structure-aware shape synthesis. arXiv: 1906.06693 (2019)
17. Lin, C., et al.: Photometric mesh optimization for video-aligned 3d object reconstruction. arXiv: 1903.08642 (2019)
18. Liu, J., Xia, Q., Li, S., Hao, A., Qin, H.: Quantitative and flexible 3d shape dataset augmentation via latent space embedding and deformation learning. Comput. Aided Geometric Design **71**, 63–76 (2019). https://doi.org/10.1016/j.cagd.2019.04.017, https://www.sciencedirect.com/science/article/pii/S0167839619300330
19. Loper, M., Mahmood, N., Romero, J., Pons-Moll, G., Black, M.J.: Smpl: a skinned multi-person linear model. ACM Trans. Graph. **34**, 248:1–248:16 (2015)
20. Mandikal, P., Babu, R.V.: Dense 3d point cloud reconstruction using a deep pyramid network. arXiv: 1901.08906 (2019)
21. Mi, Z., Luo, Y., Tao, W.: Tsrnet: Scalable 3d surface reconstruction network for point clouds using tangent convolution. arXiv: 1911.07401 (2019)
22. Oddo, C.M., Controzzi, M., Beccai, L., Cipriani, C., Carrozza, M.C.: Roughness encoding for discrimination of surfaces in artificial active-touch. IEEE Trans. Rob. **27**(3), 522–533 (2011)
23. Paszke, A., et al.: Pytorch: An imperative style, high-performance deep learning library. arXiv: 1912.01703 (2019)
24. Qi, C.R., Su, H., Mo, K., Guibas, L.J.: Pointnet: Deep learning on point sets for 3d classification and segmentation. arXiv: 1612.00593 (2016)

25. Qi, C.R., Yi, L., Su, H., Guibas, L.J.: Pointnet++: Deep hierarchical feature learning on point sets in a metric space. arXiv: 1706.02413 (2017)
26. R., M.B., Tewari, A., Seidel, H., Elgharib, M., Theobalt, C.: Learning complete 3d morphable face models from images and videos. arXiv: 2010.01679 (2020).
27. Richardson, B.A., Kuchenbecker, K.J.: Improving haptic adjective recognition with unsupervised feature learning. In: 2019 International Conference on Robotics and Automation (ICRA), pp. 3804–3810. IEEE (2019)
28. Saito, S., Huang, Z., Natsume, R., Morishima, S., Li, H., Kanazawa, A.: Pifu: Pixel-aligned implicit function for high-resolution clothed human digitization. In: 2019 IEEE/CVF International Conference on Computer Vision (ICCV), pp. 2304–2314 (2019). https://doi.org/10.1109/ICCV.2019.00239
29. Sundaram, S., Kellnhofer, P., Li, Y., Zhu, J.Y., Torralba, A., Matusik, W.: Learning the signatures of the human grasp using a scalable tactile glove. Nature **569**, 698–702 (2019). https://doi.org/10.1038/s41586-019-1234-z
30. Tulbure, A., Bäuml, B.: Superhuman performance in tactile material classification and differentiation with a flexible pressure-sensitive skin. In: 2018 IEEE-RAS 18th International Conference on Humanoid Robots (Humanoids), pp. 1–9. IEEE (2018)
31. Varley, J., DeChant, C., Richardson, A., Nair, A., Ruales, J., Allen, P.K.: Shape completion enabled robotic grasping. arXiv: 1609.08546 (2016)
32. Varley, J., Watkins-Valls, D., Allen, P.K.: Multi-modal geometric learning for grasping and manipulation. arXiv: 1803.07671 (2018)
33. Wang, K., Chen, K., Jia, K.: Deep cascade generation on point sets. In: Proceedings of the 28th International Joint Conference on Artificial Intelligence, IJCAI 2019, pp. 3726–3732. AAAI Press (2019)
34. Wang, Y., Sun, Y., Liu, Z., Sarma, S.E., Bronstein, M.M., Solomon, J.M.: Dynamic graph CNN for learning on point clouds. arXiv: 1801.07829 (2018)
35. Wen, Y., Liu, W., Raj, B., Singh, R.: Self-supervised 3d face reconstruction via conditional estimation. In: 2021 IEEE/CVF International Conference on Computer Vision (ICCV), pp. 13269–13278 (2021). https://doi.org/10.1109/ICCV48922.2021.01304
36. Windau, J., Shen, W.M.: An inertia-based surface identification system. In: 2010 IEEE International Conference on Robotics and Automation, pp. 2330–2335. IEEE (2010)
37. Wu, Z., Song, S., Khosla, A., Tang, X., Xiao, J.: 3d shapenets for 2.5d object recognition and next-best-view prediction. arXiv: 1406.5670 (2014)
38. Xie, H., Yao, H., Sun, X., Zhou, S., Zhang, S., Tong, X.: Pix2vox: Context-aware 3d reconstruction from single and multi-view images. arXiv: 1901.11153 (2019)
39. Xie, H., Yao, H., Zhou, S., Mao, J., Zhang, S., Sun, W.: Grnet: Gridding residual network for dense point cloud completion. arXiv: 2006.03761 (2020)
40. Yi, Z., et al.: Active tactile object exploration with gaussian processes. In: 2016 IEEE/RSJ International Conference on Intelligent Robots and Systems (IROS), pp. 4925–4930 (2016). https://doi.org/10.1109/IROS.2016.7759723
41. Yin, K., Huang, H., Cohen-Or, D., Zhang, H.R.: P2P-NET: bidirectional point displacement network for shape transform. arXiv: 1803.09263 (2018)

Conditional GAN for Point Cloud Generation

Zhulun Yang[1,2] (iD), Yijun Chen[1,2], Xianwei Zheng[3](✉), Yadong Chang[1,2],
and Xutao Li[1,2]

[1] Key Lab of Digital Signal and Image Processing of Guangdong Province,
Shantou University, Shantou 515063, China
{15zlyang3,21yjchen1,21ydchang,lixt}@stu.edu.cn
[2] Department of Electronic Engineering, Shantou University,
Shantou 515063, China
[3] School of Mathematics and Big Data, Foshan University, Foshan 52800, China
alex.w.zheng@hotmail.com

Abstract. Recently, 3D data generation problems have attracted more
and more research attention and have been addressed through various
approaches. However, most of them fail to generate objects with given
desired categories and tend to produce hybrids of multiple types. Thus,
this paper proposes a generative model for synthesizing high-quality
point clouds with conditional information, which is called Point Cloud
conditional Generative Adversarial Network (PC-cGAN). The generative model of the proposed PC-cGAN consists of two main components:
a pre-generator to generate rough point clouds and a conditional modifier to refine the last outputs with specific categories. To improve the
performance for multi-class conditional generation for point clouds, an
improved tree-structured graph convolution network, called BranchGCN,
is adopted to aggregate information from both ancestor and neighbor features. Experimental results demonstrate that the proposed PC-cGAN
outperforms state-of-the-art GANs in terms of conventional distance
metrics and novel latent metric, Frechet Point Distance, and avoids the
intra-category hybridization problem and the unbalanced issue in generated sample distribution effectively. The results also show that PC-cGAN
enables us to gain explicit control over the object category while maintaining good generation quality and diversity. The implementation of
PC-cGAN is available at https://github.com/zlyang3/PC-cGAN.

1 Introduction

In recent years, point clouds, a popular representation for 3D realistic objects
data, are adopted in various applications (e.g., object classification [7,9,12,18,
21,23,32,36], semantic segmentation [12,21,23,32,36], and shape completion [28,
33,35]) and have become increasingly attractive in computer vision application,
such as augmented reality [19,20,25] and virtual reality [2,29]. As each point in
raw point clouds consists of a Cartesian coordinate, along with other additional
information such as a surface normal estimate and RGB color value, point clouds

L. Wang et al. (Eds.): ACCV 2022, LNCS 13847, pp. 117–133, 2023.
https://doi.org/10.1007/978-3-031-26293-7_8

can represent a 3D object to capture intricate details by an unordered set of irregular points collected from the surface of the objects.

With huge demand for data to train models for the aforementioned applications, generative models, including generative adversarial networks (GANs) [10] and variational autoencoders (VAEs) [15], have draw significant attention to generate point cloud data with high quality and diversity. Some researchers [1] stacked fully connected layers to convert random latent codes into 3D point clouds, which were borrowed from the image generative domain. Additionally, some generative models [26,31] have utilized graph convolution networks and k-nearest neighbor techniques to generate point clouds. Recently, SP-GAN [17] succeeds in synthesize diverse and high-quality shapes and promote controllability for part-aware generation and manipulation without any part annotations. And a novel GAN was proposed to generate 3D objects conditioned on a continuous parameter in [30].

However, part of these generative models need to train separate models for each categories, leading to poor reusability of them. Worse, most of the generative model for point clouds fail to actively synthesize objects with specific categories, which causes the intra-category hybridization problem. Since these models are trained to simulate the distribution of the training data with all kind of point clouds and generate objects in random category, they are likely to generate point clouds composed of parts of several different classes of objects as Fig. 1 shown. In addition, as the datasets are unbalanced with regard to categories, aforementioned generative models also tend to generate point cloud with unbalanced distribution. Therefore, this paper proposes a novel framework, called PC-cGAN that can generate 3D point clouds from random latent codes with categories as auxiliary information to avoid the intra-category hybridization problem and data unbalance problem. Besides, to enhance our performance in terms of high quality and diversity, graph convolution layer, called BranchGCN, to aggregate information from ancestors and neighbors in feature spaces.

The main contributions of this paper are listed as follows.

- We present a BranchGCN, which passes messages from not only ancestors, but also neighbors.
- We propose a generative model, consisting of a pre-generator and a conditional modifier, to generate 3D point clouds based on the given category information in a supervised way.
- Using the conditional generation framework, the intra-category hybridization problem and the data unbalance problem in generated distribution can be solved directly.

The remainder of this paper is organized as follows. We give a summary about related studies in Sect. 2. In Sect. 3, we introduce the basic related concepts about GANs and graph convolution networks. And the details of our model are provided in Sect. 4. Next, We present the experimental setup and results in Sect. 5. A conclusion of our work is given in Sect. 6.

Fig. 1. An example of the intra-category hybridization phenomenon: The point cloud is combination of a chair and an airplane.

2   Related Works

Graph Convolutional Networks. During the past years, many works have focused on the generalization of neural networks for graph structures. [5] proposed graph convolution network for classification tasks. [16] introduced the Chebyshev approximation of spectral graph convolutions for semi-supervised graph learning. Even GCN [34] can be adopted to extract spatial-temporal features for time series data. And dynamic graph convolution network [32] was designed to extract feature for point clouds using the connectivity of pre-defined graphs. Similarly, TreeGCN [27] was introduced to represent the diverse typologies of point clouds.

GANs for Point Cloud. Although GANs for image generation takes have been comprehensively studies with success [3,8,10,11,14,24,37], but GANs for 3D point clouds have seldom been studies in computer vision domain. Recently, [1] proposed a GAN for 3D point clouds called r-GAN, the generator of which is based on fully connected layers. Since these layers failed to utilize the structural information of point clouds, the r-GAN met difficulty to synthesize realistic objects with diversity. In order to utilize the structural information in point clouds, [31] used graph convolutions in the generator for the first time. However, the computational cost for the construction of adjacency matrices is $O(V^2)$, which leads to lengthy training period. Therefore, tree-GAN [27] saved computational cost and time without construction of adjacency matrices. Instead, its generator used ancestor information from the tree to exploit the connectivity of a graph, in which only a list of tree structure is needed. But tree-GAN lacked attention on neighbor information of each point clouds, which leaded to slow convergence in its training process.

Conditional Generation. In contrast to the aforementioned generative model that learned to directly generate point clouds without regard to category, several generative models in image processing received additional conditional parameters to construct specific objects or styles. Conditional GAN [22] used explicit condition to generate hand-written digit images by an additional auxiliary classifier, which formed the basis of many other conditional generative models. StyleGAN [14] and others [8,37] investigated how to enhance desirable attributes of generated images selectively. However, in 3D computer vision, conditional generative models are rarely adopted to synthesize point cloud data. Recently, the generator of [4] introduced a progressive generative network that created both geometry and color for point clouds based on the given class labels. Their work focus on generating dense and colored point clouds and struggled with generating objects that have fewer samples in the training data. And SP-GAN [17] is able to promote controllability for part-aware shape generation and manipulation without any part annotations, while SP-GAN requires huge amount of training epochs. Moreover, the work in [30] proposed conditioning point cloud generation using continuous physical parameters, but not category information.

3   Preliminaries

3.1   GAN

In general, a generative adversarial network (GAN) is designed to train a generative sub-network that transfers Gaussian random vectors $z \in \mathbf{Z}$ ($\mathbf{Z}$ is a Gaussian distribution by default) into data in a real sample space $x \in \mathbf{X}$ ($\mathbf{X}$ is the set of training samples) to foolish the discriminator, and the discriminator that learns to judge samples from the given dataset and generator as real or fake. Therefore, the task of GAN could be formulated as the minimax objective:

$$\min_{G} \max_{D} \mathbb{E}_{x \sim \mathbb{P}_r}[\log(D((x))] + \mathbb{E}_{\tilde{x} \sim \mathbb{P}_g}[\log(1 - D(\tilde{x}))], \qquad (1)$$

where $\mathbb{P}_r$ is the data distribution from the given dataset and $\mathbb{P}_g$ is the generative distribution defined by $\tilde{x} = G(z), z \sim \mathcal{N}(\mathbf{0}, \mathbf{I})$. Additionally, in order to increase diversity and stability of GAN, a gradient penalty that penalize the norm of gradient of the critic with respect to its input is utilized in improved WGAN, so that it can satisfy the 1-Lipschitz condition.

3.2   Graph Convolution Network

The classical graph convolution network can be defined as a network with multiple message passing layers of the form

$$\mathbf{x}_i^{l+1} = \sigma \left(W^l \mathbf{x}_i^l + \sum_{j \in \mathcal{N}(i)} U^l \mathbf{x}_j^l + b^l \right), \qquad (2)$$

where $\sigma(\cdot)$ is the activation unit, $\mathbf{x}_i^l$ denotes the feature (e.g., 3D coordinate of a point cloud at the first layer) of i-th vertex in the graph at l-th layer, while $\mathcal{N}(i)$ denotes the set of all neighbor vertices of the i-th vertex. And W_l, U_l is learnable weights at l-th layer, with b_l as learnable bias at l-th layer. The first and second term in (2) are called the self-loop term and the neighbors term, respectively.

In tree-GAN [27], a novel GCN, called TreeGCN, is defined with tree structure, passing information from ancestors to descendants of points. TreeGCN can be formulated as

$$ p_i^{l+1} = \sigma \left(W^l p_i^l + \sum_{q_j^l \in \mathcal{A}(p_i^l)} U^l q_j^l + b^l \right), \tag{3} $$

where p_i^l is the i-th point in the point cloud at the l-th layer, q_j^l is the j-th neighbor of p_i^l, and $\mathcal{A}(p_i^l)$ denotes the set of all ancestors of p_i^l. That means the second term in Eq. 3 updates a current point based on all its ancestors but not its neighbors.

4   Methodology

We propose an approach inspired by cGAN [22] in image generation and tree-structured graph convolution operations in [27]. Instead of synthesizing samples without any instruction like previous GAN models for point clouds, we design a supervised generative model by restricting the output samples to a specific category and train a discriminator to judge whether the given samples are real or fake, and which categories the samples belong to.

Unlike previous conditional GANs in image processing, our generative model does not take a random vector and the given condition information as the direct input. The model generates an artificial sample without any restriction in the first step, which accelerates the convergence speed. Then the unconditioned sample is modified by the rest of the model with the given category condition. After modification, the conditional input forces the model to transform the generated sample to the givn category. We present the overview of our GAN framework in Fig. 2. BrachGCN, the basic module of PC-cGAN, is introduced in Sect. 4.1, while in Sect. 4.2, we discuss the auxiliary supervised module to help GAN to generate category-special objects. And the loss functions for training procedure is explained in Sect. 4.3.

4.1   BranchGCN

To improve TreeGCN adopted in [27], we propose a new GCN combining TreeGCN with Dynamic Graph Convolution [32] and define it as BranchGCN. Since a set of point clouds is unable to be converted into a fixed graph before being processed, and those traditional GCNs require information about connectivity of a graph, tree-GAN [27] introduced TreeGCN to avoid use prior

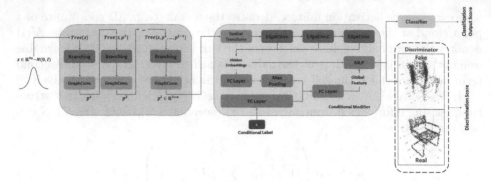

Fig. 2. Overview of model architecture: Our model consists of four parts: 1) Pre-Generator, 2) Conditional Modifier, 3) Discriminator, and 4) Classifier. The Pre-Generator generates the rough point clouds without any auxiliary information. The Conditional Modifier modifies the output point clouds of the Pre-Generator based on the given category labels. The Discriminator, meanwhile, tries to discriminate between the real point clouds and the generated point clouds. The classifier learns to classify the given point clouds, no matter whether they are real or artificial.

information about connectivity among vertices. However, TreeGCN only considers information from ancestors, but does not take horizontal connection into account. Therefore, we remain the ancestor term in TreeGCN, but add an extract term: neighbor term. The additional neighbor term is able to aggregate the information of vertices from their nearest neighbors. The neighbor term is beneficial to recompute the graph structure by using nearest neighbors in the embedded feature spaces produced by each layer. By adding the neighbor term, the receptive field can be expanded to the diameter of the point cloud. Specifically, we compute a pairwise Euclidean distance matrix in the feature spaces and then take closest k vertices for each single vertex in our experimental implementation. Figure 3 illustrates the differences between TreeGCN and BranchGCN.

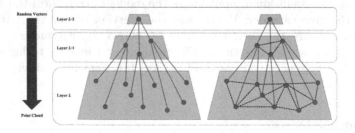

Fig. 3. TreeGCN (left) and BranchGCN (right).

To assist the neighbor term, we construct $\mathcal{G}^l$ as the k-nearest neighbor (k-NN) graph of points at l-th layer based on the pairwise distance between their features, so that the connectivity $\mathcal{E}^l \subseteq \mathcal{V} \times \mathcal{V}$ between each point could be obtained. Accordingly, the output of this term could be formulated as

$$\sum_{k:(i,k)\in\mathcal{E}} H^l p_k^l, \tag{4}$$

where H^l is also learnable weights at l-th layer. It is noticeable that the graph structure is recomputed using nearest neighbors in the feature space built in each layer. By adding this neighbors term, the Branching plays an effective role to aggregate local features during generative procedure.

4.2 Conditional Generative Model

Since there is a lack of control on traditional GANs to generate artificial samples of specific categories, these models tend to learn one-to-one mappings from a random distribution to the distribution of the real dataset, which suffer from the risk to produce mixtures of different kinds of samples and tend to produce a unbalanced distribution just like the real dataset. We attribute these two issues to the lack of explicit control on generated samples' category. On the contrary, the conditional version of GAN takes the combination of a random vector and a specific label about the desired class at the input layer to generate samples with the specific shape. Therefore, to address these issues, the proposed GAN adopts another strategy to utilize conditional information in our work. We divide the generative model into two part: a pre-generator similar to the existing GANs, and a conditional modifier that enables the generative model to employ category information to generated point clouds. As for the pre-generator, we take the generator of tree-GAN as our backbone but replace the TreeGCN module with our BranchGCN, as Fig. 2 shown. Meanwhile, we keep the Branching module in [27] to upsample the generated points from low-dimension random vectors. On the another hand, we extend the architecture of DGCNN to receive the output point clouds from the pre-generator, and an additional input that is represented as a one-hot code about category information. These codes and the outputs from the pre-generator are fed into the sub-network in our modifier together to synthesize point clouds based on the given labels.

With regard to our discriminator, since we add auxiliary information into generator, the corresponding discriminator should have the ability to distinguish which category the given point clouds belong to, and whether they are real or artificial. Consequently, an extra classifier is employed to recognize them. And our discriminator has two outputs: one for the adversarial feedback $D(x)$, and another to determine the classification results $\hat{y} = C(x)$. To adapt the new strategy, we train the overall model by Algorithm 1. Different from traditional GANs, we need to train the generator, the discriminator, and classifier simultaneously in each iteration.

Algorithm 1: Training the PC-cGAN model

---
1 $\theta_{(Gen,Mod)}, \theta_{Disc}, \theta_{Cls} \leftarrow$ initialize parameters
2 **repeat**
3 | | $(\mathbf{X}, c) \leftarrow$ random point clouds and their category labels from dataset;
4 | | $\mathbf{Z} \leftarrow$ samples from prior $\mathcal{N}(\mathbf{0}, \mathbf{I})$;
5 | | $\widetilde{\mathbf{X}} \leftarrow Mod(Gen(\mathbf{X}, c))$;
6 | | $\hat{\mathbf{X}} \leftarrow$ samples from line segments between $\mathbf{X}$ and $\widetilde{\mathbf{X}}$;
7 | | $\mathcal{L}_{(Gen,Mod)} \leftarrow -Disc(\widetilde{\mathbf{X}})$;
8 | | $\mathcal{L}_{Disc} \leftarrow Disc(\widetilde{\mathbf{X}}) - Disc(\mathbf{X}) + \lambda_{gp}(\|\nabla Disc(\hat{\mathbf{X}})\|_2 - 1)^2$;
9 | | $\mathcal{L}_{Cls} \leftarrow CrossEntropy(Cls(\mathbf{X}), c) + CrossEntropy(Cls(\widetilde{\mathbf{X}}), c)$;
   | | // Update parameters according to gradients
10 | | $\theta_{(Gen,Mod)} \xleftarrow{+} -\nabla_{\theta_{(Gen,Mod)}}(\mathcal{L}_{(Gen,Mod)})$;
11 | | $\theta_{Disc} \xleftarrow{+} -\nabla_{\theta_{Disc}}(\mathcal{L}_{Disc})$;
12 | | $\theta_{Cls} \xleftarrow{+} -\nabla_{\theta_{Cls}}(\mathcal{L}_{Cls})$;
13 **until** *deadline*;

4.3   Loss Functions

The optimization objective for general GANs is composed of two components, the generative loss, adversarial loss with a gradient penalty, just like the improved WGAN [11]. However, since the category information is added into the proposed GAN, there are some differences inside the loss functions. In PC-cGAN, the loss function of the generator, $\mathcal{L}_{gen}$, is defined as

$$\mathcal{L}_{gen} = -\mathbb{E}_{z\in\mathcal{Z}, c\in\mathcal{C}}[D(G(z, c))] + \mathcal{L}_c(G(z, c), c), \qquad (5)$$

where G and D represent the generator and discriminator, respectively. $\mathcal{Z}$ and $\mathcal{C}$ denote a latent Gaussian distribution, and the set of all the class labels. Besides, the last term in Eq. 5 represents the loss for multi-class classification of generated point clouds, defined as $\mathcal{L}_c(G(z, c), c) = CrossEntropy(C(G(z, c)), c)$, where C denotes the multi-class classifier. On the another hand, the loss of the discriminator $\mathcal{L}_{disc}$ is formulated as

$$\mathcal{L}_{disc} = \mathbb{E}_{z\in\mathcal{Z}, c\in\mathcal{C}}[D(G(z, c))] - \mathbb{E}_{x\in\mathcal{R}}[D(x)] + \lambda_{gp}\mathcal{L}_{gp} + \mathcal{L}_c(x, c), \qquad (6)$$

where x denotes real point clouds belonging to a real data distribution $\mathcal{R}$. Accordingly, in our model, we adopt a gradient penalty from WGAN, $\mathcal{L}_{gp} = \mathbb{E}_{\hat{x}}[(\|\nabla_{\hat{x}}D(\hat{x})\|_2 - 1)^2]$, where $\hat{x}$ is sampled uniformly along straight lines between pairs of points samples from the data distribution $\mathcal{R}$ and the generated distribution $G(z)$ ($z \in \mathcal{Z}$), and λ_{gp} is a weighting parameter that we set to 10.

For better convergence of our model, we add another discriminator to distinguish the generated point clouds from the pre-generator, which is illustrated in Fig. 4. Therefore, the loss function of the generative model is refined as

$$\mathcal{L}'_{gen} = \mathcal{L}_{gen} - \mathbb{E}_{z\in\mathcal{Z}}[D'(G'(z))], \qquad (7)$$

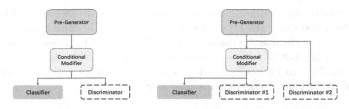

Fig. 4. The difference between the frameworks of the original PC-cGAN (left) and the improved PC-cGAN (right).

where G' and D' denote the pre-generator and the extra discriminator. Since the output point clouds from the pre-generator are lack of category constraint, the additional loss for multi-class classification $\mathcal{L}'_c$ is unnecessary. Similarly, the loss function of discriminator part is modified as

$$\mathcal{L}'_{disc} = \mathcal{L}_{disc} + \mathbb{E}_{z \in \mathcal{Z}}[D'(G'(z))] - \mathbb{E}_{x \in \mathcal{R}}[D'(x)] + \lambda_{gp}\mathcal{L}'_{gp}, \qquad (8)$$

where $\mathcal{L}'_{gp}$ is the gradient penalty term for the extra discriminator D'.

5   Experimental Results

In this section, we demonstrate the point cloud generation using PC-cGAN on ShapeNetBenchmark [6] in two tasks: point cloud generation, and point cloud modification.

5.1   Generation Point Cloud Assessment

As for experimental details, an Adam optimizer is adopted to train both our generator and discriminator sub-networks with a learning rate of $\alpha = 10^{-4}$, $\beta_1 = 0$ and $\beta_2 = 0.99$. The discriminator with a classifier was updated ten times per iteration, while the generator was updated one time per iteration. After setting these hyper-parameters, a latent vector $z \in \mathbb{R}^{96}$ sampled from a Gaussian distribution $\mathcal{N}(0, \mathbf{I})$, and the total number of generated point clouds was set to $n = 2048$.

The conventional quantitative metrics proposed by [1] are used to evaluated the quality of generated point clouds by measuring matching distances between real and artificial point clouds. Besides, we also adopt Fréchet dynamic distance (FPD) [27], a nontrivial extension of Fréchet inception distance (FID) [13], to quantitatively evaluate generated point clouds. We generated 1000 random samples for each class and performed an evaluation using the aforementioned matrices. Table 1 presents the results along with quantitative comparison to previous studies [1] on conventional metrics, and Table 2 presents the FPD score. Note that separated models were trained in all the comparative methods to generated point clouds for different classes except [4] used the same model to generate point clouds for five classes, while the proposed model is able to generate point clouds of any category (16 classes) simultaneously on ShapeNetBenchmark dataset.

Table 1. A quantitative evaluation of the Jensen-Shannon divergence (JSD), the minimum matching distance (MMD), coverage (COV) with the Earth mover's distance (EMD), and the pseudo-chamfer distance (CD). Please refer to [1] for details regarding the metrics. For the GANs with ⋆, we adopted the results from [4]. The red and blue values indicate the best and the second best results for each metric, respectively. The resolution of the evaluated point clouds was 2048 × 3.

Class	Model	JSD ↓	MMD ↓		COV ↑	
			CD	EMD	CD	EMD
Airplane	r-GAN (dense)⋆	0.182	0.0009	0.094	31	9
	r-GAN (conv)⋆	0.350	0.0008	0.101	26	7
	Valsesia et al. (no up.)	0.164	0.0010	0.102	24	13
	Valsesia et al. (up.)	0.083	0.0008	0.071	31	14
	tree-GAN⋆	0.097	0.0004	**0.068**	61	**20**
	PC-cGAN (ours)	**0.086**	**0.0006**	0.061	**53**	23
Chair	r-GAN (dense)⋆	0.235	0.0029	0.136	33	13
	r-GAN (conv)⋆	0.517	0.0030	0.223	23	4
	Valsesia et al. (no up.)	0.119	0.0033	0.104	26	20
	Valsesia et al. (up.)	**0.100**	0.0029	**0.097**	30	26
	tree-GAN⋆	0.119	0.0016	0.101	58	**30**
	PCGAN⋆	0.089	0.0027	0.093	30	33
	PC-cGAN (ours)	0.119	**0.0026**	0.109	**38**	24
Table	PCGAN⋆	0.250	0.0016	0.097	10	9
	tree-GAN	0.074	0.0032	0.115	46	35
	PC-cGAN (ours)	0.048	0.0030	0.096	50	47
Motorbike	PCGAN⋆	0.093	0.0035	0.089	45	43
	tree-GAN	0.116	0.0015	0.056	18	35
	PC-cGAN (ours)	0.062	0.0013	0.069	25	38
Car	tree-GAN	0.080	0.0014	0.089	35	18
	PC-cGAN (ours)	0.070	0.0014	0.073	38	19
Guitar	tree-GAN	0.046	0.0008	0.051	40	18
	PC-cGAN (ours)	0.083	0.0006	0.061	32	23
All (16 classes)	r-GAN (dense)	0.171	0.0021	0.155	58	29
	tree-GAN	0.105	0.0018	0.107	66	39
	PC-cGAN (ours)	0.034	0.0034	0.106	47	44

Table 2. The FPD score for point cloud samples generated by generative models. Notice that the score for real point clouds are almost zero. The point clouds were evaluated at a resolution of 2048 × 3. The bold values denote the best results.

Class	r-GAN	tree-GAN	PC-cGAN (ours)
Airplane	1.860	**0.439**	0.747
Chair	1.016	**0.809**	1.948
All (16 classes)	4.726	3.600	**2.120**

According to the quantitative results from the given tables, we have achieved comparable results in synthesizing high-quality point clouds for each category. Even though our model fails to achieve the best in terms of some metrics for some classes, the model outperforms other GANs on the most metrics (*i.e.*, JSD, MMD-EMD, COV-EMD, and FPD), which demonstrates the effectiveness of the proposed method for multi-class generation. Again, we point out that all the methods train a separate network for each single class except that PCGAN [4] trains the single model for only five classes because of their lack of ability for conditional generation. By contrast, the proposed model uses the same network to generate samples of every single category (16 classes in total) to avoid the hybridization problem and data unbalance issue. Therefore, the quantitative results demonstrate that the proposed method can achieve the comparable performance alongside conditional generation.

In addition to the quantitative results, Fig. 5 and Fig. 6 show point clouds from the real dataset, and ones generated by the baseline in [27], and our PC-cGAN after 1000 epochs of training. In Fig. 5, the generated samples are more realistic due to their large sample size. As for Fig. 6, even though the generative results are not as realistic as Fig. 5 owing to lack of samples, PC-cGAN can still generate point clouds whose shapes look like the given categories.

5.2   Point Cloud Modification

Since the conditional modifier in the proposed generative model could modify the unclassified point clouds from the pre-generator based on the given classes, it also can generate point clouds of the specified categories from any outputs of pre-generator. Figure 7 shows that during the former generative process, the pre-generator produce point clouds randomly, without specified category information, while the latter part modify the random point clouds into ones of the specified categories. And even the point clouds from the pre-generator seem to fall into the specified categories, the modifier also can enhance their realness by decreasing the distance between the distribution of generated point clouds and real point clouds further.

5.3   Ablation Study

To verify the effectiveness of the modules proposed in the previous section, including BranchGCN in Sect. 4.1 and the improved loss functions in Sect. 4.3, we further perform several ablation studies on PC-cGAN. For all experiments in this section, the models are trained in the same experimental configuration as Sect. 5.1 on the table class. Note that although the table class is chosen as an example, all the models in this section are trained on complete ShapeNetBenchmark dataset (16 classes at all) to keep consistency with Sect. 5.1. The results of our ablation studies are presented in Table 3. Applying BranchGCN module to aggregate not only ancestors' information but also features from neighbors brings considerable performance improvement in point cloud generation.

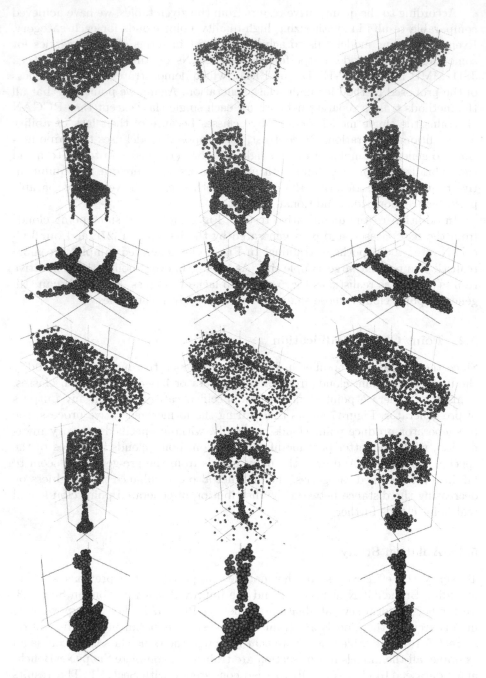

Fig. 5. Real point clouds from ShapeNetBenchmark [6] (left), and 3D point clouds generated by the baseline (middle), *i.e.*, tree-GAN [27], and our PC-cGAN (right).

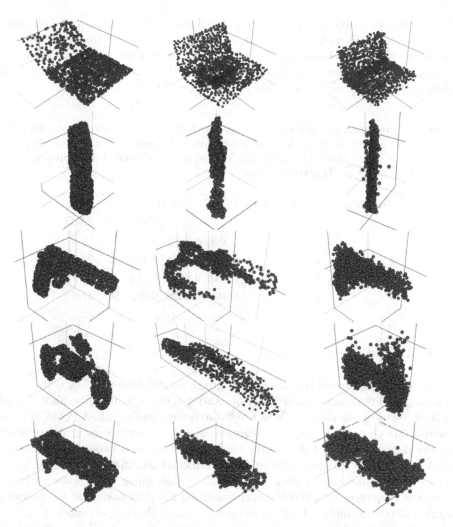

Fig. 6. A supplement for point clouds generation of other categories lacking sample data. The same order is adopted as Fig. 5

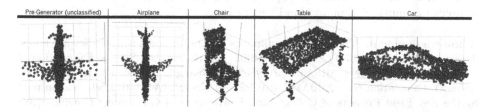

Fig. 7. Point cloud modification. The leftmost point cloud is fixed as a random object generated by the pre-generator, whose shape is closed to a coarse airplane in the implementation. The following four objects are generated by modifying the leftmost point cloud according to the given classes.

The results also demonstrate that the effectiveness of the improved loss functions. Although PC-cGAN with the losses requires more computational resources due to the extra discriminator, the additional discriminator can guarantee the pre-generator produces more realistic samples without category constraint, and enhance the quality of final outputs as mentioned in Sect. 4.3.

Table 3. Supplemented ablation study on the table class. Improved loss used the framework of improved PC-cGAN. We try replacing BranchGCN module with previous TreeGCN in [27] [**BranchGCN**], and removing the additional discriminator with improved loss function [**Improved Loss**].

BranchGCN	Improved loss	JSD↓	MMD↓		COV↑	
			CD	EMD	CD	EMD
✗	✗	0.584	0.0215	0.178	20	15
✗	✔	0.174	0.0164	0.156	22	20
✔	✗	0.080	0.0150	0.113	46	35
✔	✔	**0.048**	**0.0030**	**0.096**	**50**	**47**

6   Conclusions

In this work, a improved tree-structured graph convolution network is proposed to aggregate information from ancestors and neighbors in feature space. Based on that, we propose a conditional GAN for point clouds, called PC-cGAN, to generate 3D objects of specific categories. To introduce the given information about categories into the generative model, we propose a two-stage generator, which consist of a pre-generator and a conditional modifier to solve the intra-category hybridization problem and the data unbalance issue. Based on the two-stage structure, we build the corresponding loss functions and the training algorithm to assist our model to converge effectively. By comparisons with recent generation approaches, we evaluate the generated point clouds by PC-cGAN on the conventional metrics, including FPD score. The quantitative and visual results show that our model is capable of simulating high-quality and diverse samples for multi-class point cloud generation.

Acknowledgement. This work was supported by the National Natural Science Foundation of China (No. 61471229 and No. 61901116), the Natural Science Foundation of Guangdong Province (No. 2019A1515011950), the Guangdong Basic and Applied Basic Research Foundation (No. 2019A1515010789 and No. 2021A1515012289), and in part by the Key Field Projects of Colleges and Universities of Guangdong Province (No. 2020ZDZX3065), and in part by Shantou University Scientific Research Foundation for Talents under Grant NTF19031.

References

1. Achlioptas, P., Diamanti, O., Mitliagkas, I., Guibas, L.: Learning representations and generative models for 3D point clouds. In: Dy, J., Krause, A. (eds.) Proceedings of the 35th International Conference on Machine Learning. Proceedings of Machine Learning Research, vol. 80, pp. 40–49. PMLR (2018). https://proceedings.mlr.press/v80/achlioptas18a.html
2. Alexiou, E., Yang, N., Ebrahimi, T.: PointXR: a toolbox for visualization and subjective evaluation of point clouds in virtual reality. In: 2020 Twelfth International Conference on Quality of Multimedia Experience (QoMEX). IEEE (2020). https://doi.org/10.1109/qomex48832.2020.9123121
3. Arjovsky, M., Chintala, S., Bottou, L.: Wasserstein GAN. arXiv e-prints arXiv:1701.07875 (2017)
4. Arshad, M.S., Beksi, W.J.: A progressive conditional generative adversarial network for generating dense and colored 3D point clouds (2020). https://doi.org/10.1109/3dv50981.2020.00081
5. Bruna, J., Zaremba, W., Szlam, A., LeCun, Y.: Spectral networks and locally connected networks on graphs. In: Bengio, Y., LeCun, Y. (eds.) 2nd International Conference on Learning Representations, ICLR 2014, Banff, AB, Canada, 14–16 April 2014, Conference Track Proceedings (2014). http://arxiv.org/abs/1312.6203
6. Chang, A.X., et al.: ShapeNet: an information-rich 3D model repository. Technical report. arXiv:1512.03012, Stanford University – Princeton University – Toyota Technological Institute at Chicago (2015)
7. Charles, R.Q., Su, H., Kaichun, M., Guibas, L.J.: PointNet: deep learning on point sets for 3D classification and segmentation. In: 2017 IEEE Conference on Computer Vision and Pattern Recognition (CVPR). IEEE (2017). https://doi.org/10.1109/cvpr.2017.16
8. Choi, Y., Choi, M., Kim, M., Ha, J., Kim, S., Choo, J.: StarGAN: unified generative adversarial networks for multi-domain image-to-image translation. In: 2018 IEEE Conference on Computer Vision and Pattern Recognition, CVPR 2018, Salt Lake City, UT, USA, 18–22 June 2018, pp. 8789–8797. Computer Vision Foundation/IEEE Computer Society (2018). https://doi.org/10.1109/CVPR.2018.00916. http://openaccess.thecvf.com/content_cvpr_2018/html/Choi_StarGAN_Unified_Generative_CVPR_2018_paper.html
9. Feng, M., Zhang, L., Lin, X., Gilani, S.Z., Mian, A.: Point attention network for semantic segmentation of 3D point clouds. Pattern Recognit. **107**, 107446 (2020). https://doi.org/10.1016/j.patcog.2020.107446
10. Goodfellow, I.J., et al.: Generative adversarial networks. CoRR abs/1406.2661 (2014). http://arxiv.org/abs/1406.2661
11. Gulrajani, I., Ahmed, F., Arjovsky, M., Dumoulin, V., Courville, A.C.: Improved training of Wasserstein GANs. In: Guyon, I., et al. (eds.) Advances in Neural Information Processing Systems, vol. 30. Curran Associates, Inc. (2017). https://proceedings.neurips.cc/paper/2017/file/892c3b1c6dccd52936e27cbd0ff683d6-Paper.pdf
12. Guo, M.-H., Cai, J.-X., Liu, Z.-N., Mu, T.-J., Martin, R.R., Hu, S.-M.: PCT: point cloud transformer. Comput. Vis. Media **7**(2), 187–199 (2021). https://doi.org/10.1007/s41095-021-0229-5

13. Heusel, M., Ramsauer, H., Unterthiner, T., Nessler, B., Hochreiter, S.: GANs trained by a two time-scale update rule converge to a local nash equilibrium. In: Guyon, I., et al. (eds.) Advances in Neural Information Processing Systems 30: Annual Conference on Neural Information Processing Systems 2017, 4–9 December 2017, Long Beach, CA, USA, pp. 6626–6637 (2017). https://proceedings.neurips.cc/paper/2017/hash/8a1d694707eb0fefe65871369074926d-Abstract.html

14. Karras, T., Laine, S., Aila, T.: A style-based generator architecture for generative adversarial networks. In: 2019 IEEE/CVF Conference on Computer Vision and Pattern Recognition (CVPR), pp. 4396–4405 (2019). https://doi.org/10.1109/CVPR.2019.00453

15. Kingma, D.P., Welling, M.: Auto-encoding variational bayes. In: 2nd International Conference on Learning Representations, ICLR 2014, Banff, AB, Canada, 14–16 April 2014, Conference Track Proceedings (2014). http://arxiv.org/abs/1312.6114

16. Kipf, T.N., Welling, M.: Semi-supervised classification with graph convolutional networks. In: 5th International Conference on Learning Representations, ICLR 2017, Toulon, France, 24–26 April 2017, Conference Track Proceedings (2017). https://openreview.net/forum?id=SJU4ayYgl

17. Li, R.; Li, X.; Hui, K.-H.; Fu, C.-W.: SP-GAN: sphere-guided 3D shape generation and manipulation. In: ACM Transactions on Graphics (Proc. SIGGRAPH), vol. 40. ACM (2021)

18. Li, Y., Bu, R., Sun, M., Wu, W., Di, X., Chen, B.: PointCNN: convolution on X-transformed points. In: Bengio, S., Wallach, H.M., Larochelle, H., Grauman, K., Cesa-Bianchi, N., Garnett, R. (eds.) Advances in Neural Information Processing Systems 31: Annual Conference on Neural Information Processing Systems 2018, NeurIPS 2018, 3–8 December 2018, Montréal, Canada, pp. 828–838 (2018). https://proceedings.neurips.cc/paper/2018/hash/f5f8590cd58a54e94377e6ae2eded4d9-Abstract.html

19. Lim, S., Shin, M., Paik, J.: Point cloud generation using deep local features for augmented and mixed reality contents. In: 2020 IEEE International Conference on Consumer Electronics (ICCE). IEEE (2020). https://doi.org/10.1109/icce46568.2020.9043081

20. Ma, K., Lu, F., Chen, X.: Robust planar surface extraction from noisy and semi-dense 3D point cloud for augmented reality. In: 2016 International Conference on Virtual Reality and Visualization (ICVRV). IEEE (2016). https://doi.org/10.1109/icvrv.2016.83

21. Ma, X., Qin, C., You, H., Ran, H., Fu, Y.: Rethinking network design and local geometry in point cloud: a simple residual MLP framework. CoRR abs/2202.07123 (2022). https://arxiv.org/abs/2202.07123

22. Mirza, M., Osindero, S.: Conditional generative adversarial nets. CoRR abs/1411.1784 (2014). http://arxiv.org/abs/1411.1784

23. Qi, C.R., Yi, L., Su, H., Guibas, L.J.: PointNet++: deep hierarchical feature learning on point sets in a metric space. In: Proceedings of the 31st International Conference on Neural Information Processing Systems, NIPS 2017, pp. 5105–5114. Curran Associates Inc., Red Hook (2017)

24. Radford, A., Metz, L., Chintala, S.: Unsupervised representation learning with deep convolutional generative adversarial networks. In: Bengio, Y., LeCun, Y. (eds.) 4th International Conference on Learning Representations, ICLR 2016, San Juan, Puerto Rico, 2–4 May 2016, Conference Track Proceedings (2016). http://arxiv.org/abs/1511.06434

25. Sagawa, H., Nagayoshi, H., Kiyomizu, H., Kurihara, T.: [POSTER] hands-free AR work support system monitoring work progress with point-cloud data processing. In: 2015 IEEE International Symposium on Mixed and Augmented Reality. IEEE (2015). https://doi.org/10.1109/ismar.2015.50

26. Sarmad, M., Lee, H.J., Kim, Y.M.: RL-GAN-net: a reinforcement learning agent controlled GAN network for real-time point cloud shape completion. In: 2019 IEEE/CVF Conference on Computer Vision and Pattern Recognition (CVPR). IEEE (2019). https://doi.org/10.1109/cvpr.2019.00605

27. Shu, D., Park, S.W., Kwon, J.: 3D point cloud generative adversarial network based on tree structured graph convolutions. In: 2019 IEEE/CVF International Conference on Computer Vision (ICCV). IEEE (2019). https://doi.org/10.1109/iccv.2019.00396

28. Singh, P., Sadekar, K., Raman, S.: TreeGCN-ED: encoding point cloud using a tree-structured graph network. CoRR abs/2110.03170 (2021)

29. Tredinnick, R., Broecker, M., Ponto, K.: Progressive feedback point cloud rendering for virtual reality display. In: 2016 IEEE Virtual Reality (VR). IEEE (2016). https://doi.org/10.1109/vr.2016.7504773

30. Triess, L.T., Bühler, A., Peter, D., Flohr, F.B., Zöllner, M.: Point cloud generation with continuous conditioning. In: International Conference on Artificial Intelligence and Statistics, AISTATS 2022, 28–30 March 2022, Virtual Event, pp. 4462–4481 (2022). https://proceedings.mlr.press/v151/triess22a.html

31. Valsesia, D., Fracastoro, G., Magli, E.: Learning localized generative models for 3D point clouds via graph convolution. In: 7th International Conference on Learning Representations, ICLR 2019, New Orleans, LA, USA, 6–9 May 2019 (2019). https://openreview.net/forum?id=SJeXSo09FQ

32. Wang, Y., Sun, Y., Liu, Z., Sarma, S.E., Bronstein, M.M., Solomon, J.M.: Dynamic graph CNN for learning on point clouds. ACM Trans. Graph. **38**(5), 1–12 (2019). https://doi.org/10.1145/3326362

33. Yu, X., Rao, Y., Wang, Z., Liu, Z., Lu, J., Zhou, J.: PoinTr: diverse point cloud completion with geometry-aware transformers. In: 2021 IEEE/CVF International Conference on Computer Vision, ICCV 2021, Montreal, QC, Canada, 10–17 October 2021, pp. 12478–12487 (2021). https://doi.org/10.1109/ICCV48922.2021.01227

34. Yu, Z., Zheng, X., Yang, Z., Lu, B., Li, X., Fu, M.: Interaction-temporal GCN: a hybrid deep framework for COVID-19 pandemic analysis. IEEE Open J. Eng. Med. Biol. **2**, 97–103 (2021). https://doi.org/10.1109/ojemb.2021.3063890

35. Yuan, W., Khot, T., Held, D., Mertz, C., Hebert, M.: PCN: point completion network. In: 2018 International Conference on 3D Vision, 3DV 2018, Verona, Italy, 5–8 September 2018, pp. 728–737. IEEE Computer Society (2018). https://doi.org/10.1109/3DV.2018.00088

36. Zhao, H., Jiang, L., Jia, J., Torr, P.H.S., Koltun, V.: Point transformer. In: 2021 IEEE/CVF International Conference on Computer Vision, ICCV 2021, Montreal, QC, Canada, 10–17 October 2021, pp. 16239–16248. IEEE (2021). https://doi.org/10.1109/ICCV48922.2021.01595

37. Zhu, J.Y., Park, T., Isola, P., Efros, A.A.: Unpaired image-to-image translation using cycle-consistent adversarial networks. In: 2017 IEEE International Conference on Computer Vision (ICCV), pp. 2242–2251 (2017). https://doi.org/10.1109/ICCV.2017.244

Progressive Attentional Manifold Alignment for Arbitrary Style Transfer

Xuan Luo[1], Zhen Han[2(✉)], and Linkang Yang[1]

[1] School of Computer Science and Technology, Xi'an Jiaotong University, Xi'an, Shaanxi, China
[2] School of Computer Science, Wuhan University, Wuhan, Hubei, China
hanzhen_2003@hotmail.com
https://github.com/luoxuan-cs/PAMA

content/style   AdaIN   WCT   SANet   AdaAttN   StyleFormer   IEC   MAST   Ours

Fig. 1. The style degradation problem. Stylization results of AdaIN [13], WCT [22], SANet [29], AdaAttN [25], StyleFormer [38], IEC [2], and MAST [14] are shown. Existing methods may fail to transfer the brushstrokes (the first row) or the color distribution of the style image (the second row).

Abstract. Arbitrary style transfer algorithms can generate stylization results with arbitrary content-style image pairs but will distort content structures and bring degraded style patterns. The content distortion problem has been well issued using high-frequency signals, salient maps, and low-level features. However, the style degradation problem is still unsolved. Since there is a considerable semantic discrepancy between content and style features, we assume they follow two different manifold distributions. The style degradation happens because existing methods cannot fully leverage the style statistics to render the content feature that lies on a different manifold. Therefore we designed the progressive attentional manifold alignment (PAMA) to align the content manifold to the style manifold. This module consists of a channel alignment module to emphasize related content and style semantics, an attention module to establish the correspondence between features, and a spatial interpolation module to adaptively align the manifolds. The proposed PAMA can alleviate the style degradation problem and produce state-of-the-art stylization results.

Keywords: Style transfer · Manifold alignment · Image synthesis

Supplementary Information The online version contains supplementary material available at https://doi.org/10.1007/978-3-031-26293-7_9.

1   Introduction

Neural style transfer aims at rendering a content image with style patterns from a style image. The pioneering style transfer algorithms rather needs online optimization [6,7,9,18], or be constrained to a few styles [16,19,21,33,34,37]. Arbitrary style transfer methods [3,13,20,22,29] enable real-time stylization with arbitrary styles by leveraging statistical information. These flexible yet efficient approaches have received widespread attention from academics and industries. However, arbitrary style transfer methods suffer from the content distortion problem and the style degradation problem. The content distortion problem can be alleviated using the high frequency components [10], salient map guidance [26], and low-level features [25]. This paper focuses on the style degradation problem, which is less studied but significantly influences stylization quality.

Figure 1 demonstrates the style degradation problem. The first row shows examples generated using the style with thick colorful lines. The AdaIN [13], AdaAttN [25], StyleFormer [38], IEC [2] stylized the content structure insufficiently. Only a limited set of style patterns are used for rendering. Although the WCT [22], SANet [29], and AdaIN [13] brings sufficient style patterns, the patterns are mixed chaotically (WCT), distorted locally (SANet), or blurred and overlapped (MAST). These methods struggle to migrate the brushstroke information. The second row demonstrates that existing methods may also damage the color distribution. The AdaIN, WCT, and SANet mix the colors of the style image, producing a tainted appearance. Moreover, the AdaAttN, StyleFormer, IEC, and MAST merely adopt a few colors for rendering.

We identified that the semantic discrepancy between content and style images brings the style degradation problem. Since the content images are natural images but the style images are artificial images, the content and style features follow different manifold distributions [14]. Utilizing statistics of the data from different distributions is inherently challenging. For instance, the AdaIN [13] use the mean and variance of the style feature to transform the content feature. However, directly applying the style feature statistics distorts the style semantics and brings the style degradation problem. The MAST [14] learns a projection matrix to align the content manifold to the style manifold. The WCT [22] adopts whitening and coloring transformations to adjust the correlation of the content feature. Nevertheless, the learning-free linear transformations of MAST and WCT have limited expressiveness. For patch-based methods like SANet [29], AdaAttN [25], and StyleFormer [38], it is difficult to measure complex relations between content and style feature vectors. The patch-based methods cannot reorganize the style feature vectors to form complicated patterns.

To alleviate the style degradation problem, we proposed the progressive attentional manifold alignment (PAMA) to align the content manifold to the style manifold. The PAMA consists of a channel alignment module, an attention module, and a spatial interpolation module. In the channel alignment module, we adopt the squeeze and excitation operation [12] to extract channel weights of the content and style features. Then the content channel weights are used to re-weight the style feature while the style channel weights are used to re-weight the content feature. This cross-manifold channel alignment emphasizes the related semantics (the related channels) of content and style features, helping the attention module parse complex relations between them. The attention module is the style attentional network [29] that computes pair-wise similarity (attention map) between content and style feature vectors. The style feature is redistributed according to the attention map, building the correspondence between the content and style feature vectors. The spatial interpolation module summarizes spatial information to adaptively interpolate between the content feature and the redistributed style feature. The content and style feature vectors with similar semantics are linearly fused, forming an intermediate manifold between the content manifold and style manifold. By repeating the whole process multiple times, the content manifold is gradually aligned to the style manifold along a geodesic between them, and thus the semantic gap between content and style features is filled.

We designed a multistage loss function with decreasing weight of content loss to train the progressive alignment procedure. The content loss is the self-similarity loss [18,23,30] to preserve the original manifold structure of the content feature. The style loss is the relaxed earth mover distance [18,23,30] which optimizes along the style manifold surfaces [30]. A momentum loss [18,23,30] is used together with the style loss to preserve the magnitude information. We also adopt a color histogram loss [1] to align the color distributions explicitly. Finally, an auto-encoder loss is used to maintain the common space for manifold alignment. Our contributions can be summarized as follows:

- We proposed a new arbitrary style transfer framework named PAMA, which gradually aligns the content manifold to the style manifold with a channel alignment module, an attention module, and a spatial interpolation module.
- A multistage loss function is designed to enable progressive manifold alignment. We also adopt an auto-encoder loss to maintain the shared space for manifold alignment.
- Experiments show that the proposed framework can generate fine-grained stylization results in real-time (100 fps for 512px images on Tesla V100 GPU). The style degradation problem is alleviated significantly.

2   Related Works

2.1   Arbitrary Style Transfer

The goal of arbitrary style transfer is to generate stylization results in real-time with arbitrary content-style pairs. The mainstream arbitrary style transfer algorithms can be divided into two groups: the global transformation based and local patch based. The global transformation based methods utilize global statistics for feature transformation. One of the representative methods is AdaIN [13], which forces the mean and variance of content features to be the same as the style features. To reduce the memory consumption, DIN [15] substitutes the

VGG [32] with the MobileNet [11] and adopts dynamic convolutions for feature transformation. Although practical, using the global statistics from another manifold distribution to transform the content feature brings degraded style patterns. There are also manifold alignment based style transfer algorithms performing global transformations in a common subspace. The WCT [22] changes the covariance matrix of content features with whitening and coloring transformation, which aligns the self-similarity structure of the content and style manifolds. The MAST [14] learns to project the content manifold to a subspace where the correspondence between the two manifolds can be found. However, the WCT and MAST use learning-free transformations for manifold alignment, limiting their ability to transfer complicated style patterns.

For the local patch based methods, they manipulate the feature patches for stylization. The style swap proposed by Chen *et al.* [3] is the earliest patch based method, which swaps the content patches with the most similar style patches. The DFR [9] and AAMS [39] further extend this method with global hints. Recently, the SANet [29] is proposed to matching content and patches using the attention mechanism. Then MANet [4] disentangles the content and style representation for the attention mechanism. AdaAttN [25] enhances the attention mechanism with multi-layer features to reduce content distortion. IEC [2] uses internal-external learning and contrastive learning to refine the feature representation of SANet. However, the semantic discrepancy between content and style features makes the nearest neighbor search and attentional matching struggle to parse complex semantic relations, triggering the style degradation problem. With the manifold alignment process, the proposed PAMA can parse complex relations like regional correspondence and high-order relations.

2.2 Manifold Alignment

Manifold alignment aims at revealing the relationship between two datasets from different manifold distributions. These algorithms learn to transform the two datasets in a shared subspace, establishing the correspondence of the two datasets while preserving their manifold structures. Existing manifold alignment methods for domain adaptation can be divided into subspace learning methods [5,8], distribution alignment methods [27,36]. Huo *et al.* [14] introduce manifold alignment methods to the style transfer community, which assumes that the content and style features follow different manifold distributions. This subspace learning method aligns the content manifolds to style manifolds with a global channel transformation. The proposed PAMA is a distribution alignment method that uses linear interpolation to align the content manifold to the style manifold. The alignment is performed along a geodesic between the two manifolds.

3 Method

3.1 Overall Framework

Figure 2 shows the architecture of the proposed progressive attentional manifold alignment (PAMA). Our method uses a pre-trained VGG [32] network to encode

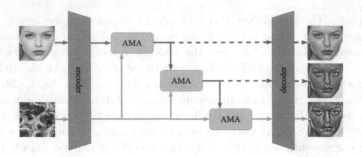

Fig. 2. The overall framework of the progressive attention manifold alignment (PAMA). The content manifold is gradually aligned to the style manifold with three independent attentional manifold alignment (AMA) blocks. The dash lines are only forwarded during training to generate the intermediate results for loss calculation.

the content image I_c and style image I_s, resulting the $ReLU4\_1$ features F_c and F_s. The features are transformed by the attentional manifold alignment (AMA) block for stylization, which consists of a channel alignment module, an attention module, and a spatial interpolation module (Fig. 3). Passed through three AMA blocks, the aligned content feature will be fed into the decoder to generate the stylized image. Following the setting of [13], the structure of the decoder is symmetric to the encoder.

3.2   Channel Alignment Module

The channel alignment module aims at emphasizing the related semantic aspects (related channels) between content and style features. The alignment is achieved by manipulating cross-manifold information to re-weight feature channels (Fig. 3). We adopted the squeeze-and-excitation operation of SENet [12]:

$$W = MLP(GAP(F)) \tag{1}$$

where a global average pooling operation (GAP) pools $F \in \mathbf{R}^{H \times W \times C}$ into $\mathbf{R}^{C}$, and a multilayer perceptron (MLP) is used to embed the channel feature to obtain the channel weights. The H, W denotes the height, width, and channels of the feature F. We applied Eq. 1 on the content feature F_c and the style feature F_s to obtain the channel weights $A_c \in \mathbf{R}^C$ and $A_s \in \mathbf{R}^C$. As demonstrated in Fig. 3, the features F_c and F_s are crossly re-weighted with A_s and A_c to, resulting the aligned features $\hat{F}_c$ and $\hat{F}_s$. The related feature channels (or related semantic aspects) between the content and style features are enhanced. This global channel re-weighting operation can help the attention module to parse cross-manifold semantics.

3.3   Attention Module

The middle part of Fig. 3 is the attention module, which redistribute the style feature according to the content structure. This module builds the spatial cor-

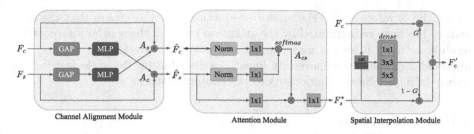

Fig. 3. The details of a single attentional manifold alignment (AMA) block. It consists of a channel alignment module, an attention module, and a spatial interpolation module.

respondence of the content and style feature vectors. The attention map is computed using normalized features:

$$A_{cs} = softmax(f(Norm(\hat{F}_c)) \otimes g(Norm(\hat{F}_s))^T) \tag{2}$$

where the $f(\cdot)$ and $g(\cdot)$ denote 1×1 convolution blocks for feature embedding, the $Norm(\cdot)$ refers to the mean-variance normalization, and the $\otimes$ is the matrix multiplication. With the attention map A_{cs} containing pair-wise similarities, the style feature vectors are redistributed according to the content feature:

$$F_s^* = \theta(A_{cs} \otimes h(\hat{F}_s)) \tag{3}$$

where the $h(\cdot)$ and $\theta(\cdot)$ are 1×1 convolution blocks for feature embedding. The F_s^* denotes the redistributed style feature. The architecture of the attention module is the same as the ones of [2, 4, 29].

3.4   Spatial Interpolation Module

The right part of Fig. 3 shows the structure of the spatial interpolation module. The spatial interpolation module summarizes spatial information to adaptively interpolate between the content feature F_c and the redistributed style feature F_s^*. Concretely, the dense operation applies multi-scale convolution kernels on the concatenated feature to compute the interpolation weights $G \in \mathbf{R}^{H \times W}$:

$$G = \frac{1}{n} \sum_{i=1}^{n} \psi_i([F_c, F_s^*]) \tag{4}$$

where $\psi_i(\cdot)$ represent the i-th convolution kernel, and the $[\cdot, \cdot]$ denotes the channel concatenation operation. The concatenated feature can help us to identify the local discrepancy between the corresponding content and style feature, figuring out the appropriate interpolation strength. The interpolation weights G is then applied for linear interpolation:

$$F_c' = G \odot F_c + (1 - G) \odot F_s^* \tag{5}$$

where the $\odot$ refers to the Hadamard production. For the reason that the style feature has been redistributed by the attention module, the spatial interpolation module actually fuses the most similar content and style feature vectors. The manifold alignment is achieved by linearly redistributing the style feature and interpolating its linear components to the content feature.

3.5 Loss Functions

Multistage losses are applied for the proposed progressive attentional manifold alignment (PAMA). For each stage, the loss is a weighted summation of the self-similarity loss [18,31] L_{ss}, the relaxed earth mover distance loss [18,23,30] L_r, the momentum loss [18,30] L_m, and the color histogram loss [1] L_h. The overall loss is the weighted sum of the multistage losses and the auto-encoder loss L_{ae}:

$$L = \sum_i (\lambda_{ss}^i L_{ss}^i + \lambda_r^i L_r^i + \lambda_m^i L_m^i + \lambda_h^i L_h^i) + L_{ae} \tag{6}$$

where the λ_x^i denotes the weight for L_x ($x \in \{ss, r, m, h\}$) of the i-th stage. In the first stage, a high initial value of the self-similar loss weight λ_{ss} is set to preserve the content manifold structure. In the following stages, the weight of the self-similarity loss decreases gradually to relax constraints and align the content manifold to the style manifold.

Our content loss is based on the L_1 distance between the self-similarity matrices of the content feature F_c and the VGG [32] feature of the stylized image F_{cs}:

$$L_{ss} = \frac{1}{H_c W_c} \sum_{i,j} \left\| \frac{D_{ij}^c}{\sum_i D_{ij}^c} - \frac{D_{ij}^{cs}}{\sum_j D_{ij}^{cs}} \right\|_1 \tag{7}$$

where D^c and D^{cs} are the pairwise cosine distance matrices of the F_c and F_{cs}, and the subscript ij denotes the element from the $i-th$ row $j-th$ column. For simplicity, the superscript of λ and L indicating the manifold alignment stage are omitted.

Following the setting of [18,30] we adapts the relaxed earth mover distance (REMD) to align the content manifold to the style manifold:

$$L_r = \max\left(\frac{1}{H_s W_s} \sum_i \min_j C_{ij}, \frac{1}{H_c W_c} \sum_j \min_i C_{ij}\right) \tag{8}$$

where the C_{ij} denotes the pair-wise cosine distance matrix between F_{cs} and F_s. We also added the moment matching loss [18,30] to regularize the magnitude of features:

$$L_m = \|\mu_{cs} - \mu_s\|_1 + \|\Sigma_{cs} - \Sigma_s\|_1 \tag{9}$$

where the μ and Σ denote the mean and covariance matrix of feature vectors. The subscript s denotes the statistic of style feature and the subscript cs denotes the statistic of the VGG feature of stylization result.

The color distribution plays a central role in solving the style degradation problem. We adopt the differentiable color histogram loss proposed in HistoGAN [1] to learn to match the color distribution explicitly:

$$L_h = \frac{1}{\sqrt{2}} \|H_s^{1/2} - H_{cs}^{1/2}\|_2 \tag{10}$$

where the H refers the color histogram feature [1], the $H^{1/2}$ denotes the element-wise square root.

The target of the auto-encoder loss is maintaining the common subspace for manifold alignment. It learns how to reconstruct the content and style images using the encoded content and style features. We optimize pixel level reconstruction and semantic level reconstruction simultaneously:

$$L_{ae} = \lambda_{ae}(\|(I_{rc} - I_c)\|_2 + \|(I_{rs} - I_s)\|_2) +$$
$$\sum_i (\|\phi_i(I_{rc}) - \phi_i(I_c)\|_2 + \|\phi_i(I_{rs}) - \phi_i(I_s)\|_2) \tag{11}$$

where I_{rc} and I_{rs} are the content and style images reconstructed from the encoded features, and the λ_{ae} is a constant weight. The $\phi_i(I)$ refers to the $ReLU\_i\_1$ layer VGG feature of image I. This common subspace is critical for the attention module to avoid matching features from different subspaces. The attention operation cannot accurately measure the similarities of features from different subspaces, leading to the shifted matching of feature vectors (Fig. 7). The common subspace also helps the spatial interpolation module interpolate between features from the same subspace, thus reducing the information loss of the pipeline.

4   Experiments

4.1   Implementation Details

Our proposed PAMA is trained with content images from COCO [24] and style images from wikiart [28]. VGG [32] features from $ReLU\_3\_1$, $ReLU\_4\_1$, and $ReLU\_5\_1$ are used to compute the self-similarity loss L_{ss}, the REMD loss L_r, and the moment matching loss L_m. For self-similarity loss, we set its weights λ_{ss}^1, λ_{ss}^2, λ_{ss}^3 to 5, 4, 3 respectively. All weights of the style loss terms L_r and L_m are set to 1. The weights of color histogram loss λ_h^1, λ_h^2, λ_h^3 are set to 1, 2, and 4. The λ_{ae} for the reconstruction loss in Eq. 11 is 50. We use the Adam [17] optimizer with learning rate of 0.0001 and momentum of 0.9. We take 8 content-style image pairs as a batch. The smaller dimension of content and style images are rescaled to 512, and then we randomly crop a 256×256 patch for efficient training. In the testing phase, our fully convolutional network can tackle images with any size.

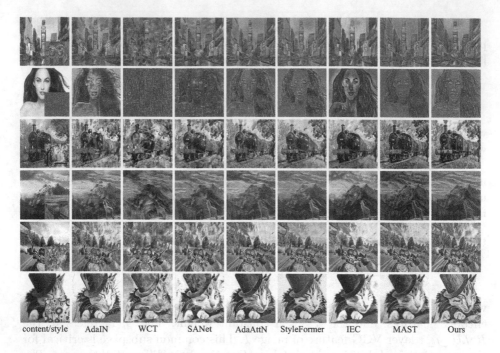

content/style   AdaIN   WCT   SANet   AdaAttN   StyleFormer   IEC   MAST   Ours

Fig. 4. Qualitative comparison.

4.2 Comparison with Prior Arts

To evaluate the proposed progressive attentional manifold alignment (PAMA), we compared it with other arbitrary style transfer methods, including global statistic based methods AdaIN [13], WCT [22], and MAST [14], local patch based methods SANet [29], AdaAttN [25], StyleFormer [38], and IEC [2].

Qualitative Comparison. Figure 4 shows the comparison results. The AdaIN [13] often renders with crack-like style patterns (the 1st, 2nd, 3rd, 5th rows) or shifted color distributions (the 1st, 4th, 5th, 6th rows). Since the content and style features lie on different manifolds, directly using the mean and variance of style features is suboptimal. The WCT [22] performs the whitening and coloring transformations to adjust the covariance of the content feature to be the same as the style feature. The covariance based transformations of WCT may mix the style patterns chaotically (the 1st, 4th, 5th, 6th rows) or distort the content structures (the 2nd, 3rd, 5th rows). The SANet [29] is a patch based method that uses the attention mechanism to match content and style patches, but mixes the style patterns (the 1st, 2nd, 3rd, 6th rows) and brings arc-shaped artifacts (1st, 5th rows). The AdaAttN [25] further introduced low-level features to improve the content preservation ability of SANet, but suffers from understylization (the 3rd, 4th, 5th, 6th rows) and color distribution shift (the 1st, 2nd, 4th, 6th rows). The StyleFormer [38] adopts a transformer [35] inspired network for stylization, but produces blurred (the 2nd, 3rd, 4th rows) and understylized (the 2nd, 5th, 6th rows) results. The IEC [2] suffers from the same problems of AdaAttN, which are the understylization problem (the 2nd, 3rd, 4th, 5th, 6th rows) and the color distribution shift problem (1st, 2nd). These patch based approaches cannot parse the complex relations between the content and style patches. The MAST [14] aligns the content manifold to the style manifold with subspace projection. This learning-free manifold alignment method has limited model capacity, thus not free from the blurry problem (the 3rd, 4th, 5th rows) and the understyliza- tion problem (the 2nd, 5th, 6th rows). The proposed PAMA aligns the content manifolds to the style manifolds, enabling the attention mechanism to establish complex relations like regional correspondence and high-order dependencies. The regional correspondence suggests that content semantic regions (e.g., the sky of the 4th row) are consistently rendered by continuous style patterns (the 1st, 4th rows). The high-order dependencies refer to the multi-level dependencies formed by progressive attention, which is essential to express complicated style patterns (the 6th, 7th rows). The proposed PAMA can alleviate the style degradation problem significantly.

Table 1. User study.

Method	Content quality	Style quality	Overall quality	Total
AdaIN [13]	273	165	252	690
WCT [22]	231	234	207	672
SANet [29]	374	351	453	1178
AdaAttN [25]	1382	541	956	2879
StyleFormer [38]	626	370	514	1510
IEC [2]	1042	589	934	2565
MAST [14]	286	242	235	763
Ours	786	2508	1449	4743

User Study. As shown in Table 1 We perform user study on eight arbitrary style transfer methods: AdaIN [13], WCT [22], SANet [29], AdaAttN [25], StyleFormer [38], IEC [2], MAST [14], and the proposed PAMA. In the study, a single sample consists of a content image, a style image, and eight corresponding stylization results generated by the eight methods. We use 25 content images and 25 style images to generate 625 samples and randomly draw 20 samples for each user. For each sample, users are asked to choose the best stylization result according to the evaluation indicators: content quality, style quality, and overall quality. Each user judges the quality according to their subjective perception. We collect 5000 votes from 250 users for each indicator. The results are shown in Table A. The result demonstrates that our method produces results with better style quality and overall performance.

Table 2. Stylization time comparison.

Method	Time (256px)	Time (512px)
AdaIN [13]	2.813 ms	2.912 ms
WCT [22]	1210 ms	4053 ms
SANet [29]	4.792 ms	6.351 ms
AdaAttN [25]	19.76 ms	22.52 ms
StyleFormer [38]	7.154 ms	9.790 ms
IEC [2]	4.936 ms	6.851 ms
MAST [14]	1488 ms	2743 ms
Ours	8.726 ms	9.937 ms

Efficiency. In Table 2, we compare the stylization time of our method with other baselines at 256px and 512px. All of the methods are evaluated on a server with an NVIDIA V100 PCIe 32G GPU. The results are the average running time of 3600 image pairs. The proposed PAMA can generate artworks in real-time (100 fps at 512px).

4.3   Ablation Study

Loss Analysis. In this part, we explore the effect of different loss weights. Firstly, we change the weight of relaxed earth mover distance loss λ_r and the weight of momentum loss λ_m simultaneously to verify the influence of stylization strength. As shown in Fig. 5(b), when the style loss is comparably low, the proposed PAMA only transfers the color information of the style image. Meanwhile, the content structure is well preserved. As the value of λ_r and λ_m increase, the delicate triangle like texture (the first row of Fig. 5) and water lily patterns (the 2nd row of Fig. 5) are introduced. In conclusion, higher λ_r and λ_m suggest more global and complicated style patterns. We also adjust the weight of the color histogram loss λ_h to demonstrate its effectiveness. The first row Fig. 6 shows that higher λ_h brings richer color patterns. The second row shows that the color histogram loss can even help the proposed PAMA match the distribution of the greyscale image. With these loss functions, the proposed PAMA can not only migrate the complicated style patterns but also render with the style palette.

Subspace Constraint. In this part, we remove the subspace constraint (Eq. 11) to explore its influences. The color distributions of the results from Fig. 7(b) and (c) are completely opposite. Without the subspace constraint, the subspace changes whenever the attention module is performed. The attention modules (except the first one) need to compute the pair-wise similarities with features from different subspaces, leading to the shifted matching of content and style feature vectors, which further influences the color distribution. The common subspace maintained by Eq. 11 is necessary for the manifold alignment process.

Channel Alignment. To verify the effectiveness of the channel alignment module, we remove the channel alignment module of PAMA for ablation study.

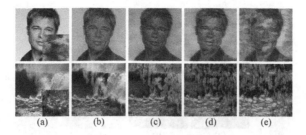

Fig. 5. The influence of the relaxed earth mover distance loss λ_r and the momentum loss λ_m. (a) the content and style images; (b) 0.5x weights; (c) 1x weights (the original weights); (d) 2x weights; (e) 4x weights.

Fig. 6. The influence of the color histogram loss λ_r. (a) the content and style images; (b) 0.5x weights; (c) 1x weights (the original weights); (d) 2x weights; (e) 4x weights.

As shown in Fig. 8(c), the style patterns generated without the channel alignment module overlap with each other (the first row) or form a blurred appearance (the second row). The attention module cannot establish a stable and consistent measurement. In the contrary, a consistent regional correspondence can be find in Fig. 8(b). The trees, buildings, and sky are consistently matched with the blue, red, and yellow patterns (the first row). The bridge, ocean, and sky are rendered with distinguishable brown, blue-green, and blue patterns(the second row). The channel alignment module is critical to parse the semantic relations between the content and style images.

Spatial Interpolation. In this section, we perform ablation study about the spatial interpolation module. Since the redistributed feature is directly passed to the next stage without interpolating with the content feature, the content structure information gradually vanishes during the manifold alignment process. Compared to Fig. 8(b), the content structures of Fig. 8(d) are blurry and distorted. By removing the spatial interpolation module, the stylized feature of the first stage is re-rendered multiple times without the guidance of content information. This makes the manifold deviates from the geodesic between the content and style manifolds, producing collapsed results. The spatial interpolation module is indispensable for the proposed PAMA.

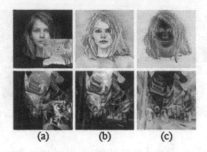

Fig. 7. The influence of the subspace constraint. (a) the content and style images; (b) the results generated with the subspace constraint; (c) the results generated without the subspace constraint.

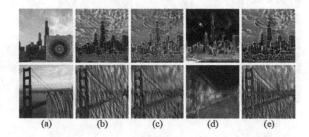

Fig. 8. Ablation studies about the architecture. (a) the content and style images; (b) the original PAMA; (c) PAMA without the channel alignment module; (d) PAMA without the spatial interpolation module; (d) PAMA with single-stage manifold alignment.

Progressive Alignment. In this part, we demonstrate the effectiveness of the progressive alignment mechanism of the proposed PAMA. Figure 8(e) shows the results generated with single-stage manifold alignment. The single-stage manifold alignment fails to establish the regional correspondence as Fig. 8(b) but renders with uniform and mixed style patterns. In the first row of Fig. 8(e), the blue patterns overlap with the yellow and orange pattern. In the second row, the result is rendered by uniform blue-green patterns without apparent variance. The attention module cannot distinguish between different content structures to render them differently. Although single-stage manifold alignment can provide acceptable results, it cannot transfer complicated style patterns and suffers from the understylization problem. Uniform and repetitive pattern are adopted for rendering. The capacity of the single-stage manifold alignment is limited.

4.4   Multi-style Transfer

Multi-style Interpolation. In this part, we linearly interpolate between stylized features of different style images. Specifically, we linearly interpolate between the stylized features from the third stage, and decode them to obtain the final

| content | style1 | 1.00:0.00 | 0.75:0.25 | 0.50:0.50 | 0.25:0.75 | 0.00:1.00 | style2 |

Fig. 9. Linear interpolation between stylized features.

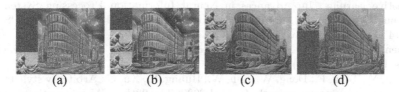

(a)   (b)   (c)   (d)

Fig. 10. Multi-style alignment. The content manifold is aligned with different style manifolds from top to bottom.

stylization results. Figure 9 illustrates the results interpolated with the five different weights. The blue and white wave patterns gradually shift to the yellow and green circuit patterns by changing the interpolation weights.

Multi-style Alignment. The proposed PAMA consists of three attentional manifold alignment (AMA) stages (Fig. 3). We align the content manifold to different style manifolds (different style feature for the three stages) to produce stylization results with textures similar to one set of styles but color distributions similar to another. Figure 10 shows the results of multi-style alignment. In Fig. 10, example (a) and (b) have circuit-like pattern in blue and grey, while example (c) and (d) consists of waves with a greenish appearance. The proposed PAMA can render the texture information and color distribution separately by aligning to multiple style manifolds.

5 Conclusion

We proposed the progressive attention manifold alignment module (PAMA) to solve the style degradation problem. The proposed PAMA consists of a channel alignment module to emphasize related semantics, an attention module to establish correspondence between features, and a spatial interpolation module to align the content manifold to the style manifold. The proposed PAMA can produce high-quality stylization results in real-time.

Acknowledgements. This work was supported in part by the National Nature Science Foundation of China under Grant 62072347.

References

1. Afifi, M., Brubaker, M.A., Brown, M.S.: HistoGAN: controlling colors of GAN-generated and real images via color histograms. In: IEEE Conference on Computer Vision and Pattern Recognition, CVPR 2021, virtual, 19–25 June 2021, pp. 7941–7950 (2021)
2. Chen, H., et al.: Artistic style transfer with internal-external learning and contrastive learning. In: Advances in Neural Information Processing Systems 34: Annual Conference on Neural Information Processing Systems 2021, NeurIPS 2021, 6–14 December 2021, virtual, pp. 26561–26573 (2021)
3. Chen, T.Q., Schmidt, M.: Fast patch-based style transfer of arbitrary style. CoRR abs/1612.04337 (2016)
4. Deng, Y., Tang, F., Dong, W., Sun, W., Huang, F., Xu, C.: Arbitrary style transfer via multi-adaptation network. In: MM 2020: The 28th ACM International Conference on Multimedia, Virtual Event/Seattle, WA, USA, 12–16 October 2020, pp. 2719–2727 (2020)
5. Fernando, B., Habrard, A., Sebban, M., Tuytelaars, T.: Unsupervised visual domain adaptation using subspace alignment. In: 2013 IEEE International Conference on Computer Vision, pp. 2960–2967 (2013)
6. Gatys, L.A., Ecker, A.S., Bethge, M.: Texture synthesis using convolutional neural networks. In: Advances in Neural Information Processing Systems 28: Annual Conference on Neural Information Processing Systems 2015, 7–12 December 2015, Montreal, Quebec, Canada, pp. 262–270 (2015)
7. Gatys, L.A., Ecker, A.S., Bethge, M.: Image style transfer using convolutional neural networks. In: 2016 IEEE Conference on Computer Vision and Pattern Recognition, CVPR 2016, Las Vegas, NV, USA, 27–30 June 2016, pp. 2414–2423 (2016)
8. Gong, B., Shi, Y., Sha, F., Grauman, K.: Geodesic flow kernel for unsupervised domain adaptation. In: 2012 IEEE Conference on Computer Vision and Pattern Recognition, pp. 2066–2073 (2012)
9. Gu, S., Chen, C., Liao, J., Yuan, L.: Arbitrary style transfer with deep feature reshuffle. In: 2018 IEEE Conference on Computer Vision and Pattern Recognition, CVPR 2018, Salt Lake City, UT, USA, 18–22 June 2018, pp. 8222–8231. Computer Vision Foundation/IEEE Computer Society (2018)
10. Hong, K., Jeon, S., Yang, H., Fu, J., Byun, H.: Domain-aware universal style transfer. In: 2021 IEEE/CVF International Conference on Computer Vision, ICCV 2021, Montreal, QC, Canada, 10–17 October 2021, pp. 14589–14597 (2021)
11. Howard, A.G., et al.: MobileNets: efficient convolutional neural networks for mobile vision applications. arXiv abs/1704.04861 (2017)
12. Hu, J., Shen, L., Sun, G.: Squeeze-and-excitation networks. In: Proceedings of the IEEE Conference on Computer Vision and Pattern Recognition, pp. 7132–7141 (2018)
13. Huang, X., Belongie, S.J.: Arbitrary style transfer in real-time with adaptive instance normalization. In: IEEE International Conference on Computer Vision, ICCV 2017, Venice, Italy, 22–29 October 2017, pp. 1510–1519 (2017)
14. Huo, J., et al.: Manifold alignment for semantically aligned style transfer. In: Proceedings of the IEEE/CVF International Conference on Computer Vision, pp. 14861–14869 (2021)

15. Jing, Y., et al.: Dynamic instance normalization for arbitrary style transfer. In: The Thirty-Fourth AAAI Conference on Artificial Intelligence, AAAI 2020, The Thirty-Second Innovative Applications of Artificial Intelligence Conference, IAAI 2020, The Tenth AAAI Symposium on Educational Advances in Artificial Intelligence, EAAI 2020, New York, NY, USA, 7–12 February 2020, pp. 4369–4376 (2020)

16. Johnson, J., Alahi, A., Fei-Fei, L.: Perceptual losses for real-time style transfer and super-resolution. In: Leibe, B., Matas, J., Sebe, N., Welling, M. (eds.) ECCV 2016, Part II. LNCS, vol. 9906, pp. 694–711. Springer, Cham (2016). https://doi.org/10.1007/978-3-319-46475-6_43

17. Kingma, D.P., Ba, J.: Adam: a method for stochastic optimization. In: 3rd International Conference on Learning Representations, ICLR 2015, San Diego, CA, USA, 7–9 May 2015, Conference Track Proceedings (2015)

18. Kolkin, N.I., Salavon, J., Shakhnarovich, G.: Style transfer by relaxed optimal transport and self-similarity. In: IEEE Conference on Computer Vision and Pattern Recognition, CVPR 2019, Long Beach, CA, USA, 16–20 June 2019, pp. 10051–10060 (2019)

19. Kotovenko, D., Sanakoyeu, A., Lang, S., Ommer, B.: Content and style disentanglement for artistic style transfer. In: 2019 IEEE/CVF International Conference on Computer Vision, ICCV 2019, Seoul, Korea (South), 27 October–2 November 2019, pp. 4421–4430 (2019)

20. Li, X., Liu, S., Kautz, J., Yang, M.: Learning linear transformations for fast image and video style transfer. In: IEEE Conference on Computer Vision and Pattern Recognition, CVPR 2019, Long Beach, CA, USA, 16–20 June 2019, pp. 3809–3817 (2019)

21. Li, Y., Fang, C., Yang, J., Wang, Z., Lu, X., Yang, M.: Diversified texture synthesis with feed-forward networks. In: 2017 IEEE Conference on Computer Vision and Pattern Recognition, CVPR 2017, Honolulu, HI, USA, 21–26 July 2017, pp. 266–274 (2017)

22. Li, Y., Fang, C., Yang, J., Wang, Z., Lu, X., Yang, M.: Universal style transfer via feature transforms. In: Advances in Neural Information Processing Systems 30: Annual Conference on Neural Information Processing Systems 2017, 4–9 December 2017, Long Beach, CA, USA, pp. 386–396 (2017)

23. Lin, T., et al.: Drafting and revision: Laplacian pyramid network for fast high-quality artistic style transfer. In: IEEE Conference on Computer Vision and Pattern Recognition, CVPR 2021, virtual, 19–25 June 2021, pp. 5141–5150 (2021)

24. Lin, T.-Y., et al.: Microsoft COCO: common objects in context. In: Fleet, D., Pajdla, T., Schiele, B., Tuytelaars, T. (eds.) ECCV 2014, Part V. LNCS, vol. 8693, pp. 740–755. Springer, Cham (2014). https://doi.org/10.1007/978-3-319-10602-1_48

25. Liu, S., et al.: AdaAttN: revisit attention mechanism in arbitrary neural style transfer. In: 2021 IEEE/CVF International Conference on Computer Vision, ICCV 2021, Montreal, QC, Canada, 10–17 October 2021, pp. 6629–6638 (2021)

26. Liu, X., Liu, Z., Zhou, X., Chen, M.: Saliency-guided image style transfer. In: IEEE International Conference on Multimedia & Expo Workshops, ICME Workshops 2019, Shanghai, China, 8–12 July 2019, pp. 66–71 (2019)

27. Long, M., Wang, J., Ding, G., Sun, J., Yu, P.S.: Transfer feature learning with joint distribution adaptation. In: 2013 IEEE International Conference on Computer Vision, pp. 2200–2207 (2013)

28. Nichol, K.: Painter by numbers (2016)

29. Park, D.Y., Lee, K.H.: Arbitrary style transfer with style-attentional networks. In: IEEE Conference on Computer Vision and Pattern Recognition, CVPR 2019, Long Beach, CA, USA, 16–20 June 2019, pp. 5880–5888 (2019)

30. Qiu, T., Ni, B., Liu, Z., Chen, X.: Fast optimal transport artistic style transfer. In: International Conference on Multimedia Modeling, pp. 37–49 (2021)

31. Shechtman, E., Irani, M.: Matching local self-similarities across images and videos. In: 2007 IEEE Conference on Computer Vision and Pattern Recognition, pp. 1–8 (2007)

32. Simonyan, K., Zisserman, A.: Very deep convolutional networks for large-scale image recognition. In: 3rd International Conference on Learning Representations, ICLR 2015, San Diego, CA, USA, 7–9 May 2015, Conference Track Proceedings (2015)

33. Ulyanov, D., Lebedev, V., Vedaldi, A., Lempitsky, V.S.: Texture networks: feed-forward synthesis of textures and stylized images. In: Proceedings of the 33nd International Conference on Machine Learning, ICML 2016, New York City, NY, USA, 19–24 June 2016, pp. 1349–1357 (2016)

34. Ulyanov, D., Vedaldi, A., Lempitsky, V.S.: Improved texture networks: maximizing quality and diversity in feed-forward stylization and texture synthesis. In: 2017 IEEE Conference on Computer Vision and Pattern Recognition, CVPR 2017, Honolulu, HI, USA, 21–26 July 2017, pp. 4105–4113 (2017)

35. Vaswani, A., et al.: Attention is all you need. In: Advances in Neural Information Processing Systems 30: Annual Conference on Neural Information Processing Systems 2017, 4–9 December 2017, Long Beach, CA, USA, pp. 5998–6008 (2017)

36. Wang, J., Chen, Y., Hao, S., Feng, W., Shen, Z.: Balanced distribution adaptation for transfer learning. In: 2017 IEEE International Conference on Data Mining (ICDM), pp. 1129–1134 (2017)

37. Wang, X., Oxholm, G., Zhang, D., Wang, Y.: Multimodal transfer: a hierarchical deep convolutional neural network for fast artistic style transfer. In: 2017 IEEE Conference on Computer Vision and Pattern Recognition, CVPR 2017, Honolulu, HI, USA, 21–26 July 2017, pp. 7178–7186 (2017)

38. Wu, X., Hu, Z., Sheng, L., Xu, D.: Styleformer: real-time arbitrary style transfer via parametric style composition. In: 2021 IEEE/CVF International Conference on Computer Vision, ICCV 2021, Montreal, QC, Canada, 10–17 October 2021, pp. 14598–14607 (2021)

39. Yao, Y., Ren, J., Xie, X., Liu, W., Liu, Y.J., Wang, J.: Attention-aware multi-stroke style transfer. In: 2019 IEEE/CVF Conference on Computer Vision and Pattern Recognition (CVPR), pp. 1467–1475 (2019)

Exp-GAN: 3D-Aware Facial Image Generation with Expression Control

Yeonkyeong Lee[1], Taeho Choi[1], Hyunsung Go[2], Hyunjoon Lee[1],
Sunghyun Cho[3], and Junho Kim[2(✉)]

[1] Kakao Brain, Seongnam, South Korea
[2] Kookmin University, Seoul, South Korea
junho@kookmin.ac.kr
[3] POSTECH, Pohang, South Korea

Abstract. This paper introduces Exp-GAN, a 3D-aware facial image generator with explicit control of facial expressions. Unlike previous 3D-aware GANs, Exp-GAN supports fine-grained control over facial shapes and expressions disentangled from poses. To this ends, we propose a novel hybrid approach that adopts a 3D morphable model (3DMM) with neural textures for the facial region and a neural radiance field (NeRF) for non-facial regions with multi-view consistency. The 3DMM allows fine-grained control over facial expressions, whereas the NeRF contains volumetric features for the non-facial regions. The two features, generated separately, are combined seamlessly with our depth-based integration method that integrates the two complementary features through volume rendering. We also propose a training scheme that encourages generated images to reflect control over shapes and expressions faithfully. Experimental results show that the proposed approach successfully synthesizes realistic view-consistent face images with fine-grained controls. Code is available at https://github.com/kakaobrain/expgan.

| generated face | pose | expression | shape | appearance |

Fig. 1. Exp-GAN generates realistic images of human faces, with explicit control of camera pose, facial expression, and facial shape. It is also capable of generating faces with different appearances keeping given facial expression and camera pose unchanged.

This work was done when the first author was with Kookmin University.

Supplementary Information The online version contains supplementary material available at https://doi.org/10.1007/978-3-031-26293-7_10.

L. Wang et al. (Eds.): ACCV 2022, LNCS 13847, pp. 151–167, 2023.
https://doi.org/10.1007/978-3-031-26293-7_10

1   Introduction

Recent years have seen a significant increase in photo-realism of synthetic images built on generative models such as generative adversarial networks (GANs) [11], variational autoencoders (VAEs) [17] and diffusion models [30]. Among them, state-of-the-art GAN models such as StyleGAN [15,16] have realized generation of extremely realistic face images of massive identities with scale-wise style control. To control over more sophisticated semantic attributes, much research has been done to explore semantically meaningful directions in the latent space [13,26, 27,34] or to learn mappings for disentangled representations [1,2,7,10,18,28,31]. However, face shapes and expressions can be controlled in a limited way because they are manipulated through attribute editing in a latent space.

For more intuitive control over semantic attributes including facial shapes and expressions, several methods that adopt 3D morphable models (3DMM) of faces to the 2D GAN framework have been proposed [2,10,31]. In [2,10], a 3D face mesh is rendered to inject various information (RGB, normal, neural features) of face shapes and expressions to the generator. However, despite using a 3D face mesh model, their results show entanglement between facial expressions and other attributes such as camera poses and identities due to the lack of multi-view consistency of the 2D GAN framework.

3D-aware GANs have been proposed to synthesize high-fidelity face images with multi-view consistency [3–5,8,12,35]. In general, 3D-aware GANs learn to generate an implicit volume feature field that can be realized as images with volume rendering. Since implicit volume features already contain 3D information, 3D-aware GANs can be successfully trained to generate face images with multi-view consistency. However, to our best knowledge, control over facial shapes and expressions in 3D-aware GANs has not yet been considered.

This paper proposes Exp-GAN, a 3D-aware facial image generator that gives us explicit controls over facial shapes and expressions with multi-view consistency. Specifically, Exp-GAN learns to synthesize a variety of facial expressions disentangled from identities and camera poses, as shown in Fig. 1. To accomplish this, Exp-GAN adopts a hybrid approach that combines the 3D morphable model (3DMM) of faces and the 3D-aware GAN into a single framework of conditional GAN that can be trained with a collection of 2D facial images. The 3DMM allows us fine-grained and intuitive control over the facial shape and expressions, while the 3D-aware GAN enables multi-view consistent photo-realistic image synthesis.

Specifically, Exp-GAN synthesizes the facial and non-facial parts separately using a neural face generator and a neural volume generator, respectively. The neural face generator adopts a 3DMM with the neural texture to synthesize features of a realistic and multi-view consistent face that fully reflect user controls over facial expression and shape given by blendshape coefficients. The neural volume generator adopts the 3D-aware GAN approach to generate volumetric features, supporting diverse and realistic image synthesis with multi-view consistency. For the seamless integration of the two separately generated features, we also introduce a feature integration method based on the volume rendering process of

NeRF [22]. Finally, we propose a training scheme based on the regression of blend-shape coefficients with discriminators for faithful image synthesis with respect to user control parameters.

We empirically show that Exp-GAN can generate various expressions, poses, and shapes of human faces. We also show that the proposed method improves the result quantitatively compared to previous works that provide expression controls with 2D StyleGANs. Our contributions can be summarized as follows:

- We propose Exp-GAN, the first 3D-aware facial image generator to achieve both multi-view consistency and fine-grained control over facial expressions.
- We propose geometrically explicit conditioning of a 3D-aware GAN with facial expression controls based on 3DMMs.
- Our hybrid approach combines the 3DMM and volumetric features for the synthesis of the facial and non-facial regions, and adopts a novel depth integration method for seamless integration of separately synthesized features.
- We also propose a novel training scheme leveraging discriminators with regression branches to train our network to faithfully reflect user controls.

2 Related Work

Expression Controls in Generative Models. Semantic editing in the latent space of GANs has been studied in [1,13,26–28,34], in which facial expression controls are handled through semantic attribute editing. SeFa [27] and GANSpace [13] discover semantically interpretable directions through latent space factorization. InterFaceGAN [26] finds linear directions in the latent space using binary-classified samples with respect to semantic attributes. StyleFlow [1] finds non-linear paths in the StyleGAN's latent space for manipulating semantic attributes using attribute-conditioned normalizing flow models. However, these approaches treat facial expression controls by means of semantic attribute editing over the pretrained StyleGAN's latent space, the diversity of expression controls is limited to simple expressions, such as smiles. To support fine-grained control over facial expressions, StyleRig [31] presents a facial attribute editing approach based on rig-like controls via 3DMMs. While StyleRig adopts the 3DMM, it still relies on the predefined StyleGAN's latent space, thus it suffers from a similar limitation to the aforementioned approaches, i.e., limited to simple expressions.

Several generative networks have been proposed that employ 3DMMs to synthesize facial images with complicated expressions [2,7,10,18]. DiscoFaceGAN [7] trains a StyleGAN-like image generator via an imitative-contrastive paradigm. GIF [10] and VariTex [2] leverage generated 3DMM face images to learn controllable face image generation. GIF [10] generates face images with expressions, with FLAME [19] conditioning signals and rendered images. VariTex [2] learns to synthesize pose-independent textures for 3DMM faces with expressions and additive features for non-facial parts to generate facial images with camera pose and expression controls. Since previous approaches with 3DMMs [2,7,10,18] rely on 2D generators such that expression information from 3DMMs is injected as projected facial features in 2D image spaces, entanglement between camera poses and expressions still exists due to the limitation of 2D approaches.

3D-Aware GANs. Recently, 3D-aware GANs have been proposed to disentangle camera poses from other attributes to achieve multi-view consistency [3–5,8,12,23,25,35]. They learn to map a random noise vector to an implicit feature field that can be consistently rendered from multiple viewpoints. SofGAN [5] decouples the latent space of portraits into a geometry and a texture space and uses the geometry space to generate a 3D geometry with a canonical pose. π-GAN [4] proposes a SIREN-based network [29] to learn a neural radiance field (NeRF)-based generator that can synthesize 3D-aware images. StyleNeRF [12] combines the NeRF and a style-based generator to improve rendering efficiency and 3D consistency for high-resolution image generation. CIPS-3D [35] learns a style-based NeRF generator with a deep 2D implicit neural representation network that efficiently generates high-resolution rendering results with partial gradient backpropagation. EG3D [3] proposes a tri-plane-based hybrid 3D representation to learn high-quality 3D geometry with fine details. GRAM [8] learns generative neural radiance fields for 3D-aware GAN by constraining point sampling and radiance field learning on 2D manifolds to generate high-quality images with fine details. Recent 3D-aware GANs successfully disentangle pose and identity to provide high-quality multi-view-consistent images. However, disentanglement of facial expression has not yet been considered in 3D-aware face image generation.

3   Framework

Figure 2 shows an overview of our framework. Our framework consists of four parts: a neural face generator, a neural volume generator, an image synthesis module, and a discrimination module. For the synthesis part of our framework, StyleGAN2-based generators [16] are used, namely G_{tex} for the neural face texture, G_{vol} for the volume feature, and G_{img} for the final image, respectively. Two StyleGAN2-based discriminators are used for the discrimination module: D_{img} for the final output and D_{aux} for the low-resolution auxiliary output.

Similar to previous generative NeRF models, we provide a camera pose $(\mathbf{R}, \mathbf{t})$ as an input to our framework to generate images from various viewpoints, where $\mathbf{R}$ is a rotation matrix, and $\mathbf{t}$ is a translation vector. We assume a fixed intrinsic camera matrix as a hyperparameter. For the explicit control of shapes and expressions of faces, we use a blendshape-based 3DMM. Specifically, we adopt DECA [9] that allows us to control the facial shape and expression using coefficient vectors $\boldsymbol{\alpha} \in \mathbb{R}^{100}$ and $\boldsymbol{\beta} \in \mathbb{R}^{50}$, respectively. To model the jaw motion, which is not supported by DECA, we introduce additional three coefficients to the expression coefficients, i.e., we use $\boldsymbol{\beta} \in \mathbb{R}^{50+3}$ as an expression coefficient vector in our framework. With the 3DMM, a face mesh that reflects user-provided coefficients $\boldsymbol{\alpha}$ and $\boldsymbol{\beta}$ can be created (Fig. 2, top-left).

Mathematically, our synthesis framework can be expressed as:

$$I = G(\mathbf{z}, \boldsymbol{\alpha}, \boldsymbol{\beta}, \mathbf{R}, \mathbf{t}), \qquad (1)$$

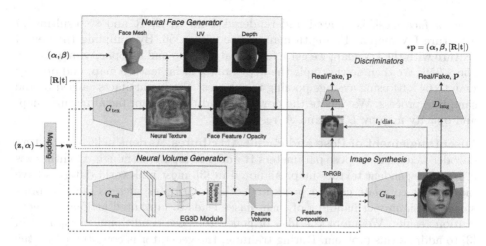

Fig. 2. Our framework. Facial feature map is generated in the *neural face generator* block, while the *neural volume generator* block synthesizes the feature volume representing non-facial regions. In the *image synthesis* block, the two features are composited by volume ray casting and upsampled to produce the output image. The synthesized output as well as a low-resolution auxiliary result are evaluated by the two discriminators in the *discriminators* block with adversarial and parameter regression losses.

where I is a 2D output image, G is our synthesis network, and $\mathbf{z} \in \mathbb{R}^{256}$ is a random latent vector. Our synthesis framework takes a form of a conditional GAN framework that produces a realistic 2D output image I from a latent vector $\mathbf{z}$ conditioned by a camera pose $(\mathbf{R}, \mathbf{t})$, a facial shape vector $\boldsymbol{\alpha}$, and an expression vector $\boldsymbol{\beta}$. The latent code $\mathbf{z}$, sampled from a multivariate unit Gaussian distribution, enables the generation of diverse identities while conditioned on the other parameters. Note that, in our framework, both $\boldsymbol{\alpha}$ and $\mathbf{z}$ control the identities of generated face images; the blendshape coefficient vector $\boldsymbol{\alpha}$ provides fine control over the facial shape, while the latent code $\mathbf{z}$ controls the appearance like hair and skin.

Following the StyleGAN architecture [15,16], our framework adopts a mapping network that transforms $\mathbf{z}$ to an intermediate latent vector $\mathbf{w} \in \mathbb{R}^{512}$ instead of directly using $\mathbf{z}$. In addition, to constrain our synthesis process on the facial shape coefficient vector $\boldsymbol{\alpha}$, we feed $\boldsymbol{\alpha}$ to the mapping network by concatenating it with $\mathbf{z}$ as shown in Fig. 2.

3.1 Neural Face Generator

The goal of the neural face generator is to generate a 2D feature map representing the facial region that fully reflects the user control parameters and the latent vector $\mathbf{z}$. Inspired by neural texture approaches [2,20,32], we generate a neural texture of size 256×256 for the facial region from the intermediate latent vector $\mathbf{w}$. Each texel within the texture has a 32-dimensional feature vector representing the appearance of the facial region and an opacity value $a \in [0, 1]$. At the same time, a face mesh is created and rendered from α, β, $\mathbf{R}$ and $\mathbf{t}$, resulting in a texture UV map and a depth map of size 256×256. By sampling the neural texture with the UV map, we get a facial feature map and an opacity map of size 256×256. We then downsample the depth map, facial feature map, and opacity map to 64×64 using average pooling to suppress aliasing artifacts caused by the sampling process. We denote the downsampled depth map, facial feature map, and opacity map by D, T, and A, respectively.

Disentanglement of Pose and Facial Attributes. As done in [3], our texture decoder takes camera pose parameters $\mathbf{R}$ and $\mathbf{t}$ as inputs, although it synthesizes a pose-independent texture map. As noted in [3], most real-world datasets have biases that correlate camera poses and other attributes such as facial expressions, and improper handling of such biases leads to unwanted artifacts in generation results. We adopt the idea of generator pose conditioning proposed in [3] to address this problem. During training, the generator is conditioned by the camera pose so that it can learn the joint distribution of pose-independent and -dependent attributes inherent in the dataset. We refer the readers to [3] for more details about the generator pose conditioning.

3.2   Neural Volume Generator

The neural volume generator generates a 3D feature volume representing non-facial regions including the hair, clothes, and background. For this purpose, we employ EG3D [3] as a backbone 3D-aware generative model because of its generation performance and architectural simplicity. Nonetheless, our method can be incorporated with other 3D-aware generative models as long as they use volumetric feature fields [4,8,12,35].

Our neural volume generator takes an intermediate latent code $\mathbf{w}$ as input and feeds it to a StyleGAN2-based generator to produce a tri-plane representation [3], a light-weight representation for volumetric features, for the non-facial region. Then, we obtain a feature volume from the generated tri-plane representation. Specifically, given a camera pose $(\mathbf{R}, \mathbf{t})$, we shoot a bundle of rays from the camera to sample features from the tri-plane representation. In our implementation, we shoot 64×64 rays and sample 48 points per ray from the feature field. We aggregate features from the tri-planes and decode them using a small multi-layer perceptron for each point. Finally, we obtain a 3D feature volume V of size $64 \times 64 \times 48$ where each voxel has a 32-dimensional feature vector and a scalar density value σ. We refer to [3] for more details about the tri-plane representation.

3.3   Image Synthesis with Feature Integration

The image synthesis module first integrates the facial feature map T and the feature volume V based on the depth map D and performs volume rendering to obtain a 2D feature map $F \in \mathbb{R}^{64 \times 64 \times 32}$. To this end, we adopt the volume ray casting algorithm with per-ray feature composition [22]. Specifically, for each spatial location in V, we have 48 features $\{\mathbf{f}_0, \ldots, \mathbf{f}_{47}\}$, density values $\{\sigma_0, \ldots, \sigma_{47}\}$,

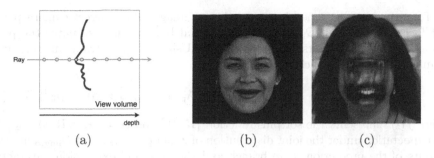

(a)   (b)   (c)

Fig. 3. Depth-based feature composition process and feature visualizations. (a) Depth-based feature composition process. Green dots represent features from V, as $\{f_0, f_1, \ldots\}$, and yellow dot represents a facial feature $\mathbf{f}^f$. (b) Visualization of a facial feature map. (c) Feature volume rendered without facial features. Non-facial regions including teeth and inner mouth are synthesized, complementing the facial feature map.

and their corresponding depths $\{d_0, d_1, \ldots, d_{47}\}$ where $d_0 \leq d_1 \leq \ldots \leq d_{47}$. For volume ray casting, we compute a set of opacity values $\{a_0, \ldots, a_{46}\}$ where $a_i = 1 - \exp(-\sigma_i(d_{i+1} - d_i))$. Then, we insert the feature $\mathbf{f}^f$ and opacity a^f from T and A into the sets of features and opacity values according to the depth d^f from D, and obtain:

$$\mathcal{F} = \left\{ \mathbf{f}'_0 = \mathbf{f}_0, \ldots, \mathbf{f}'_i = \mathbf{f}_i, \mathbf{f}'_{i+1} = \mathbf{f}^f, \mathbf{f}'_{i+2} = \mathbf{f}_{i+1}, \ldots \mathbf{f}'_N = \mathbf{f}_{N-1} \right\}, \text{ and} \quad (2)$$

$$\mathcal{A} = \left\{ a'_0 = a_0, \ldots, a'_i = a_i, a'_{i+1} = a_t, a'_{i+2} = a_{i+1}, \ldots a'_N = a_{N-1} \right\}, \quad (3)$$

where $d_i \leq d^f \leq d_{i+1}$. We then perform volume ray casting as:

$$\mathbf{f} = \sum_{i=0}^{N} T_i(1 - a'_i)\mathbf{f}'_i, \text{ where } T_i = \prod_{j=0}^{i-1} a'_j, \quad (4)$$

where $N = 48$ and $\mathbf{f}$ is an integrated feature vector. Collecting $\mathbf{f}$, we construct a 2D feature map F. Figure 3 illustrates the composition process.

The feature map F is then fed to a StyleGAN2-based superresolution network G_{img} to produce a high-resolution final RGB image. G_{img} also takes the intermediate latent vector $\mathbf{w}$ to synthesize realistic-looking high-resolution details for the final output image.

3.4 Training

We train our entire network in an end-to-end fashion, as our framework is composed of differentiable modules except for the morphing and rendering steps for the face mesh, which do not have learnable parameters. To synthesize novel images, we use adversarial learning using only 2D real images. Specifically, we attach a discriminator D_{img} to the output of the superresolution network to predict whether the final output looks real or fake.

To encourage our generator to synthesize images with correct camera poses and facial expressions, D_{img} has an additional branch that estimates the pose and expression coefficients of an input image. Using D_{img}, we train our generator by minimizing $\mathcal{L}_{\text{img}}^{\text{gen}}$, which is defined as:

$$\mathcal{L}_{\text{img}}^{\text{gen}} = \mathbb{E}_{\mathbf{z},\mathbf{p}} \left[f(-D_{\text{img},s}(G(\mathbf{z},\mathbf{p}))) + \lambda_{\mathbf{p}} \| D_{\text{img},\mathbf{p}}(G(\mathbf{z},\mathbf{p})) - \mathbf{p} \|^2 \right] \quad (5)$$

where $f(\cdot)$ represents the softplus function [15,16] and $\mathbf{p} = (\alpha, \beta, \mathbf{R}, \mathbf{t})$. $\mathbb{E}_{\mathbf{z},\mathbf{p}}$ is the expectation under the joint distribution of $\mathbf{z}$ and $\mathbf{p}$. $D_{\text{img},s}$ and $D_{\text{img},\mathbf{p}}$ are the outputs of the prediction score branch and the pose and expression parameter branch, respectively. $\lambda_{\mathbf{p}}$ is a weight to balance the loss terms.

The discriminator D_{img} is trained by $\mathcal{L}_{\text{img}}^{\text{disc}}$, which is defined as:

$$\mathcal{L}_{\text{img}}^{\text{disc}} = \mathbb{E}_{\mathbf{z},\mathbf{p}} \left[f(D_{\text{img},s}(G(\mathbf{z},\mathbf{p}))) \right] \quad (6)$$
$$+ \mathbb{E}_I \left[f(-D_{\text{img},s}(I)) + \lambda_{\mathbf{p}} \| D_{\text{img},\mathbf{p}}(I) - \mathbf{p}_{gt} \|^2 + \lambda_{r1} \| \nabla D_{\text{img},s}(I) \|^2 \right]$$

where I is a real image sample, $\mathbb{E}_I$ is the expectation under the distribution of real images, and $\mathbf{p}_{gt}$ is the ground-truth (GT) label for the camera pose and expression parameters of I. We obtain $\mathbf{p}_{gt}$ by applying a pretrained DECA encoder [9] to the real image samples before training. The last term in $\mathbb{E}_I$ is an R1 regularization term [21] to stabilize GAN training.

Training our generator with only D_{img} may converge to a low-quality local minimum as the superresolution network can be trained to synthesize images with an average pose and facial expression regardless of its input F. To resolve this, we let our network to produce an auxiliary low-resolution RGB image I_{aux} directly from F, and introduce another discriminator D_{aux}, which predicts whether F looks realistic. For this purpose, we assume the first three channels of F as the low-resolution RGB output; a similar technique is also used in DNR [32]. We then train the generator minimizing a loss function $\mathcal{L}_{\text{aux}}^{\text{gen}}$, defined as:

$$\mathcal{L}_{\text{aux}}^{\text{gen}} = \mathbb{E}_{\mathbf{z},\mathbf{p}} \left[-D_{\text{aux},s}(F_{1,2,3}(\mathbf{z},\mathbf{p})) + \lambda_{\mathbf{p}} \| D_{\text{aux},\mathbf{p}}(F_{1,2,3}(\mathbf{z},\mathbf{p})) - \mathbf{p} \|^2 \right] \quad (7)$$

where $F_{1,2,3}(\mathbf{z},\mathbf{p})$ represents the first three channels of F as a function of $\mathbf{z}$ and $\mathbf{p}$. The loss $\mathcal{L}_{\text{aux}}^{\text{disc}}$ for the discriminator D_{aux} is also defined similarly to $\mathcal{L}_{\text{img}}^{\text{disc}}$. Specifically, $\mathcal{L}_{\text{aux}}^{\text{disc}}$ is defined as:

$$\mathcal{L}_{\text{aux}}^{\text{disc}} = \mathbb{E}_{\mathbf{z},\mathbf{p}} \left[D_{\text{aux},s}(F_{1,2,3}(\mathbf{z},\mathbf{p})) \right] \quad (8)$$
$$+ \mathbb{E}_I \left[-D_{\text{aux},s}(I \downarrow) + \lambda_{\mathbf{p}} \| D_{\text{aux},\mathbf{p}}(I \downarrow) - \mathbf{p}_{gt} \|^2 + \lambda_{r1} \| \nabla D_{\text{aux},s}(I \downarrow) \|^2 \right]$$

where $I \downarrow$ is a downsampled version of I. For both D_{img} and D_{aux}, we employ the network architecture of the discriminator of StyleGAN2 [16].

In addition, as similarly done in [3,12], we employ an MSE loss to make I and I_{aux} similar to each other, defined as:

$$\mathcal{L}_{\text{MSE}} = \frac{1}{N_{I_{\text{aux}}}} \sum_i^{N_{I_{\text{aux}}}} (I_{\text{aux}}(j) - I \downarrow (j))^2, \quad (9)$$

where j is a pixel index and $N_{I_{\mathrm{aux}}}$ is the number of pixels in I_{aux}. Lastly, based on the observation that the facial region is opaque, we introduce an opacity loss

$$\mathcal{L}_{\mathrm{opacity}} = \frac{1}{N_A} \sum_j^{N_A} -\log\left(1 - A(j)\right), \tag{10}$$

where N_A is the number of pixels in A. Our final loss for the generator is then

$$\mathcal{L}^{\mathrm{gen}} = \mathcal{L}_{\mathrm{img}}^{\mathrm{gen}} + \mathcal{L}_{\mathrm{aux}}^{\mathrm{gen}} + \lambda_{\mathrm{MSE}}\mathcal{L}_{\mathrm{MSE}} + \lambda_{\mathrm{opacity}}\mathcal{L}_{\mathrm{opacity}} \tag{11}$$

where λ_{MSE} and $\lambda_{\mathrm{opacity}}$ are weights for $\mathcal{L}_{\mathrm{MSE}}$ and $\mathcal{L}_{\mathrm{opacity}}$, respectively.

4   Experiments

We train and evaluate our network with the FFHQ dataset [15]. In our training stage, we randomly sample parameters from the GT labels $\mathbf{g}_{gt}$ for each training sample to expand the sampling space, i.e., for each training sample, we randomly sample image indices i, j, and k and use the GT labels $\boldsymbol{\alpha}_i$, $\boldsymbol{\beta}_j$, and $(\mathbf{R}_k, \mathbf{t}_k)$ of the i-, j-, and k-th training images in the FFHQ dataset, respectively. The GT labels $\mathbf{g}_{gt}$ are obtained by applying a pretrained DECA [9] encoder to the FFHQ dataset. All of our qualitative results are generated with the truncation trick [15], with truncation $\psi = 0.5$. For quantitative comparisons truncation trick is not used except for the multi-view consistency score. Qualitative comparisons are conducted with images of different identities, due to the nature of GAN training. Further training details and more results, including additional comparisons and ablations, are presented in the supplementary material.

4.1   Comparison

We compare qualitative and quantitative results between our Exp-GAN and several baseline methods. We first compare our method against 3DMM-based 2D generative models, such as DiscoFaceGAN [7], GIF [10], and VariTex [2], to show how well our Exp-GAN reflects facial expression conditions in the final images with all the attributes disentangled. We also compare our method against π-GAN [4] as a baseline 3D-aware GAN to show how accurately Exp-GAN disentangles all the attributes. For DiscoFaceGAN [7] and GIF [10], we use pretrained models provided the authors. For VariTex [2], we change its 3DMM model parameters to match ours and train the model from scratch using the FFHQ dataset and the authors' implementation. For π-GAN [4] we train the model from scratch on the FFHQ dataset using the authors' implementation.

Qualitative Results. Figure 4 shows a qualitative comparison where we fix the identity $(\mathbf{z}, \boldsymbol{\alpha})$ but vary the camera pose $(\mathbf{R}, \mathbf{t})$ and expression $\boldsymbol{\beta}$. As shown in the figure, π-GAN [4] generates view-consistent facial images but has no expression control. On the other hand, all the 2D generative models incorporating 3DMM provide control over the pose and expression, but they do not guarantee

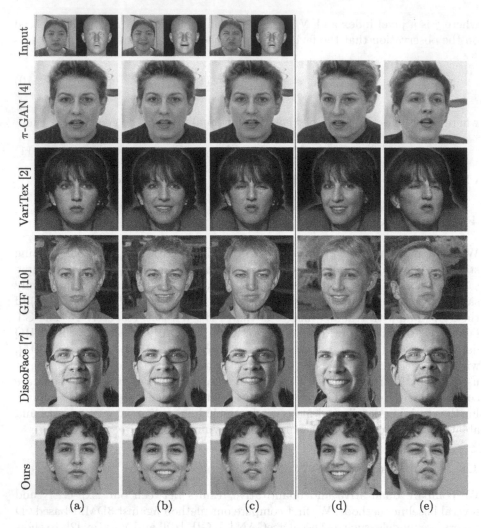

Fig. 4. Qualitative results with varying expression coefficients β and camera poses. In each method, an identity is fixed with a random seed. (top row) Three reference images from which β are extracted and corresponding face meshes are also rendered. In (a-c), we generate samples with respect to β, except for π-GAN which cannot control facial expressions. For (d) and (e), we use β from (b) and (c), respectively, with different camera poses.

view consistency. Also, despite using the 3DMM, DiscoFaceGAN [7] shows only slight facial expression changes. Compared to all the other methods, our method generates high-fidelity 3D-aware face images that faithfully reflect input facial expressions while keeping all the other attributes unchanged.

Quantitative Results. Table 1 provides quantitative comparisons in Frechét Inception Distance (FID) [14], blendshape metric (BS), multi-view consistency score (MV), and identity consistency score (ID). As DiscoFaceGAN [7] uses the

Table 1. Quantitative comparisons. Bold is the best result, and underscore is the second-best result. Refer the manuscript for details of the comparison protocols specific to algorithms and metrics.

	FID ↓	BS ↓	MV ↑	ID ↑
π-GAN [4] (128^2)	16.91	—	**24.58**	—
DiscoFaceGAN [7]	<u>15.57</u>	0.147	23.28	**0.699**
GIF [10]	28.0	<u>0.1</u>	16.56	0.435
Ours	**7.44**	**0.05**	<u>23.84</u>	<u>0.622</u>

Basel Face Model (BFM) [24] for 3DMM differently from others, for a fair comparison, we estimate BFM blendshape coefficients from all the images in FFHQ and use them in place of DECA blendshape coefficients to generate images with DiscoFaceGAN [7]. We exclude VariTex [2] from our quantitative comparisons due to the domain difference caused by background masking.

We first evaluate image quality with FID between 50K real images from FFHQ and 50K generated images. For DiscoFaceGAN and GIF, we generate 50K facial images with random latent vectors and GT parameters, similarly to ours. For π-GAN, 50K images are generated with random latent vectors and sampled GT camera poses. As shown in the FID column in Table 1, our Exp-GAN generates higher-quality images than the other methods. The FID score of Exp-GAN is comparable to 4.8 of EG3D, which is reported in [3], while Exp-GAN also provides explicitly control over shape and expression unlike EG3D.

Next, we evaluate how well input conditional facial expressions are reflected in generated images with the BS metric. For this, we generate 50K images and re-estimate blendshape coefficients from them. The BS metric is measured by the mean squared distance between the input blendshape coefficients and re-estimated ones. As shown in the BS column in Table 1, our Exp-GAN achieves a better result than previous 2D-based generative models, validating that our method can faithfully reflect the input facial expression.

To evaluate multi-view consistency of our results, we measure MV scores as proposed in StyleNeRF [12]. From the given parameters $(\mathbf{z}, \boldsymbol{\alpha}, \boldsymbol{\beta})$, we generate 9 images by varying camera poses from left to right, changing the yaw value in $[-0.5, 0.5]$ radian. Among 9 generated images, we use 5 of them as reference images and reconstruct the remaining ones with IBRNet [33]. Then we compute the MV score in terms of PSNR for the reconstructed images with the generated images as references. For evaluation, we generate 1K test cases, each with 9 images, and measure the average MV score of the test cases. Our Exp-GAN shows comparable results with that of π-GAN. DiscoFaceGAN, interestingly, shows good multi-view consistency even though it is a 2D-based approach, but it achieves relatively low facial expression accuracy as discussed earlier.

We evaluate ID score that measures how well the facial identity is preserved in various camera poses and expressions. We evaluate ID score by computing the average ArcFace [6] cosine similarity from 50K pairs of generated images.

Fig. 5. Limitation of a naïve baseline. All the results are generated using a single baseline model with the same code for **z** but different values for β. Despite the fixed **z**, the images show different identities due to the entanglement between **z** and β.

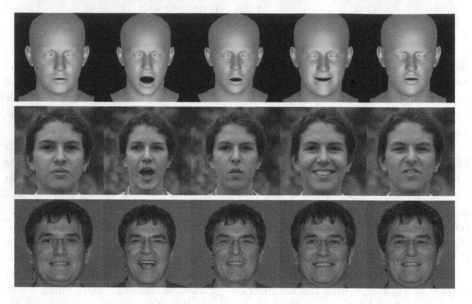

Fig. 6. Effect of the expression coefficient regression. (top) input expressions, as neutral, open mouth, half-open mouth, smile, and frown, respectively. (middle) our result, (bottom) without regression loss of expression coefficients.

For each pair, we generate images by fixing **z** and α and changing β, **R** and **t**. ID score is evaluated only on the models that allow explicit control of expression. The 'ID' column of Table 1 shows that DiscoFaceGAN performs the best, followed by ours. However, DiscoFaceGAN generates images where facial expressions are not changed as expected.

4.2 Ablation Study

Comparison with a Naïve Baseline. As a naïve baseline, we train an EG3D network with a simple modification to control facial expressions. The mapping network is modified to get not only latent vectors but also expression coefficients, as the form of a concatenated vector $[\mathbf{z}, \beta]$. Here, the 3DMM-related components are ablated, i.e., the neural face generator and the feature composition in Fig. 2. As shown in Fig. 5, changing facial expressions affects identities, evidencing entanglement between the blendshape coefficients and latent vectors.

Table 2. Ablation study of the loss terms.

	FID ↓	BS ↓	MV ↑	ID ↑
Ours (full)	7.44	0.05	23.84	0.622
w/o blendshape coeff. reg	8.90	0.10	25.93	0.723
w/o $\mathcal{L}_{\mathrm{MSE}}$	12.08	0.05	23.30	0.628
w/o $\mathcal{L}_{\mathrm{opacity}}$	10.53	0.05	23.96	0.617

Feature Integration Scheme. Similar to [2,10], facial and non-facial parts may be integrated with feature concatenation and then fed to the image synthesis module by ablating the depth composition in Sect. 3.3. We train with feature concatenation in place of our composition method for ablation and evaluate with the FID metric. The FID score is 13.19, which is worse than that of ours, 7.44. Additional qualitative comparisons are provided in the supplementary material.

Impact of Loss Terms. We conduct an ablation study to study the impact of each loss term. The quantitative results are reported in Table 2. Figure 6 shows the impact of the expression coefficient regression loss in our discriminator. Here, we ablate the blendshape coefficient β in $\mathbf{p}$ estimated by the discriminators D_{aux} and D_{img} in Fig. 2. Thanks to the face mesh, which guides the generation process, it is still possible for the network to reflect expressions without the regression loss to some extent. Still, its expressiveness is limited compared to that of our full model. Next, We ablate each term from Eq. (11) to evaluate its impact on the generator. Without $\mathcal{L}_{\mathrm{aux}}^{\mathrm{gen}}$, we cannot generate plausible images. Without $\mathcal{L}_{\mathrm{MSE}}$ or $\mathcal{L}_{\mathrm{opacity}}$, FID scores are far inferior to that of our full model. As shown in Table 2, our final model can achieve the best performance, especially in both FID and BS scores.

4.3 Additional Results

Figure 7 shows the effect of changing shape coefficients α. Although α is entangled with the latent vector $\mathbf{z}$ in our framework, it is shown that changing α results in natural-looking results with similar appearances. Figure 8 shows various facial expressions while the camera pose changes. Lastly, as shown in Fig. 9, our Exp-GAN successfully synthesizes asymmetrical facial expressions that are rare in the FFHQ dataset [16]. See the supplementary material for more results, including examples of GAN inversion and facial reenactment.

Fig. 7. Example of changing the face shape coefficients. (left) reference images from which α is extracted; (right) our synthesis results. Fixing the latent code $\mathbf{z}$ and changing only α, we can obtain images with similar appearances but different face shapes.

Fig. 8. While the camera pose changes, facial expressions of each example are controlled as neutral, smile, open mouth, half-open mouth, and frown, respectively. See uncurated results in the supplementary material.

Fig. 9. Asymmetrical facial expressions generated by our method. Our method can generate asymmetrical facial expressions that are rare in the FFHQ dataset [16] thanks to the explicit modeling of facial expressions using a 3DMM model.

5   Conclusion

We presented Exp-GAN, a novel 3D-aware GAN that can explicitly control camera poses, facial shapes and expressions. Leveraging the advantages of 3DMM and NeRF, our Exp-GAN generates features for facial and non-facial parts separately with appropriate neural approaches and seamlessly combines them to synthesize high-fidelity images via neural rendering. We showed that the depth-based feature integration in our generator and blendshape coefficient regressions in our discriminator play essential roles in the training of Exp-GAN for synthesizing images that faithfully reflect input shape and expression parameters.

Although Exp-GAN successfully disentangled several attributes as a 3D-aware GAN, it still lacks control over gaze and placement of accessories (e.g., glasses, earrings, etc.). Furthermore, Exp-GAN shows limited rendering qualities for the inside of the mouth mainly due to the lack of examples containing such a region in the dataset. We plan to address these issues for future work.

Acknowledgements. This was supported by the National Research Foundation of Korea (NRF) grant funded by the Korea government (MSIT) (2022R1F1A1074628, 2022R1A5A7000765) and Institute of Information & communications Technology Planning & Evaluation (IITP) grant funded by the Korea government (MSIT) (No. 2020-0-01826, Problem-Based Learning Program for Researchers to Proactively Solve Practical AI Problems (Kookmin University) and No. 2019-0-01906, Artificial Intelligence Graduate School Program (POSTECH)).

References

1. Abdal, R., Zhu, P., Mitra, N.J., Wonka, P.: StyleFlow: attribute-conditioned exploration of StyleGAN-generated images using conditional continuous normalizing flows. ACM Trans. Graph. (Proc. SIGGRAPH 2021) **40**(3) (2021)
2. Bühler, M.C., Meka, A., Li, G., Beeler, T., Hilliges, O.: VariTex: variational neural face textures. In: Proceedings of ICCV, pp. 13890–13899 (2021)
3. Chan, E.R., et al.: Efficient geometry-aware 3D generative adversarial networks. In: Proceedings of CVPR, pp. 16123–16133 (2022)
4. Chan, E.R., Monteiro, M., Kellnhofer, P., Wu, J., Wetzstein, G.: pi-GAN: periodic implicit generative adversarial networks for 3D-aware image synthesis. In: Proceedings of CVPR, pp. 5799–5809 (2021)
5. Chen, A., Liu, R., Xie, L., Chen, Z., Su, H., Yu, J.: SofGAN: a portrait image generator with dynamic styling. ACM Trans. Graph. **42**(1), 1–26 (2022)
6. Deng, J., Guo, J., Xue, N., Zafeiriou, S.: ArcFace: additive angular margin loss for deep face recognition. In: Proceedings of CVPR, pp. 4690–4699 (2019)
7. Deng, Y., Yang, J., Chen, D., Wen, F., Tong, X.: Disentangled and controllable face image generation via 3D imitative-contrastive learning. In: Proceedings of CVPR, pp. 5154–5163 (2020)
8. Deng, Y., Yang, J., Xiang, J., Tong, X.: GRAM: generative radiance manifolds for 3D-aware image generation. In: Proceedings of CVPR, pp. 10673–10683 (2022)
9. Feng, Y., Feng, H., Black, M.J., Bolkart, T.: Learning an animatable detailed 3D face model from in-the-wild images. ACM Trans. Graph. (Proc. SIGGRAPH 2021) **40**(8), Article No. 88 (2021)
10. Ghosh, P., Gupta, P.S., Uziel, R., Ranjan, A., Black, M., Bolkart, T.: GIF: generative interpretable faces. In: Proceedings of 3DV, pp. 868–878 (2020)
11. Goodfellow, I.J., et al.: Generative adversarial nets. In: Proceedings of NIPS, pp. 2672–2680 (2014)
12. Gu, J., Liu, L., Wang, P., Theobalt, C.: StyleNeRF: a style-based 3D-aware generator for high-resolution image synthesis. In: Proceedings of ICLR (2022)
13. Härkönen, E., Hertzmann, A., Lehtinen, J., Paris, S.: GANSpace: discovering interpretable GAN controls. In: Proceedings of NeurIPS (2020)

14. Heusel, M., Ramsauer, H., Unterthiner, T., Nessler, B., Hochreiter, S.: GANs trained by a two time-scale update rule converge to a local nash equilibrium. In: Proceedings of NIPS, pp. 6629–6640 (2017)
15. Karras, T., Laine, S., Aila, T.: A style-based generator architecture for generative adversarial networks. In: Proceedings of CVPR, pp. 4401–4410 (2019)
16. Karras, T., Laine, S., Aittala, M., Hellsten, J., Lehtinen, J., Aila, T.: Analyzing and improving the image quality of StyleGAN. In: Proceedings of CVPR, pp. 8110–8119 (2020)
17. Kingma, D.P., Welling, M.: Auto-encoding variational bayes. In: Proceedings of ICLR (2014)
18. Kowalski, M., Garbin, S.J., Estellers, V., Baltrušaitis, T., Johnson, M., Shotton, J.: CONFIG: controllable neural face image generation. In: Vedaldi, A., Bischof, H., Brox, T., Frahm, J.-M. (eds.) ECCV 2020. LNCS, vol. 12356, pp. 299–315. Springer, Cham (2020). https://doi.org/10.1007/978-3-030-58621-8_18
19. Li, T., Bolkart, T., Black, M.J., Li, H., Romero, J.: Learning a model of facial shape and expression from 4D scans. ACM Trans. Graph. (Proc. SIGGRAPH Asia 2017) **36**(4), Article No. 194 (2017)
20. Ma, S., et al.: Pixel codec avatars. In: Proceedings of CVPR, pp. 64–73 (2021)
21. Mescheder, L., Geiger, A., Nowozin, S.: Which Training Methods for GANs do actually Converge? arXiv preprint arXiv:1801.04406 (2018)
22. Mildenhall, B., Srinivasan, P.P., Tancik, M., Barron, J.T., Ramamoorthi, R., Ng, R.: NeRF: representing scenes as neural radiance fields for view synthesis. In: Vedaldi, A., Bischof, H., Brox, T., Frahm, J.-M. (eds.) ECCV 2020. LNCS, vol. 12346, pp. 405–421. Springer, Cham (2020). https://doi.org/10.1007/978-3-030-58452-8_24
23. Nguyen-Phuoc, T., Li, C., Theis, L., Richardt, C., Yang, Y.L.: HoloGAN: unsupervised learning of 3D representations from natural images. In: Proceedings of ICCV, pp. 7588–7597 (2019)
24. Paysan, P., Knothe, R., Amberg, B., Romdhani, S., Vetter, T.: A 3D face model for pose and illumination invariant face recognition. In: IEEE International Conference on Advanced Video and Signal Based Surveillance, pp. 296–301 (2009)
25. Schwarz, K., Liao, Y., Niemeyer, M., Geiger, A.: GRAF: generative radiance fields for 3D-aware image synthesis. In: Proceedings of NeurIPS (2020)
26. Shen, Y., Gu, J., Tang, X., Zhou, B.: Interpreting the latent space of GANs for semantic face editing. In: Proceedings of CVPR, pp. 9243–9252 (2020)
27. Shen, Y., Zhou, B.: Closed-form factorization of latent semantics in GANs. In: Proceedings of CVPR, pp. 1532–1540 (2021)
28. Shoshan, A., Bhonker, N., Kviatkovsky, I., Medioni, G.: GAN-control: explicitly controllable GANs. In: Proceedings of ICCV, pp. 14083–14093 (2021)
29. Sitzmann, V., Martel, J.N.P., Bergman, A., Lindell, D.B., Wetzstein, G.: Implicit neural representations with periodic activation functions. In: Proceedings of NeurIPS (2020)
30. Sohl-Dickstein, J., Weiss, E.A., Maheswaranathan, N., Ganguli, S.: Deep unsupervised learning using nonequilibrium thermodynamics. In: Proceedings of ICML (2015)
31. Tewari, A., et al.: StyleRig: rigging StyleGAN for 3D control over portrait images. In: Proceedings of CVPR, pp. 6142–6151 (2020)
32. Thies, J., Zollhöfer, M., Nießner, M.: Deferred neural rendering: image synthesis using neural textures. ACM Trans. Graph. (Proc. SIGGRAPH 2019) **38**(4) (2019)

33. Wang, Q., et al.: IBRNet: learning multi-view image-based rendering. In: Proceedings of CVPR, pp. 4690–4699 (2021)
34. Wu, Z., Lischinski, D., Shechtman, E.: StyleSpace analysis: disentangled controls for StyleGAN image generation. In: Proceedings of CVPR, pp. 12863–12872 (2021)
35. Zhou, P., Xie, L., Ni, B., Tian, Q.: CIPS-3D: A 3D-Aware Generator of GANs Based on Conditionally-Independent Pixel Synthesis. arXiv preprint arXiv:2110.09788 (2021)

DHG-GAN: Diverse Image Outpainting via Decoupled High Frequency Semantics

Yiwen Xu⬤, Maurice Pagnucco⬤, and Yang Song$^{(\boxtimes)}$⬤

School of Computer Science and Engineering, University of New South Wales,
Sydney, Australia
yiwen.xu1@student.unsw.edu.au, {morri,yang.song1}@unsw.edu.au

Abstract. Diverse image outpainting aims to restore large missing regions surrounding a known region while generating multiple plausible results. Although existing outpainting methods have demonstrated promising quality of image reconstruction, they are ineffective for providing both diverse and realistic content. This paper proposes a Decoupled High-frequency semantic Guidance-based GAN (DHG-GAN) for diverse image outpainting with the following contributions. 1) We propose a two-stage method, in which the first stage generates high-frequency semantic images for guidance of structural and textural information in the outpainting region and the second stage is a semantic completion network for completing the image outpainting based on this semantic guidance. 2) We design spatially varying stylemaps to enable targeted editing of high-frequency semantics in the outpainting region to generate diverse and realistic results. We evaluate the photorealism and quality of the diverse results generated by our model on CelebA-HQ, Place2 and Oxford Flower102 datasets. The experimental results demonstrate large improvement over state-of-the-art approaches.

Keywords: Diverse image outpainting · GAN · Image reconstruction

1 Introduction

Image outpainting (as shown in Fig. 1) aims to reconstruct large missing regions and synthesise visually realistic and semantically convincing content from a limited input content [8,22,33,37]. This is a challenging task because it utilises less neighbouring reference information to extrapolate unseen areas and the regions that are outpainted should look aesthetically genuine to the human eye. Image outpainting has gained considerable interest in recent years and has broadly novel applications, including image and video-based rendering [21], image reconstruction [37] and image modification [23].

Early outpainting methods [2,30,35,46] are usually patch-based, matching and stitching known pixel blocks or semantic vectors in the input to outpaint

Supplementary Information The online version contains supplementary material available at https://doi.org/10.1007/978-3-031-26293-7_11.

L. Wang et al. (Eds.): ACCV 2022, LNCS 13847, pp. 168–184, 2023.
https://doi.org/10.1007/978-3-031-26293-7_11

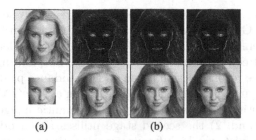

(a)   (b)

Fig. 1. Diverse image outpainting results with cropped images as inputs. (a) Original image and input image with an outpainting mask. (b) Generated high-frequency semantic image (top) and final outpainting result (down).

images. Afterwards, using the image reconstruction methods to infer semantics is proven effective for image outpainting. Wang *et al.* [37] proposed a progressive semantic inference method to expand complex structures. Lin *et al.* [22] presented a coarse-to-fine network that utilises edge maps to guide the generation of structures and textures in the outpainting region. [42] and [36] achieved a large-scale expansion of landscape images and ensured semantic coherence through a long short-term memory (LSTM) [10] encoder. However, current outpainting methods are only focused on enhancing the reconstruction quality of outpainting regions. Methods for *diverse outpainting* need to be able to infer the missing area from a small area of known pixels as well as provide diverse outputs with plausible visual and semantic information.

Currently, StyleGAN [15] has made significant progress in generating diverse images. The inverse mapping method also enables StyleGAN-based models to modify the semantics of real images. [16] and [1] demonstrate that StyleGAN can modify local semantic information based on spatially varying stylemaps. In addition, a series of variational auto-encoder (VAE) [19] based methods have been proposed to generate diverse reconstruction results. Zheng *et al.* [48] combines two VAE pipelines to trade-off output diversity and reconstruction quality based on probabilistic principles. However, this also leads to a gradual deterioration in the reconstruction quality when the similarity between the reconstruction results and the ground truth decreases. Peng *et al.* [27] proposed a model based on VQ-VAE [28], which generates a variety of structural guidance via a conditional autoregressive network PixelCNN [34]. Nevertheless, due to the randomness of the generated results of PixelCNN, it is difficult to control the structural information in the output. To introduce explicit control in the reconstruction region, an intuitive approach is to utilise artificial sketches to modify the texture and structure details [23,36].

In this paper, we focus on using *decoupled high-frequency information* to guide the diverse generation of outpainted images. There are two main challenges in this task. First, previous studies have utilised sketches as guidance for generating diverse structures, however, providing modification guidance for complex structural and texture information becomes challenging. Furthermore, it is difficult to ensure the quality, diversity and controllability of the results for outpainting.

To solve these issues, we propose Decoupled High-frequency semantic Guidance-based GAN (DHG-GAN), a diverse image outpainting network built upon a spatially varying stylemap. Our DHG-GAN is a two-stage model that starts from generating a high-frequency semantic image for the outpainting region and then completes the image reconstruction based on the guidance provided by the high-frequency semantic image. Specifically, 1) we design a StyleGAN-based model to generate for the entire image a high-frequency semantic image from a spatially varying stylemap; and, 2) the second stage utilises an encoder-decoder structure network to complete the high-frequency semantic image with low-frequency information to generate realistic outputs.

Previous research shows that high-frequency information can improve image reconstruction performance [22,25,40]. In our method, we decouple the high-frequency semantic information through Fourier high-pass filtering, which becomes the ground truth for the first stage. This decoupled high-frequency semantics can provide rich texture details and structural information for the outpainting region. By interpolating the spatially varying stylemaps, it is feasible to generate a variety of high-frequency semantic images for the outpainting region that allow the semantic completion network to synthesise diverse results (as shown in Fig. 1). We compare with Canny, Sobel, Prewitt and Laplacian edge maps and determine that our decoupled high-frequency semantic image provides the best performance in terms of quality and diversity of image outpainting. There are three main contributions in this paper:

- We present the first diverse image outpainting model utilising decoupled high-frequency semantic information, which demonstrates state-of-the-art performance on CelebA-HQ [24], Places2 [49] and Oxford Flower102 [26] datasets which are commonly used in outpainting studies.
- We propose a two-stage diverse outpainting model DHG-GAN that consists of a high-frequency semantic generator and semantic completion network. The first stage generates images to provide guidance of high-frequency semantics for the outpainting region. The second stage performs semantic completion in the outpainting region to generate realistic images.
- We design a StyleGAN-based high-frequency semantic image generator for modifying the structure and texture information in the outpainting region via a spatially varying stylemap. Ablation experiments show that our method can achieve editing of complicated structures and textures.

2   Related Work

2.1   Image Outpainting

Early outpainting models expand input images by retrieving appropriate patches from a predefined pool of patch candidates [20,32,35,46]. The performance of such methods depends on the retrieval ability and the quality and quantity of candidate images. Later inspired by generative adversarial networks (GANs) [6], semantic regeneration network (SRN) [37] incorporates a relative spatial variant loss to gradually infer semantics, which improves the performance for repairing contours and textures. In addition, Yang *et al.* [41] proposed an outpainting method that can synthesis association relationships for scene datasets. Besides, [22] and [17] utilise an edge map generation module to provide richer textural and structural guidance, thereby improving the outpainting performance. However, these GAN-based outpainting methods rely on pixel-wise reconstruction loss and do not employ random variables or control information, and hence the outpainted images are limited in diversity. Wang *et al.* [36] developed a method to generate controllable semantics in the outpainted images based on artificial sketches. However, it is difficult to use sketches to provide complex textural and structural information such as hair and petal texture. Our method is based on a high-frequency semantic image that can present more detailed guidance information for outpainting regions.

2.2 Diverse Image Reconstruction

To generate diverse reconstruction results on cropped images, some methods train VAE-based encoder-decoder networks to condition a Gaussian distribution and sample the diversity result at test time [47,48]. Zheng *et al.* [48] proposed a framework with two parallel paths; one reconstructs the original image to provide the prior distribution for the cropped regions and coupling with the conditional distribution in the other path. Peng *et al.* [27] proposed to utilise the hierarchical architecture of VQ-VAE [28] to disentangle the structure and texture information and use vector quantisation to model the discrete distribution of the structural features through auto-regression to generate a variety of structures. However, the sampled distribution constrains the diversity of outputs generated by these methods. In contrast to these methods, our model trains high and low frequency features independently in two stages and modifies the structure and texture details using an encoded stylemap to produce diverse and realistic outputs.

2.3 GAN-Based Image Editing

In order to make the reconstruction results controllable, various methods inject sketches as guidance to edit the content of the image. DeFLOCNet [23] and SC-FEGAN [13] perform sketch line refinement and colour propagation in the convolutional neural network (CNN) feature space for injected sketches. In addition, recent studies have demonstrated that GANs are capable of learning rich semantics from latent space and manipulating the corresponding features of the output images by modifying the latent code. For instance, BicycleGAN [50] constructs an invertible relationship between the latent space and the generated image, which assists in decoupling the semantic information contained in the latent code in order to achieve semantic editing. Kim *et al.* [16] and Alharbi *et al.* [1] proposed spatially varying stylemaps that enable semantic modification of generated images locally. Nevertheless, because these approaches are not designed for image reconstruction, the output images usually contain artifacts.

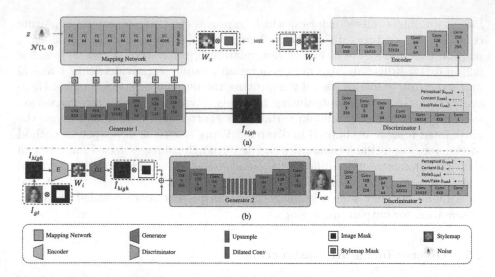

Fig. 2. Our framework illustration. (a) The high-frequency semantic image generator (HFG) consists of a mapping network, encoder, generator, and discriminator. The generator synthesises high-frequency semantic images (HSIs) based on the stylemap of generated by the mapping network. The encoder constructs an accurate inverse mapping of the synthesised HSIs through MSE supervision between the encoder and the mapping network. (b) The semantic completion network (SCN) consists of an encoder-decoder structure generator and an SN-PatchGAN discriminator. The input of the generator is a combination of the input image and a generated HSI. The semantic competition network outputs a realistic image after completing semantics.

In addition, Cheng *et al.* [4] proposed a method that searches multiple suitable latent codes through the inversion process of GAN to generate diverse outpainting results. However, this method is only designed for landscape images.

3   Methodology

In this paper, we propose the DHG-GAN method for reconstructing images with diverse structures and textures in the outpainting region. Our proposed model consists of two stages: a high frequency semantic generator (HFG) and a semantic completion network (SCN). Given an original (ground truth) image I_{gt} of size 256×256 pixels, we first obtain a cropped image $\bar{I}$ of size 128×128 pixels by $\bar{I} = I_{gt} \times M_i$, where M_i is a binary mask indicating the known region to be kept in I_{gt}. The outpainting process is to reconstruct I_{gt} from $\bar{I}$ and diversity is introduced via the HFG module. The overall framework is shown in Fig. 2.

The first stage is inspired by StyleGAN [15] and the goal of this stage is to generate a high-frequency semantic image (HSI) to provide guidance for the second-stage semantic completion. Our generated HSI contains rich textural and structural information, which is sufficiently similar to the decoupled high-frequency information from the ground truth. Accordingly, this enables the second stage to establish a strong correlation between the completed low-frequency semantic components and the associated HSI when reconstructing the image. Moreover, in the first stage, we construct an accurate mapping between the spatially varying stylemap and HSI, and then modifying the stylemap for the outpainting region enables the generator to synthesise diverse HSIs.

Existing research has shown that high-frequency information such as edge maps can be used to improve the quality of reconstructed images, particularly to sharpen contours and textures [22,25,40]. However, common edge maps, such as Canny, Sobel and Prewitt, lose a lot of texture information, whereas Laplacian edge maps contain grain noise. Therefore, we choose to decouple the high-frequency information of the RGB the channels of the ground truth image through a Fourier high-pass filter bank, which is used as the ground truth for learning the HSI. Such decoupled high-frequency information contains more comprehensive structural and textural information and less noise.

3.1 Revisting StyleGAN

We first briefly introduce StyleGAN, on which our proposed DHG-GAN is based. In the first stage, the aim of using a StyleGAN-based network is to encode samples to obtain a mapping of stylemap-to-image. Then, we can change the style code in the stylemap to get diverse images. StyleGAN proposes a style-based generator, which consists of a mapping network and synthesis network. To generate images using the generator, StyleGAN first randomly samples a latent vector Z with a Gaussian distribution $\mathcal{N}(1,0)$, then projects it to a stylemap W by the mapping network. The stylemap W is then fed into each convolutional layer in the synthesis network through an affine transformation and Adaptive Instance Normalization (AdaIN) [11]. The discriminator then distinguishes the authenticity of the images.

Unlike StyleGAN, we provide an encoder to create the inverse mapping of HSIs to stylemaps. The mean squared error (MSE) is used to minimize the difference between HSIs and stylemaps in order to construct a more precise inverse mapping. Our encoder and mapping network outputs are 3D stylemaps, which allow direct editing of structural information and textural detials for the outpainting regions. The encoder network structure is similar to the discriminator in StyleGAN.

3.2 DHG-GAN

High Frequency Semantic Generator (HFG). To generate diverse structures and textures for the outpainting region, we design an HFG module. HFG generates a high-frequency semantic image (HSI) that provides texture and structure information to guide the second stage to complete and improve the quality of image outpainting. In addition, modifying the style code in the stylemap for the outpainting region can enable HFG to synthesise various HSIs, thus providing diverse guidance for the second stage to generate diverse outputs.

Our HFG utilises a similar network structure to StyleGAN. As shown in Fig. 2, HFG consists of a mapping network F, encoder E, generator G_1 and discriminator D_1. We map the input latent code Z onto a spatially varying stylemap W_z via the mapping network. Such additional mapping has been demonstrated to learn more disentangled features [15,31]. The high-frequency information I_{high} is decoupled from the original image I_{gt} using Fourier high-pass filters, which is used as the ground truth of HFG. The generator learns to synthesise I_{high} from W_z and output the HSI $\hat{I}_{high}$. Besides, we use the encoder E to build a reverse mapping from HSI to spatially varying stylemaps W_z. This inverse mapping is to enable semantic modification for outpainting regions.

Specifically, given a latent code Z with Gaussian distribution, the mapping network $F : Z \rightarrow W_z$ produces a 3D stylemap W_z. Then, we use AdaIN operations to transfer the encoded stylemap W_z to each convolution layer in the generator. Here, AdaIN is defined as:

$$\mathbf{x}_{i+1} = a_i \frac{\mathbf{x}_i - \mu(\mathbf{x}_i)}{\sigma(\mathbf{x}_i)} + b_i \qquad (1)$$

where μ and σ compute the mean and standard deviation of each feature map $\mathbf{x}_i$. a_i and b_i are the style code computed from W_z. This enables stylemaps to be added to every synthesis module in the generator. As shown in Fig. 3, the generator contains synthesis modules of different resolution scales, and the last image-scale synthesis module generates the HSI $\hat{I}_{high}$. Then, the discriminator distinguishes I_{high} and $\hat{I}_{high}$.

The encoder E is used to establish an inverse mapping from HSI $\hat{I}_{high}$ to stylemap W_i. Then we crop W_i and W_z to keep the outpainted regions $\bar{W}_i = W_i \times (1 - M_s)$ and $\bar{W}_z = W_z \times (1 - M_s)$, where M_s is a binary mask corresponding to the outpainting region. We minimise the difference between $\bar{W}_i$ and $\bar{W}_z$ by MSE loss to train the encoder. This supervision aims to make the stylemaps $\bar{W}_i$ and $\bar{W}_z$ close in the latent space, so that E can generate stylemaps that are more suitable for semantic modification of the outpainting region. This MSE loss at the stylemap level is formulated as:

$$L_{mse_s} = \|\bar{W}_i - \bar{W}_s\|_2^2 \qquad (2)$$

We also use a combination of MSE, learned perceptual image patch similarity (LPIPS) and hinge adversarial losses to train the generator and discriminator. MSE loss measures the pixel-level similarity which can be formulated as:

$$L_{mse_i} = \|(I_{high} - \hat{I}_{high}\|_2^2 \qquad (3)$$

LPIPS loss [45] is used to measure perceptual differences and improve the perceptual quality of the HSI. Inspired by [7,12], we observe that images generated with LPIPS loss have less noise and richer textures than using perceptual loss [14]. This is due to the VGG-based perceptual loss being trained for image classification, whereas LPIPS is trained to score image patches based on human perceptual similarity judgments. LPIPS loss is defined as:

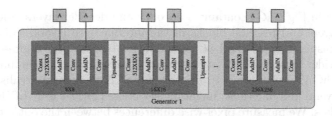

Fig. 3. Generator G_1 in HFG. The generator consists of affine transformation modules, adaptive instance normalization (AdaIN) operation and multiple scale convolution layers. After obtaining the spatially varying stylemaps, we add it to the generator using affine transformations and AdaIN.

$$L_{lpips1} = \sum_l \tau^l \left(\phi^l \left(\hat{I}_{high} \right) - \phi^l \left(I_{high} \right) \right) \tag{4}$$

where ϕ is a feature extractor based on a pre-trained AlexNet. τ is a transformation from embeddings to a scalar LPIPS score, which is computed from l layers and averaged.

In addition, the hinge adversarial loss for generator G_1 and discriminator D_1 in this stage is defined as:

$$L_{G_1} = -\mathbb{E}_{I_{high}} \left[D_1 \left(G_1 \left(A(W_z) \right), I_{high} \right) \right] \tag{5}$$

$$L_{D_1} = \mathbb{E}_{(I_{gt}, I_{high})} \left[\max \left(0, 1 - D_1 \left(I_{gt}, I_{high} \right) \right) \right] \\ + \mathbb{E}_{I_{high}} \left[\max \left(0, 1 + D_1 \left(G(A(W_z), I_{high}) \right) \right) \right] \tag{6}$$

where A is the affine transformation function. The overall objective function of HFG can be written as:

$$L_{total_1} = L_{lpips1} + L_{G_1} + L_{mse_s} + 0.1 \cdot L_{mse_i} \tag{7}$$

The pixel-level MSE loss improves the outpainting quality but also affects the diversity of HSI. Here we assign a smaller weight to L_{mse_i} based on our empirical evaluations.

Semantic Completion Network (SCN). In the second stage, we design a SCN to utilise the HSI generated in the first stage to complete the outpainting region semantically and create a realistic image. SCN has an encoder-decoder structure with a generator G_2 and discriminator D_2. To train SCN, first, based on the ground truth image I_{gt}, we obtain the high-frequency information I_{high} and generate the stylemap W_i by inverse mapping I_{high} through the first stage encoder E. The generator G_1 then uses W_i to synthesise a HSI $\hat{I}_{high}$. We then combine the cropped input image $\hat{I}$ (to be outpainted) and $\hat{I}_{high}$ into $\bar{I} = \hat{I} + \hat{I}_{high} \times (1 - M_i)$ as the input to the generator G_2. The generator G_2 then performs semantic completion on this input to generate realistic results I_{out}.

As shown in Fig. 2, G_2 contains 3 encoder and decoder layers, and there are 8 dilation layers and residual blocks in the intermediate layers. This generator structure is inspired by [25], and we using dilation blocks in the middle in order to promote a wider receptive field at the output neuron. The discriminator follows the SN-PatchGAN structure [44]. In order for SCN to generate realistic results, we utilise several loss functions, including style loss [29], $L1$ loss, LPIPS loss and adversarial loss. We measure pixel-wise differences between the output of G_2 and I_{gt} by $L1$ loss and take into account high-level feature representation and human perception by LPIPS loss. Besides, style loss compares the difference between the deep feature maps of I_{out} and the I_{gt} from pre-trained and fixed VGG-19, and has shown effectiveness in counteracting "checkerboard" artifacts produced by the transpose convolution layers. Style loss is formulated as follows:

$$L_{\text{style}} = \mathbb{E}\left[\sum_j \left\|Gram_j^\phi\left(I_{gt}\right) - Gram_j^\phi\left(I_{out}\right)\right\|_1\right] \tag{8}$$

where the $Gram_j^\phi$ is the Gram matrix of the j-th feature layer. The adversarial loss over the generator G_2 and discriminator D_2 are defined as:

$$L_{G_2} = -\mathbb{E}_{I_{gt}}\left[D_2\left(I_{out}, I_{gt}\right)\right] \tag{9}$$

$$L_{D_2} = \mathbb{E}_{I_{gt}}\left[\text{ReLU}\left(1 - D_2\left(I_{gt}\right)\right)\right] + \mathbb{E}_{I_{out}}\left[\text{ReLU}\left(1 + D_2\left(I_{out}\right)\right)\right] \tag{10}$$

Finally, the total function is defined as a weighted sum of different losses with the following coefficients:

$$L_{total_2} = L_1 + 250 \cdot L_{style} + 0.1 \cdot L_{lpips2} + 0.1 \cdot L_{G_2} \tag{11}$$

4   Experiments

4.1   Datasets and Implementation Details

We evaluate our model on three datasets, including CelebA-HQ [24], Places2 [49] and Oxford Flower102 [26]. The CelebA-HQ dataset includes 30000 celebrity face images at 1024×1024 pixels. Places2 contains 1,803,460 images with 434 different scene categories. Oxford Flower102 comprises 102 flower categories and a total of 8189 images. The official training, testing and validation splits are used for these three datasets. We resize the images to 256×256 pixels with data augmentation and use the center outpainting mask to reserve the middle region of images for testing.

Our model is implemented in Pytorch v1.4 and trained on an NVIDIA 2080 Ti GPU. The model is optimized using the Adam optimizer [18] with $\beta_1 = 0$

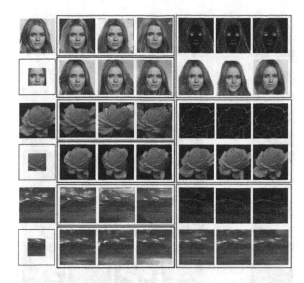

Fig. 4. Qualitative comparison results with diverse-solution methods on CelebA-HQ [24], Places2 [49] and Oxford Flower102 [26] datasets. For each dataset, from top to bottom, from left to right, the images are: original image, input image, results of PIC [48] (with green box), results of DSI [27] (with blue box), results of our method showing both HSIs and the generated images (with red box). (Color figure online)

and $\beta_2 = 0.9$. The two stages are trained with learning rates of 0.001 and 0.0001, respectively, until the loss plateaus, with a batch size of 4. In addition, in order to train the HFG, we adjusted the size scale of the stylemap. Through our experiments, we found that stylemaps of $64 \times 8 \times 8$ can produce satisfactory HSIs.

During the testing phase, we traverse the images $\{I_{gt}\}$ in the validation set to obtain a set of HSIs $\{I_{high}\}$ using Fourier high-pass filtering and utilise the encoder E to map these HSIs to obtain a variety of stylemaps. We use W_i to represent the stylemap of the current test image and $\{W_{ref}\}$ to represent the stylemaps of other images in the validation set. In addition, W_z is generated by the mapping network F through random vectors Z. We crop $\{W_{ref}\}$ and W_z through the outpainting mask, retaining only the style code for the outpainting region. For the test image I, we crop its stylemap W_i to retain the style code only for the centre kept region, corresponding to the input image that will be outpainted, so that ground truth information is not used during outpainting. Then, we can get various complete stylemap $\hat{W}_i$ by combine W_i, W_{ref} and W_z with the following operations:

$$\hat{W}_i = (W_i \times M_s) + (W_{ref} \times (1 - M_s)) \cdot 0.2 + (W_z \times (1 - M_s)) \cdot 0.8 \qquad (12)$$

where $W_i \times M_s$ denotes the cropped stylemaps of the test image, $W_{ref} \times (1 - M_s)$ and $W_z \times (1 - M_s)$ denote the cropped stylemap of a sampled reference image in the validation set and cropped stylemap generated from the random vector,

Fig. 5. Qualitative comparison results with single-solution methods on CelebA-HQ (top), Place2 (middle) and Oxford Flower102 (down) datasets. For each dataset, from left to right, the image are: (a) Original image. (b) Input images. (c) Results of CA [43]. (d) Results of EdgeConnect [25]. (e) Results of SRN [37]. (f) Results of our method.

respectively. Then, by passing the various $\hat{W}_i$ to the generator G_1 in HFG, we can obtain diverse $\hat{I}_{high}$, so that SCN can produce various outpainting results I_{out}. The coefficients assigned to the stylemaps force HSIs to trade-off between being similar to the original image or the other reference images. Following previous diverse-solution methods [47,48], we also select the top samples closest to the original image through the discriminator D_2 for qualitative and quantitative evaluation. For compared approaches, we use their official implementations and train on the same dataset with identical configurations to ensure a fair comparison. We use the centre mask to compare the results, which has the advantage of showing the generation effect of different regions in a balanced manner.

4.2 Qualitative Results

Figure 4 shows the qualitative comparison results of our model and the state-of-the-art diverse solution reconstruction models: pluralistic image completion (PIC) [48] and diverse structure inpainting (DSI) [27], on CelebA-HQ, Places2 and Oxford Flower102 datasets for the outpainting task. We did not choose models based on artifact-injected edge maps for comparison, as these models are strongly influenced by artifact input [13,23,44]. PIC and DSI are able to generate diverse results without human intervention, which are thus more directly comparable with our method. PIC is based on a probabilistic principle with

Table 1. Comparison of different methods on CelebA-HQ, Oxford Flower102 and Place2. S denotes the single-solution methods, D denotes the diverse-solution methods.

		CelebA-HQ				Place2				Flower102			
		PSNR	SSIM	MIS	FID	PSNR	SSIM	MIS	FID	PSNR	SSIM	MIS	FID
S	CA	15.22	0.52	0.017	31.62	14.29	0.51	0.018	29.09	14.16	0.49	0.018	33.46
	EC	15.26	0.55	**0.024**	27.31	16.84	0.57	0.019	29.17	16.49	0.57	0.021	29.43
	SRN	16.65	0.56	0.023	28.29	16.94	0.60	**0.023**	**27.12**	15.97	0.50	0.021	29.73
	Ours w/o HSI	15.24	0.51	0.019	30.18	16.13	0.53	0.018	29.19	15.37	0.51	0.020	32.17
D	PIC	14.77	0.45	0.019	33.12	15.10	0.51	0.020	29.19	13.68	0.48	0.017	39.11
	DSI	15.00	0.51	0.021	31.45	15.33	0.53	0.021	29.71	16.68	0.56	0.020	29.38
	Ours w/ HSI	**16.71**	**0.59**	0.023	**27.14**	**16.97**	**0.61**	**0.023**	28.59	**17.05**	**0.62**	**0.024**	**28.76**

two parallel reconstruction paths, while DSI uses an auto-regressive network to model a conditional distribution of discrete structure and texture features and samples from this distribution to generate diverse results. It can be seen that PIC generates reasonable results on CelebA-HQ but fails to achieve inner-class diversity in multi-class datasets (Oxford Flower102). DSI performs better on inner-class diversity due to the distribution obtained through an auto-regressive model. The results also show that our method can generate more realistic outputs, e.g., the generated shapes of flowers and mountains are more plausible. In addition, since our results are generated with guidance of a high-frequency semantics image, they show finer texture and structural details.

Table 2. Classification result on the Oxford Flower102 dataset.

Method	Original	CA	EdgeConnect	SRN	PIC	DSI	Ours
VGG-S	**0.9213**	0.7945	0.8285	0.8317	0.7807	0.8310	**0.8458**
Inception-v3	**0.9437**	0.7419	0.8230	0.8593	0.7719	0.8683	**0.8731**
EffNet-L2	**0.9814**	0.7385	0.8194	0.8494	0.7355	0.8511	**0.8770**

For comparison with single-solution reconstruction methods, we select the top-1 results by discriminator D_2. As shown in Fig. 5, due to the lack of prior structural information, CA [43] has difficulties generating reasonable structures, whereas EdgeConnect [25] generates better textures and structures due to the use of edge maps as semantic guidance. SRN [37] infers semantics gradually via the relative spatial variant loss [37], enabling it to generate adequate structural and textural information. Since our results are guided by richer structural and semantic information in the outpainted regions, the semantic completion network is able to better infer and complete the images.

4.3   Quantitative Results

Following the previous image reconstruction methods, we utilise the common evaluation metrics including Peak Signal-to-Noise Ratio (PSNR) and Structural

Similarity (SSIM) [38] to determine the similarity between the outpainting results and ground truth. Additionally, we use Modified Inception Score (MIS) [47] and Fréchet Inception Distance (FID) [9] as perceptual quality indicators. Furthermore, FID is capable of detecting the GAN model's mode collapse and mode dropping [24]. For multiple-solution methods, we sample 100 output images for each input image and report the average result of top 10 samples. Table 1 shows that, although our method is lower than SRN and EdgeConnect in some metrics, it still outperforms other single and diverse solution reconstruction methods. Moreover, the MIS and FID measurements demonstrate that our method is more effective in generating visually coherent content.

Furthermore, we analyse the quality of outpainted images on the Oxford Flower102 dataset by classifying the generated images for the official validation set into 102 categories as labelled in the dataset. We use pre-trained VGG-S [3], Inception-v3 [39], EffNet-L2 [5] to evaluate our generated results, which are state-of-the-art classification methods on the Oxford Flower102 dataset. As shown in Table 2, our method has a higher classification accuracy than other methods and is closer to the classification result on original images. This can be attributed to the fact that our method can generate more reasonable and realistic semantics.

Table 3. Comparison of diversity scores on the CelebA-HQ dataset.

	LPIPS - I_{out}	LPIPS - $I_{out(m)}$
PIC [48]	0.032	0.085
DSI [27]	0.028	0.073
Ours	**0.035**	**0.089**

Also, we calculate the diversity score for comparisons with diverse-solution methods using the LPIPS [51] metric on the CelebA-HQ dataset. The average score is calculated based on 50K output images generated from 1K input images. We compute the LPIPS scores for the complete images I_{out} and then only the outpainting regions $I_{out(m)}$. As shown in Table 3, our method achieves relatively higher diversity scores than other existing methods.

4.4   Ablation Study

Alternative High-Frequency Semantic Feature. We conduct an ablation study on CelebA-HQ to show the impact of using various types of high-frequency semantic features as a guide for outpainting. As shown in Fig. 6, the facial contours generated without using the high-frequency semantic map as a guide show some shrinkage and the texture is not clear enough. The high-frequency semantic information provided by utilizing Canny, Prewitt and Sobel edge detection operators can solve the contour shrinkage to an extent. However, we observed that such high-frequency semantic features cannot provide specific texture information. Although the Laplacian edge map can provide more texture details,

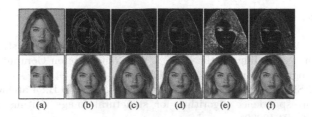

Fig. 6. Results of image outpainting using different high-frequency information. (a) Original image and input image. (b) Results of Canny. (c) Results of Sobel. (d) Results of Prewitt. (e) Results of Laplacian. (f) Results of our method.

it contains a lot of grain noise, which adversely affects the quality of image restoration and texture modification. Compared to these techniques, our method can generate overall more detailed structural and texture information.

Quality of High-Frequency Semantic Image for Diverse Outpainting Quality. We notice that the dimension of stylemaps affects the capacity of decoupling latent semantics in the stylemap and thus the fineness of diverse outpainting. To choose a suitable size for stylemaps, we consider that for 256×256 sized images, we need at least a size of $4 \times 4 \times 1$ for stylemaps to identify the outpainting region, thus performing semantic editing for the outpainting area. Furthermore, as illustrated in the supplementary material (Fig. 1), we determined that, if the latent vector dimension in the stylemap is small, the encoder will fail to establish a high quality mapping of HSI. We found that using an $8 \times 8 \times 64$ stylemap provided a sufficiently accurate mapping to avoid compromising the quality of the images generated.

5   Conclusion and Future Work

In this paper, we propose a diverse-solution outpainting method, DHG-GAN, for generating diverse and high-quality images guided by decoupled high-frequency semantic information. Our method first generates diverse high-frequency semantic images and then complete their semantics to produce the outpainted images. The proposed high-frequency semantic generator based on spatial varying stylemaps helps introduce diversity in the outpainted images. Extensive qualitative and quantitative comparisons show the superiority of our method in terms of diversity and quality of outpainting. However, a limitation of our method is that when the generated HSI is considerably different from the ground truth HSI of the outpainting region, in a few cases, the regions around the object contours can be blurry, and the structural and textural information in the generated HSI can be quite different from the ground truth in these cases. This makes it challenging for the second-stage semantic completion network to generate reasonable semantics. In our future work, we plan to improve the robustness of our method to generate diverse results. Also, a promising future direction could be exploiting HSI for outpainting complex scene images.

References

1. Alharbi, Y., Wonka, P.: Disentangled image generation through structured noise injection. In: Proceedings of the IEEE/CVF Conference on Computer Vision and Pattern Recognition, pp. 5134–5142 (2020)
2. Barnes, C., Shechtman, E., Finkelstein, A., Goldman, D.B.: PatchMatch: a randomized correspondence algorithm for structural image editing. ACM Trans. Graph. **28**(3), 24 (2009)
3. Chatfield, K., Simonyan, K., Vedaldi, A., Zisserman, A.: Return of the devil in the details: delving deep into convolutional nets. In: British Machine Vision Conference, pp. 1–12 (2014)
4. Cheng, Y.C., Lin, C.H., Lee, H.Y., Ren, J., Tulyakov, S., Yang, M.H.: Inout: diverse image outpainting via GAN inversion. In: Proceedings of the IEEE/CVF Conference on Computer Vision and Pattern Recognition, pp. 11431–11440 (2022)
5. Foret, P., Kleiner, A., Mobahi, H., Neyshabur, B.: Sharpness-aware minimization for efficiently improving generalization. arXiv preprint arXiv:2010.01412 (2020)
6. Goodfellow, I.J., et al.: Generative adversarial nets. In: Proceedings of the 27th International Conference on Neural Information Processing Systems, pp. 2672–2680 (2014)
7. Guan, S., Tai, Y., Ni, B., Zhu, F., Huang, F., Yang, X.: Collaborative learning for faster StyleGAN embedding. arXiv preprint arXiv:2007.01758 (2020)
8. Guo, D., et al.: Spiral generative network for image extrapolation. In: Vedaldi, A., Bischof, H., Brox, T., Frahm, J.-M. (eds.) ECCV 2020. LNCS, vol. 12364, pp. 701–717. Springer, Cham (2020). https://doi.org/10.1007/978-3-030-58529-7_41
9. Heusel, M., Ramsauer, H., Unterthiner, T., Nessler, B., Hochreiter, S.: GANs trained by a two time-scale update rule converge to a local Nash equilibrium. In: Advances in neural Information Processing Systems, vol. 30 (2017)
10. Hochreiter, S., Schmidhuber, J.: Long short-term memory. Neural Comput. **9**(8), 1735–1780 (1997)
11. Huang, X., Belongie, S.: Arbitrary style transfer in real-time with adaptive instance normalization. In: Proceedings of the IEEE International Conference on Computer Vision, pp. 1501–1510 (2017)
12. Jo, Y., Yang, S., Kim, S.J.: Investigating loss functions for extreme super-resolution. In: Proceedings of the IEEE/CVF Conference on Computer Vision and Pattern Recognition Workshops, pp. 424–425 (2020)
13. Jo, Y., Park, J.: SC-FEGAN: Face editing generative adversarial network with user's sketch and color. In: 2019 IEEE/CVF International Conference on Computer Vision, pp. 1745–1753 (2019). https://doi.org/10.1109/ICCV.2019.00183
14. Johnson, J., Alahi, A., Fei-Fei, L.: Perceptual losses for real-time style transfer and super-resolution. In: Leibe, B., Matas, J., Sebe, N., Welling, M. (eds.) ECCV 2016. LNCS, vol. 9906, pp. 694–711. Springer, Cham (2016). https://doi.org/10.1007/978-3-319-46475-6_43
15. Karras, T., Laine, S., Aila, T.: A style-based generator architecture for generative adversarial networks. In: Proceedings of the IEEE/CVF Conference on Computer Vision and Pattern Recognition, pp. 4401–4410 (2019)
16. Kim, H., Choi, Y., Kim, J., Yoo, S., Uh, Y.: Exploiting spatial dimensions of latent in GAN for real-time image editing. In: Proceedings of the IEEE/CVF Conference on Computer Vision and Pattern Recognition, pp. 852–861 (2021)

17. Kim, K., Yun, Y., Kang, K.W., Kong, K., Lee, S., Kang, S.J.: Painting outside as inside: edge guided image outpainting via bidirectional rearrangement with progressive step learning. In: Proceedings of the IEEE/CVF Winter Conference on Applications of Computer Vision, pp. 2122–2130 (2021)
18. Kingma, D.P., Ba, J.: Adam: A method for stochastic optimization. arXiv preprint arXiv:1412.6980 (2014)
19. Kingma, D.P., Welling, M.: Auto-encoding variational Bayes. arXiv preprint arXiv:1312.6114 (2013)
20. Kopf, J., Kienzle, W., Drucker, S., Kang, S.B.: Quality prediction for image completion. ACM Trans. Graph. (ToG) **31**(6), 1–8 (2012)
21. Lin, C.H., Lee, H.Y., Cheng, Y.C., Tulyakov, S., Yang, M.H.: InfinityGAN: Towards infinite-resolution image synthesis. arXiv preprint arXiv:2104.03963 (2021)
22. Lin, H., Pagnucco, M., Song, Y.: Edge guided progressively generative image outpainting. In: Proceedings of the IEEE/CVF Conference on Computer Vision and Pattern Recognition, pp. 806–815 (2021)
23. Liu, H., et al.: Deflocnet: deep image editing via flexible low-level controls. In: Proceedings of the IEEE/CVF Conference on Computer Vision and Pattern Recognition, pp. 10765–10774 (2021)
24. Lucic, M., Kurach, K., Michalski, M., Gelly, S., Bousquet, O.: Are GANs created equal? A large-scale study. arXiv preprint arXiv:1711.10337 (2017)
25. Nazeri, K., Ng, E., Joseph, T., Qureshi, F.Z., Ebrahimi, M.: EdgeConnect: Generative image inpainting with adversarial edge learning. arXiv preprint arXiv:1901.00212 (2019)
26. Nilsback, M.E., Zisserman, A.: Automated flower classification over a large number of classes. In: 6th Indian Conference on Computer Vision, Graphics & Image Processing, pp. 722–729. IEEE (2008)
27. Peng, J., Liu, D., Xu, S., Li, H.: Generating diverse structure for image inpainting with hierarchical VQ-VAE. In: Proceedings of the IEEE/CVF Conference on Computer Vision and Pattern Recognition, pp. 10775–10784 (2021)
28. Razavi, A., Van den Oord, A., Vinyals, O.: Generating diverse high-fidelity images with VQ-VAE-2. In: Advances in Neural Information Processing Systems, vol. 32 (2019)
29. Sajjadi, M.S., Scholkopf, B., Hirsch, M.: EnhanceNet: single image super-resolution through automated texture synthesis. In: Proceedings of the IEEE International Conference on Computer Vision, pp. 4491–4500 (2017)
30. Shan, Q., Curless, B., Furukawa, Y., Hernandez, C., Seitz, S.M.: Photo uncrop. In: Fleet, D., Pajdla, T., Schiele, B., Tuytelaars, T. (eds.) ECCV 2014. LNCS, vol. 8694, pp. 16–31. Springer, Cham (2014). https://doi.org/10.1007/978-3-319-10599-4_2
31. Shen, Y., Gu, J., Tang, X., Zhou, B.: Interpreting the latent space of GANs for semantic face editing. In: Proceedings of the IEEE/CVF Conference on Computer Vision and Pattern Recognition, pp. 9243–9252 (2020)
32. Sivic, J., Kaneva, B., Torralba, A., Avidan, S., Freeman, W.T.: Creating and exploring a large photorealistic virtual space. In: 2008 IEEE Computer Society Conference on Computer Vision and Pattern Recognition Workshops, pp. 1–8. IEEE (2008)
33. Teterwak, P., et al.: Boundless: Generative adversarial networks for image extension. In: Proceedings of the IEEE/CVF International Conference on Computer Vision, pp. 10521–10530 (2019)

34. Van Oord, A., Kalchbrenner, N., Kavukcuoglu, K.: Pixel recurrent neural networks. In: International Conference on Machine Learning, pp. 1747–1756. PMLR (2016)

35. Wang, M., Lai, Y.K., Liang, Y., Martin, R.R., Hu, S.M.: BiggerPicture: data-driven image extrapolation using graph matching. ACM Trans. Graph. **33**(6), 1–14 (2014)

36. Wang, Y., Wei, Y., Qian, X., Zhu, L., Yang, Y.: Sketch-guided scenery image outpainting. IEEE Trans. Image Process. **30**, 2643–2655 (2021)

37. Wang, Y., Tao, X., Shen, X., Jia, J.: Wide-context semantic image extrapolation. In: Proceedings of the IEEE/CVF Conference on Computer Vision and Pattern Recognition, pp. 1399–1408 (2019)

38. Wang, Z., Bovik, A.C., Sheikh, H.R., Simoncelli, E.P.: Image quality assessment: from error visibility to structural similarity. IEEE Trans. Image Process. **13**(4), 600–612 (2004)

39. Xia, X., Xu, C., Nan, B.: Inception-V3 for flower classification. In: 2nd International Conference on Image, Vision and Computing, pp. 783–787 (2017)

40. Xiong, W.,et al.: Foreground-aware image inpainting. In: Proceedings of the IEEE/CVF Conference on Computer Vision and Pattern Recognition, pp. 5840–5848 (2019)

41. Yang, C.A., Tan, C.Y., Fan, W.C., Yang, C.F., Wu, M.L., Wang, Y.C.F.: Scene graph expansion for semantics-guided image outpainting. In: Proceedings of the IEEE/CVF Conference on Computer Vision and Pattern Recognition, pp. 15617–15626 (2022)

42. Yang, Z., Dong, J., Liu, P., Yang, Y., Yan, S.: Very long natural scenery image prediction by outpainting. In: Proceedings of the IEEE/CVF International Conference on Computer Vision, pp. 10561–10570 (2019)

43. Yu, J., Lin, Z., Yang, J., Shen, X., Lu, X., Huang, T.S.: Generative image inpainting with contextual attention. In: Proceedings of the IEEE Conference on Computer Vision and Pattern Recognition, pp. 5505–5514 (2018)

44. Yu, J., Lin, Z., Yang, J., Shen, X., Lu, X., Huang, T.S.: Free-form image inpainting with gated convolution. In: Proceedings of the IEEE/CVF International Conference on Computer Vision, pp. 4471–4480 (2019)

45. Zhang, R., Isola, P., Efros, A.A., Shechtman, E., Wang, O.: The unreasonable effectiveness of deep features as a perceptual metric. In: Proceedings of the IEEE Conference on Computer Vision and Pattern Recognition, pp. 586–595 (2018)

46. Zhang, Y., Xiao, J., Hays, J., Tan, P.: FrameBreak: dramatic image extrapolation by guided shift-maps. In: Proceedings of the IEEE Conference on Computer Vision and Pattern Recognition, pp. 1171–1178 (2013)

47. Zhao, L., et al.: UCTGAN: diverse image inpainting based on unsupervised cross-space translation. In: Proceedings of the IEEE/CVF Conference on Computer Vision and Pattern Recognition, pp. 5741–5750 (2020)

48. Zheng, C., Cham, T.J., Cai, J.: Pluralistic image completion. In: Proceedings of the IEEE/CVF Conference on Computer Vision and Pattern Recognition. pp. 1438–1447 (2019)

49. Zhou, B., Lapedriza, A., Khosla, A., Oliva, A., Torralba, A.: Places: a 10 million image database for scene recognition. IEEE Trans. Pattern Anal. Mach. Intell. **40**(6), 1452–1464 (2017)

50. Zhu, J.Y., et al.: Multimodal image-to-image translation by enforcing bi-cycle consistency. In: Advances in Neural Information Processing Systems, pp. 465–476 (2017)

51. Zhu, J.Y., et al.: Toward multimodal image-to-image translation. In: Advances in Neural Information Processing Systems, vol. 30 (2017)

MUSH: Multi-scale Hierarchical Feature Extraction for Semantic Image Synthesis

Zicong Wang[1,2], Qiang Ren[1,2], Junli Wang[1,2], Chungang Yan[1,2], and Changjun Jiang[1,2(✉)]

[1] Key Laboratory of Embedded System and Service Computing (Tongji University), Ministry of Education, Shanghai 201804, China
{wangzicong,rqfzpy,junliwang,yanchungang,cjjiang}@tongji.edu.cn
[2] National (Province-Ministry Joint) Collaborative Innovation Center for Financial Network Security, Tongji University, Shanghai 201804, China

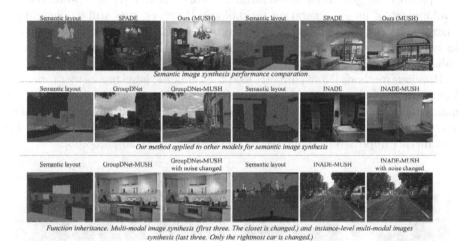

Semantic image synthesis performance comparison

Our method applied to other models for semantic image synthesis

Function inheritance. Multi-modal image synthesis (first three. The closet is changed.) and instance-level multi-modal images synthesis (last three. Only the rightmost car is changed.)

Fig. 1. Our method improves image synthesis performance by utilizing multi-scale information. It can be applied to many models and get better results.

Abstract. Semantic image synthesis aims to translate semantic label masks to photo-realistic images. Previous methods have limitations that extract semantic features with limited convolutional kernels and ignores some crucial information, such as relative positions of pixels. To address these issues, we propose MUSH, a novel semantic image synthesis model that utilizes multi-scale information. In the generative network stage, a multi-scale hierarchical architecture is proposed for feature extraction and merged successfully with guided sampling operation to enhance semantic image synthesis. Meanwhile, in the discriminative network stage, the

Supplementary Information The online version contains supplementary material available at https://doi.org/10.1007/978-3-031-26293-7_12.

model contains two different modules for feature extraction of semantic masks and real images, respectively, which helps use semantic masks information more effectively. Furthermore, our proposed model achieves the state-of-the-art qualitative evaluation and quantitative metrics on some challenging datasets. Experimental results show that our method can be generalized to various models for semantic image synthesis. Our code is available at https://github.com/WangZC525/MUSH.

Keywords: Image synthesis · Semantic information · Multi-scale hierarchical architecture

1   Introduction

The progress of generative adversarial network (GAN) [7] has promoted the development of image generation technology. However, it is still challenging to generate photorealistic images from input data. In this paper, we focus on semantic image synthesis, which aims to translate semantic label masks to photorealistic images.

Previous methods [13,39] directly feed the semantic label masks to an encoder-decoder network. Inspired by AdaIN [10], SPADE [27] uses spatially-adaptive normalization to make semantic information control the generation in normalization layers. This method achieved great success in semantic image synthesis and has been further improved by many recent methods [4,14,22,26,35,38,50].

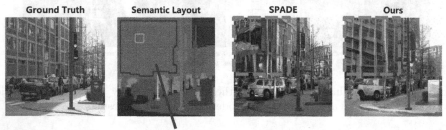

All pixels inside the blue frame have the same value after feature extraction

Fig. 2. An example of consequences of restricted receptive fields in SPADE-based models. We take a semantic label map downsampled to 64×64 in the figure for example. The SPADE-based models have a receptive field of 5×5 (marked with green frame) for feature extraction of semantic layouts. Then in the layout of the building, all pixels inside each blue frame have exactly the same value after convolution, and the model cannot distinguish the relative positions of these pixels in the corresponding objects, leading to bad performance of image synthesis. (Color figure online)

However, the receptive fields for semantic feature extraction in these methods are limited, which fails to effectively capture multi-scale information and is unable to distinguish the relative position of each pixel inside its category area, leading to bad performance especially on classes with large areas. CLADE [34,35] gets similar performance by replacing the semantic feature extraction module of SPADE by directly mapping each label of the semantic map to the corresponding parameters, which further proves that the semantic feature extraction module in SPADE cannot effectively extract the multi-scale spatial information of the relative positions of pixels inside their categories(see Fig. 2).

In addition, the discriminator in previous work [14,22,26,38,50] takes the concatenation of the image and the semantic label map in channel direction as the input. However, the features of them are added after the first layer of convolution, which can no longer be separated. Moreover, the label map is input only at the beginning of the network in the way of channel concatenation, so that its information can be easily lost in normalization layers, the above problems make it difficult for the network to work according to the label map.

To address above issues, in this paper, we propose a model that utilizes multi-scale hierarchical architecture to extract semantic features, named MUSH. The architecture is used in both the generator and the discriminator. In the generator, it helps capture specific features of each pixel according to its relative position to improve the performance on classes with large areas. Meanwhile, we merge a method in which semantic features are extracted by using individual parameters for each class with it, so as to achieve good performance on classes with small areas. In the discriminator, we use two different networks to extract semantic mask features and real images respectively. Thus, the discriminator can make better use of the semantic information to distinguish between real and generated images. Because of these strategies, the quality of generated images is significantly improved (see Fig. 1).

The contributions of this paper are summarized as follows: (1) We apply a network that can effectively extract the multi-scale features of the semantic map to the generator, so that the generator can recognize the relative positions of all pixels in each object and refine the structure of the generated objects. (2) We propose a novel approach to merge two methods of semantic feature extraction in the generator. (3) We propose a discriminator that extracts the features of the semantic map and the image separately, so as to better discriminate between real and generated images according to the semantic map. (4) With the proposed MUSH, we have achieved better experimental results than the state-of-the-art methods. And we apply our method to GroupDNet and INADE to verify that it can be generalized to various models and improve their performance.

2   Related Work

Generative Adversarial Networks (GANs) [7] have achieved great success in image generation. It contains two parts: generator and discriminator, which are respectively used to generate images and distinguish between real and fake images. CGAN [23] is proposed based on GAN. It generates images according to restricted input. Our work focuses on CGANs that do semantic image synthesis, where the input is the semantic label map.

Semantic Image Synthesis takes semantic label maps as inputs, and synthesize photorealistic images according to the maps. Many GAN-based semantic image synthesis models [11,16,17,20,44,47,48] have been proposed recently. Pix2pix [13] and pix2pixHD [39] used an encoder-decoder generator and took semantic maps as inputs directly. CRN [4] refined the output image by a deep cascaded network. SPADE [27] replaced batch normalization [12] in pix2pixHD with spatially adaptive normalization, which has achieved great success in semantic image synthesis. Many approaches have been proposed based on SPADE, such as GroupDNet [50], LGGAN [38], TSIT [14], SEAN [49], INADE [33], OASIS [30], SESAME [26] etc. GroupDNet used group convolution in the generator to implement semantically multi-modal image synthesis. LGGAN proposed a network to generate areas of each category separately. TSIT used a stream to extract features of semantic label maps. SEAN found a better way to insert style information to the network based on SPADE. INADE made the model be adaptive to instances. OASIS trained the network with only adversarial supervision. SESAME helped add, manipulate or erase objects. Different from the above models, our work focuses on the extension of the receptive field for semantic layout feature extraction. Compared with problems that the above models targeted on, the limitation of the receptive field for semantic feature extraction is more serious and easy to be ignored.

Encoder-Decoder is a popular structure for deep learning in recent years. It has been applied to many tasks of computer vision. Our semantic map feature extraction network is based on U-Net [29], which is a special encoder-decoder network. Feature maps in its decoder concatenate those the same size as them in encoder by channel, so that the decoder does not forget the encoding information. In addition, due to the basic structure of the encoder-decoder, decoded feature maps contain multi-scale features. The above advantages are what we need for semantic feature extraction. So we propose a semantic feature extraction network based on U-Net.

3   Method

We propose a novel semantic image synthesis model, MUSH. The model is trained adversarially and contains modules with multi-scale hierarchical architectures for semantic feature extraction. In the generator, the module helps image synthesis according to semantic features in various scales. A method using individual parameters for each class is merged to it. In the discriminator, the module helps distinguish between real and generated images according to semantic maps.

3.1   Overall Structure of MUSH Generator

The generator takes noise as input, and transforms it into an image by convolution, normalization and upsampling. Semantic information are added in the normalization layers. To avoid the limitation of the receptive field for semantic feature extraction and get better feature extraction abilities especially for classes with large areas, we propose a multi-scale hierarchical semantic feature extraction network and multi-scale feature adaptive normalizations (MSFA-Norm).

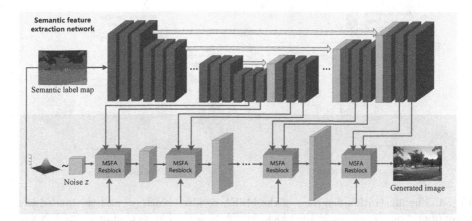

Fig. 3. The overall framework of MUSH generator. The network transforms noise into an image through MSFA-Norm residual blocks. In MSFA residual blocks, semantic information controls the procedure of image generation. The semantic feature extraction network is built based on encoder-decoder structure. Feature maps in the decoder will be input into MSFA residual blocks.

We adopt the architecture of resnet blocks [8], pack convolution layers and MSFA-Norm layers into MSFA residual blocks as the main components of the generator (see Fig. 3).

The multi-scale hierarchical architecture for semantic feature extraction is based on U-Net, which is an encoder-decoder structure. In encoding stage, the network uses multiple convolutional layers and downsampling layers to get a small-sized encoded feature map, which makes each pixel in the map contain information of a large area. In decoding stage, the network takes the encoded map as input, and outputs multi-scale feature maps from different layers. Therefore the network contains different levels. Each level processes feature maps in a specific size, which builds a multi-scale hierarchical architecture. Additionally, feature maps in the encoder are concatenated to those at the corresponding locations in the decoder with the same channel number to retain the encoding information. Feature maps of each level in the decoder are fed to MSFA-Norm.

3.2 MSFA Residual Block and MSFA-Norm

The generator uses MSFA residual blocks (see Fig. 4 (a)) and upsampling alternatively to transform noise into image. MSFA-Norms in MSFA residual blocks do normalization and control image generation with semantic information, so appropriate form of semantic information should be input here. Since feature maps obtained from the feature extraction network already contains the information of the semantic label map, it seems reasonable to input them into MSFA-Norms without the semantic label map itself. However, although the encoder-decoder extracts features with large receptive fields to achieve good feature extraction results on semantic categories with large areas, it can hardly perform well on

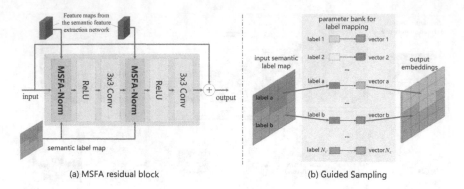

(a) MSFA residual block   (b) Guided Sampling

Fig. 4. The illustration diagrams of the MSFA residual block (a) and the guided sampling operation (b) we use in the block. MSFA residual blocks contains MSFA-Norms, ReLU layers and convolutional layers. In addition to the output of the last layer, the MSFA-Norm also takes the semantic label map and the feature map from the semantic feature extraction network as input. The guided sampling module contains a trainable parameter bank to store a corresponding vector for each label. It replaces each pixel in the semantic label map with the vector corresponding to it during calculation.

semantic categories with small areas. Because of the imbalance of sample quantities for different classes, the network training will be dominated by semantic classes with large areas, leading to classes with small areas being ignored.

Therefore, in the MSFA-Norm (see Fig. 5), we add another method to extract semantic features by using individual parameters for each class and combine it to the above mentioned method. Referring to the guided sampling (see Fig. 4 (b)) operation in CLADE [34], we maintain a parameter bank which contains a trainable vector for each category. To do the guided sampling operation, we map each category in the semantic label map to the corresponding vector to get the feature map of the input so that the operation is adaptive to different semantic categories. In this approach, we obtain features by individual parameters for each class and avoid training preference for classes with large areas.

However, how to combine these two feature extraction methods? We propose a novel method based on attention values to solve this problem. We obtain a weighted average of the two methods by taking the calculated attention values as the weights of them. For the calculation of the attention values, since performance of these methods on each pixel mainly depends on the category the pixel belongs to, we also use guided sampling to map the semantic layouts to get the attention maps of the two methods. These two maps contains all attention values of the two methods on all pixels. Attention values of different categories vary on these maps.

For the ways in which the two methods influence the input data, we use affine transformation. The multi-scale semantic feature extraction results will be convoluted twice to obtain γ_1 and β_1, which are multiplied and added to the input respectively. For the other method, γ_2 and β_2 are obtained through guided

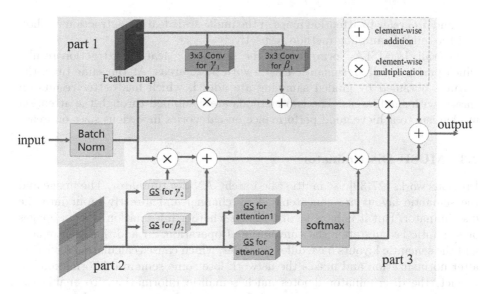

Fig. 5. Structure of MSFA-Norm. GS refers to the guided sampling operation in Fig. 4 (b). The feature map in part 1 refers to features obtained from the feature extraction network. Part 1 illustrates how multi-scale features influence the input, and part 2 illustrates how semantic features obtained by guided sampling influence the input. Part 3 shows the process of mergence of results from part 1 and part 2.

sampling, and the affine transformation calculation is the same as above. In addition, before all of these, the input of MSFA-Norms will be batch normalized first.

Let W, H, C and N be the width, height, the number of channels of the feature map to be fed into a MSFA-Norm and the batch size respectively. Value at pixel $p^{n,c,w,h}$ ($n \in N, c \in C, w \in W, h \in H$) of the feature map will be transformed into $p'^{n,c,w,h}$ after MSFA-Norms, which is expressed as follows:

$$
\begin{aligned}
p'^{n,c,w,h} = att_1^{n,w,h} \cdot \left(\gamma_1^{n,c,w,h} \cdot \frac{p^{n,c,w,h} - \mu^c}{\sigma^c} + \beta_1^{n,c,w,h} \right) + \\
att_2^{n,w,h} \cdot \left(\gamma_2^{n,c,w,h} \cdot \frac{p^{n,c,w,h} - \mu^c}{\sigma^c} + \beta_2^{n,c,w,h} \right)
\end{aligned}
\tag{1}
$$

where μ^c and σ^c are the mean and standard deviation of the values in channel c. They are used for batch normalization and expressed as:

$$
\mu^c = \frac{1}{NHW} \sum_{nhw} p^{n,c,w,h},
\tag{2}
$$

$$
\sigma^c = \sqrt{\frac{1}{NHW} \sum_{nhw} (p^{n,c,w,h})^2 - (\mu^c)^2}.
\tag{3}
$$

att_1 and att_2 refer to attention maps of the multi-scale feature extraction method and the guided sampling method respectively.

Overall, In MSFA-Norm, we use the multi-scale feature extraction results which perform well on semantic classes with large areas. At the same time, the features obtained by guided sampling are added, which has better results on classes with small areas. The two methods are combined through the attention mechanism to achieve good performance on categories in various sizes of areas.

3.3   MUSH Discriminator

Previous works [27,39] use multi-scale PatchGAN discriminator. The image and the semantic layout are concatenated by channel and directly input into the discriminator. But it is hard to distinguish whether information is from images or semantic layouts after the convolutional operation. In addition, the images and the semantic layouts have different scales, which causes calculation deviation after normalization and makes the network lose some semantic information. So in fact, the discriminator ignores much semantic information. To enable the discriminator to separate and extract features of the image and the semantic map while retaining the semantic information in the deep network, we propose a new discriminator.

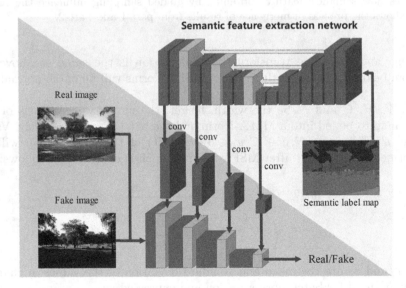

Fig. 6. Structure of MUSH discriminator. The semantic label map and image are input separately into the discriminator. We use a network similar to the semantic feature extraction network in the generator here to produce multi-scale feature maps. The decoded semantic features of each layer are concatenated to the features at all levels of the image, so that input of all convolution layers for images also contains features of the semantic label map.

The discriminator no longer uses the concatenation of image and condition as input like CGAN, but extracts the features of the semantic map separately(see Fig. 6). The extracted features at different levels are concatenated to the feature maps at the corresponding levels of image feature extraction, so that each convolutional layer of the discriminator takes both image features and semantic features as input. Therefore the deep network will not forget or discard the semantic information. Finally, the discriminative result will be more related to the semantic information.

The feature extraction network of semantic map in the discriminator is similar to the network described in 3.1. But it uses fewer layers and convolutional channels.

3.4 Training Scheme and Loss Functions

Similar to the original GAN training, we train the discriminator and generator alternately. For the discriminator, we use hinge loss [19, 43] to train referring to previous works [27]. For the generator, most previous works [21, 27, 38, 49, 50] use adversarial loss, GAN feature matching loss and perceptual loss for training. The loss function we use in generator is similar to the above, but the GAN feature matching loss is removed. GAN feature matching loss sets the difference between features of the generated image and the real image as the loss. However, our generator uses multi-scale hierarchical architecture for semantic feature extraction to generate images with finer details, while the detailed representation of each object is diverse and not unique. Training by feature difference may make the model parameters vibrate among different samples, so that the model can be difficult to converge. The training loss functions for the discriminator L_D and for the generator L_G are as follows,

$$L_D = -\mathbb{E}[min(-1 + D(x), 0)] - \mathbb{E}[min(-1 - D(G(z, m)), 0)], \qquad (4)$$

$$L_G = -\mathbb{E}[D(G(z, m))] + \lambda_p L_p(G(z, m), x) \qquad (5)$$

Where m, x and z refer to the semantic label map, a real image and the input noise of generator respectively. $G(z, m)$ denotes the image synthesized by the generator with noise z and the semantic map as input. $L_p(G(z, m), x)$ denotes the perceptual loss [15], which is used to minimize the VGG19 [32] feature difference between the generated image and the real image. λ_p refers to the weight of perceptual loss.

4   Experiments

4.1   Implementation Details

We apply the spectral norm [24] to layers in both generator and discriminator. The learning rates of generator and discriminator are 0.0001 and 0.0004 respectively. We adopt Adam optimizer [18] with $\beta 1 = 0$ and $\beta 2 = 0.999$. We train our model on V100 GPUs. The weight of perceptual loss λ_p in the loss function is 10.

4.2   Datasets and Metrics

We conduct experiments on COCO-Stuff [2], ADE20K [45], ADE20K-outdoor and Cityscapes [5]. Many previous works [1,4,6,22,26,28,36,37,47] have experimented on these datasets. COCO-Stuff has 182 semantic categories, containing 118,000 training images and 5,000 validation images. Its huge and diverse content makes it very challenging. ADE20K contains 20,210 training images and 2,000 validation images, 150 categories in total. It is a challenging dataset, too. ADE20K-outdoor is a subset of ADE20K. Cityscapes is a dataset of street scene images. Its training set and validation set contain 3,000 and 500 images respectively. All images are in high resolution, which makes it suitable for testing high resolution image synthesis of models. We use the Fréchet Inception Distance (FID) [9,31] to measure the distribution distance between the model generated images and the real images. Additionally, we use mean Intersection-over-Union (mIOU) and pixel accuracy (accu) measured by the state-of-the-art image segmentation networks for each dataset: DeepLabV2 [3,25] for COCO-Stuff, Uper-Net101 [40,46] for ADE20K, and DRN-D-105 [41,42] for Cityscapes.

4.3   Qualitative Results

Fig. 7. Comparison of MUSH with other methods on ADE20K

As shown in Fig. 7, we compare our results with the state-of-the-art approach SPADE and two popular SPADE-based approaches LGGAN and INADE. Ours have finer details, clearer eages and less generation failure (For example, the furniture in the first row has straight edges. The building in the second and the fourth row are less blurred. The wall in the third row is more clear). In addition, the model is able to consider the internal representation differentiation of different locations of complex objects, which makes the objects better generated in all parts.

4.4 Quantitative Results

We use popular semantic image synthesis models in recent years as baselines, including CRN, pix2pixHD, SPADE, GroupDNet, LGGAN, TSIT and INADE. SPADE is the state-of-the-art approach.

Table 1. Quantitative comparison of MUSH with other methods. Bold denotes the best performance. For the mIoU and accu, higher is better. For the FID, lower is better. Results of GroupDNet, LGGAN, TSIT and INADE are collected by running the evaluation on our machine.

	ADE20K			ADE20K-outdoor			Cityscapes			COCO-Stuff		
	mIoU	Accu	FID	mIoU	Accu	FID	mIoU	Accu	FID	mIoU	Accu	FID
CRN	22.4	68.8	73.3	16.5	68.6	99.0	52.4	77.1	104.7	23.7	40.4	70.4
pix2pixHD	20.3	69.2	81.8	17.4	71.6	97.8	58.3	81.4	95.0	14.6	45.8	111.5
SPADE	38.5	79.9	33.9	30.8	82.9	63.3	62.3	81.9	71.8	37.4	67.9	22.6
GroupDNet	28.3	74.7	42.0	n/a	n/a	n/a	62.5	82.2	50.2	n/a	n/a	n/a
LGGAN	38.8	80.6	32.2	n/a	n/a	n/a	64.4	82.9	55.8	n/a	n/a	n/a
TSIT	37.2	80.3	33.4	n/a	n/a	n/a	64.0	83.7	55.5	n/a	n/a	n/a
INADE	37.7	79.9	33.5	n/a	n/a	n/a	61.2	81.9	49.9	n/a	n/a	n/a
OASIS	45.0	83.6	30.7	**36.2**	85.8	55.5	**67.7**	86.3	49.3	**44.5**	**71.4**	**16.8**
SESAME	**45.7**	**84.9**	31.4	n/a	n/a	n/a	65.8	84.2	51.6	n/a	n/a	n/a
Ours	39.0	82.5	**30.3**	33.7	**86.7**	**55.2**	66.3	**87.2**	**48.0**	36.7	69.3	21.6

As shown in Table 1, our method outperforms most baselines expect OASIS and SESAME in almost all metrics on each dataset. It achieves great results especially on small datasets such as Cityscapes and ADE20K-outdoor. However, it does not get better results than OASIS and SESAME on COCO-Stuff or ADE20K. Because we used a relatively lightweight semantic feature extraction network in the experiment so that for big datasets, it does not have sufficient parameters to extract all global features of semantic label maps with so many categories. An increase in its scale can help achieve better results. In spite of these, our model in this scale gets best FID on most datasets except COCO-Stuff and is still competitive with these models. In addition, we use SPADE as the backbone of our method in the experiment. A better backbone will help improve the performance.

4.5 Generalization Ability to SPADE-Based Methods

We also carried out the improvement experiments of our methods on recent semantic image synthesis models with other functions. We choose two representative method: GroupDNet [50] and INADE [33]. GroupDNet replaces convolution

Fig. 8. After being applied to other models, our method can inherit their functions. MUSH has great performance on multi-modal image synthesis (first row, various sofas can be synthesized) by being applied to GroupDNet, and can implement instance-level multi-model image synthesis (second row) by being applied to INADE.

layers in SPADE encoder and generator with group convolution layers, so as to implement semantically multi-modal image synthesis. INADE uses random values from parametric probability distributions in denormalization to implement instance-level multi-model image synthesis. Both of them achieve good performance and implement new functions.

Table 2. Generalization ability test of MUSH on GroupDNet and INADE. Bold denotes the best performance. All results are collected by running the evaluation on our machine.

	ADE20K			Cityscapes		
	mIoU↑	Accu↑	FID↓	mIoU↑	Accu↑	FID↓
GroupDNet	28.3	74.7	42.0	62.5	82.2	50.2
GroupDNet-MUSH	**36.7**	**78.8**	**34.8**	**63.0**	**82.7**	**49.0**
INADE	37.7	79.9	33.5	61.2	81.9	49.9
INADE-MUSH	**38.6**	**81.5**	**30.9**	**64.5**	**85.3**	**49.5**

We apply MUSH's multi-scale hierarchical semantic feature extraction modules to GroupDNet and INADE. Similar to what is described in Sect. 3.2, we use guided sampling to calculate attention values of multi-scale feature extraction methods and their original denormalization methods to obtain the output of normalization layers. The new models not only inherit their functions (see Fig. 8), but also achieve better performance (see Table 2), especially for GroupDNet on ADE20K. The results show that our method has great generalization ability.

4.6   Ablations

We conduct the ablation experiments on ADE20K and Cityscapes. These are ablation configurations on generator architecture and discriminator architecture. The experimental results and our analysis are as follows.

Table 3. Ablation on generator architecture. Bold denotes the best performance.

	ADE20K		Cityscapes	
Generator architecture	mIoU↑	FID↓	mIoU↑	FID↓
MUSH	**39.0**	**30.3**	**66.3**	**48.0**
MUSH w/o guided sampling	37.0	30.9	63.4	49.5
SPADE	37.2	32.5	62.8	58.6

Ablation on the Generator Architecture. We train some alternative generators. The results are shown in Table 3. Compared to SPADE generator, MUSH generator performs better on both mIoU and FID. We also train a generator without guided sampling, which means that the guided sampling part for semantic feature extraction and the attention calculation part are eliminated, so that the MSFA controls image generation only by multi-scale semantic features. We find that the network performs worse, especially on mIoU. This shows that the multi-scale hierarchical architecture fails to extract features of classes with small areas because mIoU calculates the average results of all classes, while most classes have small areas in images of both two datasets, leading to low mIoU of its results on them. This also verifies that the method of guided sampling and the approach we add it here improve the performance.

Table 4. Ablation on discriminator architecture. Bold denotes the best performance.

	ADE20K		Cityscapes	
Discriminator architecture	mIoU↑	FID↓	mIoU↑	FID↓
MUSH	**39.0**	**30.3**	**66.3**	**48.0**
MUSH w/ GAN feature match loss	38.0	31.4	63.2	52.7
MUSH w/o VGG loss	33.2	40.5	59.3	63.4
SPADE	38.5	31.2	64.7	52.1

Ablation on the Discriminator Architecture. As shown in Table 4, compared to the MUSH generator, MUSH discriminator improves less performance but still performs better than SPADE discriminator. The experimental results also show that the GAN feature matching loss in MUSH will degrade the performance. When GAN feature matching loss is used, different samples of images with diversity will make the generator network converge in different directions in training, resulting in difficulty in model convergence and blurred areas in generated images. We also get the results that VGG loss is essential in our network. It is difficult to generate so complex images with adversarial training only, so it is reasonable that we use features from a pretrained model to guide learning.

5   Conclusions

In this paper, we propose MUSH, a semantic image synthesis method that extracts semantic features with a multi-scale hierarchical architecture. The feature extraction network for semantic label maps can calculate an unique value for each pixel, which benefits generation of classes with large areas. We also merge a semantic feature extraction method in which individual parameters are used for each class with it in order to get better results on classes with small areas. Because of these, MUSH generator performs well on various classes. The MUSH discriminator extracts features of the semantic label map and the image separately, which makes it better discriminate between real and fake images according to the semantic label map. MUSH achieves better results than the state-of-the-art approaches and can be generalized to various models to improve their performance. However, for large datasets like COCO-Stuff, it is hard for the proposed network to extract all features of semantic maps. We believe this is a promising research area.

Acknowledgements. This work was supported by the National Key Research and Development Program of China under Grant 2018YFB2100801.

References

1. Almahairi, A., Rajeshwar, S., Sordoni, A., Bachman, P., Courville, A.: Augmented cycleGAN: learning many-to-many mappings from unpaired data. In: International Conference on Machine Learning, pp. 195–204. PMLR (2018)
2. Caesar, H., Uijlings, J., Ferrari, V.: Coco-stuff: thing and stuff classes in context. In: Proceedings of the IEEE Conference on Computer Vision and Pattern Recognition, pp. 1209–1218 (2018)
3. Chen, L.C., Papandreou, G., Kokkinos, I., Murphy, K., Yuille, A.L.: DeepLab: semantic image segmentation with deep convolutional nets, Atrous convolution, and fully connected CRFs. IEEE Trans. Pattern Anal. Mach. Intell. **40**(4), 834–848 (2017)
4. Chen, Q., Koltun, V.: Photographic image synthesis with cascaded refinement networks. In: Proceedings of the IEEE International Conference on Computer Vision, pp. 1511–1520 (2017)
5. Cordts, M., et al.: The cityscapes dataset for semantic urban scene understanding. In: Proceedings of the IEEE Conference on Computer Vision and Pattern Recognition, pp. 3213–3223 (2016)
6. Dundar, A., Sapra, K., Liu, G., Tao, A., Catanzaro, B.: Panoptic-based image synthesis. In: Proceedings of the IEEE/CVF Conference on Computer Vision and Pattern Recognition, pp. 8070–8079 (2020)

7. Goodfellow, I., et al.: Generative adversarial nets. In: Advances in Neural Information Processing Systems, vol. 27 (2014)

8. He, K., Zhang, X., Ren, S., Sun, J.: Deep residual learning for image recognition. In: Proceedings of the IEEE Conference on Computer Vision and Pattern Recognition, pp. 770–778 (2016)

9. Heusel, M., Ramsauer, H., Unterthiner, T., Nessler, B., Hochreiter, S.: GANs trained by a two time-scale update rule converge to a local Nash equilibrium. In: Advances in Neural Information Processing Systems, vol. 30 (2017)

10. Huang, X., Belongie, S.: Arbitrary style transfer in real-time with adaptive instance normalization. In: Proceedings of the IEEE International Conference on Computer Vision, pp. 1501–1510 (2017)

11. Huang, X., Liu, M.-Y., Belongie, S., Kautz, J.: Multimodal unsupervised image-to-image translation. In: Ferrari, V., Hebert, M., Sminchisescu, C., Weiss, Y. (eds.) ECCV 2018. LNCS, vol. 11207, pp. 179–196. Springer, Cham (2018). https://doi.org/10.1007/978-3-030-01219-9_11

12. Ioffe, S., Szegedy, C.: Batch normalization: accelerating deep network training by reducing internal covariate shift. In: International Conference on Machine Learning, pp. 448–456. PMLR (2015)

13. Isola, P., Zhu, J.Y., Zhou, T., Efros, A.A.: Image-to-image translation with conditional adversarial networks. In: Proceedings of the IEEE Conference on Computer Vision and Pattern Recognition, pp. 1125–1134 (2017)

14. Jiang, L., Zhang, C., Huang, M., Liu, C., Shi, J., Loy, C.C.: TSIT: a simple and versatile framework for image-to-image translation. In: Vedaldi, A., Bischof, H., Brox, T., Frahm, J.-M. (eds.) ECCV 2020. LNCS, vol. 12348, pp. 206–222. Springer, Cham (2020). https://doi.org/10.1007/978-3-030-58580-8_13

15. Johnson, J., Alahi, A., Fei-Fei, L.: Perceptual losses for real-time style transfer and super-resolution. In: Leibe, B., Matas, J., Sebe, N., Welling, M. (eds.) ECCV 2016. LNCS, vol. 9906, pp. 694–711. Springer, Cham (2016). https://doi.org/10.1007/978-3-319-46475-6_43

16. Karacan, L., Akata, Z., Erdem, A., Erdem, E.: Learning to generate images of outdoor scenes from attributes and semantic layouts. arXiv preprint arXiv:1612.00215 (2016)

17. Karacan, L., Akata, Z., Erdem, A., Erdem, E.: Manipulating attributes of natural scenes via hallucination. ACM Trans. Graph. (TOG) 39(1), 1–17 (2019)

18. Kingma, D.P., Ba, J.: Adam: a method for stochastic optimization. In: ICLR (Poster) (2015)

19. Lim, J.H., Ye, J.C.: Geometric GAN. arXiv preprint arXiv:1705.02894 (2017)

20. Liu, M.Y., Breuel, T., Kautz, J.: Unsupervised image-to-image translation networks. In: Advances in Neural Information Processing Systems, vol. 30 (2017)

21. Liu, X., Yin, G., Shao, J., Wang, X., et al.: Learning to predict layout-to-image conditional convolutions for semantic image synthesis. In: Advances in Neural Information Processing Systems, vol. 32 (2019)

22. Long, J., Lu, H.: Generative adversarial networks with bi-directional normalization for semantic image synthesis. In: Proceedings of the 2021 International Conference on Multimedia Retrieval, pp. 219–226 (2021)

23. Mirza, M., Osindero, S.: Conditional generative adversarial nets. arXiv preprint arXiv:1411.1784 (2014)

24. Miyato, T., Kataoka, T., Koyama, M., Yoshida, Y.: Spectral normalization for generative adversarial networks. In: International Conference on Learning Representations (2018)

25. Nakashima, K.: Deeplab-pytorch. https://github.com/kazuto1011/deeplab-pytorch (2018)
26. Ntavelis, E., Romero, A., Kastanis, I., Van Gool, L., Timofte, R.: SESAME: semantic editing of scenes by adding, manipulating or erasing objects. In: Vedaldi, A., Bischof, H., Brox, T., Frahm, J.-M. (eds.) ECCV 2020. LNCS, vol. 12367, pp. 394–411. Springer, Cham (2020). https://doi.org/10.1007/978-3-030-58542-6_24
27. Park, T., Liu, M.Y., Wang, T.C., Zhu, J.Y.: Semantic image synthesis with spatially-adaptive normalization. In: Proceedings of the IEEE/CVF Conference on Computer Vision and Pattern Recognition, pp. 2337–2346 (2019)
28. Qi, X., Chen, Q., Jia, J., Koltun, V.: Semi-parametric image synthesis. In: Proceedings of the IEEE Conference on Computer Vision and Pattern Recognition, pp. 8808–8816 (2018)
29. Ronneberger, O., Fischer, P., Brox, T.: U-net: convolutional networks for biomedical image segmentation. In: Navab, N., Hornegger, J., Wells, W.M., Frangi, A.F. (eds.) MICCAI 2015. LNCS, vol. 9351, pp. 234–241. Springer, Cham (2015). https://doi.org/10.1007/978-3-319-24574-4_28
30. Schonfeld, E., Sushko, V., Zhang, D., Gall, J., Schiele, B., Khoreva, A.: You only need adversarial supervision for semantic image synthesis. In: International Conference on Learning Representations (2020)
31. Seitzer, M.: Pytorch-fid: FID score for PyTorch, August 2020. https://github.com/mseitzer/pytorch-fid. version 0.2.1
32. Simonyan, K., Zisserman, A.: Very deep convolutional networks for large-scale image recognition. arXiv preprint arXiv:1409.1556 (2014)
33. Tan, Z., et al.: Diverse semantic image synthesis via probability distribution modeling. In: Proceedings of the IEEE/CVF Conference on Computer Vision and Pattern Recognition, pp. 7962–7971 (2021)
34. Tan, Z., et al.: Efficient semantic image synthesis via class-adaptive normalization. IEEE Trans. Pattern Anal. Mach. Intell. **44**, 4852–4866 (2021)
35. Tan, Z., et al.: Rethinking spatially-adaptive normalization. arXiv preprint arXiv:2004.02867 (2020)
36. Tang, H., Bai, S., Sebe, N.: Dual attention GANs for semantic image synthesis. In: Proceedings of the 28th ACM International Conference on Multimedia, pp. 1994–2002 (2020)
37. Tang, H., Qi, X., Xu, D., Torr, P.H., Sebe, N.: Edge guided GANs with semantic preserving for semantic image synthesis. arXiv preprint arXiv:2003.13898 (2020)
38. Tang, H., Xu, D., Yan, Y., Torr, P.H., Sebe, N.: Local class-specific and global image-level generative adversarial networks for semantic-guided scene generation. In: Proceedings of the IEEE/CVF Conference on Computer Vision and Pattern Recognition, pp. 7870–7879 (2020)
39. Wang, T.C., Liu, M.Y., Zhu, J.Y., Tao, A., Kautz, J., Catanzaro, B.: High-resolution image synthesis and semantic manipulation with conditional GANs. In: Proceedings of the IEEE Conference on Computer Vision and Pattern Recognition, pp. 8798–8807 (2018)
40. Xiao, Tete, Liu, Yingcheng, Zhou, Bolei, Jiang, Yuning, Sun, Jian: Unified perceptual parsing for scene understanding. In: Ferrari, Vittorio, Hebert, Martial, Sminchisescu, Cristian, Weiss, Yair (eds.) ECCV 2018. LNCS, vol. 11209, pp. 432–448. Springer, Cham (2018). https://doi.org/10.1007/978-3-030-01228-1_26
41. Yu, F., Koltun, V.: Multi-scale context aggregation by dilated convolutions. In: International Conference on Learning Representations (ICLR) (2016)
42. Yu, F., Koltun, V., Funkhouser, T.: Dilated residual networks. In: Computer Vision and Pattern Recognition (CVPR) (2017)

43. Zhang, H., Goodfellow, I., Metaxas, D., Odena, A.: Self-attention generative adversarial networks. In: International Conference on Machine Learning, pp. 7354–7363. PMLR (2019)
44. Zhao, B., Meng, L., Yin, W., Sigal, L.: Image generation from layout. In: Proceedings of the IEEE/CVF Conference on Computer Vision and Pattern Recognition, pp. 8584–8593 (2019)
45. Zhou, B., Zhao, H., Puig, X., Fidler, S., Barriuso, A., Torralba, A.: Scene parsing through ade20k dataset. In: Proceedings of the IEEE Conference on Computer Vision and Pattern Recognition, pp. 633–641 (2017)
46. Zhou, B., et al.: Semantic understanding of scenes through the ade20k dataset. Int. J. Comput. Vision. **127**, 302–321 (2018)
47. Zhu, J.Y., Park, T., Isola, P., Efros, A.A.: Unpaired image-to-image translation using cycle-consistent adversarial networks. In: Proceedings of the IEEE International Conference on Computer Vision, pp. 2223–2232 (2017)
48. Zhu, J.Y., Zhang, R., Pathak, D., Darrell, T., Efros, A.A., Wang, O., Shechtman, E.: Toward multimodal image-to-image translation. In: Advances in Neural Information Processing Systems, vol. 30 (2017)
49. Zhu, P., Abdal, R., Qin, Y., Wonka, P.: Sean: image synthesis with semantic region-adaptive normalization. In: Proceedings of the IEEE/CVF Conference on Computer Vision and Pattern Recognition, pp. 5104–5113 (2020)
50. Zhu, Z., Xu, Z., You, A., Bai, X.: Semantically multi-modal image synthesis. In: Proceedings of the IEEE/CVF Conference on Computer Vision and Pattern Recognition, pp. 5467–5476 (2020)

SAC-GAN: Face Image Inpainting with Spatial-Aware Attribute Controllable GAN

Dongmin Cha[iD], Taehun Kim[✉], Joonyeong Lee[✉], and Dajin Kim[✉]

Department of CSE, Pohang University of Science and Technology (POSTECH),
Pohang, Korea
{cardongmin,taehoon1018,joonyeonglee,dkim}@postech.ac.kr
https://github.com/easternCar/spatial_attribute_control_inpaint.git

Abstract. The objective of image inpainting is refilling the masked area with semantically appropriate pixels and producing visually realistic images as an output. After the introduction of generative adversarial networks (GAN), many inpainting approaches are showing promising development. Several attempts have been recently made to control reconstructed output with the desired attribute on face images using exemplar images and style vectors. Nevertheless, conventional style vector has the limitation that to project style attribute representation onto linear vector without preserving dimensional information. We introduce spatial-aware attribute controllable GAN (SAC-GAN) for face image inpainting, which is effective for reconstructing masked images with desired controllable facial attributes with advantage of utilizing style tensors as spatial forms. Various experiments to control over facial characteristics demonstrate the superiority of our method compared with previous image inpainting methods.

1 Introduction

Image inpainting is a task about image generation which has long been studied and dealt with in computer vision. Given masked images with missing regions, the main objective of image inpainting is understanding the masked images through neural networks and refilling the hole pixels with appropriate contents to produce the final reconstructed output. Image inpainting has been mainly applied to restore damaged images or refill appropriate content after removing some specific object in a photo (Fig. 1).

Although conventional inpainting works [1,2] have demonstrated considerable reconstruction ability, image inpainting has achieved notable development since the introduction of generative adversarial networks (GAN) [3], which facilitate image synthesis ensuring plausible quality. Inpainting approaches with GAN-based methods [4–6] were able to show more visually realistic results than

Supplementary Information The online version contains supplementary material available at https://doi.org/10.1007/978-3-031-26293-7_13.

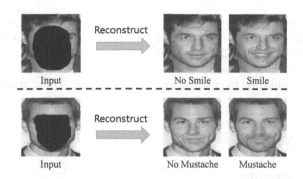

Fig. 1. Examples of face image inpainting with attribute manipulation ('smile', 'mustache') using our proposed model. We can reconstruct intended image by giving a condition, which is considered a domain of attribute.

traditional approaches due to competitive training driven by adversarial networks. Recently, many efforts are focusing on user-controllable image inpainting by refilling masked region with desired contents beyond the limitation of traditional deterministic image inpainting. For example, providing an exemplar image for the masked region [7], giving a guideline based on edge image [8,9], or facial landmarks [10].

After StarGAN [11] demonstrated GAN-based image translation beyond multiple domains, StyleGAN [12] showed the potential of image generation with user preference by mapping styles into features. Based on these translation methods, COMOD-GAN [13] adopted StyleGAN based style codes for controllable image inpainting with conditional input. Additionally, [14] made a trial to reconstruct images with style codes using AdaIN [12,15] or weight modulation [16].

Nevertheless, StyleMapGAN [17] pointed out that existing style-aware image generation or reconstruction projects style attributes into a linear vector with ignorance of spatial information for facial attributes. This means that facial attributes have information regarding shapes and dimensions, and projecting those style attributes onto linear vectors without preserving spatial information may reduce image generation quality. They proposed handling style codes as spatial tensors, called style maps.

Motivated by these previous works, we further propose spatial-aware attribute controllable GAN for image inpainting (SAC-GAN). Given input masked image, the target attribute is converted to spatial style map through convolutional mapping network $\mathcal{M}$. Then, we adopted a cross attention module to style maps for enhancement of contextual consistency in feature space to achieve long-range dependency between image feature and spatial style map. Finally, obtained style maps go through upsampling networks to produce multi-scale style maps, which are modulated to each layer of the decoder to reconstruct masked areas with proper contents with target attributes. To confirm the advantages of the proposed model, we conducted comparative experiments with other approaches and ablation studies with and without various loss conditions. The main contributions of this study are summarized as follows: (1) Attribute-controllable inpainting model with user-guided condition input and high-quality image reconstruction based on GAN.

(2) Convolutional mapper network based on modulation which preserves dimensional information for spatial style maps with cross attention module for global consistency between feature from masked image and spatial style maps. (3) Multiscale spatial style maps obtained using an upsampling network are applied to the decoder using modulation to ensure a higher quality than conventional linear style vectors.

2   Related Works

2.1   Image Inpainting

Conventional image inpainting is categorized into diffusion-based and patch-based approaches. Patch-based approaches [1] usually relied on dividing image as small patches and refilling occluded areas with outer patches by computing scores, such as cosine similarity.

However, since deep-learning became a trend in computer vision, the capability of generative adversarial networks (GAN) [3] has shown remarkable performance in image generation. Many recent researches about image inpainting have been based on GAN and showed high quality in the reconstructed image [4–6,18]. Some approaches attempted to imitate the traditional approach of patch-based inpainting with the GAN-based model [19]. Other methods [20,21] tried to utilize binary masks to emphasize pixels to be reconstructed.

Because of the deterministic property of conventional image inpainting approaches, several researches focus on diverse image restoration to produce multiple possible predictions for damaged images. For example, considering distributions [22,23], or adopting visual transformers for the prediction of prior possibilities [24]. COMOD-GAN [13] showed visually excellent inpainting outputs by using stochastic learning and style codes [12] for large scale image restoration. It showed the possibility of applying modulation in image inpainting tasks. Inspired by attention-based [25] works like Self-Attention GAN [26], UCT-GAN [7] applied feature-cross attention map for image inpainting from two features from different images to ensure high consistency between pixels from the reconstructed image by combining two features semantically.

2.2   Facial Attribute Manipulation

StarGAN [11] demonstrated face image translation for multiple domains. The improved StarGAN v2 [27] showed the possibility of generating diverse results within one domain using the additional mapping network and feature-level style encoder. Styles for image generation are usually utilized for editing facial attributes or mixing several facial images with high quality. Unlike pixel-wise

style translation [28, 29], AdaIN [15] presented style editing with the generative model using instance normalization and showed remarkable possibility in
manipulating a specific style. PA-GAN [30] focused on feature disentanglement
using pairs of generators and discriminators for progressive image generation
with attributes. StyleGAN [12] showed how a latent-space aware network facilitates multi-domain style transfer with AdaIN. StyleGAN v2 [16] proposed weight
modulation which demonstrates better image generation quality than AdaIN.

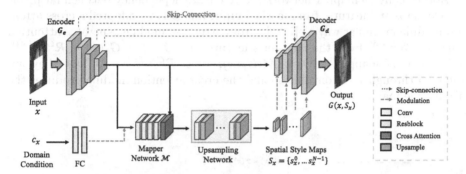

Fig. 2. Overall architecture of SAC-GAN consisting of the encoder G_e and the decoder
G_d as a generative network. The mapper $\mathcal{M}$ extracts spatial style maps [17] from
feature from the encoder and mapper. The style maps S_x are applied to decoder's
layer by weight modulation [16].

3   Proposed Methods

As shown in Fig. 2, overall network is based on an encoder-decoder based generative architecture with weight modulation [16] to handle facial attribute features
as style codes. However, beyond previous approaches dealing with styles [12, 13],
our model handles style codes as spatial tensors form containing spatial dimension information like StyleMapGAN [17].

3.1   Facial Attributes as Domains

Similar to L2M-GAN [31], we define a set representing a particular facial
attribute as a domain, which can be considered as male/female, wearing glasses,
mustache, or any other possible attributes. Then, we can consider a style code
S_k for a specific domain $c_k \rightarrow \mathcal{K}$. In implementation, we consider domain $\mathcal{K}$
as a condition value c_k. Then synthesized output from input image x will be
described as $G(x, S_k)$, where G denotes the encoder-decoder structure generator $G = \{G_e, G_d\}$. The proposed network for extracting a style maps S_k from
condition input c_k and its detailed process are explained in the following sections.

3.2   Convolutional Mapper Network

Because of projection style information to linear vectors, some significant dimensional information could be lost during the modulation process. StyleMapGAN [17] suggests handling style code as a spatial form called style maps. With this motivation, we propose a conditional mapper based on co-modulation [13]. Instead of linear vector and fully-connected layers, our mapper network $\mathcal{M}$ produces a spatial style map s_k as output.

Additionally, in mapper network $\mathcal{M}$, to ensure dependency between far pixels in style maps and features from masked input image x, we adopted a cross attention module in similar way as [7,26]. Cross attention module generates attention map $A \in \mathcal{R}^{H \times W}$ from the encoder's feature map $\mathcal{F}(x) = G_e(x) \in \mathcal{R}^{C \times H \times W}$ and output of mapper network $\mathcal{M}(G_e(x), c_k) \in \mathcal{R}^{C \times H \times W}$. At first they are reshaped through 1×1 convolution and the cross attention module produces the initial style map s_k as:

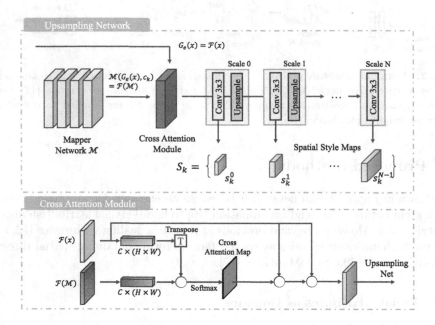

Fig. 3. Illustrations of conditional mapper network $\mathcal{M}$, upsampling layers, and cross attention module. The initial style map is generated from $\mathcal{M}$ and cross attention module gradually upsampled to produce pyramid-like spatial style maps [17].

$$A = \sigma(tr(\mathcal{F}(x)) \cdot \mathcal{F}(\mathcal{M})), \qquad s_k = (A \cdot \mathcal{F}(x)) \oplus \mathcal{F}(x) \qquad (1)$$

Then, the refined style map s_k is sent to the upsampling network to produce style maps with various scales. Where input masked image x is given, style maps for a specific attribute k are denoted as:

$$S_k = \{s_k^0, \dots, s_k^{N-1}\} = \mathcal{M}(G_e(x), c_k) \qquad (2)$$

where c_k is condition value for attribute domain $\mathcal{K}$ and $G_e(x)$ is the output feature from encoder G_e from input masked image x. The pyramid-like set S_k includes style maps with various scales. As shown in Fig. 3, the condition value c_k is converted to a one-hot vector and concatenated to the feature $G_e(x)$. The condition value is converted to a one-hot vector and passed through the mapping network. The mapped vector is reshaped to the same size as the feature $G_e(x)$ and the concatenation of the vector and $G_e(x)$ is passed through spatial mapping layers.

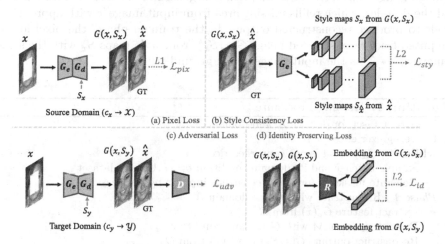

Fig. 4. Overview of our four loss functions for training our proposed models. First pixel loss ($\mathcal{L}_{pix}$) and style consistency loss ($\mathcal{L}_{sty}$) are computed from $G(x, S_x)$ (source domain $\mathcal{X}$). Then adversarial loss ($\mathcal{L}_{adv}$) and identity preserving loss ($\mathcal{L}_{id}$) are computed from $G(x, S_y)$ (target domain $\mathcal{Y}$).

3.3 Generator with Co-modulation

The encoder G_e takes images as input to represent the image as a feature map. We denoted masks pixels by concatenating binary masks to the input image. The encoder output feature $G_e(x)$ obtained from input image x and mask is reconstructed in the decoder G_d containing skip-connection [32] and style modulation [16] from the set of style codes S_i. The modulated convolutional from each decoder layer applies the intended style (S_i) to the feature map during the reconstruction process (Fig. 4).

3.4 Training Objectives

Unlike previous style-aware image translation methods that use complete images as inputs, we have to perform style modulation and image reconstruction simultaneously from masked input images that lack visual information. To achieve these objectives, we reconstruct input image x with style maps S_x from domain attribute $c_x \to \mathcal{X}$ which is in the same domain as input image $x \in \mathcal{X}$. After optimizing pixel reconstruction loss and style consistency loss in source-to-source inpainting, then we optimize adversarial loss and face identity loss in source-to-target inpainting. The training scheme is operated by four main loss functions including pixel-wise loss ($\mathcal{L}_{pix}$), adversarial loss ($\mathcal{L}_{adv}$), style consistency loss ($\mathcal{L}_{sty}$), and identity preserving loss ($\mathcal{L}_{id}$).

Pixel-Wise Reconstruction Loss. Basically, a generative model for reconstructing images requires a pixel-wise loss to refill approximate content in occluded regions. Because L2-norm loss has a drawback of making the reconstructed image blurry, L1 loss is adopted. Our generator G consists of the encoder and the decoder, which refills missing area from input image x with appropriate pixels to produce reconstructed output. In the training phase, the pixel loss is computed from reconstructed image $G(x, S_x)$ from style maps S_x with the same domain $c_x \to \mathcal{X}$ as the input image x and ground-truth $\hat{x}$.

Algorithm 1. Training Procedure

1: Prepare dataset for domain $\mathcal{X}$ and domain $\mathcal{Y}$
2: Fix recognizer encoder R
3: **while** $G_e, G_d, D, \mathcal{M}$ is not converged **do**
4:     Sample batch $\hat{x}$ and its domain condition $c_x \to \mathcal{X}$ from dataset
5:     Generate masks m for $\hat{x}$ and construct input x by $x \leftarrow \hat{x} \odot M$
      Phase 1 : Reconstruct with source domain $\mathcal{X}$
6:     Extract feature $G_e(x)$ from G_e
7:     Extract S_x from $\mathcal{M}$ with $G_e(x)$ and condition c_x
8:     Reconstruct outputs $G(x, S_x) \leftarrow G_e(x)$ from G_d
9:     Compute losses $\mathcal{L}_{re}$ and $\mathcal{L}_{sty}$
10:     Update G_e, G_d and $\mathcal{M}$ with $\mathcal{L}_{re}$ and $\mathcal{L}_{sty}$
      Phase 2 : Reconstruct with target domain $\mathcal{Y}$
11:     Pick target domain condition $c_y \to \mathcal{Y}$
12:     Extract S_y from $\mathcal{M}$ with $G_e(x)$ and condition c_y
13:     Reconstruct outputs $G(x, S_y) \leftarrow G_e(x)$ from G_d
14:     Compute losses $\mathcal{L}_{adv}$ and $\mathcal{L}_{id}$
15:     Update G_e, G_d and $\mathcal{M}$ with $\mathcal{L}_{id}$ and $\mathcal{L}_{adv}$
16:     Update D with $\mathcal{L}_{adv}$
17: **end while**

$$\mathcal{L}_{pix} = |G(x, S_x) - \hat{x}|_1 \qquad (3)$$

where S_x denotes style maps obtained from mapper with source domain condition c_x, $G(x, S_x)$ denotes the reconstructed output from x using S_x with the decoder's weight modulation.

Adversarial Loss. The synthesized image should be realistic enough to be comparable to the original image. To achieve realistic reconstruction output beyond multiple attributes, we employed a multi-domain discriminator [31,33] based on Wasserstein GAN [34]. We also applied the R1-regularization gradient penalty [35] to the adversarial loss for the discriminator's stable training with high convergence. The adversarial loss is defined as follows:

$$\mathcal{L}_{adv} = E_{\hat{x}}[-\log D(\hat{x}, c_x)] + E_{\hat{x}, c_y}[\log(1 - D(G(x, S_y), c_y))] \\ + E_{\hat{x}}[\triangledown D(\hat{x}, c_x)],$$

where S_y indicates style maps from target domain $c_y \rightarrow \mathcal{Y}$, $D(G(x, S_y), c_y)$ is the output of the discriminator for the fake image with target domain c_y; and $D(\hat{x}, c_x)$ denotes the real image description with source domain c_x, which is the same as the domain of ground-truth image $\hat{x}$.

Style Consistency Loss. We define style consistency loss [29,36] in source domain c_x to guarantee that the encoder G_e and mapper $\mathcal{M}$ extract identical style maps from reconstructed image $G(x, S_x)$ and ground-truth $\hat{x}$.

$$\mathcal{L}_{sty} = \sum_{i}^{N-1} (|s_{\hat{x}}^i - s_x^i|_2) \qquad (4)$$

where $S_x = \{s_x^0, s_x^1, ..., s_x^{N-1}\}$ are N style maps obtained from the encoded feature of the reconstructed image $G(x, S_x)$ and mapper $\mathcal{M}$, $S_{\hat{x}} = \{s_{\hat{x}}^0, s_{\hat{x}}^1, ..., s_{\hat{x}}^{N-1}\}$ denotes style maps from the feature of ground-truth image $\hat{x}$ and mapper $\mathcal{M}$ so that the style consistency loss can be computed from the summation of L-2 distances between S_x and $S_{\hat{x}}$.

Identity Preserving Loss. After computing the above three losses, we have reconstructed images with source attribute domain $\mathcal{X}$ and target attribute domain $\mathcal{Y}$. From those images we introduced identity preserving loss, which guarantees that reconstructed outputs from an image x preserve the same identity. Similar to [37], we adopted the face identity recognizer network ψ, which is the pre-trained ArcFace [38] with CASIA-WebFace [39]. The identity loss for our model is defined as:

$$\mathcal{L}_{id} = |\psi(G(x, S_x)) - \psi(G(x, S_y))|_2 \qquad (5)$$

Full Objective. Finally, the total objective for training our proposed SAC-GAN can be described as a combination of aforementioned losses:

$$\mathcal{L}_{total} = \lambda_{pix} \cdot \mathcal{L}_{pix} + \lambda_{adv} \cdot \mathcal{L}_{adv} + \lambda_{sty} \cdot \mathcal{L}_{sty} + \lambda_{id} \cdot \mathcal{L}_{id} \qquad (6)$$

where $\{\lambda_{pix}, \lambda_{adv}, \lambda_{sty}$ and $\lambda_{id}\}$ denote hyper-parameters for controlling the importance of each component. Experimentally, we conducted training and test with the hyper-parameters conditions of $\lambda_{pix} = 100$, $\lambda_{adv} = 1$, $\lambda_{sty} = 1$ and $\lambda_{id} = 0.1$. Detailed training strategy is shown in Algorithm 1.

4 Experiments

4.1 Implementation and Datasets

Datasets. Aligned face datasets have the possibility that they basically have spatial information because their significant facial components are fixed. In order to confirm that our spatial-aware method has an effect even in the unaligned face images, we prepared wild CelebA dataset [41] using only face detection based on MTCNN [42]. We cropped the face images with 25 margins and 128×128 size. CelebA contains more than 200,000 face images including various facial attributes, which are mainly used for quantitative evaluation for restoring facial images in image inpainting. For quantitative experiments, we mainly used 'smiling' attribute class data because the amount of data in each class is well balanced compared to other attributes.

Table 1. Quantitative comparison on CelebA with inpainting task for 'non-smile → non-smile' and 'smile → smile' with fixed and free form masks. The highest performances are marked in bold.

	Fixed mask			Free mask		
	FID↓	PSNR↑	SSIM↑	FID↓	PSNR↑	SSIM↑
CE [4]	13.83	19.46	0.641	11.86	20.09	0.759
CA [19]	10.35	21.87	0.694	9.13	21.87	0.803
Partial [20]	7.98	21.02	0.710	6.05	22.54	0.786
Gated [21]	6.43	22.65	0.748	4.23	23.09	0.801
Ours	**4.55**	**23.39**	**0.769**	**3.18**	**24.68**	**0.814**

Table 2. Quantitative comparison on CelebA with attribute manipulation inpainting for 'non-smile → smile' with fixed and free form masks. The highest performances are marked in bold.

	Fixed mask				Free mask			
	FID↓	PSNR↑	SSIM↑	Acc↑	FID↓	PSNR↑	SSIM↑	Acc↑
C-CE [4,40]	21.49	19.28	0.621	88.54%	13.49	0.784	21.92	91.15%
COMOD [13]	10.28	20.65	0.665	92.60%	8.18	0.811	22.57	93.28%
Ours	**10.04**	**21.30**	**0.701**	**93.82%**	**7.70**	**0.823**	**22.84**	**95.28%**

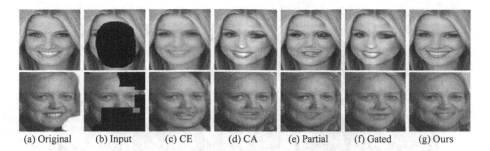

(a) Original   (b) Input   (c) CE   (d) CA   (e) Partial   (f) Gated   (g) Ours

Fig. 5. Comparison of inpainting result with CelebA dataset. (a) Ground-truth. (b) Input. (c) context encoders (CE) [4]. (d) contextual attention (CA) [19]. (e) partial convolution [20]. (f) gated convolution [21]. (g) Ours.

Implementation Details.   Our network is implemented in PyTorch and trained for 100,000 iterations using Adam optimizer [43] with a batch size of 20. Training and experiments are conducted on the NVIDIA TITAN RTX GPU. For occlusion masks, we adopted two strategies: fixed masks and irregular masks. The fixed masks cover all main facial components, including the eyes, nose, and mouth. The irregular masks are same used in gated convolution (free-form) [21], which creates random brushes, circles, and rectangles.

In order to compare the quantitative performance with other methods, we employed the facial attribute classifier based on ResNet-50 [44] to compare the visual certainty of specific attributes from the reconstructed image and FID [45] to measure the quality of generated images using Inception-V3 [46] network. We exploit several facial attributes for qualitative experiments, including 'smile', 'gender', 'glasses' and 'mustache' which are visually evident, and location information was expected to be important because those attributes tend to appear in certain areas of the face. For each attribute, we conducted training and test as two-class domains and used 90% of images for training and 10% for evaluation.

For comparative experiments with previous other inpainting models, we conducted two types of comparisons. 1) Comparing our model's inpainting performance with previous image inpainting models like context encoders (CE) [4], contextual attention (CA) [19], partial convolution [20], and gated convolution [21]. To compare our SAC-GAN with these models, we reconstructed the masked image by giving the same domain condition value as ground-truth. 2) Comparing with other models with conditional-based inpainting. In this experiment, we consider the condition as attribute domains such as wearing glasses, smiling and gender. Besides COMOD-GAN [13], we combined CGAN [40] and CE [4] as a baseline inpainting model for facial attribute manipulation without modulation. We denoted conditional CE as 'C-CE' which takes additional input condition value and reshapes it to tensor in a similar way to CGAN [40].

Smile ↔ Non-smile

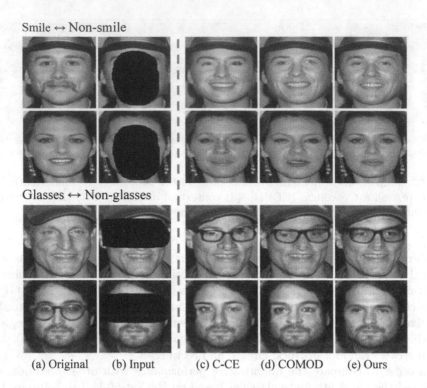

Glasses ↔ Non-glasses

    (a) Original   (b) Input   (c) C-CE   (d) COMOD   (e) Ours

Fig. 6. Examples of attribute controllable face inpainting with given domain attribute condition. From left to right are: (a) Ground-truth. (b) Input masked image. (c) Conditional CE [4,40]. (d) COMOD-GAN [13]. (e) Ours.

4.2   Quantitative Comparisons

As mentioned above, we evaluated the inpainting task of our proposed model by providing the same condition attribute with the ground-truth of the input masked image. For example, we split test set into 'smile' and 'non-smile' groups and evaluated the average accuracy in 'smile → smile' and 'non-smile → non-smile' reconstruction. Table 1 presents the quantitative comparative results for the inpainting task with various generative models.

Next, we conducted an inpainting test with facial attribute manipulation, considering an attribute as a domain. Because facial attribute translation is a hard task to evaluate using only with visual metrics, we deployed the ResNet-50 based facial attribute classifier to verify numerically that the attributes we expected were applied well during the reconstruction process. Figure 7 shows the two-class classification for specific attributes, including 'smile' and 'eye-glasses'. Our model produced higher performance and accuracy than other methods. Results of quantitative comparison in attribute manipulating experiments are shown in Table 2. In this table, we reported the results of test for 'non-smile → smile' inpainting task. Acc denotes accuracy of attribute classification about two classes: smile and non-smile.

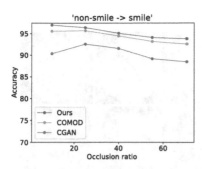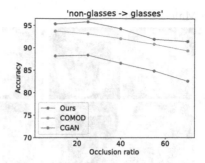

Fig. 7. Accuracy of facial attribute classification about smiling(left) and eye-glasses(right) from reconstructed images using various facial manipulation inpainting methods. Input images are occluded by irregular masks and the x-axis denotes proportion of occluded pixels from mask.

Table 3. Quantitative comparison for the ablation study with various losses on CelebA with fixed mask and facial attribute 'non-smile → smile'.

$\mathcal{L}_{pix}$	$\mathcal{L}_{adv}$	$\mathcal{L}_{sty}$	$\mathcal{L}_{id}$	FID↓	PSNR↑	SSIM↑	Acc↑
✓	✓			14.27	20.68	0.681	90.29%
✓	✓	✓		10.81	20.48	0.663	92.10%
✓	✓	✓	✓	10.28	20.86	0.685	93.01%
All losses + Cross Attention				**10.04**	**21.30**	**0.701**	**93.82%**

4.3 Qualitative Comparisons

We also conducted qualitative comparisons with the same condition as experiments for quantitative results. Reconstructed images from inpainting task are shown in Fig. 5. Although other methods produce slightly distorted outputs, applying weight modulation synthesized output images with better quality visually.

In Fig. 6, we present results of image inpainting with face attribute manipulation. From top to bottom: 'non-smile → smile', 'smile → non-smile', 'non-glasses → glasses', and 'glasses → non-glasses'. Our model synthesized visually natural output by filling masked areas with intended condition. Furthermore, our SAC-GAN generated higher quality output with spatial style maps compared to previous linear-based style modulation like COMOD-GAN [13]. Additionally, we presented another results with 'mustache' and 'gender' attributes in Fig. 8.

Male ↔ Female

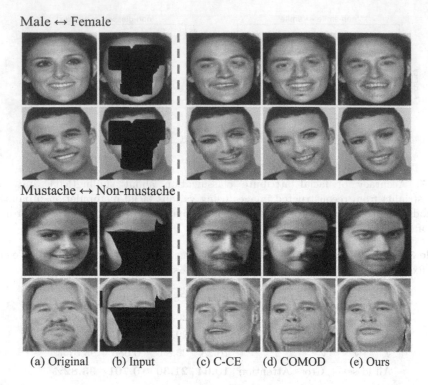

Mustache ↔ Non-mustache

(a) Original   (b) Input   (c) C-CE   (d) COMOD   (e) Ours

Fig. 8. Examples of attribute controllable face inpainting with given domain attribute condition. From left to right are: (a) Ground-truth. (b) Input masked image. (c) Conditional CE [4,40]. (d) COMOD-GAN [13]. (e) Ours.

4.4 Ablation Study

We trained our model on the auxiliary losses or module to check the effect of our loss terms. We conducted our ablation study with fixed mask and various loss conditions in 'non-smile → smile' reconstruction task. As shown in Table 3. it demonstrates preserving identity loss term produce more stable facial attribute-aware inpainting by maintaining overall identity information beyond domain conditions. The result in the bottom achieves better output, showing the benefits of the cross attention module. The visual examples are shown in Fig. 9. We can check that our cross attention improves the quality of synthesized images in visual metrics and attribute accuracy. Since style consistency loss is excluded in model (c), the 'smile' attribute was not well applied to the reconstructed output. In (d), although the expected attribute was well applied to output, it showed a limitation that the identity is not maintained.

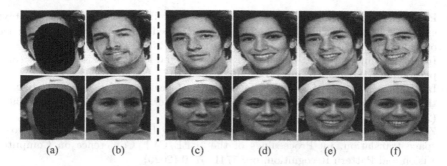

(a)   (b)   (c)   (d)   (e)   (f)

Fig. 9. Ablation study results for our model on the various losses and cross attention in inpainting task of 'non-smile → smile'. (a) Input image. (b) Ground-truth. (c) Using $\mathcal{L}_{pix} + \mathcal{L}_{adv}$. (d) Using $\mathcal{L}_{pix} + \mathcal{L}_{adv} + \mathcal{L}_{sty}$. (e) Using $\mathcal{L}_{pix} + \mathcal{L}_{adv} + \mathcal{L}_{sty} + \mathcal{L}_{id}$ and (f) Using all losses and cross attention module.

5   Conclusions

We presented SAC-GAN: Spatial-aware attribute controllable GAN for image inpainting in this paper. Our network is able to restore masked image with appropriate contents and intended attribute by using style tensors preserving spatial dimension instead of conventional linear style vectors. Through extensive experiments, we demonstrated that our proposed SAC-GAN generating visually remarkable outputs using spatial style maps. Additionally, our proposed cross attention module achieved the advantage of long-range dependency between feature from image and style map, which enhanced the performance of style-aware image inpainting. Moving forward, we expect our proposed model to reconstruct images with high quality using more complex and extensive facial attributes.

Acknowledgements. This work was supported by Institute of Information & communications Technology Planning & Evaluation(IITP) grant funded by the Korea government(MSIT) (No. 2017-0-00897, Development of Object Detection and Recognition for Intelligent Vehicles) and (No. B0101-15-0266, Development of High Performance Visual BigData Discovery Platform for Large-Scale Realtime Data Analysis)

References

1. Barnes, C., Shechtman, E., Finkelstein, A., Goldman, D.B.: PatchMatch: a randomized correspondence algorithm for structural image editing. ACM Trans. Graph. **28**, 24 (2009)
2. Wilczkowiak, M., Brostow, G.J., Tordoff, B., Cipolla, R.: Hole filling through photomontage. In: BMVC 2005-Proceedings of the British Machine Vision Conference 2005 (2005)
3. Goodfellow, I., et al.: Generative adversarial nets. In: Advances in Neural Information Processing Systems, vol. 27 (2014)

4. Pathak, D., Krahenbuhl, P., Donahue, J., Darrell, T., Efros, A.A.: Context encoders: feature learning by inpainting. In: Proceedings of the IEEE Conference on Computer Vision and Pattern Recognition, pp. 2536–2544 (2016)
5. Iizuka, S., Simo-Serra, E., Ishikawa, H.: Globally and locally consistent image completion. ACM Trans. Graph. (ToG) **36**, 1–14 (2017)
6. Li, Y., Liu, S., Yang, J., Yang, M.H.: Generative face completion. In: Proceedings of the IEEE Conference on Computer Vision and Pattern Recognition, pp. 3911–3919 (2017)
7. Zhao, L., et al.: Uctgan: Diverse image inpainting based on unsupervised cross-space translation. In: Proceedings of the IEEE/CVF Conference on Computer Vision and Pattern Recognition, pp. 5741–5750 (2020)
8. Jo, Y., Park, J.: SC-FEGAN: face editing generative adversarial network with user's sketch and color. In: Proceedings of the IEEE/CVF International Conference on Computer Vision, pp. 1745–1753 (2019)
9. Nazeri, K., Ng, E., Joseph, T., Qureshi, F., Ebrahimi, M.: EdgeConnect: structure guided image inpainting using edge prediction. In: Proceedings of the IEEE/CVF International Conference on Computer Vision Workshops (2019)
10. Yang, Y., Guo, X., Ma, J., Ma, L., Ling, H.: LAFIN: generative landmark guided face inpainting. arXiv preprint arXiv:1911.11394 (2019)
11. Choi, Y., Choi, M., Kim, M., Ha, J.W., Kim, S., Choo, J.: StarGAN: unified generative adversarial networks for multi-domain image-to-image translation. In: Proceedings of the IEEE Conference on Computer Vision and Pattern Recognition, pp. 8789–8797 (2018)
12. Karras, T., Laine, S., Aila, T.: A style-based generator architecture for generative adversarial networks. In: Proceedings of the IEEE/CVF Conference on Computer Vision and Pattern Recognition, pp. 4401–4410 (2019)
13. Zhao, S., et al.: Large scale image completion via co-modulated generative adversarial networks. arXiv preprint arXiv:2103.10428 (2021)
14. Lu, W., et al.: Diverse facial inpainting guided by exemplars. arXiv preprint arXiv:2202.06358 (2022)
15. Huang, X., Belongie, S.: Arbitrary style transfer in real-time with adaptive instance normalization. In: Proceedings of the IEEE International Conference on Computer Vision, pp. 1501–1510 (2017)
16. Karras, T., Laine, S., Aittala, M., Hellsten, J., Lehtinen, J., Aila, T.: Analyzing and improving the image quality of StyleGAN. In: Proceedings of the IEEE/CVF Conference on Computer Vision and Pattern Recognition, pp. 8110–8119 (2020)
17. Kim, H., Choi, Y., Kim, J., Yoo, S., Uh, Y.: Exploiting spatial dimensions of latent in GAN for real-time image editing. In: Proceedings of the IEEE/CVF Conference on Computer Vision and Pattern Recognition, pp. 852–861 (2021)
18. Wang, Q., Fan, H., Sun, G., Cong, Y., Tang, Y.: Laplacian pyramid adversarial network for face completion. Pattern Recogn. **88**, 493–505 (2019)
19. Yu, J., Lin, Z., Yang, J., Shen, X., Lu, X., Huang, T.S.: Generative image inpainting with contextual attention. In: Proceedings of the IEEE Conference on Computer Vision and Pattern Recognition, pp. 5505–5514 (2018)
20. Liu, G., Reda, F.A., Shih, K.J., Wang, T.-C., Tao, A., Catanzaro, B.: Image inpainting for irregular holes using partial convolutions. In: Ferrari, V., Hebert, M., Sminchisescu, C., Weiss, Y. (eds.) ECCV 2018. LNCS, vol. 11215, pp. 89–105. Springer, Cham (2018). https://doi.org/10.1007/978-3-030-01252-6_6
21. Yu, J., Lin, Z., Yang, J., Shen, X., Lu, X., Huang, T.S.: Free-form image inpainting with gated convolution. In: Proceedings of the IEEE/CVF International Conference on Computer Vision, pp. 4471–4480 (2019)

22. Zheng, C., Cham, T.J., Cai, J.: Pluralistic image completion. In: Proceedings of the IEEE/CVF Conference on Computer Vision and Pattern Recognition, pp. 1438–1447 (2019)

23. Liu, H., Wan, Z., Huang, W., Song, Y., Han, X., Liao, J.: PD-GAN: probabilistic diverse GAN for image inpainting. In: Proceedings of the IEEE/CVF Conference on Computer Vision and Pattern Recognition (2021)

24. Yu, Y., et al.: Diverse image inpainting with bidirectional and autoregressive transformers. In: Proceedings of the 29th ACM International Conference on Multimedia, pp. 69–78 (2021)

25. Vaswani, A., et al.: Attention is all you need. In: Advances in Neural Information Processing Systems, vol. 30 (2017)

26. Zhang, H., Goodfellow, I., Metaxas, D., Odena, A.: Self-attention generative adversarial networks. In: International Conference on Machine Learning, pp. 7354–7363. PMLR (2019)

27. Choi, Y., Uh, Y., Yoo, J., Ha, J.W.: Stargan v2: diverse image synthesis for multiple domains. In: Proceedings of the IEEE/CVF Conference on Computer Vision and Pattern Recognition, pp. 8188–8197(2020)

28. Isola, P., Zhu, J.Y., Zhou, T., Efros, A.A.: Image-to-image translation with conditional adversarial networks. In: Proceedings of the IEEE Conference on Computer Vision and Pattern Recognition, pp. 1125–1134 (2017)

29. Zhu, J.Y., Park, T., Isola, P., Efros, A.A.: Unpaired image-to-image translation using cycle-consistent adversarial networks. In: Proceedings of the IEEE International Conference on Computer Vision, pp. 2223–2232 (2017)

30. He, Z., Kan, M., Zhang, J., Shan, S.: Pa-gan: Progressive attention generative adversarial network for facial attribute editing. arXiv preprint arXiv:2007.05892 (2020)

31. Yang, G., Fei, N., Ding, M., Liu, G., Lu, Z., Xiang, T.: L2M-GAN: learning to manipulate latent space semantics for facial attribute editing. In: Proceedings of the IEEE/CVF Conference on Computer Vision and Pattern Recognition, pp. 2951–2960 (2021)

32. Ronneberger, O., Fischer, P., Brox, T.: U-net: convolutional networks for biomedical image segmentation. In: Navab, N., Hornegger, J., Wells, W.M., Frangi, A.F. (eds.) MICCAI 2015. LNCS, vol. 9351, pp. 234–241. Springer, Cham (2015). https://doi.org/10.1007/978-3-319-24574-4_28

33. Liu, M.Y., et al.: Few-shot unsupervised image-to-image translation. In: Proceedings of the IEEE/CVF International Conference on Computer Vision, pp. 10551–10560 (2019)

34. Arjovsky, M., Chintala, S., Bottou, L.: Wasserstein generative adversarial networks. In: International Conference on Machine Learning, pp. 214–223. PMLR (2017)

35. Mescheder, L., Geiger, A., Nowozin, S.: Which training methods for GANs do actually converge? In: International Conference on Machine Learning, pp. 3481–3490. PMLR (2018)

36. Zhou, T., Krahenbuhl, P., Aubry, M., Huang, Q., Efros, A.A.: Learning dense correspondence via 3d-guided cycle consistency. In: Proceedings of the IEEE Conference on Computer Vision and Pattern Recognition, pp. 117–126 (2016)

37. Richardson, E., et al.: Encoding in style: a StyleGAN encoder for image-to-image translation. In: Proceedings of the IEEE/CVF Conference on Computer Vision and Pattern Recognition, pp. 2287–2296 (2021)

38. Deng, J., Guo, J., Xue, N., Zafeiriou, S.: ArcFace: additive angular margin loss for deep face recognition. In: Proceedings of the IEEE/CVF Conference on Computer Vision and Pattern Recognition, pp. 4690–4699 (2019)

39. Yi, D., Lei, Z., Liao, S., Li, S.Z.: Learning face representation from scratch. arXiv preprint arXiv:1411.7923 (2014)
40. Mirza, M., Osindero, S.: Conditional generative adversarial nets. arXiv preprint arXiv:1411.1784 (2014)
41. Liu, Z., Luo, P., Wang, X., Tang, X.: Deep learning face attributes in the wild. In: Proceedings of the IEEE International Conference on Computer Vision, pp. 3730–3738 (2015)
42. Zhang, K., Zhang, Z., Li, Z., Qiao, Y.: Joint face detection and alignment using multitask cascaded convolutional networks. IEEE Signal Process. Lett. **23**, 1499–1503 (2016)
43. Kingma, D.P., Ba, J.: Adam: a method for stochastic optimization. arXiv preprint arXiv:1412.6980 (2014)
44. He, K., Zhang, X., Ren, S., Sun, J.: Deep residual learning for image recognition. In: Proceedings of the IEEE Conference on Computer Vision and Pattern Recognition, pp. 770–778 (2016)
45. Heusel, M., Ramsauer, H., Unterthiner, T., Nessler, B., Hochreiter, S.: GANs trained by a two time-scale update rule converge to a local Nash equilibrium. In: Advances in Neural Information Processing Systems, vol. 30 (2017)
46. Szegedy, C., Vanhoucke, V., Ioffe, S., Shlens, J., Wojna, Z.: Rethinking the inception architecture for computer vision. In: Proceedings of the IEEE Conference on Computer Vision and Pattern Recognition, pp. 2818–2826 (2016)

Diffusion Models for Counterfactual Explanations

Guillaume Jeanneret[(✉)], Loïc Simon, and Frédéric Jurie

Normandy University, ENSICAEN, UNICAEN, CNRS, GREYC, Caen, France
`guillaume.jeanneret-sanmiguel@unicaen.fr`

Abstract. Counterfactual explanations have shown promising results as a post-hoc framework to make image classifiers more explainable. In this paper, we propose DiME, a method allowing the generation of counterfactual images using the recent diffusion models. By leveraging the guided generative diffusion process, our proposed methodology shows how to use the gradients of the target classifier to generate counterfactual explanations of input instances. Further, we analyze current approaches to evaluate spurious correlations and extend the evaluation measurements by proposing a new metric: Correlation Difference. Our experimental validations show that the proposed algorithm surpasses previous state-of-the-art results on 5 out of 6 metrics on CelebA.

1 Introduction

Convolutional neural networks (CNNs) reached performances unimaginable a few decades ago, thanks to the adoption of very large and deep models with hundreds of layers and nearly billions of trainable parameters. Yet, it is difficult to explain their decisions because they are highly non-linear and over-parametrized. Moreover, for real-life applications, if a model exploits spurious correlations of data to forecast a prediction, the end-user will doubt the validity of the decision. Particularly, in high-stake scenarios like medicine or critical systems, ML must guarantee the usage of correct features to compute a prediction and prevent counterfeit associations. For this reason, the Explainable Artificial Intelligence (XAI) research field has been growing in recent years to progress towards understanding the decision-making mechanisms in black-box models.

In this paper, we focus on <u>post-hoc</u> explanation methods. Notably, we concentrate on the growing branch of Counterfactual Explanations (CE) [63]. CE aim to create minimal but meaningful perturbations of an input sample to change the original decision given by a fixed pretrained model. Although the objective between CE and adversarial examples share some similarities [44], the CE' perturbations must be understandable and plausible. In contrast, adversarial examples [37] contain high-frequency noise *indistinguishable* to the human eye. Overall, CE target four goals: *(i)* the explanations must flip the input's forecast using *(ii)* sparse modifications, *i.e.* instances with the smallest perturbation. Additionally, *(iii)* the explanations must be realistic and understandable by a

L. Wang et al. (Eds.): ACCV 2022, LNCS 13847, pp. 219–237, 2023.
https://doi.org/10.1007/978-3-031-26293-7_14

human. Lastly, *(iv)* the counterfactual generation method must create diverse instances. In general, counterfactual explanations seek to reveal the learned correlations related to the model's decisions.

Multiple works on CE use generative models to create tangible changes in the image [27,48,51]. Further, these architectures recognize the factors to generate images near the image-manifold [4]. Given the recent advances within image synthesis community, we propose DiME: Diffusion Models for counterfactual Explanations. DiME harnesses the denoising diffusion probabilistic models [19] to produce CE. For simplicity, we will refer to these models as diffusion models or DDPMs. To the best of our knowledge, we are the first to exploit these new synthesis methods in the context of CE.

Diffusion models offer several advantages compared to alternate generative models, such as GANs. First of all, DDPMs have several latent spaces; each one controls coarse and fine-grained details. We take advantage of low-level noise latent spaces to generate semantically-meaningfully changes in the input image. These spaces only have been recently studied by [38] for inpainting. Secondly, due to their probabilistic nature, they produce diverse sets of images. Stochasticity is ideal for CE because multiple explanations may explain a classifier's error modes. Third, Nichol and Dhariwal [42] results suggest that DDPMs cover a broader range of the target image distribution. Indeed, they noticed that for similar FID, the recall is much higher on the improved precision-recall metrics [32]. Finally, DDPMs' training is more stable than the state-of-the-art synthesis models, notably GANs. Due to their relatively new development, DDPMs are under-studied, and multiple aspects are yet to be deciphered.

We contribute a small step into the XAI community by studying the low-level noised latent spaces of DDPMs in the context of counterfactual explanations. We summarize our contributions on three axes:

- **Methodology:** (i) DiME uses the recent diffusion models to generate counterfactual examples. Our algorithm relies on a single unconditional DDPM to achieve instance counterfactual generation. To accomplish this, (ii) we derive a new way to leverage an existing (target) classifier to guide the generation process instead of using one trained on noisy instances, such as in [11]. Additionally, (iii) to reduce the computational burden, we take advantage of the forward and backward diffusion chains to transfer the gradients of the classifier under observation.
- **Evaluation:** We show that the standard MNAC metric is misleading because it does not account for possible spurious correlations. Consequently, we introduce a new metric, dubbed Correlation Difference, to evaluate subtle spurious correlations on a CE setting.
- **Performance:** We set a new state-of-the-art result on CelebA, surpassing the previous works on CE on the FID, FVA, and MNAC metrics for the *Smile* attribute and the FID and MNAC for the *Young* feature.

To further boost research on counterfactual explanations, our code and models are publicly available on Github.

2   Related Work

Our work contributes to the field of XAI, within which two families can be distinguished: *interpretable-by-design* and *post-hoc* approaches. The former includes, at the design stage, human interpretable mechanisms [2,3,6,9,22,40,71]. The latter aims at understanding the behavior of existing ML models without modifying their internal structure. Our method belongs in this second family. The two have different objectives and advantages; one benefit of *post-hoc* methods is that they rely on existing models that are known to have good performance, whereas XAI by design often leads to a performance trade-off.

Post-hoc Methods: In the field of *post-hoc* methods, there are several explored directions. Model Distillation strategies [13,58] approach explainability through fitting an interpretable model on the black-box models' predictions. In a different vein, some methods generate explanation in textual form [17,43,68]. When it comes to explaining visual information, feature importance is arguably the most common approach, often implemented in the form of saliency maps computed either using the gradients within the network [8,26,33,53,64,74] or using the perturbations on the image [45,46,62,70]. Concept attribution methods seek the most recurrent traits that describe a particular class or instance. Intuitively, concept attribution algorithms use [29] or search [13,14,69,75] for human-interpretable notions such as textures or shapes.

Counterfactual Explanations (CE): CE is a branch of *post-hoc* explanations. They are relevant to legally justify decisions made automatically by algorithms [63]. In a nutshell, a CE is the smallest meaningful change to an input sample to obtain a desirable outcome of the algorithm. Some recent methods [15,65] exploit the query image's regions and a different classified picture to interchange semantic appearances, creating counterfactual examples. Despite using the same terminology, this line of work [15,61,66] is diverging towards a task where it merely highlights regions that explains the discrepancy of the decision between the two real images, significantly differing from our evaluation protocol setup. Other works [52,63] leverage the input image's gradients with respect to the target label to create meaningful perturbations. Conversely, [1] find patterns via prototypes that the image must contain to alter its prediction. Similarly, [36,47] follow a prototype-based algorithm to generate the explanations. Even Deep Image Priors [59] and Invertible CNNs [23] have shown the capacity to produce counterfactual examples. Furthermore, theoretical analyses [24] found similarities between counterfactual explanations and adversarial attacks.

Due to the nature of the problem, the generation technique used is the key element to produce data near the image manifold. For instance, [12] optimizes the residual of the image directly using an autoencoder as a regularizer. Other works propose to use generative networks to create the CE, either unconditional [27,41,48,54,73] or conditional [34,55,60]. In this paper, we adopt more recent generation approaches, namely <u>diffusion models</u>; an attempt never considered in the past for counterfactual generation.

Diffusion Models: Diffusion models have recently gained popularity in the image generation research field [19,56]. For instance, DDPMs approached inpainting [49], conditional and unconditional image synthesis [10,19,42], super-resolution [50], even fundamental tasks such as segmentation [5], providing performance similar or even better than State-of-the-Art generative models. Further, studies like [20,57] show score-based approaches and diffusion are alternative formulations to denoise the reverse sampling for data generation. Due to the recursive generation process, DDPMs sampling is expensive. Many works have studied alternative approaches to accelerate the generation process [31,67].

The recent method of [11] targets conditional image generation with diffusion models, which they do by training a specific classifier on noisy instances to bias the generation process. Our work bears some similarities to this method, but, in our case, explaining an existing classifier trained uniquely in clean instances poses additional challenges. In addition, unlike past diffusion methods, we perform the image editing process from an intermediate step rather than the final one. To the best of our knowledge, no former study has considered diffusion models to explain a neural network counterfactually.

3 Methodology

3.1 Diffusion Model Preliminaries

We begin by introducing the generation process of diffusion models. They rely on two Markov chain sampling schemes that are inverse of one another. In the forward direction, the sampling starts from a natural image x and iteratively sample $z_1, \cdots, z_T$ by replacing part of the signal with white Gaussian noise. More precisely, letting β_t be a prescribed variance, the forward process follows the recursive expression:

$$z_t \sim \mathcal{N}(\sqrt{1 - \beta_t}\, z_{t-1}, \beta_t\, I), \tag{1}$$

where $\mathcal{N}$ is the normal distribution, I the identity matrix, and $z_0 = x$. In fact, this process can be simulated directly from the original sample with

$$z_t \sim \mathcal{N}(\sqrt{\alpha_t}x, (1 - \alpha_t)I), \tag{2}$$

where $\alpha_t := \prod_{k=1}^{t}(1 - \beta_k)$. For clarification, through the rest of the paper, we will refer to clean images with an x, while noisy ones with a z.

In the reverse process, a neural network recurrently denoises z_T to recover the previous samples $z_{T-1}, \cdots, z_0$. This network takes the current time step t and a noisy sample z_t as inputs, and produces an average sample $\mu(t, z_t)$ and a covariance matrix $\Sigma(t, z_t)$, shorthanded as $\mu(z_t)$ and $\Sigma(z_t)$, respectively. Then z_{t-1} is sampled with

$$z_{t-1} \sim \mathcal{N}(\mu(z_t), \Sigma(z_t)). \tag{3}$$

So, the DDPM algorithm iteratively employs Eq. 3 to generate an image z_0 with zero variance, *i.e.* a clean image. Some diffusion models use external information, such as labels, to condition the denoising process. However, in this paper, we employ an unconditional DDPM.

In practice, the variances β_t in Eq. 1 are chosen such that $z_T \sim \mathcal{N}(0, I)$. Further, the DDPM's trainable parameters are fitted so that the reverse and forward processes share the same distribution. For a thorough understanding on the DDPM training, we recommend the studies of Ho et al. [19] and Nichol and Dhariwal [42] to the reader. Once the network is trained, one can rely on the reverse Markov chain process to generate a clean image from a random noise image z_T. Besides, the sampling procedure can be adapted to optimize some properties following the so-called <u>guided diffusion</u> scheme proposed in [11][1]:

$$z_{t-1} \sim \mathcal{N}(\mu(z_t) - \Sigma(z_t) \nabla_{z_t} L(z_t; y), \Sigma(z_t)), \tag{4}$$

where L is a loss function using z_t to specify the wanted property of the generated image, for example, to condition the generation on a prescribed label y.

3.2 DiME: Diffusion Models for Counterfactual Explanations

We take an image editing standpoint on CE generation, as illustrated Fig. 1. We start from a query image x. Initially, we rely on the forward process starting from $x_\tau = x$ to compute a noisy version z_τ, with $1 \leq \tau \leq T$. Then we go back in the reverse Markov chain using the guided diffusion (Eq. 4) to recover a counterfactual (hence altered) version of the query sample. Building upon previous approaches for CE based on other generative models [25,55,63], we rely on a loss function composed of two components to steer the diffusion process: a classification loss L_{class}, and a perceptual loss L_{perc}. The former guides the image edition into imposing the target label, and the latter drives the optimization in terms of proximity.

In the original implementation of the guided diffusion [11], the loss function uses a classifier applied directly to the current noisy image z_t. In their context, this approach is appropriate since the considered classifier can make robust predictions under noisy observations, i.e. it was trained on noisy images. Regardless, such an assumption on the classifier under scrutiny would imply a substantial limitation in the context of counterfactual examples. We circumvent this obstacle by adapting the guided diffusion mechanism. To simplify the notations, let x_t be the clean image produced by the iterative unconditional generation on Eq. 3 using as the initial condition z_t. In fact, this makes x_t a *function* of z_t because we denoise z_t recursively with the diffusion model t times to obtain x_t. Luckily, we can safely apply the classifier to x_t since it is not noisy. So, we express our loss as:

$$L(z_t; y, x) = \mathbb{E}\left[\lambda_c L_{class}(C(y|x_t)) + \lambda_p L_{perc}(x_t, x)\right] := \mathbb{E}\left[\tilde{L}(x_t; y, x)\right], \tag{5}$$

where $C(y|x_t)$ is the posterior probability of the category y given x_t, and λ_c and λ_p are constants. Note that an expectation is present due to the stochastic nature of x_t. In practice, computing the loss gradient would require sampling several

---

[1] In [11], the guided diffusion is restricted to a specific classification loss. Still, for the sake of generality and conciseness, we provide its extension to an arbitrary loss.

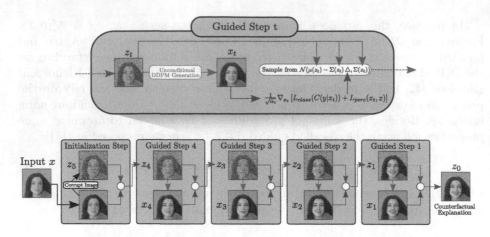

Fig. 1. DiME: Diffusion Models for Counterfactual Explanations. Given an input instance x, we perturb it following Eq. 2 to get z_τ (here $\tau = 5$). At time step t, we use the DDPM model to generate a clean image x_t to obtain the clean gradient L_{class} and L_{perc} with respect to x_t. Finally, we sample z_{t-1} using the guiding optimization process on Eq. 4, using the previously extracted clean gradients.

realizations of x_t and taking an empirical average. We restrict ourselves to a single realization per step t for computational reasons and argue that this is not an issue. Indeed, we can partly count on an averaging effect along the time steps to cope with the lack of individual empirical averaging. Besides, the stochastic nature of our implementation is, in fact, an advantage because it introduces more diversity in the produced CE, a desirable feature as advocated by [48].

Using this strategy, the dependence of the loss on x_t, rather than directly from z_t, renders the gradient computation more challenging. Indeed, formally it would require to apply back-propagation from x_t back to z_t:

$$\nabla_{z_t} L(z_t; y, x) = \left(\frac{D x_t}{D z_t} \right)^T \cdot \nabla_{x_t} \tilde{L}(x_t; y, x). \tag{6}$$

Unfortunately, this computation requires retaining Jacobian information across the entire computation graph, which is very deep when t is close to τ. As a result, backpropagation is too memory intensive to be considered an option. To bypass this pitfall, we shall rely on the forward sampling process, which operates in a single stage (Eq. 2). Using the re-parametrization trick [30], one obtains

$$z_t = \sqrt{\alpha_t} x_t + \sqrt{1 - \alpha_t} \epsilon, \ \epsilon \sim \mathcal{N}(0, \mathbf{I}). \tag{7}$$

Thus, by solving x_t from z_t, we can leverage the gradients of the loss function with respect to the noisy input, a consequence of the chain rule. Henceforth, the gradients of L with respect to the noisy image become

$$\nabla_{z_t} L(z_t; y, x) = \frac{1}{\sqrt{\alpha_t}} \nabla_{x_t} \tilde{L}(x_t; y, x). \tag{8}$$

This approximation is possible since the DDPM estimates the reverse Markov chain to fit the forward corruption process. Thereby, both processes are similar.

To sum up, Fig. 1 depicts the generation of a counterfactual explanation with our algorithm: DiME. We start by corrupting the input instance $x = x_\tau$ following Eq. 2 up to the noise level $t = \tau$. Then, we iterate the following two stages until $t = 0$: *(i)* First, using the gradients of the previous clean instance x_{t-1}, we guide the diffusion process to obtain z_{t-1} using Eq. 4 with the gradients computed in Eq. 8. *(ii)* Next, we estimate the clean image x_t for the current time step z_{t-1} with the unconditional generation pipeline of DDPMs. The final instance is the counterfactual explanation. If we do not find an explanation that fools the classifier under observation, we increase the constant λ_c and repeat the process.

Implementation Details. In practice, we incorporate additionally an ℓ_1 loss, $\eta||z_t - x||_1$, between the noisy image z_t and the input x to improve the ℓ_1 metric on the pixel space. We empirically set η small to avoid any significant impact on the quality of the explanations. Our diffusion model generates faces using 500 diffusion steps from the normal distribution. We re-spaced the sampling process to boost inference speed to generate images with 200 time-steps at test time. We use the following hyperparameters settings: $\lambda_p = 30$, $\eta = 0.05$, and $\tau = 60$. Finally, we set $\lambda_c \in \{8, 10, 15\}$ to iteratively find the counterfactuals. We consider that our method failed if we do not find any explanation after exhausting the values of λ_c. To train the unconditional DDPM model, we used the publicly available code of [11]. Our model has the same architecture as the ImageNet's Unconditional DDPM, but we used 500 sampling steps. Furthermore, the inner number of channels was set to 128 instead of 256 given CelebA's lower complexity. For training, we completed 270,000 iteration with a batch size of 75 with a learning rate of 1×10^{-4} with a weight decay of 0.05.

4 Experiments

Dataset. In this paper, we study the CelebA dataset [35]. Following standard practices, we preprocess all images to a 128 × 128 resolution. CelebA contains 200k images, labeled with 40 binary attributes. Previous works validate their methods on the smile and young binary attributes, ignoring all other attributes. Finally, the architecture to explain is a DenseNet121 [21] classifier. Given the binary nature of the task, the target label is always the opposite of the prediction. If the model correctly estimates an instance's label, we flip the model's forecast. Otherwise, we modify the input image to classify the image correctly.

Experimental Goals. In this section, we evaluate our CE approach using standard metrics. Also, we develop new tools to go beyond the current evaluation practices. Let us recap the principles of current evaluation metrics, following previous works [48,55]. The first goal of CE is to create realistic explanations that flip the classifier under observation. The capacity to change the classifier decision is typically exposed as a flip ratio (FR). Following the image synthesis research literature, the Frechet Inception Distance [18] (FID) measures the fidelity of the image distribution. The second goal of CE methods is to create

Table 1. State-of-the-Art results. We compare our model performance against the State-of-the-Art on the FID, FVA and MNAC metrics. The values in **bold** are the best results. All metrics were extracted from [48]. Our model has a 10 fold improvement on the FID metric. We extracted all results from Rodriguez *et al.*' work [48].

Method	Smile			Young		
	FID ($\downarrow$)	FVA ($\uparrow$)	MNAC ($\downarrow$)	FID ($\downarrow$)	FVA ($\uparrow$)	MNAC ($\downarrow$)
xGEM+ [27]	66.9	91.2	–	59.5	97.5	6.70
PE [55]	35.8	85.3	–	53.4	72.2	3.74
DiVE [48]	29.4	97.3	–	33.8	**98.2**	4.58
DiVE^{100}	36.8	73.4	4.63	39.9	52.2	4.27
DiME	**3.17**	**98.3**	**3.72**	**4.15**	95.3	**3.13**

proximal and sparse images. Among other tools, the XAI community adopted the Face Verification Accuracy [7] (FVA) and Mean Number of Attributes Changed (MNAC) [48]. On the one hand, the MNAC metric looks at the face attributes that changed between the input image and its counterfactual explanation, disregarding if the individual's identity changed. Finally, the FVA looks at the individual's identity without considering the difference of attributes.

As a quick caveat, let us mention that this set of standard metrics displays several pitfalls that we shall address in more detail later. First, these metric do not evaluate the diversity of the produced explanations, whereas this is an important factor. Besides, some of the metrics are at odd with a crucial purpose of CE, namely the detection of potential spurious correlations.

4.1 Realism, Proximity and Sparsity Evaluation

To begin with, the FVA is the standard metric for face recognition. To measure this value, we used the cosine similarity between the input image and its produced counterfactual on the feature space of a ResNet50 [16] pretrained model on VGGFace2 [7]. This metric considers that two instances share identity if the similarity is higher than 0.5. So, the FVA is the mean number of faces sharing the same identity with their corresponding CE. Secondly, to compute the MNAC, we fine-tuned the VGGFace2 model on the CelebA dataset. We refer to the fine-tuned model as the <u>oracle</u>. Thus, the MNAC is the mean number of attributes for which the oracle switch decision under the action of the CE. For a fair comparison with the state-of-the-art, we trained all classifiers, including the fine-tuned ResNet50 for the MNAC assessment, using the DiVE's [48] available code. Finally, previous studies [48,55] compute the FID, the FVA, and the MNAC metrics considering only those successful counterfactual examples.

DiVE do not report their flip rate (FR). This raises a concern over the fairness comparing against our method. Since some metrics depend highly on the number of samples, especially FID, we recomputed their CE. To our surprise, their flip ratio was relatively low: 44.6% for the smile category. In contrast, we

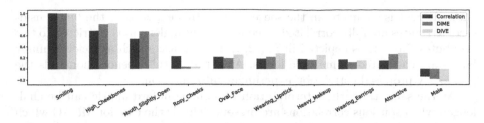

Fig. 2. Spurious Correlation Detection. We show the top 9 most correlated attributes in the label space with "smile". We obtained the Pearson Correlation Coefficient from the ground truth on the training set. Albeit the difference in the MNAC measure, DiME and DiVE achieve to detect the correlations similarly.

achieved a success rate of 97.6 and 98.9 for the smile and young attributes, respectively. Therefore, we calculated the explanations with 100 optimization steps and reported the results as DiVE^{100}. The new success rates are 92.0% for smile and 93.4% for young.

We show DiME's performance in Table 1. Our method beats the previous literature in five out of six metrics. For instance, we have a ~10-fold improvement on the FID metric for the smile category, while the young attribute has an ~8 fold improvement. We credit these gains to our generation process since it does not require entirely corrupting the input instance; hence, the coarse details of the image remain. The other methods rely on latent space-based architectures. Thus, they require to compact essential information removing outlier data. Consequently, the generated CE cannot reconstruct the missing information, losing significant visual components of the image statistics.

Despite the previous advantages, we cannot fail to notice that DiME is less effective in targeting the young attribute than the smile. The smile and young attributes have distinct features. The former is delineated by localized regions, while the latter scatters throughout the entire face. Thus, the gradients produced by the classifier differ between the attributes of choice; for the smile attribute, the gradients are centralized while they are outspread for the young attribute. We believe that this subtle difference underpins the slight drop of performance (especially with respect to FVA) in the young attribute case. This hypothetical explanation should be confirmed by a more systematic study of various attributes, though this phenomenon is out of scope of the paper.

4.2 Discovering Spurious Correlations

The end goal of CE is to uncover the modes of error of a target model, in particular its reliance on spurious correlations. Current evaluation protocols [55] search to assess the counterfeit dependencies by inducing artificial entanglements between two supposedly uncorrelated traits such as the smile and gender attributes. In our opinion, such an extreme experiment does not shed light on the ability to reveal spurious correlations for two reasons. First, the introduced entanglement is complete, in the sense that in this experiment the two considered attributes are fully correlated. Second, the entanglement is restricted to two attributes. In fact, as depicted in Fig. 2, in real datasets such as CelebA, many labels are correlated at multiple levels. As a result, this phenomenon calls the previously proposed correlation experiment into question.

At the same time, the interpretation of some standard metric can be challenged when spurious correlations are present. This is the case for MNAC which corresponds to the mean number of attributes that change under the action of a CE method. Arguably, the classical interpretation is that attributes being unrelated, a CE method that change fewer attributes (in addition to the target) is preferable. In other words between two CE methods, the one displaying the smaller MNAC is reckoned as the better one. This interpretation is at odd with the fact that the alternative method may display a higher MNAC because it actually reveals existing spurious correlations.

Consequently, we design a new metric called Correlation Difference (CD), verifying the following principles: (i) it quantifies how well a counterfactual routine captures spurious correlations. In other words, it estimates correlations between two attributes after applying the counterfactual algorithm and compare these estimates to the true dataset correlations. (ii) It should apply an oracle to predict the (unknown) attributes of counterfactual examples. (iii) The metric should preferably rely on attribute prediction changes between the original example and its explanation to mitigate potential errors of the oracle, rather than solely on the prediction made on the counterfactual. Principle (i) actually amends the failure of MNAC, while (ii) and (iii) maintain its desirable features.

To do so, we start from the definition of the Pearson correlation coefficient $c_{q,a}$ between the target attribute q and any other attribute a. Denoting X a random image sample, along with its two associated binary attribute labels Y_q and Y_a, then $c_{q,a} = \mathrm{PCC}(Y_q, Y_a)$, where PCC is the Pearson correlation coefficient operator. To cope with principle (i) we would like to estimate correlations between attributes q and a and we would like our estimation to rely on the CE method M targeting the attribute q. The main issue is that we do not know the actual attributes for the CE, $M(X, q)$, obtained from an image X. Yet, following principle (ii), we may rely on an oracle to predict these attributes. More precisely, letting $O_a(X)$ be the oracle prediction for a given image X and for the label a, we could simply compute the correlation coefficient between $O_q(M(X, q))$ and $O_a(M(X, q))$. Such an estimate would be prone to potential errors of the oracle, and following principle (iii) we would prefer to rely on attribute changes $\delta_{q,a}^M(X) = O_a(M(X, q)) - O_a(X)$.

Interestingly, $c_{q,a}$ can be reformulated as follows:

$$c_{q,a} = PCC(\delta_q, \delta_a), \tag{9}$$

where $\delta_a = Y_a - Y_a'$ (resp. $\delta_q = Y_q - Y_q'$), with (X, Y_q, Y_a) and (X', Y_q', Y_a') two independent samples. In other words, $c_{q,a}$ can be interpreted as the correlation between changes in attributes q and a among random pairs of samples. Accordingly, we use $\delta_{q,q}^M$ and $\delta_{q,a}^M$ as drop-in replacements for δ_q and δ_a in Eq. 9 to

Input   Counterfactual Explanations

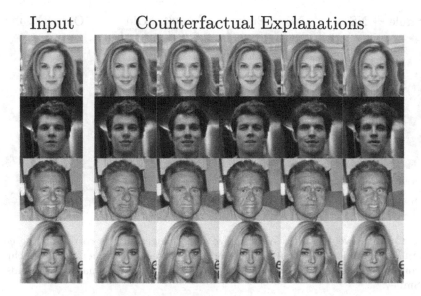

Fig. 3. Diversity Counterfactual examples. The classifier predicts first two input images as non-smiley and the last two as smiley. In this example all explanations fool the classifier. Our CE pipeline is capable of synthesising diverse counterfactuals without any additional mechanism.

obtain the estimate $c_{q,a}^M$ of $c_{q,a}$ that relies on the label changes produced by the counterfactual method M. Finally, CD for label q is merely:

$$CD_q = \sum_a |c_{q,a} - c_{q,a}^M|. \tag{10}$$

We apply our proposed metric on DiME and DiVE^{100}'s explanations. We got a CD of 2.30 while DiVE^{100} 2.33 on CelebA's validation set, meaning that DiVE^{100} lags behind DiME. However, the margin between the two approaches is only slender. This reveals our suspicions: the MNAC results presented in Table 1 give a misleading impression of a robust superiority of DiME over DiVE^{100}.

4.3   Diversity Assessment

One of the most crucial traits of counterfactual explanations methodologies is the ability to create multiple and diverse examples [39, 48]. As stated in the methodology section, DiME's stochastic properties enable the sampling of diverse counterfactuals. To measure the capabilities of different algorithms to produce multiple explanations, we computed the mean pair-wise LPIPS [72] metric between five independent runs. Formally, setting N as the length of the dataset and $n = 5$ as the number of samples, the Diversity metric σ_L is:

$$\sigma_L = \frac{1}{N} \sum_{i=1}^{N} \frac{2}{n(n+1)} \sum_{j=1}^{n} \sum_{k=j+1}^{n} LPIPS(x_j^i, x_k^i), \tag{11}$$

Smile → NS   NS → Smile   Young → Old   Old → Young

Fig. 4. Qualitative Results. We visualize some images and its corresponding counterfactual explanation produced by our proposed approach. Our methodology achieves to incorporate small but perceptually tangible changes in the image. NS stands for Non-Smiley.

A higher σ_L means increased perceptual dissimilarities between the explanations, hence, more diversity. To compute the evaluation metric, we use all counterfactual examples, even the unsuccessful instances, because we search the capacity of exploring different traits. Note that we exclude the input instance to compute the metric since we search for the dissimilarities between the counterfactuals. We compared DiME's performance with $DiVE^{100}$ and its Fisher Spectral variant on a small partition of the validation subset.

We visualize some examples in Fig. 3. All runs achieve similar performances making DiME insensible to the initial random seed. We achieved a σ_L of 0.213. In contrast, DiVE [48] and its Spectral Fisher variant obtained much lower LPIPS diversity of 0.044 and 0.086, respectively. Recall that DiME does not have an explicit mechanism to create diverse counterfactuals. Its only mechanism is the stochasticity within the sampling process (Eqs. 3 and 4). In contrast, DiVE relies on a diversity loss when optimizing the eight explanations. Yet, our methodology achieves higher σ_L metric even without an explicit mechanism.

4.4 Qualitative Results

We visualize some inputs (left) and the counterfactual examples (right) produced by DiME in Fig. 4. We show visualizations for the attributes smile and young. At first glance, the results reveal that the model performs semantical editings into the input image. In addition, uncorrelated features and coarse structure remain almost unaltered. We observe slight variations on some items, such as the pendants, or out-of-distribution shapes such as hands. DiME fails to reconstruct the exact shape of these objects, but the essential aspect remains the same.

New ICMI Studies Series

Volume 4

Published under the auspices of The International
Commission on Mathematical Instruction under the
general editorship of

Miguel de Guzmán, President   Mogens Niss, Secretary

The titles published in this series are listed at the end of Book 2.

Table 2. DiME vs variations. This table shows the advantages of the proposed adjustment to incorporate the classifier under observation. Including the clean gradients benefits DiME on all metrics, especially the FR. § FID$^+$ and ℓ_1 are computed with the same number of samples as the rest, but without filtering out unsuccessful CEs.

Method	FR ($\uparrow$)	FID$^+$($\downarrow$)	ℓ_1($\downarrow$)
Direct	19.7	**50.51**	0.0454
Naive	70.0	98.93 $\pm$ 2.36	0.0624
Early Stopping	97.3	51.97 $\pm$ 0.77	0.0467
Unconditional§	8.6	53.22 $\pm$ 0.98	0.0492
DiME	**97.9**	**50.20** $\pm$ 1.00	**0.0430**

4.5   Ablation Study: Impact of the Noise-Free Input of the Classifier

As a major contribution, we have proposed an adjustment over the guided diffusion process. It consists in applying the classifier on noise-free images x_t rather than on the current noisy version z_t to obtain a robust gradient direction. One can rightly wonder how important a role is played by this adjustment. To assess this matter, we consider several alternatives to our approach. The first alternative, dubbed <u>Direct</u>, uses the gradient (without the factor $1/\sqrt{\alpha_t}$) of the classifier applied directly to the noisy instance z_t. The second alternative, called <u>Naive</u>, uses the gradient of the original input image at each time step to guide the optimization process. Therefore, it is not subject to noise issues, but it disregards the guidance that was already applied until time step t. The last variation is a near duplicate of DiME except for the fact that it ends the guided diffusion process as soon as x_t fools the classifier. We name this approach <u>Early Stopping</u>. Eventually, we will also evaluate the DDPM generation without any guiding and beginning from the corrupted image at time-step τ to mark a reference of the performance of the DDPM model. We will refer to this variant as the <u>Unconditional</u> one.

To validate all distinct variants, we created a small and randomly selected mini-val to evaluate the various metrics. To make FID values more comparable amongst all variants, we condition its computation only on the successful CE and keep the same number of samples for all methods to mitigate the bias in FID with respect to the number of samples. We denote this fair FID as FID$^+$. Likewise to the FR and FID$^+$, we evaluate the ℓ_1 metric on successful CE.

We show the results of the different variations in Table 2. The most striking point is that when compared to the Naive and Direct approaches, the unimpaired version of DiME is the most effective in terms of FR by a large margin. This observation validates the need for our adjustment of the guided diffusion process. Further, our approach is also superior to all other variations in terms of the other metrics. At first glance, we expected the unconditional generation to have better FID than DiME and the ablated methods. However, we believe that the perceptual component of our loss is beneficial in terms of FID. Therefore, the unconditional FID is higher. Based on the same rationale, on can explain the slightly higher FID displayed by the early stopping variant. Moreover, we noticed that most instances merely shifted the decision boundary, reporting a low confidence of the posterior probability. These instances are semifactual [28] and contain features from both attributes, making them hard to analyze in the context of explainability, in our opinion.

4.6　Limitations

Although we show the benefits of using our model to generate CE, we are far from accomplishing all aspects crucial for the XAI community. For instance, we observe that DiME has two limitations. On the one hand, we adopt the most problematic aspect of DDPMs: the inference time. Namely, DiME uses ∼1800 times the DDPM model to generate a single explanation. This aspect is undesired whenever the user requires an explanation on the fly. Regardless, DiME can haste its generation process at cost of image quality since diffusion models enjoy from different strategies to boost inference time. On the other hand, we require access to the training data; a limitation shared by many studies. However, this aspect is vital in fields with sensible data. Although access to the data is permitted in many cases, we restrict ourselves to using the data without any labels.

5　Conclusion

In this paper, we explore the novel diffusion models in the context of counterfactual explanations. By harnessing the conditional generation of the guided diffusion, we achieve successful counterfactual explanations through DiME. These explanations follow the requirements given by the XAI community: a small but tangible change in the image while remaining realistic. The performance of DiME is confirmed based on a battery of standard metrics. We show that the current approach to validate the sparsity of CE has significant conflicts with the assessment of spurious correlation detection. Our proposed metric, Correlation Difference, correctly measures the impact of measuring the subtle correlation between labels. Further, DiME also exhibits strong diversity in the produced explanation. This is partly inherited from the intrinsic features of diffusion models, but it also results from a careful design of our approach. Finally, we hope that our work opens new ways to compute and evaluate counterfactual explanations.

Acknowledgements. Research reported in this publication was supported by the Agence Nationale pour la Recherche (ANR) under award number ANR-19-CHIA-0017.

References

1. Akula, A., Wang, S., Zhu, S.C.: CoCoX: generating conceptual and counterfactual explanations via fault-lines. In: Proceedings of the AAAI Conference on Artificial Intelligence, vol. 34, no. 03, pp. 2594–2601 (2020)
2. Alaniz, S., Marcos, D., Schiele, B., Akata, Z.: Learning decision trees recurrently through communication. In: Proceedings of the IEEE/CVF Conference on Computer Vision and Pattern Recognition (CVPR), pp. 13518–13527, June 2021

3. Alvarez-Melis, D., Jaakkola, T.S.: Towards robust interpretability with self-explaining neural networks. In: Advances in Neural Information Processing Systems (NeurIPS) (2018)
4. Arora, S., Risteski, A., Zhang, Y.: Do GANs learn the distribution? Some theory and empirics. In: International Conference on Learning Representations (2018). https://openreview.net/forum?id=BJehNfW0-
5. Baranchuk, D., Voynov, A., Rubachev, I., Khrulkov, V., Babenko, A.: Label-efficient semantic segmentation with diffusion models. In: International Conference on Learning Representations (2022)
6. Bohle, M., Fritz, M., Schiele, B.: Convolutional dynamic alignment networks for interpretable classifications. In: Proceedings of the IEEE/CVF Conference on Computer Vision and Pattern Recognition (CVPR), pp. 10029–10038, June 2021
7. Cao, Q., Shen, L., Xie, W., Parkhi, O.M., Zisserman, A.: VGGFace2: a dataset for recognising faces across pose and age. In: 2018 13th IEEE International Conference on Automatic Face Gesture Recognition (FG 2018), pp. 67–74 (2018). https://doi.org/10.1109/FG.2018.00020
8. Chattopadhay, A., Sarkar, A., Howlader, P., Balasubramanian, V.N.: Grad-CAM++: generalized gradient-based visual explanations for deep convolutional networks. In: 2018 IEEE Winter Conference on Applications of Computer Vision (WACV), pp. 839–847 (2018). https://doi.org/10.1109/WACV.2018.00097
9. Chen, C., Li, O., Tao, D., Barnett, A., Rudin, C., Su, J.K.: This looks like that: deep learning for interpretable image recognition. In: Wallach, H., Larochelle, H., Beygelzimer, A., d' Alché-Buc, F., Fox, E., Garnett, R. (eds.) Advances in Neural Information Processing Systems, vol. 32. Curran Associates, Inc. (2019)
10. Choi, J., Kim, S., Jeong, Y., Gwon, Y., Yoon, S.: ILVR: conditioning method for denoising diffusion probabilistic models. In: Proceedings of the IEEE/CVF International Conference on Computer Vision (ICCV), pp. 14367–14376, October 2021
11. Dhariwal, P., Nichol, A.Q.: Diffusion models beat GANs on image synthesis. In: Thirty-Fifth Conference on Neural Information Processing Systems (2021)
12. Dhurandhar, A., et al.: Explanations based on the missing: towards contrastive explanations with pertinent negatives. In: Bengio, S., Wallach, H., Larochelle, H., Grauman, K., Cesa-Bianchi, N., Garnett, R. (eds.) Advances in Neural Information Processing Systems, vol. 31. Curran Associates, Inc. (2018)
13. Ge, Y., et al.: A peek into the reasoning of neural networks: interpreting with structural visual concepts. In: Proceedings of the IEEE/CVF Conference on Computer Vision and Pattern Recognition (CVPR), pp. 2195–2204, June 2021
14. Ghorbani, A., Wexler, J., Zou, J.Y., Kim, B.: Towards automatic concept-based explanations. In: Wallach, H., Larochelle, H., Beygelzimer, A., d' Alché-Buc, F., Fox, E., Garnett, R. (eds.) Advances in Neural Information Processing Systems, vol. 32. Curran Associates, Inc. (2019)
15. Goyal, Y., Wu, Z., Ernst, J., Batra, D., Parikh, D., Lee, S.: Counterfactual visual explanations. In: ICML, pp. 2376–2384 (2019)
16. He, K., Zhang, X., Ren, S., Sun, J.: Deep residual learning for image recognition. In: 2016 IEEE Conference on Computer Vision and Pattern Recognition (CVPR), pp. 770–778 (2016)
17. Hendricks, L.A., Akata, Z., Rohrbach, M., Donahue, J., Schiele, B., Darrell, T.: Generating visual explanations. In: Leibe, B., Matas, J., Sebe, N., Welling, M. (eds.) ECCV 2016. LNCS, vol. 9908, pp. 3–19. Springer, Cham (2016). https://doi.org/10.1007/978-3-319-46493-0_1

18. Heusel, M., Ramsauer, H., Unterthiner, T., Nessler, B., Hochreiter, S.: GANs trained by a two time-scale update rule converge to a local Nash equilibrium. In: Guyon, I., et al. (eds.) Advances in Neural Information Processing Systems, vol. 30. Curran Associates, Inc. (2017). https://proceedings.neurips.cc/paper/2017/file/8a1d694707eb0fefe65871369074926d-Paper.pdf

19. Ho, J., Jain, A., Abbeel, P.: Denoising diffusion probabilistic models. In: Larochelle, H., Ranzato, M., Hadsell, R., Balcan, M.F., Lin, H. (eds.) Advances in Neural Information Processing Systems, vol. 33, pp. 6840–6851. Curran Associates, Inc. (2020)

20. Huang, C.W., Lim, J.H., Courville, A.: A variational perspective on diffusion-based generative models and score matching. In: ICML Workshop on Invertible Neural Networks, Normalizing Flows, and Explicit Likelihood Models (2021)

21. Huang, G., Liu, Z., van der Maaten, L., Weinberger, K.Q.: Densely connected convolutional networks. In: Proceedings of the IEEE Conference on Computer Vision and Pattern Recognition (CVPR), July 2017

22. Huang, Z., Li, Y.: Interpretable and accurate fine-grained recognition via region grouping. In: Proceedings of the IEEE/CVF Conference on Computer Vision and Pattern Recognition (CVPR), June 2020

23. Hvilshøj, F., Iosifidis, A., Assent, I.: ECINN: efficient counterfactuals from invertible neural networks. In: British Machine Vision Conference 2021, BMVC 2021 (2021)

24. Ignatiev, A., Narodytska, N., Marques-Silva, J.: On relating explanations and adversarial examples. In: Wallach, H., Larochelle, H., Beygelzimer, A., d' Alché-Buc, F., Fox, E., Garnett, R. (eds.) Advances in Neural Information Processing Systems, vol. 32. Curran Associates, Inc. (2019)

25. Jacob, P., Zablocki, É., Ben-Younes, H., Chen, M., Pérez, P., Cord, M.: STEEX: steering counterfactual explanations with semantics (2021)

26. Jalwana, M.A.A.K., Akhtar, N., Bennamoun, M., Mian, A.: CAMERAS: enhanced resolution and sanity preserving class activation mapping for image saliency. In: Proceedings of the IEEE/CVF Conference on Computer Vision and Pattern Recognition (CVPR), pp. 16327–16336, June 2021

27. Joshi, S., Koyejo, O., Kim, B., Ghosh, J.: xGEMs: generating examplars to explain black-box models. ArXiv abs/1806.08867 (2018)

28. Kenny, E.M., Keane, M.T.: On generating plausible counterfactual and semi-factual explanations for deep learning. In: Proceedings of the AAAI Conference on Artificial Intelligence, vol. 35, no. 13, pp. 11575–11585, May 2021. https://ojs.aaai.org/index.php/AAAI/article/view/17377

29. Kim, B., et al.: Interpretability beyond feature attribution: quantitative testing with concept activation vectors (TCAV). In: Dy, J., Krause, A. (eds.) Proceedings of the 35th International Conference on Machine Learning. Proceedings of Machine Learning Research, vol. 80, pp. 2668–2677. PMLR, 10–15 July 2018

30. Kingma, D.P., Welling, M.: Auto-encoding variational bayes. In: 2nd International Conference on Learning Representations, ICLR 2014, Banff, AB, Canada, 14–16 April 2014, Conference Track Proceedings (2014)

31. Kong, Z., Ping, W.: On fast sampling of diffusion probabilistic models. In: ICML Workshop on Invertible Neural Networks, Normalizing Flows, and Explicit Likelihood Models (2021)

32. Kynkäänniemi, T., Karras, T., Laine, S., Lehtinen, J., Aila, T.: Improved precision and recall metric for assessing generative models. CoRR abs/1904.06991 (2019)

33. Lee, J.R., Kim, S., Park, I., Eo, T., Hwang, D.: Relevance-CAM: your model already knows where to look. In: Proceedings of the IEEE/CVF Conference on Computer Vision and Pattern Recognition (CVPR), pp. 14944–14953, June 2021
34. Liu, S., Kailkhura, B., Loveland, D., Han, Y.: Generative counterfactual introspection for explainable deep learning. In: 2019 IEEE Global Conference on Signal and Information Processing (GlobalSIP), pp. 1–5 (2019)
35. Liu, Z., Luo, P., Wang, X., Tang, X.: Deep learning face attributes in the wild. In: Proceedings of International Conference on Computer Vision (ICCV), December 2015
36. Van Looveren, A., Klaise, J.: Interpretable counterfactual explanations guided by prototypes. In: Oliver, N., Pérez-Cruz, F., Kramer, S., Read, J., Lozano, J.A. (eds.) ECML PKDD 2021. LNCS (LNAI), vol. 12976, pp. 650–665. Springer, Cham (2021). https://doi.org/10.1007/978-3-030-86520-7_40
37. Madry, A., Makelov, A., Schmidt, L., Tsipras, D., Vladu, A.: Towards deep learning models resistant to adversarial attacks. In: International Conference on Learning Representations (2018)
38. Meng, C., et al.: SDEdit: guided image synthesis and editing with stochastic differential equations. In: International Conference on Learning Representations (2022)
39. Mothilal, R.K., Sharma, A., Tan, C.: Explaining machine learning classifiers through diverse counterfactual explanations. In: Proceedings of the 2020 Conference on Fairness, Accountability, and Transparency (2020)
40. Nauta, M., van Bree, R., Seifert, C.: Neural prototype trees for interpretable fine-grained image recognition. In: Proceedings of the IEEE/CVF Conference on Computer Vision and Pattern Recognition (CVPR), pp. 14933–14943, June 2021
41. Nemirovsky, D., Thiebaut, N., Xu, Y., Gupta, A.: CounteRGAN: generating realistic counterfactuals with residual generative adversarial nets. arXiv preprint arXiv:2009.05199 (2020)
42. Nichol, A.Q., Dhariwal, P.: Improved denoising diffusion probabilistic models (2021)
43. Park, D.H., et al.: Multimodal explanations: justifying decisions and pointing to the evidence. In: Proceedings of the IEEE Conference on Computer Vision and Pattern Recognition (CVPR), June 2018
44. Pawelczyk, M., Agarwal, C., Joshi, S., Upadhyay, S., Lakkaraju, H.: Exploring counterfactual explanations through the lens of adversarial examples: a theoretical and empirical analysis. arXiv:2106.09992 [cs], June 2021
45. Petsiuk, V., Das, A., Saenko, K.: RISE: randomized input sampling for explanation of black-box models. In: British Machine Vision Conference 2018, BMVC 2018, Newcastle, UK, 3–6 September 2018, p. 151. BMVA Press (2018)
46. Petsiuk, V., et al.: Black-box explanation of object detectors via saliency maps. In: IEEE Conference on Computer Vision and Pattern Recognition, CVPR 2021, Virtual, 19–25 June 2021, pp. 11443–11452. Computer Vision Foundation/IEEE (2021)
47. Poyiadzi, R., Sokol, K., Santos-Rodríguez, R., Bie, T.D., Flach, P.A.: FACE: feasible and actionable counterfactual explanations. In: Proceedings of the AAAI/ACM Conference on AI, Ethics, and Society (2020)
48. Rodríguez, P., et al.: Beyond trivial counterfactual explanations with diverse valuable explanations. In: Proceedings of the IEEE/CVF International Conference on Computer Vision (ICCV), pp. 1056–1065, October 2021
49. Saharia, C., et al.: Palette: image-to-image diffusion models. In: NeurIPS 2021 Workshop on Deep Generative Models and Downstream Applications (2021)

50. Saharia, C., Ho, J., Chan, W., Salimans, T., Fleet, D., Norouzi, M.: Image super-resolution via iterative refinement. ArXiv abs/2104.07636 (2021)
51. Sauer, A., Geiger, A.: Counterfactual generative networks. In: 9th International Conference on Learning Representations, ICLR 2021, Virtual Event, Austria, 3–7 May 2021. OpenReview.net (2021)
52. Schut, L., et al.: Generating interpretable counterfactual explanations by implicit minimisation of epistemic and aleatoric uncertainties. In: Banerjee, A., Fukumizu, K. (eds.) Proceedings of the 24th International Conference on Artificial Intelligence and Statistics. Proceedings of Machine Learning Research, vol. 130, pp. 1756–1764. PMLR, 13–15 April 2021
53. Selvaraju, R.R., Cogswell, M., Das, A., Vedantam, R., Parikh, D., Batra, D.: Grad-CAM: visual explanations from deep networks via gradient-based localization. In: Proceedings of the IEEE International Conference on Computer Vision (ICCV), October 2017
54. Shih, S.M., Tien, P.J., Karnin, Z.: GANMEX: one-vs-one attributions using GAN-based model explainability. In: Meila, M., Zhang, T. (eds.) Proceedings of the 38th International Conference on Machine Learning, ICML 2021, 18–24 July 2021, Virtual Event. Proceedings of Machine Learning Research, vol. 139, pp. 9592–9602. PMLR (2021)
55. Singla, S., Pollack, B., Chen, J., Batmanghelich, K.: Explanation by progressive exaggeration. In: International Conference on Learning Representations (2020)
56. Song, J., Meng, C., Ermon, S.: Denoising diffusion implicit models. In: International Conference on Learning Representations (2021)
57. Song, Y., Sohl-Dickstein, J., Kingma, D.P., Kumar, A., Ermon, S., Poole, B.: Score-based generative modeling through stochastic differential equations. In: International Conference on Learning Representations (2021)
58. Tan, S., Caruana, R., Hooker, G., Koch, P., Gordo, A.: Learning global additive explanations for neural nets using model distillation (2018)
59. Thiagarajan, J.J., Narayanaswamy, V., Rajan, D., Liang, J., Chaudhari, A., Spanias, A.: Designing counterfactual generators using deep model inversion. In: Beygelzimer, A., Dauphin, Y., Liang, P., Vaughan, J.W. (eds.) Advances in Neural Information Processing Systems (2021). https://openreview.net/forum?id=iHisgL7PFj2
60. Van Looveren, A., Klaise, J., Vacanti, G., Cobb, O.: Conditional generative models for counterfactual explanations. arXiv preprint arXiv:2101.10123 (2021)
61. Vandenhende, S., Mahajan, D., Radenovic, F., Ghadiyaram, D.: Making heads or tails: towards semantically consistent visual counterfactuals. arXiv preprint arXiv:2203.12892 (2022)
62. Vasu, B., Long, C.: Iterative and adaptive sampling with spatial attention for black-box model explanations. In: Proceedings of the IEEE/CVF Winter Conference on Applications of Computer Vision (WACV), March 2020
63. Wachter, S., Mittelstadt, B., Russell, C.: Counterfactual explanations without opening the black box: automated decisions and the GDPR. Harvard J. Law Technol. **31**(2), 841–887 (2018). https://doi.org/10.2139/ssrn.3063289
64. Wang, H., et al.: Score-CAM: Score-weighted visual explanations for convolutional neural networks. In: Proceedings of the IEEE/CVF Conference on Computer Vision and Pattern Recognition (CVPR) Workshops, June 2020
65. Wang, P., Li, Y., Singh, K.K., Lu, J., Vasconcelos, N.: IMAGINE: image synthesis by image-guided model inversion. In: Proceedings of the IEEE/CVF Conference on Computer Vision and Pattern Recognition (CVPR), pp. 3681–3690, June 2021

66. Wang, P., Vasconcelos, N.: SCOUT: self-aware discriminant counterfactual explanations. In: Proceedings of the IEEE/CVF Conference on Computer Vision and Pattern Recognition, pp. 8981–8990 (2020)
67. Watson, D., Ho, J., Norouzi, M., Chan, W.: Learning to efficiently sample from diffusion probabilistic models. CoRR abs/2106.03802 (2021)
68. Xian, Y., Sharma, S., Schiele, B., Akata, Z.: F-VAEGAN-D2: a feature generating framework for any-shot learning. In: Proceedings of the IEEE/CVF Conference on Computer Vision and Pattern Recognition (CVPR), June 2019
69. Yeh, C.K., Kim, B., Arik, S.Ö., Li, C.L., Pfister, T., Ravikumar, P.: On completeness-aware concept-based explanations in deep neural networks. arXiv: Learning (2020)
70. Hatakeyama, Y., Sakuma, H., Konishi, Y., Suenaga, K.: Visualizing color-wise saliency of black-box image classification models. In: Ishikawa, H., Liu, C.-L., Pajdla, T., Shi, J. (eds.) ACCV 2020. LNCS, vol. 12624, pp. 189–205. Springer, Cham (2021). https://doi.org/10.1007/978-3-030-69535-4_12
71. Zhang, Q., Wu, Y.N., Zhu, S.C.: Interpretable convolutional neural networks. In: Proceedings of the IEEE Conference on Computer Vision and Pattern Recognition (CVPR), June 2018
72. Zhang, R., Isola, P., Efros, A.A., Shechtman, E., Wang, O.: The unreasonable effectiveness of deep features as a perceptual metric. In: Proceedings of the IEEE Conference on Computer Vision and Pattern Recognition (CVPR), June 2018
73. Zhao, Z., Dua, D., Singh, S.: Generating natural adversarial examples. In: 6th International Conference on Learning Representations, ICLR 2018, Vancouver, BC, Canada, 30 April–3 May 2018, Conference Track Proceedings. OpenReview.net (2018)
74. Zhou, B., Khosla, A., Lapedriza, A., Oliva, A., Torralba, A.: Learning deep features for discriminative localization. In: Proceedings of the IEEE Conference on Computer Vision and Pattern Recognition (CVPR), June 2016
75. Zhou, B., Sun, Y., Bau, D., Torralba, A.: Interpretable basis decomposition for visual explanation. In: Ferrari, V., Hebert, M., Sminchisescu, C., Weiss, Y. (eds.) ECCV 2018. LNCS, vol. 11212, pp. 122–138. Springer, Cham (2018). https://doi.org/10.1007/978-3-030-01237-3_8

Style Image Harmonization via Global-Local Style Mutual Guided

Xiao Yan, Yang Lu, Juncheng Shuai, and Sanyuan Zhang[✉]

College of Computer Science and Technology, Zhejiang University, Hangzhou, China
csyanxiao@zju.edu.cn, syzhang@cs.zju.edu.cn

Abstract. The process of style image harmonization is attaching an area of the source image to the target style image to form a harmonious new image. Existing methods generally have problems such as distorted foreground, missing content, and semantic inconsistencies caused by the excessive transfer of local style. In this paper, we present a framework for style image harmonization via global and local styles mutual guided to ameliorate these problems. Specifically, we learn to extract global and local information from the Vision Transformer and Convolutional Neural Networks, and adaptively fuse the two kinds of information under a multi-scale fusion structure to ameliorate disharmony between foreground and background styles. Then we train the blending network GradGAN to smooth the image gradient. Finally, we take both style and gradient into consideration to solve the sudden change in the blended boundary gradient. In addition, supervision is unnecessary in our training process. Our experimental results show that our algorithm can balance global and local styles in the foreground stylization, retaining the original information of the object while keeping the boundary gradient smooth, which is more advanced than other methods.

1 Introduction

Style image harmonization is a kind of image synthesis technique. It allows artists to create new artworks with existing materials. When pasting keying footage with different styles onto the background image, style image harmonization helps pasted materials to mix the style of the background image and make the overall image harmonious. Artistic image editing is a time-consuming process and is difficult to edit under style images. Due to the sensitivity of the human visual system [1], this disharmony of the synthetic image can cause visual discomfort. These discords mainly stem from (1) the inconsistency between the foreground and the background styles, (2) and a sudden change of gradient at the boundary of foreground and background. The problems of inconsistent styles between the foreground and the background, the incoordination of factors such as color and texture between the two, and the sudden change in gradient of the boundary are remaining to be solved.

Supplementary Information The online version contains supplementary material available at https://doi.org/10.1007/978-3-031-26293-7_15.

Fig. 1. With the mask of the source image and the composite image, our algorithm can transfer the style of the background to the foreground pasted object and smooth the boundary to make the overall image harmonious.

Recently, some deep learning methods can be applied to the style image harmonization, but there are still some other problems. Wu et al. [2] proposed to combine Generative Adversarial Networks (GANs) [3] and traditional Poisson blending to synthesize real-looking images, but its training required a well-blended Ground Truth as supervision. And there was still a background color seeping into the foreground, causing the foreground to lose its semantics. In addition, artifacts were produced in some areas. Zhang et al. [4] proposed to jointly optimize the Poisson blending, content and style calculated from the deep network to iteratively update the image blend area, but it had obvious foreground distortion caused by excessive style transfer. In conclusion, the main reason for the existing problem caused by the current methods is the use of Convolutional Neural Networks (CNNs) [5–8], which cause the foreground to be affected by the corresponding regional style of the background, resulting in distortion and artifacts. Lately, the transformer-based style transfer proposed by Deng et al. [9] solved the problem that CNNs have difficulty obtaining global information and content leak of input images on style transfer, but it still cannot be applied to the local style transfer, and having a global style on the foreground makes it incompatible with the surroundings.

Style image harmonization consists of two parts: stylization and harmonization. There are some excellent methods of style transfer and image harmonization, but simple combinations cannot bring desirable results. Cong et al. [10,11] found that converting the foreground domain to the background domain helps to guide the harmonization of the foreground with background information. Sofiiuk et al. [12] used an encoder-decoder framework, which combines pre-trained foreground-aware deep high-resolution network to obtain composite images with semantic visuals. However, combining the state-of-the-art methods mentioned above with style transfer models, will still have problems such as distortion of the foreground leading to bad visuals.

In this paper, we present a framework for style images harmonization via global and local styles mutual guided, which solves the problems of foreground distortion, content loss, and semantic inconsistencies caused by the excessive transfer of local style in the existing methods, and realize a better integration of foreground pasted object into the background style image. For the harmonization of foreground and background styles, we come up with a method for style transfer that blends global and local styles. Learn from the Vision Transformer(ViT) to extract global information and CNNs to extract local information, and adaptively fuse them using a multi-scale fusion structure. For the discord caused by the sudden change of gradient at the boundary of foreground and background, we train the blending network GradGAN to recover the gradient smoothness of the blended images, and then fuse style and gradient result in a harmonious deep image. In addition, we improved the blending loss so that the training process does not require any supervision.

Our main contributions are as follows:

- We propose a novel Blending Decoder that learns to extract global and local information from the ViT and CNNs, and blends this information to make the pasted foreground have a more reasonable style.
- Motivated by Liu et al. [13], we propose a multi-scale adaptive fusion structure that bridges ViT and CNNs.
- Different from Wu et al. [2], we improve the blending loss so that the training process only needs the foreground object, background image, and mask without supervision.
- Our experimental results show that our algorithm can balance global and local styles in the foreground stylization, retaining the original information of the object while keeping the boundary gradient smooth, which is more advanced than other methods. Some parts of the results as shown in Fig. 1.

2   Relative Work

2.1   Style Transfer

The task of style transfer is to transfer the style of one drawing into another [5, 6,14,15]. Early style transfer was achieved by histogram matching [16] or global image statistics transfer [17]. Gatys et al. [5] designed the first style transfer algorithm with neural networks, using the convolutional features of VGG19 [18] and its Gram matrix to represent content and style. Huang et al. [6] proposed model-based arbitrary style transfer by making the mean and standard deviation of each channel of the content images the same as those of the style images, which is commonly used in various generation tasks [19–22]. Li et al. [7] utilized the idea of de-stylization and stylization, making the multi-layer stylization modules enable the styles of rendered images to be transferred on multiple feature scales. Recently, Vision Transformer (ViT) [22] has been widely used in the field of vision [23–26]. In order to solve the locality and spatial invariance of CNNs, Deng et al. [9] proposed transformer-based style transfer and Content-aware Positional Encoding (CAPE) to adapt to different input sizes. However, there are few methods specifically addressing the problem of local style transfer in the harmonization of style images.

2.2   Image Blending

The task of image blending is to paste an area of the cropped source image onto the target image and make the image looks harmonious as a whole. Traditional blending methods use low-level appearance statistics [27–30] to adjust the foreground image. Alpha Blend [31] uses alpha channel values of the foreground and background to blend images. Recent blending techniques primarily use gradient smoothing [32–34], which targets the smooth transition of gradients at blend boundaries due to human sensitivity to regions of the sudden change of gradient. The earliest work [35] reconstructs pixels in blended regions through gradient-domain consistency. A large number of methods are used for realistic image blending [12,36,37], only a few methods are for style images. Luan et al. [38] proposed the use of iterative stylistic transfers and refining them with adjacent pixels, which was the first method of blending paintings. Recent methods [2,4] combined neural networks and Poisson blending to generate realistic images. Jiang et al. [39] used cropping perturbed images to handle the stylistic images blending, which uses 3D color lookup tables (LUTs) to find information such as hue, brightness, and contrast. All of these approaches distort the foreground and lose its semantics. Our algorithm blends global and local styles, then smooths the blended boundary, and improves the deficiencies in the existing methods.

3   Algorithm

3.1   Overview

Our model achieves style image harmonization followed by StyTr^2 [9], WCT [7], and GP-GAN [2]. Our training data pair is $(x, y, m), x, y, m \in R^{W \times H \times 3}$. x is a composite image containing the foreground, y is the corresponding background image, and m is mask of the foreground. The goal is to blend the foreground of x into the entire image, maintain its texture and semantics while transferring style, and smoothly transitioning the paste boundary gradient with the surrounding gradient. We learn from a generator with an encoder-decoder structure to turn the source image into the target image. The objective function is shown in Eq. 1.

$$target = style(BD(TE(p), CE(p))) + poisson(GradGAN(p)), p = (x, y, m). \quad (1)$$

TE is Transformer Encoder, CE is CNN Encoder, BD is Blending Decoder, $GradGAN$ is a network for image preliminary blending, $style$ and $poisson$ are style constraint and poisson constraint for images, and *target* is the generation target.

Our framework contains five main components: A global style transfer based on the transformer, a local style transfer based on CNN, a global and local blending module, a gradient smoothing module, and a style-gradient fusion module. Figure 2 provides an overview of our framework. First, we will introduce global and local mutual guided style transfer (Sect. 3.2), so that the source cropped area has the target style while more in line with its semantics. Second, gradient-guided image fusion (Sect. 3.3) to make the stylized areas smoother in the gradient at the paste boundary. Finally, we detail the objective function (Sect. 3.4).

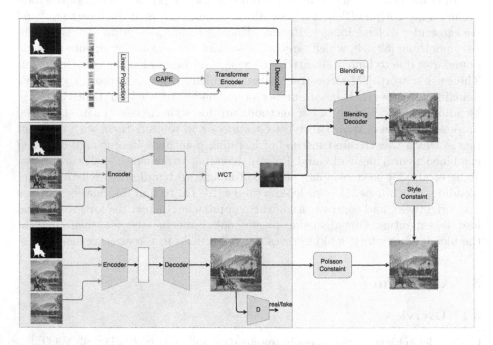

Fig. 2. Our framework contains five main components. Global style transfer based on the transformer (green part) and local style transfer based on CNN (pink part) encode style and content image separately. The global and local blending module Blending Decoder (blue part) decodes the latent code to get an image of the stylized foreground. GradGAN (orange part) generates a gradient smooth image of the boundary of the pasted area, blending styles and gradients (yellow part), and so on getting the final stylistic image harmonization output. (Color figure online)

3.2   Global and Local Mutual Guided Style Transfer

To solve the discord caused by excessively style transfer, we propose a global and local mutual guided style transfer, using a multi-scale fusion structure to bridge transformer and CNN. Blend the extracted global and local style to make the foreground has a more reasonable style. The framework consists of two encoders and one decoder.

Encoder. We utilize WCT decoders, which extract features from high-level (relu_5) to low-level (relu_1) via VGG19 [18] for Whitening and Coloring Transform (WCT).

We use transformer encoder [9] as another encoder, which contains a style encoder and a content encoder. As shown in Fig. 2, the composition and background images are used to obtain the patch sequence of images through linear projection respectively, and Content-aware Positional Encoding (CAPE) is used only for the content sequence. The input sequence is encoded as Q (query) K (key) V (value), giving the sequence outputs of style and content respectively.

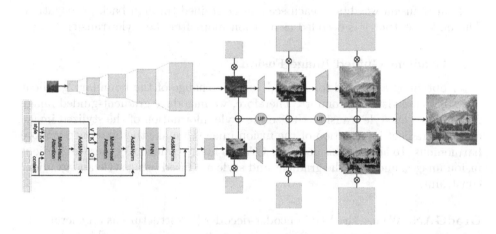

Fig. 3. We propose an adaptive multi-scale Blending Decoder. Using a multi-scale fusion structure to connect the equivalent feature maps of transformers and CNNs, bridge transformer decoders and CNN decoders, which blends global and local styles.

Adaptive Multi-scale Blending Decoder. Inspired by ASFF [13], we propose an adaptive multi-scale fusion structure that bridges the WCT decoder and the transformer decoder.

As shown in Fig. 3, in the CNN decoder path, we change the VGG19 decoder's structure in the WCT to extract three feature maps of different scales. In the transformer decoder path, the style sequence is represented as K and V, and the content sequence is represented as Q. The transformer decoder layer contains multi-head attention and a Factorization Machine supported Neural Network (FNN). The output sequence of the transformer is in the shape of $\frac{W \times H \times C}{64}$. Then use the decoder to obtain three feature maps that are equivalent to the CNN decoder output size.

Multi-scale fusion structure connects the equivalent feature maps, adaptively learning fusion spatial weights, specifically shown in Eq. 2.

$$y_{ij} = \alpha \cdot up(\alpha_{ij}^1 \cdot x_{ij}^{C1} + \alpha_{ij}^2 \cdot x_{ij}^{T1}) + \beta \cdot up(\beta_{ij}^1 \cdot x_{ij}^{C2} + \beta_{ij}^2 \cdot x_{ij}^{T2}) + \gamma \cdot up(\gamma_{ij}^1 \cdot x_{ij}^{C3} + \gamma_{ij}^2 \cdot x_{ij}^{T3}), \quad (2)$$

where y_{ij} represents the pixel value at (i, j) of the output image, $x_{ij}^{C1}, x_{ij}^{C2}, x_{ij}^{C3}$ represent pixel values at (i, j) of the feature maps of three different scales of CNN path, $x_{ij}^{T1}, x_{ij}^{T2}, x_{ij}^{T3}$ represent the pixel values at (i, j) of the feature map of three different scales of the transformer path, $\alpha_{ij} \in \alpha, \beta_{ij} \in \beta, \gamma_{ij} \in \gamma$ represent the spatial weight values at different scales, up represents upsampling, and α, β, γ represent the fusion weights of different scales. At the meantime, make $\alpha_{ij}^1 + \alpha_{ij}^2 = 1, \beta_{ij}^1 + \beta_{ij}^2 = 1, \gamma_{ij}^1 + \gamma_{ij}^2 = 1$, where α, β, γ is defined by Eq. 3.

$$\alpha_{ij}^l = \frac{e^{\lambda_{\alpha_{ij}^l}}}{e^{\lambda_{\alpha_{ij}^1}} + e^{\lambda_{\alpha_{ij}^2}}}, \beta_{ij}^l = \frac{e^{\lambda_{\beta_{ij}^l}}}{e^{\lambda_{\beta_{ij}^1}} + e^{\lambda_{\beta_{ij}^2}}}, \gamma_{ij}^l = \frac{e^{\lambda_{\gamma_{ij}^l}}}{e^{\lambda_{\gamma_{ij}^1}} + e^{\lambda_{\alpha_{ij}^2}}}. \quad (3)$$

Spatial fusion weights at each scale are obtained through back propagation. The middle output y is decoded as a fusion map after the style transfer.

3.3   Gradient-Guided Image Fusion

After obtaining a stylized foreground, sudden change of the boundary gradient can still cause visual discomfort. Therefore, we introduce gradient-guided image fusion based on style transfer. Utilize the style information of the stylized image and the gradient information of the fusion image to make the final output more harmonious. To achieve this goal, the GAN is firstly trained to generate gradient fusion images, and then the gradient and style are fused using style and gradient constraint.

GradGAN. We use the VGG encoder-decoder [18] structure as a generator to fuse the composite image to obtain an image after gradient smoothing. And use the patch discriminator to train against the generator to get a more realistic one.

Fusion. We smooth the gradient of the foreground boundary of the composite image using GradGAN, but it is still not harmonious. Different from GP-GAN [2] which needs the gradient of the original image, we fuse the gradient and the style of the generated image to obtain the final deep blending image. The low-level style and the high-level gradient are used for fusion, and the underlying features are extracted as the style constraint and the high-level features are extracted as gradient constraint.

$$S(x, x^{style}) = \sum_{ij} ||F(x_{ij}) - x_{ij}^{style}||_2,$$

$$G(x, x^{grad}) = \sum_{ij} ||P(x_{ij}^{grad}) - L(x_{ij})||_2. \quad (4)$$

The style constraint and gradient constraint for the target are shown in Eq. 4. x^{style} is the image generated by style transfer, x^{grad} is GradGAN blending image, P is gradient operator, F is the filter and L is the Laplace operator that extracts the low-level style and the high-level gradient of the image respectively.

$$T(x) = \omega \cdot S(x, x^{style}) + \varphi \cdot G(x, x^{grad}). \tag{5}$$

The final optimization goal is shown in Eq. 5. ω is style reserved parameters and φ is gradient reserved parameters.

3.4 Optimization

Style Transfer Loss. The result of the style transfer should have the same content as the composite image and the same style as the background image [7]. Therefore, the style transfer loss consists of three parts: the content perception loss, the style perception loss, and the reconstruction loss. This loss is defined by Eq. 6.

$$
\begin{aligned}
L_c &= \frac{1}{N_c} \sum_{i=1}^{N_c} ||\phi_i[x] - \phi_i[x^{compose}]||_2, \\
L_s &= \frac{1}{N_s} \sum_{i=1}^{N_s} ||Gram_i[x] - Gram_i[x^{background}]||_2, Gram_i = \phi_i[\cdot]\phi_i[\cdot]^T,
\end{aligned}
\tag{6}
$$

where x is output image, $x^{compose}$ is the composite image, $x^{background}$ is the background image, and ϕ_i is the feature map of middle layer of VGG19.

Self-supervision can help network training [40]. Therefore, we use the reconstruction loss $L_{rec} = L_c(x, I) + L_s(x, I)$ to learn more accurate content and style representations. Input two identical images I, both as content images and as style images.

$$L_{transfer} = \lambda_c L_c + \lambda_s L_s + \lambda_{rec} L_{rec}. \tag{7}$$

The total style transfer loss is shown in Eq. 7. $\lambda_c, \lambda_s, \lambda_{rec}$ is set separately for the weights of each loss.

Blending Loss. Different from GP-GAN which requires Ground Truth to recover low-resolution coarse images, our model only uses composite images and background images to recover preliminary fusion image. Specifically, blending loss consists of three parts: the generating adversarial loss, the perceptual loss, and the gradient loss.

$$
\begin{aligned}
L_{adv}^G &= E_{x' \sim P_{data}(x')}[\log D(x')] + E_{I \sim P_{data}(I)}(\log(1 - D(decoder(encoder(I))))), \\
L_{adv}^D &= -E_{x' \sim P_{data}(x')}[\log D(x')] - E_{x \sim P_G}[\log(1 - D(x))].
\end{aligned}
\tag{8}
$$

Generative adversarial loss is defined by Eq. 8, where I is the source composite image, x' is the real background image, x is the generated image, and D is the discriminator.

| Copy and Paste | Mask | Mask_dilated | Copy and Paste | Mask | Mask_dilated |

Fig. 4. The source composite image and *mask* and *mask_dilated* corresponding to the foreground. We dilate the *mask* and calculate only the gradient loss inside the *mask_dilated*.

The perceptual loss is L2 loss [41], which accelerates training and produces sharp images compared to the L1 loss [42]. We use L2 loss to make the output content the same as the foreground content of the composite image, and the output style the same as the background style.

$$L_{grad} = \sum_{ij}(|x_{i,j} - x_{i-1,j}| + |x_{i,j-1} - x_{i,j}|). \qquad (9)$$

The gradient loss penalizes the output gradient. Since only the gradient of the composite image boundary needs to be smoothed, we dilate the mask and only calculate the gradient loss inside the dilated mask. Mask and dilated mask are shown in Fig. 4. This loss is defined by Eq. 9, where x is the generated image.

$$L_{harmony} = \lambda_{adv}L_{adv} + \lambda_2 L_2 + \lambda_{grad}L_{grad}. \qquad (10)$$

The total fusion loss is shown in Eq. 10. $\lambda_{adv}, \lambda_2, \lambda_{grad}$ is set separately for the weights of each loss.

4   Experiments

4.1   Implementation Details

This section describes the implementation details of our method. For style transfer branches, we use StyTr^2 [9] and VGG19 [18] as pre-trained models, adopt Adam [43] optimizer, and employ warm-up training strategies [44]. The initial learning rate is set to 5×10^{-4}, and the decays to 10^{-5}. The $conv1\_1$, $conv2\_1$, $conv3\_1$, $conv4\_1$ of VGG19 are chosen as style representation and $conv4\_1$ as content representation. λ_c is set to 7, λ_s is set to 10, and λ_{rec} is set to 10 in Eq. 7. For GradGAN branches, we adopt Adam optimizer, where α is set to 10^{-4}, β_1 is set to 0.9, and β_2 is set to 0.999. ω is set to 1, φ is set to 1, λ_2 is set to 10 in Eq. 10. All images are reshaped into 256×256, and the datasets we used are from [2,38].

4.2 Experimental Results

In this section, we compare our method with the existing methods. Qualitative and quantitative comparisons were made, including ablation experiments, comparative experiments, and user studies.

Ablation Experiments. To fully illustrate the need for global and local blending, and gradient fusion, example results with different degrees of texture information richness are shown. As shown in the Fig. 5, the full model is superior to other baseline models. The result of w/o global (WCT [7] combine with GradGAN) is that the local style around the foreground transfer to the object, which causes the object to lose its original semantics and produce distortion. The result of w/o global&grad (WCT [7]) seems to be more harmonious with the entire image in the gradient, but the original information in the foreground is seriously missing due to its local style transfer. The result of w/o local (stytr^2 [9] combined with GradGAN) is that this position-independent global style makes the foreground incompatible with the surrounding. The result of w/o local&grad (StyTr^2 [9]) retains the original information, but it is still relatively abrupt in the entire image because it does not transfer local style and may have a big change in the gradient. The result of w/o grad (only style transfer) is more reasonable than the first two in style processing, but there is still a certain degree of texture loss. In contrast, The full model shows the best results, with the foreground area retaining its original texture and semantics while transferring the background style, and aligning with the surrounding style, while its boundary and background gradients are smoother.

Comparative Experiment. We compare our method with three others (See Fig. 6): GP-GAN [2], SSH [39], and Deep Image Blending [4]. GP-GAN is an image fusion algorithm that combines Poisson fusion and GAN, and is trained in a supervised manner. However, the color is transferred to the foreground from around, making it inconsistent with the original semantics, and it also creates artifacts in some areas. SSH uses dual data enhancement to crop perturbed images, and uses 3D color lookup tables to find information such as hue, brightness, contrast, etc., to process both real and stylistic images. However, the pasted boundary of the processed style image is very obtrusive. Deep Image Blending is an improved method based on GP-GAN, using VGG19 for style transfer, and joint optimization of Poisson loss and content style loss to blend deep images. But it distorts and produces artifacts more heavily in the foreground. Due to using VGG19 [18] as a style transfer network, the foreground has a distinctly localized style, which is unrealistic. Our model shows the best results, balancing global and local styles in the foreground stylization, maintaining the original information of the object, and making the surrounding gradient smoother.

Figure 7 shows the results of stylizing and harmonizing the image foreground using the mainstream styles transfer models [7,9], harmonization models [10,11], and stylization and harmonization of combination models separately.

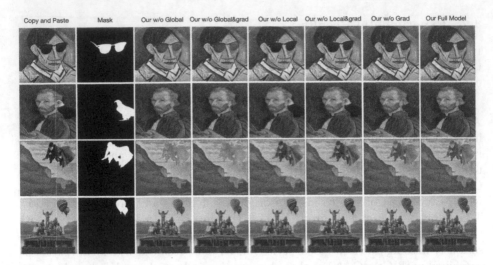

Fig. 5. Experimental study of ablation of global, local, and gradient fusion. The full model shows the best result, however other baseline models either distort the original information or are not in harmony with the surrounding, or there are problems such as a sudden change in the gradient.

As shown in the Fig. 7, the style transfer models of the traditional CNN places more emphasis on local style, and produces content distortion, such as the result of the WCT that distorts the eye part of the foreground character. Recent style transfer using transformers places more emphasis on global style, but is less coordinated with the surroundings, and the variation of gradient makes the visual effect more obtrusive. The harmonization models do not handle stylistic images very well. BargainNet [10] and Dovenet [11] converted the foreground domain to a background domain, with background information guiding the harmonization of the foreground. But we can see that the brightness of the foreground is relatively obtrusive relative to the surrounding pixels. There are translation failures in some images: the foreground lacks style information and does not fit well into the background. The result of stylization and harmonization of combination models still has foreground distortion or boundary pixel obtrusion, and the style image harmonization is not well handled. We blend global and local styles, smooth the boundary, and achieve good results.

User Studies. We conducted user studies to quantitatively evaluate the experimental results. The first experiment verifies the quality of the image generated by judging whether the provided image has been edited by the user. The second experiment compares the quality of ours and others by selecting the optimal one from the images generated by different methods. At the same time, we also measured the user reaction time to further verify the effectiveness and robustness of our method.

Copy and Paste	Mask	GP-GAN	SSH	Deep Image Blending	Ours

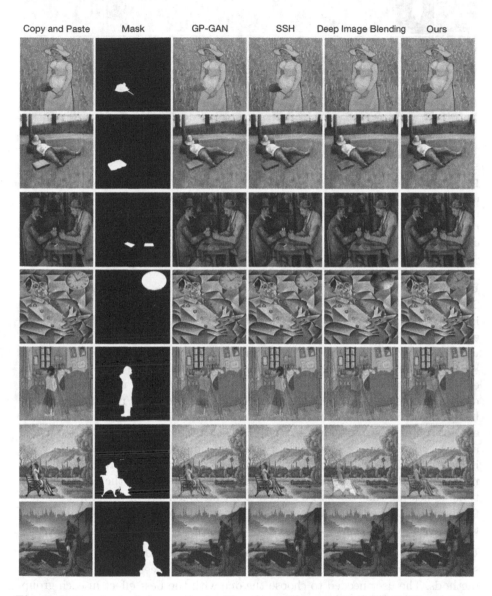

Fig. 6. Comparison of our method with others. The results of GP-GAN and Deep Image Blending show that the foreground is distorted and artifacts are produced. The gradient of pasted boundary processed by SSH is very obtrusive. Our model shows the best results, balancing global and local styles, maintaining the original information of the object, and smoother with the surrounding gradients.

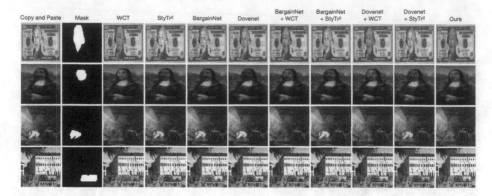

Fig. 7. Either direct use of mainstream or the latest stylized and harmonized combination models to stylize and harmonize the foreground can handle the harmonization task of style images very well. And our model blends global and local styles, smooths the boundary, and achieves the best results. (See the supplementary materials for a clearer version.)

User Study 1: Whether to Edit. We invited 30 users, and randomly selected 20 images generated by four different methods (Sect. 4.2), and unprocessed images. The user needs to answer whether it has been edited and click on the edited part. We recorded the response time of the user's answer. We asked the user in advance if they were familiar with the image in case of the impact of prior knowledge. And we only think that the correct answer is the sample if both the edit and the part click are correct, in case other parts that may be edited will cause interference. We counted the response time and error rate of each image being answered, and the unedited image statistics were answered correctly for ease of comparison. As shown in Fig. 8, it is clear that our method has a higher answer error rate and a longer user response time than other methods, which is closest to the unedited image. The high rate of answer errors indicates that most users believe that this is an unedited image. The longer the user's reaction time, indicating that the user's observation time is longer, the more difficult it is to distinguish. The larger the area that intersects the coordinate axis, the better the result obtained by the method in general.

User Study 2: Quality Comparison. We invited 30 users, and randomly selected 10 groups of images, and each group was processed by four different methods. The user needed to choose the one with the best effect in each group, and we recorded the reaction time selected by the user. The faster the reaction time, the better the method is than the others. As shown in Fig. 8, ours are considered by most users to be the best, and some images are far better than others.

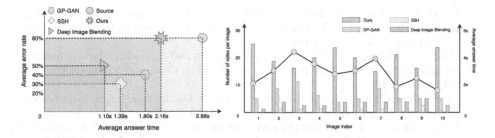

Fig. 8. Statistical results of "User Study 1 - Whether to Edit" and "User Study 2 - Quality Comparison". The higher the answer error rate, the longer the average reaction time, and the larger the area that intersects the coordinate axis, the better the result. Our method has the best results and is closer to the unprocessed image. And ours is the most selected by the user, and some images are far better than others.

5 Conclusion

Style image harmonization is the process of pasting the cropped areas from the source image into the background style image and harmonizing two as a whole. We propose a style image harmonization in which global and local information guide mutually, and solve the problems of foreground distortion, content loss, and semantic inconsistencies caused by the excessive transfer of local styles in the existing methods. Firstly, global and local styles are extracted by the transformer and CNNs separately, and an adaptive multi-scale fusion structure bridges the transformer decoder and CNNs decoder to fuse global and local styles. Secondly, the blending network GradGAN smooths the image gradient. Finally, the fusion style and gradient result in a harmonious deep image.

To evaluate the method presented, we made quantitative and qualitative comparisons. Compared to the existing methods, our model shows the best results, balancing the global and local styles on the foreground stylization, maintaining the original information of the object, and smoother with the surrounding gradient. User studies have shown that the images processed by our model are often considered unedited ones, which is superior to the results of other methods. We believe that our approach assists artists in editing their work, providing more possibilities for users to create works of art.

References

1. Xue, S., Agarwala, A., Dorsey, J., Rushmeier, H.: Understanding and improving the realism of image composites. ACM Trans. Graph. **31**, 1–10 (2012)
2. Wu, H., Zheng, S., Zhang, J., Huang, K.: GP-GAN: towards realistic high-resolution image blending. In: ACM International Conference on Multimedia, pp. 2487–2495 (2019)
3. Goodfellow, I., Pouget-Abadie, J., et al.: Generative adversarial nets. In: Advances in Neural Information Processing Systems, vol. 27 (2014)
4. Zhang, L., Wen, T., Shi, J.: Deep image blending. In: IEEE/CVF Winter Conference on Applications of Computer Vision, pp. 231–240 (2020)

5. Gatys, L.A., Ecker, A.S., Bethge, M.: Image style transfer using convolutional neural networks. In: IEEE Conference on Computer Vision and Pattern Recognition (CVPR), pp. 2414–2423 (2016)

6. Huang, X., Belongie, S.: Arbitrary style transfer in real-time with adaptive instance normalization. In: IEEE International Conference on Computer Vision (ICCV), pp. 1501–1510 (2017)

7. Li, Y., Fang, C., et al.: Universal style transfer via feature transforms. In: Advances in Neural Information Processing Systems, vol. 30 (2017)

8. Liu, S., Lin, T., et al.: AdaAttN: revisit attention mechanism in arbitrary neural style transfer. In: IEEE International Conference on Computer Vision (ICCV), pp. 6649–6658 (2021)

9. Deng, Y., Tang, F., et al.: StyTr^2: image style transfer with transformers. In: IEEE Conference on Computer Vision and Pattern Recognition (CVPR), pp. 11326–11336 (2022)

10. Cong, W., Niu, L., Zhang, J., Liang, J., Zhang, L.: Bargainnet: background-guided domain translation for image harmonization. In: IEEE International Conference on Multimedia and Expo (ICME), pp. 1–6 (2021)

11. Cong, W., Niu, L., et al.: Dovenet: deep image harmonization via domain verification. In: IEEE Conference on Computer Vision and Pattern Recognition (CVPR), pp. 8394–8403 (2020)

12. Sofiiuk, K., Popenova, P., Konushin, A.: Foreground-aware semantic representations for image harmonization. In: IEEE/CVF Winter Conference on Applications of Computer Vision, pp. 1620–1629 (2021)

13. Liu, S., Huang, D., Wang, Y.: Learning spatial fusion for single-shot object detection. arXiv preprint (2019)

14. Jing, Y., Liu, X., et al.: Dynamic instance normalization for arbitrary style transfer. In: AAAI Conference on Artificial Intelligence, vol. 34, pp. 4369–4376 (2020)

15. An, J., Huang, S., et al.: Unbiased image style transfer via reversible neural flows. In: IEEE Conference on Computer Vision and Pattern Recognition (CVPR), pp. 862–871 (2021)

16. Pitie, F., Kokaram, A.C., Dahyot, R.: N-dimensional probability density function transfer and its application to color transfer. In: IEEE International Conference on Computer Vision (ICCV), pp. 1434–1439 (2005)

17. Reinhard, E., Adhikhmin, M., Gooch, B., Shirley, P.: Color transfer between images. IEEE Comput. Graph. Appl. **21**, 34–41 (2001)

18. Sengupta, A., Ye, Y., et al.: Going deeper in spiking neural networks: VGG and residual architectures. Front. Neurosci. **13**, 95 (2019)

19. Xia, X., et al.: Joint bilateral learning for real-time universal photorealistic style transfer. In: Vedaldi, A., Bischof, H., Brox, T., Frahm, J.-M. (eds.) ECCV 2020. LNCS, vol. 12353, pp. 327–342. Springer, Cham (2020). https://doi.org/10.1007/978-3-030-58598-3_20

20. Gu, J., Ye, J.C.: AdaIN-based tunable CycleGAN for efficient unsupervised low-dose CT denoising. IEEE Trans. Comput. Imaging **7**, 73–85 (2021)

21. Karras, T., Laine, S., et al.: Analyzing and improving the image quality of StyleGAN. In: IEEE Conference on Computer Vision and Pattern Recognition (CVPR), pp. 8110–8119 (2020)

22. Dosovitskiy, A., Beyer, L., et al.: An image is worth 16x16 words: transformers for image recognition at scale. arXiv preprint (2020)

23. Yuan, L., Chen, Y., et al.: Tokens-to-token ViT: training vision transformers from scratch on imagenet. In: IEEE International Conference on Computer Vision (ICCV), pp. 558–567 (2021)

24. Arnab, A., Dehghani, M., et al.: ViViT: a video vision transformer. In: IEEE International Conference on Computer Vision (ICCV), pp. 6836–6846 (2021)
25. Wang, W., Xie, E., et al.: PVT v2: improved baselines with pyramid vision transformer. Comput. Vis. Media **8**, 1–10 (2022)
26. Zhang, P., Dai, X., et al.: Multi-scale vision longformer: a new vision transformer for high-resolution image encoding. In: IEEE International Conference on Computer Vision (ICCV), pp. 2998–3008 (2021)
27. Grundland, M., Vohra, R., et al.: Cross dissolve without cross fade: preserving contrast, color and salience in image compositing. In: Computer Graphics Forum, vol. 25, pp. 557–586 (2006)
28. Sunkavalli, K., Johnson, M.K., et al.: Multi-scale image harmonization. ACM Trans. Graph. (TOG) **29**, 1–10 (2010)
29. Tao, M.W., Johnson, M.K., Paris, S.: Error-tolerant image compositing. In: Daniilidis, K., Maragos, P., Paragios, N. (eds.) ECCV 2010. LNCS, vol. 6311, pp. 31–44. Springer, Heidelberg (2010). https://doi.org/10.1007/978-3-642-15549-9_3
30. Jia, J., Sun, J., et al.: Drag-and-drop pasting. ACM Trans. Graph. (TOG) **25**, 631–637 (2006)
31. Porter, T., Duff, T.: Compositing digital images. In: Annual Conference on Computer Graphics and Interactive Techniques, pp. 253–259 (1984)
32. Fattal, R., Lischinski, D., Werman, M.: Gradient domain high dynamic range compression. In: Annual Conference on Computer Graphics and Interactive Techniques, pp. 249–256 (2002)
33. Levin, A., Zomet, A., Peleg, S., Weiss, Y.: Seamless image stitching in the gradient domain. In: Pajdla, T., Matas, J. (eds.) ECCV 2004. LNCS, vol. 3024, pp. 377–389. Springer, Heidelberg (2004). https://doi.org/10.1007/978-3-540-24673-2_31
34. Szeliski, R., Uyttendaele, M., et al.: Fast poisson blending using multi-splines. In: IEEE International Conference on Computational Photography (ICCP), pp. 1–8 (2011)
35. Pérez, P., Gangnet, M., et al.: Poisson image editing. In: ACM SIGGRAPH 2003 Papers, pp. 313–318 (2003)
36. Ling, J., Xue, H., et al.: Region-aware adaptive instance normalization for image harmonization. In: IEEE Conference on Computer Vision and Pattern Recognition (CVPR), pp. 9361–9370 (2021)
37. Guo, Z., Zheng, H., et al.: Intrinsic image harmonization. In: IEEE Conference on Computer Vision and Pattern Recognition (CVPR), pp. 16367–16376 (2021)
38. Luan, F., Paris, S., et al.: Deep painterly harmonization. In: Computer Graphics Forum, vol. 37, pp. 95–106 (2018)
39. Jiang, Y., Zhang, H., et al.: SSH: a self-supervised framework for image harmonization. In: IEEE International Conference on Computer Vision (ICCV), pp. 4832–4841 (2021)
40. Jing, L., Tian, Y.: Self-supervised visual feature learning with deep neural networks: a survey. IEEE Trans. Pattern Anal. Mach. Intell. **43**, 4037–40581 (2020)
41. Johnson, J., Alahi, A., Fei-Fei, L.: Perceptual losses for real-time style transfer and super-resolution. In: Leibe, B., Matas, J., Sebe, N., Welling, M. (eds.) ECCV 2016. LNCS, vol. 9906, pp. 694–711. Springer, Cham (2016). https://doi.org/10.1007/978-3-319-46475-6_43
42. Zhao, H., Gallo, O., et al.: Loss functions for image restoration with neural networks. IEEE Trans. Comput. Imaging **3**, 47–40581 (2016)

43. Kingma, D.P., Ba, J.: Adam: a method for stochastic optimization. arXiv preprint (2014)
44. Xiong, R., Yang, Y., et al.: On layer normalization in the transformer architecture. In: International Conference on Machine Learning (PMLR), pp. 10524–10533 (2020)

Segmentation and Grouping

Revisiting Image Pyramid Structure for High Resolution Salient Object Detection

Taehun Kim$^{(\boxtimes)}$ [iD], Kunhee Kim, Joonyeong Lee, Dongmin Cha, Jiho Lee, and Daijin Kim

Department of CSE, Pohang University of Science and Technology (POSTECH), Pohang, Korea
{taehoon1018,kunkim,joonyeonglee,cardongmin,jiholee,dkim}@postech.ac.kr
https://github.com/plemeri/InSPyReNet.git

Abstract. Salient object detection (SOD) has been in the spotlight recently, yet has been studied less for high-resolution (HR) images. Unfortunately, HR images and their pixel-level annotations are certainly more labor-intensive and time-consuming compared to low-resolution (LR) images and annotations. Therefore, we propose an image pyramid-based SOD framework, Inverse Saliency Pyramid Reconstruction Network (InSPyReNet), for HR prediction without any of HR datasets. We design InSPyReNet to produce a strict image pyramid structure of saliency map, which enables to ensemble multiple results with pyramid-based image blending. For HR prediction, we design a pyramid blending method which synthesizes two different image pyramids from a pair of LR and HR scale from the same image to overcome effective receptive field (ERF) discrepancy. Our extensive evaluations on public LR and HR SOD benchmarks demonstrate that InSPyReNet surpasses the *State of-the-Art* (SotA) methods on various SOD metrics and boundary accuracy.

1 Introduction

While there are many successful works for SOD in low-resolution (LR) images, there are many demands on high-resolution (HR) images. One can argue that methods trained with LR datasets produce decent results on HR images by resizing the input size (Fig. 1a), but the quality in terms of the high-frequency details of prediction still remains poor in that way. Moreover, previous studies on HR prediction have been working on developing complex architectures and proposing laborious annotations on HR images [1–4] (Fig. 1b, c).

In this paper, we focus on only using LR datasets for training to produce high-quality HR prediction. To do so, we mainly focus on the structure of saliency prediction, which enables to provide high-frequency details from the image regardless of the size of the input. However, there is still another problem to be solved where the effective receptive fields (ERFs) [5] of HR images are different from

---

Supplementary Information The online version contains supplementary material available at https://doi.org/10.1007/978-3-031-26293-7_16.

the LR images in most cases. To alleviate the aforementioned issues, we propose two solid solutions which are mutually connected to each other.

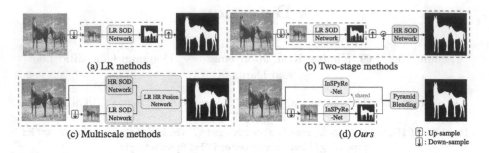

Fig. 1. Different approaches for HR SOD prediction. Areas denoted as a dashed box are trained with supervision. (a): Resizing HR input to LR, then up-sample. Works for any methods, lack of details. (b): Requires multiple training sessions, and HR datasets [1,2]. (c): Can overcome ERF discrepancy, but the architecture is complex, requires HR datasets [4]. (d): Works without HR dataset training. We predict multiscale results with single network and synthesize HR prediction with pyramid blending.

First is to design a network architecture which enables to merge multiple results regardless of the size of the input. Therefore, we propose Inverse Saliency Pyramid Reconstruction Network (InSPyReNet), which predicts the image pyramid of the saliency map. Image pyramid is a simple yet straightforward method for image blending [6], so we design InSPyReNet to produce the image pyramid of the saliency map directly. Previous works have already used image pyramid prediction, but results did not strictly follow the structure, and hence unable to use for the blending (Fig. 2). Therefore, we suggest new architecture, and new supervision techniques to ensure the image pyramid structure which enables stable image blending for HR prediction.

Second, to solve the problem of ERF discrepancy between LR and HR images, we design a pyramid blending technique for the inference time to overlap two image pyramids of saliency maps from different scales. Recent studies of HR SOD methods use two different scales of the same image, by resizing HR image to LR, to alleviate such problem [3,4], but the network should be complicated and large (Fig. 1c). Simply forwarding HR images to the InSPyReNet, or other LR SOD networks fail to predict salient region since they are not trained with HR images. Nevertheless, we notice the potential of enhancing details for high-quality details from HR prediction, even the result shows a lot of False Positives (HR prediction in Fig. 3). To combine the robust saliency prediction and details from LR and HR predictions, we blend the two image pyramids of saliency maps.

InSPyReNet *does not require HR training and datasets*, yet produces high-quality results on HR benchmarks. A series of quantitative and qualitative results on HR and LR SOD benchmarks show that our method shows SotA performance, yet more efficient than previous HR SOD methods in terms of training resources, annotation quality, and architecture engineering.

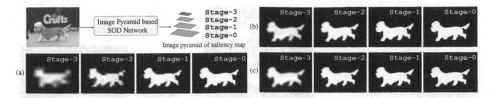

Fig. 2. Comparison of image pyramid based saliency map between (a) Chen *et al.* [7], (b) InSPyReNet, and image pyramid of (c) ground truth. Compared to the image pyramid of ground truth saliency map, Chen *et al.* shows distorted results especially for the higher stages (*e.g.*, `Stage-3`). However, our InSPyReNet shows almost identical results compared to the ground truth across each stage.

2 Related Works

Salient Object Detection. Edge-Based Models are studied in SOD for better understanding of the structure of the salient object by explicitly modeling the contour of the saliency map. Methods with auxiliary edge estimator require additional edge GT, or extra training process with extra edge datasets. For instance, EGNet [8] has an additional edge estimation branch which is supervised with additional edge-only dataset. However, the effect of edge branch is limited to the encoder network (backbone), expecting better representation with robust edge information, because the estimated edge from the edge branch is not directly used to the detection. Also, LDF [9] designed an alternative representation for the edge information. They divided the saliency map into 'body' and 'detail', which corresponds to the edge part. Unlike EGNet, they utilized both 'body' and 'detail' for the saliency prediction in the inference stage. However, to achieve the disentanglement of 'body' and 'detail' components, it requires multiple training stages and ground truth generation.

Unlike auxiliary edge models, we embedded the image pyramid structure to the network for saliency prediction, which does not require additional training process nor extra datasets, and the decoder network is implicitly trained to predict the Laplacian of the saliency map, high-frequency details of the larger scales, which implicitly includes edge information. Thanks to this simple structure, we also do not require additional training stages.

Image Segmentation for HR Images. Pixel-wise prediction tasks such as SOD resize input images into a pre-defined shape (*e.g.*, 384 × 384) for batched and memory efficient training. This is plausible since the average resolution of training datasets are usually around 300 to 400 for both width and height. For example, the average resolution is 378 × 469 for ImageNet [10], and 322 × 372 for DUTS [11]. After training, resizing input images into a pre-defined shape is often required, especially when the input image is relatively larger than the pre-defined shape (Fig. 1a). However, down-sampling large images causes severe information loss, particularly for high-frequency details. We can overcome this

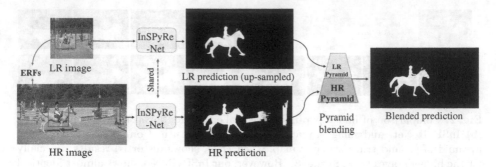

Fig. 3. Illustration of effective receptive field (ERF) [5] discrepancy between LR and HR images from InSPyReNet. LR prediction shows successful saliency prediction, but lack of details. While HR prediction shows better details but due to the ERF discrepancy (red boxes), it over detects objects. With pyramid blending, we can capture the global dependency from LR prediction while enhance local details from HR prediction at the same time. *Best viewed by zooming in.*

problem by not resizing images, but current SotA SOD methods fail to predict appropriate saliency map because they are neither trained with HR dataset nor designed to produce HR prediction. Most likely, the problem is the discrepancy between the effective receptive fields [5] of the same corresponding pixel from the original and resized images (Fig. 3).

CascadePSP [12] first tackled this problem in semantic segmentation by approaching HR segmentation by a refinement process. They trained their model with coarse segmentation masks as an input with a set of augmentation techniques, and used the model to refine an initial segmentation mask with multiple global steps and local steps in a recursive manner. However, they need an initial prediction mask from standalone models [13,14], which is definitely not resource-friendly. Zeng *et al.* [1] first proposed HR dataset for SOD task with a baseline model which consists of separate LR, HR and fusion networks. They combined global (GLFN) and local (LRN) information by two separate networks dedicated for each of them. Tang *et al.* [2] also designed LR and HR networks separately, where the branch for the HR (HRRN) gets an image and a predicted saliency map from the LR branch (LRSCN). PGNet [4] first proposed a standalone, end-to-end network for HR prediction by combining features from LR and HR images with multiple backbone networks.

Aforementioned methods require HR datasets for training, complex model architecture, multiple training sessions for submodules (Fig. 1b, c). Unlike previous methods, InSPyReNet does not require HR datasets for training, yet predicts fine details especially on object boundary.

Image Pyramid in Deep Learning Era. Studies of pixel-level prediction tasks have shown successful application of image pyramid prediction. Lap-SRN [15] first applied a Laplacian image prediction for Super Resolution task and since then, most end-to-end supervised super resolution methods adopt their structure. LRR [16] first applied a Laplacian image pyramid for the semantic segmentation task in the prediction reconstruction process. Then, Chen *et al.* [7] adopted LRR prediction strategy for the SOD with reverse attention mechanism, and UACANet [17] extended self-attention mechanism with uncertainty area for the polyp segmentation. As the above methods have already proved that without any training strategy, we can expect the network to implicitly predict the image pyramid by designing the architecture. However, without extra regularization strategy for the supervision to follow the image pyramid structure rigorously, we cannot make sure that the Laplacian images from each stage truly contains high-frequency detail (Fig. 2).

We revisit this image pyramid scheme for prediction, and improve the performance by setting optimal stage design for image pyramid and regularization methods to follow pyramidal structure. Also, to the best of our knowledge, InSPyReNet is the *first attempt to extend image pyramid prediction for multiple prediction ensembling by image blending technique.* This is because previous methods' Laplacian images did not strictly follow actual high-frequency detail. Rather, they focus more on correcting errors from the higher stages (Fig. 2). Unlike previous methods, we adopt scale-wise supervision (Fig. 4b), Stop-Gradient and pyramidal consistently loss (Sect. 3.2) for regularization which enables consistent prediction, and hence we are able to use blending technique by utilizing multiple results to facilitate more accurate results on HR benchmarks.

3 Methodology

3.1 Model Architecture

Overall Architecture. We use Res2Net [18] or Swin Transformer [19] for the backbone network, but for HR prediction, we only use Swin as a backbone. We provide a thorough discussion (Sect. 5) for the reason why we use only Swin Transformer for HR prediction.

From UACANet [17], we use Parallel Axial Attention encoder (PAA-e) for the multiscale encoder to reduce the number of channels of backbone feature maps and Parallel Axial Attention decoder (PAA-d) to predict an initial saliency map on the smallest stage (*i.e.*, Stage-3). We adopt both modules because they capture global context with non-local operation, and it is efficient thanks to the axial attention mechanism [20,21].

Refer to the stage design in Fig. 4, previous pyramid-based methods for pixellevel prediction [7,16] started with Stage-5, and ended at Stage-2. However, there are still two remaining stages to reconstruct for previous methods, which makes the reconstruction process incomplete in terms of the boundary quality. Thus, we claim that starting image pyramid from Stage-3 is sufficient, and should reconstruct until we encounter the lowest stage, Stage-0 for HR results. To recover the scale of non-existing stages (Stage-1, Stage-0), we use bi-linear interpolation in appropriate locations (Fig. 4).

We locate a self-attention-based decoder, Scale Invariant Context Attention (SICA), on each stage to predict a Laplacian image of the saliency map

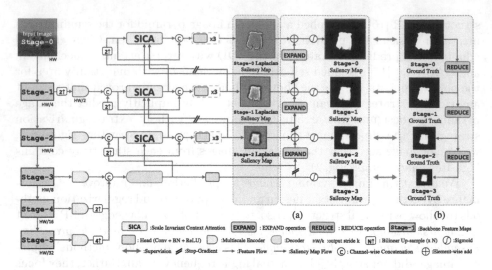

Fig. 4. The architecture of proposed InSPyReNet. (a) The initial saliency map from Stage-3 and Laplacian saliency maps from higher-stages are combined with EXPAND operation to be reconstructed to the original input size. (b) The ground-truth is deconstructed to the smaller stages for predicted saliency maps from each stage by REDUCE operation.

(Laplacian saliency map). From the predicted Laplacian saliency maps, we reconstruct saliency maps from higher-stages to the lower-stages (Fig. 4a).

Scale Invariant Context Attention. Attention-based decoder for pixel-wise prediction shows great performance due to its non-local operation with respect to the spatial dimension [22,23]. However, when the size of the input image gets larger than the training setting (*e.g.*, 384 × 384), it usually fails to produce an appropriate result for the following reason. Because as the size of input image is large enough, there exist a train-inference discrepancies for a non-local operation which flattens the feature map according to the spatial dimension and does a matrix multiplication. For instance, the magnitude of the result from the non-local operation varies depending on the spatial dimension of the input image. Moreover, the complexity of non-local operation increases quadratically as the input size increases.

To this end, we propose SICA, a scale invariant context attention module for robust Laplacian saliency prediction. As shown in Fig. 5, the overall operation of SICA follows OCRNet [23]. We found that computing object region representation causes train-inference discrepancy, so we resize input feature maps **x** and context maps **c** according to the shape from training time (h, w). Because in the training step, images are already reshaped to the fixed shape, we do not have to resize them. For context maps, unlike OCRNet, we can only access to the saliency map which is insufficient, so we generate several context maps following [17]. Further details of the context map selection and equations may be found in the supplementary material. With SICA, we can compute Laplacian saliency

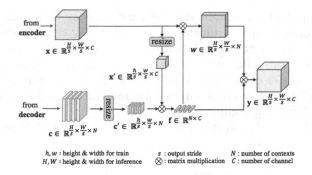

h, w : height & width for train   s : output stride   N : number of contexts
H, W : height & width for inference   ⊗ : matrix multiplication C : number of channel

Fig. 5. Illustration of Scale Invariant Context Attention (SICA)

maps more precisely for HR images, and hence can apply pyramid blending for HR prediction (Sect. 3.3).

Inverse Saliency Pyramid Reconstruction. Laplacian pyramid [24] is an image compression technique that stores the difference between the low-pass filtered image and the original image for each scale. We can interpret the Laplacian image as a remainder from the low-pass filtered signal or, in other words, high-frequency details. Inspired by this technique, we revisit the image pyramid structure by designing our network to construct a Laplacian pyramid to concentrate on the boundary details and reconstruct the saliency map from the smallest stage to its original size. We start with the saliency map from the uppermost stage (Stage-3) for the initial saliency map and aggregate high-frequency details from the Laplacian saliency maps.

Formally, we denote the saliency map and Laplacian saliency map of the jth stage as S^j and U^j, respectively. To reconstruct the saliency map from the $j + 1$th stage to the jth stage, we apply EXPAND operation [24] as follows,

$$S_e^j(x,y) = 4 \sum_{m=-3}^{3} \sum_{n=-3}^{3} g(m,n) \cdot S^{j+1}(\frac{x-m}{2}, \frac{y-n}{2}) \qquad (1)$$

where $(x,y) \in \mathcal{I}^j$ are pixel coordinates and $\mathcal{I}^j$ is a lattice domain of Stage-j. Also, $g(m,n)$ is a Gaussian filter where the kernel size and standard deviation are empirically set to 7 and 1 respectively. To restore the saliency details, we add Laplacian saliency map from SICA as follows,

$$S^j = S_e^j + U^j. \qquad (2)$$

We repeat this process until we obtain the lowest stage, Stage-0 as shown in Fig. 4a, and use it as a final prediction.

3.2   Supervision Strategy and Loss Functions

A typical way to supervise a network with multi-stage side outputs is to use bi-linear interpolation for each stage's prediction and compute the loss function with the ground-truth. However, the predicted saliency map from higher-stage is small regarding its spatial dimension, and this may cause stage-scale inconsistency, especially for the boundary area of salient objects. Instead, we focus on *"Do what you can with what you have where you are"*. In fact, the saliency output from Stage-3 cannot physically surpass the details from Stage-2, so we choose to provide each stage a suitable ground-truth. To do so, we create an image pyramid of the ground-truth (Fig. 4b).

First, we obtain the ground-truth G^j for Stage-j from G^{j-1} with REDUCE operation [24] as follows,

$$G^j(x, y) = \sum_{m=-3}^{3} \sum_{n=-3}^{3} g(m, n) \cdot G^{j-1}(2x + m, 2y + n). \tag{3}$$

From the largest scale, we deconstruct the ground-truth until we get ground-truths for each stage of our network.

For loss function, we utilize binary cross entropy (BCE) loss with pixel position aware weighting strategy $\mathcal{L}^{wbce}$ [25]. Moreover, to encourage the generated Laplacian saliency maps to follow the pyramid structure, we deconstruct S^{j-1} to the jth stage, $\tilde{S}^j$ by REDUCE operation. Then, we reinforce the similarity between S^j and reduced saliency map $\tilde{S}^j$ with pyramidal consistency loss $\mathcal{L}^{pc}$ as follows,

$$\mathcal{L}^{pc}(S^j, \tilde{S}^j) = \sum_{(x,y) \in \mathcal{I}^j} ||S^j(x, y) - \tilde{S}^j(x, y)||_1. \tag{4}$$

$\mathcal{L}^{pc}$ regularizes the lower-stage saliency maps to follow the structure of the image pyramid through the training process. We define the total loss function $\mathcal{L}$ as follows,

$$\mathcal{L}(S, G) = \sum_{j=0}^{3} \lambda_j \mathcal{L}^{wbce}(S^j, G^j) + \eta \sum_{j=0}^{2} \lambda_j \mathcal{L}^{pc}(S^j, \tilde{S}^j) \tag{5}$$

where η is set to 10^{-4} and $\lambda_j = 4^j$ for balancing the magnitude of loss across stages.

Finally, we include Stop-Gradient for the saliency map input of SICA and reconstruction process from higher-stages to force each stage saliency output to focus on each scale during training time and only affect each other in the inference time (Fig. 4). This strategy encourages the stage-wise ground-truth scheme by explicitly preventing the gradient flow from lower-stages affecting the higher-stages. Thus, supervisions with high-frequency details will not affect higher-stage decoder, which are intended only to have the abstract shape of the salient objects. While this strategy might affect the performance in terms of the multiscale scheme, we use feature maps from the different stages for multiscale encoder and SICA to compensate for this issue.

3.3 Pyramid Blending

While SICA enables saliency prediction for various image sizes, when the image gets larger, there still exists ERF discrepancies (Fig. 3). Thankfully, one very

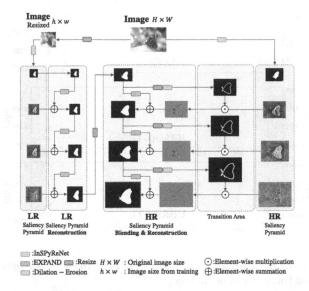

Fig. 6. Illustration of pyramid blending of InSPyReNet for HR prediction.

straightforward application for our saliency pyramid outputs is assembling multiple saliency pyramids from different inputs. We first generate saliency pyramids with InSPyReNet for original and resized images as shown in Fig. 6, namely LR and HR saliency pyramids. Then, instead of reconstructing the saliency map from the HR pyramid, we start from the lowest stage of the LR pyramid. Intuitively speaking, the LR pyramid is extended with the HR pyramid, so they construct a 7 stage saliency pyramid.

For the HR pyramid reconstruction, similar to [16], we compute the dilation and erosion operation to the previous stage's saliency map and subtract them to obtain the transition area for and multiply with the Laplacian saliency map. Transition area is used to filter out the unwanted noises from the HR pyramid, since the boundary details we need to apply should exist only around the boundary area. Unlike [16], it is unnecessary for the LR branch since we train InSPyReNet with methods in Sect. 3.2, results in the saliency pyramid are guaranteed to be consistent.

4 Experiments and Results

4.1 Experimental Settings

Implementation Details. We train our method with widely used DUTS-TR, a subset of DUTS [11] for training. We use Res2Net [18] or Swin Transformer [19] backbone which is pre-trained with ImageNet-1K or ImageNet-22K [10] respectively. Images are resized to 384×384 for training, and we use a random scale in a range of [0.75, 1.25] and crop to the original size, random rotation from -10

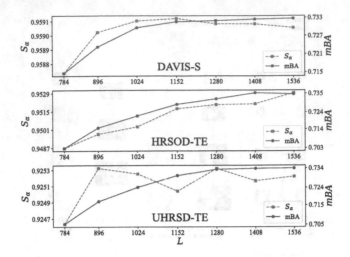

Fig. 7. Performance measure (S_α and mBA) of InSPyReNet with pyramid blending by changing L on three HR benchmarks.

to 10 degrees, and random image enhancement (contrast, sharpness, brightness) for the data augmentation. We set the batch size to 6 and maximum epochs to 60. We use Adam optimizer [26] with initial learning rate 1e−5, and follow the default PyTorch settings. Finally, we use poly learning rate decay for scheduling [13] with a factor of $(1 - (\frac{iter}{iter_{max}})^{0.9})$ and linear warm-up for the first 12000 iterations.

Evaluation Datasets and Metrics. We evaluate our method on five LR benchmarks, DUTS-TE, a subset of DUTS for evaluation, DUT-OMRON [27], ECSSD [28], HKU-IS [29], and PASCAL-S [30]. Furthermore, we evaluate our method on three HR benchmarks, DAVIS-S [31], HRSOD-TE [1], and UHRSD-TE [4]. From [32], we report S-measure (S_α) [33], maximum F-measure (F_{max}) [34], and Mean Absolute Error (MAE) [35]. Since F-measure requires a binary map, it is computed with thresholds in a range of [0, 255] and the maximum value is used for the evaluation. With the above metrics, we also report mean boundary accuracy (mBA) [36] for boundary quality measure.

4.2   Ablation Studies

Resizing Factor L. We use the resizing method for HR images from [36] for pyramid blending since current GPUs cannot deal with large sizes such as 4K images as is. So, we choose a maximum length of the shorter side of the image as L. For instance, if an input size is 1920×1080 and $L = 810$, then we resize the input into 1440×810. Moreover, we do not deploy pyramid blending process for inputs where the shorter side length is less than 512 because the difference between LR and HR pyramid is not enough for blending. We compare S_α and

Table 1. Ablation study of InSPyReNet (SwinB) with and without SICA and pyramid blending on HR benchmarks.

Resolution	Setting	DAVIS-S				HRSOD-TE				UHRSD-TE			
		$S_\alpha \uparrow$	$F_{max} \uparrow$	MAE↓	mBA↑	$S_\alpha \uparrow$	$F_{max} \uparrow$	MAE↓	mBA↑	$S_\alpha \uparrow$	$F_{max} \uparrow$	MAE↓	mBA↑
w/o pyramid blending													
384×384	–	0.953	0.949	0.013	0.705	0.945	0.941	0.019	0.700	0.927	0.932	0.032	0.724
$L = 1280$	w/o SICA	0.396	0.602	0.497	0.504	0.373	0.416	0.530	0.512	0.242	0.395	0.645	0.506
$L = 1280$	w/ SICA	0.873	0.821	0.037	**0.774**	0.886	0.873	0.043	**0.750**	0.809	0.819	0.092	**0.751**
w/ pyramid blending													
$L = 1280$	w/o SICA	0.860	0.883	0.023	0.537	0.863	0.869	0.029	0.531	0.834	0.863	0.052	0.521
$L = 1280$	w/ SICA	**0.962**	**0.959**	**0.009**	0.743	**0.952**	**0.949**	**0.016**	0.738	**0.932**	**0.938**	**0.029**	0.741

mBA on three HR datasets by varying L from 784 to 1536 (Fig. 7). We choose $L = 1280$ since mBA almost converges after that.

SICA and Pyramid Blending. To demonstrate the necessity of SICA, we evaluate InSPyReNet with and without SICA. Since SICA only takes place when the input image is large enough to make train-inference discrepancy, we demonstrate results only on HR benchmarks. Please note that all evaluation is done with resizing method mentioned above, except for the LR resolution (Table 1).

In Table 1, InSPyReNet without SICA shows the worst performance, especially for the mBA. Since mBA only considers boundary quality, InSPyReNet with SICA and without pyramid blending shows the best performance in terms of mBA measure, yet shows poor results on other SOD metrics[1]. This is because even with SICA, InSPyReNet cannot overcome the discrepancy in effective receptive fields [5] between HR and LR images. For the setting without SICA, InSPyReNet with pyramid blending shows inferior results compared to the InSPyReNet without pyramid blending, meaning that the pyramid blending technique is meaningless without SICA since it worsen the results. Thus, SICA is crucial to be included in InSPyReNet, especially for the HR pyramid in the pyramid blending. Compared to the LR setting (*i.e.*, resizing into 384×384), using both SICA and pyramid blending shows better performance for all four metrics.

4.3   Comparison with State-of-the-Art Methods

Quantitative Comparison. First, we compare InSPyReNet with 12 SotA LR SOD methods. In this experiment, we resize images same as for training. We either download pre-computed saliency maps or run an official implementation with pre-trained model parameters provided by the authors to evaluate with the same evaluation code for a fair comparison. Moreover, we re-implement Chen *et al.* [7], F³Net [25], LDF [9], MINet [40], and PA-KRN [41] with same backbones we use to demonstrate how much the training settings affects the performance

---

[1] This phenomenon shows that mBA itself cannot measure the performance of saliency detection, rather it only measures the quality of boundary itself.

Table 2. Quantitative results on five LR benchmarks. The first and the second best results for each metric are colored red and blue. ↑ indicates larger the better, and ↓ indicates smaller the better. † indicates our re-implementation.

Algorithms	Backbones	DUTS-TE			DUT-OMRON			ECSSD			HKU-IS			PASCAL-S		
		S_α ↑	F_{max} ↑	MAE↓	S_α ↑	F_{max} ↑	MAE↓	S_α ↑	F_{max} ↑	MAE↓	S_α ↑	F_{max} ↑	MAE↓	S_α ↑	F_{max} ↑	MAE↓
CNN backbone Models (ResNet, ResNext, Res2Net)																
PoolNet [37]	ResNet50	0.887	0.865	0.037	0.831	0.763	0.054	0.926	0.937	0.035	0.909	0.912	0.034	0.865	0.858	0.065
BASNet [38]	ResNet34	0.866	0.838	0.048	0.836	0.779	0.056	0.916	0.931	0.037	0.909	0.919	0.032	0.838	0.835	0.076
EGNet [8]	ResNet50	0.874	0.848	0.045	0.836	0.773	0.057	0.918	0.928	0.041	0.915	0.920	0.032	0.848	0.836	0.075
CPD [12]	ResNet50	0.869	0.840	0.043	0.825	0.754	0.056	0.918	0.926	0.037	0.905	0.911	0.034	0.848	0.833	0.071
GateNet [39]	ResNeXt101	0.897	0.880	0.035	0.849	0.794	0.051	0.929	0.940	0.035	0.925	0.932	0.029	0.865	0.855	0.064
†Chen et al. [7]	Res2Net50	0.890	0.869	0.040	0.834	0.769	0.061	0.931	0.943	0.035	0.921	0.927	0.034	0.871	0.862	0.060
†F³Net [25]	Res2Net50	0.892	0.876	0.033	0.839	0.771	0.048	0.915	0.925	0.040	0.915	0.925	0.030	0.856	0.842	0.065
†LDF [9]	Res2Net50	0.897	0.885	0.032	0.848	0.788	0.045	0.928	0.943	0.033	0.924	0.935	0.027	0.868	0.863	0.059
†MINet [40]	Res2Net50	0.896	0.883	0.034	0.843	0.787	0.055	0.931	0.942	0.031	0.923	0.931	0.028	0.865	0.858	0.060
†PA-KRN [41]	Res2Net50	0.898	0.888	0.034	0.853	0.808	0.050	0.930	0.943	0.032	0.922	0.935	0.027	0.863	0.859	0.063
Ours	Res2Net50	0.904	0.892	0.035	0.845	0.791	0.059	0.936	0.949	0.031	0.929	0.938	0.028	0.876	0.869	0.056
Transformer backbone Models (Swin, T2T-ViT)																
VST [42]	T2T-ViT-14	0.896	0.878	0.037	0.850	0.800	0.058	0.932	0.944	0.033	0.928	0.937	0.029	0.872	0.864	0.061
Mao et al. [43]	SwinB	0.917	0.911	0.025	0.862	0.818	0.048	0.943	0.956	0.022	0.934	0.945	0.022	0.883	0.883	0.050
†Chen et al. [7]	SwinB	0.901	0.883	0.034	0.860	0.810	0.052	0.937	0.948	0.030	0.928	0.935	0.029	0.876	0.868	0.058
†F³Net [25]	SwinB	0.902	0.895	0.033	0.860	0.826	0.053	0.937	0.951	0.027	0.932	0.944	0.023	0.868	0.864	0.059
†LDF [9]	SwinB	0.896	0.881	0.036	0.854	0.809	0.052	0.931	0.942	0.032	0.933	0.941	0.024	0.861	0.851	0.065
†MINet [40]	SwinB	0.906	0.893	0.029	0.852	0.798	0.047	0.935	0.949	0.028	0.930	0.938	0.025	0.875	0.870	0.054
†PA-KRN [41]	SwinB	0.913	0.906	0.028	0.874	0.838	0.042	0.941	0.956	0.025	0.933	0.944	0.023	0.873	0.872	0.056
Ours	SwinB	0.931	0.927	0.024	0.875	0.832	0.045	0.949	0.960	0.023	0.944	0.955	0.021	0.893	0.893	0.048

for other methods compared to InSPyReNet. We choose the above methods since they provided source code with great reproducibility and consistent results. As shown in Table 2, our SwinB backbone model consistently shows outstanding performance across three metrics. Moreover, our Res2Net50 backbone model shows competitive results regarding its number of parameters.

Moreover, to verify the effectiveness of pyramid blending, we compare our method with SotA methods on HR and LR benchmarks (Table 3). Among HR methods, our method shows great performance among other methods, even though we use only DUTS-TR for training. Note that previous SotA HR methods show inferior results on LR datasets and vice versa, meaning that generalizing for both scales is difficult, while our method is robust for both scales. For instance, while PGNet trained with HR datasets (H, U) shows great performance on HR benchmarks, but shows more inferior results than other methods and even LR methods on LR benchmarks, while our method shows consistent results on both benchmarks. This is because LR datasets do not provide high-quality boundary details, while HR datasets lack of global object saliency.

Qualitative Comparison. We provide a visual comparison of our method in Fig. 8 and Fig. 9 on HR benchmarks. Overall, previous SotA methods are sufficient for detecting salient objects, but shows degraded results for complex scenes. Results show that InSPyReNet can produce accurate saliency prediction for the complex, fine details thanks to the pyramid blending. Moreover, even though we train our method only with LR dataset, DUTS-TR, InSPyReNet consistently shows accurate results compared to other methods.

Table 3. Quantitative results on three HR and two LR benchmarks. Backbones; V: VGG16, R18: ResNet18, R50: ResNet50, S: SwinB. Datasets; D: DUTS-TR, H: HRSOD-TR, U: UHRSD-TR. The first and the second best results for each metric are colored red and blue. ↑ indicates larger the better, and ↓ indicates smaller the better. † indicates our re-implementation.

Algorithms	Backbone	Train datasets	HR benchmarks												LR benchmarks					
			DAVIS-S				HRSOD-TE				UHRSD-TE				DUTS-TE			DUT-OMRON		
			S_α ↑	F_{max} ↑	MAE↓	mBA↑	S_α ↑	F_{max} ↑	MAE↓	mBA↑	S_α ↑	F_{max} ↑	MAE↓	mBA↑	S_α ↑	F_{max} ↑	MAE↓	S_α ↑	F_{max} ↑	MAE↓
†Chen et al. [7]	S	D	0.934	0.925	0.018	0.697	0.915	0.907	0.032	0.684	0.915	0.919	0.034	0.712	0.901	0.883	0.034	0.860	0.810	0.052
†F³Net [25]	S	D	0.931	0.922	0.017	0.681	0.912	0.902	0.034	0.674	0.920	0.922	0.033	0.708	0.902	0.895	0.033	0.860	0.826	0.053
†LDF [9]	S	D	0.928	0.918	0.019	0.682	0.905	0.888	0.036	0.672	0.911	0.913	0.038	0.702	0.896	0.881	0.036	0.854	0.809	0.052
†MINet [40]	S	D	0.933	0.930	0.017	0.673	0.927	0.917	0.025	0.670	0.915	0.917	0.035	0.694	0.906	0.893	0.029	0.852	0.798	0.047
†PA-KRN [41]	S	D	0.944	0.935	0.014	0.668	0.927	0.918	0.026	0.653	0.919	0.926	0.034	0.673	0.913	0.906	0.028	0.874	0.838	0.042
PGNet [4]	S+R18	D	0.935	0.931	0.015	0.707	0.930	0.922	0.021	0.693	0.912	0.914	0.037	0.715	0.911	0.903	0.027	0.855	0.803	0.045
Zeng et al. [1]	V	D,H	0.876	0.889	0.026	0.618	0.897	0.892	0.030	0.623	–	–	–	–	0.824	0.835	0.051	0.762	0.743	0.065
Tang et al. [2]	R50	D,H	0.920	0.935	0.012	0.716	0.920	0.915	0.022	0.693	–	–	–	–	0.895	0.888	0.031	0.843	0.796	0.048
PGNet [4]	S+R18	D,H	0.947	0.948	0.012	0.716	0.935	0.929	0.020	0.714	0.912	0.915	0.036	0.735	0.912	0.905	0.028	0.858	0.803	0.046
PGNet [4]	S+R18	H,U	0.954	0.956	0.010	0.730	0.938	0.939	0.020	0.727	0.935	0.930	0.026	0.765	0.861	0.828	0.038	0.790	0.727	0.059
Ours	S	D	0.962	0.959	0.009	0.743	0.952	0.949	0.016	0.738	0.932	0.938	0.029	0.741	0.931	0.927	0.024	0.875	0.832	0.045

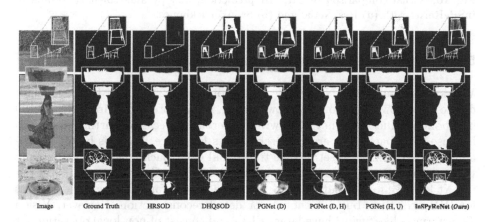

Fig. 8. Qualitative results of InSPyReNet (SwinB) compared to SotA HR methods on HRSOD-TE. *Best viewed by zooming in.*

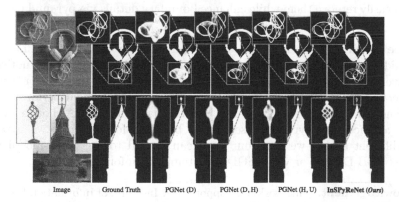

Fig. 9. Qualitative results of InSPyReNet (SwinB) compared to PGNet on UHRSD-TE. *Best viewed by zooming in.*

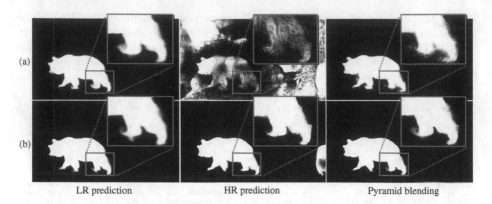

LR prediction   HR prediction   Pyramid blending

Fig. 10. Visual comparison of LR, HR prediction, and pyramid blended results of InSPyReNet with (a) Res2Net50 and (b) SwinB backbones.

5   Discussion and Conclusion

Weakness: Backbone Network. We do not use Res2Net50 for HR benchmark due to the following reason. As shown in Fig. 10, HR prediction from Res2Net50 backbone produces saliency map with numerous unnecessary artifacts. This is because CNN backbones are vulnerable to its ERF size, which is highly dependent on its training dataset. Unlike traditional CNN backbones, there are many works to minimize the above issue such as Fast Fourier Convolution [44], or ConvNeXt [45]. We found that those methods are helpful for reducing such artifacts for HR prediction, but not enough for detail reconstruction. However, Vision Transformers like SwinB have larger ERFs and consist of non-local operation for regarding global dependencies, which are suitable for our method. Thus, even the HR prediction shows some False Positives (second column, second row in Fig. 10), we can easily remove them while enhance boundary details via pyramid blending.

Future Work and Conclusion. Starting from previous works with Laplacian pyramid prediction [7,16], we have shown that InSPyReNet shows noticeable improvements on HR prediction without any HR training datasets or complex architecture. In a series of experiments, our method shows great performance on HR benchmarks while robust again LR benchmarks as well. Although we only utilize a concept of pyramid-based image blending for merging two pyramids with different scales, we hope our work can extend to the multi-modal input such as RGB-D SOD or video SOD with temporal information.

Acknowledgement. This work was supported by Institute of Information & communications Technology Planning & Evaluation (IITP) grant funded by the Korea government (MSIT) (No. 2017-0-00897, Development of Object Detection and Recognition for Intelligent Vehicles) and (No. B0101-15-0266, Development of High Performance Visual BigData Discovery Platform for Large-Scale Realtime Data Analysis).

References

1. Zeng, Y., Zhang, P., Zhang, J., Lin, Z., Lu, H.: Towards high-resolution salient object detection. In: ICCV, pp. 7234–7243 (2019)
2. Tang, L., Li, B., Zhong, Y., Ding, S., Song, M.: Disentangled high quality salient object detection. In: ICCV, pp. 3580–3590 (2021)
3. Zhang, P., Liu, W., Zeng, Y., Lei, Y., Lu, H.: Looking for the detail and context devils: high-resolution salient object detection. IEEE TIP **30**, 3204–3216 (2021)
4. Xie, C., Xia, C., Ma, M., Zhao, Z., Chen, X., Li, J.: Pyramid grafting network for one-stage high resolution saliency detection. arXiv preprint arXiv:2204.05041 (2022)
5. Luo, W., Li, Y., Urtasun, R., Zemel, R.: Understanding the effective receptive field in deep convolutional neural networks. In: NeurIPS, vol. 29 (2016)
6. Burt, P.J., Adelson, E.H.: A multiresolution spline with application to image mosaics. ACM TOG **2**, 217–236 (1983)
7. Chen, S., Tan, X., Wang, B., Hu, X.: Reverse attention for salient object detection. In: Ferrari, V., Hebert, M., Sminchisescu, C., Weiss, Y. (eds.) ECCV 2018. LNCS, vol. 11213, pp. 236–252. Springer, Cham (2018). https://doi.org/10.1007/978-3-030-01240-3_15
8. Zhao, J.X., Liu, J.J., Fan, D.P., Cao, Y., Yang, J., Cheng, M.M.: EGNet: edge guidance network for salient object detection. In: ICCV, pp. 8779–8788 (2019)
9. Wei, J., Wang, S., Wu, Z., Su, C., Huang, Q., Tian, Q.: Label decoupling framework for salient object detection. In: CVPR, pp. 13025–13034 (2020)
10. Russakovsky, O., et al.: ImageNet large scale visual recognition challenge. IJCV **115**, 211–252 (2015)
11. Wang, L., et al.: Learning to detect salient objects with image-level supervision. In: CVPR, pp. 136–145 (2017)
12. Wu, Z., Su, L., Huang, Q.: Cascaded partial decoder for fast and accurate salient object detection. In: CVPR, pp. 3907–3916 (2019)
13. Zhao, H., Shi, J., Qi, X., Wang, X., Jia, J.: Pyramid scene parsing network. In: CVPR, pp. 2881–2890 (2017)
14. Chen, L.C., Papandreou, G., Schroff, F., Adam, H.: Rethinking atrous convolution for semantic image segmentation. arXiv preprint arXiv:1706.05587 (2017)
15. Lai, W.S., Huang, J.B., Ahuja, N., Yang, M.H.: Deep Laplacian pyramid networks for fast and accurate super-resolution. In: CVPR, pp. 624–632 (2017)
16. Ghiasi, G., Fowlkes, C.C.: Laplacian pyramid reconstruction and refinement for semantic segmentation. In: Leibe, B., Matas, J., Sebe, N., Welling, M. (eds.) ECCV 2016. LNCS, vol. 9907, pp. 519–534. Springer, Cham (2016). https://doi.org/10.1007/978-3-319-46487-9_32
17. Kim, T., Lee, H., Kim, D.: UACANet: uncertainty augmented context attention for polyp segmentation. In: ACM MM, pp. 2167–2175 (2021)
18. Gao, S.H., Cheng, M.M., Zhao, K., Zhang, X.Y., Yang, M.H., Torr, P.: Res2Net: a new multi-scale backbone architecture. IEEE TPAMI **43**, 652–662 (2021)
19. Liu, Z., et al.: Swin transformer: hierarchical vision transformer using shifted windows. arXiv preprint arXiv:2103.14030 (2021)
20. Ho, J., Kalchbrenner, N., Weissenborn, D., Salimans, T.: Axial attention in multi-dimensional transformers. arXiv preprint arXiv:1912.12180 (2019)
21. Wang, H., Zhu, Y., Green, B., Adam, H., Yuille, A., Chen, L.-C.: Axial-DeepLab: stand-alone axial-attention for panoptic segmentation. In: Vedaldi, A., Bischof, H., Brox, T., Frahm, J.-M. (eds.) ECCV 2020. LNCS, vol. 12349, pp. 108–126. Springer, Cham (2020). https://doi.org/10.1007/978-3-030-58548-8_7

22. Fu, J., et al.: Dual attention network for scene segmentation. In: CVPR, pp. 3146–3154 (2019)
23. Yuan, Y., Chen, X., Wang, J.: Object-contextual representations for semantic segmentation. In: Vedaldi, A., Bischof, H., Brox, T., Frahm, J.-M. (eds.) ECCV 2020. LNCS, vol. 12351, pp. 173–190. Springer, Cham (2020). https://doi.org/10.1007/978-3-030-58539-6_11
24. Burt, P., Adelson, E.: The Laplacian pyramid as a compact image code. IEEE Trans. Commun. **31**, 532–540 (1983)
25. Wei, J., Wang, S., Huang, Q.: F^3Net: fusion, feedback and focus for salient object detection. AAAI **34**, 12321–12328 (2020)
26. Kingma, D.P., Ba, J.: Adam: a method for stochastic optimization. In: Bengio, Y., LeCun, Y. (eds.) ICLR (2015)
27. Yang, C., Zhang, L., Lu, H., Ruan, X., Yang, M.H.: Saliency detection via graph-based manifold ranking. In: CVPR, pp. 3166–3173 (2013)
28. Shi, J., Yan, Q., Xu, L., Jia, J.: Hierarchical image saliency detection on extended CSSD. IEEE TPAMI **38**, 717–729 (2015)
29. Li, G., Yu, Y.: Visual saliency based on multiscale deep features. In: CVPR, pp. 5455–5463 (2015)
30. Li, Y., Hou, X., Koch, C., Rehg, J.M., Yuille, A.L.: The secrets of salient object segmentation. In: CVPR, pp. 280–287 (2014)
31. Perazzi, F., Pont-Tuset, J., McWilliams, B., Van Gool, L., Gross, M., Sorkine-Hornung, A.: A benchmark dataset and evaluation methodology for video object segmentation. In: CVPR, pp. 724–732 (2016)
32. Wang, W., Lai, Q., Fu, H., Shen, J., Ling, H., Yang, R.: Salient object detection in the deep learning era: an in-depth survey. IEEE TPAM **I**, 1–1 (2021)
33. Fan, D.P., Cheng, M.M., Liu, Y., Li, T., Borji, A.: Structure-measure: a new way to evaluate foreground maps. In: ICCV, pp. 4548–4557 (2017)
34. Achanta, R., Hemami, S., Estrada, F., Susstrunk, S.: Frequency-tuned salient region detection. In: CVPR, pp. 1597–1604 (2009)
35. Perazzi, F., Krähenbühl, P., Pritch, Y., Hornung, A.: Saliency filters: contrast based filtering for salient region detection. In: CVPR, pp. 733–740 (2012)
36. Cheng, H.K., Chung, J., Tai, Y.W., Tang, C.K.: CascadePSP: toward class-agnostic and very high-resolution segmentation via global and local refinement. In: CVPR, pp. 8890–8899 (2020)
37. Liu, J.J., Hou, Q., Cheng, M.M., Feng, J., Jiang, J.: A simple pooling-based design for real-time salient object detection. In: CVPR, pp. 3917–3926 (2019)
38. Qin, X., Zhang, Z., Huang, C., Gao, C., Dehghan, M., Jagersand, M.: BASNet: boundary-aware salient object detection. In: CVPR, pp. 7479–7489 (2019)
39. Zhao, X., Pang, Y., Zhang, L., Lu, H., Zhang, L.: Suppress and balance: a simple gated network for salient object detection. In: Vedaldi, A., Bischof, H., Brox, T., Frahm, J.-M. (eds.) ECCV 2020. LNCS, vol. 12347, pp. 35–51. Springer, Cham (2020). https://doi.org/10.1007/978-3-030-58536-5_3
40. Pang, Y., Zhao, X., Zhang, L., Lu, H.: Multi-scale interactive network for salient object detection. In: CVPR, pp. 9413–9422 (2020)
41. Xu, B., Liang, H., Liang, R., Chen, P.: Locate globally, segment locally: a progressive architecture with knowledge review network for salient object detection. AAA **I**, 3004–3012 (2021)
42. Liu, N., Zhang, N., Wan, K., Shao, L., Han, J.: Visual saliency transformer. In: ICCV, pp. 4722–4732 (2021)

43. Mao, Y., et al.: Transformer transforms salient object detection and camouflaged object detection. arXiv preprint arXiv:2104.10127 (2021)
44. Chi, L., Jiang, B., Mu, Y.: Fast Fourier convolution. In: NeurIPS, vol. 33, pp. 4479–4488 (2020)
45. Liu, Z., Mao, H., Wu, C.Y., Feichtenhofer, C., Darrell, T., Xie, S.: A ConvNet for the 2020s. arXiv preprint arXiv:2201.03545 (2022)

BOREx: Bayesian-Optimization-Based Refinement of Saliency Map for Image- and Video-Classification Models

Atsushi Kikuchi, Kotaro Uchida, Masaki Waga[ID], and Kohei Suenaga[✉][ID]

Kyoto University, Kyoto, Japan
ksuenaga@fos.kuis.kyoto-u.ac.jp

Abstract. Explaining a classification result produced by an image- and video-classification model is one of the important but challenging issues in computer vision. Many methods have been proposed for producing heat-map–based explanations for this purpose, including ones based on the white-box approach that uses the internal information of a model (e.g., LRP, Grad-CAM, and Grad-CAM++) and ones based on the black-box approach that does not use any internal information (e.g., LIME, SHAP, and RISE).

We propose a new black-box method *BOREx* (**B**ayesian **O**ptimization for **R**efinement of visual model **Ex**planation) to refine a heat map produced by any method. Our observation is that a heat-map–based explanation can be seen as a prior for an explanation method based on Bayesian optimization. Based on this observation, BOREx conducts Gaussian process regression (GPR) to estimate the saliency of each pixel in a given image starting from the one produced by another explanation method. Our experiments statistically demonstrate that the refinement by BOREx improves low-quality heat maps for image- and video-classification results.

1 Introduction

Many image- and video-classification methods based on machine learning have been developed and are widely used. However, many of these methods (e.g., DNN-based ones) are not interpretable to humans. The lack of interpretability is sometimes problematic in using an ML-based classifier under a safety-critical system such as autonomous driving.

To address this problem, various methods to explain the result of image and video classification in the form of a heatmap called *saliency map* [5, 10, 16, 17, 21, 22, 28] have been studied. Figure 1 shows examples of saliency maps synthesized

We thank Atsushi Nakazawa for his fruitful comments on this work. KS is partially supported by JST, CREST Grant Number JPMJCR2012, Japan. MW is partially supported by JST, ACT-X Grant Number JPMJAX200U, Japan.

Supplementary Information The online version contains supplementary material available at https://doi.org/10.1007/978-3-031-26293-7_17.

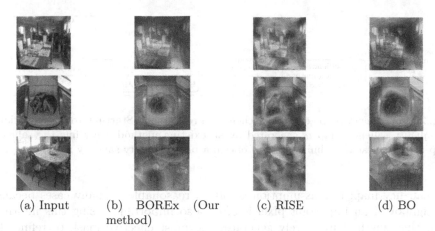

(a) Input   (b) BOREx (Our   (c) RISE   (d) BO
              method)

Fig. 1. Example of the saliency maps generated by our method BOREx (in column (b)), RISE [17] (in column (c)), and the Bayesian-optimization-based method [16] (in column (d)); the input images to each method are presented in column (a). The classification label used in the saliency maps in each row is "dining table", "Labrador retriever", and "folding chair" from the first row.

by several methods, including ours. A saliency map for an image-classification result is an image of the same size as the input image. Each pixel in the saliency map shows the contribution of the corresponding pixel in the input image to the classification result. In each saliency map, the part that positively contributes to the classification result is shown in red, whereas the negatively-contributing parts are shown in blue. The notion of saliency maps is extended to explain the results produced by a video-classification model, e.g., in [5] and [23].

These saliency-map generation techniques can be classified into two groups: the *white-box approach* and the *black-box approach*. A technique in the former group uses internal information (e.g., gradient computed inside DNN) to generate a saliency map; Grad-CAM [21] and Grad-CAM++ [5] are representative examples of this group. A technique in the latter group does not use internal information. Instead, it repeatedly perturbs the input image by occluding several parts randomly and synthesizes a saliency map based on the change in the outputs of the model to the masked images from that of the original one. The representative examples of this group are LIME [20], SHAP [15], and RISE [17].

Although these methods provide valuable information to interpret many classification results, the generated saliency maps sometimes do not correctly localize the regions that contribute to a classification result [3,9,25]. Such a low-quality saliency map cannot be used to interpret a classification result correctly.

Mokuwe et al. [16] recently proposed another black-box saliency map generation method using *Bayesian optimization* based on the theory of *Gaussian processes regression (GPR)* [19]. Their method maintains (1) the estimated saliency value of each pixel and (2) the estimated variance of the saliency values during an execution of their procedure, assuming that a Gaussian process can approximate

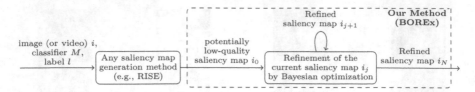

Fig. 2. Our saliency map generation scheme via refinement. Starting from a potentially low-quality saliency map i_0 generated by an existing method, we refine the saliency map using Bayesian optimization and obtain a better-quality saliency map i_N.

the saliency map; this assumption is indeed reasonable in many cases because a neighbor of an important pixel is often also important. Using this information, their method iteratively generates the most effective mask to refine the estimations and observes the saliency value using the generated mask instead of randomly generating masks. Then, the estimations are updated with the observation using the theory of Gaussian processes.

Inspired by the method by Mokuwe et al., we propose a method to *refine* the quality of a (potentially low-quality) saliency map. Our idea is that the GPR-based optimization using a low-quality saliency map i_0 as prior can be seen as a procedure to iteratively refine i_0. Furthermore, even if a saliency map i_0 generated by certain method is of low quality, it often captures the characteristic of the real saliency of the input image; therefore, using i_0 as prior is helpful to guide the optimization.

Based on this idea, we extend their approach so that it uses i_0 as prior information for their Bayesian optimization; see Fig. 2 for an overview of our saliency map generation scheme via refinement. Our method can be applied to a saliency map i_0 generated by *any* method; by the iterative refinement conducted by GPR, i_0 is refined to a better-quality saliency map as Fig. 1 presents. Each saliency map in Fig. 1b is generated by refining the one generated by RISE [17] presented in Fig. 1c; each saliency map in Fig. 1 generated by our method localizes important parts better than that by RISE.

In addition to this extension, we improve their method to generate better saliency maps in a nontrivial way; these improvements include the way a saliency value is observed using a mask and the way a saliency map is generated from the final estimation of GPR. With these extensions, our method *BOREx* (**B**ayesian **O**ptimization for **R**efinement of visual model **Ex**planation) can generate better-quality saliency maps as presented in Fig. 1.

We also present an extension of BOREx to video-classification models. Given a video-classification result, the resulting extension produces a video that indicates the saliency of each pixel in each frame using colors. Combined with a naively extended RISE for video-classification models, BOREx can generate a saliency map for a video-classification result without using any internal information of the classification model.

We implemented BOREx and experimentally evaluated the effectiveness of BOREx. The result confirms that BOREx effectively improves the quality of low-quality saliency maps, both for images and for videos, in terms of several standard metrics for evaluating saliency maps with statistical significance ($p < 0.001$). We also conducted an ablation study, which demonstrates that the additional improvements to the method by Mokuwe et al. [16] mentioned above are paramount for this effectiveness.

Our contribution can be summarized as follows.

- We propose a new black-box method to refine a saliency map generated by any method. BOREx is an extension of the method by Mokuwe et al. [16] so that it uses a saliency map to be refined as prior in its Bayesian-optimization phase. Besides the extension to take a saliency map as a prior, BOREx also enhances Mokuwe et al. [16] in several features, including how saliency values are evaluated using masks and how a saliency map is calculated from the final estimation obtained by the Bayesian optimization.
- We present an extension of BOREx to explain video-classification results. The resulting extended BOREx produces a saliency map in the form of a video in a black-box manner.
- We implemented BOREx and empirically evaluated its effectiveness. The experimental results statistically confirm the effectiveness of BOREx as a method for refining saliency-map–based explanation for image and video classifiers. We also conducted an ablation study, which demonstrates that the enhancement added to the method by Mokuwe et al. [16] is essential for the effectiveness.

Related Work. For both *white-box* and *black-box* approaches, various techniques have been proposed to explain a classification result of an image classifier by generating a saliency map. The white-box approach exploits the internal information of the classifier, e.g., the network architecture and the parameters, and generates a saliency map, typically without using the inference result. Zhou et al. [28] introduce *class activation maps (CAM)* that generate a saliency map exploiting the global average pooling layer in the classification model. Grad-CAM [21] and Grad-CAM++ [5] generalize CAM by focusing on the gradient during back propagation to relax the requirements on the architecture of the classification model. Zoom-CAM [22] is a variant of Grad-CAM that utilizes the feature map of the intermediate convolutional layers as well as the last convolutional layer. Although these techniques are efficient since an inference is not necessary, gradient-based methods do not always generate a faithful explanation because the inference result is ignored in a saliency-map generation [1,7,12,24].

In contrast, the black-box approach treats a classifier as a black-box function without using its internal information. These techniques typically perturb the given image and explain the classifier utilizing the difference in inference results between the original and the perturbed images. For example, RISE [17] and PN-RISE [10] randomly generate a mask by the Monte-Carlo method and perturb the image by occluding the pixels using the mask. Although these techniques can be applied to a model whose internal information is not available, it requires many inferences to obtain a high-quality saliency map.

As shown in Fig. 2, our technique, saliency map refinement by Bayesian optimization, requires an initial saliency map i_0 generated by an explanation technique mentioned above and refines it to improve its quality. Thus, our technique allows combining one of the techniques above and the Bayesian optimization to balance various tradeoffs. Typically, one can balance the tradeoff between the number of inferences and quality by feeding a saliency map that is not necessarily of high quality but requires less number of inferences.

Saliency-based explanation methods have also been investigated for video classifiers. Stergiou et al. [23] propose an explanation of a 3D CNN model for video classification by generating a saliency *tube* that is a 3D generalization of a saliency map. They use the white-box approach based on the idea of CAM [28]. Chattopadhyay et al. [5] show that Grad-CAM++ outperforms in the explanation of a 3D CNN model for action recognition compared to Grad-CAM. Bargal et al. [2] propose an explanation technique for recurrent neural networks (RNNs) with convolutional layers utilizing excitation backpropagation [27]. Perturbation-based black-box approaches have also been investigated to explain a video classifier by presenting salient frames [18] or a 3D generalization of a saliency map [14]. Same as the explanation of image classifiers, our technique allows combining the techniques above and the Bayesian optimization to balance various tradeoffs.

The rest of the paper is organized as follows. Section 2 defines saliency maps and reviews the saliency-map generation method by Mokuwe et al. [16]; Sect. 3 introduces BOREx and an extension for video classifiers; Sect. 4 explains the experiments; Sect. 5 concludes.

We write Λ for a set of *pixels*; we write λ for an element of Λ. An *image* is a map from Λ to $\mathbb{N}^3$; we write i for an image and $\mathcal{I}$ for the set of images. The value $i(\lambda)$ represents the RGB value of pixel λ in image i. We write $\mathcal{L}$ for the finite set of *labels*. A *classification model* is a function from $\mathcal{I}$ to a probability distribution over $\mathcal{L}$; we write M for a model. For a model M and an image i, the distribution $M(i)$ represents the confidence of M in classifying i to each label. We write $M(i, l)$ for the confidence of M classifying i to l.

2 Background

2.1 Saliency

Petsiuk et al. [17] define the saliency of each part in an image i based on the following idea: *A part in i is important for a model M classifying i as l if the confidence remains high even the other part in i is masked.* This intuition is formulated as follows by using the notion of *masks*. A mask m is a function $m \colon \Lambda \to \{0, 1\}$ that expresses how the value of each pixel of an image i is diminished; the value of pixel λ in the masked image—written $i \odot m$—is obtained by occluding the pixel λ if $m(\lambda) = 0$. Then, given a model M, an image i, and a label l, the *saliency* $S_{i,l}(\lambda)$ of pixel λ in image i in M classifying i to l is defined as follows:

$$S_{i,l}(\lambda) := \mathbb{E}[M(i \odot m, l) \mid m(\lambda) = 1]. \tag{1}$$

In the above definition and in the following, the expectation $\mathbb{E}[M(i \odot m, l) \mid m(\lambda) = 1]$ is taken over a given distribution $\mathcal{M}$ of masks. Notice that the above formula defines saliency only by the input-output relation of M. We call $S_{i,l}$ a *saliency map*.

In (1), m is randomly taken from a distribution $\mathcal{M}$ over masks that models the assumption on how a salient part tends to distribute in an image. $\mathcal{M}$ is typically designed so that it gives higher probabilities to a mask in which masked regions form lumps, rather than the one in which masked pixels are scattered around the image; this design reflects that if a pixel is salient in an image, then the neighborhoods of the pixel are often also salient.

The definition of saliency we use in this paper is the refinement of $S_{i,l}$ by Hatakeyama et al. [10] so that it takes *negative saliency* into account. Concretely, their definition of saliency $S_{i,l}^{\mathrm{PN}}$ is as follows.

$$S_{i,l}^{\mathrm{P}}(\lambda) := S_{i,l}(\lambda). \tag{2}$$

$$S_{i,l}^{\mathrm{N}}(\lambda) := \mathbb{E}[M(i \odot m, l) \mid m(\lambda) = 0]. \tag{3}$$

$$S_{i,l}^{\mathrm{PN}}(\lambda) := S_{i,l}^{\mathrm{P}}(\lambda) - S_{i,l}^{\mathrm{N}}(\lambda). \tag{4}$$

Their saliency $S_{i,l}^{\mathrm{PN}}(\lambda)$ is defined as the difference between the positive saliency $S_{i,l}^{\mathrm{P}}(\lambda)$ and the negative saliency $S_{i,l}^{\mathrm{N}}(\lambda)$. The latter is the expected confidence $M(i \odot m, l)$ conditioned by $m(\lambda) = 0$; therefore, a pixel λ is negatively salient if masking out λ contributes to increasing confidence in classifying the image as l. Hatakeyama et al. [10] show that the saliency of an irreverent pixel calculated by $S_{i,l}^{\mathrm{PN}}(\lambda)$ is close to 0, making the generated saliency map easier to interpret.

Evaluating $S_{i,l}$ and $S_{i,l}^{\mathrm{PN}}$ requires exhausting all masks, which is prohibitively expensive. Petsuik et al. [17] and Hatakeyama et al. [10] propose a method to approximate these saliency values using the Monte-Carlo method. Their implementations draw masks $\{m_1, \ldots, m_N\}$ from $\mathcal{M}$ and approximate $S_{i,l}$ and $S_{i,l}^{\mathrm{PN}}$ using the following formulas, which are derived from the definitions of $S_{i,l}$ and $S_{i,l}^{\mathrm{PN}}$ [10,17] where $p = P[m(\lambda) = 1]$:

$$S_{i,l}(\lambda) \approx \frac{1}{N} \sum_n \frac{m_n(\lambda)}{p} M(i \odot m_n, l) \tag{5}$$

$$S_{i,l}^{\mathrm{PN}}(\lambda) \approx \frac{1}{N} \sum_n \frac{m_n(\lambda) - p}{p(1-p)} M(i \odot m_n, l). \tag{6}$$

2.2 Saliency Map Generation Using Gaussian Process Regression

Mokuwe et al. [16] propose another approach to generate saliency maps for black-box classification models. Their approach uses Bayesian optimization, in particular *Gaussian process regression (GPR)* [19] for this purpose. We summarize the theory of GPR and how it serves for saliency-map generation in this section; for a detailed exposition, see [19].

In general, a Gaussian process is a set of random variables, any finite number of which constitute a joint Gaussian distribution. In our context, Gaussian

---

Algorithm 1. GPR-based saliency-map generation [16]. The function k is used in Line 9, which is kept implicit there.

Input: Model M; Image i; Label l; Function k; Upperbound of iterations N; Set of mask size $L := \{r_1, \ldots, r_q\}$.
Output: Saliency map that explains the classification of i to l by M.
1: $D \leftarrow []$
2: Set $\mu(\lambda, r) \leftarrow 0$ for every pixel λ and $r \in L$
3: $j \leftarrow 0$
4: **while** $j < N$ **do**
5:      $(\lambda, r) \leftarrow \arg\max u_{\mu, D}$
6:      Set m to a square mask whose center is λ, whose side length is r, and $m(\lambda') = 0$ if λ' is in the square
7:      $s \leftarrow M(i, l) - M(i \odot m, l)$
8:      Add (λ, s) at the end of D
9:      Update μ using Bayes' law
10:     $j \leftarrow j + 1$
11: **end while**
12: $i_{sal}(\lambda) \leftarrow \frac{1}{q} \sum_i \mu(\lambda, r_i)$ for every λ.
13: **return** i_{sal}

---

process is a distribution over functions; each f drawn from a Gaussian process maps $(\lambda, \mathbf{r})$ to a saliency value $f(\lambda, \mathbf{r}) \in \mathbb{R}$, where $\mathbf{r} \in \mathbb{R}^p$ is a vector of auxiliary parameters for determining a mask. The $\mathbf{r}$ expresses, for example, the position and the size of a generated mask. A Gaussian process is completely determined by specifying (1) a mean function $\mu(\lambda, \mathbf{r})$ that maps a pixel λ and mask parameters $\mathbf{r}$ to their expected value $\mathbb{E}[f(\lambda, \mathbf{r})]$ and (2) a covariance function $k((\lambda, \mathbf{r}), (\lambda', \mathbf{r}'))$ that maps $(\lambda, \mathbf{r})$ and $(\lambda', \mathbf{r}')$ to their covariance $\mathbb{E}[(f(\lambda, \mathbf{r}) - \mu(\lambda, \mathbf{r}))(f(\lambda', \mathbf{r}') - \mu(\lambda', \mathbf{r}'))]$. We write $\mathcal{GP}(\mu, k)$ for the Gaussian process with μ and k.

GPR is a method to use Gaussian processes for regression. Suppose we observe the saliency at several points in an image as $D := \{((\lambda_1, \mathbf{r}_1), s_1), \ldots, ((\lambda_n, \mathbf{r_n}), s_n)\}$. For an unseen $(\lambda, \mathbf{r})$, its saliency conditioned by D is obtained as a Gaussian distribution whose mean and variance can be computed by D, μ, and k. Furthermore, once a new observation is obtained, the optimization procedure can update μ using the Bayes' law. These properties allow Gaussian processes to explore new observations and predict the saliency at unseen points.

Using these properties of GPs, Mokuwe et al. [16] propose Algorithm 1 for saliency-map generation. Their method models a saliency map as a Gaussian process with mean function μ and covariance function k. Under this model, Algorithm 1 iteratively chooses (λ, r) (Line 5), observe the saliency evaluated with (λ, r) by using a mask whose center is at λ and with side length r (Lines 6 and 7), and update μ using Bayes' law (Line 9). To detect the most positively salient part with a small number of inferences, Algorithm 1 uses an *acquisition function* $u_{\mu, D}(\lambda, r)$. This function is designed to evaluate to a larger value if (1) $|\mu(\lambda, r)|$ or (2) the expected variance of the saliency at λ estimated from D is high; therefore, choosing λ and r such that $u_{\mu, D}(\lambda, r)$ is large balances exploiting the current estimation of the saliency value $\mu(\lambda, r)$ and exploring pixels whose saliency values are uncertain. To keep the search space reasonably small, we keep the shape of the generated masks simple; in Algorithm 1, to a finite set of square masks.

Algorithm 2. GPR-based refinement of a saliency map.

Input: Model M; Image i; Initial saliency map i_0; Label l; Function $k((\lambda, r), (\lambda', r'))$; Upperbound of iterations N; List of the side length of a mask $L := \{r_1, \ldots, r_p\}$.
Output: Refined saliency map obtained with GP.

1: $D \leftarrow []$
2: Set $\mu(\lambda, r) \leftarrow i_0(\lambda)$ for every pixel λ and side size $r \in L$.
3: $j \leftarrow 0$
4: **while** $j < N$ **do**
5:     $(\lambda, r) \leftarrow \arg\max u_{\mu, D}$
6:     Set m to a square mask with side length r, whose center is λ, and $m(\lambda') = 0$ if λ' is inside the rectangle
7:     $s \leftarrow M(i \odot \overline{m}, l) - M(i \odot m, l)$
8:     Add $((\lambda, r), s)$ at the end of D
9:     Update μ using Bayes' law
10:    $j \leftarrow j + 1$
11: **end while**
12: $i_{sal}(\lambda) \leftarrow \frac{1}{p} \sum_i \frac{1}{r_i^2} \mu(\lambda, r_i)$ for every λ.

13: **return** i_{sal}

Various functions that can be used as a covariance function k have been proposed; see [19] for detail. Mokuwe et al. [16] use *Matérn kernel* [19].

Algorithm 1 returns the saliency map i_{sal} by $i_{sal}(\lambda) := \frac{1}{q} \sum_i \mu(\lambda, r_i)$. The value of i_{sal} at λ is the average of $\mu(\lambda, r)$ over $r \in L$.

3   BOREx

3.1   GPR-Based Refinement of Saliency Map

Algorithm 2 is the definition of BOREx. The overall structure of the procedure is the same as that of Algorithm 1. The major differences are the following: (1) the input given to the procedures; (2) how the saliency is evaluated; and (3) how a saliency map is produced from the resulting μ. We explain each difference in the following.

Input to the Algorithm. Algorithm 2 takes the initial saliency map i_0, which is used as prior information for GPR. Concretely, this i_0 is used to initialize $\mu(\lambda, r)$ in Line 2. To generate i_0, one can use *any* saliency-map generation methods, including ones based on black-box approach [10,15,17,20] and ones based on white-box approach [5,21].

Saliency Evaluation. Algorithm 1 evaluates the saliency by calculating $M(i, l) - M(i \odot m, l)$. This value corresponds to the value of $-S_{i,l}^{N}$ around the pixel λ defined in Sect. 2.1 since it computes how much the confidence *drops* if a neighborhood of λ is masked out.

To estimate $S_{i,l}^{PN}$ instead of $-S_{i,l}^{N}$, Algorithm 2 calculates $M(i \odot \overline{m}, l) - M(i \odot m, l)$ in Line 7, where $\overline{m}$ is the *flipped* mask obtained by inverting the value at each pixel (i.e., $\overline{m}(\lambda') = 1 - m(\lambda')$ for any λ'). Since $\overline{m}(\lambda) = 1$ if and only if $m(\lambda) = 0$, the value of $M(i \odot \overline{m}, l) - M(i \odot m, l)$ is expected to be close to $S_{i,l}^{PN}(\lambda')$ if λ' is near λ.

(a) Image of goldfish. 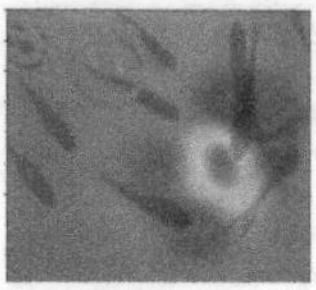   (b) Saliency map generated   (c) Saliency map generated
by Algorithm 1.   by Algorithm 2.

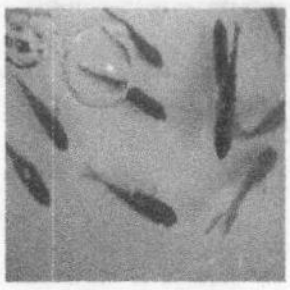

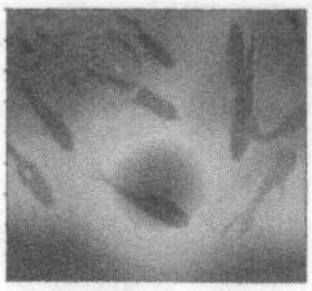

Fig. 3. Example of an image with multiple salient regions.

Another reason of using a flipped mask in the saliency observations of Algorithm 2 is to handle images in which there are multiple salient parts. For example, the image of goldfish in Fig. 3a has multiple salient regions, namely, multiple goldfish. If we apply Algorithm 1, which does not use flipped masks, to this image, we obtain the saliency map in Fig. 3b; obviously, the saliency map does not capture the salient parts in the image. This is because the value of $M(i, l) - M(i \odot m, l)$ in Line 7 of Algorithm 1 is almost same everywhere; this value becomes high for this image only if m hides *every* goldfish in the image, which is difficult using only a single mask. Our method generates the saliency map in Fig. 3c; an observed saliency value $M(i \odot \overline{m}, l) - M(i \odot m, l)$ in Algorithm 2 is higher if m hides *at least* one goldfish than if m does not hide any goldfish.

Generating Saliency Map from Resulting μ. Algorithm 2 returns the saliency map i_{sal} defined by $i_{sal}(\lambda) = \frac{1}{p} \sum_i \frac{1}{r_i^2} \mu(\lambda, r_i)$. Instead of the saliency map computed by taking the simple average over every mask in Algorithm 1, the saliency map map returned by Algorithm 2 is the average weighted by the inverse of the area $\frac{1}{r_i^2}$ of each mask with the side size r. This weighted average gives more weight to the saliency values obtained by smaller masks. Using the weighted average helps a saliency map produced by Algorithm 2 localizes salient parts better than Algorithm 1.

3.2   Extension for Video-Classification Models

Algorithm 2 can be naturally extended for a video classifier with the following changes.

– The set of masks is extended, from 2D squares specified by their side length, to 3D rectangles specified by the side length of the square in a frame, and the number of frames that they hide. Suppose a mask m with side length r and the number of frames t is applied to the pixel λ at coordinate (x, y) and at n-th frame of a video i. Then, $i \odot m$ is obtained by hiding the pixel at (x, y) in each of the n-th to $(n + t - 1)$-th frame with the 2D square mask specified by r.

- The type of functions drawn from the Gaussian process is changed to $f(\lambda, r, t)$ from $f(\lambda, r)$ in Algorithm 2 reflecting the change of the definition of masks.
- The algorithm takes $T := \{t_1, \ldots, t_k\}$ in addition to L; the set T expresses the allowed variation of parameter t of a mask.
- The expression to update i_{sal} in Line 12 of Algorithm 2 is changed to $\frac{1}{pk} \sum_i \sum_s \frac{1}{r_i^2 t_s} \mu(\lambda, r_i, t_s)$; the weight is changed to the reciprocal of the volume of each mask.

4 Experiments

We implemented Algorithm 2 and conducted experiments to evaluate the effectiveness of BOREx. Due to the limited space, we report a part of the experimental results.

See the supplementary material for the experimental environment and more results and discussions, particularly on video classification.

The research questions that we are addressing are the following.

RQ1: Does BOREx improve the quality of an input saliency map? This is to evaluate that BOREx is useful to refine a potentially low-quality saliency map, which is the main claim of this paper.

RQ2: Does Algorithm 2 produce a better saliency map than one produced by Algorithm 1 by Mokuwe et al. [16]? This is to demonstrate the merit of BOREx over the algorithm by Mokuwe et al.

RQ3: Does the extension in Sect. 3.2 useful as a saliency-map generation for video classifiers? This is to evaluate the competency of BOREx to explain a video-classification result.

Evaluation Metrics. To quantitatively evaluate the quality of a saliency map, we used the following three measures.

Insertion: For a saliency map i_{sal} explaining a classification of an image i to label l, the *insertion* metric is defined as $\sum_k M(i^{(k)}, l)$, where $i^{(k)}$ is the image obtained by masking all the pixels other than those with top-k saliency values in i_{sal} to black.

Deletion: The *deletion* metric is defined as $\sum_k M(i^{(-k)}, l)$, where $i^{(-k)}$ is the image obtained by masking all the pixels with top-k saliency values in i_{sal} to black.

F-measure: The *F-measure* in our experiments is defined as $\sum_k F(i^{(k)}, l, B_{i,l})$, where $F(i^{(k)}, l, B_{i,l})$ is the F-measure calculated from the *recall* and the *precision* of the pixels in $i^{(k)}$ against the human-annotated bounded region $B_{i,l}$ in i that indicates an object of label l.

The insertion and the deletion metrics are introduced by [17] to quantitatively evaluate how well a saliency map localizes a region that is important for a decision by a model. The higher value of the insertion metric is better; the lower value of the deletion metric is better. The higher insertion implies that

i_{sal} localizes regions in i that are enough for classifying i to l. The lower deletion implies that i_{sal} localizes regions that are indispensable for classifying i to l. The F-measure is an extension of their pointing-game metric also to consider recall, not only the precision. The higher value of F-measure is better, implying i_{sal} points out more of an important region correctly.

In what follows, we use a statistical hypothesis test called the *one-sided Wilcoxon signed-rank test* [26] (or, simply *Wilcoxon test*). This test is applied to matched pairs of values $\{(a_1, b_1), \ldots, (a_n, b_n)\}$ sampled from a distribution and can be used to check whether the median of $\{a_1, \ldots, a_n\}$ can be said to be larger or smaller than that of $\{b_1, \ldots, b_n\}$ with significance. To compare saliency generation methods X and Y, we calculate the pairs of the values of metrics evaluated with a certain dataset, the first of each are of the method X and the second are of Y; then, we apply the Wilcoxon test to check the difference in the metrics. For further details, see [26].

To address these RQs, we conducted the following experiments:

RQ1: We used RISE [17] and GRAD-CAM++ [5] to generate saliency maps for the images in PascalVOC dataset [8]; we write D_{RISE} and $D_{\text{GradCAM}++}$ for the set of saliency maps generated by RISE and GRAD-CAM++, respectively. Then, we applied BOREx with these saliency maps as input; we write $D_{\text{RISE}}^{\text{BOREx}}$ (resp., $D_{\text{GradCAM}++}^{\text{BOREx}}$) for the saliency maps generated using D_{RISE} (resp., $D_{\text{GradCAM}++}$) as input. We check whether the quality of the saliency maps in D_-^{BOREx} is better than D_- by the one-sided Wilcoxon signed-rank test. If so, we can conclude that BOREx indeed improves the saliency map generated by other methods.

RQ2: We generated saliency maps for the PascalVOC dataset using Mokuwe et al. [16] presented in Algorithm 1; we write D_{BO} for the generated saliency maps. We check if the quality of the saliency maps in $D_{\text{RISE}}^{\text{BOREx}}$ is better than D_{BO} by one-sided Wilcoxon signed-rank test. If so, we conclude the merit of BOREx over the method by Mokuwe et al.

RQ3: We generated saliency maps for the dataset in Kinetics-400 using an extension of GradCAM++ and RISE for video classification implemented by us; let the set of saliency maps $D_{M,\text{GradCAM}++}$ and $D_{M,\text{RISE}}$, respectively. Then, we applied BOREx with these saliency maps as input; we write $D_{M,\text{RISE}}^{\text{BOREx}}$ (resp., $D_{M,\text{GradCAM}++}^{\text{BOREx}}$) for the saliency maps generated using $D_{M,\text{RISE}}$ (resp., $D_{M,\text{GradCAM}++}$) as input. We check whether the quality of the saliency maps in $D_{M,-}^{\text{BOREx}}$ is better than $D_{M,-}$ by one-sided Wilcoxon signed-rank test. If so, we can conclude the merit of BOREx as an explanation method for a video-classification result.

As the model whose classification behavior to be explained, we used ResNet-152 [11] obtained from torchvision.models[1], which is pre-trained with ImageNet [6], for RQ1 and RQ2; and i3D [4] obtained from TensorFlow Hub[2], which is pre-trained with Kinetics-400 [13]. Notice that the datasets PascalVOC and

---

[1] https://pytorch.org/vision/stable/models.html.
[2] https://tfhub.dev/deepmind/i3d-kinetics-400/1.

| (a) Input | (b) BOREx | (c) RISE |

Fig. 4. Image of chairs and saliency maps to explain it.

Kinetics-400 provide human-annotated bounding regions for each label and each image, enabling computation of the F-measure.

4.1 Results and Discussion

RQ1. Table 1 shows that BOREx improved the quality of the saliency maps generated by RISE and Grad-CAM++ in several metrics with statistical significance ($p < 0.001$). Therefore, we conclude that **BOREx successfully refines an input saliency map**. This improvement is thanks to the Gaussian process regression that successfully captured the locality of the salient pixels. For example, the saliency maps in Fig. 1 suggest that BOREx is better at generalizing the salient pixels to the surrounding areas than RISE.

The time spent for GPR-based optimization was 9.26 ± 0.26 seconds in average for each image. We believe this computation time pays off if we selectively apply BOREx to saliency maps whose quality needs to be improved.

To investigate the effect of the features of BOREx presented in Sect. 3.1 (i.e., flipped masks and the saliency-map computation from the result of GPR by weighted average in its performance), we conducted an ablation study; the result is shown in Table 2. We compared BOREx with (1) a variant that does not use flipped masks (no-flip), (2) a variant that uses simple average instead of the average weighted by the inverse of the area of masks (simple-avg), and (3) a variant that does not use prior (no-prior). The statistical test demonstrates that flipped masks and weighted averages are effective in the performance of BOREx. However, the effectiveness over the no-prior variant is not confirmed. This is mainly because, if the quality of a given prior is already high, the effectiveness of BOREx is limited. Indeed, BOREx is confirmed to be effective over the no-prior case if the insertion metric of the priors is less than 0.6; see the row "no-prior (base insertion < 0.6)" in Table 2.

The statistical test did not demonstrate the improvement in the deletion metric for a saliency map generated by RISE and the F-measure for a saliency map generated by Grad-CAM++. Investigation of several images for which BOREx degrades the metrics reveals that this is partly because the current BOREx allows only square-shaped masks; this limitation degrades the deletion metric for an image with multiple objects with the target label l. For example, a single square-shaped mask cannot focus on both chairs simultaneously in the image in Fig. 4a. For such an image, BOREx often focuses on only one of the objects,

Table 1. Result of the experiments. "Image/Video": The kind of the classifier; "Compared with": the baseline method; "Metric": evaluation metric; "p-value": the p-value. The null hypothesis of each test expresses that the average of the metric of BOREx is not better than that of baseline. One asterisk indicates $p < 0.05$; two asterisks indicates $p < 0.001$.

Image/Video	Compared with	Metric	p-value
Image	RISE	F-measure	**8.307e-21****
		Insertion	**1.016e-23****
		Deletion	8.874e-01
	Grad-CAM++	F-measure	1.000
		Insertion	**5.090e-08****
		Deletion	**6.790e-04****
	BO	F-measure	**1.800e-05****
		Insertion	**6.630e-11****
		Deletion	3.111e-01
Video	RISE	F-measure	**4.988e-07****
		Insertion	8.974e-01
		Deletion	**8.161e-18****
	Grad-CAM++	F-measure	9.9980e-01
		Insertion	3.636e-01
		Deletion	**2.983e-07****

generating the saliency map in Fig. 4b. Even if we mask the right chair in Fig. 4a, we still have the left chair, and the confidence of the label "chair" does not significantly decrease, which degrades the deletion metric of the BOREx-generated saliency map.

RQ2. The last three rows of Table 1 show that the use of an initial saliency map improved the quality of the saliency maps generated by Bayesian optimization in terms of several metrics with statistical significance compared to the case where the initial saliency map is not given ($p < 0.001$). Therefore, we conclude that **BOREx produces a better saliency map than the one produced by Mokuwe et al. in terms of the insertion metric and F-measure.**

The improvement was not concluded in terms of the deletion metrics. Investigation of the generated saliency maps suggests that such degradation is observed when the quality of a given initial saliency map is too low; if such a saliency map is given, it misleads an execution of BOREx, which returns a premature saliency map at the end of the prespecified number of iterations.

RQ3. Table 1 shows the result of the experiment for RQ3. It shows that the saliency maps generated by the extensions of RISE and Grad-CAM++ for video classifiers are successfully refined by BOREx in terms of at least one metric with statistical significance ($p < 0.001$). Therefore, we conclude that **a saliency map**

Table 2. The result of ablation study.

Compared with	Metric	p-value
No-flip	F-measure	**2.987e-29****
	Insertion	**1.416e-04****
	Deletion	**2.024e-03***
Simple-avg	F-measure	**2.026e-03***
	Insertion	**1.184e-46****
	Deletion	**4.871e-03***
No-prior	F-measure	**4.514e-02***
	Insertion	3.84624e-01
	Deletion	2.2194e-01
No-prior (base insertion < 0.6)	F-measure	**2.4825e-02***
	Insertion	**3.219e-03***
	Deletion	6.47929e-01

produced by BOREx points out regions in a video that are indispensable to explain the classification result better than the other methods.

The improvement in the insertion metric over RISE and Grad-CAM++, and in F-measure over Grad-CAM++ were not concluded. The investigation of saliency maps whose quality is degraded by BOREx reveals that the issue is essentially the same as that of the images with multiple objects discussed above. A mask used by BOREx occludes the same position across several frames; therefore, for a video in which an object with the target label moves around, it is difficult to occlude all occurrences of the object in different frames. This limitation leads to a saliency map generated by BOREx that tends to point out salient regions only in a part of the frames, which causes the degradation in the insertion metric. The improvement in deletion metric seems to be due to the mask shape of BOREx. To improve the deletion metric for a video-classifier explanation, a saliency map must point out a salient region across several frames. The current mask shape of BOREx is advantageous, at least for a video in which there is a single salient object that does not move around, to cover the salient object over several frames.

5 Conclusion

This paper has presented BOREx, a method to refine a potentially low-quality saliency map that explains a classification result of image and video classifiers. Our refinement of a saliency map with Bayesian optimization applies to any existing saliency-map generation method. The experiment results demonstrate that BOREx improves the quality of the saliency maps, especially when the quality of the given saliency map is neither too high nor too low.

We are currently looking at enhancing BOREx by investigating the optimal shape of masks to improve performance. Another important research task is making BOREx more robust to an input saliency map with very low quality.

References

1. Adebayo, J., Gilmer, J., Muelly, M., Goodfellow, I.J., Hardt, M., Kim, B.: Sanity checks for saliency maps. In: Advances in Neural Information Processing Systems 31: Annual Conference on Neural Information Processing Systems 2018, NeurIPS 2018, 3–8 December 2018, Montréal, Canada, pp. 9525–9536 (2018). https://proceedings.neurips.cc/paper/2018/hash/294a8ed24b1ad22ec2e7efea049b8737-Abstract.html
2. Bargal, S.A., Zunino, A., Kim, D., Zhang, J., Murino, V., Sclaroff, S.: Excitation backprop for RNNs. In: 2018 IEEE Conference on Computer Vision and Pattern Recognition, CVPR 2018, Salt Lake City, UT, USA, June 18-22, 2018. pp. 1440–1449 (2018). https://doi.org/10.1109/CVPR.2018.00156, http://openaccess.thecvf.com/content_cvpr_2018/html/Bargal_Excitation_Backprop_for_CVPR_2018_paper.html
3. Brunke, L., Agrawal, P., George, N.: Evaluating input perturbation methods for interpreting CNNs and saliency map comparison. In: Bartoli, A., Fusiello, A. (eds.) ECCV 2020, Part I. LNCS, vol. 12535, pp. 120–134. Springer, Cham (2020). https://doi.org/10.1007/978-3-030-66415-2_8
4. Carreira, J., Zisserman, A.: Quo vadis, action recognition? A new model and the kinetics dataset. In: 2017 IEEE Conference on Computer Vision and Pattern Recognition, CVPR 2017, Honolulu, HI, USA, 21–26 July 2017, pp. 4724–4733 (2017). https://doi.org/10.1109/CVPR.2017.502
5. Chattopadhyay, A., Sarkar, A., Howlader, P., Balasubramanian, V.N.: Grad-CAM++: generalized gradient-based visual explanations for deep convolutional networks. In: 2018 IEEE Winter Conference on Applications of Computer Vision, WACV 2018, Lake Tahoe, NV, USA, 12–15 March 2018, pp. 839–847. IEEE Computer Society (2018). https://doi.org/10.1109/WACV.2018.00097
6. Deng, J., Dong, W., Socher, R., Li, L., Li, K., Fei-Fei, L.: ImageNet: a large-scale hierarchical image database. In: 2009 IEEE Computer Society Conference on Computer Vision and Pattern Recognition (CVPR 2009), 20–25 June 2009, Miami, Florida, USA, pp. 248–255 (2009). https://doi.org/10.1109/CVPR.2009.5206848
7. Dombrowski, A., Alber, M., Anders, C.J., Ackermann, M., Müller, K., Kessel, P.: Explanations can be manipulated and geometry is to blame. In: Advances in Neural Information Processing Systems 32: Annual Conference on Neural Information Processing Systems 2019, NeurIPS 2019, 8–14 December 2019, Vancouver, BC, Canada, pp. 13567–13578 (2019). https://proceedings.neurips.cc/paper/2019/hash/bb836c01cdc9120a9c984c525e4b1a4a-Abstract.html
8. Everingham, M., Van Gool, L., Williams, C.K.I., Winn, J., Zisserman, A.: The pascal visual object classes (VOC) challenge. Int. J. Comput. Vision **88**(2), 303–338 (2010)
9. Ghorbani, A., Abid, A., Zou, J.Y.: Interpretation of neural networks is fragile. In: The Thirty-Third AAAI Conference on Artificial Intelligence, AAAI 2019, The Thirty-First Innovative Applications of Artificial Intelligence Conference, IAAI 2019, The Ninth AAAI Symposium on Educational Advances in Artificial Intelligence, EAAI 2019, Honolulu, Hawaii, USA, 27 January–1 February 2019, pp. 3681–3688. AAAI Press (2019). https://doi.org/10.1609/aaai.v33i01.33013681

10. Hatakeyama, Y., Sakuma, H., Konishi, Y., Suenaga, K.: Visualizing color-wise saliency of black-box image classification models. In: Ishikawa, H., Liu, C.-L., Pajdla, T., Shi, J. (eds.) ACCV 2020, Part III. LNCS, vol. 12624, pp. 189–205. Springer, Cham (2021). https://doi.org/10.1007/978-3-030-69535-4_12
11. He, K., Zhang, X., Ren, S., Sun, J.: Deep residual learning for image recognition. In: 2016 IEEE Conference on Computer Vision and Pattern Recognition, CVPR 2016, Las Vegas, NV, USA, 27–30 June 2016, pp. 770–778 (2016). https://doi.org/10.1109/CVPR.2016.90
12. Heo, J., Joo, S., Moon, T.: Fooling neural network interpretations via adversarial model manipulation. In: Advances in Neural Information Processing Systems 32: Annual Conference on Neural Information Processing Systems 2019, NeurIPS 2019, 8–14 December 2019, Vancouver, BC, Canada, pp. 2921–2932 (2019). https://proceedings.neurips.cc/paper/2019/hash/7fea637fd6d02b8f0adf6f7dc36aed93-Abstract.html
13. Kay, W., et al.: The kinetics human action video dataset. CoRR abs/1705.06950 (2017). http://arxiv.org/abs/1705.06950
14. Li, Z., Wang, W., Li, Z., Huang, Y., Sato, Y.: Towards visually explaining video understanding networks with perturbation. In: Proceedings of WACV 2021, pp. 1119–1128. IEEE (2021). https://doi.org/10.1109/WACV48630.2021.00116
15. Lundberg, S.M., Lee, S.: A unified approach to interpreting model predictions. In: Advances in Neural Information Processing Systems 30: Annual Conference on Neural Information Processing Systems 2017, 4–9 December 2017, Long Beach, CA, USA, pp. 4765–4774 (2017). https://proceedings.neurips.cc/paper/2017/hash/8a20a8621978632d76c43dfd28b67767-Abstract.html
16. Mokuwe, M., Burke, M., Bosman, A.S.: Black-box saliency map generation using Bayesian optimisation. In: 2020 International Joint Conference on Neural Networks, IJCNN 2020, Glasgow, United Kingdom, 19–24 July 2020, pp. 1–8. IEEE (2020). https://doi.org/10.1109/IJCNN48605.2020.9207343
17. Petsiuk, V., Das, A., Saenko, K.: RISE: randomized input sampling for explanation of black box models. In: British Machine Vision Conference 2018, BMVC 2018, Newcastle, UK, 3–6 September 2018, p. 151. BMVA Press (2018). http://bmvc2018.org/contents/papers/1064.pdf
18. Price, W., Damen, D.: Play fair: frame attributions in video models. In: Ishikawa, H., Liu, C.-L., Pajdla, T., Shi, J. (eds.) ACCV 2020, Part V. LNCS, vol. 12626, pp. 480–497. Springer, Cham (2021). https://doi.org/10.1007/978-3-030-69541-5_29
19. Rasmussen, C.E., Williams, C.K.I.: Gaussian Process for Machine Learning. The MIT Press, Cambridge (2006)
20. Ribeiro, M.T., Singh, S., Guestrin, C.: "Why should I trust you?": explaining the predictions of any classifier. In: Proceedings of the 22nd ACM SIGKDD International Conference on Knowledge Discovery and Data Mining, San Francisco, CA, USA, 13–17 August 2016, pp. 1135–1144 (2016). https://doi.org/10.1145/2939672.2939778
21. Selvaraju, R.R., Cogswell, M., Das, A., Vedantam, R., Parikh, D., Batra, D.: Grad-CAM: visual explanations from deep networks via gradient-based localization. Int. J. Comput. Vis. **128**(2), 336–359 (2020). https://doi.org/10.1007/s11263-019-01228-7
22. Shi, X., Khademi, S., Li, Y., van Gemert, J.: Zoom-CAM: generating fine-grained pixel annotations from image labels. In: 25th International Conference on Pattern Recognition, ICPR 2020, Virtual Event/Milan, Italy, 10–15 January 2021, pp. 10289–10296 (2020). https://doi.org/10.1109/ICPR48806.2021.9412980

23. Stergiou, A., Kapidis, G., Kalliatakis, G., Chrysoulas, C., Veltkamp, R.C., Poppe, R.: Saliency tubes: visual explanations for spatio-temporal convolutions. In: 2019 IEEE International Conference on Image Processing, ICIP 2019, Taipei, Taiwan, 22–25 September 2019, pp. 1830–1834 (2019). https://doi.org/10.1109/ICIP.2019.8803153

24. Subramanya, A., Pillai, V., Pirsiavash, H.: Fooling network interpretation in image classification. In: 2019 IEEE/CVF International Conference on Computer Vision, ICCV 2019, Seoul, Korea (South), 27 October—2 November 2019, pp. 2020–2029 (2019). https://doi.org/10.1109/ICCV.2019.00211

25. Tomsett, R., Harborne, D., Chakraborty, S., Gurram, P., Preece, A.D.: Sanity checks for saliency metrics. In: The Thirty-Fourth AAAI Conference on Artificial Intelligence, AAAI 2020, The Thirty-Second Innovative Applications of Artificial Intelligence Conference, IAAI 2020, The Tenth AAAI Symposium on Educational Advances in Artificial Intelligence, EAAI 2020, New York, NY, USA, 7–12 February 2020, pp. 6021–6029. AAAI Press (2020). https://aaai.org/ojs/index.php/AAAI/article/view/6064

26. Woolson, R.F.: Wilcoxon signed-rank test. Wiley encyclopedia of clinical trials, pp. 1–3 (2007)

27. Zhang, J., Bargal, S.A., Lin, Z., Brandt, J., Shen, X., Sclaroff, S.: Top-down neural attention by excitation backprop. Int. J. Comput. Vis. **126**(10), 1084–1102 (2018). https://doi.org/10.1007/s11263-017-1059-x

28. Zhou, B., Khosla, A., Lapedriza, À., Oliva, A., Torralba, A.: Learning deep features for discriminative localization. In: 2016 IEEE Conference on Computer Vision and Pattern Recognition, CVPR 2016, Las Vegas, NV, USA, 27–30 June 2016, pp. 2921–2929 (2016). https://doi.org/10.1109/CVPR.2016.319

A General Divergence Modeling Strategy for Salient Object Detection

Xinyu Tian[1], Jing Zhang[2], and Yuchao Dai[1(✉)]

[1] Northwestern Polytechnical University, Xi'an, China
`daiyuchao@nwpu.edu.cn`
[2] Australian National University, Canberra, Australia

Abstract. Salient object detection is subjective in nature, which implies that multiple estimations should be related to the same input image. Most existing salient object detection models are deterministic following a point to point estimation learning pipeline, making them incapable of estimating the predictive distribution. Although latent variable model based stochastic prediction networks exist to model the prediction variants, the latent space based on the single clean saliency annotation is less reliable in exploring the subjective nature of saliency, leading to less effective saliency "divergence modeling". Given multiple saliency annotations, we introduce a general divergence modeling strategy via random sampling, and apply our strategy to an ensemble based framework and three latent variable model based solutions to explore the "subjective nature" of saliency. Experimental results prove the superior performance of our general divergence modeling strategy.

Keywords: Salient object detection · Divergence modeling

1 Introduction

When viewing a scene, human visual system has the ability to selectively locate attention [1–5] on the informative contents, which locally stand out from their surroundings. This selection is usually performed in the form of a spatial circumscribed region, leading to the so-called "focus of attention" [4]. Itti *et al.* [1] introduced a general attention model to explain the human visual search strategies [6]. Specifically, the visual input is first decomposed into a group of topographic feature maps which they defined as the early representations. Then, different spatial locations compete for saliency within each topographic feature map, such that locations that locally stand out from their surrounding persist. Lastly, all the feature maps are fed into a master "saliency map", indicating the topographical codes for saliency over the visual scene [4].

This work was supported in part by the NSFC (61871325). Code is available at https://npucvr.github.io/Divergence_SOD/.

Supplementary Information The online version contains supplementary material available at https://doi.org/10.1007/978-3-031-26293-7_18.

| Image | Multiple annotations | Our generated uncertainty maps |

Fig. 1. Given "Multiple Annotations" of each training image, the proposed "Random Sampling" strategy based solutions aim to generate diverse predictions representing the subjective nature of saliency with reliable "generated uncertainty maps".

Following the above process of saliency selection, early saliency detection models focus on detecting the informative locations, leading to the eye fixation prediction [7] task. [8,9] then extended the salient locations driven methods [1,4] and introduced the salient object detection task, which is a binary segmentation task aiming to identify the full scope of salient objects. In this way, "salient object" is defined as any item that is distinct from those around it. Many factors can lead something to be "salient", including the stimulus itself that makes the item distinct, *i.e.* color, texture, direction of movement, and the internal cognitive state of observers, leading to his/her understanding of saliency.

As an important computer vision task, salient object detection is intrinsic to various tasks such as image cropping [10], object detection [11], semantic segmentation [12], video segmentation [13], action recognition [14], image caption generation [15] and semantic image labeling [16], where saliency models can be used to extract class-agnostic important areas in an image or a video sequence. Most of the existing salient object detection models intend to achieve the point estimation from input RGB image (or RGB-D image pair) to the corresponding ground truth (GT) saliency map, neglecting the less consistent saliency regions discarded while generating the binary ground truth maps via majority voting. Although [17] presented a generative model via conditional variational auto-encoder [18] to produce stochastic prediction at test time, the generated stochastic saliency maps have low diversity, where the difference of those stochastic predictions mainly along the object boundaries, making it less effective in explaining the "subjective nature" of saliency as shown in Fig. 3.

In this paper, we study "divergence modeling" for salient object detection, and propose a general strategy to achieve effective divergence modeling (see Fig. 1), representing the "subjective nature" of saliency. Given multiple saliency labels [19] of each image, we aim to generate one majority saliency map, reflecting the majority saliency attribute, and an uncertainty map, explaining the discarded less salient regions. Specifically, we adopt a separate "majority voting" module to regress the majority saliency map for deterministic performance evaluation. We also need to generate stochastic predictions given the multiple annotations. This can be achieved with multiple model parameters (*e.g.* deep ensemble [20]) or with extra latent variable to model the predictive distribution [17], leading to our "ensemble"-based framework and "latent variable model"-based solutions.

We observe that, training simultaneously with all the annotations participating into model updating is less effective in modeling the divergence of saliency, as the model will be dominated by the majority annotation. The main reason is that there exists much less samples containing diverse annotations (around 20%) compared with samples with consistent annotations. To solve the above issue, we introduce a simple "random sampling" strategy. Specifically, within the "ensemble" framework, the network is randomly updated with one annotation. For the "latent variable model" based networks, the generated stochastic prediction is randomly compared with one annotation from the multiple annotations pool. The model is then updated with loss function based on both the deterministic prediction from the majority voting branch and one stochastic prediction from the latent variable branch. We have carried out extensive experiments and find that this simple "random sampling" based strategy works superbly in modeling the "divergence" of saliency, leading to meaningful uncertainty maps representing the "subjective nature" of saliency (see Fig. 1). We also verified the importance of multiple annotations in modeling saliency divergence, and further applied our random sampling strategy to other SOTA SOD models, which can also achieve diverse saliency predictions, demonstrating the flexibility of our approach.

Aiming at discovering the discarded less salient regions for human visual system exploration, we work on a general divergence modeling strategy for saliency detection. Our main contributions are: **1**) we introduce the first "random sampling" based divergence modeling strategy for salient object detection to produce reliable "object" level uncertainty map explaining human visual attention; **2**) we present an ensemble based framework and three latent variable model based solutions to validate our divergence modeling strategy; **3**) to maintain the deterministic performance, we also design an efficient majority voting branch to generate majority voting saliency maps without sampling at test time.

2 Related Work

Salient Object Detection: The main stream of salient object detection (SOD) models are fully-supervised [21–29], where most of them focus on effective high/low feature aggregation or structure-aware prediction [23,30–32]. With extra depth data, RGB-D SOD [33,34] models mainly focus on effective multimodal learning. In addition to above fully-supervised models, semi-supervised [35], weakly-supervised [36,37] SOD models have also been explored. We argue that most of the existing SOD models define saliency detection as a binary segmentation task in general, without exploring the subjective nature of saliency. Although [17] has taken a step further to model the predictive distribution via a conditional variational auto-encoder (CVAE) [18], the diversity of the stochastic predictions mainly distributes along object boundaries, making it less effective in discovering the discarded less salient regions.

Uncertainty Estimation: The goal of uncertainty estimation is to measure the confidence of model predictions. According to [38], the uncertainty of deep neural networks is usually divided into aleatoric uncertainty (or data uncertainty) and epistemic uncertainty (or model uncertainty). Aleatoric uncertainty describes the inherent randomness of the task (*e.g.* dataset annotation error or sensor noise), which cannot be avoided. Epistemic uncertainty is caused by our limited knowledge about the latent true model, which can be reduced by using a larger or more diverse training dataset. The aleatoric uncertainty is usually estimated via a dual-head framework [39], which produces both task related prediction and the corresponding aleatoric uncertainty. Epistemic uncertainty aims to represent the model bias, which can be obtained via a Bayesian Neural Network (BNN) [40]. In practice, Monte Carlo Dropout (MC Dropout) [41] and Deep Ensemble [20,42,43] are two widely studied epistemic uncertainty estimation techniques.

Latent Variable Model Based Stochastic Models: With extra latent variable involved, the latent variable models [18,44–46] can be used to achieve predictive distribution estimation. Currently, latent variable models have been used in numerous tasks, such as semantic segmentation [47], natural language processing [48], depth estimation [49], deraining [50], image deblurring [51], saliency detection [17,52], etc.. In this paper, we explore latent variable models for saliency divergence modeling with "random sampling" strategy.

Uniqueness of Our Solution: Different from the existing point estimation based salient object detection networks [30,53] which produce deterministic saliency maps, our method aims to generate diverse predictions representing human diverse perceptions towards the same scene. Although the existing generative model based saliency prediction networks such as [17] can produce stochastic predictions, those methods can mainly highlight the labeling noise, and are moderately effective in modeling predictive distribution, because the posterior of the latent variable z only depends on the single ground truth after majority voting. Building upon the newly released saliency dataset with multiple annotations [19] for each input image, we present a general "divergence modeling" strategy based on "random sampling", where only the randomly selected annotation can participate into model updating, leading to reliable "object-level" uncertainty map (see Fig. 1) representing the "subjective nature" of saliency.

3   Our Method

Our training dataset is $D = \{x_i, \{y_i^j\}_{j=0}^M\}_{i=1}^N$, where x_i is the input RGB image, $\{y_i^j\}_{j=1}^M$ are the multiple saliency annotations for x_i, and y_i^0 (with $j = 0$) is the saliency map after majority voting. M and N are number of annotators and the size of the training dataset, respectively. i indexes the images, and j indexes the multiple annotations [19]. We aim to model the "subjective nature" of saliency with a general divergence modeling strategy as shown in Fig. 2.

Given the multiple annotations $\{y^j\}_{j=1}^M$ for image x (we omit image index i when there is no ambiguity), we introduce a random sampling based strategy to randomly select annotations for model updating. The main reason is that we have limited samples containing diverse annotations (around 20%), and training directly with all the annotations participating into model updating will generate a biased model, where samples with consistent annotations will dominate the training process (see Table 1 and Table 2), leading to less effective divergence

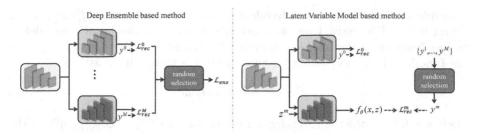

Fig. 2. The proposed strategy within the ensemble based framework (left) and the latent variable model based solutions (right). By randomly selecting one ground truth from the multiple annotations for model updating, the proposed strategy can better explore the contribution of multiple annotations for human visual system exploration.

modeling. Our "random sampling" strategy encourages the diverge annotations to fully contribute to model updating. To extensively analyse the proposed strategy, we design an ensemble based framework and three latent variable based networks with the same saliency generator f_θ, except that the extra latent variable is used in the latent variable models.

Saliency Generator: Our "saliency generator" f_θ is built upon the ResNet50 backbone [54], and we define the backbone features as $\{s_k\}_{k=1}^4$ of channel size 256, 512, 1024 and 2048 respectively. To relief the huge memory requirement and also obtain larger receptive field, we feed the backbone features to four different multi-scale dilated convolutional blocks [55] and obtain new backbone features $f_\theta^b(x) = \{s_k'\}_{k=1}^4$ of the same channel size $C = 32$. We then feed $\{s_k'\}_{k=1}^4$ to decoder from [56] $f_\theta^d(f_\theta^b(x))$ (or $f_\theta^d(f_\theta^b(x), z)$ for the latent variable model solutions) to generate saliency prediction.

Task-Related Loss Function: The widely used loss functions for saliency detection include: 1) binary cross-entropy loss, 2) boundary IOU loss [57], and 3) structure-aware loss [30], which is a weighted combination of the above two loss functions. In this paper, for all the four related divergence modeling models, we adopt structure-aware loss function as the saliency reconstruction loss $\mathcal{L}_{\mathrm{rec}}$.

3.1 Ensemble Framework

Deep ensemble [20] is a widely used method to generate multiple predictions, which uses different network structures for training to achieve multiple predictions with one single forward pass. In our case, instead of having multiple copies of mapping function from input image x to the diverse annotation y^j, we design $M + 1$ decoders with the same structure to achieve divergence modeling, where each output of the decoders is supervised by y^j (including the majority voting annotation y^0). The conventional loss definition for deep ensemble framework is computing the sum of the multiple predictions, or define the minimal loss of each decoder as the final loss to update model parameters. However, we find that the above strategies lead to less diverse predictions, making them less effective in modeling the subjective nature of saliency. In this paper, we first design an ensemble structure with $M + 1$ decoders of the same structure $f_\theta^d(f_\theta^b(x))$, which are initialized differently. Then, we randomly select a decoder branch, and define its loss function as the final loss function for model updating (see Fig. 2). The loss function of our deep ensemble based model is then defined as:

$$\mathcal{L}_{ens} = \underset{j}{\mathrm{RS}}[\{\mathcal{L}_{rec}^j\}_{j=0}^M], \tag{1}$$

where the RS[.] is the random selection operation and $\mathcal{L}_{rec}^m = \mathcal{L}(y_{pred}^m, y^m)$ is the reconstruction loss of prediction from the m_{th} decoder supervised by y^m.

3.2 Latent Variable Model Based Networks

With extra latent variable z, the latent variable models [18,44–46], e.g. variational auto-encoder (VAE) [18,44], generative adversarial net (GAN) [45], alternating back-propagation (ABP) [46], are capable of modeling the predictive distribution by sampling z from the latent space. We then design three latent variable models with the proposed "random sampling" strategy to achieve effective divergence modeling. Note that, with extra conditional variable x, our VAE based framework is indeed a conditional variational auto-encoder (CVAE) [18].

In general, given latent variable z, our latent variable model (see Fig. 2) based networks include two main parts, namely a deterministic prediction network $f_\theta(x)$ supervised by y^0 and a latent variable model $f_\theta(x, z)$ supervised by $\{y^j\}_{j\neq 0}$. The former is the same as the saliency generator with single head in the ensemble framework. For the latter, we first tile z in spatial dimension, and then concatenate it with s_4' channel-wise, leading to sz_4' of channel size $C + K$, where K is the dimension of the latent variable. We set it as $K = 32$ for all the three latent variable models. To adapt to the saliency decoder, we feed sz_4' to another 3×3 convolutional layer to obtain a new feature nz_4' of channel size C. In this way, the stochastic prediction $f_\theta(x, z)$ is obtained via: $f_\theta^d(\{s_k'\}_{k=1}^3, nz_4')$.

CVAE Solution: A CVAE [18] is a conditional directed graph model, which includes three variables, the input x or conditional variable that modulates the prior on Gaussian latent variable z and generates the output prediction y. Two main modules are included in a conventional CVAE based framework: a generator model $f_\theta(x, z)$, which is a saliency generator in this paper, and an inference model $q_\theta(z|x, y)$, which infers the latent variable z with image x and annotation y as input. Learning a CVAE framework involves approximation of the true posterior distribution of z with an inference model $q_\theta(z|x, y)$, with the loss function as:

$$\mathcal{L}_{cvae} = \underbrace{\mathbb{E}_{z \sim q_\theta(z|x,y)}[-\log p_\theta(y|x, z)]}_{\mathcal{L}_{rec}} + D_{KL}(q_\theta(z|x, y) \parallel p_\theta(z|x)), \tag{2}$$

where the first term is the reconstruction loss and the second is the Kullback-Leibler divergence of prior distribution $p_\theta(z|x)$ and posterior distribution $q_\theta(z|x, y)$. The saliency generator has been discussed, and we then introduce the details about the prior and posterior distribution models.

Prior Net and Posterior Net Construction: We have almost the same structure for the prior net and the posterior net, where the only difference is that the input channel of the first convolutional layer within the prior net is 3 (channel of the RGB image x) and that within the posterior net is 4 (channel of the concatenation of x and y). We construct our prior and posterior net with five convolution layers of the same kernel size (4×4) and two fully connected layers. The channel size of the convolutional layers are $C' = C, 2*C, 4*C, 8*C, 8*C$ ($C = 32$, which is the same as the channel reduction layers for the backbone features), and we have a batch normalization layer and a ReLU activation layer after each convolutional layer. The two fully connected layers are used to map the feature map to vectors representing the mean μ and standard deviation σ of the latent variable z. The latent variable z is then obtained via the reparameterization trick: $z = \mu + \sigma \odot \epsilon$, where $\epsilon \sim \mathcal{N}(0, \mathbf{I})$.

Divergence Modeling with CVAE: Recall that the posterior distribution is defined as $q_\theta(z|x, y)$. Given the multiple annotations $\{y^j\}_{j \neq 0}$, we randomly select one annotation y^m, and use it to define the posterior as $q_\theta(z|x, y^m)$. In this way, loss function for the CVAE based divergence modeling model is defined as:

$$\mathcal{L}_{cvae} = \mathcal{L}_{rec}^m + D_{KL}(q_\theta(z|x, y^m) \parallel p_\theta(z|x)). \tag{3}$$

GAN Solution: Within the GAN based framework, we design an extra fully convolutional discriminator g_β following [58], where β is the parameter set of the discriminator. Two different modules (the saliency generator f_θ and the discriminator g_β in our case) play the minimax game in GAN based framework:

$$\min_{f_\theta} \max_{g_\beta} V(g_\beta, f_\theta) = E_{(x,y) \sim p_{data}(x,y)}[\log g_\beta(y|x)] \\ + E_{z \sim p(z)}[\log(1 - g_\beta(f_\theta(x, z)))], \tag{4}$$

where $p_{data}(x, y)$ is the joint distribution of training data, $p(z)$ is the prior distribution of the latent variable z, which is usually defined as $p(z) = \mathcal{N}(0, \mathbf{I})$. In practice, we define loss function for the generator as the sum of a reconstruction loss $\mathcal{L}_{rec}$, and an adversarial loss $\mathcal{L}_{adv} = \mathcal{L}_{ce}(g_\beta(f_\theta(x, z)), 1)$, which is $\mathcal{L}_{gen} = \mathcal{L}_{rec} + \lambda \mathcal{L}_{adv}$, where the hyper-parameter λ is tuned, and empirically we set $\lambda = 0.1$ for stable training. $\mathcal{L}_{ce}$ is the binary cross-entropy loss. The discriminator g_β is trained via loss function as: $\mathcal{L}_{dis} = \mathcal{L}_{ce}(g_\beta(f_\theta(x, z)), 0) + \mathcal{L}_{ce}(g_\beta(y), 1)$. Similar to CVAE, for each iteration of training, we randomly select an annotation y^m, which will be treated as y for both generator and discriminator updating. In this way, with the proposed random sampling strategy for divergence modeling, generator loss and discriminator loss can be defined as:

$$\mathcal{L}_{gen} = \mathcal{L}_{rec}^m + \lambda \mathcal{L}_{adv}, \\ \mathcal{L}_{dis} = \mathcal{L}_{ce}(g_\beta(f_\theta(x, z)), 0) + \mathcal{L}_{ce}(g_\beta(y^m), 1). \tag{5}$$

ABP Solution: Alternating back-propagation [46] was introduced for learning the generator network model. It updates the latent variable and network parameters in an EM-manner. Firstly, given the network prediction with the current parameter set, it infers the latent variable by Langevin dynamics based MCMC [59], which is called "Inferential back-propagation" in [46]. Secondly, given the updated latent variable z, the network parameter set is updated with gradient descent. [46] calls it "Learning back-propagation". Following the previous variable definitions, given the training example (x, y), we intend to infer z and learn the network parameter θ to minimize the reconstruction loss as well as a regularization term that corresponds to the prior on z.

Different from the CVAE solution, where extra inference model $q_\theta(z|x, y)$ is used to approximate the posterior distribution of the latent variable z. ABP [46] samples z directly from its posterior distribution with a gradient-based Monte Carlo method, namely Langevin Dynamics [59]:

$$z_{t+1} = z_t + \frac{s^2}{2}\left[\frac{\partial}{\partial z}\log p_\theta(y, z_t|x)\right] + s\mathcal{N}(0, \mathbf{I}), \qquad (6)$$

where $z_0 \sim \mathcal{N}(0, \mathbf{I})$, and the gradient term is defined as:

$$\frac{\partial}{\partial z}\log p_\theta(y, z|x) = \frac{1}{\sigma^2}(y - f_\theta(x, z))\frac{\partial}{\partial z}f_\theta(x, z) - z. \qquad (7)$$

t is the time step for Langevin sampling, s is the step size, σ^2 is variance of the inherent labeling noise. Empirically we set $s = 0.1$ and $\sigma^2 = 0.3$ in this paper.

In practice, we sample z_0 from $\mathcal{N}(0, \mathbf{I})$, and update z via Eq. 6 by running $T = 5$ steps of Langevin sampling [59]. The final z_T is then used to generate saliency prediction in our case. Similar to the other two latent variable models, we randomly select an annotation y^m from $\{y^j\}_{j\neq 0}$, and use it to infer the latent variable z. The loss function for ABP with our random sampling strategy for saliency detection is then defined as: $\mathcal{L}_{abp} = \mathcal{L}_{rec}^m$.

3.3 Prediction Generation

Majority Voting Regression: To simulate the majority voting in the saliency labeling process, we add a deterministic majority voting prediction branch to the latent variable models, which share the same structure as our single ensemble network. We define its loss function as $\mathcal{L}_{mj} = \mathcal{L}_{rec}^0$. In this way, in addition to the stochastic prediction based loss functions for each latent variable model ($\mathcal{L}_{cvae}$, $\mathcal{L}_{gen}$, $\mathcal{L}_{abp}$), we further add the deterministic loss function $\mathcal{L}_{mj}$ to each of them. At test time, prediction from the majority voting regression branch is defined as our deterministic prediction for performance evaluation.

Obtaining the Uncertainty: Given the ensemble based framework, the predictive uncertainty is defined as entropy of the mean prediction [39]: $U_p = \mathbb{H}[\text{mean}_{j\in[1,M]}\{f_\theta^{dj}(f_\theta^b(x))\}]$, where f_θ^{dj} is the j_{th} decoder. For the latent variable models, the predictive uncertainty is defined as $U_p = \mathbb{H}[\mathbb{E}_{p_\theta(z|x)}(p_\theta(y|x, z))]$. For both models, the aleatoric uncertainty is defined as the mean entropy of multiple predictions. In practice, we usually use Monte Carlo integration as approximation of the intractable expectation operation. The epistemic uncertainty is then defined as $U_e = U_p - U_a$.

4   Experimental Results

Dataset: We use the COME dataset [19] for training as it's the only large training dataset (with 8,025 images) containing multiple annotations for each image. For each image, there are five annotations ($\{y^j\}_{j=1}^{M}$ with $M = 5$) from five different annotators. The ground truth after majority voting is y^0. The testing images include 1) DUTS testing dataset [60], 2) ECSSD [61], 3) DUT [62], 4) HKU-IS [63], 5) COME-E [19], 6) COME-H [19], where each image in COME-E [19] and COME-H [19] datasets has 5 annotations from 5 different annotators.

Evaluation Metrics: We use Mean Absolute Error ($\mathcal{M}$), mean F-measure (F_β) and mean E-measure (E_ξ) [64] for deterministic performance evaluation. We also present an uncertainty based mean absolute error to estimate the divergence modeling ability of each model. Details about these metrics will be introduced in the supplementary material.

Implementation Details: We train our models in Pytorch with ResNet-50 backbone, which is initialized with weights trained on ImageNet, and other newly added layers are initialized by default. We resize all the images and ground truth to 352×352 for both training. The maximum epoch is 50. The initial learning rate is 2.5×10^{-5}. The whole training takes 9 h, 11 h with batch size 8 on one NVIDIA GTX 2080Ti GPUs for deep ensemble model and latent variable models respectively. The inference speed is around 20 images/second.

4.1   Divergence Modeling Performance

As a divergence modeling strategy, we aim to produce both accurate deterministic predictions and reliable uncertainty. We design the following experiments to explain the divergence modeling ability of our proposed strategy.

Firstly, we train two baseline models with only the majority voting ground truth y^0 ("Base_M") and with all the annotations ("Base_A"). The results are shown in Table 1. For the former, the training dataset is $D1 = \{x_i, y_i^0\}_{i=1}^{N}$, and for the latter, the training dataset is $D2 = \{\{x_i^j, y_i^j\}_{j=0}^{M}\}_{i=1}^{N}$, where $\{x_i^j\}_{j=0}^{M} = x_i$. Then, we train all four frameworks with dataset $D2$, and report their performance as "DE_A", "GAN_A", "VAE_A" and "ABP_A" in Table 1 respectively. With the same network structures, we change the learning strategy by randomly select an annotation to learn our divergence modeling models, and the results are shown as "DE_R", "GAN_R", "VAE_R" and "ABP_R" in Table 1 respectively. In this way, for each iteration, we only use one divergent annotation to update model parameters of the stochastic prediction branch. Specifically, as shown in Fig. 2, for "DE_R", we randomly pick up one decoder branch and an annotation from $\{y^j\}_{j=0}^{M}$, and use their loss function to update the whole network. For the latent variable model based frameworks, the deterministic majority voting branch is updated with the majority voting ground truth, and the stochastic prediction branch is updated with randomly picked annotation $\{y^j\}_{j=1}^{M}$ as supervision.

The better performance of training with the randomly selected annotations (the "_R" models) compared with training with all the annotation simultaneously (the "_A" models) illustrates the effectiveness of our strategy. We also

Table 1. Performance of the proposed strategy within the ensemble and the latent variable model based frameworks.

	DUTS [60]			ECSSD [61]			DUT [62]			HKU-IS [63]			COME-E [19]			COME-H [19]		
	F_β ↑	E_ξ ↑	M ↓	F_β ↑	E_ξ ↑	M ↓	F_β ↑	E_ξ ↑	M ↓	F_β ↑	E_ξ ↑	M ↓	F_β ↑	E_ξ ↑	M ↓	F_β ↑	E_ξ ↑	M ↓
Base_A	.811	.897	.045	.903	.931	.044	.741	.846	.059	.887	.935	.036	.852	.905	.049	.807	.859	.080
Base_M	.824	.911	.041	.916	.946	.038	.754	.864	.055	.900	.950	.032	.877	.927	.040	.838	.882	.070
DE_A	.819	.906	.042	.910	.939	.040	.744	.855	.057	.895	.940	.034	.870	.920	.043	.831	.876	.073
DE_R	.816	.900	.043	.916	.944	.038	.738	.843	.060	.892	.939	.036	.869	.920	.043	.832	.880	.072
GAN_A	.803	.895	.045	.906	.939	.040	.736	.850	.060	.879	.931	.038	.855	.912	.046	.816	.867	.076
GAN_R	.817	.906	.041	.914	.944	.039	.741	.849	.059	.895	.946	.033	.873	.926	.040	.834	.882	.070
VAE_A	.798	.888	.049	.910	.940	.039	.727	.843	.068	.890	.937	.036	.853	.909	.048	.813	.865	.078
VAE_R	.822	.899	.042	.913	.936	.042	.723	.824	.062	.895	.937	.036	.880	.924	.041	.838	.879	.071
ABP_A	.777	.866	.052	.888	.916	.049	.720	.829	.063	.853	.900	.046	.835	.886	.054	.794	.844	.085
ABP_R	.805	.896	.047	.911	.938	.044	.734	.846	.060	.885	.937	.039	.864	.911	.050	.824	.863	.083

Table 2. Mean absolute error of the generated uncertainty maps. The results in each block from left to right are models trained with all the annotations simultaneously ("_A" in Table 1), only the majority annotation ("_M" in Table 3), random sampling based on three (Table 4) and five annotations pool ("_R" in Table 1).

	DE				GAN				VAE				ABP			
COME-E	.097	–	.094	.092	.085	.095	.086	.082	.089	.093	.088	.086	.108	.112	.110	.105
COME-H	.128	–	.125	.124	.118	.124	.119	.114	.123	.124	.121	.119	.142	.146	.140	.136

show the generated uncertainty maps in Fig. 1 as "Our generated uncertainty maps", where from left to right are uncertainty map from our ensemble solution, GAN solution, VAE solution and ABP solution. We observe reliable uncertainty maps with all our solutions, where uncertainty of the consistent salient regions usually distributes along object boundaries, and we produce uniform uncertainty activation for the regions with less consistent saliency agreement.

To quantitatively analyse the effectiveness of uncertainty maps with the proposed strategy, we compute mean absolute error of the estimated predictive uncertainty and the predictive uncertainty of the corresponding testing dataset, and show performance in Table 2. Given the multiple annotations, we first compute entropy of the mean annotation, and normalize it to the range of [0,1]. We then define it as the ground truth predictive uncertainty. The predictive uncertainty of the four frameworks can be obtained similarly. Table 2 shows that our random sampling based strategy leads to lower uncertainty based mean absolute error compared with updating the model with all the multiple annotations, which quantitatively validate our strategy.

4.2 Uncertainty Comparison with Single Annotation Based Models

Existing generative model based saliency detection models [17] are based on single version ground truth. Although [17] generated multiple pseudo ground truth, the training pipeline is the same as our experiment "_A" in Table 1, where the model simply learns from more pair of samples with dataset $D2 = \{\{x_i^j, y_i^j\}_{j=0}^M\}_{i=1}^N$. To compare uncertainty maps based on the single annotation (y^0 in our case) and multiple annotations ($\{y^j\}_{j=1}^M$), we train the three latent

Table 3. Performance of the latent variable models for divergence modeling with only majority voting ground truth as supervision.

	DUTS [60]			ECSSD [61]			DUT [62]			HKU-IS [63]			COME-E [19]			COME-H [19]		
	F_β ↑	E_ξ ↑	$\mathcal{M}$ ↓	F_β ↑	E_ξ ↑	$\mathcal{M}$ ↓	F_β ↑	E_ξ ↑	$\mathcal{M}$ ↓	F_β ↑	E_ξ ↑	$\mathcal{M}$ ↓	F_β ↑	E_ξ ↑	$\mathcal{M}$ ↓	F_β ↑	E_ξ ↑	$\mathcal{M}$ ↓
Base_M	.824	.911	.041	.916	.946	.038	.754	.864	.055	.900	.950	.032	.877	.927	.040	.838	.882	.070
GAN_M	.817	.905	.042	.913	.945	.038	.747	.855	.058	.893	.944	.035	.870	.923	.042	.831	.881	.071
VAE_M	.817	.905	.042	.913	.945	.038	.744	.852	.058	.893	.943	.035	.873	.924	.042	.831	.880	.072
ABP_M	.801	.893	.045	.908	.942	.040	.732	.840	.060	.886	.937	.038	.862	.916	.045	.825	.874	.074

    Image   GT   GAN_M   GAN_R   VAE_M   VAE_R   ABP_M   ABP_R

Fig. 3. Uncertainty maps of latent variable models with single majority voting GT ("_M") and multiple($M = 5$)diverse GT using our divergence modeling strategy("_R").

variable models with single majority voting ground truth y^0, and show model performance in Table 3. Similarly, we find that the majority voting ground truth based latent variable models lead to comparable performance as the base model "Base_M" trained with y^0. Then, we compare the uncertainty map difference of the two settings in Fig. 3. Different from our strategy with uniform uncertainty activation within the less confident region, the majority voting ground truth based model usually generate uncertainty distributing along object boundaries, making it less effective in modeling the "subjective nature" of saliency, which explains the necessity of multiple annotations for saliency divergence modeling.

4.3 Uncertainty w.r.t.Diversity of Annotations

In this paper, we investigate in diversity modeling ability of ensemble based model and latent variable models with multiple annotations ($M = 5$) for each input image. We further analyse the uncertainty estimation techniques with less annotations, and show model performance in Table 4, where we train our models with three different annotations. Compared with training with five different annotations, we observe similar deterministic performance when training with three annotations. Then, we visualize the produced uncertainty maps in Fig. 4. The more reliable uncertainty map of using more annotations shows that better latent space exploration can be achieved with more diverse annotations. The main reason is more annotations can provide network with more diverse targets that can be learned, making the predictions more diverse and uncertainty more focus on potentially less salient objects. We also show the mean absolute error of the estimated predictive uncertainty of using three annotations as supervision in Table 2. The slightly increased mean absolute uncertainty error further explains the necessity of more diverse annotations for saliency divergence modeling.

Table 4. Performance of our strategy based models with three diverse annotations.

	DUTS [60]			ECSSD [61]			DUT [62]			HKU-IS [63]			COME-E [19]			COME-H [19]		
	$F_\beta\uparrow$	$E_\xi\uparrow$	$\mathcal{M}\downarrow$	$F_\beta\uparrow$	$E_\xi\uparrow$	$\mathcal{M}\downarrow$	$F_\beta\uparrow$	$E_\xi\uparrow$	$\mathcal{M}\downarrow$	$F_\beta\uparrow$	$E_\xi\uparrow$	$\mathcal{M}\downarrow$	$F_\beta\uparrow$	$E_\xi\uparrow$	$\mathcal{M}\downarrow$	$F_\beta\uparrow$	$E_\xi\uparrow$	$\mathcal{M}\downarrow$
DE3	.803	.896	.045	.913	.946	.037	.731	.841	.059	.891	.942	.035	.862	.918	.043	.827	.878	.072
GAN3	.821	.905	.041	.919	.946	.037	.738	.843	.059	.897	.942	.034	.875	.925	.041	.835	.880	.071
VAE3	.824	.909	.040	.920	.948	.036	.747	.854	.057	.898	.946	.034	.876	.926	.040	.836	.881	.070
ABP3	.781	.880	.051	.906	.938	.041	.714	.835	.066	.878	.931	.040	.852	.909	.048	.815	.865	.079

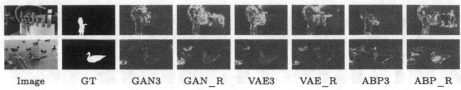

Image   GT   GAN3   GAN_R   VAE3   VAE_R   ABP3   ABP_R

Fig. 4. Uncertainty map comparison w.r.t. the number of annotations for each image.

Table 5. Performance comparison with benchmark SOD models trained on the COME training dataset. Red and blue indicate the best and the second best respectively.

Method	DUTS [60]			ECSSD [61]			DUT [62]			HKU-IS [63]			COME-E [19]			COME-H [19]		
	F_β	E_ξ	$\mathcal{M}$	F_β	E_ξ	$\mathcal{M}$	F_β	E_ξ	$\mathcal{M}$	F_β	E_ξ	$\mathcal{M}$	F_β	E_ξ	$\mathcal{M}$	F_β	E_ξ	$\mathcal{M}$
SCRN [65]	.805	.883	.047	.911	.937	.040	.738	.836	.059	.887	.929	.038	.871	.918	.043	.830	.873	.073
ITSD [66]	.824	.907	.042	.916	.945	.038	.772	.872	.053	.895	.944	.033	.870	.922	.043	.835	.880	.072
MINet [23]	.820	.902	.041	.920	.947	.037	.751	.852	.053	.898	.945	.032	.877	.924	.040	.840	.884	.068
LDF [53]	.818	.902	.043	.919	.947	.036	.754	.856	.055	.902	.948	.031	.874	.924	.041	.836	.882	.070
GateNet [24]	.820	.898	.042	.920	.943	.038	.749	.847	.056	.897	.939	.035	.876	.922	.041	.839	.880	.070
PAKRN [67]	.836	.908	.042	.920	.947	.036	.765	.863	.055	.895	.937	.037	.893	.932	.040	.849	.886	.070
SAMNet [28]	.700	.816	.076	.844	.883	.069	.689	.812	.072	.817	.879	.060	.796	.858	.073	.766	.818	.104
DCN [68]	.829	.904	.041	.923	.946	.037	.765	.861	.051	.905	.945	.032	.885	.927	.039	.843	.880	.071
PFSNet [69]	.795	.885	.053	.919	.947	.037	.739	.847	.065	.897	.944	.034	.867	.917	.044	.831	.879	.072
Ours	.824	.911	.041	.916	.946	.038	.754	.864	.055	.900	.950	.032	.877	.927	.040	.838	.882	.070

4.4   Performance Comparison with SOTA SOD Models

We compare performance of our baseline model ("Base_M" in Table 1) with state-of-the-art (SOTA) SOD models and show model performance in Table 5, where "Ours" represents our baseline model trained with the majority ground truth. As existing salient object detection models are usually trained with DUT-S [60] training dataset, we re-train those models with COME training dataset [19], where the majority voting ground truth is used as supervision. Table 5 shows comparable performance of our baseline model compared with the SOTA SOD models. Note that, the focus of our paper is achieving "divergence modeling" with informative uncertainty maps (see Fig. 1), discovering the "less salient regions" that are discarded while preparing the majority ground truth maps. Although existing models [17] can produce stochastic predictions, the generated uncertainty maps fail to discover the "object level" uncertainty (see Fig. 3). Differently, our solution can generate more reliable uncertainty maps (see Fig. 3 and Table 2), which is more informative in explaining human visual system.

Table 6. Performance of applying the proposed strategy to SOTA SOD model, where "-E", "-G", "-V" and "-A" indicate the corresponding deep ensemble, GAN, VAE and ABP based model. Note that, we perform random sampling within the models.

Method	DUTS [60]			ECSSD [61]			DUT [62]			HKU-IS [63]			COME-E [19]			COME-H [19]		
	F_β	E_ξ	$\mathcal{M}$	F_β	E_ξ	$\mathcal{M}$	F_β	E_ξ	$\mathcal{M}$	F_β	E_ξ	$\mathcal{M}$	F_β	E_ξ	$\mathcal{M}$	F_β	E_ξ	$\mathcal{M}$
PFSNet [69]	.795	.885	.053	.919	.947	.037	.739	.847	.065	.897	.944	.034	.867	.917	.044	.831	.879	.072
-E	.791	.890	.047	.904	.936	.039	.741	.858	.062	.892	.946	.035	.865	.928	.042	.839	.882	.071
-G	.830	.908	.040	.924	.948	.036	.755	.855	.055	.900	.945	.033	.889	.932	.037	.847	.887	.066
-V	.829	.902	.040	.919	.943	.038	.733	.832	.058	.897	.939	.034	.891	.930	.038	.845	.882	.069
-A	.791	.883	.053	.915	.948	.037	.739	.856	.061	.898	.951	.034	.866	.924	.043	.831	.886	.072

| Image/GT | PFSNet [69] | PFSNet-E | PFSNet-G | PFSNet-V | PFSNet-A |

Fig. 5. Uncertainty maps of SOTA SOD model (PFSNet [69] in particular) with the proposed general divergence modeling strategy, where "-E", "-G", "-V" and "-A" are introduced in Table 6. The first column show image and the majority voting ground truth, and from the second column to the last one, we show prediction (top) and the corresponding predictive uncertainty (bottom).

Applying Our Solution to SOTA SOD Model: We apply our strategy to one SOTA SOD model, namely PFSNet [69], and show the deterministic performance in Table 6, and the uncertainty maps in Fig. 5. Table 6 shows that the proposed divergence modeling strategy can keep the deterministic performance unchanged, which is consistent with our experiments in Table 3. Besides it, with the proposed "random sampling" based divergence modeling strategy, we observe more reliable "object level" uncertainty maps ("PFSNet-{E,G,V,A}" in Fig. 5) compared with the original Softmax based uncertainty map[1] (second column of Fig. 5), which further explains the superiority our strategy in applying to existing SOTA SOD models. We have also applied our strategy to other SOTA SOD models [23,24] and achieved comparable deterministic performance with reliable uncertainty estimation, which further explains superiority of our approach.

4.5 Comparison with Relative Saliency Ranking

Similar to our divergence modeling strategy, relative saliency ranking [70–73] aims to explore human perception system for better saliency understanding, while differently, it achieves this via inferring the saliency levels, and we present a general divergence modeling strategy to discover the less salient objects.

---

[1] To obtain the "Softmax uncertainty" of deterministic prediction, we compute the entropy of the prediction, which is then defined as the corresponding uncertainty.

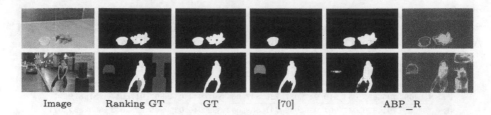

Image   Ranking GT   GT   [70]   ABP_R

Fig. 6. Performance comparison with saliency ranking model [70], where "GT" is the binary saliency GT after majority voting, and the two predictions of "ABP_R" represent the regressed majority saliency maps and the generated uncertainty maps.

To clearly explain the difference of saliency ranking and our divergence modeling strategy, we visualize the ranking ground truth maps, the ranking predictions of [70], and our generated uncertainty maps via divergence modeling in Fig. 6. Note that we run the provided model of [70] to generate the ranking predictions.

Due to the inconsistent labeling of different ranking dataset and the difficulty in generating precise ranking annotations, we argue that the ranking predictions may not be always reliable (see [70] in the fourth column of Fig. 6). Differently, our general divergence modeling strategy suffers no such issue, which is general and more effective in discovering the less salient regions (see "ABP_R" in Fig. 6). However, the explicit saliency level exploration of the relative saliency task can be benificial for our general divergence modeling strategy to have a better understanding about human perception, which will be our future work.

5   Conclusion

As a one-to-many mapping task, we argue that salient object detection deserves better exploration with effective divergence modeling strategies. Our proposed "divergence modeling" strategy aims to explore those "discarded" salient regions for better human visual system understanding. Specifically, given multiple saliency annotations for each training image, we aim to generate one majority saliency map, and an uncertainty map, explaining the discarded less salient regions. To achieve this, we propose a general random sampling based strategy and apply it to an ensemble based framework and three latent variable models. Our extensive experimental results show that multiple annotations are necessary for reliable saliency divergence modeling, and our simple random sampling based strategy works effectively in modeling the divergence of saliency. We also apply the proposed strategy to other SOTA SOD models to further explain the flexibility of our model and effectiveness of our strategy. Note that, our saliency divergence modeling approach relies on multiple annotations, which can lead to expensive labeling costs. We will further explore fixation data [74] from multiple observers to achieve both cheaper annotation and more reliable saliency divergence supervision.

References

1. Itti, L., Koch, C., Niebur, E.: A model of saliency-based visual attention for rapid scene analysis. IEEE Trans. Pattern Anal. Mach. Intell. (TPAMI) **20**, 1254–1259 (1998)
2. Tsotsos, J.K., Culhane, S.M., Wai, W.Y.K., Lai, Y., Davis, N., Nuflo, F.: Modeling visual attention via selective tuning. Artif. Intell. **78**, 507–545 (1995)
3. Olshausen, B.A., Anderson, C.H., Van Essen, D.C.: A neurobiological model of visual attention and invariant pattern recognition based on dynamic routing of information. J. Neurosci. **13**, 4700–4719 (1993)
4. Koch, C., Ullman, S.: Shifts in selective visual attention: towards the underlying neural circuitry. Hum. Neurobiol. **44**, 219–27 (1985)
5. Robinson, D., Petersen, S.: The pulvinar and visual salience. Trends Neurosci. **15**, 127–132 (1992)
6. Treisman, A.M., Gelade, G.: A feature-integration theory of attention. Cogn. Psychol. **12**, 97–136 (1980)
7. Huang, X., Shen, C., Boix, X., Zhao, Q.: Salicon: reducing the semantic gap in saliency prediction by adapting deep neural networks. In: IEEE International Conference on Computer Vision (ICCV), pp. 262–270 (2015)
8. Liu, T., et al.: Learning to detect a salient object. IEEE Trans. Pattern Anal. Mach. Intell. (TPAMI) **33**, 353–367 (2010)
9. Achanta, R., Hemami, S., Estrada, F., Susstrunk, S.: Frequency-tuned salient region detection. In: IEEE Conference on Computer Vision and Pattern Recognition (CVPR), pp. 1597–1604 (2009)
10. Wang, W., Shen, J., Ling, H.: A deep network solution for attention and aesthetics aware photo cropping. IEEE Trans. Pattern Anal. Mach. Intell. (TPAMI) **41**, 1531–1544 (2018)
11. Zhang, D., Meng, D., Zhao, L., Han, J.: Bridging saliency detection to weakly supervised object detection based on self-paced curriculum learning. In: International Joint Conference on Artificial Intelligence (IJCAI), pp. 3538–3544 (2016)
12. Wang, W., Zhou, T., Yu, F., Dai, J., Konukoglu, E., Van Gool, L.: Exploring cross-image pixel contrast for semantic segmentation. In: IEEE International Conference on Computer Vision (ICCV), pp. 7303–7313 (2021)
13. Wang, W., Shen, J., Yang, R., Porikli, F.: Saliency-aware video object segmentation. IEEE Trans. Pattern Anal. Mach. Intell. (TPAMI) **40**, 20–33 (2017)
14. Sharma, S., Kiros, R., Salakhutdinov, R.: Action recognition using visual attention. In: Conference on Neural Information Processing Systems (NeurIPS) Workshop (2015)
15. Xu, K., et al.: Show, attend and tell: Neural image caption generation with visual attention. In: International Conference on Machine Learning (ICML), vol. 37, pp. 2048–2057 (2015)
16. Oh, S.J., Benenson, R., Khoreva, A., Akata, Z., Fritz, M., Schiele, B.: Exploiting saliency for object segmentation from image level labels. In: IEEE Conference on Computer Vision and Pattern Recognition (CVPR), pp. 4410–4419 (2017)
17. Zhang, J., et al.: Uc-net: uncertainty inspired RGB-D saliency detection via conditional variational autoencoders. In: IEEE Conference on Computer Vision and Pattern Recognition (CVPR), pp. 8582–8591 (2020)
18. Sohn, K., Lee, H., Yan, X.: Learning structured output representation using deep conditional generative models. In: Conference on Neural Information Processing Systems (NeurIPS), vol. 28, pp. 3483–3491 (2015)

19. Zhang, J., et al.: Rgb-d saliency detection via cascaded mutual information minimization. In: IEEE International Conference on Computer Vision (ICCV), pp. 4338–4347 (2021)
20. Lakshminarayanan, B., Pritzel, A., Blundell, C.: Simple and scalable predictive uncertainty estimation using deep ensembles. In: Conference on Neural Information Processing Systems (NeurIPS), vol. 30 (2017)
21. Qin, X., Zhang, Z., Huang, C., Gao, C., Dehghan, M., Jagersand, M.: Basnet: boundary-aware salient object detection. In: IEEE Conference on Computer Vision and Pattern Recognition (CVPR), pp. 7479–7489 (2019)
22. Wang, W., Zhao, S., Shen, J., Hoi, S.C., Borji, A.: Salient object detection with pyramid attention and salient edges. In: IEEE Conference on Computer Vision and Pattern Recognition (CVPR), pp. 1448–1457 (2019)
23. Pang, Y., Zhao, X., Zhang, L., Lu, H.: Multi-scale interactive network for salient object detection. In: IEEE Conference on Computer Vision and Pattern Recognition (CVPR), pp. 9413–9422 (2020)
24. Zhao, X., Pang, Y., Zhang, L., Lu, H., Zhang, L.: Suppress and balance: a simple gated network for salient object detection. In: Vedaldi, A., Bischof, H., Brox, T., Frahm, J.-M. (eds.) ECCV 2020. LNCS, vol. 12347, pp. 35–51. Springer, Cham (2020). https://doi.org/10.1007/978-3-030-58536-5_3
25. Qin, X., Zhang, Z., Huang, C., Dehghan, M., Zaiane, O.R., Jagersand, M.: U2-Net: going deeper with nested u-structure for salient object detection. Pattern Recogn. (PR) **106**, 107404 (2020)
26. Tang, L., Li, B., Zhong, Y., Ding, S., Song, M.: Disentangled high quality salient object detection. In: IEEE International Conference on Computer Vision (ICCV), pp. 3580–3590 (2021)
27. Li, A., Zhang, J., Lv, Y., Liu, B., Zhang, T., Dai, Y.: Uncertainty-aware joint salient object and camouflaged object detection. In: IEEE Conference on Computer Vision and Pattern Recognition (CVPR), pp. 10071–10081 (2021)
28. Liu, Y., Zhang, X.Y., Bian, J.W., Zhang, L., Cheng, M.M.: Samnet: stereoscopically attentive multi-scale network for lightweight salient object detection. IEEE Trans. Image Process. (TIP) **30**, 3804–3814 (2021)
29. Siris, A., Jiao, J., Tam, G.K., Xie, X., Lau, R.W.: Scene context-aware salient object detection. In: IEEE International Conference on Computer Vision (ICCV), pp. 4156–4166 (2021)
30. Wei, J., Wang, S., Huang, Q.: F^3net: fusion, feedback and focus for salient object detection. In: AAAI Conference on Artificial Intelligence (AAAI), pp. 12321–12328 (2020)
31. Chen, Z., Xu, Q., Cong, R., Huang, Q.: Global context-aware progressive aggregation network for salient object detection. In: AAAI Conference on Artificial Intelligence (AAAI), pp. 10599–10606 (2020)
32. Liu, J.J., Hou, Q., Cheng, M.M., Feng, J., Jiang, J.: A simple pooling-based design for real-time salient object detection. In: IEEE Conference on Computer Vision and Pattern Recognition (CVPR), pp. 3917–3926 (2019)
33. Zhou, T., Fu, H., Chen, G., Zhou, Y., Fan, D.P., Shao, L.: Specificity-preserving rgb-d saliency detection. In: IEEE International Conference on Computer Vision (ICCV), pp. 4681–4691 (2021)
34. Sun, P., Zhang, W., Wang, H., Li, S., Li, X.: Deep rgb-d saliency detection with depth-sensitive attention and automatic multi-modal fusion. In: IEEE Conference on Computer Vision and Pattern Recognition (CVPR), pp. 1407–1417 (2021)
35. Lv, Y., Liu, B., Zhang, J., Dai, Y., Li, A., Zhang, T.: Semi-supervised active salient object detection. Pattern Recogn. (PR) **123**, 108364 (2021)

36. Piao, Y., Wang, J., Zhang, M., Lu, H.: Mfnet: multi-filter directive network for weakly supervised salient object detection. In: IEEE International Conference on Computer Vision (ICCV), pp. 4136–4145 (2021)
37. Zhang, J., Yu, X., Li, A., Song, P., Liu, B., Dai, Y.: Weakly-supervised salient object detection via scribble annotations. In: IEEE Conference on Computer Vision and Pattern Recognition (CVPR), pp. 12546–12555 (2020)
38. Zhang, J., et al.: Dense uncertainty estimation. arXiv preprint arXiv:2110.06427 (2021)
39. Kendall, A., Gal, Y.: What uncertainties do we need in bayesian deep learning for computer vision? In: Conference on Neural Information Processing Systems (NeurIPS), vol. 30 (2017)
40. Depeweg, S., Hernández-Lobato, J.M., Doshi-Velez, F., Udluft, S.: Decomposition of uncertainty in bayesian deep learning for efficient and risk-sensitive learning. In: International Conference on Machine Learning (ICML), pp. 1184–1193 (2018)
41. Gal, Y., Ghahramani, Z.: Dropout as a bayesian approximation: representing model uncertainty in deep learning. In: International Conference on Machine Learning (ICML), pp. 1050–1059 (2016)
42. Chitta, K., Feng, J., Hebert, M.: Adaptive semantic segmentation with a strategic curriculum of proxy labels. arXiv preprint arXiv:1811.03542 (2018)
43. Osband, I., Blundell, C., Pritzel, A., Van Roy, B.: Deep exploration via bootstrapped dqn. In: Conference on Neural Information Processing Systems (NeurIPS), vol. 29, pp. 4026–4034 (2016)
44. Kingma, D.P., Welling, M.: Auto-encoding variational bayes. In: International Conference on Learning Representations (ICLR) (2014)
45. Goodfellow, I., et al.: Generative adversarial nets. In: Conference on Neural Information Processing Systems (NeurIPS), vol. 27, pp. 2672–2680 (2014)
46. Han, T., Lu, Y., Zhu, S.C., Wu, Y.N.: Alternating back-propagation for generator network. In: AAAI Conference on Artificial Intelligence (AAAI) (2017)
47. Li, Z., Togo, R., Ogawa, T., Haseyama, M.: Variational autoencoder based unsupervised domain adaptation for semantic segmentation. In: IEEE International Conference on Image Processing (ICIP), pp. 2426–2430 (2020)
48. Cotterell, R., Eisner, J.: Probabilistic typology: deep generative models of vowel inventories. In: Proceedings of the 55th Annual Meeting of the Association for Computational Linguistics, pp. 1182–1192 (2017)
49. Puscas, M.M., Xu, D., Pilzer, A., Sebe, N.: Structured coupled generative adversarial networks for unsupervised monocular depth estimation. In: 2019 International Conference on 3D Vision (3DV), pp. 18–26 (2019)
50. Du, Y., Xu, J., Zhen, X., Cheng, M.M., Shao, L.: Conditional variational image deraining. IEEE Trans. Image Process. (TIP) **29**, 6288–6301 (2020)
51. Ramakrishnan, S., Pachori, S., Gangopadhyay, A., Raman, S.: Deep generative filter for motion deblurring. In: IEEE Conference on Computer Vision and Pattern Recognition (CVPR) Workshop, pp. 2993–3000 (2017)
52. Liu, Z., Tang, J., Xiang, Q., Zhao, P.: Salient object detection for rgb-d images by generative adversarial network. Multimedia Tools Appl. **79**, 25403–25425 (2020)
53. Wei, J., Wang, S., Wu, Z., Su, C., Huang, Q., Tian, Q.: Label decoupling framework for salient object detection. In: IEEE Conference on Computer Vision and Pattern Recognition (CVPR), pp. 13025–13034 (2020)
54. He, K., Zhang, X., Ren, S., Sun, J.: Deep residual learning for image recognition. In: IEEE Conference on Computer Vision and Pattern Recognition (CVPR), pp. 770–778 (2016)

55. Yang, M., Yu, K., Zhang, C., Li, Z., Yang, K.: Denseaspp for semantic segmentation in street scenes. In: IEEE Conference on Computer Vision and Pattern Recognition (CVPR), pp. 3684–3692 (2018)
56. Ranftl, R., Lasinger, K., Hafner, D., Schindler, K., Koltun, V.: Towards robust monocular depth estimation: mixing datasets for zero-shot cross-dataset transfer. IEEE Trans. Pattern Anal. Mach. Intell. (TPAMI) **44**, 1623–1637 (2020)
57. Luo, Z., Mishra, A., Achkar, A., Eichel, J., Li, S., Jodoin, P.M.: Non-local deep features for salient object detection. In: IEEE Conference on Computer Vision and Pattern Recognition (CVPR), pp. 6609–6617 (2017)
58. Hung, W.C., Tsai, Y.H., Liou, Y.T., Lin, Y.Y., Yang, M.H.: Adversarial learning for semi-supervised semantic segmentation. In: British Machine Vision Conference (BMVC). (2018)
59. Neal, R.: MCMC using hamiltonian dynamics. In: Handbook of Markov Chain Monte Carlo (2011)
60. Wang, L., et al.: Learning to detect salient objects with image-level supervision. In: IEEE Conference on Computer Vision and Pattern Recognition (CVPR), pp. 136–145 (2017)
61. Yan, Q., Xu, L., Shi, J., Jia, J.: Hierarchical saliency detection. In: IEEE Conference on Computer Vision and Pattern Recognition (CVPR), pp. 1155–1162 (2013)
62. Yang, C., Zhang, L., Lu, H., Ruan, X., Yang, M.H.: Saliency detection via graph-based manifold ranking. In: IEEE Conference on Computer Vision and Pattern Recognition (CVPR), pp. 3166–3173 (2013)
63. Li, G., Yu, Y.: Visual saliency based on multiscale deep features. In: IEEE Conference on Computer Vision and Pattern Recognition (CVPR), pp. 5455–5463 (2015)
64. Fan, D.P., Gong, C., Cao, Y., Ren, B., Cheng, M.M., Borji, A.: Enhanced-alignment measure for binary foreground map evaluation. In: International Joint Conference on Artificial Intelligence (IJCAI), pp. 698–704 (2018)
65. Wu, Z., Su, L., Huang, Q.: Stacked cross refinement network for edge-aware salient object detection. In: IEEE International Conference on Computer Vision (ICCV), pp. 7264–7273 (2019)
66. Zhou, H., Xie, X., Lai, J.H., Chen, Z., Yang, L.: Interactive two-stream decoder for accurate and fast saliency detection. In: IEEE Conference on Computer Vision and Pattern Recognition (CVPR), pp. 9141–9150 (2020)
67. Xu, B., Liang, H., Liang, R., Chen, P.: Locate globally, segment locally: a progressive architecture with knowledge review network for salient object detection. In: AAAI Conference on Artificial Intelligence (AAAI), pp. 3004–3012 (2021)
68. Wu, Z., Su, L., Huang, Q.: Decomposition and completion network for salient object detection. IEEE Trans. Image Process. (TIP) **30**, 6226–6239 (2021)
69. Ma, M., Xia, C., Li, J.: Pyramidal feature shrinking for salient object detection. In: AAAI Conference on Artificial Intelligence (AAAI), pp. 2311–2318 (2021)
70. Liu, N., Li, L., Zhao, W., Han, J., Shao, L.: Instance-level relative saliency ranking with graph reasoning. IEEE Trans. Pattern Anal. Mach. Intell. (TPAMI) **44**, 8321–8337 (2021)
71. Islam, M.A., Kalash, M., Bruce, N.D.: Revisiting salient object detection: simultaneous detection, ranking, and subitizing of multiple salient objects. In: IEEE Conference on Computer Vision and Pattern Recognition (CVPR), pp. 7142–7150 (2018)
72. Siris, A., Jiao, J., Tam, G.K., Xie, X., Lau, R.W.: Inferring attention shift ranks of objects for image saliency. In: IEEE Conference on Computer Vision and Pattern Recognition (CVPR), pp. 12133–12143 (2020)

73. Tian, X., Xu, K., Yang, X., Du, L., Yin, B., Lau, R.W.: Bi-directional object-context prioritization learning for saliency ranking. In: IEEE Conference on Computer Vision and Pattern Recognition (CVPR), pp. 5882–5891 (2022)
74. Xu, J., Jiang, M., Wang, S., Kankanhalli, M.S., Zhao, Q.: Predicting human gaze beyond pixels. J. Vision **14**, 28 (2014)

Full-Scale Selective Transformer
for Semantic Segmentation

Fangjian Lin[1,2,3], Sitong Wu[2], Yizhe Ma[1], and Shengwei Tian[1(✉)]

[1] School of Software, Xinjiang University, Urumqi, China
`tianshengwei@163.com`
[2] Baidu VIS, Beijing, China
[3] Institute of Deep Learning, Baidu Research, Beijing, China

Abstract. In this paper, we rethink the multi-scale feature fusion from two perspectives (scale-level and spatial-level) and propose a full-scale selective fusion strategy for semantic segmentation. Based on such strategy, we design a novel segmentation network, named Full-scale Selective Transformer (FSFormer). Specifically, our FSFormer adaptively selects partial tokens from all tokens at all scales to construct a token subset of interest for each scale. Therefore, each token only interacts with the tokens within its corresponding token subset of interest. The proposed full-scale selective fusion strategy can not only filter out the noisy information propagation but also reduce the computational costs to some extent. We evaluate our FSFormer on four challenging semantic segmentation benchmarks, including PASCAL Context, ADE20K, COCO-Stuff 10K, and Cityscapes, outperforming the state-of-the-art methods. We evaluate our FSFormer on four challenging semantic segmentation benchmarks, including PASCAL Context, ADE20K, COCO-Stuff 10K, and Cityscapes, outperforming the state-of-the-art methods.

Keywords: Semantic segmentation · Transformer · Full-scale feature fusion

1 Introduction

Semantic segmentation aims to predict a semantic label for each pixel in the image, which plays an important role for various applications such as autonomous driving [10] and medical analysis [27]. However, precisely recognize every pixel is still challenging as the objects vary across a wide range of scales. Since FPN [21], a typical and natural solution for this problem is to leverage both high-resolution feature maps with more detail information in shallow layers and high-level feature maps with richer semantics in deep layers via multi-scale feature fusion.

Many works [5,17,19–21,31,32,35,38,40,41,44] have explored how to fuse multi-scale features. We rethink multi-scale feature fusion from two perspectives,

---

F. Lin and S. Wu—Equal contributions.

Supplementary Information The online version contains supplementary material available at https://doi.org/10.1007/978-3-031-26293-7_19.

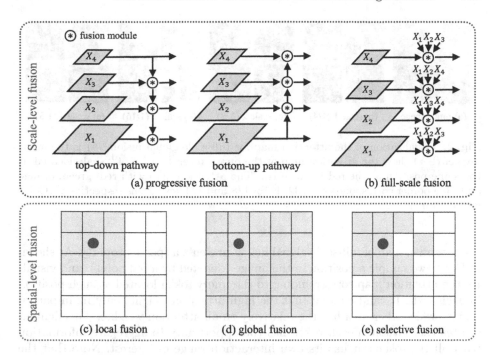

Fig. 1. Comparisons with different scale-level and spatial level feature fusion strategies. The gray shadow area in (c) (d) (e) represents the allowed interaction region of the query (denoted by red point). (Color figure online)

scale-level and spatial-level. The former refers to the fusion strategy across different scales, for example, for one scale, which scales can it interact with. And the latter refers to the interaction range of each token in the spatial dimension. On the one hand, previous scale-level fusion involves two main strategies, that is, progressive fusion and full-scale fusion. As shown in Fig. 1(a), progressive fusion has two typical pathway (top-down and bottom-up), where the token at one scale can only interact with the tokens at its adjacent scale. By contrast, in full-scale fusion (Fig. 1(b)), each token at one scale can interact with all the tokens at any scale. It has been proved that full-scale fusion has more advantages [18]. On the other hand, spatial-level fusion is a more popular topic. Benefited from the development of convolution, local fusion has been dominant for a long time. As shown in Fig. 1(c), each token can only aggregate information from its neighbourhoods. Since the attention mechanism and Transformer architecture become show promising prospects, global fusion achieves more and more attention. As shown in Fig. 1(d), each token can exchange information with all the tokens.

In order to accommodate both scale-level fusion and spatial-level fusion, we explore the full-scale global fusion using attention mechanism. Specifically, each token can interact with all the tokens at any scale. Although full-scale and global fusion strategies provide larger interaction range, they introduce more computation burden. Therefore, how to balance the trade-off between performance and computational costs is a valuable problem. Through visualization, we found that

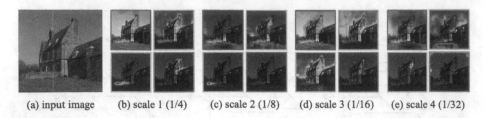

| (a) input image | (b) scale 1 (1/4) | (c) scale 2 (1/8) | (d) scale 3 (1/16) | (e) scale 4 (1/32) |

Fig. 2. Visualization of the attention map for full-scale global fusion. (a) is the input image. (b–e) show the attention map of the query token at $1^{st} \sim 4^{th}$ scale located at the same position as the red point in (a). The attention map with red, green, orange and blue border corresponds to 1/32, 1/16, 1/8 and 1/4 key tokens respectively. (Color figure online)

the attention map of full-scale global fusion presents a sparse property. As shown in Fig. 2, we sample a position in the image (denoted by a red point), and visualize the attention map corresponding to the query token located at such position in each scale. It can be found that the high attention weights only lie in partial region, while other area has the relatively small attention weights. This demonstrates that although each token has the opportunity to aggregate information with all the tokens, it has its own interaction range of interest. Note that the effective interaction range is not just simply local or global. In addition, compared with the last four columns of Fig. 2, the token at the same spatial location of the feature map at different scales present a different interaction region of interest pattern. Based on this observation, we believe that the range of feature fusion is the more accurate the better, rather than the larger the better. Making each token only interact with other tokens within its region of interest may be a breakthrough to filter out the noisy information and reduce the computational costs.

In this paper, we propose a Full-scale Selective Transformer (FSFormer) for semantic segmentation. The core idea is to perform interaction among tokens via the proposed full-scale selective fusion strategy. Specifically, for each scale, our FSFormer adaptively select partial tokens from all tokens at all the scales to construct a token subset of interest. Each token only interact with the tokens within its corresponding token subset of interest, which is shared by the tokens belonging to the same scale. Such full-scale selective fusion strategy can not only filter out the noisy information propagation but also reduce the computational costs to some extent. To verify the effectiveness, we evaluate our FSFormer on four widely-used semantic segmentation benchmarks, including PASCAL Context [24], ADE20K [43], COCO-Stuff 10K [3], and Cityscapes [10], achieving 58.91%, 54.43%, 49.80%, and 84.46% mIoU respectively, outperforming the state-of-the-art methods.

2   Related Work

Multi-scale Features Fusion. There are various works exploring how to fuse multi-scale features for semantic segmentation. Inspired by FPN [21] that employed a top-down pathway and lateral connections for progressively fusing multi-scale features for object detection, Semantic-FPN [17] and SETR-MLA [41] extended this architecture to fuse multi-scale features for semantic segmentation. Based on this top-down fusion, ZigZagNet [20] proposed top-down and bottom-up propagations to aggregate multi-scale features, while FTN [31] proposed Feature Pyramid Transformer for multi-scale feature fusion. Differently, PSPNet [40] and DeepLab series [4–6] fused multi-scale features via concatenation at the channel dimension. Different from these methods that fused features on the local region, ANN [44] proposed an Asymmetric Fusion Non-local Block for fusing all features at one scale for each feature (position) on another scale, while FPT [38] proposed Grounding Transformer to ground the "concept" of the higher-level features to every pixel on the lower-level ones. Different from these methods that fuse features from preset subset for queries, we explore how to dynamically select informative subset from the whole multi-scale feature set and fuse them for each query feature.

Transformer-Based Semantic Segmentation. Since Alexey *et al.* [11] introduced Visual Transformer (ViT) for image classification, it has attracted more and more attentions to explore how to use Transformer for semantic segmentation. These methods focused on exploring the various usages of Transformer, including extracting features [26,33,41] from input image, learning class embedding [30,34], or learning mask embedding [9]. For example, SETR [41] treated semantic segmentation as a sequence-to-sequence prediction task and deployed a pure transformer (i.e., without convolution and resolution reduction) to encode an image as a sequence of patches for feature extraction. DPT [26] reassembled the bag-of-words representation provided by ViT into image-like features at various resolutions, and progressively combined them into final predictions. Differently, Trans2Seg [34] formulated semantic segmentation as a problem of dictionary look-up, and designed a set of learnable prototypes as the query of Transformer decoder, where each prototype learns the statistics of one category. SegFormer [33] used Transformer-based encoder to extract features and the lightweight MLP-decoder to predict pixel by pixel. Segmenter [30] employed a Mask Transformer to learn a set of class embedding, which was used to generate class masks. Recent MaskFormer [9] proposed a simple mask classification model to predict a set of binary masks, where a transformer decoder was used to learn mask embedding. Different from these works, we explore how to use Transformer to fuse multi-scale features.

3   Method

3.1   Overview

The overall framework of our FSFormer is shown in Fig. 3. Given the input image $\mathcal{I} \in \mathbb{R}^{3 \times H \times W}$, the backbone first maps it into multi-scale features $\{X_i\}_{i=1}^{4}$, where

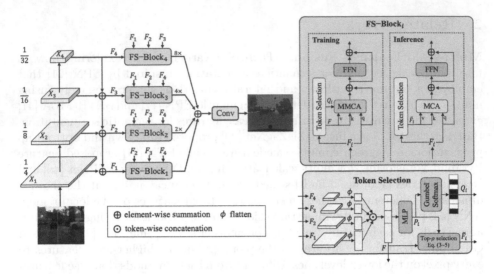

Fig. 3. The overall framework of our FSFormer, whose core component is Full-scale Selective Fusion Block (FS-Block).

$X_i \in \mathbb{R}^{2^{i-1}C \times \frac{H}{2^{i+1}} \times \frac{W}{2^{i+1}}}$. H and W denotes the height and width respectively. i indicates the scale index and C is the basic channel number. Then, a top-down pathway injects the high-level semantics into all scales to produce enhanced multi-scale representations $\{F_i\}_{i=1}^{4}$, where $F_i \in \mathbb{R}^{D \times \frac{H}{2^{i+1}} \times \frac{W}{2^{i+1}}}$, and D is the channel number of decoder. Next, a full-scale selective block is employed on each scale feature $\{F_i\}$ for context modeling. Finally, we up-sample multi-scale features to the same resolution, followed by an element-wise summation and a simple convolution head (including a 3×3 and a 1×1 convolution) to predict final segmentation result.

3.2 Full-Scale Selective Fusion Block

Previous works [18,19,21,31,38,40] have shown that fusing multi-scale features from multiple scales are critical for improving semantic segmentation, since the objects in the scene often present a variety of scales. High-resolution features in shallow layers contain more spatial details than low-resolution ones in deeper layers, while the latter contains richer semantics. Besides, small-scale objects have no precise locations in the higher-level since the multiple down-sample operations, while large-scale objects have weak semantics at the lower-level since the insufficient receptive fields.

With regard to the scale-level fusion, fully-scale fusion [18], where each token at one scale has the ability to aggregate information from all the tokens at any scale, shows more advantages than the progressive fusion [21] (each scale can only fuse information from its adjacent scale). According to the spatial-level fusion, convolution-based local fusion takes the dominant position, before the superior global fusion implemented by recent popular attention mechanisms [12] and Transformer architectures [11,31]. Based on our attention map visualization under full-scale fusion (Fig. 2), we found that although the global attention allows each token to fuse information from all the tokens, each token has its own interaction token subset of interest. Thus, the selective spatial-level fusion strategy may provide a better trade-off between performance and computational costs.

As the core component of our FSFormer, full-scale selective fusion block (FS-Block) aims to combine the full-scale fusion and selective spatial fusion. Specifically, as shown in Fig. 3, for each token, it first predicts a token subset of interest from all the tokens at all scales through a token selection module. Note that the tokens at each scale share the same subset. Then, each token only aggregates information from the tokens within its corresponding token subset of interest via a transformer-based module.

Token Selection Module. The token selection module is designed to adaptively select a token subset of interest for each scale according to the image content. Figure 3 shows the pipeline of the token selection module in i-th FS-Block. Given the multi-scale features $\{F_i\}_{i=1}^4$, it first concatenate them along token dimension, after a flatten operation,

$$F = \text{Concat}\Big(\phi(F_1), \phi(F_2), \phi(F_3), \phi(F_4)\Big) \in \mathbb{R}^{L \times D}, \tag{1}$$

where $L = \sum_{i=1}^4 \frac{HW}{2^{2i+2}}$, and ϕ denotes the flatten operation upon the spatial dimension. Then, we employ a MLP module to dynamically predict the interest scores $P_i \in [0,1]^L$ of all tokens for scale i,

$$P_i = \text{Softmax}\big(\text{MLP}(F)\big), \tag{2}$$

where $P_i^j, j \in [0, 1, ..., L-1]$ represents the interest score of the j-th token $F_j \in \mathbb{R}^D$ to the tokens F_i at scale i. Next, given a pre-defined selection ratio $\rho \in (0,1]$, we select the ρL tokens with top-ρ interest scores P_i from the whole token set F, resulting the token subset of interest $\widetilde{F}_i \in \mathbb{R}^{\rho L \times D}$ for scale i. The process can be formulated as follows:

$$\theta_i = \Theta(P_i), \tag{3}$$

$$Q_i^{\theta_i^j} = \begin{cases} 1, & 0 \le j < \rho L \\ 0, & \rho L \le j < L \end{cases} \tag{4}$$

$$\widetilde{F}_i = F[Q_i = 1], \tag{5}$$

where the Argsort operation Θ (in descending order) is first employed on P_i to obtain the sorted indexes $\theta_i \in [0, L-1]$. θ_i is further used to generate a binary mask Q_i, which indicate which tokens are selected. $Q_i^j = 1$ means the j-th token is selected into the token subset of interest for scale i, otherwise not selected. $[\delta]$ in Eq. (5) means fetching operation by the given condition δ.

However, such hard selection process is non-differentiable. To overcome this problem, we apply the gumbel-softmax technique [15] to generate the binary mask Q_i from the soft probability distribution P_i,

$$Q_i = \text{Gumbel-softmax}(P_i) \in \{0,1\}^L. \tag{6}$$

The gumbel-softmax is differentiable, thus enables the end-to-end optimization during training.

Full-Scale Selective Fusion. Inspired by the success of Transformer architecture [11], we utilize a transformer layer (including an attention module and feed-forward network (FFN)) for the context modeling. Specifically, we extend the multi-head self attention into multi-head cross attention (MCA) to enable the different sources of query, key and value, which is required for our full-scale selective fusion. MCA is responsible for token-wise interaction, whose forward pass can be formulated as follows:

$$X^{\text{MCA}} = \textbf{MCA}(X_q, X_k, X_v) = \frac{X_q X_k^T}{\sqrt{D}} \cdot X_v, \tag{7}$$

where X_q, X_k and X_v denote query, key and value embedding respectively. D is the channel number of X_q, X_k and X_v. FFN is in charge of channel-wise projection. We use the same structure of FFN as [11], which contains a layer normalization [2] and a multi-layer perceptron (MLP) module.

Figure 3 illustrates the detailed structure of the i-th FS-Block for inference and training, respectively. During inference, MCA takes the i-th scale tokens F_i as query, and the selected token subset of interest $\widetilde{F}_i$ generated by the token selection module as key and value, i.e., $F_i^{\text{MCA}} = \textbf{MCA}(F_i, \widetilde{F}_i, \widetilde{F}_i) \in \mathbb{R}^{H_i W_i \times L}$. Thus, each token at scale i has the ability to interact with all the tokens within its corresponding interested token subset $\widetilde{F}_i$, ranging from all the scales.

However, during training, the token subset of interest is sampled by gumbel-softmax, resulting in a non-uniform number of tokens for samples within a batch, which prevents the parallel computing. To overcome this issue, we introduce a masked attention mechanism, named masked multi-head cross attention (MMCA), to not only parallelize the computation but also cut down the interactions between each query token and its uninterested tokens. The MMCA takes the i-th scale tokens F_i, full-scale tokens F and selection mask Q_i as inputs, and output F_i^{MMCA} with the same size as F_i.

$$F_i^{\text{MMCA}} = \textbf{MMCA}(F_i, F, Q_i) \in \mathbb{R}^{H_i W_i \times D}. \tag{8}$$

Specifically, it first compute the non-selective full-scale fusion via the multi-head cross attention between F_i and F,

$$A = \frac{F_i F^T}{\sqrt{D}} \in \mathbb{R}^{H_i W_i \times L}. \tag{9}$$

Then, we generate the binary selection mask $M_i \in \{0, 1\}^{H_i W_i \times L}$ for all tokens at scale i by repeating $Q_i \in \{0, 1\}^L$ $H_i W_i$ times, since all the tokens belonging to i-th scale share the same token subset of interest. Note that the mask M_i is shared by all heads. Next, the effects of uninterested tokens in the attention map are filtered out by the following masking mechanism,

$$\tilde{A}_{ij} = \frac{\exp(A_{ij})M_{ij}}{\sum_{k=1}^{L} M_{ik}}. \tag{10}$$

Note that the mask M_i is shared by all heads. Equation (10) does not change the size of attention map, thus $\tilde{A}$ has the same size with A. Finally, such masked attention map $\tilde{A}$ is multiplied with the whole token set F to generate the final tokens,

$$F_i^{\mathrm{MMCA}} = \tilde{A}F \in \mathbb{R}^{H_i W_i \times D}. \tag{11}$$

Token Reduction for Efficiency. According to Eq. (9), the computational complexity of our MMCA is $O(H_i W_i L)$, which causes heavy computation burden when token number is large (*i.e.*, high-resolution feature maps). In order to improve its efficiency, we further design a meta-learning based projection mechanism to squeeze the query token sequence to a shorter one. Specifically, we perform a projection matrix $R_i \in \mathbb{R}^{N_i \times N_i'}$ on query tokens $F_i \in \mathbb{R}^{N_i \times D}$ to compress the sequence length of query embedding,

$$\widehat{F}_i = R_i^T F_i \in \mathbb{R}^{N_i' \times D}, \tag{12}$$

where $N_i = H_i W_i$ is the original sequence length of F_i. $N_i' = \frac{N_i}{r}$, where r is the reduction ratio. Considering the projection matrix requires the ability to perceive the image content, we dynamically generate R_i through a MLP layer Φ conditioned on the query tokens F_i,

$$R_i = \Phi(F_i). \tag{13}$$

Then, the squeezed query $\widehat{F}_i$ and full-scale tokens F are passed through MMCA as Eq. (8).

$$\widehat{F}_i^{\mathrm{MMCA}} = \mathbf{MMCA}(\widehat{F}_i, F, Q_i) \in \mathbb{R}^{N_i' \times D}. \tag{14}$$

Finally, we re-project the $\widehat{F}_i^{\mathrm{MMCA}}$ back to the original sequence length N_i,

$$F_i^{\mathrm{MMCA}} = R_i \widehat{F}_i^{\mathrm{MMCA}} \in \mathbb{R}^{N_i \times D}. \tag{15}$$

3.3   Loss Function

We now describe the training objectives of our FSFormer. We adopt the widely-used cross-entropy loss for the final predicted probability of each pixel,

$$\mathcal{L}_{\mathrm{ce}} = \sum_{n=1}^{N} \mathrm{CrossEntropy}(y_n, \widehat{y}_n), \tag{16}$$

where y_n and $\widehat{y}_n$ denote the ground-truth one-hot label and predicted probability distribution of n-th pixel.

Similar to previous works [22,32], we also apply a lightweight segmentation head (1×1 convolution) on the stage 3 output of backbone to project the channel dimension to class number. An auxiliary loss $\mathcal{L}_{\mathrm{aux}}$ is employed on the output of such segmentation head. $\mathcal{L}_{\mathrm{aux}}$ is also implemented by cross-entropy loss.

In addition, in order to constrain the ratio of the selected tokens of interest to a predefined value $\rho \in (0, 1]$, we utilize an MSE loss to regularize the predicted interest scores $\widehat{P}_i$ in Eq. (2),

$$\mathcal{L}_{\mathrm{reg}} = \frac{1}{S} \sum_{i=1}^{S} \|\rho - \frac{1}{L} \sum_{j=1}^{L} (P_i^j)\|^2, \tag{17}$$

where i is the scale index, and S equals to 4 in our experiments.

Overall, the total loss function consists of three terms:

$$\mathcal{L} = \mathcal{L}_{\mathrm{ce}} + \alpha \mathcal{L}_{\mathrm{reg}} + \beta \mathcal{L}_{\mathrm{aux}}, \tag{18}$$

where α and β are hyper-parameters. Following previous work [36,39,40], we set the weight β of auxiliary loss to 0.4. We ablate the α in the experiment section.

4   Experiments

4.1   Experimental Setup

Datasets. We conduct experiments on four widely-used public benchmarks: *ADE20K* [43] is a very challenging benchmark including 150 categories, which is split into 20000 and 2000 images for training and validation. *Cityscapes* [10] carefully annotates 19 object categories of urban scene images. It contains 5K finely annotated images, split into 2975 and 500 for training and validation. *COCO-Stuff 10K* [3] is a large scene parsing benchmark, which has 9000 training images and 1000 testing images with 182 categories (80 objects and 91 stuffs). *PASCAL Context* [24] is an extension of the PASCAL VOC 2010 detection challenge. It contains 4998 and 5105 images for training and validation, respectively. Following previous works, we evaluate the most frequent 60 classes (59 categories with background).

Backbone. For fair comparisons with other methods, we employ the well-known ResNet-101 [13] and Swin Transformer [22] as backbone. All the backbones are pre-trained on ImageNet-1K [28].

Hyper-parameters. The channel D of features F_i is set to 256, the weight α of is set to 0.4, and the target ratio ρ is set to 0.6. The head number of MCA is 8.

Training. We follow the previous works [30,31,36,41] to set up the training strategies for fair comparisons. The data augmentation consists of three steps: (i) random horizontal flip, (ii) random scale with the ratio between 0.5 and 2, (iii) random crop (480×480 for PASCAL Context, 512×512 for ADE20K

Table 1. Comparison with the state-of-the-art methods on ADE20K-*val*, Cityscapes *val* COCO-Stuff 10K-*test* and PASCAL Context-*val*. m.s.: multi-scale inference. "†" means that larger input resolution is used (640 × 640 for ADE20K and 1024 × 1024 for Cityscapes). "R101" is short for ResNet-101.

Method	Encoder	mIoU (m.s.)			
		ADE20K	Cityscapes	COCO-Stuff 10K	PASCAL
PSPNet [40]	R101	45.35	80.04	38.86	47.15
DeepLabV3+ [7]	R101	46.35	82.03	–	48.26
EncNet [39]	R101	44.65	76.97	–	–
ANN [44]	R101	45.24	81.30	–	52.80
OCRNet [37]	R101	–	81.87	39.50	54.80
DANet [12]	R101	45.02	82.02	39.70	52.60
CCNet [14]	R101	45.04	80.66	–	–
GFFNet [18]	R101	45.33	81.80	39.20	54.20
FPT [38]	R101	45.90	82.20	–	–
RecoNet [8]	R101	45.54	–	41.50	54.80
MaskFormer [9]	R101	47.20	81.40	39.80	–
FSFormer (ours)	R101	46.56	82.13	41.73	55.23
SETR [42]	ViT-L	50.28	82.15	45.80	55.83
MCIBI [16]	ViT-L	50.80	–	44.89	–
Segmenter [30]†	ViT-L	53.60	81.30	–	59.00
SegFormer [33]†	MiT-B5	51.80	84.00	46.70	–
UperNet [32]	Swin-L	51.17	-	47.71	57.29
UperNet [32]†	Swin-L	53.50	–	–	–
FSFormer (ours)	Swin-L	53.33	83.64	49.80	58.91
FSFormer (ours)†	Swin-L	54.43	84.46	–	–

and COCO-Stuff 10K, and 768 × 768 for Cityscapes). We use AdamW [23] as the optimizer with 0.01 weight decay. The initial learning rate is 0.00006 for ADE20K and Cityscapes, and 0.00002 on PASCAL Context and COCO-Stuff 10K. The training process contains 160k iterations for ADE20K, 60k iterations for COCO-Stuff 10k, and 80k iterations for Cityscapes and PASCAL Context. The batch size is set to 8 for Cityscapes, and 16 for other datasets. We initialize the encoder by the ImageNet-1K [28] pre-trained parameters, and other parts randomly. Synchronized BN [25] is used to synchronize the mean and standard-deviation of BN [29] across multiple GPUs. All the experiments are implemented with PyTorch [1] and conducted on 8 NVIDIA V100 GPUs.

Evaluation. The performance is measured by the widely-used mean intersection of union (mIoU) for all experiments. For the multi-scale inference, we follow previous works [22,41] to average the predictions of our model at multiple scales [0.5, 0.75, 1.0, 1.25, 1.5, 1.75].

4.2   Comparisons with the State-of-the-Arts

ADE20K val. Table 1 reports the comparison with the state-of-the-art methods on the ADE20K validation set. Equipped with Swin-L as backbone, our FSFormer is +2.16% mIoU higher (53.33% vs. 51.17%) than UperNet. Recent methods [22,30] show that using a larger resolution (640 × 640) can bring more improvements. When a larger resolution (640 × 640) is adopted, our FSFormer outperforms UperNet by +0.93% (54.43% vs. 53.50%) under the same Swin-L backbone. In addition, our FSFormer (Swin-L) is +2.63% mIoU higher than SegFormer (MiT-B5) (54.43% vs. 51.80%). Although Segmenter [30] uses the stronger ViT-L [11] backbone than Swin-L, our FSFormer also show a +0.73% mIoU advantage than Segmenter. These results demonstrate that the effectiveness of our method.

Cityscapes val. Table 1 shows the comparative results on Cityscapes validation set. Our FSFormer is +1.49% superior than SETR [41] (83.64% vs. 82.15%). According to [33], a higher input resolution of 1024 × 1024 can bring further performance gain. Thus, we also train our model under such resolution. It can be seen that our FSFormer (Swin-L) outperforms SegFormer (MiT-B5) and Segmenter (ViT-L) by 0.46% and 3.16% mIoU. When using the widely-used ResNet-101 as backbone, our FSFormer achieves 82.13% mIoU, which is +0.26% and +0.73% higher than the well-known OCRNet [36] and the promising MaskFormer [9], respectively.

COCO-Stuff 10K Test. As shown in Table 1, our FSFormer achieves 49.80% mIoU, outperforming UperNet by 2.09% under Swin-L backbone. Compared with MCIBI with a stronger ViT-L backbone, our FSFormer presents a +4.91% mIoU superiority. Besides, equipped with ResNet-101 as backbone, our FSFormer achieves 41.73% mIoU, which is +0.23% higher than the previous best RecoNet.

PASCAL Context val. Table 1 compares our method with the state-of-the-arts on PASCAL Context validation set. Our FSFormer is +0.43% mIoU higher than RecoNet with ResNet-101 as backbone (55.23% vs. 54.80%). With Swin-L as backbone, our FSFormer achieves 58.91% mIoU, outperforming UperNet by +1.62%. Compared with the methods with using stronger ViT-L as backbone, our FSFormer (Swin-L) is +3.08% mIoU higher than SETR and achieves comparable performance with Segmenter (58.91% vs. 59.00%).

4.3   Ablation Study

In this sub-section, we study the effect of key designs and hyper-parameters of our approach. All the ablation studies are conducted under Swin-T [22] backbone on PASCAL Context dataset.

Effect of Key Designs. We ablate the effect of two key designs (full-scale fusion and token selection) in Table 2. The baseline model denotes the single-scale fusion without token selection, *i.e.*, each token at one scale can only interact

Table 2. Ablation on the effect of full-scale fusion and token selection. s.s.: single-scale inference.

Baseline	Full-scale fusion	Token selection	FLOPs	Params	mIoU (s.s.)
✓			54.0G	33M	46.75
✓	✓		75.5G	38M	48.70
✓	✓	✓	73.8G	39M	49.33

with tokens within the same scale, which achieves only 46.75% mIoU. The full-scale fusion brings an obvious improvement (+1.95%), reaching 48.70% mIoU. Benefited from the token selection operation, the FLOPs is reduced by 1.7G and the performance further increase by +0.63%, achieving 49.33% mIoU.

Compare with Different Token Selection Manners. To verify the effectiveness of our adaptive token selection strategy, we compare it with two simple and intuitive manners: (i) random selection, (ii) uniform selection. As shown in Fig. 4 (a), these two fixed selection manners lead to about 1.5% performance decrease, and approximately 2% lower than our adaptive token selection manner. This shows that the token subset of interest need to be adaptively selected according to the image content.

Token Selection Ratio. The token selection ratio ρ represents the proportion of tokens of interest selected from tokens at all scales, that is an indicator of the size of the token subset of interest. A larger token selection ratio means each token can interact with more tokens, while also cause more computation burden and may introduce noise information. Where, $\rho = 1.0$ means that no selection is performed, that is, each token can interact with all the tokens at any scale. Figure 4 (d) shows the performance when the token selection ratio ρ varies within [0.1, 1.0]. It can be seen that the performance changes with token selection ratio in a unimodal pattern within a range of about 1%. Specifically, the mIoU increases from 48.40% over $\rho = 0.1$–0.6, peaking at 49.33%, and then falls back to 48.70% at $\rho = 1.0$. Note that the token selection ratio is not the larger the better, which may attribute to the noisy information aggregation caused by excessive interaction range. The best mIoU is achieved at a token selection ratio of 0.6, thus we set ρ to 0.6 by default. The results demonstrates the necessity of selecting a token subset of interest during feature fusion.

Weight of Regularization Loss for Token Selection Ratio. As mentioned in Sect. 3.3, we apply a regularization loss to constrain the ratio of selected tokens of interest to a predefined value ρ. Figure 4 (c) shows the effect of different weights (ranged between 0 and 1) for this regularization loss. It can be seen that $\alpha = 0.4$ outperforms its counterparts, achieving the best performance with 49.33% mIoU. Thus, we set $\alpha = 0.4$ by default.

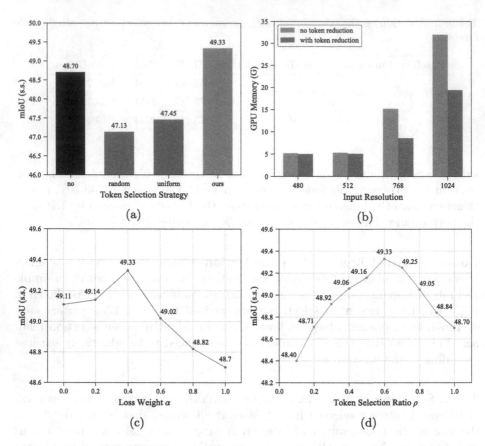

Fig. 4. Effect of (a) different token selection strategies, (b) token reduction, (c) different weights of regularization loss for token selection ratio, and (d) different token selection ratio. "s.s." denotes single-scale inference.

Effect of Token Reduction. Here, we study the effect of our token reduction strategy on both performance and GPU memory. As shown in Fig. 4(b), our token reduction can effectively relieve the GPU memory burden. Note that the larger the input resolution, the more obviously the memory burden will be reduced. Furthermore, such token reduction strategy can bring a slight +0.27% mIoU gain (49.33% vs. 49.06%).

4.4   Further Analysis

Comparisons with Other Multi-scale Fusion Decoders. To further verify the effectiveness of our full-scale selective fusion, we compare our FSFormer with other decoders with different multi-scale fusion strategies in Table 3. The results demonstrate the superiority of our method. Compared with other progressive local fusion methods (SETR, Semantic FPN and UperNet), our FSFormer

Table 3. Comparisons with other decoders with multi-scale feature fusion on (a) performance under different backbones on PASCAL Context *val* set, and (b) computational costs. We report the FLOPs and Params of decoders, relative to the backbone. The input resolution is set to 480 × 480. "s.s." denotes single-scale inference.

(a)

Encoder	SETR [41]	Semantic FPN [17]	UperNet [32]	GFFNet [18]	FTN [31]	FSFormer
Swin-S	50.48	50.49	51.67	51.76	52.14	**52.58**
Swin-B	51.51	51.48	52.52	52.58	52.81	**53.12**
Swin-L	56.62	56.78	56.87	56.90	57.29	**57.63**

(b)

	SETR [41]	Semantic FPN [17]	UperNet [32]	GFFNet [18]	FTN [31]	FSFormer
FLOPs	13G	112G	187G	85G	39G	**52G**
Params	3M	54M	37M	17M	25M	**12M**

outperforms the best one (*i.e.*, UperNet) among them by +0.91% and +0.76% under Swin-T and Swin-L. Compared with the full-scale local fusion decoder, GFFNet, our FSFormer has +0.82%, +0.54% and +0.73% gains in mIoU with Swin-T, Swin-B and Swin-L as backbone respectively. Compared with FTN, a transformer-based progressive global fusion decoder, our FSFormer is +0.44%, +0.31% and +0.34% higher than FTN under different Swin Transformer backbones.

5   Conclusion

In this paper, we first rethink the multi-scale feature fusion from two perspectives (scale-level and spatial-level), and then propose a full-scale selective fusion strategy for semantic segmentation. Based on the proposed fusion mechanism, we design a Full-scale Selective Transformer (FSFormer) for semantic segmentation. Specifically, our FSFormer adaptively select partial tokens from all tokens at all the scales to construct a token subset of interest for each scale. Therefore, each token only interact with the tokens within its corresponding token subset of interest. The proposed full-scale selective fusion strategy can not only filter out the noisy information propagation but also reduce the computational costs to some extent. Extensive experiments on PASCAL Context, ADE20K, COCO-Stuff 10K, and Cityscapes have shown that our FSFormer can outperform the state-of-the-art methods in semantic image segmentation, demonstrating that our FSFormer can achieve better results than previous multi-scale feature fusion methods.

References

1. Paszke, A., et al.: Automatic differentiation in PyTorch. In: Advances in Neural Information Processing Systems Workshop (2017)
2. Ba, J.L., Kiros, J.R., Hinton, G.E.: Layer normalization. arXiv preprint arXiv:1607.06450 (2016)
3. Caesar, H., Uijlings, J., Ferrari, V.: Coco-stuff: thing and stuff classes in context. In: Proceedings of the IEEE Conference on Computer Vision and Pattern Recognition, pp. 1209–1218 (2018)
4. Chen, L.C., Papandreou, G., Kokkinos, I., Murphy, K., Yuille, A.L.: DeepLab: semantic image segmentation with deep convolutional nets, atrous convolution, and fully connected CRFs. IEEE Trans. Pattern Anal. Mach. Intell. **40**(4), 834–848 (2017)
5. Chen, L.C., Papandreou, G., Kokkinos, I., Murphy, K., Yuille, A.L.: DeepLab: semantic image segmentation with deep convolutional nets, atrous convolution, and fully connected CRFs. IEEE Trans. Pattern Anal. Mach. Intell. **40**(4), 834–848 (2018). https://doi.org/10.1109/TPAMI.2017.2699184
6. Chen, L.C., Papandreou, G., Schroff, F., Adam, H.: Rethinking atrous convolution for semantic image segmentation. arXiv preprint arXiv:1706.05587 (2017)
7. Chen, L.-C., Zhu, Y., Papandreou, G., Schroff, F., Adam, H.: Encoder-decoder with atrous separable convolution for semantic image segmentation. In: Ferrari, V., Hebert, M., Sminchisescu, C., Weiss, Y. (eds.) ECCV 2018. LNCS, vol. 11211, pp. 833–851. Springer, Cham (2018). https://doi.org/10.1007/978-3-030-01234-2_49
8. Chen, W., et al.: Tensor low-rank reconstruction for semantic segmentation. In: Vedaldi, A., Bischof, H., Brox, T., Frahm, J.-M. (eds.) ECCV 2020. LNCS, vol. 12362, pp. 52–69. Springer, Cham (2020). https://doi.org/10.1007/978-3-030-58520-4_4
9. Cheng, B., Schwing, A.G., Kirillov, A.: Per-pixel classification is not all you need for semantic segmentation. arXiv preprint arXiv:2107.06278 (2021)
10. Cordts, M., et al.: The cityscapes dataset for semantic urban scene understanding. In: Proceedings of the IEEE Conference on Computer Vision and Pattern Recognition, pp. 3213–3223 (2016)
11. Dosovitskiy, A., et al.: An image is worth 16×16 words: transformers for image recognition at scale. In: ICLR (2021)
12. Fu, J., et al.: Dual attention network for scene segmentation. In: Proceedings of the IEEE/CVF Conference on Computer Vision and Pattern Recognition, pp. 3146–3154 (2019)
13. He, K., Zhang, X., Ren, S., Sun, J.: Deep residual learning for image recognition. In: Proceedings of the IEEE Conference on Computer Vision and Pattern Recognition, pp. 770–778 (2016)
14. Huang, Z., Wang, X., Huang, L., Huang, C., Wei, Y., Liu, W.: CCNet: criss-cross attention for semantic segmentation. In: Proceedings of the IEEE/CVF International Conference on Computer Vision, pp. 603–612 (2019)
15. Jang, E., Gu, S., Poole, B.: Categorical reparameterization with Gumbel-Softmax. arXiv preprint arXiv:1611.01144 (2016)
16. Jin, Z., et al.: Mining contextual information beyond image for semantic segmentation. In: Proceedings of the IEEE/CVF International Conference on Computer Vision, pp. 7231–7241 (2021)
17. Kirillov, A., Girshick, R., He, K., Dollár, P.: Panoptic feature pyramid networks. In: Proceedings of the IEEE/CVF Conference on Computer Vision and Pattern Recognition, pp. 6399–6408 (2019)

18. Li, X., Zhao, H., Han, L., Tong, Y., Tan, S., Yang, K.: Gated fully fusion for semantic segmentation. In: AAAI (2020)
19. Lin, D., Ji, Y., Lischinski, D., Cohen-Or, D., Huang, H.: Multi-scale context intertwining for semantic segmentation. In: Ferrari, V., Hebert, M., Sminchisescu, C., Weiss, Y. (eds.) ECCV 2018. LNCS, vol. 11207, pp. 622–638. Springer, Cham (2018). https://doi.org/10.1007/978-3-030-01219-9_37
20. Lin, D., et al.: ZigZagNet: fusing top-down and bottom-up context for object segmentation. In: Proceedings of the IEEE/CVF Conference on Computer Vision and Pattern Recognition, pp. 7490–7499 (2019)
21. Lin, T.Y., Dollár, P., Girshick, R., He, K., Hariharan, B., Belongie, S.: Feature pyramid networks for object detection. In: Proceedings of the IEEE Conference on Computer Vision and Pattern Recognition, pp. 2117–2125 (2017)
22. Liu, Z., et al.: Swin transformer: hierarchical vision transformer using shifted windows. arXiv preprint arXiv:2103.14030 (2021)
23. Loshchilov, I., Hutter, F.: Decoupled weight decay regularization. arXiv preprint arXiv:1711.05101 (2017)
24. Mottaghi, R., et al.: The role of context for object detection and semantic segmentation in the wild. In: Proceedings of the IEEE Conference on Computer Vision and Pattern Recognition, pp. 891–898 (2014)
25. Peng, C., et al.: MegDet: a large mini-batch object detector. In: Proceedings of the IEEE Conference on Computer Vision and Pattern Recognition, pp. 6181–6189 (2018)
26. Ranftl, R., Bochkovskiy, A., Koltun, V.: Vision transformers for dense prediction. arXiv preprint arXiv:2103.13413 (2021)
27. Ronneberger, O., Fischer, P., Brox, T.: U-Net: convolutional networks for biomedical image segmentation. In: Navab, N., Hornegger, J., Wells, W.M., Frangi, A.F. (eds.) MICCAI 2015. LNCS, vol. 9351, pp. 234–241. Springer, Cham (2015). https://doi.org/10.1007/978-3-319-24574-4_28
28. Russakovsky, O., et al.: ImageNet large scale visual recognition challenge. Int. J. Comput. Vision 115(3), 211–252 (2015)
29. Ioffe, S., Szegedy, C.: Batch normalization: accelerating deep network training by reducing internal covariate shift. In: Proceedings of the 32nd International Conference on International Conference on Machine Learning, vol. 37, pp. 448–456 (2015)
30. Strudel, R., Garcia, R., Laptev, I., Schmid, C.: Segmenter: transformer for semantic segmentation. In: Proceedings of the IEEE/CVF International Conference on Computer Vision (ICCV), pp. 7262–7272, October 2021
31. Wu, S., Wu, T., Lin, F., Tian, S., Guo, G.: Fully transformer networks for semantic image segmentation. arXiv preprint arXiv:2106.04108 (2021)
32. Xiao, T., Liu, Y., Zhou, B., Jiang, Y., Sun, J.: Unified perceptual parsing for scene understanding. In: Ferrari, V., Hebert, M., Sminchisescu, C., Weiss, Y. (eds.) ECCV 2018. LNCS, vol. 11209, pp. 432–448. Springer, Cham (2018). https://doi.org/10.1007/978-3-030-01228-1_26
33. Xie, E., Wang, W., Yu, Z., Anandkumar, A., Alvarez, J.M., Luo, P.: SegFormer: simple and efficient design for semantic segmentation with transformers. arXiv preprint arXiv:2105.15203 (2021)
34. Xie, E., et al.: Segmenting transparent object in the wild with transformer. arXiv preprint arXiv:2101.08461 (2021)
35. Yu, F., Koltun, V.: Multi-scale context aggregation by dilated convolutions. arXiv preprint arXiv:1511.07122 (2015)

36. Yuan, Y., Chen, X., Wang, J.: Object-contextual representations for semantic segmentation. arXiv preprint arXiv:1909.11065 (2019)
37. Yuan, Y., Chen, X., Wang, J.: Object-contextual representations for semantic segmentation. In: Vedaldi, A., Bischof, H., Brox, T., Frahm, J.-M. (eds.) ECCV 2020. LNCS, vol. 12351, pp. 173–190. Springer, Cham (2020). https://doi.org/10.1007/978-3-030-58539-6_11
38. Zhang, D., Zhang, H., Tang, J., Wang, M., Hua, X., Sun, Q.: Feature pyramid transformer. In: Vedaldi, A., Bischof, H., Brox, T., Frahm, J.-M. (eds.) ECCV 2020. LNCS, vol. 12373, pp. 323–339. Springer, Cham (2020). https://doi.org/10.1007/978-3-030-58604-1_20
39. Zhang, H., et al.: Context encoding for semantic segmentation. In: Proceedings of the IEEE Conference on Computer Vision and Pattern Recognition, pp. 7151–7160 (2018)
40. Zhao, H., Shi, J., Qi, X., Wang, X., Jia, J.: Pyramid scene parsing network. In: Proceedings of the IEEE Conference on Computer Vision and Pattern Recognition, pp. 2881–2890 (2017)
41. Zheng, S., et al.: Rethinking semantic segmentation from a sequence-to-sequence perspective with transformers. arXiv preprint arXiv:2012.15840 (2020)
42. Zheng, S., et al.: Rethinking semantic segmentation from a sequence-to-sequence perspective with transformers. In: Proceedings of the IEEE/CVF Conference on Computer Vision and Pattern Recognition, pp. 6881–6890 (2021)
43. Zhou, B., et al.: Semantic understanding of scenes through the ADE20K dataset. Int. J. Comput. Vision **127**(3), 302–321 (2019)
44. Zhu, Z., Xu, M., Bai, S., Huang, T., Bai, X.: Asymmetric non-local neural networks for semantic segmentation. In: Proceedings of the IEEE/CVF International Conference on Computer Vision, pp. 593–602 (2019)

Semi-supervised Semantic Segmentation with Uncertainty-Guided Self Cross Supervision

Yunyang Zhang[1], Zhiqiang Gong[1], Xiaoyu Zhao[1], Xiaohu Zheng[2],
and Wen Yao[1(✉)]

[1] Defense Innovation Institute, Chinese Academy of Military Science, Beijing, China
zhangyunyang17@csu.ac.cn, wendy0782@126.com
[2] College of Aerospace Science and Engineering, National University of Defense
Technology, Changsha, China

Abstract. As a powerful way of realizing semi-supervised segmentation, the cross supervision method learns cross consistency based on independent ensemble models using abundant unlabeled images. In this work, we propose a novel cross supervision method, namely uncertainty-guided self cross supervision (USCS). To avoid multiplying the cost of computation resources caused by ensemble models, we first design a multi-input multi-output (MIMO) segmentation model which can generate multiple outputs with the shared model. The self cross supervision is imposed over the results from one MIMO model, heavily saving the cost of parameters and calculations. On the other hand, to further alleviate the large noise in pseudo labels caused by insufficient representation ability of the MIMO model, we employ uncertainty as guided information to encourage the model to focus on the high confident regions of pseudo labels and mitigate the effects of wrong pseudo labeling in self cross supervision, improving the performance of the segmentation model. Extensive experiments show that our method achieves state-of-the-art performance while saving 40.5% and 49.1% cost on parameters and calculations.

Keywords: Semi-supervised semantic segmentation · Consistency regularization · Multi-input multi-output · Uncertainty

1 Introduction

Semantic segmentation is a significant fundamental task in computer vision and has achieved great advances in recent years. Compared with other vision tasks, the labeling process for semantic segmentation is much more time and labor consuming. Generally, tens of thousands of samples with pixel-wise labels are essential to guarantee good performance for such a known data-hungry task. However, the high dependence of large amounts of labeled data for training would undoubtedly restrict the development of semantic segmentation. Semi-supervised semantic segmentation, employing limited labeled data as well as abundant unlabeled data for training segmentation models, is regarded as an effective approach to tackle this problem, and has achieved remarkable success for the task [10,11,17,21,28].

Advanced semi-supervised semantic segmentation methods are mainly based on consistent regularization. It is under the assumption that the prediction for the same object with different perturbations, such as data augmentation for input images [9,17], noise interference for feature maps [28] and the perturbations from ensemble models [6,14], should be consistent. Among these perturbations, the one through ensemble models usually provide better performance since it can learn the consistent correlation from each other adaptively. The earnings of a single model acquired from unlabeled images can be improved by cross supervision between models achieved by forcing consistency of the predictions.

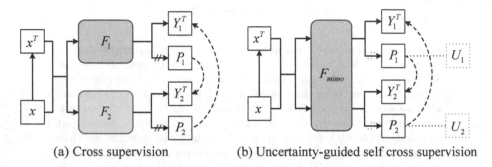

(a) Cross supervision   (b) Uncertainty-guided self cross supervision

Fig. 1. Illustrating the architectures for (a) cross supervision and (b) our method uncertainty-guided self cross supervision. In our approach, x^T means the transformed image of x; Y_1^T and Y_2^T mean the predictions from a multi-input multi-output model F_{mimo}; P_1 and P_2 mean the pseudo labels for Y_1^T and Y_2^T; U_1 and U_2 mean the uncertainty of P_1 and P_2, respectively.

Despite impressive performance, the cost of time and memory for cross supervision is usually multiplicatively increased due to the parallel training of ensemble models with different model architectures or different initializations. To break this limitation, we propose a Self Cross Supervision method, which build only one model to obtain different views and significantly reduces the computation cost while achieving high performance.

Specifically, we impose cross supervision based on a multi-input multi-output (MIMO) model rather than multiple independent models. Commonly, one model hard to produce diversity. Thus we implicitly fit two subnetworks within single basic network utilizing the over-parameterization of neural network, achieving diversity with single model. Through MIMO, multiple predictions can be obtained under a single forward pass, and then the purpose can be achieved almost "free" [13]. In our method, instead of ensemble models, cross supervision is realized by one MIMO model, which is called Self Cross Supervision.

Compared with multiple independent models, the performance of each subnetwork in one MIMO is compromised when the capacity of the MIMO is limited. The subnetwork with poor representation ability generate pseudo labels with large noise, further confusing the training process and making false propagation from one subnetwork to others [46–48]. To suppress the noisy pseudo labels, we propose employing uncertainty guided the process of learning with wrong pseudo labels. Uncertainty is used to evaluate the quality of predictions without ground truth. Generally, regions with large uncertainty represent poor prediction and vice versa [1,38]. For the task at hand, uncertainty can be used as the guided information to indicate the confidence of the pseudo labels of unlabelled samples and supervise the self cross supervision process by reducing the effects of wrong pseudo labeling, and such proposed method is called Uncertainty-guided Self Cross Supervision (USCS). The comparisons between cross supervision and our USCS is shown in Fig. 1.

In conclusion, our contributions are:

1. We firstly propose a self cross supervision method with a multi-input multi-output (MIMO) model. Our method realizes cross supervision through enforcing the consistency between the outputs of MIMO, and greatly reduces the training cost of the model.
2. We propose uncertainty-guided learning for self cross supervision to improve the performance of the model, which uses the uncertainty information as the confidence of the pseudo labels and supervises the learning process by reducing the effects of wrong pseudo labeling.
3. Experiments demonstrate that our proposed model surpasses most of the current state-of-the-art methods. Moreover, compared with cross supervision, our method can achieve competitive performance while greatly reducing training costs.

2 Related Work

2.1 Semantic Segmentation

Semantic segmentation is a pixel-wise classification task, which marks each pixel of the image with the corresponding class. Most of the current semantic segmentation models are based on the encoder-decoder structure [2,26,30]. The encoder reduces the spatial resolution generating a high-level feature map, and the decoder gradually restores spatial dimension and details. Fully convolutional neural networks (FCN) [22] is the first encoder-decoder-based segmentation model. The subsequent works improve the context dependence by dilated convolutions [4,42], maintaining high resolution [33,37], pyramid pooling [41,44], and self-attention mechanism [36]. DeepLabv3+ [5] is one of the state-of-the-art methods, which is employed as the segmentation model in this work.

2.2 Semi-supervised Learning

Semi-supervised learning focuses on high performance using abundant unlabeled data under limited labeled data, so as to alleviate the training dependence on labels [15,19,45]. Most of the current semi-supervised learning methods are based on empirical assumptions of the image itself, such as smoothness assumption, and low-density assumption [35].

Based on the smoothness assumption, prior works use the consistent regularization semi-supervised method, which encourage the model to predict the similar output for the perturbed input. This kind of works tries to minimize the difference between perturbed samples generated by data augmentations, e.g., Mean Teacher [34], VAT [24] and UDA [39]. As for the low-density assumption, the pseudo label based semi-supervised learning [19, 29, 39] is the representative method, which realizes the low-density separation by minimizing the conditional entropy of class probability for the unlabeled data. In order to utilize the merits of different assumptions, prior works also propose effective methods based on both or more. Among these methods, joint learning with the pseudo label and consistent regularization is a successful one and has achieved impressive performance, such as MixMatch [3], FixMatch [32] and DivideMix [20]. Our approach utilizes consistent regularization and the pseudo label to construct semi-supervised learning.

2.3   Semi-supervised Semantic Segmentation

As a dense prediction task, semantic segmentation is laborious and time-consuming in manual annotations. Therefore, using unlabeled images to improve model performance is an effective way for cost reduction. Most of the semi-supervised semantic segmentation approaches are based on the consistent regularization [9, 17, 27, 49]. For example, PseudoSeg [49] enforces the consistency of the predictions with weak and strong data augmentations, similar to FixMatch [32]. CAC [17] utilizes contextual information to maintain the consistency between features of the same identity under different environments. CCT [28] maintains the agreement between the predictions from the features with various perturbations. GCT [14] and CPS [6] adopt different model structures or model initializations to generate the perturbations of predictions and achieve state-of-the-art performance. However, the training cost of time and memory for ensemble models is expensive in GCT and CPS. Different from prior works, our approach enforces the consistency of predictions from a multi-input multi-output network and greatly reduces the training costs.

3   Method

In the following sections, we first introduce the overview of USCS in Sect. 3.1. The self cross supervision with MIMO model is proposed in Sect. 3.2. To ameliorate pseudo label quality, we propose the uncertainty-guided learning in Sect. 3.3.

As a common semi-supervised learning task, a dataset $\mathcal{X}$ consisting of labeled images $\mathcal{X}_l$ with labels $\mathcal{Y}$ and unlabeled images $\mathcal{X}_{ul}$ is employed to train a segmentation network. In our USCS, we extra applied transformation T on unlabeled images $\mathcal{X}_{ul}$ got the transformed images $\mathcal{X}_{ul}^T = T(\mathcal{X}_{ul})$. Both unlabeled images $\mathcal{X}_{ul}$ and transformed images $\mathcal{X}_{ul}^T$ are employed to construct self cross supervision.

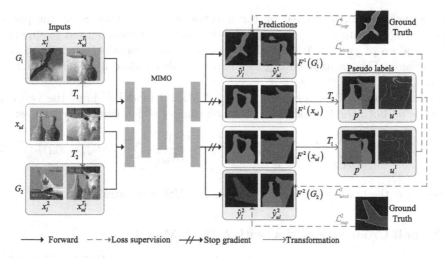

Fig. 2. The USCS framework. We aim to maintain the consistency between the predictions from a multi-input multi-output (MIMO) model. Since MIMO accepts two different group images, we adopted transformation consistency to realize the purpose.

3.1 Overview of USCS

The USCS framework is shown in Fig. 2. In contrast to the general cross supervision method using several independent models, we instead employ a multi-input multi-output (MIMO) model. Specifically, the MIMO model F has two input and output branches which can be seen as the subnetworks with shared parameters, accepting two groups independently sampled data G_k ($k \in \{1, 2\}$) and output corresponding segmentation results. In USCS, each group data is denoted as $G_k = \left\{ x_l^k, x_{ul}^{T_k} \right\}$ ($x_l^k \in \mathcal{X}_l, x_{ul}^{T_k} \in \mathcal{X}_{ul}^T$). For $k \in \{1, 2\}$, x_l^1 and x_l^2 are the labeled images with different batch sampling order, $x_{ul}^{T_1}$ and $x_{ul}^{T_2}$ are the transformed images with distinct transformation T_k.

Given an image $x^k \in G_k$, the MIMO model F first predicts $\hat{y}^k = \{\hat{y}_l^k, \hat{y}_{ul}^k\}$, where $\hat{y}_l^k = F(x_l^k)$ and $\hat{y}_{ul}^k = F(x_{ul}^{T_k})$. As common semantic segmentation models, the prediction $\hat{y}_l^k$ is supervised by its corresponding ground-truth $y \in \mathcal{Y}$ as:

$$\mathcal{L}_{sup}^k \left(x_l^k, y \right) = \frac{1}{|\Omega|} \sum_{i \in \Omega} \ell_{ce}(\hat{y}_l^k(i), y(i)), \tag{1}$$

where $\ell_{ce}(*)$ is the standard Cross Entropy loss, and Ω is the region of image with size $H \times W$.

To explore the unlabeled images, we repeat the original unlabeled images x_{ul} twice, the MIMO model F makes two groups independent predictions $F^1(x_{ul})$ and $F^2(x_{ul})$ on the same images x_{ul} as shown at the bottom of Fig. 2. Then the same transformation T_1 and T_2 are respectively performing on $F^2(x_{ul})$ and $F^1(x_{ul})$, obtaining $p^1 = T_1(F^2(x_{ul}))$, $p^2 = T_2(F^1(x_{ul}))$. Besides, the uncertainties u^1 and u^2 are estimated for two transformed predictions p^1 and p^2, respectively. Then, p^1 guided by the uncertainty u^1 is regarded as the pseudo labels of $x_{ul}^{T_1}$ to supervise $\hat{y}_{ul}^1$. Similarly, the same operation is used to supervise $\hat{y}_{ul}^2$ based on p^2 and u^2. We call the above process uncertainty-guided self cross supervision. The constraint $\mathcal{L}_{uscs}$ and more details are described in Sect. 3.2 and Sect. 3.3.

Finally, our method for the training of MIMO model F joint the two constraints on both the labeled and unlabeled images which can be written as:

$$\mathcal{L}(\mathcal{X}, \mathcal{Y}) = \sum_{k=1,2} (\frac{1}{|\mathcal{X}_l|} \sum_{x_l^k \in \mathcal{X}_l} \mathcal{L}_{sup}^k (x_l^k, y) + \frac{1}{|\mathcal{X}_{ul}|} \sum_{x_{ul} \in \mathcal{X}_{ul}} \lambda \mathcal{L}_{uscs}^k (x_{ul})), \qquad (2)$$

where λ is the trade-off weight to balance the USCS constraint.

3.2   Self Cross Supervision with MIMO Model

The proposed self cross supervision is implemented over the MIMO model. Before presenting self cross supervision, the MIMO model used in USCS is firstly introduced. Based on the fact that neural networks are heavily overparameterized models [13], we can train a MIMO model containing multiple independent subnetworks and acquire multiple predictions of one input under a single forward pass of the model. Different from the single neural network archtiecture, the MIMO model replaces the single input layer by N input layers, which can receive N datapoint as inputs. And N output layers are added to make N predictions based on the feature before output layers. Compared with a single model, the MIMO model obtains the performance of ensembling with the cost of only a few increased parameters and calculations.

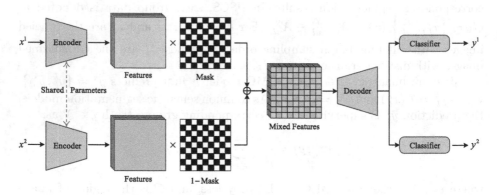

Fig. 3. The structure of MIMO segmentation model. The features after the encoder are fused by the grid mix.

In USCS, we construct a MIMO model with two inputs and outputs, whose structure is shown in Fig. 3. For better extract object features, the entire encoder part is utilized as the input layer of the model (the original MIMO model employs the first convolutions layers of the model as the input layer). However, two independent encoders (the input layer) increase the model parameters and computation. We share the parameters of two encoders to avoid this problem. The features of two inputs extracted by the encoder must be fused before entering the decoder. To effective combines inputs into a shared representation, the grid mix is adopted to replace the original summing method [13] in MIMO as:

$$\mathcal{M}_{gridmix}(f_1, f_2) = \mathbb{1}_{\mathcal{M}} \odot f_1 + (\mathbb{1} - \mathbb{1}_{\mathcal{M}}) \odot f_2, \tag{3}$$

where f_1 and f_2 are the features of two inputs, respectively; $\mathbb{1}_{\mathcal{M}}$ is a binary grid mask with grid size g.

The self cross supervision enforces two predictions of MIMO learn from each other. The output y^1 is considered the pseudo label to supervise the output y^2, vice versa. As mentioned previously, two inputs of MIMO are different, while the self cross supervision is feasible only when the inputs are the same. We overcome this issue by introducing the transformation consistency regularization [25], which assumes that the prediction $F(T(x))$ of the transformed image $T(x)$ must be equal to the transformed prediction $T(F(x))$ of the original image x.

As shown in Fig. 2, the MIMO model F predicts two transformed unlabeled images $x_{ul}^{T_1}$ and $x_{ul}^{T_2}$, obtaining $\hat{y}_{ul}^1$ and $\hat{y}_{ul}^2$. Self cross supervision expects two outputs of the MIMO model to supervise each other. However, the semantics of the outputs $\hat{y}_{ul}^1$ and $\hat{y}_{ul}^2$ are different. To achieve the self cross supervision, we input the original unlabeled image x_{ul} to the MIMO model, getting two individual predictions $F^1(x_{ul})$ and $F^2(x_{ul})$ without gradient. We further obtain two transformed predictions $p^1 = T_1(F^2(x_{ul}))$ and $p^2 = T_2(F^1(x_{ul}))$ by performing the transformation T_1 and T_2, respectively. The transformed predictions p^1 should have the similar semantics with $\hat{y}_{ul}^1$, thus we regard p^1 as the pseudo label of $\hat{y}_{ul}^1$. Similarly, the transformed prediction p^2 is considered as the pseudo label to supervise $\hat{y}_{ul}^2$.

Through the above process, the MIMO model F can realize cross supervision by itself. The self cross supervision constraint on unlabeled data is defined as:

$$\mathcal{L}_{scs}^k(x_{ul}) = \frac{1}{|\Omega|} \sum_{i \in \Omega} \ell_{ce}(\hat{y}_{ul}^k(i), p^k(i)). \tag{4}$$

3.3 Uncertainty-Guided Learning

The pseudo label obtained from the prediction exists noise, especially when the capacity of subnetworks in MIMO is limited. The poor model representation leads to plenty of inaccurate pseudo labels. The noisy pseudo label will mislead the model and interfere with the optimization direction in self cross supervision. In addition, the noise caused by one model is likely to propagate to another model through self cross supervision, resulting in the accumulation and propagation of errors and hindering the performance. It is necessary to filter the pseudo label with inferior quality to improve the overall performance of the model.

Uncertainty estimation is an effective method to evaluate noise in prediction [18]. Noise often exists in regions with large uncertainties. Figure 4 shows the uncertainty visualization. Based on this observation, we propose to employ uncertainty to guide the pseudo label with noise in cross supervision. Firstly, we estimate the uncertainty of pseudo label through the Shannon Entropy [31], which is defined as:

$$U = -\sum_{c=1}^{C} p(c) \log p(c), \tag{5}$$

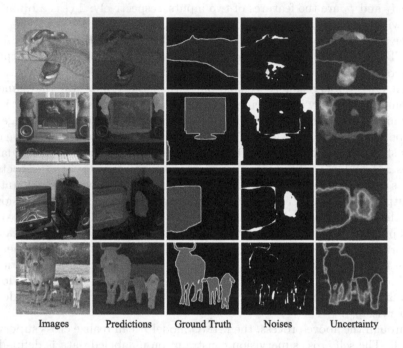

| Images | Predictions | Ground Truth | Noises | Uncertainty |

Fig. 4. Uncertainty visualization. Highly bright regions represent large uncertainties in the uncertainty map.

where C is the softmax predicted class related to the category of the dataset, p is the softmax predicted vector with C. We normalize U into range $(0,1)$, and set $\widehat{U} = 1 - U$. Then, the pseudo label can be divided into confident and uncertain regions by setting a threshold γ. We fully receive the pixels in the confident region, which are regarded as the true label. As for the uncertain regions, we assign low loss weights to high uncertain pixels. Thus the model can also learn from the pixels in the uncertain regions, which avoids the loss of useful information. We define the uncertainty weight mask as:

$$W = \begin{cases} 1 & \widehat{U} \geq \gamma \\ \widehat{U}/\gamma & \widehat{U} < \gamma \end{cases} \tag{6}$$

In the end, we multiply the weight mask W to the self cross supervision constraint and rewrite the Eq. 4, getting the uncertainty-guided self cross supervision constraint:

$$\mathcal{L}_{uscs}^{k}(x_{ul}) = \frac{1}{|\Omega|\,\|W\|_{1,1}} \sum_{i\in\Omega} W(i) \cdot \ell_{ce}(\hat{y}_{ul}^{k}(i), p^{k}(i)), \qquad (7)$$

where $\|W\|_{1,1}$ means the $L_{p,q}$ norm of matrix W.

4 Expermients

4.1 Experimental Setup

Datasets. PASCAL VOC 2012 [8] is the most prevalent benchmark for semi-supervised semantic segmentation with 20 object classes and one background class. The standard dataset contains 1464 images for training, 1449 for validation, and 1456 for testing. Following previous works [6], we adopt the augmented set provided from SBD [12] as our entire training set, which contains 10582 images.

Implementation Details. The results are obtained by training the MIMO model, modified on the basis of Deeplabv3+ [5]. We regard the backbone of the segmentation model as the encoder, whose weights are initialized with the pretrained model on ImageNet [7]. The other components except the final classifier are considerd as the decoder which are initialized randomly.

Following the previous works [6], we utilize "poly" learning rate decay policy where the base learning rate is scaled by $(1 - iter/max\_iter)^{0.9}$. Mini-batch SGD optimizer is adopted with the momentum and weight decay set to 0.9 and 10^{-4} respectively. During the training, images are randomly cropped to 320×320, random horizontal flipping with a probability of 0.5, and random scaling with a ratio from 0.5 to 2.0 are adopted as data augmentation. We train PASCAL VOC 2012 for 3×10^{4} iters with batch size set to 16 for both labeled and unlabeled images. The base learning rates are 0.01 for backbone parameters and 0.001 for others. The trade-off weight λ is set to 1 after adjustment.

Besides, we found that the MIMO model based on Deeplabv3+ cannot accommodate two independent subnetworks due to the limited capacity. Thus, we relax independence same as [13] by sampling two same inputs from the training set with probability ρ, i.e., the input x_2 of the MIMO model is set to be equal to x_1 with probability ρ. During the training, we employ CutMix [43] as transformation, same as [6]. We average two outputs of the MIMO model to generate the final results for evaluation.

Evaluation. We use the mean Intersection-over-Union (mIoU) as the evaluation metric as a common practice. To evaluate training time and memory cost reduction in USCS, Multiply-Accumulate Operations (MACs) and the number of parameters are adopted as the metric. Besides, we employ the non overlap ratio for the outputs of the MIMO model as metric to measure the diversity of subnetworks. The low non overlap ratio means poor diversity.

4.2   Results

In this section, we report the results compared with supervised baselines and other SOTA methods in different partition protocols, i.e., the full training set is split with 1/16, 1/8, 1/4, and 1/2 ratios for labeled images and the remainder as unlabeled images.

Improvements Over Supervised Baselines.   Figure 5 illustrates the improvements of our approach compared with full supervised learning (trained with the same partition protocol). Specifically, our method outperforms the supervised baseline by 5.70%, 4.23%, 2.63%, and 1.30% under 1/16, 1/8, 1/4, and 1/2 partition protocols separately with Resnet-50. On the other settings, the gains obtained by our approach are also stably: 5.20%, 3.32%, 2.03%, and 1.50% under 1/16, 1/8, 1/4, and 1/2 partition protocols separately with Resnet-101.

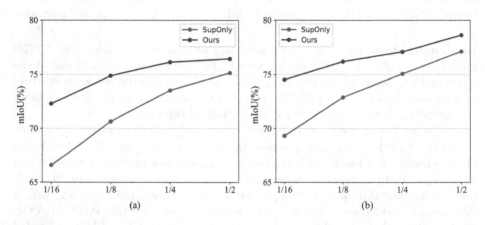

Fig. 5. Improvements over the supervised baseline on PACAL VOC 2012 with (a) Resnet-50 and (b) Resnet-101

Comparison with SOTA.   The results compared with other semi-supervised approaches are shown in Table 1. Our method performs better than most methods under different partition protocols with Resnet-50 and Resnet-101 as backbones. Compared with CAC [17], our approach improves by 2.2%, 2.48%, 2.15% under 1/16, 1/8, and 1/4 partition protocols separately with Resnet-50. Compared with CPS [6], the advantage of our method is a great reduction in the number of parameters and calculations as shown in Table 2. We acquired 40.5% and 49% economization on MACs and parameters with Resnet-50, which signify the cost decrease of training time and memory. Besides, our method only needs twice forward pass, while CPS needs four times. As for accuracy, our method achieves around 1% improvement in all cases with Resnet-50.

Table 1. Comparison with SOTA on PASCAL VOC 2012. All the approachs are based on Deeplabv3+. The * indicates the approaches re-implemented by [6]. Best results are in bold; suboptimal results are in italics.

Methods	Network	1/16	1/8	1/4	1/2
MT* [34]	Deeplabv3+ Resnet50	66.77	70.78	73.22	75.41
CutMix-Seg* [9]		68.90	70.70	72.46	74.49
CCT* [28]		65.22	70.87	73.43	74.75
GCT* [14]		64.05	70.47	73.45	75.20
CAC [17]		70.10	72.40	74.00	-
ECS [23]		-	70.20	72.60	-
CPS [6]		*71.98*	*73.67*	*74.90*	*76.15*
Ours		**72.30**	**74.88**	**76.15**	**76.45**
MT* [34]	Deeplabv3+ Resnet101	70.59	73.20	76.62	77.61
CutMix-Seg* [9]		72.56	72.69	74.25	75.89
CCT* [28]		67.94	73.00	76.17	77.56
GCT* [14]		69.77	73.30	75.25	77.14
CAC [17]		72.40	74.60	76.30	-
ELN [16]		-	75.10	76.58	-
ST++ [40]		*74.50*	**76.30**	*76.60*	-
Ours		**74.52**	*76.20*	**77.09**	**78.63**

4.3 Ablation Study

This section conducts the ablation study to exhibit the roles of self cross supervision (SCS) and uncertainty-guided learning (UL) in our method. Besides, the influences of uncertainty threshold γ, feature fusion methods, and input repetition probability ρ are reported, respectively. All the experiments are run based on 1/8 partition protocols on PASCAL VOC 2012.

Table 2. Training cost comparison with CPS [6] and SupOnly in the backbone of Resnet-50 and Resnet-101.

Methods	Resnet-50		Resnet-101		Forwardings
	MACs(G)↓	Params(M)↓	MACs(G)↓	Params(M)↓	
SupOnly	23.84	39.78	31.45	58.77	1
CPS	95.36	79.56	125.80	117.54	4
Ours	56.74	40.49	71.94	59.48	2

Uncertainty Guided Self Cross Supervision. The contribution of self cross supervision and uncertainty-guided learning are shown in Table 3. It is important to note that we adopt the result of CPS with CutMix augmentation [6] as a baseline to ensure fairness. We report a slight decline in performance after replacing CPS with SCS. While, the improvements yielded by UL are 1.23% with the Resnet-50. We can see that SCS heavily reduces training costs of time and memory, and UL improves the performance without extra cost.

Table 3. Ablation study of different components under 1/8 partition protocols on PASCAL VOC 2012.

CPS	SCS	UL	Deeplabv3+ with Resnet-50		
			mIoU(%)↑	MACs(G)↓	Params(M)↓
✓			73.67	95.36	79.56
	✓		73.65	56.74	40.49
	✓	✓	**74.88**	**56.74**	**40.49**

Uncertainty Threshold γ. We investigate the influence of threshold γ used to control the uncertain weight mask as shown in Eq. 6. The results in Fig. 6(a) show that: with the increase of γ, the model reduces the weight of learning for noisy pixels in pseudo label and performs best when $\gamma = 0.5$. When the continuous increase of γ, the performance degrades due to the model tends to regard all pixels in pseudo label as noise, reducing the weight of confident pixels in pseudo label. We visualize the effect of threshold γ to uncertainty in Fig. 7.

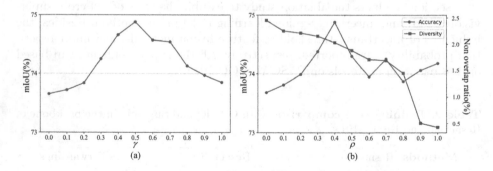

Fig. 6. The ablation study on (a) uncertainty threshold γ and (b) input repetition probability ρ.

Input Repetition Probability ρ. We show the influence of probability ρ on both accuracy and diversity Table 5. When $\rho = 0$, the training images are sam-

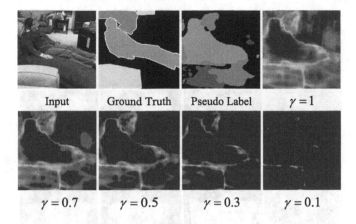

Fig. 7. Visual comparison with different uncertainty threshold γ.

pled independently for both subnetworks, and the MIMO model acquired great diversity but poor accuracy as it can not contain two independent subnetworks. As ρ grew, the diversity of the MIMO model gradually decayed, the independence of the subnetwork is relaxed to release the limited model capacity. The performance reaches the peak at $\rho = 0.4$, where get a trade-off between the diversity and the capacity of the MIMO model.

Table 4. The affects of feature fusion methods on mIoU(%) and non overlap ratio(%).

Fusion methods	Grid mix				Summing
	1	3	5	7	
mIoU(%)	74.88	74.10	73.45	73.64	73.32
Non overlap ratio(%)	2.03	2.53	1.64	1.40	0.94

Feature Fusion. We show the influence of feature fusion methods, summing and grid mix, on both accuracy and diversity in Table 4. The block size g of the grid mix is set as 1, 3, 5, and 7. We can see that the grid mix surpasses the summing feature fusion method on both mIoU scores and non overlap ratios. The accuracy of the MIMO model decreases as g increases, while the diversity reaches the top at $g = 3$. We use $g = 1$ in our method for all the experiments.

Qualitative Results. Figure 8 visualizes some segmentation results on PAS-CAL VOC 2012. The supervised results display the bad accuracy caused by the limited labeled training samples. For example, in the 2-nd row, the supervised

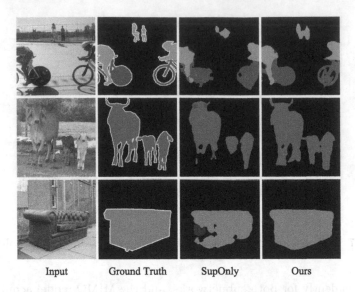

Input   Ground Truth   SupOnly   Ours

Fig. 8. Qualitative results from Pascal VOC 2012.

baseline mislabels the cow as the horse in many pixels. While our method successfully corrected the wrong annotation. Besides, the segmentation labeled by our method is more exquisite than the supervised-only method.

5   Conclusions

In this paper, we propose a new cross supervision based semi-supervised semantic segmentation approach, Uncertainty-guided Self Cross Supervision (USCS). Our method achieves self cross supervision by imposing the consistency between the subnetworks of a multi-input multi-out (MIMO) model, avoiding high computations from ensemble training. Limited by the model capacity, the subnetwork representation ability of MIMO is poor, resulting in large pseudo label noise. In order to alleviate the problem of noise accumulation and propagation in the pseudo label, we proposed uncertainty-guided learning, utilizing the uncertainty as guided information to reduce the effects of wrong pseudo labeling. Experiments show our approach dramatically reduces training costs and achieves powerful competitive performance.

Acknowledgements. This work was supported by the Natural Science Foundation of China under Grant 62001502.

References

1. Abdar, M., et al.: A review of uncertainty quantification in deep learning: techniques, applications and challenges. Inf. Fus. **76**, 243–297 (2021)
2. Badrinarayanan, V., Kendall, A., Cipolla, R.: SegNet: a deep convolutional encoder-decoder architecture for image segmentation. IEEE Trans. Pattern Anal. Mach. Intell. **39**(12), 2481–2495 (2017)
3. Berthelot, D., et al.: Mixmatch: a holistic approach to semi-supervised learning. In: 32nd Proceedings Conference on Advances in Neural Information Processing Systems (2019)
4. Chen, L.-C., Papandreou, G., Kokkinos, J., Murphy, K., Yuille, A.L.: DeepLab: semantic image segmentation with deep convolutional nets, atrous convolution, and fully connected CRFs. IEEE Trans. Pattern Anal. Mach. Intell., **40**(4), 834–848 (2017)
5. Chen, L.-C., Zhu, Y., Papandreou, G., Schroff, F., Adam, H.: Encoder-decoder with atrous separable convolution for semantic image segmentation. In: Ferrari, V., Hebert, M., Sminchisescu, C., Weiss, Y. (eds.) ECCV 2018. LNCS, vol. 11211, pp. 833–851. Springer, Cham (2018). https://doi.org/10.1007/978-3-030-01234-2_49
6. Chen, X., Yuan, Y., Zeng, C., Wang, J.: Semi-supervised semantic segmentation with cross pseudo supervision. In Proceedings of the IEEE/CVF Conference on Computer Vision and Pattern Recognition, pp. 2613–2622 (2021)
7. MMSegmentation Contributors. MMSegmentation: Openmmlab semantic segmentation toolbox and benchmark (2020). https://github.com/open-mmlab/mmsegmentation
8. Everingham, M., et al.: The pascal visual object classes challenge: a retrospective. Int. J. Comput. Vis. **111**(1), 98–136 (2015)
9. French, G., Aila, T., Laine, S., Mackiewicz, M., Finlayson, G.: Semi-supervised semantic segmentation needs strong, high-dimensional perturbations. In: ICLR (2019)
10. Gong, Z., Zhong, P., Weidong, H.: Statistical loss and analysis for deep learning in hyperspectral image classification. IEEE Trans. Neural Netw. Learn. Syst. **32**(1), 322–333 (2020)
11. Gong, Z., Zhong, P., Yang, Yu., Weidong, H., Li, S.: A CNN with multiscale convolution and diversified metric for hyperspectral image classification. IEEE Trans. Geosci. Remote Sens. **57**(6), 3599–3618 (2019)
12. Hariharan, B., Arbeláez, P., Bourdev, L., Maji, S., Malik, J.: Semantic contours from inverse detectors. In: 2011Iinternational Conference on Computer Vision, pp. 991–998. IEEE (2011)
13. Havasi, M., et al.: Training independent subnetworks for robust prediction. arXiv preprint arXiv:2010.06610 (2020)
14. Ke, Z., Qiu, D., Li, K., Yan, Q., Lau, R.W.H.: Guided collaborative training for pixel-wise semi-supervised learning. In: Vedaldi, A., Bischof, H., Brox, T., Frahm, J.-M. (eds.) ECCV 2020. LNCS, vol. 12358, pp. 429–445. Springer, Cham (2020). https://doi.org/10.1007/978-3-030-58601-0_26
15. Kipf, T.N., Welling, M.: Semi-supervised classification with graph convolutional networks. arXiv preprint arXiv:1609.02907 (2016)
16. Kwon, D., Kwak, S.: Semi-supervised semantic segmentation with error localization network. In: Proceedings of the IEEE/CVF Conference on Computer Vision and Pattern Recognition, pp. 9957–9967 (2022)

17. Lai, X., et al.: Semi-supervised semantic segmentation with directional context-aware consistency. In: Proceedings of the IEEE/CVF Conference on Computer Vision and Pattern Recognition, pp. 1205–1214 (2021)
18. Lakshminarayanan, B., Pritzel, A., Blundell, C.: Simple and scalable predictive uncertainty estimation using deep ensembles. In: 30th Proceedings on Advances in Neural Information :Processing Systems, vol. 30 (2017)
19. Lee, D.-H., et al.: Pseudo-label: The simple and efficient semi-supervised learning method for deep neural networks. In Workshop on Challenges in Representation Learning, ICML, p. 896 (2013)
20. Li, J., Socher, R., Hoi, S.C.H.: Dividemix: learning with noisy labels as semi-supervised learning. arXiv preprint arXiv:2002.07394 (2020)
21. Liu, Y., Tian, Y., Chen, Y., Liu, F., Belagiannis, M., Carneiro, C.: Perturbed and strict mean teachers for semi-supervised semantic segmentation. arXiv preprint arXiv:2111.12903 (2021)
22. Long, J., Shelhamer, E., Darrell., T.: Fully convolutional networks for semantic segmentation. In Proceedings of the IEEE Conference on Computer Vision and Pattern Recognition, pp. 3431–3440 (2015)
23. Mendel, R., de Souza, L.A., Rauber, D., Papa, J.P., Palm, C.: Semi-supervised segmentation based on error-correcting supervision. In: Vedaldi, A., Bischof, H., Brox, T., Frahm, J.-M. (eds.) ECCV 2020. LNCS, vol. 12374, pp. 141–157. Springer, Cham (2020). https://doi.org/10.1007/978-3-030-58526-6_9
24. Miyato, T., Maeda, S., Koyama, M., Ishii, S.: Virtual adversarial training: a regularization method for supervised and semi-supervised learning. IEEE Trans. Pattern Anal. Mach. Intell. **41**(8), 1979–1993 (2018)
25. Mustafa, A., Mantiuk, R.K.: Transformation consistency regularization – a semi-supervised paradigm for image-to-image translation. In: Vedaldi, A., Bischof, H., Brox, T., Frahm, J.-M. (eds.) ECCV 2020. LNCS, vol. 12363, pp. 599–615. Springer, Cham (2020). https://doi.org/10.1007/978-3-030-58523-5_35
26. Noh, H., Hong, S., Han, B.: Learning deconvolution network for semantic segmentation. In: Proceedings of the IEEE International Conference on Computer Vision, pp. 1520–1528 (2015)
27. Olsson, V., Tranheden, W., Pinto, J., Svensson, L.: Classmix: segmentation-based data augmentation for semi-supervised learning. In Proceedings of the IEEE/CVF Winter Conference on Applications of Computer Vision, pp. 1369–1378 (2021)
28. Ouali, Y., Hudelot, C., Tami, M.: Semi-supervised semantic segmentation with cross-consistency training. In: Proceedings of the IEEE/CVF Conference on Computer Vision and Pattern Recognition, pp. 12674–12684 (2020)
29. Pham, H., Dai, Z., Xie, Q., Le, Q.V.: Meta pseudo labels. In: Proceedings of the IEEE/CVF Conference on Computer Vision and Pattern Recognition, pp. 11557–11568 (2021)
30. Ronneberger, O., Fischer, P., Brox, T.: U-Net: convolutional networks for biomedical image segmentation. In: Medical Image Computing and Computer-Assisted Intervention – MICCAI 2015, pp. 234–241 (2015). https://doi.org/10.1007/978-3-319-24574-4_28
31. Claude Elwood Shannon: A mathematical theory of communication. ACM SIGMOBILE Mob. Ccomput. Commun. Rev. **5**(1), 3–55 (2001)
32. Sohn, K., et al.: FixMatch: simplifying semi-supervised learning with consistency and confidence. Adv. Neural. Inf. Process. Syst. **33**, 596–608 (2020)
33. Sun, K., Xiao, B., Liu, D., Wang, J.: Deep high-resolution representation learning for human pose estimation. In Proceedings of the IEEE/CVF Conference on Computer Vision and Pattern Recognition, pp. 5693–5703 (2019)

34. Tarvainen, A., Valpola, H.: Mean teachers are better role models: weight-averaged consistency targets improve semi-supervised deep learning results. In: 30th Advances in Neural Information Processing Systems (2017)
35. Van Engelen, J.E., Hoos, H.H.: A survey on semi-supervised learning. Machi. Learn. **109**(2), 373–440 (2020)
36. Vaswani, A., et al.: Attention is all you need. In: 30th Proceedings on Advances in Neural Information Processing Systems (NIPS 2017) (2017)
37. Wang, J., et al.: Deep high-resolution representation learning for visual recognition. IEEE Trans. Pattern Anal. Mach. Intell. **43**(10), 3349–3364 (2020)
38. Wang, Y., Peng, J., Zhang, X.: Uncertainty-aware pseudo label refinery for domain adaptive semantic segmentation. In: Proceedings of the IEEE/CVF International Conference on Computer Vision, pp. 9092–9101 (2021)
39. Xie, Q., Dai, Z., Hovy, E., Luong, T., Le, Q.: Unsupervised data augmentation for consistency training. Adv. Neural. Inf. Process. Syst. **33**, 6256–6268 (2020)
40. Yang, L., Zhuo, W., Qi, L., Shi, Y., Gao, Y.: St++: make self-training work better for semi-supervised semantic segmentation. In: Proceedings of the IEEE/CVF Conference on Computer Vision and Pattern Recognition, pp. 4268–4277 (2022)
41. Yang, M., Yu, K., Zhang, C., Li, Z., Yang, K.: Denseaspp for semantic segmentation in street scenes. In: Proceedings of the IEEE Conference on Computer Vision and Pattern Recognition, pp. 3684–3692 (2018)
42. Yu, F., Koltun, Y.: Multi-scale context aggregation by dilated convolutions. arXiv preprint arXiv:1511.07122 (2015)
43. Yun, S., Han, D., Oh, S.J., Chun, S., Choe, J., Yoo. Y.: Cutmix: regularization strategy to train strong classifiers with localizable features. In: Proceedings of the IEEE/CVF International Conference on Computer Vision, pp. 6023–6032 (2019)
44. Zhao, H., Shi, J., Qi, X., Wang, X., Jia, J.: Pyramid scene parsing network. In: Proceedings of the IEEE Conference on Computer Vision and Pattern Recognition, pp. 2881–2890 (2017)
45. Zhu, X.J.: Semi-supervised learning literature survey. Mach. Learn. **109**, 373–440 (2005)
46. Zhu, Y., et al.: Improving semantic segmentation via efficient self-training. IEEE Trans. Pattern Anal. Mach. Intell. (2021)
47. Zou, Y., Yu, Z., Vijaya Kumar, B.V.K., Wang, J.: Unsupervised domain adaptation for semantic segmentation via class-balanced self-training. In: Ferrari, V., Hebert, M., Sminchisescu, C., Weiss, Y. (eds.) ECCV 2018. LNCS, vol. 11207, pp. 297–313. Springer, Cham (2018). https://doi.org/10.1007/978-3-030-01219-9_18
48. Zou, Y., Yu, Z., Liu, X., Kumar, B.V.K., Wang, J.: Confidence regularized self-training. In: Proceedings of the IEEE/CVF International Conference on Computer Vision, pp. 5982–5991 (2019)
49. Zou, Y., et al.: Pseudoseg: designing pseudo labels for semantic segmentation. arXiv preprint arXiv:2010.09713 (2020)

A Cylindrical Convolution Network for Dense Top-View Semantic Segmentation with LiDAR Point Clouds

Jiacheng Lu[1], Shuo Gu[1(✉)], Cheng-Zhong Xu[2], and Hui Kong[2(✉)]

[1] PCA Lab, Key Lab of Intelligent Perception and Systems for High-Dimensional Information of Ministry of Education, and Jiangsu Key Lab of Image and Video Understanding for Social Security, School of Computer Science and Engineering, Nanjing University of Science and Technology, Nanjing, China
{lujiacheng,shuogu}@njust.edu.cn
[2] University of Macau, Macau, China
{czxu,huikong}@um.edu.mo

Abstract. Accurate semantic scene understanding of the surrounding environment is a challenge for autonomous driving systems. Recent LiDAR-based semantic segmentation methods mainly focus on predicting point-wise semantic classes, which cannot be directly used before the further densification process. In this paper, we propose a cylindrical convolution network for dense semantic understanding in the top-view LiDAR data representation. 3D LiDAR point clouds are divided into cylindrical partitions before feeding to the network, where semantic segmentation is conducted in the cylindrical representation. Then a cylinder-to-BEV transformation module is introduced to obtain sparse semantic feature maps in the top view. In the end, we propose a modified encoder-decoder network to get the dense semantic estimations. Experimental results on the SemanticKITTI and nuScenes-LidarSeg datasets show that our method outperforms the state-of-the-art methods with a large margin.

1 Introduction

Semantic perception of the surrounding environments is important for autonomous driving systems. In order to achieve reliable semantic estimations in top-view representation, autonomous vehicles are usually equipped with camera and LiDAR sensors. Benefiting from the rapid development of convolutional neural networks (CNNs), a large number of camera-based semantic segmentation networks, like Fully Convolutional Network (FCN) [1], ERFNet [2], U-Net [3], etc., have been proposed and proved to be effective. However, most of these methods are only applicable to the segmentation in perspective view, and accurate transformation from perspective to top-view is still a challenge. The camera sensor

---

Supplementary Information The online version contains supplementary material available at https://doi.org/10.1007/978-3-031-26293-7_21.

lacks effective geometric perception of the environment. In recent years, with the release of large-scale 3D LiDAR semantic segmentation datasets (SemanticKITTI [4] and nuScenes-LidarSeg [5]), the LiDAR-based semantic segmentation performance has been significantly increased. Because these datasets only provide point level semantic labels, most works only perform sparse semantic segmentation and predict point-wise semantic classes. The obtained sparse results still need further processing before used. Therefore, some works conduct dense top-view semantic segmentation with sparse inputs. Compared with sparse predictions, the dense top-view segmentation results are more valuable for some upper-level tasks such as the navigation and path planning of autonomous driving vehicles.

In this paper, we focus on the dense semantic segmentation of LiDAR point clouds in top-view representation. Compared with 2D camera images, the 3D LiDAR data can retain precise and complete spatial geometric information of the surroundings. Therefore, we can generate accurate top-view maps in a simpler way. However, the main problem is that the LiDAR data is sparsely distributed, and the generated top-view maps are sparse. In order to deal with the sparsity of 3D LiDAR data, some LiDAR-based segmentation approaches [6] project the 3D point clouds onto 2D bird's-eye-view (BEV[1]) images, and conduct dense semantic segmentation with 2D convolution networks. However, the projection process inevitably leads to a certain degree of information loss. Some methods [7–9] use pillar-level representation and point-wise convolution to retain and obtain more information in the height direction. These approaches still focus more on 2D convolution, neglecting the rich geometric relationships between precise 3D point cloud data.

To solve the problems mentioned above, we make use of the cylinder representation and 3D sparse convolution networks in our work. Compared with 2D images or pillar representation, the cylinder representation can maintain the 3D geometric information. The cylindrical partition divides the LiDAR point cloud dynamically according to the distances in cylindrical coordinates, and provides a more balanced distribution than 3D voxelization. The 3D sparse convolution networks can effectively integrate the geometric relationships of LiDAR point clouds, extract informative 3D features and save significant memory at the same time.

After the 3D sparse convolution networks, we introduce a cylinder-to-BEV module to convert the obtained semantic features in cylindrical representation to BEV maps. The cylinder-to-BEV module uses the coordinate information of 3D points to establish corresponding relationships, and transfers features between the two representations. The transformed feature maps in top-view are sparse, so we further propose a modified U-Net network to get the final dense segmentation results. We use groups of dilated convolutions with different receptive field sizes in different stages of downsampling and upsampling to capture more descriptive spatial features, and use grouped convolutions to reduce the FLOPs while maintaining an acceptable level of accuracy.

---

[1] BEV is another expression for top view.

The main contributions of this work lie in three aspects:

- We propose an end-to-end cylindrical convolution network that can generate accurate semantic segmentation results in top-view. The combination of cylinder representation and 3D sparse convolution greatly improves the segmentation performance.
- We propose a cylinder-to-BEV module and a modified U-Net to efficiently use 3D features to enhance the dense semantic segmentation in top-view.
- The proposed method outperforms the state-of-the-art methods on the SemanticKITTI and nuScenes-LidarSeg datasets, which demonstrates the effectiveness of the model.

2   Related Work

2.1   Image Semantic Segmentation in Bird's Eye View

Understanding the surrounding environment is an essential part of an autonomous driving system. To accomplish this, many previous works created a semantic map in Bird's Eye View (BEV) that can distinguish drivable regions, sidewalks, cars, bicycles and so on [10–14]. Image semantic segmentation in BEV usually consists of following components: an encoder for encoding features in the image view, a view transformer for converting the features from the image view to BEV, an encoder for further encoding the features in BEV and a semantic head for label classification [10–12]. Thomas Roddick et al. [13] chose feature pyramid networks like in [15] when extracting the image-view features. Weixiang Yang et al. [16] implemented cross-view transformation module that consists of the cycled view projection and the cycled view transformer in order to enrich the features getting from front-view image.

2.2   Semantic Segmentation of LiDAR Input

Since the recent launch of large-scale datasets, such as SemanticKITTI [4] and nuScenes [5], there have been an increasing number of studies on semantic segmentation of LiDAR point clouds. However, due to the sparse, irregular, and disorderly LiDAR point cloud data, it is still challenging to be processed and applied to semantic segmentation. The methods are mostly divided into three branches: point-based methods, grid-based methods, and projection-based methods.

Point-based methods directly process raw LiDAR point clouds. PointNet [17] is a pioneer and a representative of point-based approaches, which learns features for each point using shared MLPs. PointNet++ [18], an upgrade to PointNet, generates point cloud subsets by clustering and employs PointNet to extract point features from each subset. RandLA-Net [19] uses a random sampling to boost processing speed and local spatial encoding and attention-based pooling. KPConv [20] develops spatial kernels to adapt convolution operations to point clouds. However, due to computational complexity and memory requirements, the performance on large-scale LiDAR point cloud datasets is limited.

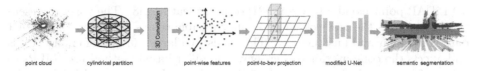

Fig. 1. Overview of the proposed dense top-view semantic segmentation network. Given a 3D LiDAR point cloud, the network first divides it into cylindrical partitions and applies a 3D sparse convolution module to obtain high-level features. Then, the cylinder-to-BEV module converts the semantic features in cylindrical representation to BEV maps. Finally, a modified U-Net is used to predict the dense top-view semantic segmentation results.

Grid-based methods convert point clouds into uniform voxels, after which, 3D convolution can be employed on voxel data. In VoxelNet [21], point clouds are quantized into uniform 3D volumetric grids, which maintains the 3D geometric information. Fully convolutional point networks [22] achieve uniform sampling in 3D space by collecting a fixed number of points around each sampled site and then apply a U-Net to extract information from multiple scales. Cyliner3D [23] recommends dividing the original point cloud into cylindrical grids to distribute the point clouds more evenly throughout the grids.

Furthermore, projection-based methods project 3D point clouds onto different 2D images. SqueezeSeg series [24,25], RangeNet series [26] and Salsanet series [6,27] deploy the spherical projection on the LiDAR data. What's more, some approaches achieve top-view semantic segmentation after projecting the point cloud into a BEV image [8,28–30]. Bieder et al. [28] turn 3D LiDAR data into a multi-layer top-view map for accurate semantic segmentation. Following this route, PillarSegNet [8] is proposed to learn features from the pillar encoding and conduct 2D dense semantic segmentation in the top view. These methods take LiDAR point cloud as input and generates dense top-view semantic grid maps, which provide a fine-grained semantic understanding that is necessary for distinguishing drivable and non-drivable areas.

3   Proposed Method

In this section, we introduce the architecture of the proposed dense top-view semantic segmentation method, as illustrated in Fig. 1. The whole network consists of three parts, a 3D cylindrical encoding, a cylinder-to-BEV module, and a 2D encoder-decoder network. The network takes sparse a LiDAR point cloud as input and generates dense semantic maps in top-view. The design of each module will be detailed in the following.

3.1   3D Cylindrical Encoding

Effectively extracting the features of 3D point cloud data is an important part of the LiDAR-based semantic segmentation. Previous methods usually convert the

3D LiDAR point cloud into voxels [21] or pillar features [8]. The voxel representation quantizes the point cloud into uniform cubic voxels. However, the LiDAR data is irregular and unstructured. This leads to a large number of empty voxels and affects the computational efficiency. The pillar-based method uses MLPs to extract features, which cannot make full use of the 3D topology and rich spatial geometric relationships.

Based on these considerations, we apply cylindrical coordinates to represent LiDAR data in this paper, which is firstly proposed by Xinge Zhu et al. [23]. The density of the 3D LiDAR point cloud usually varies, and the density in nearby areas is significantly higher than that in remote areas. Uniformly dividing the LiDAR data with different densities will lead to an unbalanced distribution of points. The cylindrical partitions can cover areas with different size of grids, which grow by distance, evenly distribute points on different cylindrical grids, and provide a more balanced representation. The cylindrical coordinate system is defined as follows,

$$\begin{cases} \rho = \sqrt{x^2 + y^2} \\ \theta = arctan(y, x) \\ z = z \end{cases} \tag{1}$$

where (x, y, z) represents the Cartesian coordinate, and (ρ, θ, z) represents its corresponding cylindrical coordinate. The radius ρ and tangent angle θ denote the distance from the origin in the x-y plane and the tangent angle between the y and x directions, respectively.

Compared with the voxel representation, although the number of empty elements is reduced, the cylindrical representation of LiDAR data is still sparse. Therefore, we apply a 3D sparse convolution network to extract features, which can efficiently process sparse data and increase the computing speed. More details of the network can be referred in [23].

3.2 Cylinder-to-BEV Module

After the 3D feature encoding module, we can obtain high-level features with rich semantic information in the form of cylindrical representation. Since the goal is to predict dense semantic categories in top-view, we need to convert the cylindrical features into BEV maps before the 2D semantic segmentation module.

Figure 2 shows two types of transformations, without and with point guidance. Without point guidance means that we use the correspondence between the cylindrical coordinate system and the BEV coordinate system to directly convert the features. However, the cylindrical grid is different from the BEV grid, which can lead to deviation problems at the boundaries. As shown in the left of Fig. 2, the cylindrical grid and the BEV grid have different shapes, in which the yellow denotes the cylindrical grid with features, and the purple represents the empty cylindrical grid without features. One BEV grid overlaps with two cylindrical grids. Transforming directly from cylinder to BEV may establish correspondence between the purple cylindrical grid and the BEV grid instead

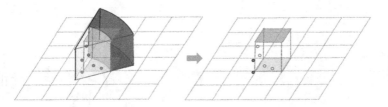

Fig. 2. Illustration of conversion from cylinder to BEV without and with point guidance. Left shows the transformation without point guidance, in which the yellow denotes cylindrical grid with features, and the purple represents empty cylindrical grid without features. Right shows the transformation with point guidance, the point serves as intermediate to connect the cylinder and BEV grids. (Color figure online)

of the yellow one, resulting in loss of information. The right of Fig. 2 shows the transformation with point guidance. The cylindrical features are first converted to point features according to Eq. 1. Then, the point features are converted to BEV features according to Eq. 2. Using point features as intermediate can preserve more useful features and cover the BEV grids more completely.

The transformation from point to BEV grid is described bellow. Given a point (x, y, z) and the feature f, its corresponding coordinates in BEV are calculated as,

$$\begin{cases} u = x/precision + W/2 \\ v = y/precision + H/2 \end{cases} \qquad (2)$$

where (u, v) represents the corresponding point in BEV, *precision* denotes the resolution of BEV. W and H represent the width and height of the BEV map, respectively.

When converting point features to BEV features, a lot of geometric information may be lost due to the many-to-one problem. Different from some methods that use maximum compression and retain only one point feature, we sort the points corresponding to the same BEV grid by height, and retain the features of the highest and lowest points. Therefore, the features of each BEV grid are as follows,

$$F_{bev} = (r_l, x_l, y_l, z_l, f_l, r_h, x_h, y_h, z_h, f_h) \qquad (3)$$

where r denotes the distance and the subscripts l and h represent the lowest and highest points, respectively. The features (z_h, f_h, z_l, f_l) of the highest and lowest points can provide the height range and spatial features of each BEV grid. For example, the spatial features of the highest and lowest points for roads, vehicles, and pedestrians vary greatly, which is very useful in determining the semantic categories.

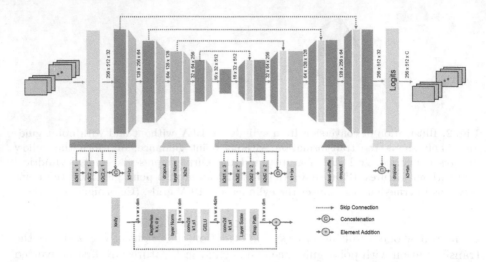

Fig. 3. Flowchart of the proposed modified U-Net. k, d, s, bn, and $\times$ represent the kernel size, dilation rate, stride, batch normalization, and block numbers, respectively. Blocks of different colors represent convolutional layers of different structures. Among them, the light blue blocks represent ordinary convolution, dropout, and normalization blocks. The pink blocks represent a designed convolution block, which is the base block of each convolution layer. (Color figure online)

3.3 2D Semantic Segmentation

After the cylinder-to-BEV process, we can get sparse BEV maps with rich semantic information. In this section, we will introduce the 2D semantic segmentation that is used to densify the semantic predictions. The network is based on U-Net structure and added various convolution designs, including dilated convolution, depth-wise convolution, inverse bottleneck, etc., as shown in Fig. 3.

Encoder-Decoder Architecture. Building upon the U-Net framework, we use convolutional blocks in both encoder and decoder, supplemented by appropriate design, to make the network more suitable for the LiDAR-based semantic segmentation task. As a characteristic of U-Net, skip-connection is also used to improve image segmentation accuracy by fusing low-level and high-level features. Considering the computational overhead, we use a separate downsampling layer and a pixel-shuffle layer instead of transpose convolution in the upsampling part.

Depth-Wise Convolution and Inverse Bottleneck. Depth-wise convolution is adapted from grouped convolution, in which the number of groups equals the number of channels. The advantage is that it greatly reduces the floating-point operations while maintaining an acceptable level of accuracy. An important design of blocks in Transformer [31], MobileNetV2 [32] and ConvNet [33] is the inverse bottleneck. The dimension of the intermediate hidden layer is larger than that of the input and output layers. We implement depth-wise convolution and combine it with two 1×1 convolutions to form an inverse bottleneck. As shown in Fig. 3, we combine these two designs and apply them as the basic block whose color is pink.

Dilated Convolution Groups. Unlike increasing the size of the convolution kernel, which greatly increases the number of parameters, dilated convolution provides a cost-effective way to extract more descriptive features. Because different stages of U-Net have different scales of information, we use convolution groups with different receptive field sizes for each stage in downsampling and upsampling as shown in Fig. 3. In each group, a 1×1 convolution is used to extract the spatial information from different receptive fields after concatenating the outputs of each dilated convolution. Meanwhile, a dropout layer and a pooling layer are added at the end. Following Transformer [31], the number of convolution blocks in different stages of downsampling and upsampling is adjusted to (3,3,9,3). As seen in the Fig. 3, the blue block represents a convolution group consisting of 3 convolution blocks, and the cyan block represents a convolution group consisting of 9 convolution blocks.

In addition, we use GeLU activation function instead of ReLU in the basic block, and reduce the number of activation functions used in each block. Similarly, we use fewer normalization layers and replace BatchNorm with LayerNorm, as in Transformer.

3.4 Loss Function

The unbalanced data distribution in the dataset can make model training difficult. Especially for the classes with fewer samples, the network predicts them with a lower frequency than that of the classes with more samples. To solve this problem, we use the weighted cross-entropy loss function, whose weight is equal to the inverse square root of the frequency of each class, as shown below:

$$L_{wce}(y, \hat{y}) = -\sum_{i=1}^{n} \lambda_i p(y_i) log(p(\hat{y}_i)) \tag{4}$$

where n denotes the number of classes, y_i and $\hat{y}_i$ represent the ground truth and the prediction, respectively. $\lambda_i = 1/\sqrt{f_i}$ and f_i denotes the frequency of the ith class.

In addition, we also incorporate the Lovász-Softmax loss in the training process. The Jaccard loss function is directly defined based on the Intersection over Union (IoU) metric. However, it is discrete and its gradient cannot be calculated directly. In [34], the lovász extension is proposed, which is derivable and can be used as the loss function to guide the training process. Specifically, the Lovász-Softmax loss can be expressed as follows:

$$L_{ls} = \frac{1}{|C|} \sum_{c \in C} \overline{\Delta_{J_c}}(m(c)),$$

$$m_i(c) = \begin{cases} 1 - x_i(c) & if \quad c = y_i(c), \\ x_i(c) & otherwise \end{cases}$$

(5)

where C denotes the class number, $\overline{\Delta_{J_c}}$ represents the lovász extension of the Jaccard index. $x_i(c)$ and $y_i(c)$ represent the predicted probability and the ground truth of pixel i for class c, respectively.

The final loss function is a linear combination of the weighted cross-entropy loss and the Lovász-Softmax loss, as shown below:

$$L = L_{wce} + L_{ls}$$

(6)

4   Experiments

In order to evaluate the segmentation performance of the proposed network, we carry out experiments on SemanticKITTI [4] and nuScenes-LidarSeg [5] datasets with raw LiDAR data, sparse semantic segmentation ground truths, and the aggregated dense semantic segmentation ground truths. The experimental results show that our network achieves state-of-the-art performance in both SemanticKITTI and nuScenes-LidarSeg datasets.

4.1   Datasets

SemanticKITTI. The SemanticKITTI is a large-scale outdoor point cloud dataset with precise pose information and semantic annotations of each LiDAR point. The training set consists of sequences 00-07 and 09-10, and the evaluation set consists of sequence 08, containing 19130 and 4071 LiDAR scans, respectively. As in [8], we merge the 19 classes into 12 classes. Specifically, The *motorcyclist* and *bicyclist* are merged to *rider*. The *bicycle* and *motorcycle* are merged to *two-wheel*. The *car*, *truck* and *other-vehicle* are merged to *vehicle*. The *traffic-sign*, *pole* and *fence* are merged to *object*. The *other-ground* and *parking* are merged to *other-ground*. The *unlabeled* pixels are not considered in the training process.

nuScenes-LidarSeg. The nuScenes-LidarSeg provides semantic annotations for each LiDAR point in the 40,000 keyframes, marking a total of 1.4 billion LiDAR points, including 32 classes. Similarly, we map the *adult*, *child*, *policeofficer*, and *constructionworker* to *pedestrian*, *bendybus* and *rigidbus* to *bus*. These class labels for *barrier*, *car*, *constructionvehicle*, *truck*, *motorcycle*, *trafficcone*, *trailer*, *driveablesurface*, *sidewalk*, *manmade*, *otherflat*, *terrain* and *vegetation* remain unchanged. The other classes are mapped to *unlabeled*. As a result, we merge 32 classes into 16 classes on the nuScenes-LidarSeg dataset.

4.2   Label Generation

Sparse Label Generation. As described in [8], we project the 3D LiDAR point cloud onto the BEV grid map and perform weighted statistical analysis on the frequency of each class in each grid to obtain the most representative grid-wise semantic label. For each grid, the weighted calculation formula of its label c_i is defined as follows:

$$c_i = argmax_{c \in [1,C]} (w_c n_{i,c}), \qquad (7)$$

where C is the number of the semantic classes, w_c denotes the weight for class c, and $n_{i,c}$ represents the number of points of class c in grid i. In addition, the weights of the traffic participant classes, such as *person*, *rider*, *two-wheel*, and *vehicle*, are chosen as 5. The weight of the *unlabeled* class is set as 0 and the weights of other classes are set as 1.

Dense Label Generation. We use the precise pose information provided by SemanticKITTI to aggregate consecutive LiDAR scans and generate dense top-view ground truths, which can provide fine-grained descriptions of the surrounding environment. As in [8], the neighboring LiDAR scans with a distance less than twice the farthest distance are selected as the supplement to the current frame. Based on the provided poses, we transform the adjacent LiDAR point clouds to the coordinate system of the current scan, and then we can get dense aggregation following Eq. 7. In addition, to avoid confusion caused by overlapping, we only aggregate static objects and ignore moving objects.

4.3   Evaluation Metrics

To evaluate the performance of the proposed dense top-view semantic segmentation method, we apply the widely used intersection-over-union (IoU) and mean intersection-over-union (mIoU) in all classes, which are defined as follows:

$$IoU_i = \frac{P_i \cap G_i}{P_i \cup G_i}, \quad mIoU = \frac{1}{C} \sum_{i=1}^{C} IoU_i, \qquad (8)$$

where P_i denotes the set of pixels whose predicted semantic labels are class i, G_i represents the set of pixels whose corresponding ground truths are class i, and C represents the total number of classes.

4.4   Implementation Details

We deploy the proposed network on a server with a single NVIDIA Geforce RTX 2080Ti-11GB GPU, running with PyTorch. The initial learning rate is 0.01, the epoch size is 30, and the batch size of 2 with an adam optimizer.

In the preprocessing step, the input LiDAR point cloud is first cropped into $[(-51.2, 51.2), (-51.2, 51.2), (-5.0, 3.0)]$ meters in the x, y, z directions,

Table 1. Quantitative results on the SemanticKITTI dataset [4]

Mode	Method	mIoU [%]	■ vehicle	■ person	■ two-wheel	■ rider	■ road	■ sidewalk	■ other-ground	■ building	■ object	■ vegetation	■ trunk	■ terrain
Sparse Train Sparse Eval	Bieder et al. [28]	39.8	69.7	0.0	0.0	0.0	85.8	60.3	25.9	72.8	15.1	68.9	9.9	69.3
	Pillar [8]	55.1	79.5	15.8	25.8	51.8	89.5	70.0	38.9	80.6	25.5	72.8	38.1	72.7
	Pillar + Occ [8]	55.3	82.7	20.3	24.5	51.3	90.0	71.2	36.5	81.3	28.3	70.4	38.5	69.0
	Pillar + Occ + P	57.5	85.1	24.7	16.9	60.1	90.7	72.9	38.3	82.9	30.1	80.4	35.4	72.8
	Pillar + Occ + LP	57.8	85.9	24.2	18.3	57.6	91.3	74.2	**39.2**	82.4	29.0	80.6	38.0	72.9
	Pillar + Occ + LGP [9]	58.8	85.8	34.2	26.8	58.5	91.3	74.0	38.1	82.2	28.7	79.5	35.7	71.3
	Our	**67.9**	**89.5**	**59.7**	**52.7**	**74.1**	**92.7**	**76.2**	36.5	**85.8**	**37.5**	**83.3**	**50.6**	**75.7**
Sparse Train Dense Eval	Bieder et al. [28]	32.8	43.3	0.0	0.0	0.0	84.3	51.4	22.9	54.7	10.8	51.0	6.3	68.6
	Pillar [8]	37.5	45.1	0.0	0.1	3.3	82.7	57.5	29.7	64.6	14.0	58.5	25.5	**68.9**
	Pillar + Occ [8]	38.4	52.5	0.0	0.2	3.0	**85.6**	60.1	29.8	**65.7**	16.1	56.7	26.2	64.5
	Pillar + Occ + P	40.9	53.3	11.3	13.1	7.0	83.6	**60.3**	30.2	63.4	15.7	**61.4**	24.6	67.2
	Pillar + Occ + LP	**41.5**	**57.3**	11.3	9.5	**10.4**	85.5	60.1	**31.2**	64.6	16.9	59.5	25.3	66.8
	Pillar + Occ + LGP [9]	40.4	55.8	10.8	14.1	9.3	84.5	58.6	26.8	62.4	15.2	59.2	26.3	62.3
	Our	38.5	53.1	21.2	26.4	4.8	72.8	52.3	22.1	52.1	20.0	47.8	31.5	57.2
Dense Train Dense Eval	Pillar [8]	42.8	70.3	5.4	6.0	8.0	89.8	65.7	34.0	65.9	16.3	61.2	23.5	67.9
	Pillar + Occ [8]	44.1	72.8	7.4	4.7	10.2	90.1	66.2	32.4	67.8	17.4	63.1	27.6	69.2
	Pillar + Occ + P	44.9	72.1	6.8	6.2	9.9	90.1	65.8	37.8	67.1	18.8	**68.1**	24.7	**71.4**
	Pillar + Occ + LP	44.8	73.0	7.8	6.1	10.6	90.6	66.5	33.7	67.6	17.7	67.6	25.5	70.4
	Pillar + Occ + LGP [9]	44.5	**73.2**	6.5	6.5	9.5	90.8	**66.5**	**34.9**	68.0	18.8	67.0	22.8	70.0
	Our	**48.8**	70.0	**25.9**	**28.0**	**22.5**	90.8	65.4	32.7	**68.3**	**20.9**	64.4	**30.6**	66.1

respectively. Then, the cropped data is divided into 3D representation $\mathbb{R} \in$ $512 \times 360 \times 32$ by cylindrical partition, where three dimensions represent radius, tangent angle, and height, respectively. After the 3D sparse convolution networks, the features are converted to a BEV map, covering the area of $[(-51.2, 51.2), (-25.6, 25.6)]$ meters in the x, y directions. The size of the BEV map is $B \times 48 \times 256 \times 512$, representing batch size, feature channels, image height and width, respectively. The resolution is $[0.2, 0.2]$ meters. The final output of the network is the semantic prediction result whose size is 256×512. Since the range of the semantic ground truth is $[(-50.0, 50.0), (-25.0, 25.0)]$ meters and the resolution is $[0.1, 0.1]$, we use linear interpolation to zoom in the network output, and then crop it to the same size as the ground truth.

4.5 Results on SemanticKITTI Dataset

We use two training modes and two evaluation modes for dense top-view semantic segmentation, following [28]: Sparse Train and Sparse Eval, Sparse Train and Dense Train, Dense Train and Dense Eval. Among them, Sparse Eval represents using the sparse top-view semantic segmentation ground truth derived from a single LiDAR scan, Dense Eval represents using the generated dense top-view ground truth.

Table 1 shows the quantitative comparison with other state-of-the-art methods. The proposed method achieves a performance improvement of **9.1**% over the current best result in the sparse evaluation mode, and **3.9**% improvement in the dense evaluation mode. In particular, our method greatly improves the performance of classes with small spatial size, including *person*, *two-wheel* and *rider*, and also performs well on other classes. In the sparse mode, the IoUs of these three classes are improved by **25.5**%, **25.9**% and **25.6**%, respectively. In the dense mode, they are increased by **18.1**%, **21.8**% and **21.9**%. This proves the effectiveness of our method in semantic segmentation.

Table 2. Quantitative results on the nuScenes-LidarSeg dataset [5].

Mode	Method	mIoU [%]	barrier	bicycle	bus	car	const-vehicle	motorcycle	pedestrian	cone	trailer	truck	drivable	other-flat	sidewalk	terrain	manmade	vegetation
Dense Train Dense Eval	Pillar [8]	22.7	10.8	0.0	5.3	1.6	6.0	0.0	0.0	0.8	19.59	0.8	83.4	35.5	45.0	52.3	48.5	54.3
	MASS [9]	32.7	28.4	0.0	24.0	35.7	16.4	2.9	4.4	0.1	29.3	21.2	87.3	46.9	51.6	56.3	56.8	61.4
	Our	33.7	25.0	3.2	26.1	46.9	15.0	11.8	10.9	6.7	22.6	25.7	85.6	40.2	48.3	58.6	62.0	51.2

4.6 Results on nuScenes-LidarSeg Dataset

In addition to SemanticKITTI dataset, we also evaluate our method on the nuScenes-LidarSeg dataset for dense top-view semantic segmentation. As shown in Table 2, our network achieves better performance than other ones. The proposed network obtains a **1.0%** performance improvement over the state-of-the-art method. Our method is superior in categories with sparse points, such as *bicycle, motorcycle, pedestrian* and *cone*. The IoU of *car* has been significantly improved by **11.2%**.

Table 3. Ablation study on the SemanticKITTI dataset. All experiments are carried out in dense mode.

Baseline	Cylinder	Cylinder-to-BEV	Modified U-Net	mIoU [%]
✓				38.9
✓	✓			45.1
✓	✓	✓		47.5
✓	✓	✓	✓	**48.8**

4.7 Ablation Studies

In this section, we conduct extensive ablation experiments to investigate the effects of different components in our method. We create several variants of our network to verify the contributions of each components Table 3 summarizes the semantic segmentation results on the SemanticKITTI evaluation dataset in dense mode. The *Baseline* represents the method of using raw point features, point-to-BEV projection and a simple encoder-decoder network with traditional convolution blocks. The *Cylinder* represents replacing point features with cylindrical features and direct cylinder-to-BEV projection without point-guidance. The *Cylinder-to-BEV* represents using cylinder-to-BEV projection with point as intermediate. The *ModifiedU-Net* means using a 2D modified U-Net in the 2D semantic segmentation part.

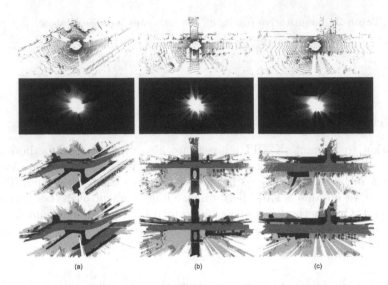

Fig. 4. Qualitative results generated by our approach on the SemanticKITTI validation set. From top to bottom in each column, we display the input point cloud, the 2D occupancy map, the ground truth and the prediction from our method. The unobserved areas were erased using the observability map as in [9]

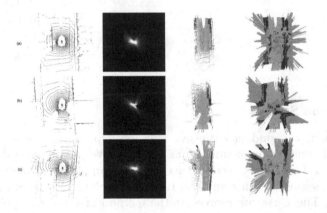

Fig. 5. Qualitative results generated by our approach on the nuScenes dataset. From left to right in each row, we display the input point cloud, the 2D occupancy map, the ground truth and the prediction from our method.

The results in Table 3 show that when dealing with outdoor sparse point clouds, the cylindrical encoding is quite successful in gathering rich characteristics from input data, and greatly improves the spatial feature extraction. Compared with methods that ignore 3D information and convert LiDAR data to 2D representation directly, we focus on investigating the spatial geometric relationships of LiDAR points, thus achieving an improvement of 6.2%. The well-designed cylinder-to-BEV module selects key characters in each grid of the

2D top-view, and further increases the performance of 2.4%. The modified U-Net with dilated convolution, depth-wise convolution and inverse bottleneck can also bring a 1.3% performance improvement.

4.8 Qualitative Analysis

As shown in Fig. 4 and Fig. 5, the proposed network can get an accurate semantic understanding of the surrounding environment. It can not only recognize large objects like roads, vehicles, and buildings, but also segment smaller objects more accurately, such as pedestrians, bicycles, motorbikes, and riders. This demonstrates that our method can effectively deal with outdoor, large-scale, sparse, and density-varying 3D point cloud data, and improve the dense semantic segmentation performance in the 2D top-view.

5 Conclusion

In this paper, we propose an end-to-end cylindrical convolution network for dense top-view semantic segmentation with LiDAR data only. We use cylindrical LiDAR representation and 3D CNNs to extract semantic and spatial information, which can effectively preserve more 3D connections and deal with the sparse density of point clouds. Moreover, we introduce an efficient cylinder-to-BEV module to transform features from cylindrical representation to BEV map and provide guidance for the proposed modified U-Net based semantic segmentation in the top-view. We perform extensive experiments and ablation studies on the SemanticKITTI and nuScenes-LidarSeg datasets, and achieve state-of-the-art performance.

Acknowledgements. This work was supported by the National Natural Science Found of China (Grant No. 62106106), the Key Laboratory of Intelligent Perception and Systems for High-Dimensional Information of Ministry of Education (Nanjing University of Science and Technology, Grant JYB202106), the National Key Research and Development Program of China (No. 2019YFB2102100), the Science and Technology Development Fund of Macau SAR (File no. 0015/2019/AKP and AGJ-2021-0046), the Guangdong-Hong Kong-Macao Joint Laboratory of Human-Machine Intelligence-Synergy Systems (No. 2019B121205007) and the startup project of Macau University (SRG2021-00022-IOTSC and SKL-IOTSC(UM)-2021-2023).

References

1. Long, J., Shelhamer, E., Darrell, T.: Fully convolutional networks for semantic segmentation. In: Proceedings of the IEEE Conference on Computer Vision and Pattern Recognition, pp. 3431–3440 (2015)
2. Romera, E., Alvarez, J.M., Bergasa, L.M., Arroyo, R.: ERFNet: efficient residual factorized convnet for real-time semantic segmentation. IEEE Trans. Intell. Transp. Syst. **19**, 263–272 (2017)

3. Ronneberger, O., Fischer, P., Brox, T.: U-Net: convolutional networks for biomedical image segmentation. In: Navab, N., Hornegger, J., Wells, W.M., Frangi, A.F. (eds.) MICCAI 2015. LNCS, vol. 9351, pp. 234–241. Springer, Cham (2015). https://doi.org/10.1007/978-3-319-24574-4_28

4. Behley, J., et al.: SemanticKITTI: a dataset for semantic scene understanding of lidar sequences. In: Proceedings of the IEEE/CVF International Conference on Computer Vision, pp. 9297–9307 (2019)

5. Caesar, H., et al.: nuScenes: a multimodal dataset for autonomous driving. In: Proceedings of the IEEE/CVF Conference on Computer Vision and Pattern Recognition, pp. 11621–11631 (2020)

6. Cortinhal, T., Tzelepis, G., Erdal Aksoy, E.: SalsaNext: fast, uncertainty-aware semantic segmentation of LiDAR point clouds. In: Bebis, G., et al. (eds.) ISVC 2020. LNCS, vol. 12510, pp. 207–222. Springer, Cham (2020). https://doi.org/10.1007/978-3-030-64559-5_16

7. Lang, A.H., Vora, S., Caesar, H., Zhou, L., Yang, J., Beijbom, O.: PointPillars: fast encoders for object detection from point clouds. In: Proceedings of the IEEE/CVF Conference on Computer Vision and Pattern Recognition, pp. 12697–12705 (2019)

8. Fei, J., Peng, K., Heidenreich, P., Bieder, F., Stiller, C.: PillarSegNet: pillar-based semantic grid map estimation using sparse lidar data. In: 2021 IEEE Intelligent Vehicles Symposium (IV), pp. 838–844. IEEE (2021)

9. Peng, K., et al.: MASS: multi-attentional semantic segmentation of lidar data for dense top-view understanding. IEEE Trans. Intell. Transp. Syst. **23**(9), 15824–15840 (2022)

10. Huang, J., Huang, G., Zhu, Z., Du, D.: BEVDet: high-performance multi-camera 3D object detection in bird-eye-view. arXiv preprint arXiv:2112.11790 (2021)

11. Ng, M.H., Radia, K., Chen, J., Wang, D., Gog, I., Gonzalez, J.E.: BEV-seg: bird's eye view semantic segmentation using geometry and semantic point cloud. arXiv preprint arXiv:2006.11436 (2020)

12. Pan, B., Sun, J., Leung, H.Y.T., Andonian, A., Zhou, B.: Cross-view semantic segmentation for sensing surroundings. IEEE Robot. Autom. Lett. **5**, 4867–4873 (2020)

13. Roddick, T., Cipolla, R.: Predicting semantic map representations from images using pyramid occupancy networks. In: Proceedings of the IEEE/CVF Conference on Computer Vision and Pattern Recognition, pp. 11138–11147 (2020)

14. Philion, J., Fidler, S.: Lift, splat, shoot: encoding images from arbitrary camera rigs by implicitly unprojecting to 3D. In: Vedaldi, A., Bischof, H., Brox, T., Frahm, J.-M. (eds.) ECCV 2020. LNCS, vol. 12359, pp. 194–210. Springer, Cham (2020). https://doi.org/10.1007/978-3-030-58568-6_12

15. Lin, T.Y., Dollár, P., Girshick, R., He, K., Hariharan, B., Belongie, S.: Feature pyramid networks for object detection. In: Proceedings of the IEEE Conference on Computer Vision and Pattern Recognition, pp. 2117–2125 (2017)

16. Yang, W., et al.: Projecting your view attentively: monocular road scene layout estimation via cross-view transformation. In: Proceedings of the IEEE/CVF Conference on Computer Vision and Pattern Recognition, pp. 15536–15545 (2021)

17. Qi, C.R., Su, H., Mo, K., Guibas, L.J.: PointNet: deep learning on point sets for 3D classification and segmentation. In: Proceedings of the IEEE Conference on Computer Vision and Pattern Recognition, pp. 652–660 (2017)

18. Qi, C.R., Yi, L., Su, H., Guibas, L.J.: PointNet++: deep hierarchical feature learning on point sets in a metric space. In: Advances in Neural Information Processing Systems, vol. 30 (2017)

19. Hu, Q., et al.: RandLA-Net: efficient semantic segmentation of large-scale point clouds. In: Proceedings of the IEEE/CVF Conference on Computer Vision and Pattern Recognition, pp. 11108–11117 (2020)

20. Thomas, H., Qi, C.R., Deschaud, J.E., Marcotegui, B., Goulette, F., Guibas, L.J.: KPConv: flexible and deformable convolution for point clouds. In: Proceedings of the IEEE/CVF International Conference on Computer Vision, pp. 6411–6420 (2019)

21. Zhou, Y., Tuzel, O.: VoxelNet: end-to-end learning for point cloud based 3D object detection. In: Proceedings of the IEEE Conference on Computer Vision and Pattern Recognition, pp. 4490–4499 (2018)

22. Rethage, D., Wald, J., Sturm, J., Navab, N., Tombari, F.: Fully-convolutional point networks for large-scale point clouds. In: Proceedings of the European Conference on Computer Vision (ECCV), pp. 596–611 (2018)

23. Zhu, X., et al.: Cylindrical and asymmetrical 3D convolution networks for lidar segmentation. In: Proceedings of the IEEE/CVF Conference on Computer Vision and Pattern Recognition, pp. 9939–9948 (2021)

24. Wu, B., Wan, A., Yue, X., Keutzer, K.: SqueezeSeg: convolutional neural nets with recurrent CRF for real-time road-object segmentation from 3D lidar point cloud. In: 2018 IEEE International Conference on Robotics and Automation (ICRA), pp. 1887–1893. IEEE (2018)

25. Wu, B., Zhou, X., Zhao, S., Yue, X., Keutzer, K.: SqueezeSegV2: improved model structure and unsupervised domain adaptation for road-object segmentation from a lidar point cloud. In: 2019 International Conference on Robotics and Automation (ICRA), pp. 4376–4382. IEEE (2019)

26. Milioto, A., Vizzo, I., Behley, J., Stachniss, C.: RangeNet++: fast and accurate lidar semantic segmentation. In: 2019 IEEE/RSJ International Conference on Intelligent Robots and Systems (IROS), pp. 4213–4220. IEEE (2019)

27. Aksoy, E.E., Baci, S., Cavdar, S.: SalsaNet: fast road and vehicle segmentation in lidar point clouds for autonomous driving. In: 2020 IEEE Intelligent Vehicles Symposium (IV), pp. 926–932. IEEE (2020)

28. Bieder, F., Wirges, S., Janosovits, J., Richter, S., Wang, Z., Stiller, C.: Exploiting multi-layer grid maps for surround-view semantic segmentation of sparse lidar data. In: 2020 IEEE Intelligent Vehicles Symposium (IV), pp. 1892–1898. IEEE (2020)

29. Paigwar, A., Erkent, Ö., Sierra-Gonzalez, D., Laugier, C.: GndNet: fast ground plane estimation and point cloud segmentation for autonomous vehicles. In: 2020 IEEE/RSJ International Conference on Intelligent Robots and Systems (IROS), pp. 2150–2156. IEEE (2020)

30. Kong, X., Zhai, G., Zhong, B., Liu, Y.: PASS3D: precise and accelerated semantic segmentation for 3D POINT cloud. In: 2019 IEEE/RSJ International Conference on Intelligent Robots and Systems (IROS), pp. 3467–3473. IEEE (2019)

31. Liu, Z., et al.: Swin transformer: hierarchical vision transformer using shifted windows. In: Proceedings of the IEEE/CVF International Conference on Computer Vision, pp. 10012–10022 (2021)

32. Sandler, M., Howard, A., Zhu, M., Zhmoginov, A., Chen, L.C.: MobileNetV2: inverted residuals and linear bottlenecks. In: Proceedings of the IEEE Conference on Computer Vision and Pattern Recognition, pp. 4510–4520 (2018)

33. Tan, M., et al.: MnasNet: platform-aware neural architecture search for mobile. In: Proceedings of the IEEE/CVF Conference on Computer Vision and Pattern Recognition, pp. 2820–2828 (2019)
34. Berman, M., Triki, A.R., Blaschko, M.B.: The lovász-softmax loss: a tractable surrogate for the optimization of the intersection-over-union measure in neural networks. In: Proceedings of the IEEE Conference on Computer Vision and Pattern Recognition, pp. 4413–4421 (2018)

Causes of Catastrophic Forgetting in Class-Incremental Semantic Segmentation

Tobias Kalb[1]([envelope]) [iD] and Jürgen Beyerer[2,3] [iD]

[1] Porsche Engineering Group GmbH, Porschestraße 911, 71287 Weissach, Germany
tobias.kalb@porsche-engineering.de
[2] Fraunhofer Institute of Optronics, Systems Technologies and Image Exploitation IOSB, Karlsruhe, Germany
juergen.beyerer@iosb.fraunhofer.de
[3] Karlsruhe Institute of Technology, 76131 Karlsruhe, Germany

Abstract. Class-incremental learning for semantic segmentation (CiSS) is presently a highly researched field which aims at updating a semantic segmentation model by sequentially learning new semantic classes. A major challenge in CiSS is overcoming the effects of catastrophic forgetting, which describes the sudden drop of accuracy on previously learned classes after the model is trained on a new set of classes. Despite latest advances in mitigating catastrophic forgetting, the underlying causes of forgetting specifically in CiSS are not well understood. Therefore, in a set of experiments and representational analyses, we demonstrate that the semantic shift of the background class and a bias towards new classes are the major causes of forgetting in CiSS. Furthermore, we show that both causes mostly manifest themselves in deeper classification layers of the network, while the early layers of the model are not affected. Finally, we demonstrate how both causes are effectively mitigated utilizing the information contained in the background, with the help of knowledge distillation and an unbiased cross-entropy loss.

Keywords: Continual learning · Semantic segmentation · Class-incremental

1 Introduction

Semantic segmentation is a long-standing problem in computer vision, which aims to assign a semantic label to each pixel of an image. However, a fundamental constraint of the traditional semantic segmentation benchmarks is that they assume that all classes are known beforehand and are all learned at once. This assumption limits the use of semantic segmentation for practical applications, as within a realistic scenario the model should be able to learn new classes

Supplementary Information The online version contains supplementary material available at https://doi.org/10.1007/978-3-031-26293-7_22.

without requiring a complete retraining of all previous ones. For this reason, the recently emerged field of class-incremental semantic segmentation (CiSS) focuses on achieving this goal of incrementally learning new classes. The two main challenges that CiSS has to overcome are catastrophic forgetting [11, 26] of old classes and the semantic shift of the background class. Recently, several methods were proposed to mitigate the limitations brought on by these challenges [3, 8, 19, 28, 29, 40]. Contrary to methods in class-incremental image classification, CiSS methods are mostly based on the concept of knowledge distillation [14]. While significant progress has been made in mitigating the effects of catastrophic forgetting and the semantic shift of the background class, there is limited understanding of how the drop in accuracy manifests itself within the CNN-based semantic segmentation model. The focus of our work is to identify the causes of forgetting in CiSS. Specifically, we aim at revealing how activation drift, inter-task-confusion and task-recency bias affect the performance in CiSS and how existing approaches overcome these effects. Our main contributions are:

1. We study the impact of the semantic shift of the background class on PascalVoc2012 [10] in 3 different task protocols: *overlapped*, *disjoint* and a novel *full disjoint* setup.
2. We analyse the degree of activation drift in various layers by stitching them with the previous task's network and reveal that semantic shift of the background class is the main cause of the catastrophic drop in performance in CiSS. Forgetting mainly happens in the decoder layers of the model, where discriminating features for old classes of the encoder are assigned to new visually similar classes or to the background class. However, the re-appearance of previous classes in the background of subsequent training tasks also reduces the internal activation drift in the encoder.
3. We introduce Decoder Retraining Accuracy to analyse the degree to which the decoder of the network contributes to inter-task confusion. We observe that methods that do not use any form of replay fail to learn discriminating features for all classes. Specifically, the model is not able to distinguish old classes from new classes that are visually closely related, e.g. *train* and *bus*.
4. In our final analysis, we apply the UNCE loss from [3] to the full disjoint setting and provide observations that support our inferences.

2   Related Work

2.1   Continual Learning

The majority of research in continual learning is focused on developing methods to overcome the effects of catastrophic forgetting. This is achieved by dedicating a subset of parameters to each task either dynamically or explicitly, by penalizing updates on important parameters for the previous task during training on a new task [1, 7, 18, 43], directly storing data from previous task for later replay [13, 34, 35, 39], as well as knowledge-distillation-based methods, which utilize the activations of the old network as regularization term during training on new data [23]. In a recent empirical survey by Masana *et al.* [25] the best performing methods on class-incremental image classification all utilize replay to cope with forgetting [25]. Other surveys and investigations confirm the notion that class-incremental learning requires some form of replay [41] and that prior regularization methods might fail at learning discriminating features in the class-incremental setting [16,22]. For a state-of-the-art overview on class-incremental image classification we refer to [25].

2.2   Class-Incremental Semantic Segmentation

When comparing the most successful approaches in class-incremental image classification [25] and CiSS [3,8,17,19,28,29,40], it is noticeable that contrary to classification in CiSS most of the recent methods build on the idea of Learning without Forgetting (LwF) [23] and utilize a knowledge distillation-based loss. In most CiSS benchmarks knowledge distillation-based approaches even outperform replay-based methods [17]. As a result, state-of-the-art CiSS approaches mostly rely on some form of knowledge distillation [14] and do not require any form of replay. These approaches have been proven to be effective on their respective benchmarks. However, they require that previously learned classes will reappear in the future training images, because otherwise they would still suffer from forgetting, as we will confirm in our experiments. Therefore, recent approaches also investigated how to integrate exemplar replay into CiSS [9,17,24] or utilizing unlabelled auxiliary data [2,42].

2.3   Studying Effects of Catastrophic Forgetting

Prior work on understanding catastrophic forgetting in deep learning for image classification used representational analysis techniques such as centered kernel alignment [20] and linear probing to conclude that deeper layers are disproportionately the cause of forgetting in CNNs [6,33]. We will confirm this is also true for CiSS. Other work analyzed the loss-landscape of continually trained models [30] or investigated how the task sequence [32] and task similarity [33] impact forgetting. Contrary to prior work, we concentrate on identifying the causes of forgetting that arise specifically in CiSS, as the semantic shift of background class and the multi-class nature of semantic segmentation introduce new challenges into the continual learning setting that are not yet well understood.

3   Problem Formulation

Before going into details about the causes of forgetting, we define the general task of CiSS. The goal of semantic segmentation is to assign a class out of a set of pre-defined classes $\mathcal{C}$ to each pixel in a given image. A training task $T = \{(x_m, y_m)\}_{m=1}^M$ consists of a set of M images $x \in \mathcal{X}$ with $\mathcal{X} = \mathbb{R}^{H \times W \times 3}$ and corresponding labels $y \in \mathcal{Y}$ with $\mathcal{Y} = \mathcal{C}^{H \times W}$. Given the task T the goal in semantic segmentation is to learn a mapping $f_\theta : \mathcal{X} \to \mathbb{R}^{H \times W \times |\mathcal{C}|}$ from the image space $\mathcal{X}$ to a posterior probability vector q. The output segmentation mask for a single pixel i of the image is obtained as $\bar{y}_i = \arg\max_{c \in \mathcal{C}} q_{i,c}$. In the class-incremental learning setting the model f_θ is not trained on a single task T but on a sequence of tasks T_t. Each task T_t extends the previous set of classes $\mathcal{C}_{t-1}$ by a set of novel classes $\mathcal{S}_t$ resulting in the new label set $\mathcal{C}_t = \mathcal{C}_{t-1} \cup \mathcal{S}_t$. In this setting, the labels of classes $\mathcal{C}_{t-1}$ are not included in the training set of T_t, with the exception of the background class b. After learning a task T_t, the model is required to correctly discriminate between all the observed classes $\mathcal{C}_t$.

3.1 Causes of Forgetting in Class-Incremental Learning

The fundamental challenge in class-incremental learning is that during optimization of the model on a new T_t, the model is optimized without regard to the previous classes $\mathcal{C}_{t-1}$, which leads to catastrophic forgetting of previous classes. Masana *et al.* [25] stated four causes of forgetting in class-incremental classification.

- **Weight Drift:** During optimization on T_t, the weights of the model that were relevant to the previous task T_{t-1} are updated without regard to the previous task, resulting in drop of performance on task T_{t-1}.
- **Activation Drift:** A change of the weights of the model directly results in a change of internal activations and to the output of the model.
- **Inter-task confusion:** The objective in class-incremental learning is to correctly discriminate between between all the observed classes $\mathcal{C}_t$. However, as the classes are never jointly trained, the learned features are not optimized to discriminate classes from different tasks, as shown in Fig. 1.
- **Task-recency bias:** In the class-incremental setting, the model is optimized to predict new classes without regarding the old classes. This leads to a strong bias for the most recently learned classes. This bias can easily be seen in the confusion matrix, as shown in Fig. 3a. Furthermore, especially in the final classification layer, the weights and biases of the classifier layer have higher magnitudes of the weights vectors for new classes [15]. In CiSS with the addition of the background class, an additional bias to the background class can be observed.

In the following we want to investigate how these causes manifest themselves in CiSS and how existing approaches mitigate these causes.

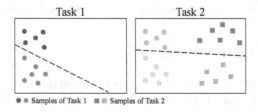

Fig. 1. Visualization of task confusion in class-incremental learning. As classes of Task 1 (circles) and classes of Task 2 (squares) are never trained at the same time, the classifier never learns to discriminate between circles and squares, which causes inter-task confusion.

4   Methods to Measure Forgetting

Purely accuracy based evaluation only allows restrictive insight into the causes
of forgetting of a model. Therefore, in this section, we present the key methods
that we use in our analysis, including layer matching [5] and decoder retraining
accuracy. We utilize these methods to measure the activation shift between a
model f_0 and f_1. The model f_0 is trained on T_0, whereas f_1 is initialized with
the parameters of f_0 and incrementally trained on T_1.

4.1   Layer Matching with Dr. Frankenstein

The Dr. Frankenstein toolset proposed by Csiszárik *et al.* [5] aims to match
the activations of two neural networks on a given layer by joining them with a
stitching layer, compare Fig. 2. The goal of the stitching layer is to transform the
activations of a specific layer of f_0 to the corresponding activations of a model f_1.
The stitching layer is initialized using least-squares-matching and is optimized
using the loss function the network was trained with. In order to measure the
similarity of the learned representations, the accuracy of the resulting Frankenstein Network is evaluated on the test set and compared to the initial accuracy
of the model f_0. The higher the resulting relative accuracy is, the closer the
learned representations of the models are to each other. In our analysis we omit
the stitching layer and directly use the activations of f_0 in f_1, as we noticed in
our experiments that the initial activations are already very similar. We attribute
this to the fact that the models are closely related because f_1 is initialized with
the parameters of f_0. Our setup can be seen on the right side in Sect. 4.1. We
denote the resulting Frankenstein Network when layer n of model f_1 is stitched
to layer $n+1$ of model f_0 as $f_{1,0}^n$, i.e. the network depicted in Sect. 4.1 would
be $f_{1,0}^2$. If the accuracy of the resulting Frankenstein network is not affected,
this is clear evidence that the internal representations of f_1 were not altered
drastically during training on T_1. This analysis will give insight into how much
the activation at a specific layer has changed after training continually, but will
give no insight into a possible positive backward transfer.

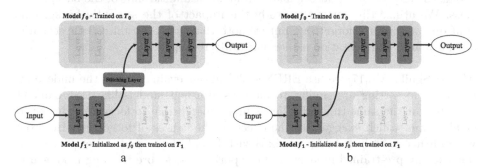

Fig. 2. Comparing a) the original Dr. Frankenstein layer matching approach of using
an additional stitching layer as proposed by [5] and b) our approach for measuring
activation drift in continual learning by directly propagating the activations of f_1 to
f_0 without a stitching layer.

4.2 Decoder Retrain Accuracy

Inspired by use of linear probing for continual learning [6], which measures representational forgetting by calculating the difference in accuracy an optimal linear classifier achieves on an old task before and after introducing a new task, we propose Decoder Retraining Accuracy. To measure the Decoder Retraining Accuracy, we freeze the encoder of the model and retrain the decoder on all classes with the same training configuration and measure the mIoU on a test set. Instead of measuring the representational drift as in [6] this will give a measure of how useful the learned representations of the encoder are to discriminate between classes of different tasks, effectively measuring the contribution of inter-task-confusion.

5 Experiments

Datasets: We conduct our experiments using the PascalVoc-2012 [10] dataset, which contains 20 object classes and a background class. We follow the established CiSS PascalVoc-15-5 split that is widely used [3,8,28,29]. The PascalVOC-15-5 split is a two step incremental learning task, which consists of learning 15 classes (1–15) in the first step T_0 and the remaining 5 classes (16–20) in the second step T_1.[1] We follow the two distinct class-incremental settings for *Disjoint* and *Overlapped* proposed by Cermelli *et al.* [3]. In both settings, only the set of current classes S_t is labelled, while the rest is labelled as background b. However, in the *Disjoint* setting, the images of the current task T_t only contain pixels of classes C_t, meaning that images that contain pixels belonging to classes of future tasks will be discarded in the training set of T_t. In the *Overlapped* setting, pixels can belong to any of the classes, but classes that do not belong to current training set will be labeled as background. Finally, in order to study the impact of the semantic shift that the background class is subjected to in the *Disjoint* and *Overlapped* setting, we introduce the *Fully Disjoint* setting. In this setting each task only contains pixels belonging to the current set of classes, which therefore avoids the interference that originates from the semantic shift of the background class. We utilize this setting to study the impact of the semantic background shift on catastrophic forgetting, but we note that this is an unrealistic scenario, as in semantic segmentation classes naturally re-appear.

Model: Similar to [17] we use ERFNet [36] in our evaluation, as the underlying effects of forgetting are similar to more established models like DeepLabV3+ [4], while at the same time ERFNet is more susceptible to forgetting due to its smaller size, which exaggerates the effect of the causes of forgetting. However, we confirm our findings in Appendix B with DeepLabv3+ and U-Net [37]. We do not use any pre-trained models in our experiments, as pre-training is known to

---

[1] As the focus of this paper is to understand the general causes of forgetting in CiSS, we leave the study of the impact of different splits, more classes and longer task sequences to future work.

increase robustness to catastrophic forgetting [12, 27]. Instead, we use the same randomly initialized weights for every method in our experiments.

Methods Compared: In our evaluation, we focus on evaluating naive fine-tuning approaches, representative regularization and replay methods as these cover the basic concepts that are considered as potential solutions for CiSS. For prior regularization methods we consider EWC [18] and MAS [1]. For data regularization methods we use LwF [23], with the modification of [19], in which the distillation loss is only applied to the parts of the image that are labelled as background. We explicitly do not consider incremental improvements of knowledge distillation based approaches of [3, 8, 24, 29], as our focus is on evaluating the underlying causes of forgetting. For replay we save 20 samples for each class in the buffer. For every experiment we also list results for the *offline* model, which is jointly trained on all classes in one step. Further information regarding the implementation details and selected hyperparameters can be found in the supplementary material.

5.1 Semantic Background Shift and Class Confusion

First, we study the impact of the semantic background shift on forgetting by comparing the results of the selected CiSS methods on the *Overlapped, Disjoint*, and *Full Disjoint* tasks. These tasks have a varying degree of semantic shift of the background class. The results are displayed in Table 1. For the *Overlapped* and *Disjoint* tasks, it can be noted that only LwF and Replay effectively learn to discriminate between all classes. EWC and MAS effectively mitigate forgetting of old classes (0–15) compared to Fine-Tuning, but they also inhibit the learning of new classes (16–20). The reason for the low mIoU for EWC and MAS on all classes can be inferred from the confusion matrix, in which EWC and MAS exhibit a strong bias to the background class and a minor bias towards a few new classes, shown in Figs. 3b and 3c. LwF and Replay reduce both biases. In the *Overlapped* and *Disjoint* setting, LwF and Replay achieve similar performance, as in this setting LwF can effectively replay old classes by discovering them in the background of new images. However, once these classes do not re-appear in the background, as is the case in the *Full Disjoint* task, LwF develops a strong bias towards selected new classes. In contrast to this, Replay benefits from the *Full Disjoint* setting as the training is no longer affected by the semantic background shift. Similarly, MAS and EWC also show significant improvement in this setting, as they benefit from the fact that old classes do not appear as background in the new task, thus not interfering with previously learned knowledge. This is especially noticeable in the confusion matrices of the *Full Disjoint* setting, in which the bias towards the background class is greatly reduced[2]. This demonstrates that the semantic shift of the background class is a significant cause of forgetting for prior regularization methods like EWC and MAS and that it has

---

[2] The confusion matrices are shown in the supplementary material.

a noticeable effect on replay as well. The results also indicate that knowledge distillation is the most effective method to combat this semantic shift.

Finally, upon a closer look at the semantics of the false positives, we see that old classes that are falsely assigned to a new class share semantic and visual properties. In this case *bus* (6), *car* (7), *boat* (5) are assigned to *train* (19), whereas *cow* (10) and *horse* (13) are classified as *sheep* (17). The remaining classes that do not share such a relationship with new classes are falsely classified as background. This confusion can only be alleviated by either Replay or LwF when old classes re-appear in the background in subsequent tasks. To clarify which layers of the model contribute the most to the bias for the background class and newly learned classes, next up we investigate the internal activation drift.

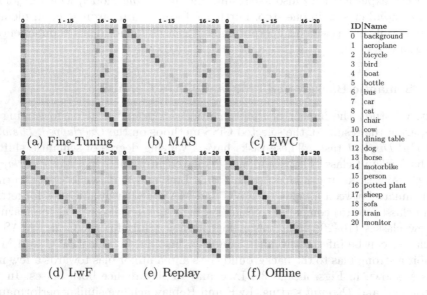

(a) Fine-Tuning   (b) MAS   (c) EWC

(d) LwF   (e) Replay   (f) Offline

ID	Name
0	background
1	aeroplane
2	bicycle
3	bird
4	boat
5	bottle
6	bus
7	car
8	cat
9	chair
10	cow
11	dining table
12	dog
13	horse
14	motorbike
15	person
16	potted plant
17	sheep
18	sofa
19	train
20	monitor

Fig. 3. Confusion matrices after training on PascalVoc-15-5 (disjoint). The confusion matrix for Fine-Tuning a) shows a severe bias to the background class and the classes of the most recent task (16–20). EWC [18] and MAS [1] decrease the bias in exchange for worse accuracy on the most recent classes. Replay and LwF [23] reduce the bias towards new classes and the background.

5.2   Measuring the Effect of Activation Drift

We use the Dr. Frankenstein tool set to measure the activation drift for each layer between the model before and after learning T_1, to investigate which layers are affected the most by the internal activation drift. We follow the setup from Fig. 2 without an additional stitching layer, meaning that the activations of the layer n under examination f_1 are directly propagated to the layer $n+1$ in f_0. The resulting Frankenstein Network $f_{1,0}^n$ is then evaluated on the data of the first

Table 1. Results of semantic segmentation on Pascal-VOC 2012 in Mean IoU (%) on the overlapped, disjoint and full disjoint settings.

PascalVoc 15-5 semantic segmentation									
Method	Overlapped			Disjoint			Full Disjoint		
	0–15	16–20	All	0–15	16–20	All	0–15	16–20	All
Fine-Tuning	4.5	22.2	8.8	4.6	23.0	9.0	5.1	16.3	7.8
MAS [1]	24.1	10.8	21.0	30.6	12.9	26.4	35.6	12.9	30.2
EWC [18]	23.8	11.8	21.0	28.1	10.1	23.8	35.2	10.9	29.4
Replay	41.3	31.5	39.0	42.2	29.1	39.1	48.2	28.8	43.6
LwF [23]	45.8	28.2	41.6	44.4	25.4	39.9	35.9	12.6	30.4
Offline	55.7	47.6	53.8	55.7	47.6	53.8	55.7	47.6	53.8

task (classes 0–15). The mIoU relative to the initial performance on the first task is shown in Fig. 4. Overall, we observe that the activations of the early layers of the network (layers 0–4) stay very stable for every approach and that later layers are disproportionately affected by activation drift, especially the layers of decoder. A similar observation was made in prior work on image classification [6,31]. This confirms that forgetting in CiSS is also mostly affecting deeper layers of the network. Furthermore, the results show that EWC and MAS effectively prevent severe activation drift in the deeper layers of the encoder, dropping only to about 90% of the initial mIoU on the disjoint task, compared to the 30% of the stitched fine-tuning model. This suggests that forgetting for EWC and MAS is less severe as accuracy in Table 1 would reveal. The reason for this could be two-fold: Firstly, the bad accuracy could be attributed to the classifier being biased towards new classes (task-recency bias) or secondly that the regularization methods fail to learn meaningful features that help to discriminate between old and new classes as they are never trained jointly. While a biased classifier is fixed more easily, inter-task confusion is a fundamental shortcoming of prior regularization methods [22].

Another striking phenomenon is the severe change of activations at the third decoder layer (layer 18) that Fine-Tuning, MAS and EWC show on the *Disjoint* task. The predictions of the specific Frankenstein Networks $f_{1,0}^{17}$ and $f_{1,0}^{18}$ in Fig. 5, show that $f_{1,0}^{17}$ is able to correctly classify old classes (*bike, person*) as such, but $f_{1,0}^{18}$ assigns the background class to these regions. Therefore, it can be concluded that the sudden activation change of MAS originates from the fact that features that were evidence for old classes in f_0 are now attributed evidence for the background class. This validates that the semantic shift of the background class is mostly affecting the later layers of the decoder and that the features for old classes are in fact not forgotten, but assigned to the background class. Similar observations can be made for EWC and Fine-Tuning. In Appendix D we measure the similarity with Centered Kernel Alignment to support our observations. When completely avoiding the semantic background shift in the

Full Disjoint task, we observe that activation drift for the fine-tuned model is much more pronounced in the middle layers of the encoder (layer 8–15), which implies that the re-appearance of old classes, even though they are labelled as background, is mitigating the activation drift in the earlier layers of the model.

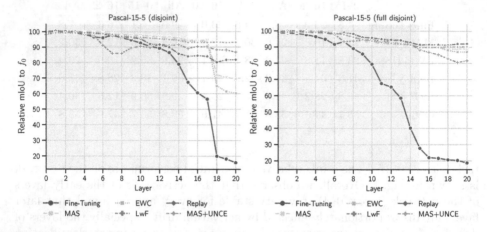

Fig. 4. Activation drift between f_1 to f_0 measured by relative mIoU on the first task of the Frankenstein Networks stitched together at specific layers (horizontal axis). The layers of the encoder are layer 0–15 (grey area), the decoder layers are 17–20 (white area). The activations in the early layers of the encoder stay very stable for all methods, whereas EWC, MAS and Fine-Tuning have a severe drift in activations in the decoder layers of the network, which is clear evidence that forgetting is mostly affecting later layers in the *Disjoint* setting.

5.3 The Impact of Inter-task Confusion on the Encoder

As we observed in Sect. 5.2 that the early layers of a model trained with a continual learning method do not suffer from severe activation drift, in this experiment we investigate how useful the learned features of the encoder of the different methods are to discriminate between all classes. Therefore, we measure Decoder Retrain Accuracy, introduced in Sect. 4.2, for which the decoder of the model is retrained on all classes and subsequently evaluated on the test set. The first observation to be made when looking at the retraining accuracy in Table 3, is that all methods improve after decoder retraining, though EWC, MAS and Fine-Tuning show bigger improvements than LwF and Replay. This again confirms that forgetting in the encoder is not as severe for Fine-Tuning, EWC and MAS as the accuracy indicates. Furthermore, it also verifies that MAS and EWC are effectively preserving important features for old classes in the encoder, but that the biased decoder layer might wrongly attribute important features for old classes to the background class or new classes, which leads to a severe amount of misclassifications.

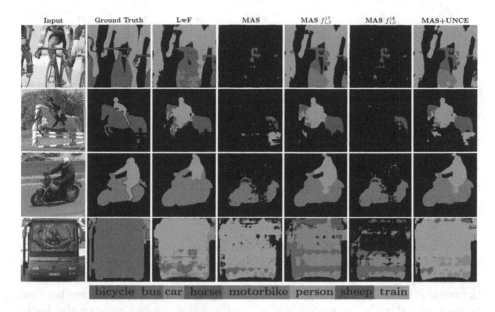

| Input | Ground Truth | LwF | MAS | MAS $f_{1,0}^{17}$ | MAS $f_{1,0}^{18}$ | MAS+UNCE |

bicycle bus car horse motorbike person sheep train

Fig. 5. Visualizations of the segmentation maps for LwF, MAS and the resulting Frankenstein Networks of MAS $f_{1,0}^{17}$ and $f_{1,0}^{18}$. The predictions of $f_{1,0}^{17}$ and $f_{1,0}^{18}$ show that up until layer 17 the information for previously learned classes *person* and *horse* is still available, but is assigned to the *background* in layer 18.

Still, as EWC, MAS and Fine-Tuning do not achieve a comparable mIoU as LwF or Replay after decoder retraining, it can be concluded that the learned features of the encoder are less useful for discriminating between all classes. Specifically the aforementioned related classes *bus* (6), *car* (7), *boat* (5), *train* (19), as well as *cow* (10), *horse* (13), *sheep* (17) cannot be effectively classified after retraining, compare Fig. 6. We hypothesize that Replay does not suffer from inter-task confusion since old classes are taken into account when optimizing for new classes, leading to more discriminative features. The same holds for LwF in the *Overlapped* and *Disjoint* setting, in which old classes are effectively replayed, by using soft-labels for old classes that are discovered in the background.

5.4 Mitigating Background Bias and the Task Recency Bias

A simple method to reduce the recency bias in the classification layer that is used in class-incremental classification is to calculate the cross-entropy loss (CE) only for classes of the current training set [25]. This enforces that errors are only back-propagated for probabilities that are related to the current set of classes:

$$\ell_{\text{ce}}(y, q) = -\frac{1}{I} \sum_{i \in \mathcal{I}} \sum_{c \in \mathcal{C}_t} y_{i,c} \log(q_{i,c}) \tag{1}$$

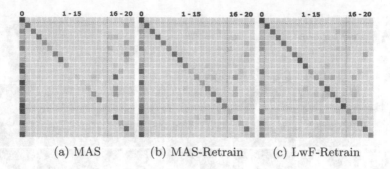

(a) MAS   (b) MAS-Retrain   (c) LwF-Retrain

Fig. 6. Confusion matrices before (a) and after b), c) retraining the decoder on all classes of PascalVoc2012.

However, in the case of CiSS, this addition has proven to be less effective than the standard cross-entropy loss [8]. Therefore, an unbiased cross-entropy loss (UNCE) is proposed in [3], which accounts for the uncertainty of the content of the background class. This is achieved by comparing the pixels that are labelled as background with the probability of having either an old class or the background predicted by the model:

$$\ell_{ce}(y, q) = -\frac{1}{I} \sum_{i \in \mathcal{I}} \sum_{c \in \mathcal{C}} y_{i,c} \log(\hat{q}_{i,c}) \tag{2}$$

$$\hat{q}_{i,c} = \begin{cases} \sum_{k \in \mathcal{C}_{t-1}} q_{i,k} & \text{if } c = b \\ q_{i,c} & \text{otherwise} \end{cases} \tag{3}$$

In addition, Weight Normalization Layers [38] were also successfully used in classification tasks to address the recency bias [21]. In the next experiment we study the impact of UNCE and UNCE combined with Weight Normalization to combat the recency and background bias in CiSS.

The results in Table 2 show that UNCE improves the accuracy for all approaches on the *Disjoint* setting. Specifically, the prior regularization methods MAS and EWC show a significantly higher accuracy compared to the basic cross-entropy loss. This can be attributed to the fact that UNCE effectively mitigates the background bias, as we can see in the confusion matrix in Fig. 3f and the segmentation maps in Fig. 5. In addition, the severe activation drift that we observed in Sect. 5.2 between layer 17 and 18 for MAS completely vanishes with the use of UNCE. Therefore, UNCE effectively resolves the confusion of the old classes with the background class. This confirms the assumption that a major cause of forgetting in Sect. 5.1 was in fact a bias of the classifier towards the background and the new classes. However, the confusion matrices shows that while the background bias is severely reduced by using UNCE, the semantic confusion of old and new classes is amplified.

In the *Full Disjoint* setting the use of UNCE does not improve the performance as much as it does in the *Disjoint* setting. The reason is that in the *Full*

Disjoint setting the pixels of old classes do not re-occur and thus the de-biasing effect of UNCE is decreased. Therefore, the content of the background class plays an important role to mitigate forgetting. Of the selected approaches, only Replay benefits from the addition of the Weight Normalization layer. Finally, we note that EWC and MAS, with the addition of UNCE, show competitive performance to the remaining approaches and more recent approaches like MiB [3], even without the use of knowledge-distillation or replay. However, we hypothesize that for longer task sequence and more classes MiB will outperform prior regularization methods, as they will not be able to learn discriminative features.

Table 2. Results on Pascal-15-5 in mIoU (%) on the disjoint and full-disjoint settings with: Cross-Entropy Loss (CE), Unbiased Cross-Entropy (UNCE) and UNCE combined with a Weight Normalization (UNCE+WN). UNCE effectively reduces forgetting for all approaches, especially for EWC and MAS.

Method	PascalVoc 15-5 (disjoint)									PascalVoc 15-5 (full disjoint)								
	CE			UNCE			UNCE+WN			CE			UNCE			UNCE+WN		
	0–15	16–20	All	0–15	16–20	All	0–15	16–20	All	0–15	16–20	All	0–15	16–20	All	0–15	16–20	All
Fine-Tuning	4.6	23.0	9.0	10.4	21.8	13.1	16.5	21.6	17.7	5.1	16.3	7.8	6.0	15.5	8.3	7.7	15.2	9.5
EWC [18]	28.1	10.1	23.8	48.2	11.6	39.4	17.0	9.5	15.2	35.2	10.9	29.4	41.1	9.8	33.6	34.8	9.7	28.8
MAS [1]	30.6	12.9	26.4	45.8	14.4	38.3	41.0	13.9	34.6	35.6	12.9	30.2	39.1	12.3	32.7	32.5	11.8	27.6
LwF [23]	44.4	25.4	39.9	45.3	22.9	40.0	46.6	19.7	40.2	35.9	12.6	30.4	38.0	13.8	32.2	38.8	13.3	32.8
Replay	42.2	29.1	39.1	47.2	31.4	43.5	48.1	31.9	44.3	48.2	28.8	43.6	47.7	28.0	43.0	48.8	28.5	44.0
MiB [3]	-	-	-	48.6	21.7	42.2	49.4	24.1	43.3	-	-	-	47.6	19.7	41.0	48.6	20.7	42.0

Table 3. Decoder retraining results on Pascal-VOC. mIoU$_I$ and mIoU$_R$ denote the mIoU (%) before and after retraining.

PascalVoc 15-5 - decoder retraining accuracy						
Method	Overlapped		Disjoint		Full Disjoint	
	mIoU$_I$	mIoU$_R$	mIoU$_I$	mIoU$_R$	mIoU$_I$	mIoU$_R$
Fine-Tuning	8.8	28.0	9.0	27.9	7.8	22.0
MAS [1]	21.0	34.3	26.4	36.2	30.2	37.3
EWC [18]	21.0	34.3	23.8	35.1	29.4	36.9
LwF [23]	41.6	45.3	39.9	43.3	30.4	38.1
Replay	39.0	42.6	39.1	42.9	43.6	45.6
Offline	53.8	54.6	53.8	54.6	53.8	54.6

Table 4. Classification results on PascalVoc-15-5.

PascalVoc 15-5 classification			
Method	Full disjoint		
	0–15	16–20	All
Fine-Tuning	13.6	27.6	17.1
MAS [1]	32.0	27.8	31.0
EWC [18]	27.2	25.4	26.8
LwF [23]	39.6	32.7	37.9
Replay	42.1	34.2	40.1
Offline	51.3	54.8	52.2

5.5 The Role of the Background Class to Overcome Forgetting

The prior observations show that in CiSS the semantic shift of the background class is a major cause of a rapid drop in performance if not addressed correctly. However, if the uncertainty of the content of the background class is taken into account by either UNCE, Knowledge Distillation or both, the appearance of old classes in the background can to some extent be used for replay. In the experiments of the *Full Disjoint* setting we see that once classes do not reoccur, these methods are less effective, whereas explicit replay of classes benefits from avoiding the semantic shift. The ranking of the methods of *Full Disjoint* setting in CiSS is also similar to the ranking of the same methods for class-incremental image classification, compare Table 4. This indicates that the discrepancy in performance of LwF and replay in image classification is due to the missing background class. Looking at it the other way around, this could also mean that introducing an out-of-set class for image classification could help to reduce forgetting in the class-incremental setting without requiring explicit replay via stored samples, as the re-appearing classes in the out-of-set-class play a similar role as explicit replay.

6   Conclusion

We studied the major causes of catastrophic forgetting in CiSS, answering how it manifests itself in the hidden representations of the network and how the background class both causes severe forgetting and decreases activation drift. Using representational similarity techniques, we demonstrated that forgetting is concentrated at deeper layers and that re-appearing classes mitigate activation drift in the encoder even when they are labelled as background. Moreover, we show that EWC and MAS are effectively reducing representational drift in the later layers of the encoder, but suffer from severe background and recency bias, which leads to the sudden drop in accuracy. These biases manifest themselves in deeper layers of the networks by assigning previous discriminating features for the previous classes to the background class or visually related classes.

The background bias can be effectively alleviated using an unbiased cross-entropy loss, which leads to a significant improvement for all methods, when classes re-appear in the background of new training data. Finally, we find that only methods that in some form replay old classes during training of new classes can learn to correctly discriminate between all classes after incremental training, as otherwise the model fails to learn to discriminate between new and old classes that share similar visual features. Overall, the results of our work provide a foundation for deeper understanding of the principles of forgetting in CiSS and open the door to future directions to explore methods for its mitigation.

Acknowledgments. The research leading to these results is funded by the German Federal Ministry for Economic Affairs and Climate Action within the project "KI Delta Learning" (Förderkennzeichen 19A19013T). The authors would like to thank the consortium for the successful cooperation.

References

1. Aljundi, R., Babiloni, F., Elhoseiny, M., Rohrbach, M., Tuytelaars, T.: Memory aware synapses: learning what (not) to forget. In: Proceedings of the European Conference on Computer Vision (ECCV) (2018)

2. Cermelli, F., Fontanel, D., Tavera, A., Ciccone, M., Caputo, B.: Incremental learning in semantic segmentation from image labels. In: Proceedings of the IEEE/CVF Conference on Computer Vision and Pattern Recognition (CVPR), pp. 4371–4381 (2022)
3. Cermelli, F., Mancini, M., Bulo, S.R., Ricci, E., Caputo, B.: Modeling the background for incremental learning in semantic segmentation. In: Proceedings of the IEEE Computer Society Conference on Computer Vision and Pattern Recognition, pp. 9230–9239 (2020). https://doi.org/10.1109/CVPR42600.2020.00925. https://arxiv.org/abs/2002.00718
4. Chen, L.C., Zhu, Y., Papandreou, G., Schroff, F., Adam, H.: Encoder-decoder with atrous separable convolution for semantic image segmentation. In: Proceedings of the European Conference on Computer Vision (ECCV) (2018)
5. Csiszárik, A., Kőrösi-Szabó, P., Matszangosz, A., Papp, G., Varga, D.: Similarity and matching of neural network representations. In: Ranzato, M., Beygelzimer, A., Dauphin, Y., Liang, P., Vaughan, J.W. (eds.) Advances in Neural Information Processing Systems, vol. 34, pp. 5656–5668. Curran Associates, Inc. (2021). https://proceedings.neurips.cc/paper/2021/file/2cb274e6ce940f47beb8011d8ecb14 62-Paper.pdf
6. Davari, M., Asadi, N., Mudur, S., Aljundi, R., Belilovsky, E.: Probing representation forgetting in supervised and unsupervised continual learning. In: Proceedings of the IEEE/CVF Conference on Computer Vision and Pattern Recognition (CVPR), pp. 16712–16721 (2022)
7. Delange, M., et al.: A continual learning survey: defying forgetting in classification tasks. IEEE Trans. Pattern Anal. Mach. Intell. **44**(7), 3366–3385 (2021). https://doi.org/10.1109/TPAMI.2021.3057446
8. Douillard, A., Chen, Y., Dapogny, A., Cord, M.: PLOP: learning without forgetting for continual semantic segmentation. In: Proceedings of the IEEE Conference on Computer Vision and Pattern Recognition (CVPR) (2021)
9. Douillard, A., Chen, Y., Dapogny, A., Cord, M.: Tackling catastrophic forgetting and background shift in continual semantic segmentation. CoRR abs/2106.15287 (2021). https://arxiv.org/abs/2106.15287
10. Everingham, M., Van Gool, L., Williams, C.K.I., Winn, J., Zisserman, A.: The PASCAL Visual Object Classes Challenge 2012 (VOC2012) Results (2012). http://www.pascal-network.org/challenges/VOC/voc2012/workshop/index.html
11. French, R.: Catastrophic forgetting in connectionist networks. Trends Cogn. Sci. **3**(4), 128–135 (1999). https://doi.org/10.1016/s1364-6613(99)01294-2
12. Gallardo, J., Hayes, T.L., Kanan, C.: Self-supervised training enhances online continual learning. In: British Machine Vision Conference (BMVC) (2021)
13. Hayes, T.L., Kafle, K., Shrestha, R., Acharya, M., Kanan, C.: REMIND your neural network to prevent catastrophic forgetting. In: Vedaldi, A., Bischof, H., Brox, T., Frahm, J.-M. (eds.) ECCV 2020. LNCS, vol. 12353, pp. 466–483. Springer, Cham (2020). https://doi.org/10.1007/978-3-030-58598-3_28
14. Hinton, G., Vinyals, O., Dean, J.: Distilling the knowledge in a neural network. In: NIPS Deep Learning and Representation Learning Workshop (2015). http://arxiv.org/abs/1503.02531
15. Hou, S., Pan, X., Loy, C.C., Wang, Z., Lin, D.: Learning a unified classifier incrementally via rebalancing. In: The IEEE Conference on Computer Vision and Pattern Recognition (CVPR) (2019)
16. Hsu, Y.C., Liu, Y.C., Ramasamy, A., Kira, Z.: Re-evaluating continual learning scenarios: a categorization and case for strong baselines. In: NeurIPS Continual learning Workshop (2018). https://arxiv.org/abs/1810.12488

17. Kalb, T., Roschani, M., Ruf, M., Beyerer, J.: Continual learning for class- and domain-incremental semantic segmentation. In: 2021 IEEE Intelligent Vehicles Symposium (IV), pp. 1345–1351 (2021). https://doi.org/10.1109/IV48863.2021. 9575493

18. Kirkpatrick, J., et al.: Overcoming catastrophic forgetting in neural networks. Proc. Natl. Acad. Sci. **114**(13), 3521–3526 (2017). https://doi.org/10.1073/pnas. 1611835114. https://www.pnas.org/doi/abs/10.1073/pnas.1611835114

19. Klingner, M., Bär, A., Donn, P., Fingscheidt, T.: Class-incremental learning for semantic segmentation re-using neither old data nor old labels. In: 2020 IEEE 23rd International Conference on Intelligent Transportation Systems (ITSC), pp. 1–8. IEEE (2020)

20. Kornblith, S., Norouzi, M., Lee, H., Hinton, G.: Similarity of neural network representations revisited. In: 36th International Conference on Machine Learning, ICML 2019, vol. 2019-June, pp. 6156–6175 (2019). https://arxiv.org/abs/1905.00414

21. Lesort, T., George, T., Rish, I.: Continual learning in deep networks: an analysis of the last layer (2021). https://doi.org/10.48550/ARXIV.2106.01834. https://arxiv. org/abs/2106.01834

22. Lesort, T., Stoian, A., Filliat, D.: Regularization shortcomings for continual learning (2020)

23. Li, Z., Hoiem, D.: Learning without forgetting. IEEE Trans. Pattern Anal. Mach. Intell. **40**(12), 2935–2947 (2018). https://doi.org/10.1109/TPAMI.2017.2773081

24. Maracani, A., Michieli, U., Toldo, M., Zanuttigh, P.: Recall: replay-based continual learning in semantic segmentation. In: Proceedings of the IEEE/CVF International Conference on Computer Vision (ICCV), pp. 7026–7035 (2021)

25. Masana, M., Liu, X., Twardowski, B., Menta, M., Bagdanov, A.D., van de Weijer, J.: Class-incremental learning: survey and performance evaluation. arXiv preprint arXiv:2010.15277 (2020)

26. McCloskey, M., Cohen, N.J.: Catastrophic interference in connectionist networks: the sequential learning problem. In: Psychology of Learning and Motivation - Advances in Research and Theory, vol. 24, pp. 109–165 (1989). https://doi.org/10. 1016/S0079-7421(08)60536-8

27. Mehta, S.V., Patil, D., Chandar, S., Strubell, E.: An empirical investigation of the role of pre-training in lifelong learning (2021). https://doi.org/10.48550/ARXIV. 2112.09153. https://arxiv.org/abs/2112.09153

28. Michieli, U., Zanuttigh, P.: Incremental learning techniques for semantic segmentation. In: Proceedings - 2019 International Conference on Computer Vision Workshop, ICCVW 2019, pp. 3205–3212 (2019). https://doi.org/10.1109/ICCVW.2019. 00400. http://arxiv.org/abs/1907.13372

29. Michieli, U., Zanuttigh, P.: Continual semantic segmentation via repulsion-attraction of sparse and disentangled latent representations. In: Proceedings of the IEEE/CVF Conference on Computer Vision and Pattern Recognition (CVPR), pp. 1114–1124 (2021)

30. Mirzadeh, S.I., Farajtabar, M., Gorur, D., Pascanu, R., Ghasemzadeh, H.: Linear mode connectivity in multitask and continual learning. In: International Conference on Learning Representations (2021). https://openreview.net/forum?id=Fmg_ fQYUejf

31. Neyshabur, B., Sedghi, H., Zhang, C.: What is being transferred in transfer learning? In: Larochelle, H., Ranzato, M., Hadsell, R., Balcan, M., Lin, H. (eds.) Advances in Neural Information Processing Systems, vol. 33, pp. 512–523. Curran Associates, Inc. (2020). https://proceedings.neurips.cc/paper/2020/file/ 0607f4c705595b911a4f3e7a127b44e0-Paper.pdf

32. Nguyen, C.V., Achille, A., Lam, M., Hassner, T., Mahadevan, V., Soatto, S.: Toward Understanding Catastrophic Forgetting in Continual Learning (2019). http://arxiv.org/abs/1908.01091

33. Ramasesh, V.V., Dyer, E., Raghu, M.: Anatomy of catastrophic forgetting: hidden representations and task semantics. In: International Conference on Learning Representations (2021). https://openreview.net/forum?id=LhY8QdUGSuw

34. Rebuffi, S.A., Kolesnikov, A., Sperl, G., Lampert, C.H.: iCaRL: incremental classifier and representation learning. In: CVPR (2017)

35. Rolnick, D., Ahuja, A., Schwarz, J., Lillicrap, T., Wayne, G.: Experience replay for continual learning. In: Wallach, H., Larochelle, H., Beygelzimer, A., d' Alché-Buc, F., Fox, E., Garnett, R. (eds.) Advances in Neural Information Processing Systems, vol. 32. Curran Associates, Inc. (2019). https://proceedings.neurips.cc/paper/2019/file/fa7cdfad1a5aaf8370ebeda47a1ff1c3-Paper.pdf

36. Romera, E., Álvarez, J.M., Bergasa, L.M., Arroyo, R.: ERFNet: efficient residual factorized convnet for real-time semantic segmentation. IEEE Trans. Intell. Transp. Syst. **19**(1), 263–272 (2018). https://doi.org/10.1109/TITS.2017.2750080

37. Ronneberger, O., et al.: U-net: convolutional networks for biomedical image segmentation (2015). https://doi.org/10.48550/ARXIV.1505.04597. https://arxiv.org/abs/1505.04597

38. Salimans, T., Kingma, D.P.: Weight normalization: a simple reparameterization to accelerate training of deep neural networks. In: Lee, D., Sugiyama, M., Luxburg, U., Guyon, I., Garnett, R. (eds.) Advances in Neural Information Processing Systems, vol. 29. Curran Associates, Inc. (2016). https://proceedings.neurips.cc/paper/2016/file/ed265bc903a5a097f61d3ec064d96d2e-Paper.pdf

39. Shin, H., Lee, J.K., Kim, J., Kim, J.: Continual learning with deep generative replay. In: Advances in Neural Information Processing Systems, vol. 30 (2017)

40. Tasar, O., Tarabalka, Y., Alliez, P.: Incremental learning for semantic segmentation of large-scale remote sensing data. IEEE J. Sel. Top. Appl. Earth Obs. Remote Sens. **12**(9), 3524–3537 (2019). https://doi.org/10.1109/JSTARS.2019.2925416

41. van de Ven, G.M., Siegelmann, H.T., Tolias, A.S.: Brain-inspired replay for continual learning with artificial neural networks. Nat. Commun. **11**(1), 4069 (2020). https://doi.org/10.1038/s41467-020-17866-2

42. Yu, L., Liu, X., van de Weijer, J.: Self-training for class-incremental semantic segmentation. IEEE Trans. Neural Netw. Learn. Syst. 1–12 (2022). https://doi.org/10.1109/TNNLS.2022.3155746

43. Zenke, F., Poole, B., Ganguli, S.: Continual learning through synaptic intelligence. In: Precup, D., Teh, Y.W. (eds.) Proceedings of the 34th International Conference on Machine Learning. Proceedings of Machine Learning Research, vol. 70, pp. 3987–3995. PMLR, International Convention Centre, Sydney, Australia (2017). http://proceedings.mlr.press/v70/zenke17a.html

Application of Multi-modal Fusion Attention Mechanism in Semantic Segmentation

Yunlong Liu[1]([⊠]) (ID), Osamu Yoshie[2] (ID), and Hiroshi Watanabe[1] (ID)

[1] Graduate School of Fundamental Science and Engineering, Waseda University, Tokyo, Japan
liuyunlong@akane.waseda.jp, hiroshi.watanabe@waseda.jp
[2] Graduate School of Information, Production and Systems, Waseda University, Kitakyushu, Japan
yoshie@waseda.jp

Abstract. The difficulty of semantic segmentation in computer vision has been reintroduced as a topic of interest for researchers thanks to the advancement of deep learning algorithms. This research aims into the logic of multi-modal semantic segmentation on images with two different modalities of RGB and Depth, which employs RGB-D images as input. For cross-modal calibration and fusion, this research presents a novel FFCA Module. It can achieve the goal of enhancing segmentation results by acquiring complementing information from several modalities. This module is plug-and-play compatible and can be used with existing neural networks. A multi-modal semantic segmentation network named FFCANet has been designed to test the validity, with a dual-branch encoder structure and a global context module developed using the classic combination of ResNet and DeepLabV3+ backbone. Compared with the baseline, the model used in this research has drastically improved the accuracy of the semantic segmentation task.

1 Introduction

Semantic segmentation has been a fundamental and critical problem in computer vision [1] for the past long time. The advent of neural networks significantly improved semantic segmentation performance [2]. CNN based on the Encoder-Decoder structures [3] has become the mainstream, and its accuracy and efficiency have greatly exceeded other methods.

However, semantic segmentation in indoor scenes [4] is still a challenging task, because its semantic information is more complex than outdoor scene [5]. Multi-modal input is a feasible solution for indoor scene [6]. The RGB-D method is becoming increasingly popular among multi-modal segmentation systems since it may gather spatial information and scene structure coding.

Previously, the biggest obstacle to RGB-D solutions was that depth images were difficult to obtain. However, with the proliferation of depth cameras [7],

L. Wang et al. (Eds.): ACCV 2022, LNCS 13847, pp. 378–397, 2023.
https://doi.org/10.1007/978-3-031-26293-7_23

collecting depth photos is no longer expensive and complicated. The popularity of civilian depth cameras [8], such as the Microsoft Kinect sensor [6], has made acquiring depth photos much easier and cheaper. Consequently, semantic segmentation datasets based on RGB-D images have started to arise sequentially [9]. RGB-D-based semantic segmentation is becoming more and more popular and common.

Further modalities bring additional spatial information, but fusing these two modalities becomes new challenge [10]. Calculating HHA [11] is a reasonable way to enhance depth in preprocessing phase for extracting features better. HHA includes three channels: horizontal disparity, height above ground, and the angle the pixel's local surface normal makes with the inferred gravity direction [12].

In the deep-learning-based RGB-D segmentation methods, to extract features from RGB and depth modalities separately, it is common to design a dual branching network structure [13]. The depth features are gradually fused into the feature map of RGB when the network goes deeper. Adding features with depth information to RGB can significantly improve the accuracy of segmentation [14]. Especially in the case of using HHA, since the HHA image has three channels, the same structure can be used to extract RGB and depth information.

In this paper, We innovatively propose a multi-modal feature fusion attention mechanism for cross-modal calibration, which calibrates the information of different modalities in both feature channels and spatial dimensions through two different attention mechanisms. The unique feature fusion module enables the network to better capture the complementary information among different modalities based on the standard features, thus improving the segmentation results.

According to its characteristics, this module is named the feature fusion cascade attention mechanism module, abbreviated as the FFCA module. The method chapter will introduce this module, including multi-modal attention and feature fusion, and show the semantic segmentation model based on this FFCA module: FFCANet.

2   Related Work

2.1   Semantic Segmentation

In computer vision, the first application of CNN is image classification [15]. Consequently, other directions also emerged with deep learning-based approaches, such as detection, generation, and segmentation. Similar to the concept of "encoding" in NLP [16], the Neural Network for semantic segmentation task usually consists of two parts: The part that extracts the feature map is called the encoder, and the part that obtains the segmentation result from the feature map is called the decoder.

Encoder is used to extract the feature map though cropping or compressing the input image, they generally have similar structure with the backbone used for classification task [17], such as AlexNet [18] and so on [19–23]. These networks focus on how to extract features from the input image more efficiently, then have gradually developed several different styles. Some networks choose to utilize convolutional kernels of different sizes to obtain multi-scale contextual features [19,24–26]; others desire to make more efficient use of the information available in the image while reducing the computational overhead DenseNet [27–29]. The most popular network is the ResNet families [21]. It adds identity connections to solve the degradation problem in deep models. This scheme the most common encoder for semantic segmentation tasks.

The word decoder first appeared in SegNet [3] through a series of deconvolution [30] or up-sampling operations. The decoder restores the feature map to the original input size to realize the pixel-by-pixel classification of the input image. Compared with the traditional CNN structure, these decoders generally have a different convolutional structure [31–33]. These kernels use multiple sizes of the kernel to capture features at different scales to solve the multi-scale problem [34–36]. A famous example is DeepLab series [37–40]. This structure is widely used in various tasks related to semantic segmentation. The most advanced DeepLabv3+ [40] in the series employs atrous convolution and spatial pyramidal pooling module and simultaneously improves speed and accuracy.

Another unique structure is the global context module. It can incorporate global information into the feature map. Before the emergence of CNN, context modules are already present in the traditional approaches [41,42]. Subsequently, this concept was introduced to deep learning [43–46]. A more general approach is to insert the context module between the encoder and the decoder. The most typical example of this structure is EncNet [47]. By introducing the context encoding module, EncNet is capable of processing global contextual information.

2.2   RGB-D Segmentation

As a kind of multi-modal approach [48], using RGB-D images as input are particularly common in the segmentation task of indoor scenes. Since RGB-D photographs contain spatial information missing from 2D images, they can reveal more relevant elements in indoor situations unrelated to lighting. Couprie et al. [14] found that using depth information can significantly improve the distinguishability of objects with a similar appearance.

Researchers discovered the effectiveness of deep information quite early, even before deep learning emerged. The traditional method of RGB-D segmentation can be considered as a relatively fixed pipeline mode [49]. Similar to traditional 2D image segmentation, pre-segmentation segments the original RGB-D image into more basic units, such as superpixels [50], blocks [51] and regions [52], then these units will be classified by traditional statistical [53–55] or machine learning methods [12,56]. The earliest attempt came from Silberman et al. [5]. They created the NYUv1 data set, which was eventually expanded to NYUv2 [4]. This approach based on Conditional Random Field (CRF) and has inspired many subsequent researches [57–59], even in some early models of deep learning also containing CRF-like structures [37,38,60].

The subsequent advent of deep learning likewise brought significant developments to RGB-D segmentation. These neural networks usually use two encoder branches to extract RGB and depth features separately. This structure was first seen in FuseNet [13]. The focus of this type of research is how to integrate RGB and depth information. Common strategies include Early Fusion [14], Middle Fusion [30] or Late Fusion [61]. These early approaches inspired later research, such as RDFNet [62] which extends RefineNet [63] with a multi-modal fusion block. LSTM-CF [64] uses a horizontal LSTM to capture RGB and depth features, then introduces a vertical LSTM to combine them. Liu et al. improved the HHA [11] encoding by integrating 2D and 3D information [61]. They also extended the VGG [14] encoder in DeepLab [37] for RGB-D semantic segmentation. Cheng et al. proposed a gated fusion method [30], which studied the paired relationship between adjacent RGB-D pixels. MTI-Net [65] discussed the importance of considering task interaction on multiple scales when extracting task information in a multi-task learning setting. ICM [66] is an ensemble classification model that proposes regularization based on variance. ESANet [67] has an encoder with two branches and uses the attention mechanism to fuse depth into the RGB encoder in several stages. ShapeConv [68] introduces a shape-aware convolution layer to process depth features.

The method used in this paper is to combine the mid-term fusion of the attention mechanism, use the attention mechanism to calibrate the features from different modalities, and fuse the features of each stage of the encoder through element-by-element addition.

3 Method

After a detailed discussion of the strengths and weaknesses of best practices for RGB-D semantic segmentation, we propose a new method for RGB-D semantic segmentation, inspired by attention mechanisms like CBAM [69] and SKNet [70]. The crucial part of this method is a Feature Fused Cascade Attention Module (FFCA Module).

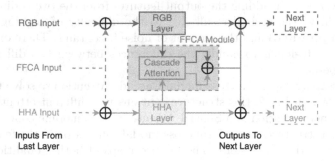

Fig. 1. FFCA module

Considering that each down-sampling occurs at the beginning of the hidden layers, the output of each hidden layer is the feature map with the highest level of abstraction. Therefore, feature fusion should occur after each hidden layer. Figure 1 shows the insertion method and internal structure of the FFCA Module proposed in this study. Different features from two different modalities

are converged into the RGB and HHA branches in an element-by-element summation after a cross-modal calibration based on the Attention mechanism to
achieve the fusion of features. HHA [11] is a generalization of Depth images that
makes it easier to apply CNN algorithms to RGB-D data, which is a considerable
improvement over depth channels alone.

3.1   FFCA Module: Feature Fused Cascade Attention Module

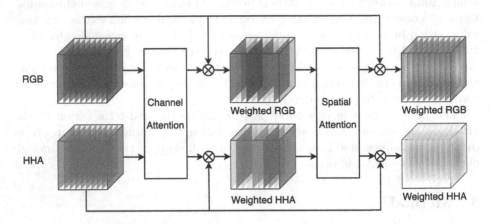

Fig. 2. Cascaded channel and spatial attention

The extra spatial information contained in RGB-D can compensate well for the
lack of RGB compared to pure RGB's traditional 2D image semantic segmentation. However, simply adding the output features from the two coding branches
may not achieve the desired result due to the difficulty of aligning the depth
information with RGB and the amount of noise it contains. Therefore, the critical of feature fusion is to handle the differences between two different image
signals properly.

This research proposes a cross-modal Cascade Attention to solve these problems. As shown in Fig. 2, this structure contains two different attention mechanisms: Channel Attention and Spatial Attention. This module concatenates the
two attention structures to perform cross-modal calibration in the feature maps'
channel and spatial dimensions. The feature maps of both modalities have the
same size. The calibration assigns a pair of weights for elements at the same
position in two modalities to facilitate subsequent feature fusion by element-by-
element addition.

3.2 Multiple Layer Channel Attention

The structure of multiple layer channel attention in cascaded attention shown in Fig. 3. This multilayer Channel Attention can be regarded as a modified solution of the multi-input attention mechanism in SKNet [70]. The difference is that the two feature maps come from two different modalities.

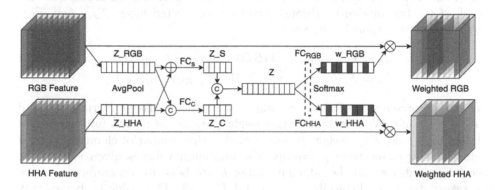

Fig. 3. Multiple layer channel attention

This channel attention has two layers: the first concatenates and sums the feature vectors, thus extracting the separated and fused features of the two modalities. The second concatenates these two features, then uses two fully connected layers and softmax to obtain the RGB and HHA attention vectors. Taking any RGB feature map $RGB \in \mathbb{R}^{H \times W \times C}$, the HHA feature map $HHA \in \mathbb{R}^{H \times W \times C}$ as input, the operation of this Channel Attention can be described in the following mathematical language:

Global Averaging Pooling: Like the Squeeze operation in SENet [71], the RGB and HHA feature maps need to undergo a global averaging pooling $\mathcal{F}_{gp}$ after being fed into the Attention module to obtain the feature vectors $z_{RGB} \in \mathbb{R}^C$ and $z_{HHA} \in \mathbb{R}^C$, enabling them to be fed into the subsequent fully connected layer. Specifically, the c-th element of both vectors is computed utilizing the c-th channel of shape H × W in the corresponding feature map:

$$z_{RGB_c} = \mathcal{F}_{gp}(RGB_c) = \frac{1}{H \times W} \sum_{i=1}^{H} \sum_{j=1}^{w} RGB_c(i,j) \tag{1}$$

$$z_{HHA_c} = \mathcal{F}_{gp}(HHA_c) = \frac{1}{H \times W} \sum_{i=1}^{H} \sum_{j=1}^{w} HHA_c(i,j) \tag{2}$$

Multi-layered Full-Connection: The role of the two full-connection layers in the Attention block is to produce a Scale vector z that fuses two different modalities. to be able to take advantage of the common features of both modalities while calibrating across modalities, the inputs to the first full-connection layer are $z_{RGB} \in \mathbb{R}^C$ and $z_{HHA} \in \mathbb{R}^C$ summing element by element the fusion vector at $z_{sum} \in \mathbb{R}^C$, and $z_{concat} \in \mathbb{R}^{2C}$ obtained by concatenating the two. The first layer extracts the fusion feature $z_s \in \mathbb{R}^d$ and the separation feature $z_c \in \mathbb{R}^d$ for each of the two modalities through two fully connected layers $\mathcal{F}_{\mathcal{FC}_s}$ and $\mathcal{F}_{\mathcal{FC}_c}$ which are at the same level:

$$z_s = \mathcal{F}_{\mathcal{FC}_s}(z_{sum}) = \delta\left(\mathcal{B}\left(W_s \times (z_{RGB} \oplus z_{HHA})\right)\right) \tag{3}$$

$$z_s = \mathcal{F}_{\mathcal{FC}_s}(z_{sum}) = \delta\left(\mathcal{B}\left(W_s \times (z_{RGB} \| z_{HHA})\right)\right) \tag{4}$$

where δ represents the ReLU activation layer, $\mathcal{B}$ is the Batch Norm layer, and $W_s \in \mathbb{R}^{d \times c}$ and $W_c \in \mathbb{R}^{d \times 2C}$ are the weight parameters for the fully connected layers $\mathcal{F}_{\mathcal{FC}_s}$ and $\mathcal{F}_{\mathcal{FC}_c}$ weight parameters. d is the number of channels reduced after a fully connected squeeze and the minimum value is given via L. The shrinkage ratio r and the minimum value L are both hyperparameters of the network structure. Typically, $r = 16$ and $L = 32$. This value is dynamically adjusted by itself in a similar way to that in SENet and proportion to a certain range:

$$d = \max(C/r, L) \tag{5}$$

The second layer of full concatenation is equivalent to the excitation operation in SENet and is used to obtain the weights of both RGB and HHA modalities in the channel dimension. The fused features z_s and separated features z_c in the first layer are stitched together into a feature vector $z \in \mathbb{R}^{2d}$ of twice the length, which contains both fused and separated features for both RGB and HHA modalities, making the fully connected layer $\mathcal{F}_{\mathcal{FC}_{RGB}}$, which is located in the second layer for the two different modalities, and $\mathcal{F}_{\mathcal{FC}_{HHA}}$ can extract the required feature weights for each. Similar to the first layer, the weights of the two full connections are $W_{RGB} \in \mathbb{R}^{C \times 2d}$ and $W_{HHA} \in \mathbb{R}^{C \times 2d}$ respectively.

$$w_{RGB} = \mathcal{F}_{\mathcal{FC}_{RGB}}(z) = W_{RGB} \times (z_S \| z_c) \tag{6}$$

$$w_{HHA} = \mathcal{F}_{\mathcal{FC}_{HHA}}(z) = W_{HHA} \times (z_s \| z_c) \tag{7}$$

Cross-modal Softmax Normalization: Multi-layer full connectivity has filtered out those feature channels from RGB and HHA that are more useful for subsequent segmentation tasks and given them higher weights. However, some of the corresponding channels of the two feature maps may be redundant or contain some information that would interfere with each other. For the subsequent feature fusion to proceed smoothly, using softmax to calibrate the feature weights jointly w_{RGB} and w_{HHA} is necessary.

$$w_{RGB}(\text{Calibrated}) = \mathcal{F}_{\text{Softmax}}(w_{RGB} \| w_{HHA})_{\dim=1}[w_{RGB}] \tag{8}$$

$$w_{RGB}(\text{Calibrated}) = \mathcal{F}_{\text{Softmax}}(w_{RGB} \| w_{HHA})_{\dim=1}[w_{HHA}] \tag{9}$$

In this process, the weight vectors w_{RGB} and w_{HHA} are no longer spliced into longer vectors, but adding a matrix of dimension $W_{RGB\|HHA} \in R^{2 \times C}$, the normalization of Softmax is performed in this new extended dimension. This normalization allows the weights of the feature channels at the corresponding positions of the two modalities to always sum to 1, enabling the maximum exploitation of the complementary features of the different modalities on the channels.

3.3   Fusion Spatial Attention

This research addresses RGB-D feature fusion problems by introducing a Fusion Spatial Attention module. Compared with the general single-mode space Attention mechanism, the multi-modal space fusion attention includes inputs from two different modalities in the final convolution process. Figure 4 shows this structure. In addition to splicing the pooled single-channel features, the spatial feature also contains an additional mixing channel.

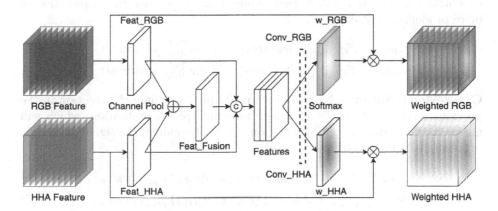

Fig. 4. Fusion spatial attention

Taking as input an arbitrary RGB feature map $RGB \in \mathbb{R}^{H \times W \times C}$ and an HHA feature map $HHA \in \mathbb{R}^{H \times W \times C}$, the following mathematical language will describe this Spatial Attention operation:

Channel Averaging Pooling: channel pooling $\mathcal{F}_{cp}$ is a compression of the feature map from the channel dimension. A feature map of dimension $H \times W \times C$ will be compressed to $H \times W$, keeping only one channel. The value of pixel (i.j) at any position in this single-channel feature map is the mean value of the pixels at the corresponding position for all channels in the original feature map:

$$\text{feat}_{RGB}(i,j) = \mathcal{F}_{cp}(RGB) = \frac{1}{C} \sum_{c=1}^{C} RGB_c(i,j) \tag{10}$$

$$\text{feat}_{HHA}(i,j) = \mathcal{F}_{cp}(HHA) = \frac{1}{C} \sum_{c=1}^{C} HHA_c(i,j) \tag{11}$$

Fusion Channels: Spatial attention generally contains two pooling operations, to obtain RGB and depth feature channel. The two feature channel are concatenated into a two-channel feature map to provide redundant information.

On this basis, a hybrid channel is also innovatively introduced in this study to obtain the fusion information between the two modalities. The spatial features from two modalities are summed pixel-by-pixel to let subsequent convolution exploit the complement information better:

$$\text{feat}_{\text{Fusion}} = \text{feat}_{\text{RGB}} \oplus \text{feat}_{\text{HHA}} \tag{12}$$

$$\text{feature} = \text{feat}_{\text{RGB}} \| \text{feat}_{\text{Fusion}} \| \text{feat}_{\text{HHA}} \tag{13}$$

Spatial Weights: This Attention module uses two convolutional layers of the same size $\mathcal{F}_{Conv_{RGB}}$ and $\mathcal{F}_{Conv_{HHA}}$ to generate the Spatial Attention weights for RGB and HHA respectively. The Sigmoid activation function is discarded here and used for subsequent cross-modal calibration. The size of both sets of convolution kernels is 7×7, cause the larger size allows for the aggregation of more prodomain features:

$$w_{\text{RGB}} = \mathcal{F}_{Conv_{RGB}} \text{ (features)} = \text{Conv}_{\text{RGB}}^{7\times7} \text{ (features)} \tag{14}$$

$$w_{\text{HHA}} = \mathcal{F}_{C_{\sigma n H_{HHA}}} \text{ (features)} = \text{Conv}_{\text{HHA}}^{7\times7} \text{ (features)} \tag{15}$$

Cross-Modal Softmax Normalization: Similar to Channel Attention in the previous section, Softmax is used here for joint spatial calibration, which will normalize the matrix of spatial attention. This is to address the signal alignment problem of RGB and HHA:

$$w_{RGB(\text{Calibrated})} = \mathcal{F}_{\text{Softmax}} (w_{RGB} \| w_{HHA})_{\dim=1} [w_{RGB}] \tag{16}$$

$$w_{RGB(\text{Calibrated})} = \mathcal{F}_{\text{Softmax}} (w_{RGB} \| w_{HHA})_{\dim=1} [w_{HHA}] \tag{17}$$

Since w_{RGB} and w_{HHA} are no longer feature vectors but one-dimensional feature maps, the stitched $w_{RGB\|HHA} \in R^{2\times W \times H}$ will have a three-dimensional shape. softmax normalization is still performed in this new extended dimension of length 2. This step allows the weights of the two sets of feature maps corresponding to spatial locations to always sum to 1, which can compensate for the different responses of the RGB and depth maps at the edges of the object and promote a better alignment of the two signals.

3.4   FFCANet: Feature Fused Cascade Attention Network

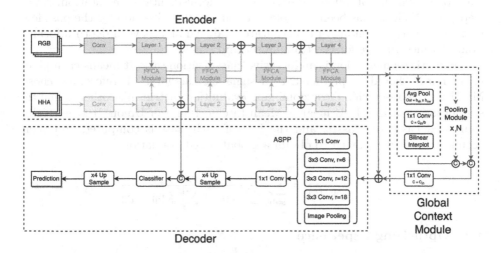

Fig. 5. Network structure of FFCANet

The FFCA Module is a plug-and-play cross-modal calibration and feature fusion module based on the Attention mechanism. Therefore, it also requires a network structure to host the module for the semantic segmentation task. We have built a network structure for the semantic segmentation task by modifying existing network components based on existing research. This network efficiently combines with the FFCA Module. We named it Feature Fused Cascade Attention Network, or FFCANet for short.

The structure of the network in this study shown in Fig. 5. The overall structure of the network consists of an encoder, context module, and decoder. The encoder part is chosen from ResNet [21], which is most commonly used in semantic segmentation tasks and is extended into two branches connected by FFCA Module. Context module is similar to pyramid pooling in PSPNet [34], refers to Seichter et al.'s scheme used in ESANet [72] and modified their approach. Since the two-branch structure of the encoder doubles the network parameters, DeepLabV3+ [40] with a smaller number of parameters was chosen for the decoder to balance the accuracy of the network with the memory overhead.

4   Experiment

To verify the validity of the innovative work made in this study, the NYUv2 [4] dataset was used as the benchmark for testing. Subsequent ablation experiments will also be conducted on this dataset.

4.1   Dataset and Metrics

Due to the scarcity of indoor RGBD datasets for semantic segmentation, NYU Depth v2 (NYUv2) has been the gold standard in this direction for the past few years. The dataset contains 1449 accurately labeled images with depth information, of which 795 are for the training set and 654 for the test set.

Semantic segmentation is an intensive classification task. It means each pixel in an image should be predicted to a semantic category. Therefore, we chose MIoU as this research's most dominant evaluation metric, like other semantic segmentation tasks. MIoU mean is Mean Intersection over Union. It is generally computed based on classes, and the IoU of each class is computed and then accumulated and averaged to obtain a global-based evaluation.

$$\text{MIoU} = \frac{1}{k+1} = \sum_{i=0}^{k} \frac{p_{ii}}{\sum_{j=0}^{k} p_{ij} + \sum_{j=0}^{k} p_{ji} - p_{ii}} \tag{18}$$

4.2   Optimizing Experience

This experiment aims to investigate the best way to use the optimal FFCA Module with the contextual module. We conduct experiments on three Backbone with two FFCA Module combining strategies: ResNet50, ResNet101, and ResNet152. The results are shown in Table 1.

Table 1. Result of optimize experience

Backbone	FFCAM	Context	P Acc	M Acc	FW Acc	MIoU
ResNet50	5	ppm-1357	76.80	62.01	63.57	50.11
	5	ppm-15	76.52	62.43	63.84	50.46
	4	ppm-1357	77.16	62.67	63.91	51.08
	4	ppm-15	76.96	62.71	63.90	51.19
ResNet101	5	ppm-1357	77.47	63.03	64.02	51.54
	5	ppm-15	77.78	62.98	64.22	51.81
	4	ppm-1357	77.92	63.28	64.77	52.32
	4	ppm-15	78.13	63.14	64.86	52.58
ResNet152	5	ppm-1357	77.32	63.87	64.89	52.53
	5	ppm-15	77.81	63.92	65.01	52.59
	4	ppm-1357	78.01	64.71	65.08	53.09
	4	ppm-15	78.39	65.31	65.72	53.30

We plug FFCA Module after each hidden layer for feature fusion. In addition to using four modules, another fusion strategy uses five modules, which the FFCA Module also inserted after the initial first convolutional layer. Two different resolution combinations have been experimented with for the contextual modules to find the best combination of pyramidal pooling sizes. One was a two-way pooling branch of 1×1 and 5×3; the other was a four-way pooling branch of 1×1, 3×2, 5×3, and 7×5.

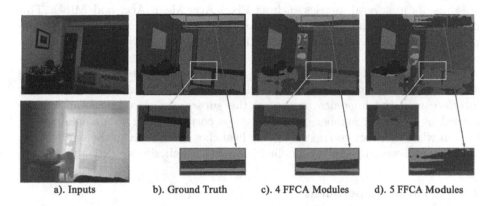

a). Inputs   b). Ground Truth   c). 4 FFCA Modules   d). 5 FFCA Modules

Fig. 6. Local edge detail of the result (Color figure online)

The encoder structure using four FFCA Modules is optimal in the optimizing experiment, with the best calibration and fusion of features between different modes. Figure 6 shows this visualized result. The network structure using 4 FFCA Module performs better at the edges of different objects, as shown in the yellow bordered area. They contain less inter-adhesion in the transition region. In contrast, at the locations marked by the red borders, the 5 FFCA Module structure classification results show many broken edge features, indicating that the RGB and depth signals are not well aligned.

Table 2. Comparison result

Method	P Acc	M Acc	MIoU
FCN [17]	65.4	46.1	34.0
CRF-RNN [53]	66.3	48.9	35.4
DeepLab [40]	68.7	46.9	36.8
ACNet [46]	–	–	48.3
MTI-Net [65]	75.3	62.9	49.0
RDFNet [62]	76	62.8	50.1
ESANet [67]	–	–	50.5
ICM [66]	75.4	–	50.7
CANet [72]	76.6	63.8	51.2
ShapeConv [68]	75.8	62.8	51.3
NANet [73]	77.9	–	52.3
SA-Gate [74]	77.9	–	52.4
FFCANet	**78.4**	**65.3**	**53.3**

After determining the network structure, we compared the performance of the best version of FFCANet with similar other work on the publicly available NYUv2 dataset. Table 2 shows the results. Notably, this work achieves remarkable results in several metrics such as Pixel Acc, Mean Acc, and MIoU. This result indicates the advantage of this cross-modal calibration and fusion mechanism in dealing with complex indoor environments with depth images containing noise.

For the pyramid pooling context, using too many combinations of pooling at different resolutions does not boost the network's accuracy. However, it may interfere with the inference process of the subsequent decoder. That may be caused by the small number of feature maps corresponding to a single pooling branch when there are too many pooling branches. Therefore, in the final version of the model, we only use two pooling branches. Their size is 1×1 and 5×3.

4.3   Ablation Experiment

Table 3. Result of ablation experience

Backbone	Encoder	Context	P Acc	M Acc	FW Acc	MIoU
ResNet50	RGB-D	No	75.88	61.43	62.33	49.49
	RGB-D	Yes	76.03	61.79	62.54	49.72
	FFCAM	No	76.34	62.28	63.58	50.74
	FFCAM	**Yes**	**76.96**	**62.71**	**63.9**	**51.19**
ResNet101	RGB-D	No	77.02	62.80	63.82	51.17
	RGB-D	Yes	77.38	62.98	64.52	51.54
	FFCAM	No	77.92	62.85	64.23	52.11
	FFCAM	**Yes**	**78.13**	**63.14**	**64.86**	**52.58**
ResNet152	RGB-D	No	77.52	62.99	63.93	51.54
	RGB-D	Yes	77.58	63.14	64.31	51.87
	FFCAM	No	77.87	64.43	65.27	53.03
	FFCAM	**Yes**	**78.40**	**65.31**	**65.72**	**53.30**

The results of the ablation experiments for the network structure are shown in Table 3, demonstrating the validity of the novel structure of the FFCA Module. A plain RGB-D two-branch segmentation network has been used as the baseline, which removed the FFCA Module between two encoder branches and used a simple element-wise adding instead. The introduction of the global context module also impacts the results, so the global context module is also a variable in the ablation experiments.

This result revealed that the performance of the network is significantly affected by the FFCA Module. When the network uses the FFCA Module as the feature fusion mechanism, there is a significant improvement in the accuracy of the model. Depending on the backbone, this difference can reach approximately 1.4% MIoU. The introduction of global pyramid pooling also contributes a slight accuracy improvement to the model, with a maximum difference of only approximately 0.3%, which is not as significant as the improvement of the FFCA Module.

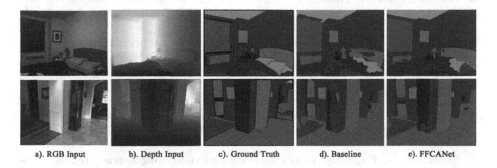

a). RGB Input   b). Depth Input   c). Ground Truth   d). Baseline   e). FFCANet

Fig. 7. Visualisation result compared with baseline

The FFCANet with the FFCA Module has obtained better segmentation results than Baseline shown in Fig. 7. As seen from the figure, the segmentation results of FFCANet have fewer category errors, more accurate object edges, and almost no shape breaking. It is due to the FFCA Module's ability to calibrate across modalities and its Squeeze-and-Excitation feature in Channel Attention. This feature allows the module to suppress defects and noise in the depth image very well, acquiring depth information while reducing the interference of harmful parts in the final segmentation result.

5   Conclusion

In this paper, we propose a neural network called FFCANet for accurately executing RGB-D semantic segmentation tasks. We have built a network structure for the semantic segmentation task by modifying the existing ResNet. This module can achieve cross-modal calibration of RGB information with depth information and fuse complementary information. Our experiments show that this ability has made FFCANet get the performance improvement in RGB-D semantic segmentation task.

As the novel structure, the role of FFCA Module is to incorporates two different modalities. This attention module is designed to be plug-and-play, can be combined with any other RGB-D semantic segmentation network have double-branch encoder structure without increasing the burden of calculation. Compared with the baseline in ablation experiment, the model used in this research has obviously improved the accuracy of the semantic segmentation task.

References

1. Thoma, M.: A survey of semantic segmentation. arXiv preprint arXiv:1602.06541 (2016)
2. Yuan, X., Shi, J., Gu, L.: A review of deep learning methods for semantic segmentation of remote sensing imagery. Expert Syst. Appl. **169**, 114417 (2021)
3. Badrinarayanan, V., Kendall, A., Cipolla, R.: Segnet: a deep convolutional encoder-decoder architecture for image segmentation. IEEE Trans. Pattern Anal. Mach. Intell. **39**(12), 2481–2495 (2017)
4. Silberman, N., Hoiem, D., Kohli, P., Fergus, R.: Indoor segmentation and support inference from RGBD images. In: Fitzgibbon, A., Lazebnik, S., Perona, P., Sato, Y., Schmid, C. (eds.) ECCV 2012. LNCS, vol. 7576, pp. 746–760. Springer, Heidelberg (2012). https://doi.org/10.1007/978-3-642-33715-4_54
5. Quattoni, A., Torralba, A.: Recognizing indoor scenes. In: 2009 IEEE Computer Society Conference on Computer Vision and Pattern Recognition (CVPR 2009), 20–25 June 2009, Miami, Florida, USA. pp. 413–420. IEEE Computer Society (2009)
6. Silberman, N., Fergus, R.: Indoor scene segmentation using a structured light sensor. In: IEEE International Conference on Computer Vision Workshops, ICCV 2011 Workshops, Barcelona, Spain, November 6–13, 2011, pp. 601–608. IEEE Computer Society (2011)
7. Webb, J., Ashley, J.: Depth Image Processing, pp. 49–83. Apress, Berkeley (2012)
8. Cai, Z., Han, J., Liu, L., Shao, L.: RGB-D datasets using microsoft kinect or similar sensors: a survey. Multim. Tools Appl. **76**(3), 4313–4355 (2017)
9. Firman, M.: RGBD datasets: Past, present and future. In: 2016 IEEE Conference on Computer Vision and Pattern Recognition Workshops, CVPR Workshops 2016, Las Vegas, NV, USA, June 26 - July 1, 2016. pp. 661–673. IEEE Computer Society (2016)
10. Zhang, Y., Funkhouser, T.A.: Deep depth completion of a single RGB-D image. In: 2018 IEEE Conference on Computer Vision and Pattern Recognition, CVPR 2018, Salt Lake City, UT, USA, June 18–22, 2018. pp. 175–185. Computer Vision Foundation / IEEE Computer Society (2018)
11. Gupta, S., Girshick, R., Arbeláez, P., Malik, J.: Learning rich features from RGB-D Images for object detection and segmentation. In: Fleet, D., Pajdla, T., Schiele, B., Tuytelaars, T. (eds.) ECCV 2014. LNCS, vol. 8695, pp. 345–360. Springer, Cham (2014). https://doi.org/10.1007/978-3-319-10584-0_23
12. Gupta, S., Arbelaez, P., Malik, J.: Perceptual organization and recognition of indoor scenes from RGB-D images. In: 2013 IEEE Conference on Computer Vision and Pattern Recognition, Portland, OR, USA, June 23–28, 2013. pp. 564–571. IEEE Computer Society (2013)
13. Hazirbas, C., Ma, L., Domokos, C., Cremers, D.: FuseNet: incorporating depth into semantic segmentation via fusion-based CNN architecture. In: Lai, S.-H., Lepetit, V., Nishino, K., Sato, Y. (eds.) ACCV 2016. LNCS, vol. 10111, pp. 213–228. Springer, Cham (2017). https://doi.org/10.1007/978-3-319-54181-5_14
14. Couprie, C., Farabet, C., Najman, L., LeCun, Y.: Indoor semantic segmentation using depth information. In: Bengio, Y., LeCun, Y. (eds.) 1st International Conference on Learning Representations, ICLR 2013, Scottsdale, Arizona, USA, May 2–4, 2013, Conference Track Proceedings (2013)
15. LeCun, Y., Bottou, L., Bengio, Y., Haffner, P.: Gradient-based learning applied to document recognition. Proc. IEEE **86**(11), 2278–2324 (1998)

16. Bahdanau, D., Cho, K., Bengio, Y.: Neural machine translation by jointly learning to align and translate. In: Bengio, Y., LeCun, Y. (eds.) 3rd International Conference on Learning Representations, ICLR 2015, San Diego, CA, USA, May 7–9, 2015, Conference Track Proceedings (2015)

17. Long, J., Shelhamer, E., Darrell, T.: Fully convolutional networks for semantic segmentation. In: IEEE Conference on Computer Vision and Pattern Recognition, CVPR 2015, Boston, MA, USA, June 7–12, 2015. pp. 3431–3440. IEEE Computer Society (2015)

18. Krizhevsky, A., Sutskever, I., Hinton, G.E.: ImageNet classification with deep convolutional neural networks. In: Bartlett, P.L., Pereira, F.C.N., Burges, C.J.C., Bottou, L., Weinberger, K.Q. (eds.) 26th Annual Conference on Neural Information Processing Systems 2012. 3–6 December 2012, Lake Tahoe, Nevada, United States, pp. 1106–1114 (2012)

19. Szegedy, C., et al.: Going deeper with convolutions. In: IEEE Conference on Computer Vision and Pattern Recognition, CVPR 2015, Boston, MA, USA, June 7–12, 2015. pp. 1–9. IEEE Computer Society (2015)

20. Simonyan, K., Zisserman, A.: Very deep convolutional networks for large-scale image recognition. In: Bengio, Y., LeCun, Y. (eds.) 3rd International Conference on Learning Representations, ICLR 2015, San Diego, CA, USA, May 7–9, 2015, Conference Track Proceedings (2015)

21. He, K., Zhang, X., Ren, S., Sun, J.: Deep residual learning for image recognition. In: 2016 IEEE Conference on Computer Vision and Pattern Recognition, CVPR 2016, Las Vegas, NV, USA, 27–30 June 2016. pp. 770–778. IEEE Computer Society (2016)

22. Howard, A.G., et al.: Mobilenets: efficient convolutional neural networks for mobile vision applications. arXiv preprint arXiv:1704.04861 (2017)

23. Zhang, X., Zhou, X., Lin, M., Sun, J.: Shufflenet: an extremely efficient convolutional neural network for mobile devices. In: 2018 IEEE Conference on Computer Vision and Pattern Recognition, CVPR 2018, Salt Lake City, UT, USA, June 18–22, 2018, pp. 6848–6856. Computer Vision Foundation/IEEE Computer Society (2018)

24. Ioffe, S., Szegedy, C.: Batch normalization: accelerating deep network training by reducing internal covariate shift. In: Bach, F.R., Blei, D.M. (eds.) Proceedings of the 32nd International Conference on Machine Learning, ICML 2015, Lille, France, 6–11 July 2015. JMLR Workshop and Conference Proceedings, vol. 37, pp. 448–456. JMLR.org (2015)

25. Szegedy, C., Vanhoucke, V., Ioffe, S., Shlens, J., Wojna, Z.: Rethinking the inception architecture for computer vision. In: 2016 IEEE Conference on Computer Vision and Pattern Recognition, CVPR 2016, Las Vegas, NV, USA, 27–30 June 2016. pp. 2818–2826. IEEE Computer Society (2016)

26. Szegedy, C., Ioffe, S., Vanhoucke, V., Alemi, A.A.: Inception-v4, inception-ResNet and the impact of residual connections on learning. In: Singh, S., Markovitch, S. (eds.) Proceedings of the Thirty-First AAAI Conference on Artificial Intelligence, 4–9 February 2017, San Francisco, California, USA, pp. 4278–4284. AAAI Press (2017)

27. Huang, G., Liu, Z., van der Maaten, L., Weinberger, K.Q.: Densely connected convolutional networks. In: 2017 IEEE Conference on Computer Vision and Pattern Recognition, CVPR 2017, Honolulu, HI, USA, July 21–26, 2017, pp. 2261–2269. IEEE Computer Society (2017)

28. Tan, M., Le, Q.V.: Efficientnet: rethinking model scaling for convolutional neural networks. In: Chaudhuri, K., Salakhutdinov, R. (eds.) Proceedings of the 36th International Conference on Machine Learning, ICML 2019, 9–15 June 2019, Long Beach, California, USA. Proceedings of Machine Learning Research, vol. 97, pp. 6105–6114. PMLR (2019)

29. Tan, M., Le, Q.V.: Efficientnetv2: Smaller models and faster training. In: Meila, M., Zhang, T. (eds.) Proceedings of the 38th International Conference on Machine Learning, ICML 2021, 18–24 July 2021, Virtual Event. Proceedings of Machine Learning Research, vol. 139, pp. 10096–10106. PMLR (2021)

30. Cheng, Y., Cai, R., Li, Z., Zhao, X., Huang, K.: Locality-sensitive deconvolution networks with gated fusion for RGB-D indoor semantic segmentation. In: 2017 IEEE Conference on Computer Vision and Pattern Recognition, CVPR 2017, Honolulu, HI, USA, July 21–26, 2017. pp. 1475–1483. IEEE Computer Society (2017)

31. Yu, F., Koltun, V.: Multi-scale context aggregation by dilated convolutions. In: Bengio, Y., LeCun, Y. (eds.) 4th International Conference on Learning Representations, ICLR 2016, San Juan, Puerto Rico, May 2–4, 2016, Conference Track Proceedings (2016)

32. Peng, C., Zhang, X., Yu, G., Luo, G., Sun, J.: Large kernel matters - improve semantic segmentation by global convolutional network. In: 2017 IEEE Conference on Computer Vision and Pattern Recognition, CVPR 2017, Honolulu, HI, USA, 21–26 July 2017. pp. 1743–1751. IEEE Computer Society (2017)

33. Chollet, F.: Xception: Deep learning with depthwise separable convolutions. In: 2017 IEEE Conference on Computer Vision and Pattern Recognition, CVPR 2017, Honolulu, HI, USA, 21–26 July 2017. pp. 1800–1807. IEEE Computer Society (2017)

34. Zhao, H., Shi, J., Qi, X., Wang, X., Jia, J.: Pyramid scene parsing network. In: 2017 IEEE Conference on Computer Vision and Pattern Recognition, CVPR 2017, Honolulu, HI, USA, 21–26 July 2017. pp. 6230–6239. IEEE Computer Society (2017)

35. He, K., Zhang, X., Ren, S., Sun, J.: Spatial pyramid pooling in deep convolutional networks for visual recognition. In: Fleet, D., Pajdla, T., Schiele, B., Tuytelaars, T. (eds.) ECCV 2014. LNCS, vol. 8691, pp. 346–361. Springer, Cham (2014). https://doi.org/10.1007/978-3-319-10578-9_23

36. Yang, M., Yu, K., Zhang, C., Li, Z., Yang, K.: Denseaspp for semantic segmentation in street scenes. In: 2018 IEEE Conference on Computer Vision and Pattern Recognition, CVPR 2018, Salt Lake City, UT, USA, June 18–22, 2018. pp. 3684–3692. Computer Vision Foundation / IEEE Computer Society (2018)

37. Chen, L., Papandreou, G., Kokkinos, I., Murphy, K., Yuille, A.L.: Semantic image segmentation with deep convolutional nets and fully connected CRFs. In: Bengio, Y., LeCun, Y. (eds.) 3rd International Conference on Learning Representations, ICLR 2015, San Diego, CA, USA, 7–9 May 2015, Conference Track Proceedings (2015)

38. Chen, L., Papandreou, G., Kokkinos, I., Murphy, K., Yuille, A.L.: DeepLab: Semantic image segmentation with deep convolutional nets, atrous convolution, and fully connected CRFs. IEEE Trans. Pattern Anal. Mach. Intell. **40**(4), 834–848 (2018)

39. Chen, L., Papandreou, G., Schroff, F., Adam, H.: Rethinking atrous convolution for semantic image segmentation. arXiv preprint arXiv:1706.05587 (2017)

40. Chen, L.-C., Zhu, Y., Papandreou, G., Schroff, F., Adam, H.: Encoder-decoder with atrous separable convolution for semantic image segmentation. In: Ferrari, V., Hebert, M., Sminchisescu, C., Weiss, Y. (eds.) ECCV 2018. LNCS, vol. 11211, pp. 833–851. Springer, Cham (2018). https://doi.org/10.1007/978-3-030-01234-2_49

41. Choi, M.J., Lim, J.J., Torralba, A., Willsky, A.S.: Exploiting hierarchical context on a large database of object categories. In: The Twenty-Third IEEE Conference on Computer Vision and Pattern Recognition, CVPR 2010, San Francisco, CA, USA, 13–18 June 2010. pp. 129–136. IEEE Computer Society (2010)
42. Mottaghi, R., et al.: The role of context for object detection and semantic segmentation in the wild. In: 2014 IEEE Conference on Computer Vision and Pattern Recognition, CVPR 2014, Columbus, OH, USA, 23–28 June 2014. pp. 891–898. IEEE Computer Society (2014)
43. Liu, W., Rabinovich, A., Berg, A.C.: Parsenet: Looking wider to see better. arXiv preprint arXiv:1506.04579 (2015)
44. Hung, W., et al.: Scene parsing with global context embedding. In: IEEE International Conference on Computer Vision, ICCV 2017, Venice, Italy, 22–29 October 2017. pp. 2650–2658. IEEE Computer Society (2017)
45. Yu, C., Wang, J., Peng, C., Gao, C., Yu, G., Sang, N.: BiSeNet: bilateral segmentation network for real-time semantic segmentation. In: Ferrari, V., Hebert, M., Sminchisescu, C., Weiss, Y. (eds.) ECCV 2018. LNCS, vol. 11217, pp. 334–349. Springer, Cham (2018). https://doi.org/10.1007/978-3-030-01261-8_20
46. Fu, J., et al.: Adaptive context network for scene parsing. In: 2019 IEEE/CVF International Conference on Computer Vision, ICCV 2019, Seoul, Korea (South), October 27 - November 2, 2019. pp. 6747–6756. IEEE (2019)
47. Zhang, H., et al.: Context encoding for semantic segmentation. In: 2018 IEEE Conference on Computer Vision and Pattern Recognition, CVPR 2018, Salt Lake City, UT, USA, 18–22 June 2018. pp. 7151–7160. Computer Vision Foundation/IEEE Computer Society (2018)
48. Baltrusaitis, T., Ahuja, C., Morency, L.: Multimodal machine learning: a survey and taxonomy. IEEE Trans. Pattern Anal. Mach. Intell. $41(2)$, 423–443 (2019)
49. Fooladgar, F., Kasaei, S.: A survey on indoor RGB-D semantic segmentation: from hand-crafted features to deep convolutional neural networks. Multim. Tools Appl. $79(7\text{--}8)$, 4499–4524 (2020)
50. Coupé, P., Manjón, J.V., Fonov, V.S., Pruessner, J.C., Robles, M., Collins, D.L.: Patch-based segmentation using expert priors: application to hippocampus and ventricle segmentation. Neuroimage $54(2)$, 940–954 (2011)
51. Wang, M., Liu, X., Gao, Y., Ma, X., Soomro, N.Q.: Superpixel segmentation: a benchmark. Signal Process. Image Commun. 56, 28–39 (2017)
52. Kaganami, H.G., Zou, B.: Region-based segmentation versus edge detection. In: Pan, J., Chen, Y., Jain, L.C. (eds.) Fifth International Conference on Intelligent Information Hiding and Multimedia Signal Processing (IIH-MSP 2009), Kyoto, Japan, 12–14 September 2009, pp. 1217–1221. IEEE Computer Society (2009)
53. Zheng, S., et al.: Conditional random fields as recurrent neural networks. In: 2015 IEEE International Conference on Computer Vision, ICCV 2015, Santiago, Chile, 7–13 December 2015. pp. 1529–1537. IEEE Computer Society (2015)
54. Banica, D., Sminchisescu, C.: Second-order constrained parametric proposals and sequential search-based structured prediction for semantic segmentation in RGB-D images. In: IEEE Conference on Computer Vision and Pattern Recognition, CVPR 2015, Boston, MA, USA, 7–12 June 2015. pp. 3517–3526. IEEE Computer Society (2015)
55. Bo, L., Ren, X., Fox, D.: Kernel descriptors for visual recognition. In: Lafferty, J.D., Williams, C.K.I., Shawe-Taylor, J., Zemel, R.S., Culotta, A. (eds.) 24th Annual Conference on Neural Information Processing Systems 2010. Proceedings of a meeting held, 6–9 December 2010, Vancouver, British Columbia, Canada. pp. 244–252. Curran Associates, Inc. (2010)

56. Hermans, A., Floros, G., Leibe, B.: Dense 3d semantic mapping of indoor scenes from RGB-D images. In: 2014 IEEE International Conference on Robotics and Automation, ICRA 2014, Hong Kong, China, May 31 - June 7, 2014. pp. 2631–2638. IEEE (2014)

57. Lerma, C.D.C., Kosecká, J.: Semantic parsing for priming object detection in indoors RGB-D scenes. Int. J. Robotics Res. **34**(4–5), 582–597 (2015)

58. Ren, X., Bo, L., Fox, D.: RGB-(D) scene labeling: Features and algorithms. In: 2012 IEEE Conference on Computer Vision and Pattern Recognition, Providence, RI, USA, 16–21 June 2012. pp. 2759–2766. IEEE Computer Society (2012)

59. Müller, A.C., Behnke, S.: Learning depth-sensitive conditional random fields for semantic segmentation of RGB-D images. In: 2014 IEEE International Conference on Robotics and Automation, ICRA 2014, Hong Kong, China, May 31 - June 7, 2014. pp. 6232–6237. IEEE (2014)

60. Wang, S., Lokhande, V.S., Singh, M., Körding, K.P., Yarkony, J.: End-to-end training of CNN-CRF via differentiable dual-decomposition. arXiv preprint arXiv:1912.02937 (2019)

61. McCormac, J., Handa, A., Leutenegger, S., Davison, A.J.: Scenenet RGB-D: can 5m synthetic images beat generic ImageNet pre-training on indoor segmentation? In: IEEE International Conference on Computer Vision, ICCV 2017, Venice, Italy, 22–29 October 2017. pp. 2697–2706. IEEE Computer Society (2017)

62. Lee, S., Park, S., Hong, K.: Rdfnet: RGB-D multi-level residual feature fusion for indoor semantic segmentation. In: IEEE International Conference on Computer Vision, ICCV 2017, Venice, Italy, 22–29 October 2017. pp. 4990–4999. IEEE Computer Society (2017)

63. Lin, G., Milan, A., Shen, C., Reid, I.D.: RefineNet: multi-path refinement networks for high-resolution semantic segmentation. In: 2017 IEEE Conference on Computer Vision and Pattern Recognition, CVPR 2017, Honolulu, HI, USA, 21–26 July 2017. pp. 5168–5177. IEEE Computer Society (2017)

64. Li, Z., Gan, Y., Liang, X., Yu, Y., Cheng, H., Lin, L.: LSTM-CF: unifying context modeling and fusion with LSTMs for RGB-d scene labeling. In: Leibe, B., Matas, J., Sebe, N., Welling, M. (eds.) ECCV 2016. LNCS, vol. 9906, pp. 541–557. Springer, Cham (2016). https://doi.org/10.1007/978-3-319-46475-6_34

65. Vandenhende, S., Georgoulis, S., Van Gool, L.: MTI-Net: multi-scale task interaction networks for multi-task learning. In: Vedaldi, A., Bischof, H., Brox, T., Frahm, J.-M. (eds.) ECCV 2020. LNCS, vol. 12349, pp. 527–543. Springer, Cham (2020). https://doi.org/10.1007/978-3-030-58548-8_31

66. Shi, H., Li, H., Wu, Q., Song, Z.: Scene parsing via integrated classification model and variance-based regularization. In: IEEE Conference on Computer Vision and Pattern Recognition, CVPR 2019, Long Beach, CA, USA, 16–20 June 2019. pp. 5307–5316. Computer Vision Foundation/IEEE (2019)

67. Seichter, D., Köhler, M., Lewandowski, B., Wengefeld, T., Gross, H.: Efficient RGB-D semantic segmentation for indoor scene analysis. In: IEEE International Conference on Robotics and Automation, ICRA 2021, Xi'an, China, May 30 - June 5, 2021. pp. 13525–13531. IEEE (2021)

68. Cao, J., Leng, H., Lischinski, D., Cohen-Or, D., Tu, C., Li, Y.: Shapeconv: shape-aware convolutional layer for indoor RGB-D semantic segmentation. In: 2021 IEEE/CVF International Conference on Computer Vision, ICCV 2021, Montreal, QC, Canada, 10–17 October 2021. pp. 7068–7077. IEEE (2021)

69. Woo, S., Park, J., Lee, J., Kweon, I.S.: CBAM: convolutional block attention module. In: Ferrari, V., Hebert, M., Sminchisescu, C., Weiss, Y. (eds.) Computer Vision - ECCV 2018–15th European Conference, Munich, Germany, September 8–14, 2018, Proceedings, Part VII. LNCS, vol. 11211, pp. 3–19. Springer, Cham (2018). https://doi.org/10.1007/978-3-030-01234-2_1

70. Li, X., Wang, W., Hu, X., Yang, J.: Selective kernel networks. In: IEEE Conference on Computer Vision and Pattern Recognition, CVPR 2019, Long Beach, CA, USA, 16–20 June 2019. pp. 510–519. Computer Vision Foundation/IEEE (2019)

71. Hu, J., Shen, L., Sun, G.: Squeeze-and-excitation networks. In: 2018 IEEE Conference on Computer Vision and Pattern Recognition, CVPR 2018, Salt Lake City, UT, USA, 18–22 June 2018. pp. 7132–7141. Computer Vision Foundation/IEEE Computer Society (2018)

72. Zhou, H., Qi, L., Huang, H., Yang, X., Wan, Z., Wen, X.: Canet: co-attention network for RGB-D semantic segmentation. Pattern Recognit. **124**, 108468 (2022)

73. Zhang, G., Xue, J., Xie, P., Yang, S., Wang, G.: Non-local aggregation for RGB-D semantic segmentation. IEEE Signal Process. Lett. **28**, 658–662 (2021)

74. Chen, X., Lin, K.-Y., Wang, J., Wu, W., Qian, C., Li, H., Zeng, G.: Bi-directional cross-modality feature propagation with separation-and-aggregation gate for RGB-D semantic segmentation. In: Vedaldi, A., Bischof, H., Brox, T., Frahm, J.-M. (eds.) ECCV 2020. LNCS, vol. 12356, pp. 561–577. Springer, Cham (2020). https://doi.org/10.1007/978-3-030-58621-8_33

Foreground-Specialized Model Imitation for Instance Segmentation

Dawei Li[1], Wenbo Li[2(✉)], and Hongxia Jin[2]

[1] Amazon Lab126, Sunnyvale, USA
daweili@amazon.com
[2] Samsung Research America AI Center, Mountain View, USA
{wenbo.li1,hongxia.jin}@samsung.com

Abstract. Instance segmentation is formulated as a multi-task learning problem. However, knowledge distillation is not well-suited to all subtasks except the multi-class object classification. Based on such a competence, we introduce a lightweight foreground-specialized (FS) teacher model, which is trained with foreground-only images and highly optimized for object classification. Yet, this leads to discrepancy between inputs to the teacher and student models. Thus, we introduce a novel Foreground-Specialized model Imitation (FSI) method with two complementary components. First, a reciprocal anchor box selection method is introduced to distill from the most informative output of the FS teacher. Second, we embed the foreground-awareness into student's feature learning via either adding a co-learned foreground segmentation branch or applying a soft feature mask. We conducted an extensive evaluation against the others on COCO and Pascal VOC.

Keywords: Knowledge distillation · Instance segmentation

1 Introduction

To deploy deep learning models on resource-constrained edge devices, researchers have been devoting efforts in four major directions: (1) model compression, i.e., quantization and pruning [10]; (2) better light-weight backbones such as MobileNet [14] and ShuffleNet [35]; (3) reduced model architecture such as YOLO [28] and SSD [24] for one-stage object detection and (4) model imitation which trains a compact and device-friendly model to imitate the behavior a more powerful yet more computationally expensive model. The focus of this work is to improve the model imitation for object instance segmentation.

A typical and widely used model imitation method is KD [13] which transfers the knowledge from a large model with stronger generalization capability to a lightweight model, while both models are trained on the same dataset. One major challenge faced by the existing KD methods is that the teacher models where the

---

D. Li and W. Li—Equal contributions.

D. Li—This work was performed at Samsung Research America.

L. Wang et al. (Eds.): ACCV 2022, LNCS 13847, pp. 398–413, 2023.
https://doi.org/10.1007/978-3-031-26293-7_24

(a) Teacher in KD   (b) Foreground-specialized Teacher

Fig. 1. The difference between teacher models in traditional knowledge distillation (KD) and our foreground-specialized teacher model. **(a)** In KD, the teacher model is normally much larger than the student model. **(b)** The foreground-specialized teacher model can be as small as the student model and thus can be efficiently trained and imitated.

knowledge is transferred from normally have many layers and parameters, e.g., using VGGNet [31] as the teacher backbone, so training and distilling from such models are time-consuming and would rely on high-end computing devices with large memories[1]. In addition, a recent work [6] shows that the final accuracy of the student model does not increase monotonically with the size of the teacher model. As the teacher model gets larger, the accuracy of the student model first increases and then decreases. This phenomenon means that substantial effort would be required to explore the optimal teacher model in order to achieve the most accurate student model.

The goal of the object instance segmentation is to detect and delineate each distinct object of interest from an input image. It is normally formulated as a multi-task learning problem including bounding box detection, object classification and mask prediction [11]. The multi-task learning formulation poses additional challenges for applying the KD to object instance segmentation, because the KD is designed exclusively for the multi-class classification problem, i.e., the object classification, which is only one out of three sub-tasks of the object instance segmentation. Therefore, it is counterproductive to apply KD to the other two sub-tasks of the object instance segmentation, i.e., bounding box detection and mask prediction, which KD is not designed for.

In order to resolve the above challenges, we focus on exploring a more effective and efficient way of model imitation for the object instance segmentation. Unlike the conventional KD in which the teacher model takes the same input as the student model, we simplify the input to the teacher model in order to relieve its learning burdens on the two sub-tasks, i.e., bounding box detection and mask prediction, but makes the teacher model dedicated to object classification which KD is designed exclusively for. Specifically, as shown in Fig. 1, we simplify the input to the teacher model by removing the background stuff from the training images while leaving only pixels for the foreground objects. In this fashion, it will

---

[1] In distillation, the memory cost can be reduced when the outputs of teacher models are pre-computed. Yet, this disables on-the-fly data augmentation, a critical component for improving the model accuracy especially when the dataset is small.

Table 1. Model accuracy (mAP@[0.5,.95]) comparison on COCO dataset by transforming the input with different backbone architectures of YOLACT. **Standard:** the model trained and validated using the complete images. **FS-Teacher:** the model trained and validated using foreground-only images.

Model	ResNet-18	ResNet-50	ResNet-101
Standard	23.76	27.97	29.73
FS-Teacher	41.08	46.69	47.86

be tremendously simpler for a teacher model on the bounding box detection and the mask prediction, because the foreground objects are salient on the vacant background.

We name the teacher model trained on the simplified input as the foreground-specialized (FS) teacher model. As shown in Table 1, we observe a significant performance gain achieved by the teacher models on the simplified input, over that achieved by the standard teacher models on the original input. One may argue the "unfairness" of the comparison between the FS teacher models and the standard ones in Table 1 because of the presence/absence of the input simplification on the validation data, i.e., foreground-only images vs. the original images. Yet, the KD method is mainly concerned about the output of a teacher model while barely favoring preprocessing tricks, so the performance gain is still quite beneficial despite the negligible "unfairness". In Table 1, note that even the FS teacher model based on the backbone ResNet-18 performs much better than the standard teacher model based on ResNet-101. As such, it can enable a significantly more efficient teacher model.

As shown in Fig. 1, compared to the standard teacher model in KD, the FS teacher model takes different input but shares the output format. To accommodate this new change, we introduce a novel Foreground-Specialized model Imitation (FSI) which includes two complementary modules that allow the student model to better imitate the teacher. First, instead of distilling knowledge from all three types of teacher output, i.e., classification, bounding boxes, and instance masks, we only distill knowledge from the teacher's classification output which is what the teacher model was designed exclusively for. In addition, to deal with the highly unbalanced positive and negative anchor boxes, a reciprocal anchor box selection method is introduced to distill knowledge based on those most informative teacher outputs. Second, though we could not filter out the background from the input of the student model with ease, the student model can be encouraged to better imitate the teacher by embedding the foreground-awareness into the feature learning. Particularly, we introduce two solutions by either applying a learned latent soft foreground mask to the intermediate convolution features (at the cost of reduced inference speed) or co-learning a foreground segmentation task by attaching a branch to the student's backbone (not affecting the inference speed). Both solutions are demonstrated to be effective in improving the student model's accuracy through extensive evaluations.

2   Related Works

It is a challenging task to deploy deep neural networks on pervasive mobile and edge computing devices that have limited computing resources. Towards this goal, model compression techniques [10] have been introduced to approximate a given deep learning model with a compact one that reduces the storage cost, and representative methods include quantization [16,27], pruning [12], and low-rankness [17]. Model compression is effective in reducing the model size, however, the accuracy is generally bounded by the model before compression and may rely on specialized hardware and/or software support for speedup [9].

More complicated vision tasks like object detection and instance segmentation rely on a multi-stage architecture [11,29] to achieve favorable accuracy at the cost of heavy computation. To enable on-device inference, researchers have come up with a more effective approach that uses a single-stage architecture [2,24,28] to substantially reduce the computational cost while still achieving satisfactory accuracy. Key technologies behind the success include multi-scale anchor boxes [24], feature pyramid networks [21], focal loss [22], etc.

On the other hand, model imitation technologies, mostly, KD [13] have been introduced to enforce additional guidance using a large teacher model's prediction in addition to the ground truth labels. The KD method was first introduced in the classification problem and the insight behind it is that a pre-trained large model on the same dataset has already learned the essential underlying relationship among different classes that can be generalized to new unseen samples, e.g., cars are close to trucks but are quite different from apples, and this essential information is reflected by the model's output. Later works have extended KD to other problems including object detection [3,33], semantic segmentation [25], and sequence learning problems [18]. However, a recent thorough evaluation [6] on the efficacy of KD indicates that larger models do not often make better teachers due to mismatched capacity of teacher and student models.

Another research topic related to our work is learning using privileged information (LUPI) [19,32], which assumes additional information or modality about the data is provided at training but may not be available at test time. Most existing LUPI works assume the extra data modalities, e.g., the depth modality, can be easily obtained or generated [5,8,20,26]. However, for many real application scenarios, the majority training data are large-scale crowd-sourced wild data which does not have such extra information. Instead, our approach generates specialized training data from the ground-truth labels as a resolution to simplify the task and train strong specialized models. In addition, only our work uses such models as the KD teachers.

3   Method

In this section, we present our method in detail. We first give a brief recap of the YOLACT one-stage object instance segmentation method. Then we describe how the FS teacher model is trained and the necessity of imitating the model.

Table 2. Misclassification errors on COCO dataset.

Model	Conf@0.5	Conf@0.7	Conf@0.9
Standard	49.71%	42.59%	37.67%
FS-Teacher	38.63%	30.87%	23.34%

After that, we delineate our model imitation approach targeting the peculiarities that are different from the existing KD method.

3.1 Recap of YOLACT

Two-stage object instance methods such as Mask-RCNN [11] are not suitable for on-device inference. Following the idea of single-stage object detection methods [24,28], researchers have recently proposed one-stage object instance segmentation methods [2,34] enabling real-time inference speed on mobile and edge devices. Among those methods, YOLACT [2] is the state-of-the-art solution considering both the accelerated inference speed and satisfactory segmentation accuracy. Compared to Mask-RCNN, it achieves about 4× speed-up with fairly close accuracy.

YOLACT is based on the RetinaNet [22] architecture which is composed of a feature pyramid network (FPN) [21] based backbone and a set of classification and box regression branches following each pyramid feature map to predict the class category and bounding box coordinates for each anchor box. On top of RetinaNet, YOLACT adds a protonet branch to infer a set of latent mask prototypes and further, for each anchor box, predicts a coefficient vector for composing the mask prototypes. The final output mask for each anchor box is the weighted sum of the latent mask prototypes given the predicted coefficients.

3.2 FS Teacher Training

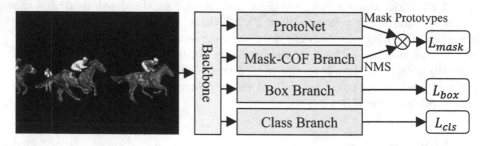

Fig. 2. The training of a teacher based on YOLACT.

The FS model is trained using exactly the same model architecture and outputs as introduced in Sect. 3.1 and is illustrated in Fig. 2. The only change is that the background of the input image is removed. This input transformation does not require additional annotation efforts since the foreground can be conveniently determined as the union of masks for each foreground object. Pixels not covered by the foreground mask is set to 0 (i.e., black) before feeding into the model.

The teacher model does not need to be larger than the student model and we demonstrate that the student model could achieve substantial accuracy improvement when both the teacher model and the student model use the same backbone network (i.e., ResNet-18). However, it must be noted that (1) the output feature map sizes from both models' backbone and (2) the number of anchor boxes are required to be consistent between the two models so that each anchor box from the student model can find a unique mapping from the teacher, and thus it knows where it can distill the knowledge.

Boosted Classification Accuracy. In this subsection, we discuss and validate that the FS teacher model has significantly improved classification performance. For instance segmentation, there could be two major types of errors. (1) Object not detected, i.e., there's no positive object detected or a detected object with IoU below a given threshold. (2) Misclassification, i.e., an object is detected with enough IoU to a ground-truth but is wrongly classified as a different class. The FS teacher model is trained without background interference which means that the training burden is greatly reduced to mostly classifying objects and identifying the object boundaries (especially for overlapped objects).

We measure the misclassification errors of the standard model which is trained and validated using the complete images and the FS model and we show the comparison in Table 2. Specifically, we calculate the misclassification error as the percentage of objects (averaged over all classes) which (1) has been detected with an IoU above 0.75 to a ground-truth object and (2) is wrongly classified. In addition, we set different confidence thresholds on each measurement, e.g., Conf@0.5 means we use the detected objects that are classified to a object class with confidence score greater than 0.5. The measured result indicates that the FS teacher model can reduce the misclassification errors by more than 14%.

Challenges on Distilling from FS Model. The challenges are mostly from the fact that the teacher and the student models have different input. For the student model, there's no trivial way of obtaining foreground masks in real-time inference. Applying foreground segmentation (using models like DeepLab [4]) as pre-processing not only incurs extra computational burden but also cannot guarantee a flawless foreground mask. Therefore, instead of manipulating on the input inference images, we propose a new distillation method to have the student model effectively imitate the FS teacher model as described in detail in the following subsection.

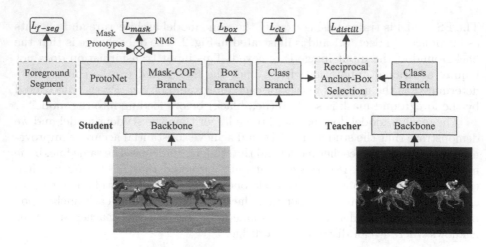

Fig. 3. Overview of the imitation training with FSI. The *Reciprocal Anchor-Box Selection* module identifies the critical knowledge from the teacher for the student to learn; the *Foreground Segment* module embeds the foreground-awareness into the student's feature learning.

3.3 FSI: Foreground-Specialized Model Imitation

As shown in Fig. 3, FSI is contains two novel modules: (1) reciprocal anchor box selection and (2) foreground segment.

Reciprocal Anchor Box Selection. The core idea is to have the student to learn from only what the teacher is specialized for. The FS teacher has been trained to concentrate on classifying the foreground objects while it hasn't been challenged much to predict the masks or to regress the bounding boxes, considering that the image background has been zeroed out. This means the distillation should only focus on teacher's classification output for each anchor box. Meanwhile, distilling the classification output for all anchor boxes is highly inefficient given the extremely unbalanced distribution of positive and negative boxes and would absorb teacher's adverse knowledge when it makes a wrong prediction.

Therefore, we propose a reciprocal anchor box selection method to effectively learn the most critical knowledge from the teacher. Concretely, the selected knowledge includes the teacher's "opinion" on two sets of anchors. The first set includes anchor boxes that the teacher gives positive opinions, i.e., correctly predicted as foreground objects. This part represents the essence of teacher's knowledge that the student should absorb. The second set includes anchor boxes that have been wrongly predicted as foreground objects by the student but correctly predicted as background by the teacher. This represents student's wrong knowledge. In Algorithm 1, we present the pseudo-code for the method and we tensorize the operations for fast training in our implementation.

Algorithm 1: Reciprocal Anchor Box Selection

Input : $\hat{Y}_{cls}^t$ and $\hat{Y}_{cls}^s$ are classification output from the *teacher* model and the *student* model of all anchor boxes S, respectively.

Output: S_{recip} is a set of selected anchor boxes for distillation.

```
1  begin
2  │  S_recip ← ∅ for  i = 1 → size(S) do
3  │  │  ŷ_i^t = Ŷ_cls^t.get(i);   t_cls = argmax(ŷ_i^t) ;        // teacher prediction
4  │  │  ŷ_i^s = Ŷ_cls^s.get(i);   s_cls = argmax(ŷ_i^s) ;        // student prediction
5  │  │  if  t_cls = gt_i and (t_cls > 0 or s_cls > 0) ;   // 0 represents background
6  │  │  then
7  │  │  │  └  S_recip.insert(i)
```

Foreground-Aware Feature Learning. FS teacher extracts features without the background interference. To imitate the teacher better, we propose two alternative approaches to embed the foreground awareness into student's learned features.

Soft Feature Mask. Inspired by the Squeeze-and-Excitation Networks [15], as shown in Fig. 4(a), we generate a latent soft feature mask for FPN features for an adaptive feature calibration based on teacher's guidance. Despite its effectiveness in improving the student's accuracy, it incurs 17% inference speed reduction with the ResNet18 backbone.

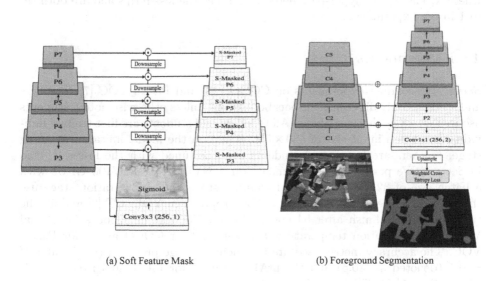

(a) Soft Feature Mask   (b) Foreground Segmentation

Fig. 4. Foreground-aware feature learning approaches.

Foreground Segmentation. Instead of adding additional operations to the feature maps, we attach a binary segmentation branch to the FPN features of the student as shown in Fig. 4(b), we apply a weighted pixel-wise binary cross-entropy loss [30] posing higher weight on the foreground pixels so that the foreground regions get more attention in the feature learning. The added layers (white layers in Fig. 4(b)) will only be used in the training and will not be touched during inference, so the student's inference speed will not be affected. We prefer to this approach because of its balanced performance on accuracy and efficiency.

Loss Function. We introduce two additional loss terms (of which the weights are both set to 1) to the loss function of YOLACT. The first is KD loss $L_{distill}$ which measures the KL-Divergence between the classification outputs of the teacher and student on a set of selected anchor boxes S_{recip}:

$$L_{distill} = \frac{1}{|S_{recip}|} \sum_{i \in S_{recip}} KL_{div}(\sigma(\frac{\hat{y}_i^t}{T}), \sigma(\frac{\hat{y}_i^s}{T})), \tag{1}$$

where T is a temperature [13] and $\sigma(\cdot)$ denotes the softmax.

The second loss term is the binary foreground segmentation loss $L_{f\text{-}seg}$ which is a weighted pixel-wise binary cross-entropy loss [30] and the weight between foreground and background pixels is set as 2:1.

Our overall learning objective is:

$$L_{YOLACT} + \lambda_1 L_{distill} + \lambda_2 L_{f\text{-}seg} \tag{2}$$

where λ_1 and λ_2 are hyper-parameter to balance the loss terms and are both set to 1 in our experiments.

4   Experiments

Setup. We conduct experiments on COCO [23] and Pascal VOC [7]. Teachers and students only differ in their backbones, and unless otherwise noted, students use ResNet-18 as their backbone. All teachers and students are first trained individually with an input size of 550×550. Then, in the model imitation training stage, teachers are frozen and students are fine-tuned with the losses defined in Sect. 3. We perform all experiments on 4 NVIDIA Tesla V100 GPUs with a batch size of 32. Both the initial training stage and the imitation stage use the SGD optimizer following the schedule proposed in YOLACT [2] and all the training sessions finish after 54 epochs for COCO and 112 epochs for Pascal VOC. The distillation temperature T is set to 3 for COCO and 1 for Pascal VOC. The accuracy metric used are the mean average precision (mAP) at IoU = 0.5 (denoted as @0.5) and the mAP averaged for IoU $\in [0.5 : 0.05 : 0.95]$ (denoted as @[.5, .95]).

Table 3. Comparison on COCO. The first row shows the performance of the student without being taught by the teacher.

Teacher backbone	Method	COCO		Pascal VOC	
		Mask@0.5	Mask@[0.5,.95]	Mask@0.5	Mask@[0.5,.95]
N/A	Student	40.44	23.76	68.98	43.74
ResNet-50	KD-Cls	42.96	25.29	70.61	45.90
	KD-Hint	43.15	25.68	70.08	45.78
	KD-FHint	43.67	25.94	70.08	45.78
	KD-All	43.17	25.65	70.12	45.56
ResNet-101	KD-Cls	42.49	25.26	69.07	44.67
	KD-Hint	42.52	25.04	69.39	44.69
	KD-FHint	43.11	25.37	69.70	44.84
	KD-All	42.72	25.12	69.81	44.98
ResNet-50	FSI-SFM	44.29	26.42	70.55	**46.27**
	FSI-FS	44.30	26.50	70.66	**46.27**
ResNet-101	FSI-SFM	43.97	26.23	69.62	44.65
	FSI-FS	43.77	26.11	68.89	44.30
ResNet-18	FSI-SFM	44.27	26.39	**71.04**	46.21
	FSI-FS	**44.43**	**26.53**	70.72	46.17

Compared Baselines. We compare our method against 4 KD baseline methods which use a large teacher that is trained on the same dataset and shares the same input/output format as the student. (1) *KD-Cls* distills exactly the same teacher knowledge as our method but uses a larger teacher model. (2) *KD-Hint* [3] distills CNN features from intermediate layers in addition to the classification output. Specifically, we add an adaptation layer (1x1 conv) to each FPN's output feature map, and the output of each adaptation layer is compared to the corresponding feature map of the teacher model by calculating the L2 distance as the hinted feature loss $L_{distill\text{-}hint}$. (3) *KD-FHint* [33] is similar to *KD-Hint*, but only distill CNN features from the foreground regions. (4) *KD-All* is built on top of *KD-Hint* via adding an additional mask prototype distillation loss $L_{distill\text{-}mask}$ (smoothed L1 loss) for the corresponding mask prototypes generated in YOLACT's protonet. The compared KD baseline methods use two different backbones for teachers, i.e., ResNet-50 and ResNet-101. Our methods are named as FSI-SFM and FSI-FS. Suffix "SFM" and "FS" represent alternative approaches to embed the foreground awareness, i.e., Soft Feature Mask (SFM) and Foreground Segmentation (FS).

Comparison with Baselines. We show results on COCO and Pascal VOC in Table 3. The proposed FSI method brings notable improvement: mAP has increased by 2.77 and 2.47 respectively for Mask@[0.5,.95]. For Pascal VOC, the result ResNet-18 student model almost achieves the accuracy of the ResNet-50 model (46.21 vs 46.87 referring to Table 7). FSI-SFM and FSI-FS have similar

Table 4. Per-class result on COCO. **BT:** Accuracy before teaching. **AT:** Accuracy after teaching.

	Person	Bike	Car	m-bile	Plane	Bus	Train	Truck	Boat	t-light
BT	28.52	8.71	22.75	19.03	40.93	51.24	54.47	21.51	10.10	14.52
AT	31.56	10.64	25.57	21.97	42.95	54.23	56.37	24.65	12.15	15.82
	hydrant	s-sign	p-meter	bench	bird	cat	dog	horse	sheep	cow
BT	50.76	53.38	32.21	8.97	16.14	55.65	47.40	27.81	23.85	28.86
AT	53.72	55.67	35.39	11.32	15.53	60.57	50.14	29.91	25.40	30.85
	elephant	bear	zebra	giraffe	b-pack	umbrella	handbag	tie	suitcase	frisbee
BT	42.35	59.34	40.37	37.86	5.12	28.81	4.96	12.21	16.51	42.29
AT	46.33	63.39	41.56	39.25	6.98	32.68	6.47	9.12	19.50	47.40
	skis	snow-b	s-ball	kite	b-bat	b-glove	skate-b	surf-b	t-racket	bottle
BT	0.65	10.30	25.31	14.65	11.44	24.67	14.89	17.63	38.85	16.38
AT	0.91	12.87	26.31	16.38	13.10	29.01	19.05	20.07	41.51	19.42
	w-glass	cup	fork	knife	spoon	bowl	banana	apple	sandwich	orange
BT	15.23	23.43	2.60	1.64	1.50	24.83	9.51	8.74	25.22	18.04
AT	17.58	26.44	4.65	2.77	3.23	26.71	11.38	12.74	29.64	19.66
	broccoli	carrot	hotdog	pizza	donut	cake	chair	couch	p-plant	bed
BT	12.65	6.61	12.68	37.80	30.00	22.26	4.74	22.67	10.34	25.72
AT	15.38	10.03	14.89	38.56	35.40	26.25	6.73	26.60	13.01	27.87
	d-table	toilet	tv	laptop	mouse	remote	k-board	c-phone	m-wave	oven
BT	9.55	47.52	45.15	43.74	45.73	12.38	34.77	19.57	41.63	22.61
AT	11.34	51.78	50.09	45.52	48.37	15.60	38.33	23.37	48.03	26.25
	toaster	sink	fridge	book	clock	vase	scissor	t-bear	h-drier	t-brush
BT	14.73	22.71	36.19	1.36	40.44	19.82	8.68	28.73	2.93	4.88
AT	18.90	26.00	41.18	1.76	42.36	22.31	16.72	31.85	4.89	5.49

Table 5. Ablation study on COCO.

RECIP	SFM	FS	NegBox	Hint	Mask@0.5	Mask@[0.5,.95]
✗	✗	✓	✗	✗	43.86	26.15
✓	✗	✗	✗	✗	43.28	25.70
✓	✗	✓	✓	✗	44.23	26.35
✓	✗	✓	✗	✓	44.08	26.34
✓	✗	✓	✗	✗	44.43	26.53
✓	✓	✗	✗	✗	44.27	26.39

performance, and both outperforms the compared KD baselines. In addition, we don't observe clear accuracy improvement by adding more loss terms to these baselines. This might be due to the fact that a large number of hyper-parameters need to be set and optimized for different datasets and teacher architectures. By contrast, our method only distills the knowledge from the teacher's classification output and thus avoids the complicated hyper-parameter tuning. Overall, we observe that a too heavy teacher (ResNet-101 vs. ResNet-50) is not necessarily

Table 6. Results of co-learned object detection on COCO and Pascal VOC. The first row shows the accuracy of the student.

T-Backbone	Method	COCO		Pascal VOC	
		Box@0.5	Box@[0.5,.95]	Box@0.5	Box@[0.5,.95]
ResNet-18	Student	44.61	25.01	72.23	44.88
ResNet-50	KD-Cls	46.80	26.86	72.71	46.83
	KD-Hint	47.33	27.14	72.50	46.75
	KD-FHint	47.67	27.24	72.60	46.71
	KD-All	47.34	27.16	72.87	46.69
ResNet-101	KD-Cls	46.83	26.67	72.76	44.25
	KD-Hint	46.45	25.86	72.44	44.12
	KD-FHint	47.21	27.07	72.46	44.19
	KD-All	46.57	25.96	72.74	44.40
ResNet-18	FSI-SFM	48.23	27.67	**73.07**	**47.61**
ResNet-18	FSI-FS	**48.54**	**28.01**	73.01	47.12

better than the lightweight one (ResNet-18) as a consequence of greatly mismatched model capacity as addressed in [6]. In addition, we present the result for each COCO class in Table 4. Among the 80 classes, 78 of them get obviously improved accuracy. The only exceptions are birds (slightly reduced accuracy) and ties (relatively small objects with a small number of validation images).

Ablation Study. To show the effect of the different components we have designed, we present the ablation study on COCO in Table 5. The abbreviated keywords in the table are explained as follows: *RECIP:* Reciprocal anchor box selection. If marked as ✗, we only select the anchor boxes that the teacher has correctly predicted as foreground objects and ignores those wrongly predicted by the student. *NegBox:* 300 random selected background boxes correctly predicted by teachers are added for distillation. *Hint:* Feature distillation loss [3]. *SFM:* Foreground awareness with soft feature mask. *FS:* Training with foreground segmentation branch. The first row of Table 5 reflects the effectiveness of the reciprocal anchor box selection. When the student only learns from teacher's essential knowledge but does not correct its own mistake, mAP drops by 0.57. Without learning the foreground-aware features, mAP drops by 1.15 as indicated at the second row.[2] Distilling on more anchor boxes or adding additional distillation loss actually reduces the performance.

Co-learned Object Detection Task. Object detection is a co-learned task for instance segmentation in YOLACT. We see the trend is similar to the instance segmentation in the evaluation. The experiment result and the comparison with baseline methods are presented in Table 6.

---

[2] Simply adding the foreground awareness (FS) to the student without a teacher can improve mAP by 0.4.

Large Student Models. The proposed FSI method can be applied to improving large students, which is not feasible for the conventional KD. We conduct experiments on students with ResNet50 and ResNet101 as backbones where teachers share the same architecture. In addition to YOLACT, we also use the recently released YOLACT++ [1] which further improves the base model accuracy by incorporating deformable convolutions, optimizing anchor box scales and aspect ratios, and adding a mask re-scoring branch. We present the results in Table 7 where considerable improvements are observed in the majority of scenarios. The only exceptions are mAP values of Mask@0.5 on Pascal VOC where only a slight improvement is achieved indicating an upper bound of the YOLACT architecture. However, we still observe an increase of 2 for mAP of Mask@[0.5,.95] showing that the proposed FSI method outputs more higher-quality masks.

Larger Teacher Model. In this experiment, we investigate whether using a larger teacher model, i.e., the model size of the teacher model is larger than the student, can further improve the model accuracy. We run experiments by replacing the small ResNet-18 backbone in the FSI teacher models with larger backbones of ResNet-50 and ResNet-101. The experiment results on MS COCO and Pascal VOC datasets are presented in Table 8. For both datasets, we observe that using a large backbone in the teacher network could not further improve the accuracy of the student models. This result can be explained as the mismatched capacity [6] that small students are unable to mimic large teachers on their classification capability. The performance even drops when ResNet-101 backbone is used due to the huge difference between the model sizes of the teacher model and the student model.

Working with Model Compression. KD and model compression are two complimentary directions. KD is to improve an originally lightweight model but compression is to reduce the model size while trying to maintain the accuracy. Therefore, a common practise is to first apply KD to improve the model's accuracy and then apply the compression to reduce the model size. We present the result of applying quantization [16] to a model distilled by our FSI method and an original base model in Table 9. Both models have similar accuracy drop after compression but our distilled model still performs better.

Training Efficiency. Because a small teacher model is used, FSI has advantages in greatly reduced training overhead compared to traditional KD with a large teacher model.

Training Speed. The proposed method also substantially reduces the training time in 3 aspects (1) *Simpler task:* Training the teacher model is much faster since the transformed input images make the training task simpler and we find that the model converges with around 40% less epochs; (2) *Smaller teacher model:* The savings come from the reduced model forward time and further speedup could come from using a larger batch size. (3) *Easier hyper-parameter tuning:* Determining the optimal hyper-parameter setting (i.e., weights among different loss terms) for the compared baseline methods requires significant amount of training/engineering efforts. Compared to our method, existing solutions have more loss terms and also need to find the optimal model size for the teacher.

Table 7. Large students'results.

Dataset	Model	Training state	Mask@0.5	Mask@[0.5,.95]
COCO	YOLACT ResNet50	Before teaching	45.92	27.97
		After teaching	49.71	30.43
	YOLACT ResNet101	Before teaching	48.01	29.73
		After teaching	52.06	31.95
	YOLACT++ ResNet50	Before teaching	52.71	33.69
		After teaching	54.49	35.03
	YOLACT++ ResNet101	Before teaching	53.17	34.46
		After teaching	54.84	35.55
Pascal	YOLACT ResNet50	Before teaching	72.34	46.87
		After teaching	72.59	48.93
	YOLACT ResNet101	Before teaching	72.72	48.26
		After teaching	73.11	50.21

Table 8. Large teachers on COCO and Pascal VOC.

T-Backbone	Method	MS COCO		Pascal VOC	
		Mask@0.5	Mask@[0.5,.95]	Mask@0.5	Mask@[0.5,.95]
ResNet-18	Student	40.44	23.76	68.98	43.74
ResNet-50	FSI-SFM	44.29	26.42	70.55	46.27
	FSI-FS	44.30	26.50	70.66	46.27
ResNet-101	FSI-SFM	43.97	26.23	69.62	44.65
	FSI-FS	43.77	26.11	68.89	44.30
ResNet-18	FSI-SFM	44.27	26.39	**71.04**	**46.21**
ResNet-18	FSI-FS	**44.43**	**26.53**	70.72	46.17

Table 9. Compressing model before and after distillation.

Model	mAP before compression	mAP after compression (4×)
ResNet-18 (Original)	23.76	22.89
ResNet-18 (Distilled)	26.53	25.62

5   Conclusion

In this paper, we introduce FSI, a foreground-specialized teacher model imitation framework for improving the accuracy of instance segmentation methods. Given that the teacher takes different input from the student, FSI incorporates two novel modules to have the student learn from the teacher better: (1) identifying the most essential teacher knowledge and (2) embedding the foreground-awareness into student's feature learning. We demonstrate the effectiveness of FSI on COCO and Pascal VOC by comparing them to KD baselines. The methodology presented in this work could have the potential to be applied to other multi-task learning problems.

References

1. Bolya, D., Zhou, C., Xiao, F., Lee, Y.J.: YOLACT++: better real-time instance segmentation. CoRR abs/1912.06218 (2019)
2. Bolya, D., Zhou, C., Xiao, F., Lee, Y.J.: YOLACT: real-time instance segmentation. In: ICCV (2019)
3. Chen, G., Choi, W., Yu, X., Han, T.X., Chandraker, M.: Learning efficient object detection models with knowledge distillation. In: NIPS (2017)
4. Chen, L., Papandreou, G., Schroff, F., Adam, H.: Rethinking atrous convolution for semantic image segmentation. CoRR abs/1706.05587 (2017)
5. Chen, S., Zhao, Q.: Attention-based autism spectrum disorder screening with privileged modality. In: ICCV (2019)
6. Cho, J.H., Hariharan, B.: On the efficacy of knowledge distillation. In: ICCV (2019)
7. Everingham, M., Eslami, S.M.A., Gool, L.V., Williams, C.K.I., Winn, J.M., Zisserman, A.: The pascal visual object classes challenge: a retrospective. IJCV **111**, 98–136 (2015)
8. Garcia, N.C., Morerio, P., Murino, V.: Modality distillation with multiple stream networks for action recognition. In: ECCV (2018)
9. Han, S., et al.: ESE: efficient speech recognition engine with sparse LSTM on FPGA. In: FPGA (2017)
10. Han, S., Mao, H., Dally, W.J.: Deep compression: compressing deep neural network with pruning, trained quantization and Huffman coding. In: ICLR (2016)
11. He, K., Gkioxari, G., Dollár, P., Girshick, R.B.: Mask R-CNN. In: ICCV (2017)
12. He, Y., Liu, P., Wang, Z., Hu, Z., Yang, Y.: Filter pruning via geometric median for deep convolutional neural networks acceleration. In: CVPR (2019)
13. Hinton, G.E., Vinyals, O., Dean, J.: Distilling the knowledge in a neural network. CoRR abs/1503.02531 (2015)
14. Howard, A.G., et al.: Mobilenets: efficient convolutional neural networks for mobile vision applications. CoRR abs/1704.04861 (2017)
15. Hu, J., Shen, L., Sun, G.: Squeeze-and-excitation networks. In: CVPR (2018)
16. Jacob, B., et al.: Quantization and training of neural networks for efficient integer-arithmetic-only inference. In: CVPR (2018)
17. Kim, Y., Park, E., Yoo, S., Choi, T., Yang, L., Shin, D.: Compression of deep convolutional neural networks for fast and low power mobile applications. In: ICLR (2016)
18. Kim, Y., Rush, A.M.: Sequence-level knowledge distillation. In: EMNLP (2016)

19. Lambert, J., Sener, O., Savarese, S.: Deep learning under privileged information using heteroscedastic dropout. In: CVPR (2018)
20. Lee, K., Ros, G., Li, J., Gaidon, A.: SPIGAN: privileged adversarial learning from simulation. In: ICLR (2019)
21. Lin, T., Dollár, P., Girshick, R.B., He, K., Hariharan, B., Belongie, S.J.: Feature pyramid networks for object detection. In: CVPR (2017)
22. Lin, T., Goyal, P., Girshick, R.B., He, K., Dollár, P.: Focal loss for dense object detection. In: ICCV (2017)
23. Lin, T.-Y., et al.: Microsoft COCO: common objects in context. In: Fleet, D., Pajdla, T., Schiele, B., Tuytelaars, T. (eds.) ECCV 2014. LNCS, vol. 8693, pp. 740–755. Springer, Cham (2014). https://doi.org/10.1007/978-3-319-10602-1_48
24. Liu, W., et al.: SSD: single shot MultiBox detector. In: Leibe, B., Matas, J., Sebe, N., Welling, M. (eds.) ECCV 2016. LNCS, vol. 9905, pp. 21–37. Springer, Cham (2016). https://doi.org/10.1007/978-3-319-46448-0_2
25. Liu, Y., Chen, K., Liu, C., Qin, Z., Luo, Z., Wang, J.: Structured knowledge distillation for semantic segmentation. In: CVPR (2019)
26. Luo, Z., Hsieh, J., Jiang, L., Niebles, J.C., Fei-Fei, L.: Graph distillation for action detection with privileged modalities. In: ECCV (2018)
27. Rastegari, M., Ordonez, V., Redmon, J., Farhadi, A.: XNOR-Net: ImageNet classification using binary convolutional neural networks. In: Leibe, B., Matas, J., Sebe, N., Welling, M. (eds.) ECCV 2016. LNCS, vol. 9908, pp. 525–542. Springer, Cham (2016). https://doi.org/10.1007/978-3-319-46493-0_32
28. Redmon, J., Divvala, S.K., Girshick, R.B., Farhadi, A.: You only look once: unified, real-time object detection. In: CVPR (2016)
29. Ren, S., He, K., Girshick, R.B., Sun, J.: Faster R-CNN: towards real-time object detection with region proposal networks. In: NIPS (2015)
30. Ronneberger, O., Fischer, P., Brox, T.: U-Net: convolutional networks for biomedical image segmentation. In: Navab, N., Hornegger, J., Wells, W.M., Frangi, A.F. (eds.) MICCAI 2015. LNCS, vol. 9351, pp. 234–241. Springer, Cham (2015). https://doi.org/10.1007/978-3-319-24574-4_28
31. Simonyan, K., Zisserman, A.: Very deep convolutional networks for large-scale image recognition. In: ICLR (2015)
32. Vapnik, V., Izmailov, R.: Learning using privileged information: similarity control and knowledge transfer. JMLR 16(1), 2023–2049 (2015)
33. Wang, T., Yuan, L., Zhang, X., Feng, J.: Distilling object detectors with fine-grained feature imitation. In: CVPR (2019)
34. Xie, E., et al.: Polarmask: single shot instance segmentation with polar representation. In: CVPR (2020)
35. Zhang, X., Zhou, X., Lin, M., Sun, J.: Shufflenet: an extremely efficient convolutional neural network for mobile devices. In: CVPR (2018)

Causal-SETR: A SEgmentation TRansformer Variant Based on Causal Intervention

Wei Li and Zhixin Li[✉]

Guangxi Key Lab of Multi-source Information Mining and Security,
Guangxi Normal University, Guilin 541004, China
lizx@gxnu.edu.cn

Abstract. We present a novel SEgmentaion TRansformer variant based on causal intervention. It serves as an improved vision encoder for semantic segmentation. Many studies have proved that vision transformers (ViT) can achieve a competitive benchmark on these downstream tasks, which shows that they can learn feature representations well. In other words, it is good at observing the instance from the image. However, in the human visual system, to recognize the objects in the scene, it is necessary to observe the objects themselves and introduce some prior knowledge for producing higher confidence results. Inspired by this, we introduced a structural causal model (SCM) to model images, category labels, and context. Beyond observing, we propose a causal intervention method by removing the confounding bias of global context and plugging it in the ViT encoder. Unlike other sequence-to-sequence prediction tasks, we use causal intervention instead of likelihood. Besides, the proxy training objective of the framework is to predict the contextual objects of a region. Finally, we combine this encoder with the segmentation decoder. Experiments show that our proposed method is flexible and effective.

Keywords: Causal intervention · Vision transformer · Semantic segmentation

1 Introduction

Semantic segmentation divides visual input into different semantically interpretable categories, which is a challenging task requiring accurate prediction of the object category, shape, and location. Both convolutional-based encoders [5, 19, 25] and transformer-based encoders [10, 17, 38] are good at telling us "**what**", but not "**why**". In particular, once the input image has been fed to the encoder, the rich

This work is supported by National Natural Science Foundation of China (Nos. 62276073, 61966004), Guangxi Natural Science Foundation (No. 2019GXNSF DA245018), Guangxi "Bagui Scholar" Teams for Innovation and Research Project, and Guangxi Collaborative Innovation Center of Multi-source Information Integration and Intelligent Processing.

L. Wang et al. (Eds.): ACCV 2022, LNCS 13847, pp. 414–430, 2023.
https://doi.org/10.1007/978-3-031-26293-7_25

and effective feature representation can be learned to provide a high confidence probability $P(Y)$. Furthermore, some empirical investigations [11, 26, 36, 37] use prior knowledge, graph neural networks and other techniques to learn the co-occurrence probability $P(Y|X)$ between objects and integrate it with the probability learned by the encoder for joint learning in order to increase the prediction probability. However, Wang et al. [30] raise questions about the validity of the co-occurrence probability learned by machine, and they think the machine usually fails to describe the exact visual relationships, or, even if the prediction is correct, the underlying visual attention is illogical. Improving the capabilities of the segmentation systems by acquiring higher co-occurrence information with a better degree of confidence is thus a crucial issue.Contextual information is crucial for semantic segmentation tasks. Some methods based on graph convolutional networks (GCN) learn the rich context to improve the feature representation capability. Moreover, Zheng et al. [38] provide a rethinking of the segmentation model and contribute a new encoder-decoder architecture built by pure transformers. This architecture does not involve spatial resolution down-sampling, but rather global context modeling at each layer of the encoder transformer. With the global context, they propose a new perspective to the semantic segmentation. As previously stated, the machine is incapable of describing the precise visual relationship. (For example, the "visual" simply conveys the "what" or "where" of a "person" or "car".) It is just a more descriptive symbol than its corresponding English word. When there is a bias, such as when more "car" areas than "human" regions co-occur with the term "road," visual attention is more likely to focus on the "car" region. These works [6, 7, 31] attempt to introduce unsupervised external features to obtain more robust co-occurrence relations, thus improving the segmentation performance. Contrary to human's recognition system, current deep learning approaches cannot yet extract or explain causality.

According to these causal theories [21–23], we intend to reconsider the segmentation based on the vision transformer model design and contribute a causal intervention attempt. In particular, we model images, category labels, and contextual information using SCM and eliminate confounder bias through causal intervention. Thus, we obtain contextual information regarding causality to drive the learning of more robust semantic features and the exploration of the deeper relationships between various objects. In addition, we design a fusion module for integrating the original feature and the causality context, so intervening in the learnt features and making the learning process more like the human learning process. It's worth noticing that the proposed module is plug and play. We can easily plug it into other downstream tasks. In summary, we make the following contributions in this paper:

- We introduced the structural casual model to model images, category labels, contextural information, and removed the observation bias by causal intervention. Thus, we get the contextural causality information, which collects more robust semantic relations.
- We incorporated external knowledge into the processing of causal intervention as well as further guided the ViT model to provide a more robust feature representation.

– We designed a fusion module for integrating the original feature and the causality context, which performs reasoning and directs the downstream task (*e.g.*, semantic segmentation) to explore more causality interconnections.

To demonstrate the efficacy of the proposed method, we duplicated these approaches into an end-to-end training network and performed extensive experiments on the benchmark of semantic segmentation. Experiments demonstrate that our proposed methods are practicable and efficient.

2 Related Work

2.1 Vision Transformer

The most related to our work is the vision transformer (ViT) [10] and its variants [28,32]. ViT treats an image as a set of fixed size (i.e., 16×16) and non-overlapping patches, then directly feeds them to a transformer architecture. Thus, it converts the dense prediction task to a sequence-to-sequence task. Compared to CNNs, it achieves a competitive speed-accuracy tradeoff on classification. However, ViT requires large-scale training datasets (i.e., JFT-300M). DeiT [29] adapts the knowledge distillation for reducing the complexity and finetuning the ViT, allowing ViT to be effective using the smaller ImageNet-1K dataset. We noticed that ViT lefts the results of image classification. However, it is still unsuitable for use as a general-purpose backbone on dense prediction tasks or handling high-resolution images due to its low-resolution feature map and the quadratic increase in complexity with image resolution. DETR [3], SETR [38] directly upsampling or deconvolution the features but with dissatisfied performance in detection and segmentation respectively. As far as we know, no one has tried to introduce the perspective of causality into ViT for semantic segmentation. Empirically, our proposed approaches are effective and flexible, achieving a new state-of-the-art in semantic segmentation task.

2.2 Causality in Vision

Due to the fact that deep learning is an effective yet unexplained black box, more and more academics are attempting to combine its complementary strengths. Causal inference [21–23] has been researched in several domains, including classification [4,20], adversarial learning [14,15], and reinforcement learning [2,9]. The most related to our work is VC R-CNN [30]. They constructed a causal region of interest (RoI) using Faster R-CNN [27] and then use this contextual RoI further to improve the performance of several multimodal downstream tasks, including image caption (IC), visual question answering (VQA), vision common-sense reasoning (VCR).

The core idea between ours and VC R-CNN is backdoor adjustment solution. However, they did not report the potential interest in semantic segmentationfield. We observed that semantic segmentation tasks also require causal contextual information for advancement. Due to the task gap, we cannot directly introduce the causal RoI information [30] as context in our investigation. Therefore, we reconstructed a structural causal model on the benchmark of semantic segmentation tasks. Despite the fact that we both intend to mine the rich contextual via a backdoor modification approach, VC R-CNN uses a backdoor adjustment method to eliminate the visual bias caused by the model's "observing" behavior. Consequently, VC R-CNN may learn "common sense" without any external monitoring. It is important to note that the most significant distinction between VC R-CNN and our approach is that we additionally intervene on external knowledge to improve the performance of semantic segmentation.

CONTA [35] proposes a contextual adjustment network to improve the semi-supervised semantic segmentation benchmark. Similar to the case with the causal RoI features [30] , we cannot directly utilize the CONTA-supplied SCM. On the one hand, the backbone we use is a ViT rather than a CNN, and on the other hand, the general paradigm for weakly supervised semantic segmentation tasks does not correspond to the paradigm used for fully supervised image recognition tasks. We aim to improve the performance of ViT via adapting the backdoor adjustment solution for semantic segmentation tasks.

Hybridization effect will result in harmful bias, mislead attention module to learn false correlation in data, and consequently reduce the model's generalizability. However, Xu et al. [34] think that confounding is unobservable. Thus, they propose a novel attention mechanism: causal attention (CATT) which can eliminate the confounding effect in existing attention-based vision-language models. Unlike CATT, we employ backdoor adjustment solution as opposed to front-door adjustment solution. Furthermore, we focus on designing a generic ViT architecture via causal intervention, not a attention mechanism.

In summary, we aim to model the semantic segmentation tasks in detail using backdoor adjustment. VC R-CNN reported the inspiring performance in IC, VQA and VCR tasks. Similarly, CONTA reported the good performance in semi-supervised semantic segmentation task. They eliminate the observation biases from within the model using backdoor adjustment. However, we extracted some common sense from an external knowledge dataset which are presented as textual data, and introduced them into the backdoor adjustment processing. For segmentation tasks, contextual information plays an important role, thus, we convert the external knowledge (textual data) to visual commonsense features using GCN. Afterthat, we notice that ViT is a unified framework for modeling language and vision, and we make an attempt to rethink the advantage of causal intervention in the ViT-based model. Different from CNN, ViT lacks some inductive bias (*e.g.*, invariance, local connectivity, weight sharing) due to the framework design. Therefore, we complement external knowledge with the strengths of ViT from a causal perspective. As a result, we further improve the benchmark in segmentation tasks.

3   Methods

We attempt to intervene in the feature representation learned by the vision transformer encoder, thus, obtaining more explanatory contextual information, which improves the performance of semantic segmentation tasks. For example, the deep encoder learns contextual information with observation bias from the dataset (*e.g.,* if there are "car", "road" and "person" co-occur in an input image, the encoder is more likely to focus on the common co-occurrence relationships in the dataset.). Perhaps the classification result is correct, but the underlying visual attention is not reasonable. To address this, we propose the causal-intervention-based framework for obtaining a more causal context. The overview of the framework for semantic segmentation is shown in Fig. 1. It is worth noting that our proposed method can be used plug and play on any transformer-based encoders and is compatible with downstream recognition tasks.

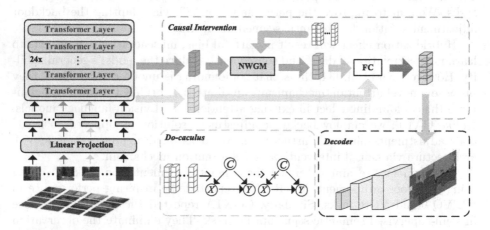

Fig. 1. Overview of our proposed semantic segmentation framework.

Visual attention is effective at learning the correlation $(P(Y|X))$ between objects. However, it is limited to the fixed input image. Therefore, it can only learn the explicit correlation in this image. In other words, visual attention is incapable of observing nonexistent objects in the input image. It disregards the implicit causality that causes the observation bias to confound the existence of objects X and Y.

To mine the implicit causality, we first build the confounder set $C \in \mathbb{R}^{N \times D}$. N represents the number of objects in the datasets,, and D is the dimension of the middle output produced by the feature encoder. Besides, we build another C' using external knowledge. Then, we "take" the objects C from other context, and "put" them around X and Y for testing if X causes the existence of Y when given C. The operators ("take" and "put") are the paradigm of intervention, implying that the probability of C depends on human intervention, but is independent on X or Y. By intervening, we force the conversion of the correlation observed only in the fixed image to a global or external causal-based context. ($P(Y|X) \rightarrow P(Y|do(X))$)

More intuitively, human will not make corresponding inferences just based on what they "see" in front of eyes. That is what we are different from the machine's visual recognition system. For example, we always keep rethinking or imaging "If there are other objects C , will object X still causes object Y?" instead of the passive observation:"If there is object X, how likely there will exist object Y? ". Thanks to intervention, we convert $P(Y|X)$ to $P(Y|do(X))$. We simulate the $do - calculus$ by "taking" non-local context that even might not be in the input image, "putting" them around pairs of objects that we want to intervene.

3.1   Structural Causal Model

As shown in Fig. 2, we intuitively demonstrate the principle of *do-calculus*. Specifically, we formulate causalities among observed objects X, confounder set C, and observed objects Y with a structural causal model. The symbol "$\rightarrow$" denotes the causalities between two nodes (*e.g.*, X causes Y).

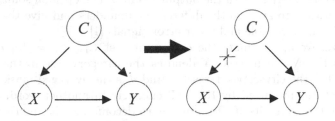

Fig. 2. Modeling the causalities by SCM

$C \rightarrow X$. It is widely known that context C affects the performance of the semantic segmentation model. In other words, C guide the model to what or where is "car", "road", and "person" in an image. We can hardly ever build a generative context for $C \rightarrow X$, but we will introduce an ingenious method to extract the links suitable for segmentation tasks

$X \rightarrow Y$. The link between X and Y denotes X causes Y (*e.g.* , if X exists in the image, what is the probability of Y exists in the same image?). It is learned by the likelihood: $P(Y|X)$. Although we have seen many successful convolutional-based methods make great progress by alleviating this likelihood, we still firmly believe that this is biased.

$X \leftarrow C \rightarrow Y$. The confounder set C can be interpreted as a generic contextual corpus. It will cause both of X and Y by "taking" the implicit context that can not be observed in the local images and "putting" it to the local receptive field. However, it might leads to spurious correlations by only learning from the likelihood which is formulated as:

$$P(Y|X) = \sum_c P(Y|X,c)P(c|X) \tag{1}$$

Where the context C introduces the observational bias $P(c|X)$. We are more concerned with the prediction of Y, so the causal link between C and X is not the major link we need to focus on. If we intervene X, the causal link between C and X is cutoff. We apply the Bayes rules again. Thus, we have:

$$P(Y|X) = \sum_c P(Y|X,c)P(c) \tag{2}$$

where $P(c)$ denotes the probability of prior label, it deliberately forces X to incorporate every c fairly. $c \in C$ is the set of the objects from the contextual corpus.

3.2 Deconfounding Bias

Where will we intervene in the human reasoning system when we find that confounding factors interfere with the output results? The optimal solution defined by mathematics is to average the different confounders and give the maximum weight to the more reliable and fewer error signals [1].

Therefore, we approximate the confounder set $C_{\text{internal}} = \{c_1, c_2, \ldots, c_n\}$, where c_n is the $N \times d$ matrix, N denotes the category size in the datasets (e.g., $N = 19$ in the cityscapes dataset) , and d is the averaged mask of the i-th category features produced by the ViT encoder. In another word, C_{internal} is produced by the model itself which means to deconfound the internal bias.

Furthermore, we acquire an **external knowledge** set $\mathcal{E}$ from Visual Genome dataset [16]. However, the Visual Genome dataset consists of 30K object categories for the specific downstream task. For the semantic segmentation task, we attempt to mine the co-occurrence relationships about different object categories that appear in Cityscapes and ADE20K. Specifically, we get the subset $\mathcal{E}^{\text{external}} \in \mathbb{R}^{C \times C}$ of $\mathcal{E}$. C is set to 150 (ADE20K contains 150 object categories and overlaps with Citiscapes). $\mathcal{E}^{\text{external}} \in \mathbb{R}^{150 \times 150}$ is a 150×150 symmetric matric which means the relation pairs are symmetric. After that, we normalized the matrix elements to obtain D. $D_{ii} = \sum_{j=1}^{C} \mathcal{E}_{ij}$ The final $\mathcal{E}^{\text{external}} \in \{\mathcal{E}_{00}, \cdots \mathcal{E}_{ij}\}$ is calculated by $\mathcal{E}_{ij} = \frac{\mathcal{E}_{ij}}{\sqrt{D_{ij}D_{jj}}}$. To maintain the consistency between semantic relations and visual features, we introduce a graph structure $\mathcal{G} = (\mathcal{N}^L, \mathcal{E}^{\text{external}})$, where $\mathcal{N}^L$ is produced by global vectors for word representation such as GloVe [24]. Besides, we feed it to two GCN layers to capture external semantic knowledge. Each GCN layer is formulated by

$$H^{(l+1)} = \sigma(\widetilde{D}^{-\frac{1}{2}} \mathcal{E}^{\text{external}} \widetilde{D}^{-\frac{1}{2}} H^{(l)} W^{(l)}) \tag{3}$$

Where $\widetilde{D}$ denotes the degree of $\mathcal{E}^{\text{external}}$. We get $H^1 \in \mathbb{R}^{C \times d}$ and $H^2 \in \mathbb{R}^{C \times D}$ through each GCN layer respectively. C and d denote the number of objects and the dimension of the representation, respectively. D denotes the depth of visual features produced by GCN.

To project $H^2 \in \mathbb{R}^{C \times D}$ and the visual features $X_{\text{visual}} \in \mathbb{R}^{D \times W \times H}$ to a common subspace, we adopt a feature mapping module. Specifically, we compressed the dimension of $X_{\text{visual}} \in \mathbb{R}^{D \times W \times H}$ to $\hat{X}_{\text{visual}} \in \mathbb{R}^{D \times (W \odot H)}$. Then, we transpose the dimension of $\hat{X}_{\text{visual}}$ to $\mathbb{R}^{(W \odot H) \times D}$. We further compressed channels of visual features with the same of the numbers of objects: $\hat{X}_{\text{visual}} \in \mathbb{R}^{(W \odot H) \times C}$ with two fully connection (FC) layers $F_1 \in \mathbb{R}^{D \times C}$. After that, we flattered the multiple of H^2 and $\hat{X}_{\text{visual}}$ and got $\hat{K} \in \mathbb{R}^{D \times W \times H}$. Furthermore, we concatenated X_{visual} and $\hat{K}$ and fed it to FC layer $F_2 \in \mathbb{R}^{2D \times D}$. In this way, we have converted the textual to the visual features.

However, despite the major challenge in trading off between annotation cost and noisy multi-modal pairs, common sense is not always recorded in text due to the reporting bias. Thus, we try to alleviate the bias with the intervention. With the same as the internal knowledge, we average mask of the i-th category features in the dimension D and got the deconfound external knowledge C_{external}. The overall logits are formulated by

$$C_{\text{external}} = \text{AVG}(F_2\{[\phi(F_1(\phi(X_{\text{visual}}))^T \odot H^2)]||X_{\text{visual}}\}^T) \qquad (4)$$

where $\phi(\cdot)$ denotes the dimension transpose function. $\odot, ||$ are matrix multiplication and matrix concatenation respectively.

3.3 Causal Intervention Module

Recalling the Fig. 1, we get X's context (see in the red arrow) x and Y (see in the yellow arrow) after the image fed to the ViT encoder. The last layer of classification tasks is the Softmax layer: $P(y^c|x,c) = Softmax(f_y(x,c))$, where $f_y(\cdot)$ calculates the logits for N categories, and y denotes that $f(\cdot)$ is parameterized by Y's context y. The overall output of logits is defined as:

$$P(Y|do(X)) := \mathbb{E}_c[Softmax(f_y(x,c))] \qquad (5)$$

We use the normalized weighted geometric mean (NWGM) [33] to move the outer expectation into the Softmax function as:

$$\mathbb{E}_c[Softmax(f(x,c))] \approx Softmax(\mathbb{E}_c[f_y(x,c)]) \qquad (6)$$

For the classification task, we use the linear model $f_y(x,c) = \mathbf{W}_1 x + \mathbf{W}_2 \mathbb{E}[g_y(c)]$, where $\mathbf{W}_1, \mathbf{W}_2$ denote the fully connected layers, it is formulated by:

$$\mathbb{E}_c[f_y(x,c)] = \mathbf{W}_1 x + \mathbf{W}_2 \mathbb{E}_c[g_y(c)] \qquad (7)$$

where $\mathbb{E}_c[g_y(c)]$ is calculated by the attention mechanism. Specifically, we are given $y \in Y$ and $c \in C$, the attention vector α is calculated by $softmax(q^T K/\sqrt{\sigma})$, then, we get $A = [\alpha; \ldots, \alpha]$ by the broadcasting operation. The most intuitive explanation is that we use attention mechanism to obtain the focus point between two objects, where [;] denotes broadcasting along the row. $q = W_3y, K = W_4C^T$. W_3, W_4 map each vector to the common subspace and σ is a constant scaling factor with the first dimension of W_3, W_4. Finally, $\mathbb{E}[g_y(c)] = \sum_c [A \odot C]P(c)$, where $\odot$ and $P(c)$ denote the element-wise product and prior statistic probability respectively. In summary, we obtain regions of interest similar to human visual system from a global perspective.

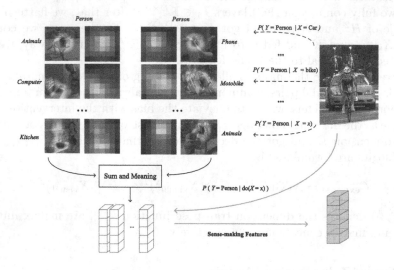

Fig. 3. "sense-making" processing.

3.4   Objective Function

Given the two features, $x \in \mathbf{X}$ and $y \in \mathbf{Y}$ from ViT encoder, x and the context C are fed to the NWGM module, and the removed confounder features are obtained. Furthermore, we adapt the fully connected layer to learned the relationship p_i between y and $do(x)$. The loss of this processing is $\frac{1}{K} \sum_i \mathcal{L}_{cxt}(p_i, y_i^c)$, Finally an enhanced feature is feed to the decoder. Our training objective is formulated by:

$$\mathcal{L} = \mathcal{L}_{seg}(p_i, \hat{y}_i^c) + \frac{1}{K} \sum_i \mathcal{L}_{cxt}(p_i, y_i^c) \tag{8}$$

where $\hat{y}_i^c$ denotes the ground-truth label provided by dataset, and y_i^c denotes the label on sub classification task. According to $P(Y|do(X))$, Y is one of the K context objects with the label y_i^c. $\mathcal{L}_{cxt}(p_i, y_i^c)$ is calculated by $-\log(p_i[y_i^c])$.

4   Experiments

We conduct experiments on semantic segmentation task semantic segmentation with Cityscapes and ADE20K. The details are as below.

4.1 Settings

Datasets: We use two commonly used datasets for semantic segmentation: Cityscapes [8], ADE20K [39]. The Cityscapes dataset contains 5000 images of driving scenes in urban environments (2975 for train, 500 for validation, 1525 for test). The resolution per image is 1024×2048 contains 19 categories of fine-grained annotations. ADE20K contains over 25K images (20k for training, 2k for validation, 3k for test). These images are densely annotated with an open dictionary label set.

Metric: We use mean intersection-over-union (mIoU) to calculate the ratio of the intersection and union of the two sets of true and predicted values. The classification task returns a true positive (TP), false positive (FP), true negative (TN) and false-negative (FN). It is formulated by

$$
\begin{aligned}
\mathrm{mIoU} &= \frac{\mathrm{TP}}{\mathrm{FP} + \mathrm{FN} + \mathrm{TP}} \\
&= \frac{1}{k+1} \sum_{i=0}^{k} \frac{p_{ii}}{\sum_{j=0}^{k} p_{ij} + \sum_{j=0}^{k} p_{ji} - p_{ii}}
\end{aligned}
\tag{9}
$$

where p_{ij} represents the total number whose true value is i but predicted to be j. p_{ii} denotes the number of TP, p_{ij} and p_{ji} denote FP and FN, respectively. $k + 1$ is the count of classes (including the background class.)

Implementation Details: We use the mmsegmentation toolbox to carry on experiments effectively. (i) the random resize with a ratio between 0.5 and 2, the random cropping $(768, 512)$ and 480 for Cityscapes, ADE20K, and the random horizontal flipping during training all the experiments. (ii) The total iteration is set to 80,000 and 160,000 for the experiments on Cityscapes and ADE20K, respectively, and both cases with batch size 16 and 8, respectively. (iii) We adopt a polynomial learning rate decay schedule and employ SGD as the optimizer. Momentum and weight decay are set to 0.9 and 0, respectively, for all the experiments on the two datasets. The initial learning rates are set to 0.001 on ADE20K and 0.01 on Cityscapes. (iv) To obtain the context C, we employed the pre-trained ViT model with the ground-truth labels as the input to extract the features for each object.

4.2 Comparision to State-of-the-Art

We conducted comparative experiments on some representative models; the results are shown in Table 1. APNB [40], CCNET [13], SPNet [12] proposed to explore the way of enhancing the ability of spatial contextual representation. SETR [38], Segmenter [28] provides us with a new perspective, that is, using transformer-based encoder [18] to capture richer and more effective global context semantic information. Swin Transformer provides an effective patch embedding method based on shift windows to reduce network size. However, all of the above models do not provide a causal explanation. It is worth noting that we only plug the causal intervention module with the internal and external context in SETR and achieve a new SOTA in both Cityscapes and ADE20K.

Table 1. Comparison with the state-of-the-art methods.

Method	Publication	Backbone	mIoU (%)	
			Cityscapes	ADE20K
APNB	ICCV'19	ResNet101	81.30	45.24
CCNET	ICCV'19	ResNct101	81.40	45.76
SPNet	CVPR'20	ResNet101	82.00	45.60
EfficientFCN	ECCV'20	ResNet101	–	45.28
KRNet	ICASSP'21	ResNet101	82.20	45.65
SETR	CVPR'21	ViT-L	82.15	50.28
Swin Transformer	ICCV'21	Swin-L	–	53.50
Segformer	NIPS'21	Seg-L-Mask/16	82.20	51.80
Segmenter	ICCV'21	ViT-L	–	53.63
Ours	–	ViT-L	**83.21**	**54.48**

Qualitative Analysis. Qualitative results are shown in Fig. 4. We use different coloured boxes to mark the differences between our model and the SETR. From the figure, we can observe that our proposed model has more accurate and more fine segmentation performance (marked with orange or yellow boxes). More objects are segmented: small, occluded, and indistinct segments. Global context information can effectively learn the co-occurrence relationship between different objects. However, some examples of objects being misclassified are circled with black boxes. It is effective for using causal intervention to remove the bias in contextual information.

4.3   Ablation Study

The Contributions of Different Module: To fairly evaluate our proposed method, we carry out different settings and report results in Table 2. Multi-scale test with random flipping (MS+Flip) is commonly used to improve semantic segmentation performance. We plugged the internal contextual information, removed confounder bias in SETR, and achieved 52.35% mIoU. It significantly increases mAP over the baseline by up to 2.07. By using MS+Filp, we achieved 53. 28% mIoU. Furthermore, we study the influence of introducing external contextual information with two settings: 1) The external knowledge without de-confounding (marked by **External**); 2) The external knowledge with de-confounding (marked by **External***). We first introduced the external knowledge without de-confounding, and got the margin improvement (0/13%mIoU).

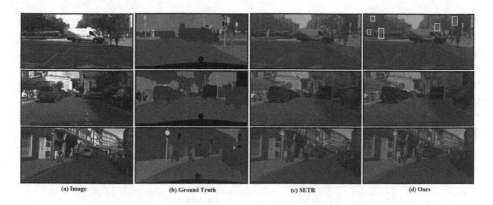

Fig. 4. Qualitative results on Cityscapes dataset.

Secondly, with de-confounded external knowledge, we achieve an improvement of 1.05%mIoU.Plugging the de-confounded external knowledge without MS+Flip, the mIoU is up to 52. 70%. Finally, we adopt MS+Flip, internal knowledge, and de-confounded external knowledge. The overall performance is over SETR by up to 8%. In summary, it is effective to improve the performance of ViT by introducing internal knowledge and external knowledge.

Table 2. Ablation study on ADE20K datasets. Internal: intervention with internal knowledge. External: The external knowledge without de-confounding. External*: The external knowledge with de-confounding

Method	MS+Filp	Internal	External	External*	mIoU (%)
SETR	✓				50.28
Ours		✓			52.35
Ours	✓	✓			53.28
Ours			✓		50.41
Ours	✓		✓		50.94
Ours				✓	51.33
Ours	✓			✓	52.70
Ours		✓	✓		52.93
Ours	✓	✓	✓		53.40
Ours		✓		✓	53.93
Ours	✓	✓		✓	54.48

The Influence of Adapting Different GCN Layers: As mentioned in Sect. 3.2, we map the natural language co-occurrence probability in the Visual Gnome dataset to a common feature space using the word embedding method.

Then, the GCN is used to extract the semantic information of the different objects from the common feature space. Therefore, it is necessary to discuss how many GCN layers are most beneficial for our method. As shown in Table.3, when the number of GCN layers increases, segmentation performance drops on both datasets. The optimal layer number of GCN is related to the sparsity degree of the adjacency matrix. When the sparsity degree of the graph is low, the over-smoothing phenomenon will soon occur. As a result, performance degradation occurs when more GCN layers are used.

Table 3. The influence of different depths of GCN in external knowledge mapping.

Layers	Encoder	Cityscapes	ADE20K	mIoU (%)
2 layers	ViT-L	✓		**81.58**
2 layers	ViT-L		✓	**52.70**
3 layers	ViT-L	✓		79.13
3 layers	ViT-L		✓	51.17
4 layers	ViT-L	✓		79.01
4 layers	ViT-L		✓	50.85
5 layers	ViT-L	✓		78.71
5 layers	ViT-L		✓	50.58

The Influence of Different Word Embedding Methods: Similar to Sect. 4.3, we use different word embedding methods to integrate external knowledge better. Thus, we investigate four different word embedding methods, including Word2vec, GoogleNews, GloVe and the FastText word embedding. Figure 5 shows the results using different word embeddings on Cityscapes and ADE20K. From the figure, we can see that when using different word embeddings as graph's nodes, the segmentation mIoU will not be affected significantly. Furthermore, using GloVe could lead to better performance. The reason is that the word embeddings learned from large text corpus maintain some implicit knowledge.

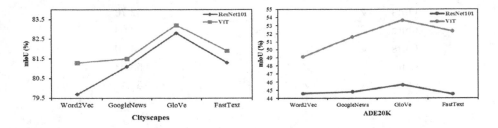

Fig. 5. Influence of word embedding methods

4.4   Plug and Play

We chose several representative ViT-based encoders and plugged our method into them to test its extensibility. For each method, we select the backbone encoder that performs best on the ADE20K benchmark. This also implies that they occupy the greatest number of parameters. For more details, we show in Table 4. We plug our method in SETR (the baseline) and achieve 54.48 % mIoU. It is up to 4.20 % higher than SETR. Swin-Large [18], Seg-L-Mask/16 [28] is the advanced version of ViT. Swin-Large introduces the W-MSA operation, which reduces computation complexity. They use windows instead of patches, which makes the computational complexity of W-MSA linear with image size. Seg-L-Mask/16 adapts image patch join processing and class embedding for re-modeling the global context. Furthermore, the Mask Transformer can perform direct segmentation rather than class embedding. By incorporating our method, the mIoU is increase to 0.68 % and 1.33 %, respectively. SegFormer contains a novel hierarchical Transformer encoder as well as a lightweight All-MLP decoder. It generates multi-scale features that do not require position coding, thereby avoiding position-coding interpolation, which leads to performance degradation when the test resolution differs from the training resolution. By plugging our method into Segformer, we achieveWe achieve 53.71 % mIoU (up to 1.46 % higher than SegFormer). It is worth noting that our method introduces extra parameters with calculating scale dot-product attention ($2 \times 512 \times 1024$), linear addition ($2 \times N \times 1024$) and feature embedding ($N \times 1024$). N denotes the count of categories. In short, we improved on several benchmarks by adjusting a few parameters. This demonstrates that our method works and that it can be applied to any other ViT-based segmentation encoder.

Table 4. Plug our method in different ViT-based Methods.

Method	Backbone	Params (M)	Original mIoU (%)	mIoU (%)
SETR	ViT-large	308	50.30	54.48
Swin-Transformer	Swin-large	234	53.53	54.21
SegFormer	MiT-B5	84	51.80	53.13
Segmenter	Seg-L-Mask/16	307	52.25	53.71

5   Conclusions

We provide a rethinking to semantic segmentation based on vision transformer model design and contribute a causal intervention attempt. Different from other tasks, we explained the model based on causal intervention. Using only feature concatenation, we improve on segmentation task, and then the model is closer to the human recognition system. Furthermore, causality can not only be explained by intervention but also many, counterfactual methods deserve further consideration. Therefore, we will further use causality to explore the next generation of artificial intelligence in the future.

References

1. Badde, S., Hong, F., Landy, M.S.: Causal inference and the evolution of opposite neurons. Proceed. Nat. Acad. Sci. **118**(36), e2112686118 (2021)
2. Bengio, Y., et al.: A meta-transfer objective for learning to disentangle causal mechanisms. arXiv preprint arXiv:1901.10912 (2019)
3. Carion, N., Massa, F., Synnaeve, G., Usunier, N., Kirillov, A., Zagoruyko, S.: End-to-end object detection with transformers. In: Vedaldi, A., Bischof, H., Brox, T., Frahm, J.-M. (eds.) ECCV 2020. LNCS, vol. 12346, pp. 213–229. Springer, Cham (2020). https://doi.org/10.1007/978-3-030-58452-8_13
4. Chalupka, K., Perona, P., Eberhardt, F.: Visual causal feature learning. arXiv preprint arXiv:1412.2309 (2014)
5. Chen, L.C., Papandreou, G., Kokkinos, I., Murphy, K., Yuille, A.L.: DeepLab: semantic image segmentation with deep convolutional nets, atrous convolution, and fully connected CRFs. IEEE Trans. Pattern Anal. Mach. Intell. **40**(4), 834–848 (2017)
6. Chen, S., Li, Z., Tang, Z.: Relation R-CNN: a graph based relation-aware network for object detection. IEEE Signal Process. Lett. **27**, 1680–1684 (2020)
7. Chen, S., Li, Z., Yang, X.: Knowledge reasoning for semantic segmentation. In: ICASSP 2021–2021 IEEE International Conference on Acoustics, Speech and Signal Processing (ICASSP), pp. 2340–2344 (2021)
8. Cordts, M., et al.: The cityscapes dataset for semantic urban scene understanding. In: Proceedings of the IEEE/CVF Conference on Computer Vision and Pattern Recognition, pp. 3213–3223 (2016)
9. Dasgupta, I., et al.: Causal reasoning from meta-reinforcement learning. arXiv preprint arXiv:1901.08162 (2019)
10. Dosovitskiy, A., et al.: An image is worth 16x16 words: Transformers for image recognition at scale. arXiv preprint arXiv:2010.11929 (2020)
11. Fu, J., et al.: Dual attention network for scene segmentation. In: Proceedings of the IEEE/CVF Conference on Computer Vision and Pattern Recognition, pp. 3146–3154 (2019)
12. Hou, Q., Zhang, L., Cheng, M.M., Feng, J.: Strip pooling: rethinking spatial pooling for scene parsing. In: Proceedings of the IEEE/CVF Conference on Computer Vision and Pattern Recognition, pp. 4003–4012 (2020)
13. Huang, Z., Wang, X., Huang, L., Huang, C., Wei, Y., Liu, W.: CCNet: Criss-Cross attention for semantic segmentation. In: Proceedings of the IEEE/CVF International Conference on Computer Vision, pp. 603–612 (2019)
14. Kalainathan, D., Goudet, O., Guyon, I., Lopez-Paz, D., Sebag, M.: Sam: structural agnostic model, causal discovery and penalized adversarial learning (2018)
15. Kocaoglu, M., Snyder, C., Dimakis, A.G., Vishwanath, S.: CausalGAN: learning causal implicit generative models with adversarial training. arXiv preprint arXiv:1709.02023 (2017)
16. Krishna, R., et al.: Visual genome: connecting language and vision using crowd-sourced dense image annotations. Int. J. Comput. Vision **123**(1), 32–73 (2017)
17. Li, Z., Sun, Y., Zhu, J., Tang, S., Zhang, C., Ma, H.: Improve relation extraction with dual attention-guided graph convolutional networks. Neural Comput. Appl. **33**(6), 1773–1784 (2021)
18. Liu, Z., et al.: Swin transformer: hierarchical vision transformer using shifted windows. arXiv preprint arXiv:2103.14030 (2021)

19. Long, J., Shelhamer, E., Darrell, T.: Fully convolutional networks for semantic segmentation. In: Proceedings of the IEEE Conference on Computer Vision and Pattern Recognition, pp. 3431–3440 (2015)
20. Lopez-Paz, D., Nishihara, R., Chintala, S., Scholkopf, B., Bottou, L.: Discovering causal signals in images. In: Proceedings of the IEEE Conference on Computer Vision and Pattern Recognition, pp. 6979–6987 (2017)
21. Pearl, J.: Theoretical impediments to machine learning with seven sparks from the causal revolution. arXiv preprint arXiv:1801.04016 (2018)
22. Pearl, J., Glymour, M., Jewell, N.P.: Causal inference in statistics: a primer. John Wiley & Sons (2016)
23. Pearl, J., Mackenzie, D.: The book of why: the new science of cause and effect. Basic books (2018)
24. Pennington, J., Socher, R., Manning, C.D.: Glove: global vectors for word representation. In: Proceedings of the 2014 Conference on Empirical Methods in Natural Language Processing, pp. 1532–1543 (2014)
25. Quan, Y., Li, Z., Chen, S., Zhang, C., Ma, H.: Joint deep separable convolution network and border regression reinforcement for object detection. Neural Comput. Appl. **33**(9), 4299–4314 (2021)
26. Redondo-Cabrera, C., Baptista-Ríos, M., López-Sastre, R.J.: Learning to exploit the prior network knowledge for weakly supervised semantic segmentation. IEEE Trans. Image Process. **28**(7), 3649–3661 (2019)
27. Ren, S., He, K., Girshick, R., Sun, J.: Faster R-CNN: towards real-time object detection with region proposal networks. Adv. Neural. Inf. Process. Syst. **28**, 91–99 (2015)
28. Strudel, R., Garcia, R., Laptev, I., Schmid, C.: Segmenter: transformer for semantic segmentation. arXiv preprint arXiv:2105.05633 (2021)
29. Touvron, H., Cord, M., Douze, M., Massa, F., Sablayrolles, A., Jégou, H.: Training data-efficient image transformers & distillation through attention. In: International Conference on Machine Learning, pp. 10347–10357. PMLR (2021)
30. Wang, T., Huang, J., Zhang, H., Sun, Q.: Visual commonsense representation learning via causal inference. In: Proceedings of the IEEE/CVF Conference on Computer Vision and Pattern Recognition (2020)
31. Wei, H., Li, Z., Huang, F., Zhang, C., Ma, H., Shi, Z.: Integrating scene semantic knowledge into image captioning. ACM Trans. Multimedia Comput. Commun. Appl. (TOMM) **17**(2), 1–22 (2021)
32. Wu, H., Xiao, B., Codella, N., Liu, M., Dai, X., Yuan, L., Zhang, L.: CvT: introducing convolutions to vision transformers. arXiv preprint arXiv:2103.15808 (2021)
33. Xu, K., et al.: Show, attend and tell: neural image caption generation with visual attention. In: International Conference on Machine Learning, pp. 2048–2057 (2015)
34. Yang, X., Zhang, H., Qi, G., Cai, J.: Causal attention for vision-language tasks. In: Proceedings of the IEEE/CVF Conference on Computer Vision and Pattern Recognition, pp. 9847–9857 (2021)
35. Zhang, D., Zhang, H., Tang, J., Hua, X.S., Sun, Q.: Causal intervention for weakly-supervised semantic segmentation. In: Advances in Neural Information Processing Systems 33 (2020)
36. Zhang, H., Zhang, H., Wang, C., Xie, J.: Co-occurrent features in semantic segmentation. In: Proceedings of the IEEE/CVF Conference on Computer Vision and Pattern Recognition, pp. 548–557 (2019)
37. Zhang, J., Li, Z., Zhang, C., Ma, H.: Stable self-attention adversarial learning for semi-supervised semantic image segmentation. J. Vis. Commun. Image Represent. **78**, 103170 (2021)

38. Zheng, S., et al.: Rethinking semantic segmentation from a sequence-to-sequence perspective with transformers. In: Proceedings of the IEEE/CVF Conference on Computer Vision and Pattern Recognition, pp. 6881–6890 (2021)
39. Zhou, B., Zhao, H., Puig, X., Fidler, S., Barriuso, A., Torralba, A.: Scene parsing through ade20k dataset. In: Proceedings of the IEEE/CVF Conference on Computer Vision and Pattern Recognition, pp. 633–641 (2017)
40. Zhu, Z., Xu, M., Bai, S., Huang, T., Bai, X.: Asymmetric non-local neural networks for semantic segmentation. In: Proceedings of the IEEE/CVF International Conference on Computer Vision, pp. 593–602 (2019)

A Joint Framework Towards Class-aware and Class-agnostic Alignment for Few-shot Segmentation

Kai Huang[1] , Mingfei Cheng[2]([✉]) , Yang Wang[1], Bochen Wang[1], Ye Xi[1],
Feigege Wang[1], and Peng Chen[1]

[1] Alibaba Group, Hangzhou, China
{zhouwan.hk,wanyuan.wy,bochen.wbc,yx150449,
feigege.wfgg,yuanshang.cp}@alibaba-inc.com
[2] Singapore Management University, Singapore, Singapore
mfcheng.2022@phdcs.smu.edu.sg

Abstract. Few-shot segmentation (FSS) aims to segment objects of unseen classes given only a few annotated support images. Most existing methods simply stitch query features with independent support prototypes and segment the query image by feeding the mixed features to a decoder. Although significant improvements have been achieved, existing methods are still face class biases due to class variants and background confusion. In this paper, we propose a joint framework that combines more valuable class-aware and class-agnostic alignment guidance to facilitate the segmentation. Specifically, we design a hybrid alignment module which establishes multi-scale query-support correspondences to mine the most relevant class-aware information for each query image from the corresponding support features. In addition, we explore utilizing base-classes knowledge to generate class-agnostic prior mask which makes a distinction between real background and foreground by highlighting all object regions, especially those of unseen classes. By jointly aggregating class-aware and class-agnostic alignment guidance, better segmentation performances are obtained on query images. Extensive experiments on PASCAL-5^i and COCO-20^i datasets demonstrate that our proposed joint framework performs better, especially on the 1-shot setting.

Keywords: Few-shot learning · Semantic segmentation · Hybrid alignment

1 Introduction

Semantic segmentation has made tremendous progress thanks to the advancement in deep convolutional neural networks. The performance of standard

---

K. Huang and M. Cheng—Equal contribution.

Supplementary Information The online version contains supplementary material available at https://doi.org/10.1007/978-3-031-26293-7_26.

L. Wang et al. (Eds.): ACCV 2022, LNCS 13847, pp. 431–447, 2023.
https://doi.org/10.1007/978-3-031-26293-7_26

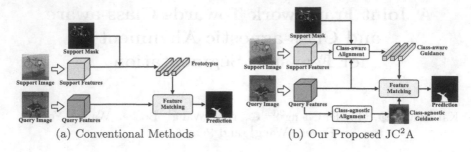

(a) Conventional Methods   (b) Our Proposed JC²A

Fig. 1. Illustration of (a) conventional methods and (b) our proposed JC^2A. JC^2A in (b) joint class-aware and class-agnostic guidance rather than only independent prototypes in (a) to guide the query image segmentation.

supervised semantic segmentation [4,23,56] heavily relies on large-scale datasets [8,24] and will drop drastically on unseen classes. However, obtaining large-scale datasets requires substantial human efforts, which is costly and infeasible in general. Inspired by the few-shot learning [35], few-shot segmentation (FSS) has been proposed to alleviate the need of huge annotated data set. Conventional FSS methods are built on meta-learning [36], which is supposed to learn a generic meta-learner from seen classes and then adopted to handle unseen classes with few annotated support samples. Specifically, as shown in Fig. 1(a), the features of query and support images are firstly extracted by a shared convolutional neural network. Then the support features within the target object regions are transferred to prototypes [6,20] which are used to guide the query image segmentation with a feature matching module, e.g., relational network [38].

Despite of the recent progresses made by these work [20,25,38,42,46], there still exist two limitations on FSS. 1) Class-aware bias: The support prototypes extracted independently from the query feature are not discriminative enough due to the variations of objects within the same class. 2) Class-agnostic bias: FSS treats objects of unseen classes as background during training, e.g., *person* in the query image of Fig. 1, which leads to model bias toward the seen classes rather than class-agnostic.

Some researches [46,47,51] have tried to address one of the above limitations. To eliminate the influence of class-aware bias, feature interaction [47,51] are adopted to fuse relative class-aware features in support samples by calculating the pixel-to-pixel relationship between query and support features, which ignore the contextual information and still suffer from the second limitation. Recently, MiningFSS [46] tries to mine latent object features by using pseudo class-agnostic labels and an extra training branch, which is more complex and does not consider the query-support relationship. To sum up, it is inspiring to study how to fully explore class-aware relationship between query and support samples and class-agnostic information as a joint guidance to improve FSS.

In this work, we propose a novel joint framework, Joint Class-aware and Class-agnostic Alignment Network (JC^2A), to address the above-mentioned problems with one stone. For each query image, as shown in Fig. 1(b), JC^2A aims to guide the segmentation by jointly aggregating most relevant class-aware information from the support image and class-agnostic information by object region mining. Specifically, JC^2A explores class-aware guidance by aligning the prototypes based on the feature relationships between query features and support features, and designs a Hybrid Prototype Alignment Module (HPAM) to build point-to-point and point-to-block correspondences. The class-aware prototypes produced by HPAM contains not only spatial details but also contextual cues of objects in the support feature. To eliminate class-agnostic bias and focus on regions of all objects well, JC^2A proposes a Class-agnostic Knowledge Mining Module (CKMM) to mine object regions of all classes in the query image, including seen and unseen classes. The CKMM provides a class-agnostic object mask by highlighting all non-background regions. By aggregating both class-aware prototypes and class-agnostic object mask as a joint guidance, better segmentation performance are obtained on query images. In addition, comprehensive experiments on PASCAL-5^i and COCO-20^i validate the effectiveness of our proposed JC^2A in comparison with ablations and alternatives.

2   Related Work

Semantic Segmentation is the task to assign a specific category label to each pixel in an image. Inspired by Fully Convolutional Network (FCN) [27], state-of-the-art segmentation methods [3,4,22,54,56] have been proposed and applied in various fields [5,21,33,57]. Recently, dilated convolution [3,4,48], pyramid features [54,55], non-local modules [12,59], vision transformer [56] and skip connections [23,34] are adopted to perceive more contextual information and preserve spatial details. However, these supervised methods heavily rely on a large amount of pixel-level labeled data. In this work, we focus on FSS which performs better on unseen classes with a handful of annotations.

Few-shot Learning is meant to efficient adapt to handle new tasks with limited empirical information available, which emphasizes on the generalization capability of a model. In order to reflect the ability of fitting to new categories given a few annotated data, episodes-based training and verification strategy [39] has been the foundation of major few-shot learning methods. Meta-based learning methods [7,9,18] maintain a meta-learner to boost the ability of fast acclimatization for new tasks. For instance, meta-manager [9,19] for parameters optimization, meta-memorizer [32,58] for storing the properties of prototypes and meta-comparator [37] for feature retrieval between query image and support set. Metric-based methods [14,15,36] aim to construct a unified similarity measure within the multi-tasks, such as embedding distance of Matching Networks [39], parameterized metric of Relation Networks [37] and structural distance of DeepEMD [50].

Few-shot Segmentation aims to give a dense prediction for the query image with only a few annotated support images. The pioneer work OSLSM [35] generates segmentation parameters based on support images by the two-branch network including a conditional branch and a segmentation branch. Later, prototype-based methods [6,20,38,43,46,49] adopting this two-branch paradigm became the mainstream solutions for FSS. The prototype was first proposed in

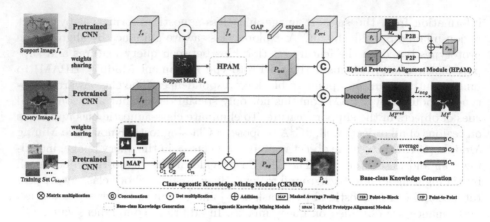

Fig. 2. Overview of Joint Class-aware and Class-agnostic Alignment Network (JC^2A). A shared encoder is adopted to extract features for support set, query set and base-classes set. Hybrid Prototype Alignment Module (HPAM) and Class-agnostic Knowledge Mining Module (CKMM) are used to generate context-aware prototypes and class-agnostic probability map respectively, which then jointly guide the segmentation.

few-shot segmentation [6], which is directly used to guide the segmentation of query images by comparison [6,38,41,46,52,53]. Some works adopt part-aware prototypes [20,26,45] to contain more diverse support features which may not be needed by the query image. Interaction-based methods [28,40,51] extract class-aware information for the query images by only considering point-level correspondence between query and support features. MiningFSS [46] mines the latent objects by a class-agnostic constraint, which needs extra annotations and parameters. Existing methods rarely focus on both class-aware and class-agnostic information for the query image. In this work, we jointly use class-aware and class-agnostic information as an aggregated guidance for FSS.

3 Method

3.1 Task Formulation

Few-shot segmentation (FSS) aims to quickly adapt to the given segmentation tasks with only a few annotated data available. For a K-shot few-shot segmentation task, K labeled samples from the same class make up a support set S of this task, and there is another unlabeled sample set named query set Q. FSS needs to segment the target of the specific class in the query set Q only with the help of the support set S.

The training and evaluation of FSS models are usually performed by the episodic paradigm [39]. Specifically, the model is trained on a base-classes set C_{base} and evaluated on a novel-classes set C_{novel} with the precondition $C_{base} \cap C_{novel} = \emptyset$. For each episode, samples of K-shot FSS episodic task with class c from C_{base} or C_{novel} are randomly sampled. The final segmentation performance is reported by averaging the results on the query set of various episodic tasks.

3.2 Overview

Figure 2 illustrates an overview of our joint framework (JC^2A) for few-shot segmentation. JC^2A mainly contains two components: Hybrid Prototype Alignment Module (HPAM) and Class-agnostic Knowledge Mining Module (CKMM). HPAM is designed to generate most relevant class-aware support prototypes for each query and support image pair by establishing multi-scale query-support relations. Specifically, HPAM builds Point-to-Point and Point-to-Block correspondences between query and support features and combines them as class-aware prototypes, which is able to provide more useful class-aware guidance with spatial details and contextual information. CKMM aims to mitigate the class-agnostic bias mainly caused by background confusion by highlighting all foreground regions in the query image. In CKMM, object features of seen classes in the training set are used to mine the class-agnostic guidance for each query image. More details are introduced in the following.

3.3 Hybrid Prototype Alignment Module (HPAM)

Inspired by [43,51], we utilize the mutual information between the support feature and the query feature to mine more class-aware information from the support feature for the query image. However, simply computing pixel-level attention which contains limited weakened semantics ignores the context. Thus, Hybrid Prototype Alignment Module (HPAM) is proposed to mine the most relevant class-aware guidance from support features by mixing Point-to-Point Alignment and Point-to-Block Alignment, which calculate different scale correspondences between the support and the query.

Point-to-Point Alignment (P2P). Considering the support image I_s and its binary mask M_s as the support instance pair $\{I_s, M_s\}$, and I_q denotes the query image, their respective features extracted by the pretrained CNN are represented as f_s and f_q. A Hadamard product is performed in support feature map with its binary mask to obtain the masked support features $\tilde{f}_s$. We adopt an attention mechanism to establish the point-to-point relation between $\tilde{f}_s$ and f_q. Formally,

$$Q_q = f_q W_Q, \quad K_s = \tilde{f}_s W_K, \quad V_s = \tilde{f}_s W_V, \tag{1}$$

where W_Q, W_K, W_V are learnable projection parameters, Q_q, K_s, $V_s \in \mathbb{R}^{C \times (H \times W)}$ are projected features. An attention map is then obtained by the dot-product operation between Q_q and K_s, which will bring heavy computation as the resolution of feature map grows. To balance computational efficiency and keeping as much target information as possible, a linear attention mechanism [16,30] is adopted to decompose the calculation of the attention map, which has linear complexity. Thus, the point-to-point class-aware information can be obtained by the linear attention:

$$P_{aw}^d = \Phi(Q_q) \times (\Psi(K_s)^T \times V_s) \in \mathbb{R}^{C \times (H \times W)}, \tag{2}$$

where $\Phi(\cdot)$ and $\Psi(\cdot)$ are the decoupling functions to approximate the attention map of normal attention mechanism, in which batch normalization function [13] and softmax function are commonly used respectively. In order to accommodate the mini-batch learning and suppress the noisy alignment of FSS, we adopt the ReLU function [1] instead of batch normalization to generate positive responses. The Eq. (2) can be rewritten as:

$$P_{aw}^d = \text{ReLU}(Q_q) \times (softmax(K_s)^T \times V_s) \in \mathbb{R}^{C \times (H \times W)}. \tag{3}$$

In this way, we obtain the class-aware information of each query point under the full support key-value point features with an dense way. The point-to-point dense alignment between query features and masked support features can offer a detailed and complete pixel level alignment, which tends to find similar grainy vision information from the intra-task targets of support set.

Point-to-Block Alignment (P2B). The Point-to-Point approach only focuses on single point alignment which contains limited weakened semantics. It is observed that the feature points with similar semantic information are always close in spatial locations, consequently, the semantic alignment can be performed in a sparse way with spatial blocked targets of support images. Specifically, We pick out several key feature blocks to represent the class-aware target, such as head-block, leg-block and tail-block for a horse. Then, a point-to-block linear attention alignment between query point-feature and masked support block-feature is formed to catch the semantic class-aware information.

It is worth mentioning that the permutation invariance [17] of attention mechanism may disrupt the topological relation of feature blocks. Take a horse as an example, it is clear that the head-block is in the front of the tail-block according to the actual spatial location relationship. The topological relation becomes chaotic in the linear attention, which will impact the expression of semantic level information. Thus, we adopt two parameterized matrices to act as the point-specific position embedding, which provides a valid signal which carries the original positional information:

$$f_q^p = f_q + p_q, \tilde{f}_s^p = \tilde{f}_s + p_s \tag{4}$$

where $p_q, p_s \in \mathbb{R}^{H \times W}$ are learnable position embedding for query and support features respectively. Similar to Eq. (1), the queries, keys and values with the position embedding are represented as Q_q^p, K_s^p and V_s^p. Assume the block area is $m \times m$ in feature pixels, the concatenated block patch of masked support features is formulated as:

$$\mathcal{B}(\tilde{f}_s^p) = Concat(\tilde{f}_s^p[(r-1)m : rm, (c-1)m : cm])_{i=1}^N \in \mathbb{R}^{C \times N \times m^2}, \tag{5}$$

where r is the row position of i-th block and can be calculated with the round down of im/W, c indicates the column position of i-th block which is obtained with $i - rW/m$, and $N = HW/m^2$ is the total number of block patches.

The importance of the i-th block patch is obtained according to the target coverage with corresponding block in its binary mask:

$$(\mathcal{I}_s)_i = \sum_{o=(r-1)m}^{rm} \sum_{j=(c-1)m}^{cm} M_s(o, j)/m^2. \tag{6}$$

The top k key feature blocks are selected by the importance ranking within these block patches as:

$$\mathcal{B}(\tilde{f}_s^p|Topk) = \mathcal{B}(\tilde{f}_s^p|Sort(\mathcal{I}_s)[: k]) \in \mathbb{R}^{C \times k \times m^2}. \tag{7}$$

It inevitably contains pure empty regions when the number of valid blocks is smaller than k. The alignment outcomes of such regions keeps zero response and is insignificant compared with the positive response of those target regions. Therefore, mining an amount of pure empty regions basically does not affect the learning. The keys and values of blocked support features can be represented as:

$$\mathcal{B}(K_s^p), \mathcal{B}(V_s^p) = \mathcal{B}(\tilde{f}_s^p W_S|Topk), \mathcal{B}(\tilde{f}_s^p W_V|Topk). \tag{8}$$

where the reformulated $\mathcal{B}(K_s^p), \mathcal{B}(V_s^p) \in \mathbb{R}^{C \times N \times m^2}$ are the sparsification of support key-value with more focused and explicit semantic information on support features. The sparse feature alignment between query features and blocked support features is further expressed as:

$$P_{aw}^s = \frac{1}{m^2} \sum_{m \times m} \text{ReLU}(Q_q^p)(softmax(\mathcal{B}(K_s^p))^T \mathcal{B}(V_s^p)) \in \mathbb{R}^{C \times H \times W}. \tag{9}$$

Compared with the Point-to-Point prototype alignment in Eq. (3), we align support prototypes with the blocked support key-value features by using P2B. Moreover, instead of using the block set with a fixed partition, the sparse feature blocks contain pivotal semantic information of targets. Thus the point-to-block prototype alignment aggregates the integrated semantic information with the key feature blocks and offers a block-level semantic alignment, which can generate more stable and smooth class-aware information.

Hybrid Prototype. Our final hybrid aligned prototype combines the aligned prototypes generated by Point-to-Point and Point-to-Block alignment respectively, which is summarized as:

$$P_{aw} = P_{aw}^d + P_{aw}^s \in \mathbb{R}^{C \times H \times W}. \tag{10}$$

As aforementioned, the P2P alignment is designed to obtain a dense alignment and search the class-aware information with the view of local vision, although the candidate targets of query image usually get positive response, it also introduce background points for the low discrimination of visual features. Thus, by considering the semantic matching between query points and blocked class-aware targets, the P2B alignment can help to filter out these semantic irrelevant points and smooth the class-aware information of P2P alignment. **Extension to Feature Pyramid.** Feature pyramid has been widely used in few-shot segmentation [20,38,49] due to its abundant multi-scale feature maps. Higher-level hierarchical feature maps possess more centralized semantic information but with lower resolution. In this context, fixed number of key feature blocks become inappropriate, which may introduce noisy information for high-level feature maps or miss essential parts of target for low-level feature maps. In order to tolerate this variation, we apply specific number of block patches corresponding to different scale support feature maps. More concretely, a top k set $\{k_l\}_{l=1}^{L}$ is prepared for L-layers feature pyramid, k_l decreases as l increases. With such a variational top k for sparse feature alignment, the blocked support features can gain semantic information of class-aware targets according to the multi-scale feature pyramid.

3.4   Class-agnostic Knowledge Mining Module (CKMM)

Due to the limitation of few-shot segmentation datasets, there is only one seen class for the effective targets, and others unseen are treated as the background. The ability of adapting to novel classes is in doubt when same or similar classes are viewed as background in the training process.

Mining latent target with base-classes set was firstly proposed by [53], which focuses on search target by part-specific attributes. However, it suffers from complicated multi-states optimization and the low-semantic target parts are more easy to match the background. Inspired by the feature prototype [36] in few-shot classification, we exploits another simple but effective way with class-specific attitudes. Specifically, we propose to mine the latent target information as well as class-agnostic information with the masked feature prototypes, which are obtained by the base-classes. Given the feature map $f^{c,i} \in \mathbb{R}^{C \times HW}$ and its binary mask $M^{c,i}$ with class c, the spacial weighted Global Average Pooling (wGAP) [49,52] for i-th instance pair $\{f^{c,i}, M^{c,i}\}$ of class c is defined as:

$$f_{wGPA}^{c,i} = \frac{\sum_{h,w} f^{c,i}(h,w) M^{c,i}(h,w)}{\sum_{h,w} M^{c,i}(h,w)} \in \mathbb{R}^{1 \times C}, \qquad (11)$$

where h and w are the height index and width index with the limitation of H and W. Feature prototype of class c is obtained by averaging over the wGAP of all instance pairs with class c, and the feature prototype of base-classes C_{base} is concatenated with each single-class feature prototype:

$$FP_{base} = Concat([\sum_{i=1}^{I_c} f_{wGPA}^{c,i}/I_c]_{c \in C_{base}}) \in \mathbb{R}^{|C_{base}| \times C}, \qquad (12)$$

where I_c is the total instance pairs of class c, and $|C_{base}|$ indicates the cardinality of base-classes set. The feature prototype FP_{base} can be regarded as the aggregation of seen base-classes, and each row of FP_{base} contains the most common feature of this class. Subsequently, the latent targets with similar or partially similar features are possibly mined to replenish the missing class-agnostic information. The class-agnostic probability map of query image I_q is calculated by the dot-product between the base feature prototype and query features:

$$\bar{P}_{ag} = \frac{1}{|C_{base}|} \sum_{|C_{base}|} FP_{base} \cdot f_q \in \mathbb{R}^{1 \times W \times H}. \qquad (13)$$

For the convenience of adapting different base-classes set, we make the average operation in $|C_{base}|$ dimension to obtain a compositive probability map of class-agnostic information for query features, which acts as a kind of prior mask. Different from the class-specific prior mask in PFENet [38], our class-agnostic probability map has the ability to search not only the class that is common with support set but also the latent class existed in other meta-tasks.

3.5 Multiple Information Aggregation (MIA)

The class-aware information generated by HPAM aims to provide more discriminative prototypes for the current meta-task. The CKMM provides the class-agnostic information to eliminate the background confusion during training. To joint these guidance, we simply combine these two sets of information by concatenating:

$$P_{multiple} = Concat([P_{aw}, \bar{P}_{ag}]) \in \mathbb{R}^{(C+1) \times H \times W}. \tag{14}$$

The concatenated information then is passed through 1×1 convolution along with the original query features and support features for further information aggregation. The predicted mask of query image is later obtained by a feature decoder with multi-scale residual layers refer to [38,42].

4 Experiments

4.1 Implementation Details

Dataset. We validate the effectiveness of our proposed method on two standard few-shot segmentation datasets: PASCAL-5^i [35] and COCO-20^i [29]. PASCAL-5^i consists of PASCAL VOC 2012 [8] with extra mask annotations from SDS [10] dataset. It contains 20 object classes which are evenly divided into 4 folds: $\{5^i, i \in \{0, 1, 2, 3\}\}$. COCO-$20^i$ is a more challenging dataset for few-shot segmentation, which is modified from MS COCO [24]. It splits 80 categories into 4 folds: $\{20^i, i \in \{0, 1, 2, 3\}\}$. Following the standard experimental settings [38], on both datasets, three folds are selected for training while the remaining fold is used for evaluation in each single experiment. During the evaluation, 1000 episodes in the target fold are randomly sampled for both datasets.

Evaluation Metrics. Following [29,52], we adopt mean intersection over union (mIoU) and foreground-background IoU (FB-IoU) as our evaluation metrics. Specifically, mIoU is computed by averaging over IoU values of all classes in a fold. FB-IoU calculates the average of foreground and background IoU in a fold (e.g., $C = 2$), which treats all object categories as a single foreground class. The average of all the folds is reported as the final mIoU/FB-IoU. For the multi-shot case, we leverage the decision-level fusion strategy [20,49,52] by averaging the predicted masks between each single support instance and the query image.

Training Details. Our proposed model is constructed on PyTorch [31] and trained on a single NVIDIA RTX 2080Ti. We build our model with the ResNet50 [11] and ResNet101 [11] as backbones. Our model is optimized by the SGD with an initial learning rate of $2.5e\text{-}3$, where momentum is 0.9 and the weight decay is set to $1e\text{-}4$. During training, the batch size is set to 4 and parameters of the backbone are not updated. All images together with the masks are all resized to 473×473 for training and tested with their original sizes. We construct four-layer feature pyramid for the mixed alignment module, the top k set is set to $\{60, 20, 5, 3\}$ as l increases, which also corresponds to 20% number of block patches with respective scales.

Table 1. Performance of 1-shot and 5-shot segmentation on PASCAL-5^i. Results in **bold** indicate the best performance and the underlined ones are the second best.

Backbone	Method	1-shot						5-shot					
		fold-0	fold-1	fold-2	fold-3	mIoU	FB-IoU	fold-0	fold-1	fold-2	fold-3	mIoU	FB-IoU
ResNet50	PGNet [51]	56.0	66.9	50.6	50.4	56.0	69.9	57.7	68.7	52.9	54.6	58.5	70.5
	SCL [49]	63.0	70.0	56.5	57.7	61.8	71.9	64.5	70.9	57.3	58.7	62.9	72.8
	SAGNN [43]	64.7	69.6	57.0	57.2	62.1	73.2	64.9	70.0	57.0	59.3	62.8	73.3
	CMN [44]	64.3	70.0	57.4	59.4	62.8	72.3	65.8	70.4	57.6	60.8	63.7	72.8
	PFENet [38]	61.7	69.5	55.4	56.3	60.8	73.3	63.1	70.7	55.8	57.9	61.9	73.9
	RePRI [2]	60.2	67.0	61.7	47.5	59.1	–	64.5	70.8	**71.7**	60.3	66.8	–
	MiningFSS [46]	59.2	71.2	**65.6**	52.5	62.1	–	63.5	71.6	71.2	58.1	66.1	–
	HSNet [28]	64.3	70.7	60.3	60.5	64.0	**76.7**	70.3	73.2	67.4	**67.1**	**69.5**	80.6
	CyCTR [53]	65.7	71.0	59.5	59.7	64.0	–	69.3	**73.5**	63.8	63.5	67.5	–
	JC²A (ours)	**67.3**	**72.4**	57.7	**60.7**	**64.5**	76.5	68.6	72.9	58.7	62.0	65.4	76.8
ResNet101	PPNet [26]	52.7	62.8	57.4	47.7	55.2	70.9	60.3	70.0	69.4	60.7	65.1	77.5
	DAN [40]	54.7	68.6	57.8	51.6	58.2	71.9	57.9	69.0	60.1	54.9	60.5	72.3
	PFENet [38]	60.5	69.4	54.4	55.9	60.1	72.9	62.8	70.4	54.9	57.6	61.4	73.5
	RePRI [2]	59.6	68.6	**62.2**	47.2	59.4	–	66.2	71.4	67.0	57.7	65.6	–
	MiningFSS [46]	60.8	71.3	61.5	56.9	62.6	–	65.8	74.9	**71.4**	63.1	68.8	–
	HSNet [28]	67.3	72.3	62.0	**63.1**	66.2	77.6	71.8	74.4	67.0	**68.3**	70.4	80.6
	CyCTR [53]	67.2	71.1	57.6	59.0	63.7	–	71.0	75.0	58.5	65.0	67.4	–
	JC²A (ours)	**68.2**	**74.4**	59.8	63.0	**66.4**	**78.8**	70.6	**75.2**	61.9	64.8	68.1	**80.6**

Table 2. Performance of 1-shot and 5-shot segmentation on COCO-20^i. Results in **bold** indicate the best performance and the underlined ones are the second best.

Backbone	Method	1-shot						5-shot					
		fold-0	fold-1	fold-2	fold-3	mIoU	FB-IoU	fold-0	fold-1	fold-2	fold-3	mIoU	FB-IoU
ResNet50	PPNet [26]	31.5	22.6	21.5	16.2	23.0	–	45.9	29.2	30.6	29.6	33.8	–
	RePRI [2]	31.2	38.1	33.3	33.0	34.0	–	38.5	46.2	40.0	43.6	42.1	–
	MMNet [42]	34.9	41.0	37.2	37.0	37.5	–	37.0	40.3	39.3	36.0	38.2	–
	CMN [44]	37.9	44.8	38.7	35.6	39.3	61.7	42.0	50.5	41.0	38.9	43.1	63.3
	CyCTR [53]	38.9	43.0	39.6	39.8	40.3	–	41.1	48.9	45.2	**47.0**	45.6	–
	MiningFSS [46]	**46.8**	35.3	26.2	27.1	33.9	–	**54.1**	41.2	34.1	33.1	40.6	–
	HSNet [28]	36.3	43.1	38.7	38.7	39.2	68.2	43.3	51.3	**48.2**	45.0	46.9	70.7
	JC²A (ours)	40.4	**47.4**	**44.5**	**43.5**	**44.0**	**70.0**	44.3	**53.5**	46.0	45.8	**47.4**	**71.5**
ResNet101	PMMs [45]	29.5	36.8	28.9	27.0	30.6	–	33.8	42.0	33.0	33.3	35.5	–
	PFENet [38]	34.3	33.0	32.3	30.1	32.4	58.6	38.5	38.6	38.2	34.3	37.4	61.9
	SCL [49]	36.4	38.6	37.5	35.4	37.0	–	38.9	40.5	41.5	38.7	39.9	–
	SAGNN [43]	36.1	41.0	38.2	33.5	37.2	60.9	40.9	48.3	42.6	38.9	42.7	63.4
	MiningFSS [46]	**50.2**	37.8	27.1	30.4	36.4	–	**57.0**	46.2	37.3	37.2	44.4	–
	HSNet [28]	37.2	44.1	42.4	41.3	41.2	69.1	45.9	53.0	51.8	47.1	**49.5**	**72.4**
	JC²A (ours)	41.5	**48.6**	**45.6**	42.9	**44.7**	**70.6**	43.7	**55.2**	47.3	**47.7**	48.5	72.0

4.2 Comparisons

To verify the effectiveness of our proposed method, we compare with alternatives on the two few-shot segmentation datasets [24, 35]. Extensive experiments with various backbones show that our model achieves the best performance as shown in Table 1 and Table 2.

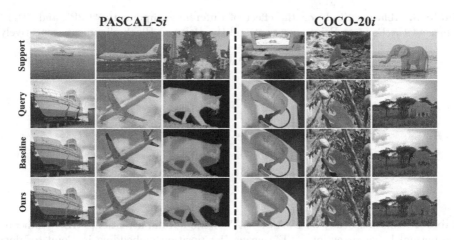

Fig. 3. Qualitative results on PASCAL-5^i and COCO-20^i. Oriented top to bottom, each row shows the ground truth of query images (yellow), the baseline results (blue) and ours results (blue), respectively. (Color figure online)

Quantitative Results. In Table 1, we show the comparative results of our JC^{2}A and alternative FSS methods on PASCAL-5^i. Although our method does not perform better on PASCAL-5^i 5-shot, our method achieves competitive performance compared with other methods on the 1 shot setting. The highest increment mIoU based metrics is around 2 points (e.g. from 72.3% to 74.4% for fold-1 with ResNet101). Table 2 presents the results of different approaches on COCO-20^i. It can be found that our JC^{2}A outperforms significantly compared with alternatives on both 1-shot and 5-shot settings. With the backbone of ResNet50, our method outperforms the second best by 3.7% mIoU and 0.5% mIoU on 1-shot setting and 5-shot setting respectively. The performance gains with different backbones further demonstrate the superiority of our JC^{2}A, particularly with the backbone of ResNet101 on COCO-20^i, which exceeds the second best model by 3.5% on 1-shot. From the above comparison, we conclude that our JC^{2}A achieves better performance. Besides, we think that JC^{2}A is more suitable for few-shot segmentation, because it obtains the SOTA on 1-shot setting, which means fewer annotated samples are required in JC^{2}A.

Qualitative Results. Figure 3 provides visual examples of JC^{2}A on PASCAL-5^i and COCO-20^i. Compared our results (the 4th row) with the baseline (the 3rd row), JC^{2}A yields fewer false predictions in base classes and background. Besides, JC^{2}A can capture more details and maintain a more complete structure of the target object. These results verify that the joint class-aware and class-agnostic guidance is effective for FSS.

Table 3. Ablation Study on the effect of different components. "P2P" and "P2B" represent the Point-to-Point alignment and the Point-to-Block alignment respectively.

P2P	P2B	CKMM	Parameters	mIoU	
				1-shot	5-shot
			34.09M	59.0	60.6
✓			34.52M	61.3	62.2
	✓		34.61M	62.7	63.3
✓	✓		34.61M	63.6	64.7
		✓	34.11M	61.7	62.8
✓	✓	✓	34.64M	**64.5**	**65.4**

Table 4. Ablation Study on HPAM. "NA" is the normal attention, "NLA" indicates the normal linear attention, "PE" means the position embedding in Point-to-Block alignment, "IS" is the inference speed on 1-shot setting.

Setting	1-shot			5-shot		
	mIoU ↑	FB-IoU ↑	IS ↑	mIoU ↑	FB-IoU ↑	IS ↑
NA	64.2	76.9	1.00×	65.5	77.2	0.19×
NLA	62.8	74.5	4.11×	63.3	74.9	0.84×
Cosine	62.3	74.4	2.86×	62.8	73.3	0.55×
Ours w/o PE	63.1	74.9	4.07×	64.0	75.2	0.80×
Ours	64.5	76.5	4.10×	65.4	76.8	0.82×

4.3   Ablation Studies

To analyze the impact of each component in JC^2A, we conduct extensive ablation studies on PASCAL-5^i. Here, our baseline model is obtained from JC^2A excluding HPAM and CKMM.

Model Effectiveness. We first conduct an ablation study to show the effectiveness of the Hybrid Prototype Alignment Module (HPAM) and Class-agnostic Knowledge Mining Module (CKMM). Results are summarized in Table 3. It is noted that the model using HPAM (P2P+P2B) outperforms the baseline (1st row) by 4.6% and 4.1% on 1-shot and 5-shot settings respectively. CKMM provides class-agnostic information to FSS by highlighting all object regions. Observing results of 1st row and 5th row in Table 3, we can see CKMM improves the results by a large margin with 2.7% mIoU on 1-shot and 2.2% mIoU on 5-shot, which shows the effectiveness of CKMM. The last row of Table 3 demonstrates that the combination of these two modules performs better than only using each of them. We can infer that HPAM and CKMM mutually benefit during meta-learning.

Hybrid Prototype Alignment Module (HPAM). HPAM contains different scale prototype alignments, P2P and P2B. The 2nd row and the 3rd row in Table 3 proves the effectiveness of combination of P2P and P2B. Table 4 studies

Table 5. Effectiveness of CKMM (Class-agnostic guidance) for different encoder-decoder based FSS methods on PASCAL-5^i and COCO-20^i.

Dataset	Method	1-shot					5-shot				
		fold-0	fold-1	fold-2	fold-3	mIoU	fold-0	fold-1	fold-2	fold-3	mIoU
PASCAL-5^i	PFENet [38]	$61.7_{+0.82}$	$69.5_{+0.59}$	$55.4_{+1.24}$	$56.3_{+1.05}$	$60.8_{+0.93}$	$63.1_{+0.65}$	$70.7_{+1.14}$	$55.8_{+0.98}$	$57.9_{+0.77}$	$61.9_{+0.89}$
	SCL [49]	$63.0_{+0.65}$	$70.0_{+1.63}$	$56.5_{+0.47}$	$57.7_{+0.70}$	$61.8_{+0.86}$	$64.5_{+0.79}$	$70.9_{+1.28}$	$57.3_{+0.54}$	$58.7_{+0.77}$	$62.9_{+0.85}$
	MM-Net [42]	$62.7_{+1.72}$	$70.2_{+0.60}$	$57.3_{+0.54}$	$57.0_{+0.98}$	$61.8_{+0.96}$	$62.2_{+1.89}$	$71.5_{+1.05}$	$57.5_{+0.66}$	$62.4_{+0.96}$	$63.4_{+1.14}$
COCO-20^i	PFENet [38]	$34.3_{+1.05}$	$33.0_{+0.89}$	$32.3_{+0.67}$	$30.1_{+0.90}$	$32.4_{+0.88}$	$38.5_{+0.94}$	$38.6_{+0.77}$	$38.2_{+0.93}$	$34.3_{+0.84}$	$37.4_{+0.87}$
	SCL [49]	$36.4_{+1.26}$	$38.6_{+1.30}$	$37.5_{+0.78}$	$35.4_{+1.03}$	$37.0_{+1.09}$	$38.9_{+1.17}$	$40.5_{+1.34}$	$41.5_{+0.88}$	$38.7_{+1.01}$	$39.9_{+1.10}$
	MM-Net [42]	$35.4_{+1.51}$	$41.7_{+0.90}$	$37.5_{+1.33}$	$40.1_{+1.06}$	$36.2_{+1.20}$	$37.8_{+1.66}$	$41.0_{+1.11}$	$40.3_{+1.28}$	$36.9_{+1.37}$	$39.0_{+1.36}$

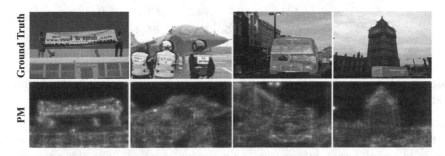

Fig. 4. Visualization of class-agnostic probability maps (PM) generated by CKMM. The 1st row shows query images with their annotations (yellow). The 2nd row shows probability maps which highlight all object regions of seen and unseen classes. (Color figure online)

the influence of operations in HPAM. Since the feature alignment is accomplished by a modified linear attention, we compare it with the normal attention (NA), normal linear attention (NLA) and cosine interaction. It is clear that our class-aware feature alignment achieves competitive performance with greatly increased inference speed. Although the normal attention way gets slight superiority in some cases, it trades off with a huge computational cost which is reflected in its slowest inference speed. For the reason of mini-batch data form in few-shot segmentation, the normalization decoupling function suffers from instability of data distribution and gets worse performance than ReLU function adopted in our method. The interaction ability of cosine similarity is weaker than the attention-based measure for its lack of nonlinear mapping and noisy suppression. Besides, as shown in the 4th row and the bottom row of Table 4, the position embedding also plays a positive role in our method to improve the results.

Class-agnostic Knowledge Mining Module (CKMM). CKMM is designed to provide class-agnostic alignment guidance for FSS by highlighting all object regions. As shown in Fig. 4, CKMM is able to successfully highlight all object regions. To further demonstrate the effectiveness of CKMM and its generated class-agnostic probability maps (PM), we apply it to several encoder-decoder based FSS methods [38,42,49]. We adopt ResNet50 and ResNet101 as the backbone of PASAL-5^i and COCO-20^i dataset respectively. The experimental results in Table 5 indicate that our CKMM can also boost other FSS approaches without

Table 6. Ablation study on different ways of hybrid prototypes and information aggregation. Three common operations are compared: Multiply, Add and Concat.

Component	Setting	1-shot		5-shot	
		mIoU	FB-IoU	mIoU	FB-IoU
HPAM	*Multiply*	60.2	72.3	61.8	73.5
	Add	**64.5**	**76.5**	**65.4**	**76.8**
	Concat	64.0	75.7	64.8	76.1
MIA	*Multiply*	58.8	70.1	60.9	73.0
	Add	62.7	74.3	63.3	75.0
	Concat	**64.5**	**76.5**	**65.4**	**76.8**

upsetting the original structures. It also proves that the class-agnostic alignment guidance is beneficial for FSS.

Hybrid Prototypes & Information Aggregation. Table 6 shows the ablation study of different information aggregation methods. For the aggregation of obtaining hybrid prototypes from P2P and P2B, the *add* operation obtains better performance than others. The recommended operation is *concat* between mixed alignment for class-specific targets and probability map for class-agnostic targets. It is reasonable that the *add* operation is more suitable to aggregate the information with similar properties and the *concat* operation prefers differentiated information with the precondition of absence of curse of dimensionality.

5　Conclusion

In this paper, we have proposed a joint framework JC^2A towards class-aware and class-agnostic alignment for few-shot segmentation. JC^2A contains two critical modules: Hybrid Prototype Alignment Module (HPAM) and Class-agnostic Knowledge Mining Module (CKMM), then combines these two modules to jointly guide the query image segmentation. HPAM aims to generate class-aware guidance for the query image by combining multi-scale aligned prototypes between query features and support features. To prevent background confusion and class-agnostic bias, CKMM uses base-classes knowledge to produce a class-agnostic probability mask for the query image, which highlights object regions of all classes especially those of unseen classes. Comparisons with FSS alternatives validate the effectiveness of joint class-aware and class-agnostic information in guiding the query image segmentation. Potential extensions of JC^2A include developing more replaceable components for each module, thus improving FSS performance.

References

1. Agarap, A.F.: Deep learning using rectified linear units (relu). arXiv preprint arXiv:1803.08375 (2018)
2. Boudiaf, M., Kervadec, H., Masud, Z.I., Piantanida, P., Ben Ayed, I., Dolz, J.: Few-shot segmentation without meta-learning: a good transductive inference is all you need? In: CVPR (2021)
3. Chen, L.C., Papandreou, G., Kokkinos, I., Murphy, K., Yuille, A.L.: Deeplab: semantic image segmentation with deep convolutional nets, atrous convolution, and fully connected crfs. TPAMI **40**, 834–848 (2017)
4. Chen, L.-C., Zhu, Y., Papandreou, G., Schroff, F., Adam, H.: Encoder-decoder with atrous separable convolution for semantic image segmentation. In: Ferrari, V., Hebert, M., Sminchisescu, C., Weiss, Y. (eds.) ECCV 2018. LNCS, vol. 11211, pp. 833–851. Springer, Cham (2018). https://doi.org/10.1007/978-3-030-01234-2_49
5. Cheng, M., Zhao, K., Guo, X., Xu, Y., Guo, J.: Joint topology-preserving and feature-refinement network for curvilinear structure segmentation. In: ICCV (2021)
6. Dong, N., Xing, E.P.: Few-shot semantic segmentation with prototype learning. In: BMVC (2018)
7. Elsken, T., Staffler, B., Metzen, J.H., Hutter, F.: Meta-learning of neural architectures for few-shot learning. In: CVPR (2020)
8. Everingham, M., Eslami, S.M.A., Gool, L.V., Williams, C.K.I., Winn, J.M., Zisserman, A.: The pascal visual object classes challenge: a retrospective. Int. J. Comput. Vision **111**, 98–136 (2015)
9. Finn, C., Abbeel, P., Levine, S.: Model-agnostic meta-learning for fast adaptation of deep networks. In: ICML. PMLR (2017)
10. Hariharan, B., Arbeláez, P., Girshick, R., Malik, J.: Simultaneous detection and segmentation. In: Fleet, D., Pajdla, T., Schiele, B., Tuytelaars, T. (eds.) ECCV 2014. LNCS, vol. 8695, pp. 297–312. Springer, Cham (2014). https://doi.org/10.1007/978-3-319-10584-0_20
11. Hu, T., Yang, P., Zhang, C., Yu, G., Mu, Y., Snoek, C.G.M.: Attention-based multi-context guiding for few-shot semantic segmentation. In: AAAI (2019)
12. Huang, Z., Wang, X., Huang, L., Huang, C., Wei, Y., Liu, W.: CCNET: criss-cross attention for semantic segmentation. In: ICCV (2019)
13. Ioffe, S., Szegedy, C.: Batch normalization: accelerating deep network training by reducing internal covariate shift. In: ICML (2015)
14. Jiang, W., Huang, K., Geng, J., Deng, X.: Multi-scale metric learning for few-shot learning. IEEE Trans. Circ. Syst. Video Technol. **31**(3), 1091–1102 (2020)
15. Karlinsky, L., et al.: Repmet: representative-based metric learning for classification and few-shot object detection. In: CVPR (2019)
16. Katharopoulos, A., Vyas, A., Pappas, N., Fleuret, F.: Transformers are rnns: fast autoregressive transformers with linear attention. In: ICML (2020)
17. Lee, J., Lee, Y., Kim, J., Kosiorek, A., Choi, S., Teh, Y.W.: Set transformer: a framework for attention-based permutation-invariant neural networks. In: ICML (2019)
18. Lee, K., Maji, S., Ravichandran, A., Soatto, S.: Meta-learning with differentiable convex optimization. In: CVPR (2019)
19. Lee, Y., Choi, S.: Gradient-based meta-learning with learned layerwise metric and subspace. In: ICML. PMLR (2018)
20. Li, G., Jampani, V., Sevilla-Lara, L., Sun, D., Kim, J., Kim, J.: Adaptive prototype learning and allocation for few-shot segmentation. In: CVPR (2021)

21. Li, Y., et al.: Fully convolutional networks for panoptic segmentation. In: CVPR (2021)
22. Lin, D., Ji, Y., Lischinski, D., Cohen-Or, D., Huang, H.: Multi-scale context intertwining for semantic segmentation. In: Ferrari, V., Hebert, M., Sminchisescu, C., Weiss, Y. (eds.) ECCV 2018. LNCS, vol. 11207, pp. 622–638. Springer, Cham (2018). https://doi.org/10.1007/978-3-030-01219-9_37
23. Lin, G., Milan, A., Shen, C., Reid, I.: Refinenet: multi-path refinement networks for high-resolution semantic segmentation. In: CVPR (2017)
24. Lin, T.-Y., et al.: Microsoft COCO: common objects in context. In: Fleet, D., Pajdla, T., Schiele, B., Tuytelaars, T. (eds.) ECCV 2014. LNCS, vol. 8693, pp. 740–755. Springer, Cham (2014). https://doi.org/10.1007/978-3-319-10602-1_48
25. Liu, B., Ding, Y., Jiao, J., Ji, X., Ye, Q.: Anti-aliasing semantic reconstruction for few-shot semantic segmentation. In: CVPR (2021)
26. Liu, Y., Zhang, X., Zhang, S., He, X.: Part-aware prototype network for few-shot semantic segmentation. In: Vedaldi, A., Bischof, H., Brox, T., Frahm, J.-M. (eds.) ECCV 2020. LNCS, vol. 12354, pp. 142–158. Springer, Cham (2020). https://doi.org/10.1007/978-3-030-58545-7_9
27. Long, J., Shelhamer, E., Darrell, T.: Fully convolutional networks for semantic segmentation. In: CVPR (2015)
28. Min, J., Kang, D., Cho, M.: Hypercorrelation squeeze for few-shot segmentation. In: ICCV (2021)
29. Nguyen, K., Todorovic, S.: Feature weighting and boosting for few-shot segmentation. In: ICCV (2019)
30. Pan, Y., Yao, T., Li, Y., Mei, T.: X-linear attention networks for image captioning. In: CVPR (2020)
31. Paszke, A., et al.: Automatic differentiation in pytorch. In: NeurIPS Autodiff Workshop (2017)
32. Ramalho, T., Garnelo, M.: Adaptive posterior learning: few-shot learning with a surprise-based memory module. In: ICLR (2018)
33. Reiss, S., Seibold, C., Freytag, A., Rodner, E., Stiefelhagen, R.: Every annotation counts: multi-label deep supervision for medical image segmentation. In: CVPR (2021)
34. Ronneberger, O., Fischer, P., Brox, T.: U-Net: convolutional networks for biomedical image segmentation. In: Navab, N., Hornegger, J., Wells, W.M., Frangi, A.F. (eds.) MICCAI 2015. LNCS, vol. 9351, pp. 234–241. Springer, Cham (2015). https://doi.org/10.1007/978-3-319-24574-4_28
35. Shaban, A., Bansal, S., Liu, Z., Essa, I., Boots, B.: One-shot learning for semantic segmentation. In: BMVC (2017)
36. Snell, J., Swersky, K., Zemel, R.: Prototypical networks for few-shot learning. In: NeurIPS (2017)
37. Sung, F., Yang, Y., Zhang, L., Xiang, T., Torr, P.H., Hospedales, T.M.: Learning to compare: relation network for few-shot learning. In: CVPR (2018)
38. Tian, Z., Zhao, H., Shu, M., Yang, Z., Li, R., Jia, J.: Prior guided feature enrichment network for few-shot segmentation. TPAMI 44, 1050–1065 (2020)
39. Vinyals, O., Blundell, C., Lillicrap, T., Wierstra, D., et al.: Matching networks for one shot learning. In: NIPS (2016)
40. Wang, H., Zhang, X., Hu, Y., Yang, Y., Cao, X., Zhen, X.: Few-shot semantic segmentation with democratic attention networks. In: Vedaldi, A., Bischof, H., Brox, T., Frahm, J.-M. (eds.) ECCV 2020. LNCS, vol. 12358, pp. 730–746. Springer, Cham (2020). https://doi.org/10.1007/978-3-030-58601-0_43

41. Wang, K., Liew, J.H., Zou, Y., Zhou, D., Feng, J.: Panet: few-shot image semantic segmentation with prototype alignment. In: ICCV (2019)
42. Wu, Z., Shi, X., Lin, G., Cai, J.: Learning meta-class memory for few-shot semantic segmentation. In: Proceedings of the IEEE/CVF International Conference on Computer Vision, pp. 517–526 (2021)
43. Xie, G.S., Liu, J., Xiong, H., Shao, L.: Scale-aware graph neural network for few-shot semantic segmentation. In: CVPR (2021)
44. Xie, G.S., Xiong, H., Liu, J., Yao, Y., Shao, L.: Few-shot semantic segmentation with cyclic memory network. In: ICCV (2021)
45. Yang, B., Liu, C., Li, B., Jiao, J., Ye, Q.: Prototype mixture models for few-shot semantic segmentation. In: Vedaldi, A., Bischof, H., Brox, T., Frahm, J.-M. (eds.) ECCV 2020. LNCS, vol. 12353, pp. 763–778. Springer, Cham (2020). https://doi.org/10.1007/978-3-030-58598-3_45
46. Yang, L., Zhuo, W., Qi, L., Shi, Y., Gao, Y.: Mining latent classes for few-shot segmentation. In: ICCV (2021)
47. Yang, X., et al.: Brinet: towards bridging the intra-class and inter-class gaps in one-shot segmentation. In: BMVC (2020)
48. Yu, F., Koltun, V.: Multi-scale context aggregation by dilated convolutions. In: ICLR (2016)
49. Zhang, B., Xiao, J., Qin, T.: Self-guided and cross-guided learning for few-shot segmentation. In: CVPR (2021)
50. Zhang, C., Cai, Y., Lin, G., Shen, C.: Deepemd: few-shot image classification with differentiable earth mover's distance and structured classifiers. In: CVPR (2020)
51. Zhang, C., Lin, G., Liu, F., Guo, J., Wu, Q., Yao, R.: Pyramid graph networks with connection attentions for region-based one-shot semantic segmentation. In: ICCV (2019)
52. Zhang, C., Lin, G., Liu, F., Yao, R., Shen, C.: Canet: class-agnostic segmentation networks with iterative refinement and attentive few-shot learning. In: CVPR (2019)
53. Zhang, G., Kang, G., Yang, Y., Wei, Y.: Few-shot segmentation via cycle-consistent transformer. In: NIPS (2021)
54. Zhao, H., Shi, J., Qi, X., Wang, X., Jia, J.: Pyramid scene parsing network. In: CVPR (2017)
55. Zhen, M., et al.: Joint semantic segmentation and boundary detection using iterative pyramid contexts. In: CVPR (2020)
56. Zheng, S., et al.: Rethinking semantic segmentation from a sequence-to-sequence perspective with transformers. In: CVPR (2021)
57. Zheng, Z., Zhong, Y., Wang, J., Ma, A.: Foreground-aware relation network for geospatial object segmentation in high spatial resolution remote sensing imagery. In: CVPR (2020)
58. Zhu, L., Yang, Y.: Compound memory networks for few-shot video classification. In: Ferrari, V., Hebert, M., Sminchisescu, C., Weiss, Y. (eds.) ECCV 2018. LNCS, vol. 11211, pp. 782–797. Springer, Cham (2018). https://doi.org/10.1007/978-3-030-01234-2_46
59. Zhu, Z., Xu, M., Bai, S., Huang, T., Bai, X.: Asymmetric non-local neural networks for semantic segmentation. In: ICCV (2019)

Active Domain Adaptation
with Multi-level Contrastive Units
for Semantic Segmentation

Hao Zhang[1,2] and Ruimao Zhang[1(✉)] (iD)

[1] Shenzhen Research Institute of Big Data, The Chinese University of Hong Kong,
Shenzhen, China
`ruimao.zhang@ieee.org`
[2] University of Illinois Urbana-Champaign, Champaign, USA
`haoz19@illinois.edu`

Abstract. To further reduce the cost of semi-supervised domain adaptation (SSDA) labeling, a more effective way is to use active learning (AL) to annotate a selected subset with specific properties. However, domain adaptation tasks are always addressed in two interactive aspects: domain transfer and the enhancement of discrimination, which requires the selected data to be both uncertain under the model and diverse in feature space. Contrary to active learning in classification tasks, it is usually challenging to select pixels that contain both the above properties in segmentation tasks, leading to the complex design of pixel selection strategy. To address such an issue, we propose a novel Active Domain Adaptation scheme with Multi-level Contrastive Units (ADA-MCU) for semantic image segmentation. A simple pixel selection strategy followed with the construction of multi-level contrastive units is introduced to optimize the model for both domain adaptation and active supervised learning. In practice, MCUs are constructed from intra-image, cross-image, and cross-domain levels by using both labeled and unlabeled pixels. At each level, we define contrastive losses from center-to-center and pixel-to-pixel manners, with the aim of jointly aligning the category centers and reducing outliers near the decision boundaries. In addition, we also introduce a categories correlation matrix to implicitly describe the relationship between categories, which are used to adjust the weights of the losses for MCUs. Extensive experimental results on standard benchmarks show that the proposed method achieves competitive performance against state-of-the-art SSDA methods with 50% fewer labeled pixels and significantly outperforms state-of-the-art with a large margin by using the same level of annotation cost. Code will be in https://github.com/haoz19/ADA-MCU.

H. Zhang—Research done when Hao Zhang was a RA at SRIBD and CUHK, SZ.

Supplementary Information The online version contains supplementary material available at https://doi.org/10.1007/978-3-031-26293-7_27.

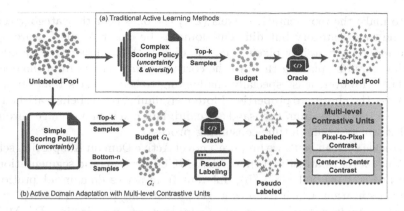

Fig. 1. Traditional active learning methods via the complex sample selection policy for domain adaptation task and (b) the pipeline of our proposed Active Domain Adaptation with Multi-level Contrastive Units. G_l and G_h denote low and high uncertainty pixel groups determined by the uncertainty score.

1 Introduction

Semantic segmentation is one of the most classic tasks in computer vision and image processing. The goal is to learn the semantic context in the image and automatically annotate each pixel a category label according to such learned information [1,2]. In practice, such a task always requires excessively numerous annotations, limiting its scalability in real applications. One way to reduce annotation cost is to leverage a large amount of virtual data that is easy to obtain labels from game engines to extend training samples (*e.g.*, GTA5, SYNTHIA, Synscapes). However, the model trained merely with virtual data performs terribly on real-world data distribution because of the domain shifts. Therefore, many domain adaption methods are raised in recent years to bridge the gap between label-rich virtual data and label-scarce real-life data.

One way to address the domain adaptation problem is to train the model under the semi-supervised manner (SSDA), which jointly leverages fully labeled source domain data and a subset of labeled target domain data in the training phase. Compared to unsupervised domain adaptation (UDA), semi-supervised methods are able to significantly improve the segmentation accuracy through a small amount of annotated samples. Although the SSDA method makes a good balance between performance and annotation cost, there is still a big gap compared with the performance of the fully supervised approaches. In practice, how to bridge this gap by effectively labeling data is still an open issue.

In the literature, active learning (AL), which aims to select a subset of samples with specific properties, is proposed to annotate the training samples in a cost-effective way [19–21]. Although it has extensive research [23] on various areas by adopting active learning, it remains a challenge to deal with domain adaptation semantic segmentation since we need to address this task by dealing with two entangled issues from the pixel level, *i.e.*, **aligning domain distribution** and **improve the model's discriminant ability**. In simple terms, we need to make the representative pixels (*i.e.,* the ones near the category center) from the same category but different domains be closer in the feature space, and to reduce uncertainty pixels (*i.e.,* the ones with low predictive confidence) of each domain by pushing them to the corresponding centers from the decision boundary. However, it is especially challenging to select pixels that simultaneously meet the above two properties. Moreover, mining pixels being satisfied with the above properties from thousands of candidate pixels in an image through a complex scheme is also a time-consuming process.

To tackle such an issue, we propose a novel Active Domain Adaptation scheme with Multi-level Contrastive Units (ADA-MCU) for semantic segmentation. As shown in Fig. 1 (a) and Fig. 1 (b), different from active domain adaptation for image classification, which requires a complex scoring policy to labeled specific samples for model supervision, we use a simple selection policy in ADA-MCU to divide each image into two subsets of pixels, (*i.e.,* actively labeled ones with low confidence scores and unlabeled ones), then adopt pixels from these two subsets to construct contrastive pairs from multiple perspectives for model optimization. Specifically, at the **cross-domain** level, we enforce the distribution of the feature representations in the target domain being aligned to the source domain. While at the **intra-image** and **cross-image** level, we enforce the instance representations belonging to the same category to be closer in the feature space and make those are belonging to different categories to be far away in both source and target domain. Additionally, for each contrastive level, we do the alignment in two perspectives by using two kinds of loss, *i.e.,* center to center contrastive loss and pixel to pixel contrastive loss. The former enforces the centers of distribution to be aligned while the latter reduces the uncertainty of pixels by pulling them closer to the category center representations. In this way, two domains would be better aligned by employing the synergy of the above two losses.

In practice, since the misclassified pixels are highly relevant to its spatial layout, *e.g.,* the `sidewalk` is more likely to be misclassified as the `road` but be rarely misclassified as `sky`, we further introduce a dynamic categories correlation matrix (DCCM) to model the implicit relationship between each pair of categories. DCCM will be updated online during the training phase, aiming to adjust the weights of contrastive losses for the MCUs. Such a categories-aware contrastive loss could further improve the discriminative feature representation learning across domains. In this way, the domain transfer and discrimination enhancement are unified into one single framework.

The main contributions of this article can be summarized as follows,

1. We propose ADA-MCU, a novel active learning scheme, which uses a simple selection policy along with the construction of MCUs to optimize the model.
2. We introduce a simple yet effective scheme to construct Multi-level Contrastive Units (MCU) to regularize model training from multiple perspectives and propose a dynamic categories correlation matrix (DCCM) to describe the implicit relationship between categories, making more effective usage of the labeled and unlabeled pixels for model training.
3. As shown in Fig. 1 (c), extensive experiments demonstrate that our proposed method can achieve similar performance on two standard synthetic-to-real

semantic segmentation benchmarks with less than 50% labeled data compared with current semi-supervised domain adaptation methods, and significantly outperforms state-of-the-art with a large margin by using the same level of annotation cost.

2 Related Work

2.1 Domain Adaptation for Semantic Segmentation

Domain adaptation task aims to explore how to transfer the knowledge that the model learned from one domain to another. In domain adaptation task, typically there are two domains, which have a certain domain shift but also share some common knowledge with each other. The domain that we train our model originally with is called source domain and the domain that we want to transfer our model to is called target domain. According to the usage of target domain annotations, we can divide domain adaptation semantic segmentation task into two main parts: 1) unsupervised domain adaptation one termed UDA and 2) semi-supervised domain adaptation termed SSDA. UDA is aimed at transferring the knowledge obtained from a labeled source domain to an unlabelled target domain. While SSDA aims at narrowing the gap with the fully supervised results by using a small set of labeled samples.

In recent years, some UDA methods for semantic segmentation [3–5] have been proposed via adversarial training which relies on a discriminator to measure the divergence between two domains' distributions. Adversarial based methods aims at aligning the feature space of source domain and target domain by confusing the discriminator. Image translation has also been widely used in the UDA methods [6,7]. As we know the most obvious gap between source domain and target domain is the color distribution. The aligning of color distribution is a very easy but efficient way to improve the performance. Recently some works [45] proposed that class-agnostic training paradigm gives model stronger generalization ability, which also pointed out a promising way of UDA. In addition, several methods [8,9] also focus on self-training that generates pseudo labels for unlabelled data in the target domain.

For the SSDA, some methods [10–12] implemented feature alignment across domain from global and semantic level. For example, Chen et al. [11] proposed a framework based on dual-level domain mixing to address the differences in the amount of the labeled data between two domains. Huang et al. [12] aligned features by employing a few labeled target samples as anchors. However, none of those methods focus on sample selection, especially from the pixel-level perspective. We introduce pixel-level active learning to make more effective use of labeled pixels compared with SSDA.

Contrastive Learning. Contrastive learning (CL) aims at pushing positive sample pairs away from the negative ones in the representation space to learn better representations. For positive pair sampling, mainstream methods [13,14] create different views of each sample using multiple perturbations. At the same time, negative pairs can be obtained by hard example mining strategies [15,16].

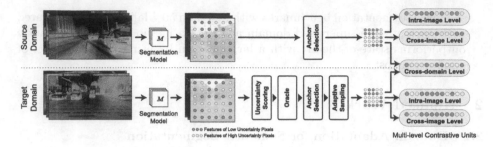

Fig. 2. Detailed illustration of construction of Multi-level Contrastive Units.

For the semantic segmentation task, CL could also be used to do intra-domain model pre-training [17]. Recently, [18] also propose a pixel-wise contrast scheme for semantic segmentation, making the representation of each category's pixel more compact.

Active Learning. Active learning (AL) aims at obtaining high performance with low annotation costs by selecting the most informative samples. Usually, there are three major views to address AL tasks. 1) uncertainty-based methods [19,20], 2) diversity-based methods [21,22], 3) methods based on expected model change [23]. The former selected the samples with the highest uncertainty, while the latter selected the samples that could better represent the whole dataset. The last one aims to select the sample which can lead to more effect on the model. Compared to the existing works in active learning, our proposed method address AL in pixel-level and we construct MCUs for assistance in order to achieve great performance without a complicated selection strategy.

3 Methodology

In this section, we will describe our proposed method ADA-MCU in detail. Firstly, we will introduce a simple pixel-level sample selection policy, dividing each image into different pixel subsets. Secondly, we will present how to construct multi-level contrastive units (MCU) using labeled samples and unlabeled ones. At last, we will discuss how to apply contrastive loss with the dynamic category correlation to optimize the segmentation model.

3.1 Problem Setting and Notation

The goal of semantic segmentation domain adaptation is to transfer the model from source domain X_s to the target domain X_t. In our setting, we have a fully labeled source domain $\{(x_s^n, y_s^n)\}_{n=1}^{N_s}$, indicating the n-th source image x_s^n with the ground truth label map y_s^n, and the target domain $\{x_t^n\}_{n=1}^{N_t}$. Here N_s and N_t denote the number of source and target domain images, respectively. For the pixel-level active domain adaptation, the n-th target image x_t^n contains two subsets in pixel-level, naming active annotated pixel $x_t^{n:(i,j)}$ with its corresponding ground-truth label $y_t^{n:(i,j)}$ and unlabeled pixel $x_t^{n:(\bar{i},\bar{j})}$, where (i,j) and $(\bar{i},\bar{j})$ denote the labeled and unlabeled pixel positions in the target image. We use M_e and M_a to represent the number of expected labeled pixels number and already labeled pixels. And M_t is the total number of pixels in the target training dataset.

3.2   Active Pixel Annotation via Uncertainty Score

In practice, we first use source domain images to train the segmentation network. Once the pre-trained model $\mathcal{F}$ is obtained, we use the output of $\mathcal{F}$ on both source and target images to obtain the calculate their predictive uncertainty scores. The uncertainty score of pixel (i,j) in the n-th image can be calculated as follows,

$$S(x^{n:(i,j)}) = E(x^{n:(i,j)}) + \gamma D_{\mathrm{KL}}(p^{n:(i,j)} \| \hat{p}^{n:(i,j)}) \qquad (1)$$

where the first part $E(\cdot)$ is the pixel-wise entropy, which is calculated by Eq. (2). The second part, $D_{\mathrm{KL}}(\cdot)$ is the pixel-wise KL divergence between the predictions of the main segmentation head and the auxiliary head.[1] γ is a hyperparameter to control the weights of two uncertainty indicators. $E(\cdot)$ and $D_{\mathrm{KL}}(\cdot)$ can be calculated as follows,

$$E(x^{n:(i,j)}) = \frac{-1}{\log(C)} \sum_{c=1}^{C} p_c^{n:(i,j)} \log p_c^{n:(i,j)}, \qquad (2)$$

$$D_{\mathrm{KL}}(p \| \hat{p}) = \sum_{c=1}^{C} p_c (\log p_c - \log \hat{p}_c), \qquad (3)$$

where C denotes the total number of categories, By calculating the entropy of the pixel and the KL divergence of two head outputs, the degree of each pixel's uncertainty is obtained for further pixel annotation.

Using the scoring function mentioned above with the threshold π_{high}, π_{low}, we can divide both source and target image pixels into three groups termed *high*, *low*, and *medium* uncertainty groups respectively. Here the higher uncertainty indicates the lower model predictive confidence. By adjusting π_{high}, we can control the annotation rate of the target domain. We will discuss how to use pixels in different groups for contrastive unit construction in the next subsection.

Category Center Generation Strategy. After we divide both source and target pixels of each training batch into three groups, we use those pixels from the low uncertainty group with high predictive confidence to generate the category center. Intuitively, high confident samples always lie in the center of category

---

[1]   The auxiliary loss is proposed in [1] to improve the accuracy. We leverage the outputs from both auxiliary and main segmentation heads of DeepLab v2 to calculate KL divergence.

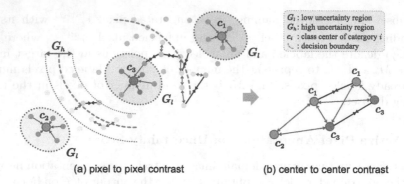

(a) pixel to pixel contrast   (b) center to center contrast

Fig. 3. Illustration of the construction of MCUs. We leverage both labeled data from G_h (when adding adaptive sampling, we also consider G_m to label) and unlabeled data from G_l to construct MCUs, and in each level of MCUs we use both pixel-to-pixel and center-to-center contrast to regularize the model in both uncertainty and diversity perspectives. Note that the representations can be from different pairs of images at different levels. High transparency means high uncertainty.

clusters, leading to a high density. According to the boundary assumption [34], the decision boundary should not across the high-density region of a cluster. In other words, high confident pixels are always reliable and can be applied as representative of the cluster. Therefore, the aggregation of those pixels in the low uncertainty group for each domain can be considered as the representation of each category's center.

Annotation via Adaptive Sampling. When labeling the target domain pixels, we introduce adaptive sampling (AS) to maximize the advantages of active learning strategy. For each training batch, after dividing each target domain image into three groups: G_h^t, G_m^t and, G_l^t, we randomly select the same number of pixels in those three groups. Then for G_h^t and G_m^t, we give ground-truth labels to those pixels, and we use the current predictions as pseudo labels to annotate those pixels selected by AS in G_l^t. Note that the size of each group changes for each training iteration. Thus the annotated pixels in G_h^t, G_m^t and G_l^t should be dynamically sampled, and we called it *annotation via adaptive sampling*. In practice, the size of G_m^t is always larger than G_h^t, therefore, in other words, we give a higher sampling rate to the group with high uncertainty for labeling (Fig. 3).

3.3   Multi-level Contrastive Unit Construction

After obtaining the active labeled target pixels, we construct the multi-level contrastive units for domain adaptation and discrimination enhancement, which is partially inspired by the pixel-wise contrastive learning for semantic segmentation [18]. For each batch size that contains the same number of source and target images, we construct the contrastive units in three levels: (1) intra-image level, (2) cross-image level, and (3) cross-domain level.

Anchor Selection Policy. Before constructing the contrastive units for each level, we first require to select the anchor pixels for each contrastive unit to construct. In practice, rather than mining informative samples, we follow the work [18], determining whether it is a hard anchor through its uncertainty and whether the model predicts it correctly. In the source domain, we compare the predictive pixels' labels and their ground truth and then select the ones which are incorrectly predicted as anchors. Similarly, in the target domain, we use active learning to annotate the pixels with high uncertainty. Then the incorrectly predicted pixels in the high uncertainty group[2] are selected as the target anchors.

Contrastive Unit Construction. In each level, we firstly extract anchors from images using the above anchor selection policy. Then the pixels with the same label as the anchor are indicated as the positive samples, while the others with different labels are identified as the negative ones. After that, we construct the contrastive units in three levels, which are illustrated in Fig. 2. For the *intra-image level*, we extract positive pixels and negative pixels from the same image according to anchors. While there is a problem that the independence of images leads to the loss of information across the whole dataset. For example, the class car and class train may never appear in the same image through the entire domain, making these two categories lack inter-actions at the intra-image level. Therefore, we introduce *cross-image level* contrast to solving this problem. For the cross-images level, we first extract anchors from one image and extract positive pixels and negatives ones from another random image in the same domain. In this way, any two categories could be able to have interactions with each other along with the training iterations. According to the above strategy, the intra-image and the cross-image contrast units encourage the feature representation to be more discriminative in a specific domain. In contrast, we introduce *cross-domain level* contrastive units to align source and target distribution for the domain transfer. Different from cross-images level contrast, for cross-domain level, we extract anchors from one image in a certain domain and extract positive samples and negative ones from another image from the other domain. Through the synergy of three levels' contrast units, we could achieve the discrimination enhancement and domain alignment in the model optimization process.

3.4 Pixel-level Active Domain Adaptation with MCUs

Segmentation Loss Function. In target domain, we use G_h^t, G_m^t, and G_l^t to denote the labeled pixels selected by adaptive sampling in high, medium, and low uncertainty groups, because there only labels for those samples selected by AL with AS in target domain. Thus the active training loss by using the labeled high uncertainty pixels can be defined as,

$$\mathcal{L}_{\text{seg}}^{G_h^t, G_m^t} = \sum_{n:(i,j) \in G_h^t, G_m^t} \mathcal{L}_{\text{CE}}(p^{n:(i,j)}, y^{n:(i,j)}) \tag{4}$$

---

[2] Note that the pixel with high uncertainty does not necessarily to have the wrong predictive result.

where $p^{n:(i,j)}$ and $y^{n:(i,j)}$ are the prediction and ground truth of pixel in position (i,j) of n-th target image, and $\mathcal{L}_{CE}$ denotes the cross-entropy loss function. On the contrary, pixels in G_l^t are with low uncertainty, which means more likely to be correctly predicted. Thus, we directly use their predictions as pseudo labels to calculate the loss,

$$\mathcal{L}_{seg}^{G_l^t} = \sum_{n:(\bar{i},\bar{j})\in G_l^t} \mathcal{L}_{CE}(p^{n:(\bar{i},\bar{j})}, \widehat{y}^{n:(\bar{i},\bar{j})}), \tag{5}$$

where $\widehat{y}^{n:(\bar{i},\bar{j})} = \text{softmax}(p^{n:(\bar{i},\bar{j})})$. Thus the overall segmentation loss can be represented as follows,

$$\mathcal{L}_{seg} = \mathcal{L}_{seg}^{G_h^t, G_m^t} + \mathcal{L}_{seg}^{G_l^t} + \mathcal{L}_{seg}^{G_h^s, G_m^s, G_l^s}. \tag{6}$$

where G_h^s, G_m^s and G_l^s denote the pixel groups of the source domain. Note that since we have all annotations for images in source domain, we just calculate cross-entropy loss as their segmentation loss.

Dynamic Categories Correlation Matrix (DCCM). In practice, the context information and spatial layout are critical for segmentation accuracy. For instance, class road and sidewalk are misclassified more frequently, while sky and road are rarely to be misclassified. To better leverage such information, we introduce DCCM to describe the implicit relationship between any two categories. Denote $M_k^{(c_u,c_v)}$ as the number of pixels being misclassified from category c_u to c_v. Let M^{c_u} indicate the number of all pixels in c_u, then the error rate $R_k^{(c_u,c_v)}$ can be calculated as follow,

$$R_k^{(c_u,c_v)} = M_k^{(c_u,c_v)} / M^{c_u} \tag{7}$$

And at each iteration, we could dynamically update each element of the correlation matrix W by using exponential moving average as,

$$w_\tau^{(u,v)} = \beta \, w_{\tau-1}^{(u,v)} + (1-\beta) \, R_k^{(c_u,c_v)}, \tag{8}$$

where $w_\tau^{(u,v)}$ denotes the correlation coefficient of u-th and v-th categories at the iteration τ. Then DCCM will further guide multi-level contrastive units by adjusting the weight of contrastive loss for each MCU.

Contrastive Loss Function. In each level of MCUs, we define loss function in two perspectives. On the one hand, we introduce pixel-to-pixel (p2p) contrastive loss based on the labeled pixels to reduce their uncertainty by pulling the same class pixel being close and pushing different class samples being apart. The loss function is defined based on InfoNCE [28], modified by using the weights in DCCM.

$$\mathcal{L}_{con}^{p2p} = \frac{1}{|P_p|} \sum_{p^+ \in P_p} -\log \mathcal{H}_p, \tag{9}$$

$$\mathcal{H}_p = \frac{\exp(w^{(u,v)} \, p \cdot p^+ / \lambda)}{\exp(w^{(u,v)} \, p \cdot p^+ / \lambda) + \sum_{p^- \in N_p} \exp(w^{(u,v)} \, p \cdot p^- / \lambda)}, \tag{10}$$

where $\cdot$ denotes dot multiplication of two vectors with the scalar as the output. P_p and N_p denote pixel-wise embedding collections of the positive and negative samples.

On the other hand, we introduce class-to-class (c2c) contrastive loss by using the category centers introduced in the former section, which are most representative for each category. Then we define c2c contrastive loss in the same form with the p2p,

$$\mathcal{L}_{con}^{c2c} = \frac{1}{|P_c|} \sum_{c^+ \in P_c} \log \mathcal{H}_c, \tag{11}$$

where P_c and N_c denote class-wise embedding collections of the positive and negative samples, and $\mathcal{H}_c$ has the same form as Eqn. 10 but uses class-wise anchors instead of pixel-wise ones. Using two kinds of contrastive losses as regular terms, we can encourage the features from the same category to be closer and from different categories to be further. And the DCCM can increase the weight between the categories which are more likely to be misclassified so as to guiding model optimization. In this way, the total loss of the proposed ADA-MCU can be presented as $\mathcal{L}_{total} = \mathcal{L}_{seg} + \mathcal{L}_{con}$, where $\mathcal{L}_{con}$ is the sum of ($\mathcal{L}_{con}^{p2p} + \mathcal{L}_{con}^{c2c}$) from three levels.

Table 1. Experimental results on GTA5-to-Cityscapes compared with current SSDA, semi-supervised learning (SSL) methods. 19-class mIoU (%) scores are reported on Cityscapes validation set by using 1.7%, 3.4%, 6.8%, 16.8% labeled pixels from whole dataset.

Type	Method	Label percentage (%)			
		1.7	3.4	6.8	16.8
Supervised	Image-wise	-	41.9	47.7	55.5
SSL	CutMix (bmvc20)	-	50.8	54.8	61.7
	DST-CBC (arxiv20)	-	48.7	54.1	60.6
SSDA	MME (cvpr19)	-	52.6	54.4	57.6
	MinEnt (cvpr19)	47.5	49.0	52.0	55.3
	AdvEnt (cvpr19)	44.9	46.9	50.2	55.4
	ASS (cvpr20)	50.1	54.2	56.0	60.2
	FDA (cvpr20)	53.1	54.1	56.2	59.2
	DDM (cvpr21)	-	61.2	60.5	64.3
	PCL (arxiv21)	54.2	55.2	57.0	60.4
	Ours	**58.7**	**61.6**	**63.9**	**65.8**

Table 2. Experimental results on SYNTHIA-to-Cityscapes compared with current SSDA and semi-supervised learning (SSL) methods. 13-class mIoU (%) scores are reported on Cityscapes validation set. Note that 1.7% pixels of 2795 images are at the same pixel number of 50 images.

Type	Method	Label percentage (%)			
		1.7	3.4	6.8	16.8
Supervised	Image-wise	-	53.0	58.9	61.0
SSL	CutMix (bmvc20)	-	61.3	66.7	71.7
	DST-CBC (arxiv20)	-	59.7	64.3	68.9
SSDA	MME (cvpr19)	-	59.6	63.2	66.7
	MinEnt (cvpr19)	52.9	56.4	57.9	62.5
	AdvEnt (cvpr19)	51.4	55.2	59.6	62.6
	ASS (cvpr20)	60.7	62.1	64.8	69.8
	FDA (cvpr20)	58.5	62.0	64.4	66.8
	DDM (cvpr21)	-	68.4	69.8	71.7
	PCL (arxiv21)	61.2	63.4	65.2	70.3
	Ours	**66.2**	**69.1**	**70.6**	**73.1**

Image   Ground Truth   Fully Supervise   Ours   AdvEnt+CycleGAN

Fig. 4. Visualization of the segmentation results. Ours stands for our proposed method with 20% active annotations.

4 Experiment

4.1 Dataset, Setting and Implementation

We evaluate our proposed method by using the two standard large-scale segmentation benchmarks for domain adaptation, GTA5-to-Cityscapes, and SYNTHIA-to-Cityscapes. Following the previous method [31], we apply 19 classes domain adaptation for the former, and 13 classes for the latter. We conduct extensive experiments and report mean Intersection-over-Union (mIoU) compared with existing domain adaptation methods. All of the methods employ Deeplab v2 [2] as the basic model, which utilizes a pre-trained ResNet-101 [25] on ImageNet as backbone network. To measure the uncertainty, we calculate the KL divergence in Eqn. (3) by using multi-level outputs coming from both `conv4` and `conv5` feature maps. All experiments are run on a single Tesla V100 GPU with 32 GB of memory. All the models are trained by the Stochastic Gradient Descent (SGD) optimizer with an initial learning rate of 2.5×10^{-4} and decreasing with the polynomial annealing procedure with the power of 0.9.

Before the domain adaptation procedure, we use translated source domain data to train the model first. Then in each iteration, we randomly select 4 images, contains 2 from the source domain and 2 from the target domain. For **source domain images**, we feed them to the model and get the predictions and pixel-wise features. After that, we use predictions and labels to calculate cross-entropy loss, execute anchor selection and calculate intra-image and cross-image contrastive loss. While for **target domain images**, since we don't have any annotations at the beginning, we directly calculate the pixel-wise uncertainty score of the model output (after `softmax` function) and divide the batch of images into three groups according to the uncertainty, where γ is set to 0.5 in Eqn. (1). We ask the Oracle to label the high uncertainty group, and use the predictions as pseudo labels for pixels in low uncertainty group. Then, we can calculate cross-entropy loss and intra-image and cross-image contrastive loss similar to source domain. Furthermore, we also calculate cross-domain contrastive loss using both source domain and target domain data (Table 3).

Table 3. Experiments setting of ablation study.

	AS	MCU-L_i	MCU-L_d	DCCM
AL(w/o AS)				
AL(w AS)	✓			
AL(w MCU$_i$)	✓	✓		
AL(w MCU$_d$)	✓	✓	✓	
Full Model	✓	✓	✓	✓

4.2 Comparison with State-of-the-Art Methods

As presented in Tables 1 and 2, we compare the proposed method with two SSL methods, *i.e.* CutMix [29], DST-CBC [30], and seven SSDA methods, *i.e.* MME [31], ASS [10], MinEnt [4], AdvEnt [4], FDA [33], PCL [32], DDM [11], in different percentage of annotation: 1.7%, 3.4%, 6.8%, 16.8%. As expected, compared with those methods, our purposed method has achieved a significant accuracy (mIoU) improvement.

From the Table 1, we can clearly see that our method can achieve comparable results with only about 50% of the annotations compared with other methods. Even compared with the state of art SSDA method DDM, we can still achieve similar performance using 30% fewer annotations. In addition, our proposed method can achieve similar performance with the fully supervised method using only 16.8% annotation of the whole target set. The visualization of the segmentation results of fully supervised method, UDA method [4] and our proposed method with 20% annotations are shown in Fig. 4. We can clearly see that our proposed scheme with 20% annotations can obviously improve the effect of some critical small areas, *e.g.* sign, rider and person.

5   Comparison with State-of-the-Art ADA Methods

Recently, one article termed Multi-Anchor Active Domain Adaptation (MADA) [37], which is about active domain adaptation for semantic segmentation, has been accepted by ICCV2021 as the oral representation. The authors claim that it is the first study to adopt active learning to assist the domain adaptation regarding the semantic segmentation tasks, which adopts multiple anchors obtained via clustering-based method to characterize the feature distribution of the source-domain and multi-anchor soft-alignment loss to push the features of the target samples towards multiple anchors leading to better latent representation. Since the DeepLab v3+ [39] is adopted as the backbone network of this method, we have not listed in the main article considering fair comparisons with previous DeepLab v2 [2] based methods.

Different from our proposed *pixel-level* group partition regarding the uncertainty, such a method still adopts an *image-level* scheme to conduct active target sample selection against source anchors. As presented in Table 5, we compare our method with this active domain adaptation (ADA) method, MADA [37].

For fairness, we follow the same setting that MADA uses. The DeepLab v3+ [39] is applied with the pre-trained ResNet-101 on ImageNet as the backbone network. We set $M_e = 5\%$ since MADA also selects 5% target-domain samples as active samples for their experiments. The other settings are the same as the former experiments described in the main article. The experiment shows that our proposed method outperform MADA by a large margin, *i.e.*, 1.6% mIoU, which demonstrates the proposed method could take little annotation workload but brings large performance gain compared with the recent SOTA method.

5.1 Ablation Study

Effectiveness of AL with Adaptive Sampling. Firstly, we want to investigate the effectiveness of our proposed active learning strategy. As shown in in Table 1, **UDA** is the performance of an unsupervised domain method AdvEnt with cycleGAN [4]. **RBA** denotes the performance of a pixel-level Region-based active learning method proposed in [35]. **AL(w/o AS)** indicates the degradation model which only uses the pixels from the high uncertainty group (*i.e.* pixels with manual labeling) and all of the pixels from the low uncertainty group (*i.e.* pixels with pseudo labels) to optimize the model. According to the results shown in Table 1, we surprisingly find out that in our work, merely adding an active learning strategy may lead to performance degradation, *e.g.* 43.2% compared with 46.3%. One acceptable reason is that our proposed scheme not only uses labeled samples but also uses some of the unlabeled ones with pseudo labels to supervise the network, and the proportion of selected labeled pixels in different images varies greatly. For instance, if we set the annotation percentage for the whole dataset to be 10%, some of the images may get 2% or less labeled data (*e.g.* images with simple scenes) yet more than 50% pseudo label to supervise the model together. In such case, the loss calculated by those ground truth labels would be overwhelmed by the loss calculated by the pseudo labels, especially when the annotation budget is extremely limited. After we introduce adaptive sampling (AS) to our pixel-level active learning scheme, we get a satisfactory result (64.3% mIoU) shown by **AL(w AS)** in Table 1. Such result also outperforms **RBA** by 2.5%, showing the superiority of our proposed active selection strategy (Table 4).

Table 4. Evaluation of different components of proposed method on GTA5-to-Cityscapes, with 20% labeled pixels except UDA.

GTA5 to cityscapes

	Road	Sidewalk	Building	Wall	Fence	Pole	Light	Sign	Vege	Terrace	Sky	Person	Rider	Car	Truck	Bus	Train	Motor	Bike	mIoU
UDA (cvpr19)	92.4	52.7	83.8	32.4	24.1	30.7	33.2	25.7	83.7	35.1	85.1	58.2	27.4	85.5	37.1	41.9	2.1	25.2	22.6	46.3
RBA (wacv19)	–	–	–	–	–	–	–	–	–	–	–	–	–	–	–	–	–	–	–	61.8
AL(w/o AS)	90.4	34.9	82.3	30.0	23.4	27.4	31.9	21.9	84.0	38.5	77.8	58.4	25.0	84.8	26.9	34.9	1.5	27.6	19.8	43.2
AL(w AS)	96.4	75.7	86.8	40.3	42.0	47.4	46.1	65.4	87.9	44.0	84.3	68.6	44.9	91.5	66.7	72.6	53.9	41.9	64.7	64.3
AL(w MCU$_1$)	96.8	76.5	86.8	40.5	43.2	47.4	48.5	66.3	88.6	50.7	80.7	69.4	48.4	91.7	67.4	73.2	54.2	45.6	66.6	65.5
AL(w MCU$_4$)	97.1	77.4	87.8	42.1	43.9	48.1	47.4	65.3	87.4	55.1	82.9	72.1	49.1	91.2	70.4	73.1	55.3	45.7	66.3	66.1
Full Model	**97.2**	**78.3**	**88.4**	**46.0**	42.9	**48.5**	**48.6**	**66.5**	**89.2**	54.9	**89.3**	70.3	**49.7**	**92.1**	**70.9**	72.2	49.0	**46.4**	**67.0**	**66.7**

Table 5. Experimental results on GTA5-to-Cityscapes compared with current Active learning domain adaptation (ADA) method [37] with 5% annotations. The segmentation network used in the above experiment is DeepLab v3+ [39], which utilizes a pre-trained ResNet-101 on ImageNet as the backbone.

GTA5 to cityscapes (DeepLab v3+)	Road	Sidewalk	Building	Wall	Fence	Pole	Light	Sign	Vege	Terrace	Sky	Person	Rider	Car	Truck	Bus	Train	Motor	Bike	mIoU
MADA (iccv21)	95.1	69.8	88.5	43.3	48.7	45.7	53.3	59.2	89.1	46.7	91.5	73.9	50.1	91.2	60.6	56.9	48.4	51.6	68.7	64.9
Full Model	97.3	78.5	88.7	50.8	44.3	49.6	49.5	64.1	89.3	55.9	91.8	68.7	37.5	91.6	65.9	74.6	58.6	41.5	65.5	66.5

6 Conclusion

In this paper, we propose ADA-MCU, a novel active learning method, which uses a simple selection policy along with the construction of MCUs to optimize the segmentation model. As shown in Fig. 1, such a scheme abandons the complex sample selection policy in previous methods, leading to a more efficient active supervised training process. To the best of our knowledge, this work is the first study to conduct pixel-level annotation-based active domain adaptation for semantic image segmentation. The multi-level contrastive units (MCU), together with dynamic categories correlation matrix (DCCM), are carefully designed for efficient active supervised model training, leading to many appealing benefits. (1) It enables the models to learn more compact feature representation for each category. (2) It could employ fewer annotations (16.8%) to achieve comparable performance with fully supervised method (65.3% mIoU). (3) It is effective for dealing with boundaries and small objects. Future work will combine the proposed scheme with more powerful architecture, *e.g.,* vision transformer, to explore more challenge tasks, such as panoptic segmentation.

Acknowledgements. The work is supported in part by the Young Scientists Fund of the National Natural Science Foundation of China under grant No. 62106154, by Natural Science Foundation of Guangdong Province, China (General Program) under grant No.2022A1515011524, by CCF-Tencent Open Fund, by Shenzhen Science and Technology Program ZDSYS20211021111415025, and by the Guangdong Provincial Key Laboratory of Big Data Computing, The Chinese University of Hong Kong (Shenzhen).

References

1. Zhao, H., Shi, J., Qi, X., Wang, X., Jia, J.: Pyramid scene parsing network. In: Proceedings of the IEEE Conference on Computer Vision and Pattern Recognition, pp. 2881–2890 (2017)
2. Chen, L.-C., Papandreou, G., Kokkinos, I., Murphy, K., Yuille, A.L.: DeepLab: semantic image segmentation with deep convolutional nets, Atrous convolution, and fully connected CRFs. .IEEE Trans. Pattern Anal. Mach. Intell. **40**(4), 834–848 (2017)
3. Hoffman, J., Wang, D., Yu, F., Darrell, T.: FCNs in the wild: pixel-level adversarial and constraint-based adaptation. arXiv preprint arXiv:1612.02649 (2016)

4. Vu, T.-H., Jain, H., Bucher, M., Cord, M., Pérez, P.: Advent: adversarial entropy minimization for domain adaptation in semantic segmentation. In: Proceedings of the IEEE/CVF Conference on Computer Vision and Pattern Recognition, pp. 2517–2526 (2019)
5. Huang, J., Guan, D., Lu, S., Xiao, A.: MLAN: multi-level adversarial network for domain adaptive semantic segmentation. arXiv preprint arXiv:2103.12991 (2021)
6. Hoffman, J., et al.: Cycada: cycle-consistent adversarial domain adaptation. In: International Conference on Machine Learning, pp. 1989–1998. PMLR (2018)
7. Huang, J., Guan, D., Xiao, A., Lu, S.: FSDR: frequency space domain randomization for domain generalization. In: Proceedings of the IEEE/CVF Conference on Computer Vision and Pattern Recognition, pp. pp. 6891–6902 (2021)
8. Zou, Y., Yu, Z., Vijaya Kumar, B.V.K., Wang, J.: Unsupervised domain adaptation for semantic segmentation via class-balanced self-training. In: Ferrari, V., Hebert, M., Sminchisescu, C., Weiss, Y. (eds.) ECCV 2018. LNCS, vol. 11207, pp. 297–313. Springer, Cham (2018). https://doi.org/10.1007/978-3-030-01219-9_18
9. Huang, J., Guan, D., Xiao, A., Lu, S.: Cross-view regularization for domain adaptive panoptic segmentation. In: Proceedings of the IEEE/CVF Conference on Computer Vision and Pattern Recognition, pp. 10:133–10:144 (2021)
10. Wang, Z., et al.: Alleviating semantic-level shift: A semi-supervised domain adaptation method for semantic segmentation. In: Proceedings of the IEEE/CVF Conference on Computer Vision and Pattern Recognition Workshops, pp. 936–937 (2020)
11. Chen, S., Jia, X., He, J., Shi, Y., Liu, J.: Semi-supervised domain adaptation based on dual-level domain mixing for semantic segmentation. In: Proceedings of the IEEE/CVF Conference on Computer Vision and Pattern Recognition, pp. 11:018–11:027 (2021)
12. Huang, J., Guan, D., Xiao, A., Lu, S.: Semi-supervised domain adaptation via adaptive and progressive feature alignment. arXiv preprint arXiv:2106.02845 (2021)
13. Wu, Z., Xiong, Y., Yu, S.X., Lin, D.: Unsupervised feature learning via non-parametric instance discrimination. In: Proceedings of the IEEE Conference on Computer Vision and Pattern Recognition, pp. 3733–3742 (2018)
14. Caron, M., Misra, I., Mairal, J., Goyal, P., Bojanowski, P., Joulin, A.: Unsupervised learning of visual features by contrasting cluster assignments. arXiv preprint arXiv:2006.09882 (2020)
15. Khosla, P., et al.: Supervised contrastive learning. arXiv preprint arXiv:2004.11362 (2020)
16. Robinson, J., Chuang, C.-Y., Sra, S., Jegelka, S.: Contrastive learning with hard negative samples. arXiv preprint arXiv:2010.04592 (2020)
17. Chaitanya, K., Erdil, E., Karani, N., Konukoglu, E.: Contrastive learning of global and local features for medical image segmentation with limited annotations. arXiv preprint arXiv:2006.10511 (2020)
18. Wang, W., Zhou, T., Yu, F., Dai, J., Konukoglu, E., Van Gool, L.: Exploring cross-image pixel contrast for semantic segmentation. arXiv preprint arXiv:2101.11939 (2021)
19. Lewis, D.D., Catlett, J.: Heterogeneous uncertainty sampling for supervised learning. In: Machine Learning Proceedings, pp. 148–156. Elsevier 1994 (1994)
20. Scheffer, T., Decomain, C., Wrobel, S.: Active hidden Markov models for information extraction. In: Hoffmann, F., Hand, D.J., Adams, N., Fisher, D., Guimaraes, G. (eds.) IDA 2001. LNCS, vol. 2189, pp. 309–318. Springer, Heidelberg (2001). https://doi.org/10.1007/3-540-44816-0_31

21. Jain, S.D., Grauman, K.: Active image segmentation propagation. In: Proceedings of the IEEE Conference on Computer Vision and Pattern Recognition, pp. 2864–2873 (2016)
22. Hoi, S.C., Jin, R., Zhu, J., Lyu, M.R.: SemiSupervised SVM batch mode active learning with applications to image retrieval. ACM Trans. Inf. Syst. (TOIS) **27**(3), 1–29 (2009)
23. Vezhnevets, A., Buhmann, J.M., Ferrari, V.: Active learning for semantic segmentation with expected change. In: IEEE Conference on Computer Vision and Pattern Recognition 2012, pp. 3162–3169. IEEE (2012)
24. Siddiqui, Y., Valentin, J., Nießner, M.: ViewAL: active learning with viewpoint entropy for semantic segmentation. In: Proceedings of the IEEE/CVF Conference on Computer Vision and Pattern Recognition, pp. 9433–9443 (2020)
25. He, K., Zhang, X., Ren, S., Sun, J.: Deep residual learning for image recognition. In: Proceedings of the IEEE Conference on Computer Vision and Pattern Recognition, pp. 770–778 (2016)
26. Paszke, A., et al.: Pytorch: an imperative style, high-performance deep learning library. In: Advances in Neural Information Processing Systems, vol. 32, pp. 8026–8037 (2019)
27. Zhu, J.-Y., Park, T., Isola, P., Efros, A.A.: Unpaired image-to-image translation using cycle-consistent adversarial networks. In: Proceedings of the IEEE International Conference on Computer Vision, pp. 2223–2232 (2017)
28. Oord, A., Li, Y., Vinyals, O.: Representation learning with contrastive predictive coding. arXiv preprint arXiv:1807.03748 (2018)
29. French, G., Aila, T., Laine, S., Mackiewicz, M., Finlayson, G.: Semi-supervised semantic segmentation needs strong, high-dimensional perturbations (2019)
30. Feng, Z., Zhou, Q., Cheng, G., Tan, X., Shi, J., Ma, L.: Semi-supervised semantic segmentation via dynamic self-training and classbalanced curriculum. arXiv preprint arXiv:2004.08514 (2020)
31. Saito, K., Kim, D., Sclaroff, S., Darrell, T., Saenko, K.: Semi-supervised domain adaptation via minimax entropy. In: Proceedings of the IEEE/CVF International Conference on Computer Vision, pp. 8050–8058 (2019)
32. Liu, W., Ferstl, D., Schulter, S., Zebedin, L., Fua, P., Leistner, C.: Domain adaptation for semantic segmentation via patch-wise contrastive learning. arXiv preprint arXiv:2104.11056 (2021)
33. Yang, Y., Soatto, S.: FDA: Fourier domain adaptation for semantic segmentation. In: Proceedings of the IEEE/CVF Conference on Computer Vision and Pattern Recognition, pp. 4085–4095 (2020)
34. Chapelle, O., Zien, A.: Semi-supervised classification by low density separation. In: International Workshop on Artificial Intelligence and Statistics, pp. 57–64. PMLR (2005)
35. Kasarla, T., Nagendar, G., Hegde, G.M., Balasubramanian, V., Jawahar, C.: Region-based active learning for efficient labeling in semantic segmentation. In: 2019 IEEE Winter Conference on Applications of Computer Vision (WACV), pp. 1109–1117. IEEE (2019)
36. He, K., Fan, H., Wu, Y., Xie, S., Girshick, R.: Momentum contrast for unsupervised visual representation learning. In: Proceedings of the IEEE/CVF Conference on Computer Vision and Pattern Recognition, pp. 9729–9738 (2020)
37. Ning, M., et al.: Multi-anchor active domain adaptation for semantic segmentation. arXiv preprint arXiv:2108.08012 (2021)
38. Chen, L.-C., Papandreou, G., Schroff, F., Adam, H.: Rethinking Atrous convolution for semantic image segmentation. arXiv preprint arXiv:1706.05587 (2017)

39. Chen, L.-C., Zhu, Y., Papandreou, G., Schroff, F., Adam, H.: Encoder-decoder with atrous separable convolution for semantic image segmentation. In: Ferrari, V., Hebert, M., Sminchisescu, C., Weiss, Y. (eds.) ECCV 2018. LNCS, vol. 11211, pp. 833–851. Springer, Cham (2018). https://doi.org/10.1007/978-3-030-01234-2_49

40. Tsai, Y.-H., Hung, W.-C., Schulter, S., Sohn, K., Yang, M.-H., Chandraker, M.: Learning to adapt structured output space for semantic segmentation. In: Proceedings of the IEEE Conference on Computer Vision and Pattern Recognition, pp. 7472–7481 (2018)

41. Ros, G., Sellart, L., Materzynska, J., Vazquez, D., Lopez, A.M.: The Synthia dataset: a large collection of synthetic images for semantic segmentation of urban scenes. In: Proceedings of the IEEE Conference on Computer Vision and Pattern Recognition, pp. 3234–3243 (2016)

42. Hjelm, R.D., et al.: Learning deep representations by mutual information estimation and maximization. arXiv preprint arXiv:1808.06670 (2018)

43. Chen, X., Fan, H., Girshick, R., He, K.: Improved baselines with momentum contrastive learning. arXiv preprint arXiv:2003.04297 (2020)

44. Chen, T., Kornblith, S., Norouzi, M., Hinton, G.: A simple framework for contrastive learning of visual representations. In: International Conference on Machine Learning, pp. 1597–1607. PMLR (2020)

45. Qi, L., et al.: Open-world entity segmentation. arXiv preprint arXiv:2107.14228 (2021)

Motion and Tracking

D³: Duplicate Detection Decontaminator for Multi-Athlete Tracking in Sports Videos

Rui He[1], Zehua Fu[1,2], Qingjie Liu[1,2]([✉]), Yunhong Wang[1],
and Xunxun Chen[3]

[1] Laboratory of Intelligent Recognition and Image Processing (IRIP Lab), Beihang University (BUAA), Xueyuan Road No. 37, Haidian District, Beijing, China
{heruihr,zehua_fu,qingjie.liu,yhwang}@buaa.edu.cn
[2] Hangzhou Innovation Institute, Beihang University, Hangzhou, China
[3] National Computer Network Emergency Response Technical Team/Coordination Center of China (CNCERT or CNCERT/CC), Beijing, China
cxx@cert.org.cn

Abstract. Tracking multiple athletes in sports videos is a very challenging Multi-Object Tracking (MOT) task, since athletes often have the same appearance and are intimately covered with each other, making a common occlusion problem becomes an abhorrent duplicate detection. In this paper, the duplicate detection is newly and precisely defined as occlusion misreporting on the same athlete by multiple detection boxes in one frame. To address this problem, we meticulously design a novel transformer-based Duplicate Detection Decontaminator (D³) for training, and a specific algorithm Rally-Hungarian (RH) for matching. Once duplicate detection occurs, D³ immediately modifies the procedure by generating enhanced box losses. RH, triggered by the team sports substitution rules, is exceedingly suitable for sports videos. Moreover, to complement the tracking dataset that without shot changes, we release a new dataset based on sports video named RallyTrack. Extensive experiments on RallyTrack show that combining D³ and RH can dramatically improve the tracking performance with 9.2 in MOTA and 4.5 in HOTA. Meanwhile, experiments on MOT-series and DanceTrack discover that D³ can accelerate convergence during training, especially saving up to 80% of the original training time on MOT17. Finally, our model, which is trained only with volleyball videos, can be applied directly to basketball and soccer videos, which shows the priority of our method. Our dataset is available at https://github.com/heruihr/rallytrack.

Keywords: Multi-Athlete Tracking · Multi-Object Tracking · Transformer

---

Supplementary Information The online version contains supplementary material available at https://doi.org/10.1007/978-3-031-26293-7_28.

L. Wang et al. (Eds.): ACCV 2022, LNCS 13847, pp. 467–484, 2023.
https://doi.org/10.1007/978-3-031-26293-7_28

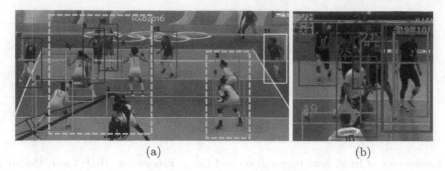

Fig. 1. A labeled sample in RallyTrack and the duplicate detection problem occurred in a TransTrack model: (a) a labeled sample with heavy occlusions, label_7 is totally covered by label_0 in the left dash box; (b) red box shows duplicate detection, the same individual is detected by two queries with two IDs. (Color figure online)

1   Introduction

Sports video analysis possesses wide application prospects and is currently receiving plenty of attention from academia and industry. Scene understanding in sports video can be utilized for data statistics [13], highlight extraction [30], tactics analysis [21]. Multi-Athlete Tracking (MAT) [20] is a basic task in sports video-based scene understanding, which occupies a pivotal position.

Unlike general Multi-Object Tracking (MOT) [9,29], in MAT, different athletes generally share a high similarity in appearance and they often have a diversity of action changes and abrupt movements. The former difficulty, from our observation, turns a common occlusion problem in MOT to duplicate detection in sports video, which is defined in this paper as occlusion misreporting on the same object by multiple predictions in the same frame. The latter one leads to objects detetion missing, which often accompany with duplicate detection. In contrast to general person-based MOT, Fig. 1(a) displays the difficulties with two yellow dash boxes. These two main difficulties make MAT a challenging task.

Then Fig. 1(b) illustrates duplicate detection with a red box, which may be caused by two possibilities. One is that all athletes are detected, and an athlete being repeatedly detected is treated as occluding an additional invisible athlete. The other is that not all athletes are detected, some detections are missing while someone is repeatedly detected. To address this issue, we design a Duplicate Detection Decontaminator (D^3), which can keep watch on the training procedure. Once a duplicate detection occurs, D^3 can generate additionally enhanced self-GIoU [35] losses during training. Then the losses will be gradually backpropagated to force the duplicate detecting boxes to keep away from each other. When duplicate detection disappears, D^3 would not produce loss anymore. We also offer a specific matching algorithm called Rally-Hungarian (RH) algorithm for MAT, which is triggered by the substitution rule in team sports like volleyball.

What is more, to make up for the lack of shot change, a new dataset namely RallyTrack is annotated, which is based on a scene of sports videos, and Fig. 1(a) is a labeled sample. Unlike videos commonly used in the scientific research of MAT, live sports videos include shot changes as sports video streaming always uses multiple cameras. Although all athletes remain on the scene after each shot change, the association becomes challenging as it is hard to predict the trajectories of athletes. However, there are a considerable amount of sports videos available online. Making use of those massive data can help improve athletes' competitiveness, e.g., using live sports videos for tactical analysis. Therefore, building a MAT dataset with shot changes is both significant academically and practically.

Intensive experimental results on RallyTrack demonstrate the efficiency of D^3 and RH. During our experiments, we discover that duplicate detection is not only a prominent problem in MAT but also an unnoticed barrier hidden in MOT, which makes a model converge slowly. Experimental results on MOT17 [29] show D^3 can save up to 80% of original training time. More experiments on MOT16 [22], MOT20 [9], and DanceTrack [37] also confirm the priority of D^3.

The main contributions of this study are as follows. (i) We design a Duplicate Detection Decontaminator (D^3) which supervises the training procedure to optimize detection and tracking boxes. (ii) We design a matching algorithm called Rally-Hungarian (RH) for MAT to further improve tracking result. (iii) We annotate a new dataset named RallyTrack, which is based on scenes of sports videos, to make up for the lack of videos without shot change. (iv) We perform extensive experiments to demonstrate and verify that the proposed method improves the tracking performance on MAT with a total enhancement of 9.2 for MOTA and 4.5 for HOTA, and D^3 can accelerate training convergence on MOT.

2 Related Work

2.1 Multiple Object Tracking Datasets

Human-Based Datasets. Concentrating on variant scenarios, a large number of multiple object tracking datasets have been collected, and human tracking datasets accounted for a big proportion. PETS [11], MOT15 [22], MOT17 [29] and MOT20 [9] datasets become popular in this community. MOT datasets mainly contain a handful of pedestrian videos, which are limited to regular movements of objects and distinguishable appearances. As a consequence of that, multiple object tracking could be easily achieved with the association by pure appearance matching [31]. More recently DanceTrack [37] is proposed as being expected to make research rely less on visual discrimination and depend more on motion analysis. However, the background in DanceTrack is usually identical to the foreground so detecting is easy and tracking is hard. Collected from real and noisy sports videos, our dataset is both challenging in detecting and tracking.

Diverse Datasets. Besides, WILDTRACK [6], Youtube-VOS [45], and MOTS [41] are proposed for diverse objectives. With the development of autonomous driving, KITTI [12] is interested in vehicles and pedestrians. Then larger scale autonomous driving datasets BDD100K [48] and Waymo [36] are published. Limiting by lanes and traffic rules, the motion patterns of objects in these datasets are even more regular than moving people. What is more, some datasets broaden their horizon on more diverse object categories, such as ImageNet-Vid [10] and TAO [8].

2.2 Object Detection in MOT

Tracking by Detection. Object detection [24,33,34] develops so vigorously that a lot of methods would like to utilize powerful detectors to pursue higher tracking performance. RetinaTrack [26] and ChainedTracker [32] apply the one-stage object detector RetinaNet [24] for tracking. For its simplicity and efficiency, CenterNet [54] becomes a popular detector adopted by CenterTrack [53] and FairMOT [51]. The YOLO series detectors [33] are also put to use by TransMOT [7] due to their excellent balance of accuracy and speed. Single image tracking is easy for these methods. However, as is pointed out by [39], when occlusion happens, missing and low-scoring detections would influence the quality of object linking. Therefore, the information of the previous frame is usually leveraged to enhance the video detection performance. Recently, Versatile Affinity Network (VAN) [23] is proposed to handle incomplete detection issues, affinity computation between target and candidates, and decision of tracking termination. [16] presents an approach injecting spatiotemporally derived information into convolutional AutoEncoder in order to produce a suitable data embedding space for multiple object tracking.

Joint-Detection-and-Tracking. Achieving detection and tracking simultaneously in a single stage is the destination of the joint-detection-and-tracking pipeline. Some early methods [7] utilize single object tracking (SOT) [3] or Kalman filter [19]to predict the location of the tracklets in the following frame and fuse the predicted boxes with the detection boxes. Then by combining the detection boxes and tracks, Integrated-Detection [52] boosts the detection performance. Recently, Tracktor [1] directly regress the previous frame tracking boxes to provide tracking boxes on the current step. From a shared backbone, JDE [42] and FairMOT [51] learn the object detection task and appearance embedding task in the meantime. Different from CenterTrack [53] localizing objects by tracking-conditioned detection and predicting their offsets to the previous frame, ChainedTracker [32] chains paired bounding boxes estimated from overlapping nodes, in which each node covers two adjacent frames. More recently, transformer-based [40] detectors like DETR [5,55] are adopted by several methods, such as TransTrack [38], TrackFormer [28], and MOTR [49]. Our method also follows this structure to utilize the similarity with tracklets to strengthen the reliability of detection boxes.

2.3 Data Association

Tracking by Matching Appearance. Appearance similarity is useful in long-range matching and serves as a linchpin in many multi-object tracking methods. DeepSORT [43] adopts a stand-alone Re-ID model to extract appearance features from the detection boxes. POI [47] achieves excellent tracking performance depending on the high-quality detection. Recently, because of their simplicity and efficiency, joint detection and Re-ID models, such as RetinaTrack [26], QuasiDense (QDTrack) [31], JDE [42], FairMOT [51], become more and more prevalent.

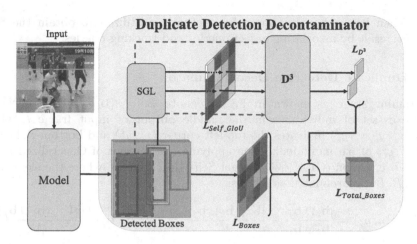

Fig. 2. In the training stage, a matrix L_{Self_GIoU} is constructed according to the detected boxes of the input frame t by Self-GIoU Loss (SGL) function. Then D^3 will set a Lower Bound (LB) to check L_{Self_GIoU}. Values lower than LB in the matrix will be regarded as duplicate detection. The values are output and added to the boxes loss L_{Boxes} as a total boxes loss L_{Total_Boxes} to be backpropagated. If there is no duplicate detection, D^3 will do nothing.

Tracking with Motion Analysis. Tracking objects by estimating their motion is a natural and intuitive idea. SORT [4] first adopts Kalman filter [19] to predict the location of the tracklets in the new frame, and then by Hungarian algorithm [46] computes the IoU between the detection boxes and the predicted boxes as the similarity for tracking. STRN [44] presents a similarity learning framework between tracks and objects. Tracking by associating almost every detection box instead of only the high score ones, for the low score detection boxes, ByteTrack [50] utilizes their similarities with tracklets to recover true objects and filter out the background detections. Recently attention mechanism [40] can directly propagate boxes between frames and perform association implicitly. TransTrack [38] is designed to learn object motions and achieves robust results in cases of large camera motion or low frame rate.

3   Duplicate Detection Decontaminator and Rally-Hungarian Algorithm

In this section, the working mechanism of Duplicate Detection Decontaminator (D^3) and Rally-Hungarian (RH) matching algorithm will be introduced in detail respectively. In a transformer-based joint-detection-and-tracking model, objects in an image are detected by harnessing learned object queries, which is a set of learnable parameters trained together with all other parameters in the network. While training a model, duplicate detection appears and D^3 will unroll its power then. Then RH, a box IoU matching method, is utilized to obtain the final tracking result by associating object queries and tracking queries.

3.1 Duplicate Detection Decontaminator

At training stage as shown in Fig. 2, denote $B = \{\mathbf{b}_i | i = 1, \cdots, N\}$ as the boxes set of individuals in the middle output of input frame t, where $\mathbf{b}_i = (x_i^1, y_i^1, x_i^2, y_i^2)$ indicates the top-left corner (x_i^1, y_i^1) and bottom-right corner (x_i^2, y_i^2) of ith individual. Then applying the concept of Generalized Intersection over Union (GIoU) [35], we get Self-GIoU from B, where the element in $GIoU(B, B)$ is formulated as follows:

$$GIoU(\mathbf{b}_i, \mathbf{b}_j) = \frac{|\mathbf{b}_i \cap \mathbf{b}_j|}{|\mathbf{b}_i \cup \mathbf{b}_j|} - \frac{|V \setminus (\mathbf{b}_i \cup \mathbf{b}_j)|}{|V|} = IoU(\mathbf{b}_i, \mathbf{b}_j) - \frac{|V \setminus (\mathbf{b}_i \cup \mathbf{b}_j)|}{|V|} \tag{1}$$

where V is the smallest convex hull that encloses both $\mathbf{b}_i$ and $\mathbf{b}_j$, IoU means Intersection over Union. Then a matrix L_{Self_GIoU} is constructed by Self-GIoU Loss (SGL) function as follows:

$$L_{Self\_GIoU} = SGL(B) = 1 - GIoU(B, B) \tag{2}$$

L_{Self_GIoU} is a symmetric matrix in which the elements on the diagonal of the matrix are all 0, as painted white in Fig. 2. Then D^3 will set a Lower Bound (LB) to check L_{Self_GIoU}. Once a value of the element in L_{Self_GIoU} is lower than the LB, which means duplicate detection happens, D^3 will output the value and add it to the detected boxes loss as a total boxes loss to be backpropagated. If there is no duplicate detection, D^3 will do nothing. The mechanism of D^3 is as follows:

$$L_{D^3} = D^3(L_{Self\_GIoU}) = \frac{1}{2} \sum_{i=1}^{N} \sum_{j=1}^{N} l_{ij}, \quad l_{ij} < LB \tag{3}$$

$$L_{Total\_Boxes} = L_{Boxes} + L_{D^3} \tag{4}$$

where l_{ij} is an element located at ith row and jth column in $L_{Self-GIoU}$. l_{ij} is equal to l_{ji} in a symmetric matrix so the output of D^3 should be divided by two. When the model is equipped D^3, duplicate detection may be within limits. However, in sports video, the quality of MOT could go a step further while making use of some special rules of sports, which pedestrian video is not in the possession of.

3.2 Rally-Hungarian Algorithm

Although the Hungarian algorithm can still work, its limitation is shown when applied to sports videos. So Rally-Hungarian (RH) algorithm is provided and its overview is shown in Fig. 3. RH models the substitution rule of team sports. The

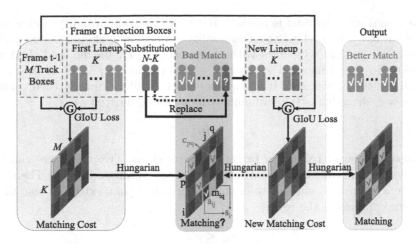

Fig. 3. Through a tracking model, detection boxes of frame t and track boxes of frame $t-1$ are acquired. Then detection boxes are split into Lineup and Substitution. A rally of constructing matching cost matrix by computing GIoU loss from Lineup and track boxes, getting matching pairs by applying Hungarian algorithm on the matrix, replacing the bad matching item with Substitution to form new Lineup is executed, which is called Rally-Hungarian (RH).

players who are on the court are called Lineup, and the others from the same team are called Substitution. Usually, if a Lineup player performs not good, he or she will be replaced by a Substitution player. Then RH is introduced in detail.

Through a tracking model, detection boxes of frame t which are denoted as $B_{det} = \{\mathbf{b}_n | n = 1, \cdots, N\}$, in which the elements are sorted in descending order by detection score, and track boxes of frame $t-1$ as $B_{track} = \{\mathbf{b}_j | j = 1, \cdots, M\}$ are acquired, as the definition of boxes set applied in Sect. 3.1. According to the substitution rule, we split detection boxes set B_{det} to B_{lineup} and B_{sub}. The top K elements in B_{det} are regarded as Lineup $B_{lineup} = \{\mathbf{b}_i | i = 1, \cdots, K\}$ and the rest elements as Substitution $B_{sub} = \{\mathbf{b}_k | k = K + 1, \cdots, N\}$. We provide a mathematical explanation of setting limitation K in RH, please refer to supplementary 1. Then we construct matching cost matrix $\mathbf{C} \in \mathbb{R}^{K \times M}$ by computing GIoU loss from B_{lineup} and B_{track} as follows:

$$\mathbf{C} = 1 - GIoU(B_{lineup}, B_{track}) \tag{5}$$

where GIoU is the same as Eq. (1). Then utilizing Hungarian Algorithm on $\mathbf{C}$, we could acquire a set of matching indices pairs $P = \{(i, j) | i \in [1, K]; j \in [1, M]\}$ as follows:

$$P = Hungarian(\mathbf{C}) \tag{6}$$

P are labeled by check marks in Fig. 3. If $\mathbf{b}_i$ and $\mathbf{b}_j$ belong to one individual, item c_{ij} in $\mathbf{C}$ should be a relatively small value as the light color square shows, which means an individual is tracked. However, if an abnormality is chosen, as marked in red, a row where the abnormality is occupying should be replaced.

Here, we explain why rows with abnormalities can be replaced. In Fig. 3, according to the Hungarian algorithm, a abnormality a_{ij} in $\mathbf{C}$ at row i and column j is the best match. It means that in $\mathbf{C}$, of all the mismatched columns $s_{i\cdot}$ in row i, the value is the minimum, which can be written as:

$$a_{ij} = min\{s_{i\cdot}\} \tag{7}$$

It is also not considered a new target in sports video. Two inferences can then be drawn. First, some duplicate detections are not eliminated by D^3. Then there must be a value m_{iq} smaller than the abnormality that exists among all the matching columns in the row i,

$$m_{iq} < a_{ij} \tag{8}$$

and the smaller value does not match. That is to say, in column q, two similar values exist in row i and another row p,

$$m_{iq} \approx c_{pq} \tag{9}$$

which means row i is duplicate detection and can be replaced. Secondly, this abnormality is exactly the minimum value of this row, indicating a bad quality of the matching. As a consequence, a row, or a detection box, with an abnormality could be replaced.

Then the bad matching detection box in B_{lineup}, regarded as B_{bad}, could be replaced by a substitution in B_{sub}, and a new lineup set B_{new} is composed as follows:

$$B_{new} = (B_{lineup} \setminus B_{bad}) \cup B_{sub} = (B_{lineup} \setminus \mathbf{b}_i) \cup \mathbf{b}_k \tag{10}$$

Looping Eqs. (5), (6), (10) as the dash arrows until each $\mathbf{C}_{ij}$ becomes acceptable or the B_{sub} is empty. In the end, we get a better match pair set. In the field of volleyball, a rally means a round will not stop until the ball touch floor, like a loop. So we name our matching strategy as Rally-Hungarian (RH) algorithm, and "R" may have a dual meaning of "Replace".

4   RallyTrack Dataset

There are plenty of MOT Datasets as we have mentioned in Sect. 2.1. However, there are few datasets for MAT. Driven by this observation, a question arises: Is anything difficult while exploring MAT? To discover the mystery in MAT, we annotate a RallyTrack dataset based primarily on sports videos for the MAT task as shown in Fig. 4. In this section, RallyTrack Dataset will be introduced in detail and our labeling method is provided in our supplementary material 2.

In RallyTrack, videos are from different views, broadcast or fixed, and different gender, men or women, of volleyball games. To guarantee training data and

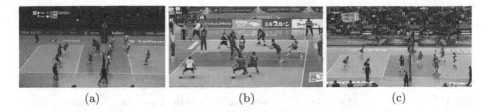

(a)　　　　　　　　(b)　　　　　　　　(c)

Fig. 4. Some samples in RallyTrack: (a) broadcast view of men's game; (b) fixed view of men's game; (c) broadcast view of women's game.

Table 1. Datasets comparison between MOT17 and RallyTrack. F/V means frames per video. O/F means objects per frame. T/V means tracks per video.

Dataset	Subset	Videos	Frames	F/V	Objects	O/F	Tracks	T/V
MOT17	Train	7	5316	759.4	85828	16.1	546	78
	Test	7	5919	845.6	–	–	–	–
	Total	14	11235	802.5	–	–	–	–
RallyTrack	Train	10	8104	810.4	68449	8.5	122	12.2
	Test	10	9757	975.7	91126	9.3	126	12.6
	Total	20	17861	893.1	159575	8.9	248	12.4

test data are not crossed, only will games from different Series be set as train and test. For example, if both games come from Rio 2016 Olympic Games, they should be put into a train set or test set together, even if each team is different. Some details of RallyTrack are then displayed in Table 1. All of our data are labeled in MOT17 annotation format. As the test set's ground truths of MOT17 are not published, only the train set is calculated. In this table, column F/V refers to the number of frames showing more frames in RallyTrack than in each MOT17 video. Column O/F means objects per frame which show that individuals in RallyTrack are less than MOT17. Column T/V means tracks per video which show that trajectories in RallyTrack are also less than MOT17. However, given the overall situation of O/F and T/V, O/F is closer to T/V in RallyTrack than in MOT17, suggesting that RallyTrack has a longer personal trajectory than MOT17.

5 Experimental Results

5.1 Experimental Setup

We evaluate D³ on benchmarks: RallyTrack, MOT17, MOT16, MOT20, and DanceTrack. Following previous practice [38,53], we split all the training sets of the MOT-series into two parts, one for training and the other for validation. The operation is samely applied on RallyTrack where half of the train set will be used and the whole test set will be tested. The widely-used MOT metrics set

Table 2. Experiments on RallyTrack. Our method makes an amazing 9.2 promotion on MOTA, 7.0 on IDF1 and 4.5 on HOTA to baseline TransTrack (TT). Best in bold.

Model	MOTA↑	IDF1↑	MOTP↑	MT↑	FP↓	FN↓	IDS↓	HOTA↑	DetA↑	AssA↑
TT [38]	59.5	28.8	77.8	70.6	15370	19489	2049	27.9	51.9	15.2
TT+RH	62.0	33.3	77.8	66.7	12557	20310	**1788**	30.2	52.3	17.7
TT+D^3	66.4	29.7	78.1	**78.6**	13676	**14848**	2107	29.2	55.8	15.5
TT+D^3+RH	**68.7**	**35.8**	**78.1**	77.0	**11359**	15350	1847	**32.4**	**56.3**	**18.9**

[2] is adopted for quantitative evaluation where multiple objects tracking accuracies (MOTA) is the primary metric to measure the overall performance. What is more, the higher order tracking accuracy (HOTA) [18,27], which explicitly balances the effect of performing accurate detection, association and localization into a single unified metric for comparing trackers, is also applied. While evaluating RH, only RallyTrack is used.

For a fair comparison, we maintain most of the settings in TransTrack [38], such as ResNet-50 [15] network backbone, Deformable DETR [55] based transformer structure, AdamW [25] optimizer, batch size 16. The initial learning rate is 2e−4 for the transformer and 2e−5 for the backbone. The weight decay is 1e−4. All transformer weights are initialized with Xavier-init [14]. The backbone model is pre-trained on ImageNet [10] with frozen batch-norm layers [17]. Data augmentation includes random horizontal, random crop, scale augmentation, and resizing the input images whose shorter side is by 480–800 pixels while the longer side is by at most 1333 pixels. When the model is trained for 150 epochs, the learning rate drops by a factor of 10 at the 100th epoch.

5.2 Experiments on RallyTrack and Others

Our models are evaluated on RallyTrack as shown in Table 2. In this table, the original TransTrack (TT) [38] model based on Deformable Transformer [55] is regarded as a baseline, Rally-Hungarian (RH) algorithm and Duplicate Detection Decontaminator (D^3) could be evaluated respectively or jointly. In this table, TT with both D^3 and RH gets a stunning 9.2 rating on MOTA, 7.0 on IDF1 and 4.5 on HOTA to baseline. The results show that our methods are not only good at detecting multiple athletes but also associating them. It is mainly caused by decontaminating the duplicate detections and many athletes are correctly detected and tracked.

MOT17 is another dataset mainly used to measure the effectiveness of D^3 as shown in Table 3. In this dataset, the main function of our approach is to reduce training time. The hyperparameters in the first column mean Lower Bound (LB) in D^3. LB is chosen according to different self-GIoU losses from different datasets. Different self-GIoU losses are caused by different resolutions of videos. In this table, by actively eliminating duplicate detection, D^3 can save 80% of TT's

Table 3. Experiments on MOT17, MOT16, MOT20, and DanceTrack. Our method converges faster. Best in bold.

D³	Epoch	MOTA↑	IDF1↑	MOTP↑	MT↑	FP↓	FN↓	IDS↓	HOTA↑	DetA↑	AssA↑
w/o	150	65.1	63.6	81.9	36.8	1918	16440	**438**	52.6	54.0	51.7
0.010	150	65.3	62.9	82.2	38.3	1849	**16358**	457	53.0	54.4	52.1
w/o	30	64.9	62.6	82.0	36.3	1862	16537	477	52.1	53.9	50.7
0.010	30	**65.3**	**63.6**	**82.2**	**38.6**	**1833**	16398	480	**53.4**	**54.5**	**52.8**
mot16											
w/o	30	64.1	61.3	81.8	40.7	**2434**	16186	**544**	50.9	53.6	48.9
0.011	30	**65.3**	**61.6**	81.8	40.7	2578	**15328**	601	**52.3**	**55.0**	**50.3**
mot20											
w/o	30	72.5	63.2	82.9	51.6	12882	153K	2978	52.4	59.4	46.3
0.017	30	**73.2**	**64.6**	82.9	**53.4**	**12831**	**149K**	**2808**	**53.6**	**60.1**	**47.9**
DT											
w/o	50	76.5	39.4	85.2	**70.7**	19087	29130	4795	**38.9**	66.8	22.9
0.012	50	76.3	37.4	84.8	68.5	19432	**28947**	5026	37.1	66.4	21.0
w/o	25	**76.6**	37.6	**85.2**	70.0	18710	29348	**4685**	38.1	67.1	21.8
0.012	25	76.5	**39.4**	84.9	67.8	**18518**	29557	4808	38.7	66.2	**22.9**

training time, making the model converge faster from 150 down to 30 epochs. Instead, too many training epochs can lead to overfitting. We then demonstrated the priority of our approach by experimenting directly with MOT16 and MOT20 in the same setting as MOT17 for only 30 epochs. DanceTrack (DT) dataset is also measured with training on train set and testing on val set. As shown in Table 3 only trained in 50 epochs could our method save 50% of the original training time and almost maintain the basic performance.

Datasets and solutions are massive for MOT after long-term development while it is not for MAT. We hope to provide a paradigm for MOT methods to easily extend to MAT. So D³ is proposed as a connection between them. D³ retains an almost complete structure of TT, allowing TT to expand for MAT (Table 2) while maintaining the original MOT capabilities (Table 3).

5.3 Ablation Study

In this section, we conduct a comprehensive ablation study for the proposed D³ and RH.

What is the Shortest Training Time and the Best Lower Bound of D ³? Training epochs and lower bound (LB) are two key factors for D³ networks. During training, short training time leads to non-convergence and long training time leads to overfitting. Then a small LB has little effect and a large LB leads to the elimination of non-duplicate detection. We verify the impact in training with

Table 4. Training time and lower bound. Best in bold.

Model	LB	Epoch	MOTA↑	IDF1↑	MT↑	FP↓	FN↓	IDS↓	HOTA↑	DetA↑	AssA↑
TT	–	150	59.4	28.7	69.8	15520	19426	2052	28.1	51.9	15.4
TT+D³	0.010	150	61.9	29.2	73.0	15393	17166	2164	28.9	53.4	15.9
TT+D³	0.011	150	66.3	29.5	78.6	13806	**14750**	2109	29.2	55.8	15.5
TT+D³	0.012	150	61.8	29.0	74.6	15954	16540	2309	28.8	53.0	16.0
TT	–	40	59.5	28.8	70.6	15370	19489	**2049**	27.9	51.9	15.2
TT+D³	0.011	40	**66.4**	**29.7**	**78.6**	**13676**	14848	2107	**29.2**	**55.8**	**15.5**

Table 5. Age L, Top K and replacement of RH.

	L	K	Replace	MOTA	IDF1	HOTA
TT+D³	32	–	No	66.4	29.7	29.2
TT+D³	80	–	No	66.3	32.2	30.3
TT+D³+RH	32	12	No	67.7	30.9	30.3
TT+D³+RH	80	12	No	67.7	33.2	31.0
TT+D³+RH	80	12	Yes	**68.7**	**35.8**	**32.4**

different training epochs and LB settings. Table 4 shows that using 40 training epochs and 0.011 LB brings the best performance in terms of MOTA, IDF1, and HOTA. On one hand, D³ makes more correct predictions shown by FP and FN with a desirable gap. On the other hand, LB in D³ should be carefully set to determine whether duplicate detection exists.

What is the Best Age L of RH? Should RH Set Top K and Conduct Replacement? First, age L means that if a tracking box is unmatched, it keeps as an "inactive" tracking box until it remains unmatched for L consecutive frames. Inactive tracking boxes can be matched to detection boxes and regain their ID. Following [38], we choose L = 32 and then lengthen L to 80. Because in sports videos individuals who are always on the court will reappear in an image. Second, we set top K = 12 as the data are based on volleyball videos. Finally, whether replace with substitution is also evaluated. Table 5 shows that using age L = 80, setting a limitation K = 12, and conducting replacement bring the best performance.

What is the Best Replacing Strategy of RH? Different replacing strategies (RS) may lead to different tracking performances. So 5 different RS of the RH algorithm are examined as shown in Table 6. In this table, we assume that there are p to-be replaced items in B_{bad} and q items in Substitution. Delete No. means the number of removed items in B_{bad}, and "1st Bad" means deleting the first item in B_{bad}. Replace No. means the number of being replaced items in B_{sub}. As the elements in B_{sub} are already sorted in descending order by detection score, the first one has the highest detection score in B_{sub}, which is marked as "1st

Table 6. Replacing strategies of RH.

	Delete No.	Replace No.	Complexity	MOTA	IDF1	HOTA	FPS
RS1	p	1 (1st Score)	$O(p)$	67.5	31.9	30.4	6.41
RS2	p	1 (1st Good)	$O(pq)$	67.5	31.8	30.2	6.43
RS3	p	$min\{p,q\}$	$O(p \cdot min\{p,q\})$	67.6	32.2	30.4	6.33
RS4	1 (1st Bad)	1 (1st Score)	$O(1)$	68.4	35.5	**32.5**	**6.52**
RS5	1 (1st Bad)	1 (1st Good)	$O(q)$	**68.7**	**35.8**	32.4	6.44

Table 7. Experiments on basketball and soccer.

Basketball	MOTA↑	IDF1↑	MOTP↑	MT↑	FP↓	FN↓	IDS↓	HOTA↑	DetA↑	AssA↑
TT [38]	39.7	12.5	72.9	20.0	3995	3396	636	14.8	43.1	5.2
TT+RH	45.1	14.4	73.0	20.0	3079	3680	593	15.9	44.6	5.7
TT+D³	53.2	12.5	74.7	20.0	2450	**3205**	605	15.2	**48.6**	4.8
TT+D³+RH	**54.4**	**16.8**	**74.7**	20.0	**2166**	3420	520	17.4	48.0	**6.3**
Soccer										
TT [38]	**60.1**	21.2	79.3	33.3	1232	**4205**	327	23.2	**51.4**	10.5
TT+RH	59.2	**24.0**	**79.4**	42.9	1287	4260	354	24.1	50.9	11.4
TT+D³	55.7	19.3	78.9	33.3	**1156**	4868	381	20.8	48.2	9.0
TT+D³+RH	57.9	21.4	78.8	28.6	1175	4907	395	22.0	47.9	10.2

Score". When a bad item is replaced by a high score substitution, it is also able to get a bad match. So all the items in B_{sub} could be replaced to find a good match, and the first item composing a good match is marked as "1st Good". Then time complexity of each strategy is also analyzed. In this table, the total q is set 3, so the FPSes are close.

5.4 Details of Basketball and Soccer Videos

Additionally, extending the RallyTrack dataset to other sports, we labeled 1484 frames of basketball and 1422 frames of soccer and tested them as shown in Table 7. 13384 objects are in basketball and 14802 objects in soccer. Results indicate that our method can be directly applied to basketball videos rather than soccer videos. The scene in basketball is more similar to volleyball than that in soccer, and occlusion is not serious in soccer mainly because the background is easily distinguishable and larger, as visualized in Fig. 5.

5.5 Visualization

We visualize two examples tracked by four different tracking models as shown in Fig. 5. In Fig. 5(a), heavy duplicate detections happen and an object is missing while using a base model TT; then in Fig. 5(b), with RH, some duplicate detections disappear; moreover in Fig. 5(c), when equipped D³, the missing object is found; finally in Fig. 5(d), combining D³ and RH could get the best and the

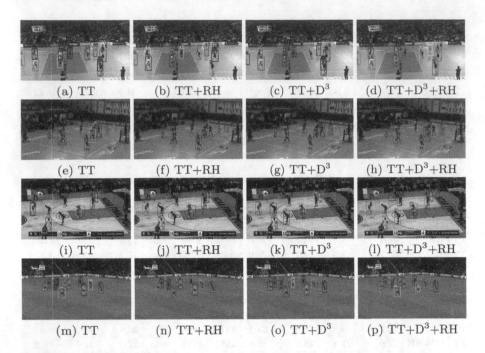

Fig. 5. Visualization of frame No. 942 in test_0100017 and frame No. 715 in validation_0160025 tracked by four different tracking models, and the models are directly applied to basketball and soccer videos.

clearest tracking result. Then the best model is directly applied to basketball by setting $N = 15$, $K = 10$, $q = 5$ in RH, and soccer by $N = 20$, $K = 15$, $q = 5$ for the court of soccer is so large that usually not all individuals are visible.

6   Conclusion

In this paper, to address duplicate detection in MAT, we design a Duplicate Detection Decontaminator (D^3) which supervises the training procedure. Then we design a Rally-Hungarian (RH) matching algorithm to go a step further on MAT. Experiments on our labeled RallyTrack show the priority of our methods. D^3 could also be utilized for saving training time on MOT17, MOT16, MOT20, and DanceTrack. Moreover, our model trained with volleyball data can be directly applied on other team sports videos like basketball or soccer, which may encourage more research exploring MAT applications.

Acknowledgements. This work was supported by the National Natural Science Foundation of China under Grant U20B2069 and the Fundamental Research Funds for the Central Universities.

References

1. Bergmann, P., Meinhardt, T., Leal-Taixé, L.: Tracking without bells and whistles. In: International Conference on Computer Vision, pp. 941–951 (2019)
2. Bernardin, K., Stiefelhagen, R.: Evaluating multiple object tracking performance: the CLEAR MOT metrics. EURASIP J. Image Video Process. **2008**, 246309 (2008)
3. Bertinetto, L., Valmadre, J., Henriques, J.F., Vedaldi, A., Torr, P.H.S.: Fully-convolutional Siamese networks for object tracking. In: Hua, G., Jégou, H. (eds.) ECCV 2016. LNCS, vol. 9914, pp. 850–865. Springer, Cham (2016). https://doi.org/10.1007/978-3-319-48881-3_56
4. Bewley, A., Ge, Z., Ott, L., Ramos, F.T., Upcroft, B.: Simple online and realtime tracking. In: International Conference on Image Processing, pp. 3464–3468 (2016)
5. Carion, N., Massa, F., Synnaeve, G., Usunier, N., Kirillov, A., Zagoruyko, S.: End-to-end object detection with transformers. In: Vedaldi, A., Bischof, H., Brox, T., Frahm, J.-M. (eds.) ECCV 2020. LNCS, vol. 12346, pp. 213–229. Springer, Cham (2020). https://doi.org/10.1007/978-3-030-58452-8_13
6. Chavdarova, T., et al.: WILDTRACK: a multi-camera HD dataset for dense unscripted pedestrian detection. In: Computer Vision and Pattern Recognition, pp. 5030–5039 (2018)
7. Chu, P., Wang, J., You, Q., Ling, H., Liu, Z.: TransMOT: spatial-temporal graph transformer for multiple object tracking. arxiv abs/2104.00194 (2021)
8. Dave, A., Khurana, T., Tokmakov, P., Schmid, C., Ramanan, D.: TAO: a large-scale benchmark for tracking any object. In: Vedaldi, A., Bischof, H., Brox, T., Frahm, J.-M. (eds.) ECCV 2020. LNCS, vol. 12350, pp. 436–454. Springer, Cham (2020). https://doi.org/10.1007/978-3-030-58558-7_26
9. Dendorfer, P., et al.: MOT20: a benchmark for multi object tracking in crowded scenes. arxiv abs/2003.09003 (2020)
10. Deng, J., Dong, W., Socher, R., Li, L., Li, K., Fei-Fei, L.: ImageNet: a large-scale hierarchical image database. In: Computer Vision and Pattern Recognition, pp. 248–255 (2009)
11. Ellis, A., Ferryman, J.M.: PETS2010 and PETS2009 evaluation of results using individual ground truthed single views. In: International Conference on Advanced Video and Signal-Based Surveillance, pp. 135–142 (2010)
12. Geiger, A., Lenz, P., Urtasun, R.: Are we ready for autonomous driving? The KITTI vision benchmark suite. In: Computer Vision and Pattern Recognition, pp. 3354–3361 (2012)
13. Giancola, S., Amine, M., Dghaily, T., Ghanem, B.: SoccerNet: a scalable dataset for action spotting in soccer videos. In: Computer Vision and Pattern Recognition Workshops, pp. 1711–1721 (2018)
14. Glorot, X., Bengio, Y.: Understanding the difficulty of training deep feedforward neural networks. In: International Conference on Artificial Intelligence and Statistics. JMLR Proceedings, vol. 9, pp. 249–256 (2010)
15. He, K., Zhang, X., Ren, S., Sun, J.: Deep residual learning for image recognition. In: Computer Vision and Pattern Recognition, pp. 770–778 (2016)
16. Ho, K., Kardoost, A., Pfreundt, F.-J., Keuper, J., Keuper, M.: A two-stage minimum cost multicut approach to self-supervised multiple person tracking. In: Ishikawa, H., Liu, C.-L., Pajdla, T., Shi, J. (eds.) ACCV 2020. LNCS, vol. 12623, pp. 539–557. Springer, Cham (2021). https://doi.org/10.1007/978-3-030-69532-3_33

17. Ioffe, S., Szegedy, C.: Batch normalization: accelerating deep network training by reducing internal covariate shift. In: International Conference on Machine Learning. JMLR Workshop and Conference Proceedings, vol. 37, pp. 448–456 (2015)
18. Jonathon Luiten, A.H.: Trackeval (2020). https://github.com/JonathonLuiten/TrackEval
19. Kalman, R.E.: A new approach to linear filtering and prediction problems. J. Basic Eng. **82D**, 35–45 (1960)
20. Kong, L., Huang, D., Wang, Y.: Long-term action dependence-based hierarchical deep association for multi-athlete tracking in sports videos. IEEE Trans. Image Process. **29**, 7957–7969 (2020)
21. Kong, L., Zhu, M., Ran, N., Liu, Q., He, R.: Online multiple athlete tracking with pose-based long-term temporal dependencies. Sensors **21**(1), 197 (2021)
22. Leal-Taixé, L., Milan, A., Reid, I.D., Roth, S., Schindler, K.: MOTChallenge 2015: towards a benchmark for multi-target tracking. arxiv abs/1504.01942 (2015)
23. Lee, H., Kim, I., Kim, D.: VAN: versatile affinity network for end-to-end online multi-object tracking. In: Ishikawa, H., Liu, C.-L., Pajdla, T., Shi, J. (eds.) ACCV 2020. LNCS, vol. 12623, pp. 576–593. Springer, Cham (2021). https://doi.org/10.1007/978-3-030-69532-3_35
24. Lin, T., Goyal, P., Girshick, R.B., He, K., Dollár, P.: Focal loss for dense object detection. In: International Conference on Computer Vision, pp. 2999–3007 (2017)
25. Loshchilov, I., Hutter, F.: Decoupled weight decay regularization. In: International Conference on Learning Representations (2019)
26. Lu, Z., Rathod, V., Votel, R., Huang, J.: RetinaTrack: online single stage joint detection and tracking. In: Computer Vision and Pattern Recognition, pp. 14656–14666 (2020)
27. Luiten, J., et al.: HOTA: a higher order metric for evaluating multi-object tracking. Int. J. Comput. Vision **129**, 548–578 (2020)
28. Meinhardt, T., Kirillov, A., Leal-Taixe, L., Feichtenhofer, C.: TrackFormer: multi-object tracking with transformers. In: The IEEE Conference on Computer Vision and Pattern Recognition (CVPR), June 2022
29. Milan, A., Leal-Taixé, L., Reid, I.D., Roth, S., Schindler, K.: MOT16: a benchmark for multi-object tracking. arxiv abs/1603.00831 (2016)
30. Niu, Z., Gao, X., Tian, Q.: Tactic analysis based on real-world ball trajectory in soccer video. Pattern Recogn. **45**(5), 1937–1947 (2012)
31. Pang, J., et al.: Quasi-dense similarity learning for multiple object tracking. In: Computer Vision and Pattern Recognition, pp. 164–173 (2021)
32. Peng, J., et al.: Chained-tracker: chaining paired attentive regression results for end-to-end joint multiple-object detection and tracking. In: Vedaldi, A., Bischof, H., Brox, T., Frahm, J.-M. (eds.) ECCV 2020. LNCS, vol. 12349, pp. 145–161. Springer, Cham (2020). https://doi.org/10.1007/978-3-030-58548-8_9
33. Redmon, J., Farhadi, A.: YoloV3: an incremental improvement. arxiv abs/1804.02767 (2018)
34. Ren, S., He, K., Girshick, R.B., Sun, J.: Faster R-CNN: towards real-time object detection with region proposal networks. In: Conference and Workshop on Neural Information Processing Systems, pp. 91–99 (2015)
35. Rezatofighi, H., Tsoi, N., Gwak, J., Sadeghian, A., Reid, I.D., Savarese, S.: Generalized intersection over union: a metric and a loss for bounding box regression. In: Computer Vision and Pattern Recognition, pp. 658–666 (2019)

36. Sun, P., et al.: Scalability in perception for autonomous driving: Waymo open dataset. In: Computer Vision and Pattern Recognition, pp. 2443–2451 (2020)
37. Sun, P., et al.: DanceTrack: multi-object tracking in uniform appearance and diverse motion. In: Proceedings of the IEEE/CVF Conference on Computer Vision and Pattern Recognition (CVPR) (2022)
38. Sun, P., et al.: TransTrack: multiple-object tracking with transformer. arxiv abs/2012.15460 (2020)
39. Tang, P., Wang, C., Wang, X., Liu, W., Zeng, W., Wang, J.: Object detection in videos by high quality object linking. IEEE Trans. Pattern Anal. Mach. Intell. **42**(5), 1272–1278 (2020)
40. Vaswani, A., et al.: Attention is all you need. In: Conference and Workshop on Neural Information Processing Systems, pp. 5998–6008 (2017)
41. Voigtlaender, P., et al.: MOTS: multi-object tracking and segmentation. In: Computer Vision and Pattern Recognition, pp. 7942–7951 (2019)
42. Wang, Z., Zheng, L., Liu, Y., Li, Y., Wang, S.: Towards real-time multi-object tracking. In: Vedaldi, A., Bischof, H., Brox, T., Frahm, J.-M. (eds.) ECCV 2020. LNCS, vol. 12356, pp. 107–122. Springer, Cham (2020). https://doi.org/10.1007/978-3-030-58621-8_7
43. Wojke, N., Bewley, A., Paulus, D.: Simple online and realtime tracking with a deep association metric. In: International Conference on Image Processing, pp. 3645–3649 (2017)
44. Xu, J., Cao, Y., Zhang, Z., Hu, H.: Spatial-temporal relation networks for multi-object tracking. In: International Conference on Computer Vision, pp. 3987–3997 (2019)
45. Xu, N., et al.: YouTube-VOS: sequence-to-sequence video object segmentation. In: Ferrari, V., Hebert, M., Sminchisescu, C., Weiss, Y. (eds.) ECCV 2018. LNCS, vol. 11209, pp. 603–619. Springer, Cham (2018). https://doi.org/10.1007/978-3-030-01228-1_36
46. Yaw, H.: The Hungarian method for the assignment problem. Naval Res. Logistics Q. **2**(1–2), 83–97 (1955)
47. Yu, F., Li, W., Li, Q., Liu, Yu., Shi, X., Yan, J.: POI: multiple object tracking with high performance detection and appearance feature. In: Hua, G., Jégou, H. (eds.) ECCV 2016. LNCS, vol. 9914, pp. 36–42. Springer, Cham (2016). https://doi.org/10.1007/978-3-319-48881-3_3
48. Yu, F., et al.: BDD100K: a diverse driving dataset for heterogeneous multitask learning. In: Computer Vision and Pattern Recognition, pp. 2633–2642 (2020)
49. Zeng, F., Dong, B., Zhang, Y., Wang, T., Zhang, X., Wei, Y.: MOTR: end-to-end multiple-object tracking with transformer. In: Avidan, S., Brostow, G., Cissé, M., Farinella, G.M., Hassner, T. (eds.) European Conference on Computer Vision, ECCV 2022. LNCS, vol. 13687. Springer, Cham (2022). https://doi.org/10.1007/978-3-031-19812-0_38
50. Zhang, Y., et al.: ByteTrack: multi-object tracking by associating every detection box (2022)
51. Zhang, Y., Wang, C., Wang, X., Zeng, W., Liu, W.: FairMOT: on the fairness of detection and re-identification in multiple object tracking. Int. J. Comput. Vis. **129**(11), 3069–3087 (2021)
52. Zhang, Z., Cheng, D., Zhu, X., Lin, S., Dai, J.: Integrated object detection and tracking with tracklet-conditioned detection. arxiv abs/1811.11167 (2018)

53. Zhou, X., Koltun, V., Krähenbühl, P.: Tracking objects as points. In: Vedaldi, A., Bischof, H., Brox, T., Frahm, J.-M. (eds.) ECCV 2020. LNCS, vol. 12349, pp. 474–490. Springer, Cham (2020). https://doi.org/10.1007/978-3-030-58548-8_28
54. Zhou, X., Wang, D., Krähenbühl, P.: Objects as points. arxiv abs/1904.07850 (2019)
55. Zhu, X., Su, W., Lu, L., Li, B., Wang, X., Dai, J.: Deformable DETR: deformable transformers for end-to-end object detection. In: International Conference on Learning Representations (2021)

Group Guided Data Association for Multiple Object Tracking

Yubin Wu[1,2], Hao Sheng[1,2,3](✉), Shuai Wang[1,2], Yang Liu[1,2],
Zhang Xiong[1,2,3], and Wei Ke[3]

[1] State Key Laboratory of Virtual Reality Technology and Systems, School of
Computer Science and Engineering, Beihang University, Beijing 100191, China
{yubin.wu,shenghao,shuaiwang,xiongz}@buaa.edu.cn, liu.yang@buaa.edu.cn
[2] Beihang Hangzhou Innovation Institute Yuhang, Xixi Octagon City, Hangzhou,
Yuhang District, China
[3] Faculty of Applied Sciences, Macao Polytechnic University, Macao SAR, China
wke@mpu.edu.mo

Abstract. Multiple Object Tracking (MOT) usually adopts the
Tracking-by-Detection paradigm, which transforms the problem into
data association. However, these methods are restricted by detector performance, especially in dense scenes. In this paper, we propose a novel
group-guided data association, which improves the robustness of MOT
to error detections and increases tracking accuracy in occlusion areas.
The tracklets are firstly clustered into groups of related motion patterns
by a graph neural network. Using the idea of grouping, the data association is divided into two stages: intra-group and inter-group. For the
intra-group, based on the structural relationship between objects, detections are recovered and associated by min-cost network flow. For inter-
group, the tracklets are associated with the proposed hypotheses to solve
long-term occlusion and reduce false positives. The experiments on the
MOTChallenge benchmark prove our method's effects, which achieves
competitive results over state-of-the-art methods.

Keywords: Multiple object tracking · Target grouping · Data
association

1 Introduction

Multi-Object Tracking (MOT) is an essential topic in computer vision and is
widely used in video understanding, intelligent transportation [30], and surveillance systems. Benefiting from the progress of detectors, MOT methods usually adopt the Tracking-by-Detection paradigm, which associates detections with
object identities. However, in the case of frequent object interaction and dense
occlusion in practical applications, detetors are often performed with errors.
Consequently, it is difficult for trackers to recover the missed detection; on the
other hand, appearance metrics are no longer reliable when objects overlapped.
These problem become the main challenge for the MOT methods.

According to the different needs of online [29,38] and batch processing [35,
36], association algorithms are generally divided into two types: the one is local

L. Wang et al. (Eds.): ACCV 2022, LNCS 13847, pp. 485–500, 2023.
https://doi.org/10.1007/978-3-031-26293-7_29

(a) A group with missed detections   (b) Two group intersect

Fig. 1. The examples of object groups, where frame numbers are colored in yellow. (a) The red boxes denotes detections while the blue dashed boxes denotes missed detections. (b) The red and green boxes represent two different groups of objects respectively. (Color figure online)

optimization, such as Bipartite Graph [3], and Heuristic Hypothesis [8]. The other is global optimization, such as Network Flow [40], and Conditional Random Field [21]. The local optimization algorithms are more robust to cumulative detection errors but at the expense of the ability to handle long trajectories. The achievements of these methods are mainly attributed to recovering missed detections by regression and searching with additional detectors. The global optimization algorithms maintain higher trajectory integrity but are disturbed by detection errors due to delayed decisions. These methods focus on feature modeling of long trajectories for the anti-occlusion association.

Motivation: Aiming at the above problems, we designed a novel data association method with group information. As the main target of MOT, pedestrians often move as different groups by companions and roads. As shown in Fig. 1 (a), a group of pedestrians have a similar movement pattern and maintain a stable relative position for a short time. In addition to entering and leaving the scene, the number of pedestrians in the group also remained stable. This inspired us to recover the missed detections (blue dashed boxes in Fig. 1 (a)) by position and number constraints when occlusion occurs. As shown in Fig. 1 (b), pedestrians in different groups continue to move independently after a short occlusion. Therefore, associating pedestrians with groups respectively can avoid the interference caused by occlusion. In summary, this requires a data association method with group granularity.

In this paper, we propose a data association method guided by group information for better tracking accuracy in dense scenes. In the MOT problem, objects association can be naturally transformed into a graph problem. Based on previous work [34], we design a graph network to cluster objects into groups. For intra-group association, the method assumes the number and relative position of objects are maintained stable. We construct dummy detections for the intra-group association to meet the constraints and use the min-cost network flow model to obtain the tracklets. Although the dummy nodes recover missing detection, it brings some false positives in tracklets. Therefore, in the inter-group association, we establish the association hypotheses for tracklets. Before solving the hypotheses, the pruning strategy is used to reduce the false positives, and then the algorithm measure appearance affinity to get complete trajectories. Experiments show that our method significant improvement in detection recovery and long-term association.

In summary, our main contributions include:

- Design a graph network to obtain tracklets group by aggregating objects motion information.
- Propose Intra-Group association by min-cost network flow, which recover missing detection by constraint in group.
- Propose Inter-Group association by hypotheses proposal of tracklets, which use pruning to reduce false positives and measure appearance affinity to solve long-term occlusion problem.

2   Related Work

The data association problem in multi-target tracking aims to distinguish multiple identity tags of detections. In association measurement, one kind of method can achieve better tracking by adopting multi-stage [13] and multi-granularity association strategy [27]. The other method uses multiple features [28] to constrain the feasible solution space in each decision window [22,41], so as to reduce the computational overhead and difficulty. We consider combining the advantages of these two types of methods. The proposed method uses group-guided two-stage association from different granularity and distinguishes different occluded targets, which reduces the interference in the calculation. To achieve higher accuracy, grouping and two-step association rely on global information, so our method is batched and not designed for real-time systems.

Motion pattern analysis based on social groups contributes to improving the accuracy of MOT in dense scenes. Pellegrini et al. [23] proposed data association by joint modeling of pedestrian trajectories and groupings. Zhao et al. [44] proposed a tracking method using motion patterns for very crowded scenes. Kratz et al. [16] proposed a tracking method using local spatio-temporal motion patterns in extremely crowded scenes. These methods mainly study the use of groups to predict the future movement of targets and the structural information of the group is not fully utilized. Chen et al. [6] proposed an online learned elementary grouping model for multi-target tracking. Chen et al. [5] proposed PSTG-based multi-label optimization for multi-target tracking. This method uses the group

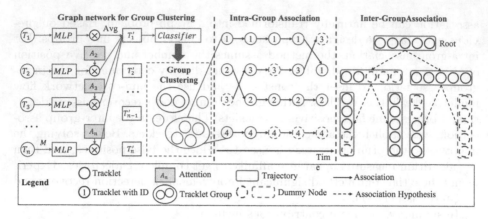

Fig. 2. The general framework of group-guided data association. (1) Initial tracklets are clustered into tracklet groups by the graph network. (2) Intra-group tracklets are associated with the min-cost network flow model. (3) Inter-group trajectories are associated by solving the hypotheses tree.

structure for tracking, but the method is only designed for pairwise target modeling. For tracking problems in more complex scenes, the grouping method of such methods is not robust enough. Sadeghian [26] using interaction model of object position for affinity measurement. Liu [18] associates objects by graph matching which considers group structure in measurement. These methods model the relative position structure and do not use the group as the unit for the association. Based on previous research, we use the group structure to recover detection and achieve reliable tracklets, and on the other hand, we use groups as a unit to conduct intra-group and inter-group association of two steps. The method makes full use of group information and achieves accurate tracking in dense scenes.

3 Proposed Method

As shown in Fig. 2, the proposed method consists of three steps. Initial tracklets $\{T_1, T_2, \cdots, T_n\}$ provide motion information for the graph network. Through embedding aggregation and classification, the network gives the grouping score between tracklets. By measuring scores with the threshold, tracklets can be clustered into groups.

To build the optimization model across multiple frames, we use the minimum cost network flow model for the intra-group association. We propose an algorithm to estimate the maximum number of target IDs and only use motion metrics to achieve better computational performance. The dummy nodes in the association graph recover a large number of candidate missed detection.

To obtain long trajectories and eliminate false positives in dummy nodes, we take each trajectory as the root node, establish the association hypothesis and solve the optimal branch. The inter-group association can further reduce the fragments of trajectories and provide long-distance modeling capability for the method.

3.1 Graph Network for Group Clustering

The structure of graph network for group clustering is shown in Fig. 2. To model data association as a graph $G = (V, E)$, vertex set V is consist of all initial tracklets $\{T_1, T_2, \cdots, T_n\}$ in intra-group association window W_{intra}. For tracklet T_j of length p, motion feature M_j is defined as follows:

$$M_j = \{m_1, m_2, \cdots, m_p\} \tag{1}$$

$$m_i = (\frac{x_i/W - \mu_x}{\sigma_x}, \frac{y_i/H - \mu_y}{\sigma_y}, \frac{w_i/W - \mu_w}{\sigma_w}, \frac{h_i/H - \mu_h}{\sigma_h}), i \in [1, p], \tag{2}$$

where m_i denotes motion feature of detection D_i in tracklet T_j. (x_i, y_i) denotes detection coordinates, w_i and h_i are width and height of detection. W and H are the width and height of the image, μ and σ represent the mean value and standard deviation. Normalization based on image size makes it easier to improve training efficiency and network performance. To obtain a fixed feature dimension, if the length $p < W_{intra}$, the algorithm performs linear interpolation to insure an equal length of M.

To improve the expression representation capability, the Multilayer Perceptron (MLP) is used to encode features to 512 dimensions, which consist of two Full Connection layers (FC) and Rectified Linear Unit (ReLU). By embedding aggregation, the graph network uses the context information between tracklets to improve the discrimination. For tracklet T_1, embeddings from all tracklets overlapped with it are averaged to update T_1'. To avoid the problem of over smoothing, we calculate the self-attention to discover the importance of other overlapped tracklets. For tracklet pair (T_i, T_j), we follow the paradigm of GAT [34], the importance E_{ij} of tracklet T_j to tracklet T_i is formulated with shared weight $\mathbf{W}$:

$$E_{ij} = A(\mathbf{W}T_i', \mathbf{W}T_j'), \tag{3}$$

where $A()$ is a single-layer feedforward neural network, which maps the high-dimensional features to a number as the attention coefficient. The attention coefficient E_{ij} is nonlinear expressed by LeakyReLU and normalized to α_{ij} by Softmax. To balance the possible deviation of the attention, the multi-head attention is used in the prediction layer. The output is averaged as follows:

$$T_i' = \frac{1}{K} \sum_{k=1}^{K} \sum_{j \in n} \alpha_{ij}^k T_j', \tag{4}$$

where K is number of multi-head attention and n are first-order neighbor vertices of j. The group score $S_{i,j}$ is cosine distance between T_i' and T_j'. The graph network gives a cluster probability between 0 and 1. Therefore, we set threshold $t_g = 0.7$, when $S_{i,j} > t_g$, tracklets T_i and T_j are clustered into the same group.

Tracklets Initialization and Training: Initial tracklets provide the basis for group clustering. Based on the method proposed in [27], we extract initial tracklets by affinity measurement of detections. Considering the camera frame rate and pedestrian moving speed, longer tracklets are needed to provide motion features for group clustering. However, the computational performance and accuracy of the baseline method deteriorate with the increase in tracklet length. To handle this problem, frames in an association window are first extracted every 5 frames to generate tracklets. Then, parallel computation is carried out on these 5-frame-long fragments for speed up.

In training, we minimize the cross-entropy loss over all tracklets between the labeled samples and the prediction. The training data are generated from MOT17 [20] and MOT20 [10] datasets. The positive samples are obtained by measuring the tracklets in the relevant spatio-temporal region. For tracklet pairs, We calculate and accumulate the relative position changes between detections frame by frame. If the cumulative deviation is less than 50% of the object's average displacement, tracklets are labeled into a group. Furthermore, we also manually marked and corrected some positive samples. We randomly shift bounding boxes and delete detections between tracklets to simulate the deviation and missing of the detector. The ratio of positive and negative samples is 1:3. We train for 5000 iterations with a learning rate $5 \cdot 10^{-4}$, weight decay term 10^{-4} and an Adam Optimizer with β_1 and β_2 set to 0.9 and 0.999, respectively. By searching the parameters in the train set, we obtained the optimal parameters. The number of attention head K is set to 4.

3.2   Intra-Group Data Association

In this section, we introduce the intra-group data association. By analyzing object behavior and training data, we propose group constraints that the relative position and number of objects in the same group remain stable for a short time. This property can provide a basis for data association. For example, as shown in Fig. 1 (a), detections are missing in frame 246, and group constraints can be used to construct dummy detection for recovery.

As shown in Fig. 2, the number of object IDs in the group remains static, which is crucial for recovering missed detection. However, there may be two kinds of errors in the initial tracklets: one is that the same object is divided into two different tracklets, and the other is that two different objects are associated into one tracklet. These problems will lead to errors in the number of object IDs in initial tracklets.

In intra-group association, the method first ensures the maximization of recall rate, so it is needed to obtain the maximum possible number N of objects in the window W_{intra} and provide motion features for dummy nodes. First, we use a 2-frame sliding window to calculate the best Perfect matchings of bipartite graph $\mathbf{G} = (\mathbf{V}, \mathbf{E}, \mathbf{C})$ and use the Kuhn-Munkres algorithm to assign ID for detections with match $\mathbf{M}$. The Affinity_measure for edges $\mathbf{E}_i$ is based on deepsort [3].

Input: $D_1, D_2, \cdots, D_{W_{intra}}$
Output: $N, D_1, D_2, \cdots, D_{W_{intra}}$

$N=0$
foreach D_i, D_{i+1}, *where* i *from* 1 *to* $W_{intra} - 1$ **do**
   | $V_i = \{D_i, D_{i+1}\}$
   | E_i =Full connection from D_i to D_{i+1}
   | C_i =Affinity_measure(E_i)
   | $G_i = (V_i, E_i, C_i)$
   | M_i =Kuhn-Munkres(G_i)
   | **if** $|D_i| > |D_{i+1}|$ **then**
   | | Add $(|D_i| - |D_{i+1}|)$ dummy nodes to D_{i+1}
   | **end**
   | **if** $|D_i| < |D_{i+1}|$ **then**
   | | Add $(|D_{i+1}| - |D_i|)$ dummy nodes to D_i
   | **end**
   | $N=$Max$(|M_i|, N)$
end

return $N, D_1, D_2, \cdots, D_{W_{intra}}$

Algorithm 1: Algorithm for maximum number N of object IDs

To meet the object number constraint, We add the dummy nodes as mismatched detections replica in the subsequent or previous frames. Afterward, N is set to the maximum number of tracklets with different IDs among all frames. The detailed algorithm is shown in Algorithm 1.

To obtain higher tracking accuracy, we use the global optimization association. Considering the group constraints, the min-cost network flow model is suitable for modeling the problem. The association edges between the detections are regarded as the path in the network, and the similarity between the detections denotes the cost of the path, solving the min-cost flow of the network between multiple frames can provide the optimal solution. The linear programming problem is as follows:

$$argmin \sum_{i,j \in D} C_i f_i + C_{i,j} f_{i,j} + C_s f_{s,i} + C_t f_{i,t} \qquad (5)$$

$$s.t. \qquad f \in \{0,1\} \qquad (6)$$

$$f_i = f_s + \sum_{j \in D} f_{j,i} = \sum_{j \in D} f_{i,j} + f_t \qquad (7)$$

$$0 \le \sum_i f_{s,i} \le 1 \qquad (8)$$

$$0 \le \sum_i f_{t,i} \le 1 \qquad (9)$$

$$\sum_{i \in D} f_i \le 1, \qquad (10)$$

where $f_{i,j}$ denotes flow from detection D_i to D_j and f_i denotes flow from detection D_i to itself. f_s, f_t denotes the virtual source and sink flow with fixed cost C_s and C_t. The data association edge is determined by whether f is activated, i.e. $f \in \{0,1\}$. Eq .(7) constrains the independence of each trajectory. Eqs. (8) and (9) constrain the flow of each target in the network is activated, so the total

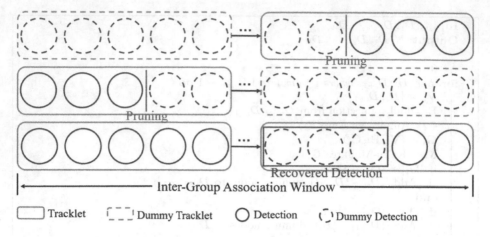

|☐ Tracklet |˻ ˺ Dummy Tracklet   ○ Detection   ˻ ˻ Dummy Detection

Fig. 3. Pruning and detection recovery example.

flow should be 1. Eq .(10) constrains that two trajectories do not cross one node. Cost C_i denotes the negative confidence of detection D_i, which means the higher the detection confidence, the more likely the detection will be included in the trajectory. Cost $C_{i,j}$ from detection D_i to D_j is as follows:

$$C_{i,j} = GIOU(D_i|D_j), \tag{11}$$

where $GIOU(D_i|D_j)$ denotes generalized intersection over union [24] between detection D_i and D_j. Compared with the traditional Kalman filter, GIOU represents the degree of position similarity and has higher efficiency. Limited by the intra-group association window, the target motion state is relatively stable. Therefore, affinity measurement without appearance feature and linear or nonlinear filter can also obtain high measurement accuracy.

3.3   Inter-Group Data Association

In this section, we introduce the inter-group data association. Due to the group size, the intra-group association cannot model trajectories for a longer time. There are interruptions and fragments caused by window division or occlusion in the trajectories. Trajectories obtained in each group are associated in window W_{inter}. With the increase in the time interval, the error of motion prediction will gradually accumulate. Appearance affinity has become an important basis for tracking associations between groups. Following the tracklet level multiple hypothesis framework [27], the appearance affinity measure S between tracklets T_i and T_j is as follows:

$$S = \frac{1}{|T_i|} \sum_{D_i \in T_i} S_{app}(D_i, D_j), \tag{12}$$

where, D_j denotes the detection in T_j that closest to T_i in time. S_{app} denotes the cosine distance of appearance feature vectors between detection D_i and D_j. S averages the appearance of all detections in the T_i which makes it more robust in the face of occlusion. The appearance features of detections are obtained by the Re-Id network [33].

As shown in Fig. 1, there are 2 kinds of nodes in the tracking graph: the original nodes and the dummy nodes. The dummy nodes are the node that is obtained by an extended Kalman filter which predicts the position of the target when it comes to long-term occlusion. Only the original nodes can be used as the root of the hypothesis tree. The association hypothesis increases exponentially with the number of targets, so effective pruning strategies are needed to improve computational efficiency.

Pruning: we take the appearance feature and the space distance as the evaluation branch of the multiple hypothesis tree. However, the correct association may be cut out because of local occlusion, so the algorithm retains the delayed decision to avoid this problem. The dummy detections used by intra-group association can recover the missed detection, but it will cause false-positive problems. We propose a pruning strategy to solve the problem, as shown in Fig. 3. By traversing the branches in the current association hypothesis tree, two kinds of pruning situations are found. One is the branch of continuous dummy detections from the beginning of the window, and the other is the branch of continuous dummy detections from somewhere in the tracklet to the end of the window. We truncate the trajectory according to the position of the red line in Fig. 3. These truncated dummy detections do not match the previous or subsequent objects in the whole calculation window, so they can be regarded as false positives. In addition, as shown in the green box in the figure, the detection of occlusion in the middle is correctly restored, which improves the recall rate of multi-target tracking results.

After pruning, we solve the multi-dimensional assignment problem for the association hypothesis with the strategy proposed in [27] to obtain the final trajectories. The whole MOT method can perform near-online and retain the delay of window size W_{inter}.

4   Experiments

4.1   Dataset and Metric

In experiments, our method is tested on MOT15 [17], MOT17 [20] and MOT20 [10] datasets, which are most widely used in MOT. MOT15 is a comprehensive data set integrating KITTI, ETH, and PETS datasets. MOT17 contains three kinds of public detectors to test the effect of the MOT method on different detectors. Videos are collected from moving and static cameras respectively. MOT20 is designed for ultra-dense scenes, which is more challenging for methods. CLEAR MOT metrics [2] is used to evaluate the method. In addition, IDF1 [25] is used to measure the ID accuracy. Higher Order Tracking Accuracy (HOTA) [19]

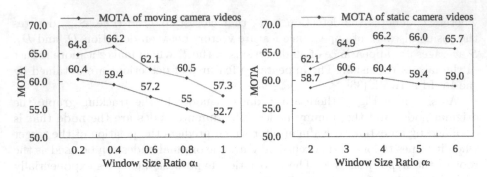

Fig. 4. Window size ratio analysis by moving and static camera videos respectively in MOT17 training set.

Table 1. Comparison of clustering methods on MOT17 validation set

Dataset	Cluster	Setting	mean_Acc ↑	MOTA ↑	IDF1 ↑
MOT17 val	k-means++	k=$\sqrt{No.\ Obj.}$	0.69	34.3	44.7
MOT17 val	agglomerative	Group average	0.84	42.1	53.6
MOT17 val	Our		**0.89**	**43.2**	**55.0**

is the geometric mean of detection accuracy and association accuracy. Averaged across localization thresholds. For a fair comparison, the tests for all methods use the public detections provided by the dataset.

4.2   Parameters Analysis

The window size of intra-group W_{intra} and inter-group W_{inter} are the main parameters in our method. We searched for the optimal setting of both window sizes on the MOT17 training set. For different window size ratios, Multi-Object Tracking Accuracy (MOTA) of moving and static camera videos in the MOT17 training set are shown in Fig. 4. The window size of the intra-group is related to the object speed and video frame rate, and the window size of the inter-group shall be an integral multiple of the intra-group. Therefore, we define $\alpha_1 = \frac{W_{intra}}{framerate}$ and $\alpha_2 = \frac{W_{inter}}{W_{intra}}$. As shown in Fig. 4, with the increase of α_1, MOTA increased slightly and then decreased because the object structure in the group is no longer stable. Especially in a video with a moving camera, the movement state of the target changes greatly, so it is necessary to use a smaller intra-group window. With the increase of α_2, MOTA increases with the inter-group window, due to the measurement information increases, and the longer trajectory is included in the correlation hypothesis. After α_2 reaches 4, considering the longer time interval of the tracklets is not effective. Therefore, to obtain the best results and achieve a balance between moving and static camera videos, we set $\alpha_1 = 1$ and $\alpha_2 = 4$ for the following experiments.

Table 2. Ablation study on MOT17 training set

Settings	MOTA ↑	IDF1 ↑	FP ↓	FN ↓	IDsw ↓
TT17 [41]	56.5	67.0	**9,116**	136,572	572
$A_{intra} + A_{inter}$	58.4	68.7	10,496	131,189	752
$A_{intra}^D + A_{inter}$	60.3	70.5	27,245	**100,275**	581
$A_{intra}^D + A_{inter}^P$	**64.2**	**75.1**	14,232	103,587	**549**

Table 3. Comparison on MOT15 benchmark

Tracker	MOTA ↑	IDF1 ↑	HOTA	FP↓	FN↓	ID Sw. ↓
KCF [7]	38.9	44.5	33.1	7,321	29,501	720
CRFTrack [37]	40.0	49.6	37.3	10,295	25,917	658
Tracktor++v2 [1]	46.6	47.6	37.6	**4,624**	26,896	1,290
Lif_TsimInt [14]	47.2	57.6	43.8	7,635	24,277	554
mfi_tst [39]	49.2	52.4	41.5	8,707	21,594	912
ApLift [15]	51.1	59.0	45.7	10,070	19,288	677
MPNTrack [4]	51.5	58.6	45.0	7,620	21,780	**375**
Our	**57.5**	**60.6**	**46.8**	7,637	**18,013**	466

4.3 Ablation Study

Considering that only group clustering is used without data association, the tracking results will be poor, so we did not use group cluster alone as baseline comparison. First, we made the comparison of cluster methods in the Table 1. Heuristic methods often need to manually design and measure feature metrics and search parameters for different scenarios, which is not robust. Graph network has become a mainstream paradigm in the field of deep clustering recently. By embedding aggregation, the context information and high-dimensional features of the graph structure are extracted, which can better model the interaction and motion affinities between trajectories. We separated the validation set from MOT17 and selected common clustering methods for comparison. Without fine-tuning for parameters, we achieve better results on both cluster and tracking.

To verify the effect of each component of the method, we used the MOT17 training set for the ablation experiment. Since our method adopts a multi-stage grouping association strategy based on tracklets, we select a similar tracklet level multi-hypothesis tracking method [41] as the baseline for comparison. As shown in second row of Table 2, A_{intra} denotes intra-group association without dummy node, and A_{inter} denotes inter-group association without pruning. We first use A_{intra} and A_{inter} instead of the tracklet generation and association in [41]. Compared with the baseline method in the first row, the basic group association reduces false positives and false negatives and slightly improves main metrics. In the third row of the Table 2, A_{intra}^D denotes intra-group association with dummy node. Compared with the previous results, the false negative is significantly reduced after the introduction of dummy nodes, indicating that the method recovers a large number of missed detection. However, the dummy

Table 4. Comparison on MOT17 benchmark

Dataset	Method	Detection	MOTA ↑	IDF1 ↑	FP+FN ↓
MOT17 Test	FairMOT [43]	Centernet [11]	73.7	72.3	144,984
MOT17 Test	FairMOT [43]+Group	Centernet [11]	**74.6**	**76.5**	**141,280**
MOT17 Test	ByteTrack [42]	YOLOX [12]	80.3	77.3	109,212
MOT17 Test	Our	YOLOX [12]	**81.0**	**80.0**	**105,668**

Table 5. Comparison on MOT20 benchmark

Method	MOTA ↑	IDF1 ↑	HOTA ↑	FP ↓	FN ↓	IDsw ↓
LPC [9]	56.3	62.5	49.0	**11,726**	213,056	1,562
MPN [4]	57.6	59.1	46.8	16,953	201,384	**126**
ApLift [15]	58.9	56.5	46.6	17,739	192,736	2,241
mfi_tst [39]	59.3	59.1	47.1	36,150	172,782	1,919
TMOH [31]	60.1	61.2	48.9	38,043	165,899	2,342
MPTC [32]	60.6	59.7	48.5	45,318	153,978	4,533
Our	**64.4**	**65.7**	**53.4**	70,976	**110,614**	2,708

introduces wrong estimates resulting in higher false positives. In the fourth row of the Table 2, A_{inter}^P denotes inter-group association with pruning. By using the pruning strategy, FP is effectively suppressed and the best result is achieved. The focus of the model affects the preferences of FP and FN, which is a trade-off problem. The bottleneck of the MOT method lies in the recall of the detector, that is FN problem (FN is always one order of magnitude larger than FP in most data sets). Therefore, our strategy is to reduce FN more and keep the sum of FN and FP lower. It is worth mentioning that ID sw. is reduced because pruning reduces unnecessary solution space for the association, thus avoiding ID Sw. caused by similarity measurement ambiguity.

4.4 Benchmark Evaluation

To compare our method with other advanced methods, we chose the classic MOT15, MOT17 benchmark and the most challenging MOT20 benchmark. To fairly compare the performance of data association algorithms, all methods and results use the same detector. In order to prove the generality of the method, we add experiments of group data association method for better detectors and tracking methods. As shown in the Table 4, using anchor-free detection and introducing our group association into the popular end to end method FairMOT [43], all the main metrics have been improved. By using same YOLOX [12] detection as ByteTrack, we have achieved SOTA results on MOT17 benchmark. As shown in Tables 3 and 5, the best-published results on the leaderboard are listed. Compared with state-of-the-art methods, our method achieves the highest result for MOTA, IDF1, and HOTA. The group-based data association strategy is conducive to achieving higher tracking and identity accuracy. In particular, our

Fig. 5. The visual tracking results of MOT20 test set.

method restores missing detection through group relationships, which has significant advantages in reducing FN. We selected the tracking results of representative frames from the MOT20 test set as a visual display, shown in Fig. 5. It can be observed that in this dense scene, our method can maintain a more stable track ID. The total results can be found on the official website of the MOTChallenge[1].

5   Conclusion

In this paper, we propose a novel group-guided data association for MOT. The graph neural network is designed to obtain the initial groups of tracklets. By analyzing the potential groups of objects, we design a two-stage data association. Intra-group associations utilize the group constraints to achieve more accurate trajectories in dense scenes and recover missed detection. The inter-group association uses the appearance features and proposes tracklets hypothesis to solve the long-term occlusion problem, which improves the trajectory integrity. In the MOT benchmark, the experiments prove the effectiveness of our algorithm, which achieve better results than previous state-of-the-art methods.

Acknowledgements. This study is partially supported by the National Key R&D Program of China (No.2019YFB2102200), the National Natural Science Foundation of

---

[1]   https://motchallenge.net/

China (No.61872025), the Science and Technology Development Fund, Macau SAR(File no.0001/2018/AFJ), and the Open Fund of the State Key Laboratory of Software Development Environment (No. SKLSDE2021ZX-03). Thank you for the support from the HAWKEYE Group.

References

1. Bergmann, P., Meinhardt, T., Leal-Taixe, L.: Tracking without bells and whistles. In: Proceedings of the IEEE/CVF International Conference on Computer Vision, pp. 941–951 (2019)
2. Bernardin, K., Stiefelhagen, R.: Evaluating multiple object tracking performance: the clear mot metrics. EURASIP J. Image Video Process. **2008**, 1–10 (2008)
3. Bewley, A., Ge, Z., Ott, L., Ramos, F., Upcroft, B.: Simple online and realtime tracking. In: 2016 IEEE International Conference on Image Processing, pp. 3464–3468. IEEE (2016)
4. Brasó, G., Leal-Taixé, L.: Learning a neural solver for multiple object tracking. In: Proceedings of the IEEE/CVF Conference on Computer Vision and Pattern Recognition, pp. 6247–6257 (2020)
5. Chen, J., Sheng, H., Li, C., Xiong, Z.: PSTG-based multi-label optimization for multi-target tracking. Comput. Vis. Image Underst. **144**, 217–227 (2016)
6. Chen, X., Qin, Z., An, L., Bhanu, B.: An online learned elementary grouping model for multi-target tracking. In: Proceedings of the IEEE Conference on Computer Vision and Pattern Recognition, pp. 1242–1249 (2014)
7. Chu, P., Fan, H., Tan, C.C., Ling, H.: Online multi-object tracking with instance-aware tracker and dynamic model refreshment. In: IEEE Winter Conference on Applications of Computer Vision, pp. 161–170. IEEE (2019)
8. Chu, P., Ling, H.: FAMNet: joint learning of feature, affinity and multi-dimensional assignment for online multiple object tracking. In: Proceedings of the IEEE/CVF International Conference on Computer Vision (2019)
9. Dai, P., Weng, R., Choi, W., Zhang, C., He, Z., Ding, W.: Learning a proposal classifier for multiple object tracking. In: Proceedings of the IEEE/CVF Conference on Computer Vision and Pattern Recognition, pp. 2443–2452 (2021)
10. Dendorfer, P., et al.: MOT20: a benchmark for multi object tracking in crowded scenes. arXiv preprint arXiv:2003.09003 (2020)
11. Duan, K., Bai, S., Xie, L., Qi, H., Huang, Q., Tian, Q.: CenterNet: keypoint triplets for object detection. In: Proceedings of the IEEE/CVF International Conference on Computer Vision, pp. 6569–6578 (2019)
12. Ge, Z., Liu, S., Wang, F., Li, Z., Sun, J.: YOLOX: exceeding yolo series in 2021. arXiv preprint arXiv:2107.08430 (2021)
13. Ho, K., Kardoost, A., Pfreundt, F.J., Keuper, J., Keuper, M.: A two-stage minimum cost multicut approach to self-supervised multiple person tracking. In: Proceedings of the Asian Conference on Computer Vision (2020)
14. Hornakova, A., Henschel, R., Rosenhahn, B., Swoboda, P.: Lifted disjoint paths with application in multiple object tracking. In: International Conference on Machine Learning, pp. 4364–4375. PMLR (2020)

15. Hornakova, A., Kaiser, T., Swoboda, P., Rolinek, M., Rosenhahn, B., Henschel, R.: Making higher order mot scalable: an efficient approximate solver for lifted disjoint paths. In: Proceedings of the IEEE/CVF International Conference on Computer Vision, pp. 6330–6340 (2021)
16. Kratz, L., Nishino, K.: Tracking pedestrians using local spatio-temporal motion patterns in extremely crowded scenes. IEEE Trans. Pattern Anal. Mach. Intell. **34**(5), 987–1002 (2011)
17. Leal-Taixé, L., Milan, A., Reid, I., Roth, S., Schindler, K.: MOTchallenge 2015: towards a benchmark for multi-target tracking. arXiv preprint arXiv:1504.01942 (2015)
18. Liu, Q., Chu, Q., Liu, B., Yu, N.: GSM: graph similarity model for multi-object tracking. In: IJCAI, pp. 530–536 (2020)
19. Luiten, J., et al.: HOTA: a higher order metric for evaluating multi-object tracking. Int. J. Comput. Vision **129**(2), 548–578 (2021)
20. Milan, A., Leal-Taixé, L., Reid, I., Roth, S., Schindler, K.: MOT16: a benchmark for multi-object tracking. arXiv preprint arXiv:1603.00831 (2016)
21. Milan, A., Schindler, K., Roth, S.: Multi-target tracking by discrete-continuous energy minimization. IEEE Trans. Pattern Anal. Mach. Intell. **38**(10), 2054–2068 (2015)
22. Mykheievskyi, D., Borysenko, D., Porokhonskyy, V.: Learning local feature descriptors for multiple object tracking. In: Proceedings of the Asian Conference on Computer Vision (2020)
23. Pellegrini, S., Ess, A., Van Gool, L.: Improving data association by joint modeling of pedestrian trajectories and groupings. In: Daniilidis, K., Maragos, P., Paragios, N. (eds.) ECCV 2010. LNCS, vol. 6311, pp. 452–465. Springer, Heidelberg (2010). https://doi.org/10.1007/978-3-642-15549-9_33
24. Rezatofighi, H., Tsoi, N., Gwak, J., Sadeghian, A., Reid, I., Savarese, S.: Generalized intersection over union: A metric and a loss for bounding box regression. In: Proceedings of the IEEE/CVF Conference on Computer Vision and Pattern Recognition, pp. 658–666 (2019)
25. Ristani, E., Solera, F., Zou, R., Cucchiara, R., Tomasi, C.: Performance measures and a data set for multi-target, multi-camera tracking. In: Hua, G., Jégou, H. (eds.) ECCV 2016. LNCS, vol. 9914, pp. 17–35. Springer, Cham (2016). https://doi.org/10.1007/978-3-319-48881-3_2
26. Sadeghian, A., Alahi, A., Savarese, S.: Tracking the untrackable: learning to track multiple cues with long-term dependencies. In: Proceedings of the IEEE/CVF International Conference on Computer Vision, pp. 300–311 (2017)
27. Sheng, H., Chen, J., Zhang, Y., Ke, W., Xiong, Z., Yu, J.: Iterative multiple hypothesis tracking with tracklet-level association. IEEE Trans. Circuits Syst. Video Technol. **29**(12), 3660–3672 (2018)
28. Sheng, H., et al.: Combining pose invariant and discriminative features for vehicle reidentification. IEEE Internet Things J. **8**(5), 3189–3200 (2020)
29. Sheng, H., et al.: Near-online tracking with co-occurrence constraints in blockchain-based edge computing. IEEE Internet Things J. **8**(4), 2193–2207 (2020)
30. Sheng, H., et al.: High confident evaluation for smart city services. Front. Environ. Sci. **10**, 1103 (2022)
31. Stadler, D., Beyerer, J.: Improving multiple pedestrian tracking by track management and occlusion handling. In: Proceedings of the IEEE/CVF Conference on Computer Vision and Pattern Recognition, pp. 10958–10967 (2021)

32. Stadler, D., Beyerer, J.: Multi-pedestrian tracking with clusters. In: IEEE International Conference on Advanced Video and Signal Based Surveillance, pp. 1–10. IEEE (2021)

33. Sun, Y., Zheng, L., Yang, Y., Tian, Q., Wang, S.: Beyond part models: person retrieval with refined part pooling (and a strong convolutional baseline). In: Ferrari, V., Hebert, M., Sminchisescu, C., Weiss, Y. (eds.) ECCV 2018. LNCS, vol. 11208, pp. 501–518. Springer, Cham (2018). https://doi.org/10.1007/978-3-030-01225-0_30

34. Veličković, P., Cucurull, G., Casanova, A., Romero, A., Lio, P., Bengio, Y.: Graph attention networks. arXiv preprint arXiv:1710.10903 (2017)

35. Wang, S., Sheng, H., Yang, D., Zhang, Y., Wu, Y., Wang, S.: Extendable multiple nodes recurrent tracking framework with RTU++. IEEE Trans. Image Process. **31**, 5257–5271 (2022)

36. Wang, S., Sheng, H., Zhang, Y., Wu, Y., Xiong, Z.: A general recurrent tracking framework without real data. In: Proceedings of the IEEE/CVF International Conference on Computer Vision, pp. 13219–13228 (2021)

37. Xiang, J., Xu, G., Ma, C., Hou, J.: End-to-end learning deep CRF models for multi-object tracking deep CRF models. IEEE Trans. Circuits Syst. Video Technol. **31**(1), 275–288 (2020)

38. Xu, Y., Chen, Y., Zhang, Y., Zhu, Q., He, Y., Sheng, H.: Bilateral association tracking with Parzen window density estimation. IET Image Processing (2022)

39. Yang, J., Ge, H., Yang, J., Tong, Y., Su, S.: Online multi-object tracking using multi-function integration and tracking simulation training. Applied Intelligence, pp. 1–21 (2021)

40. Zhang, L., Li, Y., Nevatia, R.: Global data association for multi-object tracking using network flows. In: Proceedings of the IEEE Conference on Computer Vision and Pattern Recognition, pp. 1–8. IEEE (2008)

41. Zhang, Y., et al.: Long-term tracking with deep tracklet association. IEEE Trans. Image Process. **29**, 6694–6706 (2020)

42. Zhang, Y., et al.: ByteTrack: multi-object tracking by associating every detection box. arXiv preprint arXiv:2110.06864 (2021)

43. Zhang, Y., Wang, C., Wang, X., Zeng, W., Liu, W.: FairMOT: on the fairness of detection and re-identification in multiple object tracking. Int. J. Comput. Vision **129**(11), 3069–3087 (2021)

44. Zhao, X., Gong, D., Medioni, G.: Tracking using motion patterns for very crowded scenes. In: Fitzgibbon, A., Lazebnik, S., Perona, P., Sato, Y., Schmid, C. (eds.) ECCV 2012. LNCS, vol. 7573, pp. 315–328. Springer, Heidelberg (2012). https://doi.org/10.1007/978-3-642-33709-3_23

Consistent Semantic Attacks
on Optical Flow

Tom Koren, Lior Talker$^{(\boxtimes)}$, Michael Dinerstein, and Ran Vitek

Samsung Israel R&D Center, Tel Aviv, Israel
{lior.talker,m.dinerstein,ran.vitek}@samsung.com

Abstract. We present a novel approach for semantically targeted adversarial attacks on Optical Flow. In such attacks the goal is to corrupt the flow predictions of a specific object category or instance. Usually, an attacker seeks to hide the adversarial perturbations in the input. However, a quick scan of the output reveals the attack. In contrast, our method helps to hide the attacker's intent in the output flow as well. We achieve this thanks to a regularization term that encourages off-target consistency. We perform extensive tests on leading optical flow models to demonstrate the benefits of our approach in both white-box and black-box settings. Also, we demonstrate the effectiveness of our attack on subsequent tasks that depend on the optical flow.

Keywords: Adversarial attacks · Optical flow · Semantic attacks

1 Introduction

Optical Flow (OF) is a crucial subtask of many safety-critical pipelines. It is especially important for Advanced Driver Assistance Systems (ADAS) and autonomous vehicles, where unreliable optical flow can be hazardous and life-threatening. For example, *Time-To-Collision* (TTC) methods often rely on optical flow [1–4], and their errors can have dangerous consequences.

In this paper we consider malicious manipulations aiming to lead the OF predictions astray. These manipulations are represented by perturbations, sometimes subtle, that are introduced into the input pixels. In the literature such perturbations are referred to as Adversarial Attacks (AA) [5–7].

The attacker's goal is to damage a system's performance and remain unnoticed. The defender's goal is to design a system that operates reliably despite such attacks. To achieve this goal, the defender can, for example, use some AA detection method to discard suspicious inputs. One approach to detect AAs [9,10], is to examine the input to the attacked model. Another approach, which we consider in this paper, is to examine the *output* of the attacked model. We show that a straightforward attack on OF may be fairly easy to detect in the output. We propose an AA method, which is more difficult to detect, but has a similar or stronger effect on OF.

Supplementary Information The online version contains supplementary material available at https://doi.org/10.1007/978-3-031-26293-7_30.

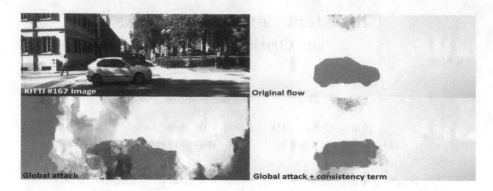

Fig. 1. An example of a targeted attack on HD3-PPAC [8] optical flow model. Top: the original predicted flow and the corresponding input image. Bottom left: the corrupted flow using a non-consistent attack. Bottom right: the corrupted flow using a consistent targeted attack on the vehicle in the scene.

We use the following observation: in the context of automotive applications some objects are more important than others. Obvious examples of such important objects are pedestrians and vehicles. For TTC systems, failing to estimate the correct flow for pedestrians and vehicles can lead to fatal accidents.

Assuming a semantic or instance segmentation of the observed scene is given, an attacker can specify a target to attack. That is, instead of targeting the entire image, only a subset, defined by its semantic segmentation, can be selected. Likewise, instead of perturbing all the pixels in the image, a subset of the pixels can be chosen at which the attacker introduces malicious perturbations.

Adversarial attacks targeting only a subset of pixels may still alter the predictions of other pixels in the image. To make the attack less detectable, it is beneficial to corrupt the target prediction without affecting the rest of the predictions. We refer to attacks that leave off-target predictions unaffected (or affected as little as possible) as "consistent attacks". In this paper we present a method to create targeted consistent adversarial attacks on optical flow. The chosen targets for the attacks are the "vehicle" and the "human" categories.

Figure 1 depicts an example of such an attack. The bottom row presents two attacked flows (encoded using Middlebury [11] color-wheel): consistent and non-consistent. Both attacks achieve their goal - the flow of the corresponding target, the vehicle, is heavily damaged. However, the difference between them is readily seen. A consistently attacked flow looks reasonable, while a non-consistent attack results in a flow which is chaotically cluttered. In this paper we show that the first attack is less detectable than the second.

To achieve the effect described above, we introduce a new optimization term. We refer to it as "consistency term". While being a relatively simple addition to the optimization loss, the consistency term leads to multiple improvements in the generated adversarial attacks when compared to the baseline non-consistent settings. First, as expected, the impact on the flow predictions of non-target scene objects is significantly reduced. Second, the effectiveness of attacks on targets increases. Third, our experiments show that in the "black-box" setting, i.e. where the attacked model is inaccessible, we observe a much better transferability [12]. Finally, we demonstrate that consistent attacks are more effective on a TTC system while being less noticeable by detection methods.

We have conducted an extensive set of evaluations using five leading optical flow methods [8,13–16]. There are three main groups of experiments where we compare attacks obtained with and without our consistency term: global, local and cross-category attacks. The target of those attacks is always the same, but the subsets of pixels that are perturbed are different. In a global attack setting, the perturbation can be distributed over all pixels. In a local attack setting, only pixels of the target object can be perturbed. Finally, in a cross-category setting, in order to corrupt the flow of some target object, we perturb pixels of some other object. We have also evaluated the effect of these attacks on a downstream TTC task. We compare consistent and non-consistent attacks in terms of the tradeoff between their effect on TTC and a AA detection score.

To summarize our contribution, we are the first to study targeted attacks on optical flow models. We introduce a new term to the optimization loss, which we name the "consistency" term, to preserve the optical flow of non-target objects. This helps to hide the attacks in the output of the optical flow. We show that the resulting consistent attacks are more effective than the non-consistent attacks. We demonstrate that these attacks are more transferable, i.e. more efficient in a black-box setting, than the non-consistent attacks. Finally, we show that under three detection methods these attacks are more effective against a downstream TTC system.

1.1 Related Work

The history of optical flow methods goes back to the early 1980 s,s, when the foundational studies of Lucas-Kanade [17] and Horn-Schunk [18] were published. Since then, hundreds of classical computer vision techniques were proposed. A substantive survey on the non-deep optical flow methods can be found in [19]. In the era of deep learning for computer vision much attention has been paid to optical flow. Deep neural network (DNN) based OF methods such as [8,13–16,20–22] have left the classical approaches far behind in terms of performance, which is reflected in the results of the KITTI'15 benchmark [23], where the leading non-deep optical flow method [24] is scored about 150-th place.

DNN based optical flow models can be divided into groups according to their architecture characteristics: encoder-decoder [20,21] and spatial pyramid [8,13, 16,22,25] networks. Some models ([8,15,16,22]) use a coarse-to-fine technique to refine their predictions. Others [14] operate with full resolution features at every stage of the model. In addition, a model can be equipped with a recurrent refinement mechanism, which is placed on top of an optical flow model, as in [26]. Finally, the RAFT model [14], which has demonstrated the *state-of-the-art* performance on KITTI'15 [23], consists of the encoder-decoder part followed by a simple recurrent module utilizing GRU [27] blocks.

Although adversarial perturbations are possible in many ML models, the rise of deep neural networks has opened the door for massive research effort in adversarial perturbations. Historically, many of the early adversarial attacks were carried out in the context of image classification tasks [6,7,28,29]. The attacker's goal was to force a model to misclassify the input image. Many attack schemes were developed and tested on such models. One of the most cited is the so-called Fast Gradient Sign Method (FGSM) [6]. In their original work, Goodfellow et al. [6] suggest a fast method to create adversarial input to a classification model. Consider an input x to a classification model M, a hyper-parameter ϵ, a loss function l and y_{true} - the target associated with x. Assume the model predicts a label y for an input x, i.e., M(x) = y. In their work they show that an adversarial example x_{adv} could then be computed by $x_{adv} = x + \epsilon Sign(\nabla(l(M(x), y_{true}))$. Shortly after [30] introduces a straightforward way to extend this method. They named this new approach the Iterative Fast Gradient Sign Method (IFGSM) [30]. They suggest to iteratively use the same update step on the input. To do so, set $x_{adv}^0 = x$ and iteratively update $x_{adv}^{i+1} = x_{adv}^i + \epsilon \cdot Sign(\nabla(l(M(x_{adv}^i), y_{true}))$. Since these attacks are thoroughly researched and well understood [6,31,32] we adopt them to our attack approach.

Later on, adversarial attack methods that target specific objects in the image were introduced against object detectors [33–35]. In [33] it is shown how to force a SOTA detection model to classify all detections of a semantic class as another class while leaving all other detections unchanged. Liao *et al* [35] proposed a local attack that only perturb a specific detection bounding box, achieving a stronger effect than a global perturbation for the same attack budget. In [34], an analysis of object detection from the viewpoint of multi-task learning leads to a method to (partially) defend object detectors against adversarial attacks.

Recently, adversarial attacks have expanded beyond image classification and object detection to include dense prediction tasks such as semantic segmentation, depth, and optical flow. Promising results are shown in each of these tasks [36–39]. Such attacks often demonstrate the ability to target specific subsets of pixels rather than the entire image. For semantic segmentation, it was shown that pixels belonging to specific instances of pedestrians can be labeled by the attacked model as a road [36]. In [39], depth prediction has been successfully manipulated in many ways, such as removing the target entirely and aligning its depth with the surrounding background.

In the past couple of years, there is a growing interest in adversarial attacks on optical flow models [38,40–43]. Ranjan *et al* [38] demonstrated the possible benefits of a patch attack against leading models. First, they showed that this attack is very successful against encoder-decoder like architectures, but less effective for spatial pyramid types of models. They also showed it to be reproducible in "real life" conditions, with a hostile patch printed on a board and displayed in front of a camera. A follow-up work [40] conjectures that a principal cause for the success of adversarial attacks on OF is the small size of their receptive field. [41,42] introduce methods to corrupt the prediction of action recognition systems by attacking the OF modules they rely on. Finally, [44] proposes to rank OF method, in addition to their prediction accuracy, by their robustness to AAs. Specifically, to quantify the robustness, they propose a strong attack against OF models which is easily bounded so the comparison between methods is valid. Differently from the above, we consider the effectiveness of adversarial attacks on OF methods from the perspective of the ability to hide them in the output, in addition to their impact on performance.

Finally, one of the most important tools in risk assessment and collision avoidance for autonomous agents, e.g., robots and autonomous vehicles, is estimating the TTC [1–4,45,46]. A popular approach to estimate the TTC is using OF [1–4]. For example, [1] fuses 2D OF vectors, and per-pixel estimated scale change, to "upgrade" 2D OF to 3D, allowing the direct computation of the TTC. We use [1] to demonstrate the impact of our consistency term on the TTC, and the benefits it has over the non-consistent attack.

2   Method

The inputs for an optical flow network f_{flow} are two $H{\times}W{\times}3$ RGB images $I_1(x,y), I_2(x,y)$ where RGB channels ranges between [0,1]. The output is an $H{\times}W{\times}2$ optical flow vector map $V(x,y)$. The goal of an attacker is to find an additive perturbation to the input that would shift the attacked optical flow map V' away from the original prediction V, as in [38].

To calculate this perturbation, we use two binary masks (see examples in Fig. 2). The first mask, M_{target}, selects target pixels with the aim to change their optical flow, where in Fig. 2, M_{target} is the vehicle's instance mask. The second mask, $M_{perturb}$, specifies the pixels that may change due to the perturbation, where in Fig. 2 we allow only the pixels of the "nature" category to change.

Consider the first mask M_{target} with N non-zero entries specifying the pixels of the object (category or instance) we aim to attack. Our attack term, $l_{attack}(V',V)$, is then defined by the $L1$ norm between the attacked V' and original V flows, averaged over M_{target}, as given by

$$l_{attack}(V',V) = \frac{1}{N} \sum_{(x,y)\in M} |V'(x,y) - V(x,y)|_1, \qquad (1)$$

where $(x,y) \in M$ iff $M_{target}(x,y) = 1$.

To encourage the flow on the remaining scene to stay unaffected by the attack, we add a consistency term, $l_{con}(V',V)$, which is the negative $L1$ norm of the difference between original and attacked flows, averaged over non-attacked pixels:

$$l_{con}(V',V) = -\frac{1}{HW - N} \sum_{(x,y)\notin M} |V'(x,y) - V(x,y)|_1, \qquad (2)$$

where $(x,y) \notin M$ iff $M_{target}(x,y) = 0$, and since N pixels are attacked, we have $HW - N$ non-attacked pixels.

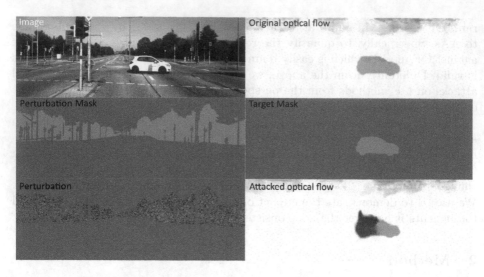

Fig. 2. Consistent attack method. In the top row, the original image (left) and OF (right) are presented. In the middle row, the perturbation mask (left) and target mask (right) are presented. In the bottom row, the perturbation on the "nature" category (left) and the attacked OF (right) are presented.

Our final loss is composed of these two terms, l_{attack} and l_{con}. The trade-off between the terms is controlled by the consistency coefficient α:

$$l_{total} = l_{attack} + \alpha l_{con}. \tag{3}$$

In order to attack semantic categories we require ground truth semantic labeling. This is only provided for the first image I_1 in the data we use for the attacks. Thus we have restricted our perturbation to the first image I_1. The second image I_2 is left unperturbed by our attack.

Let us denote the first image after the i-th perturbation as $I_1^{(i)}$, the i'th perturbation as $\delta I_1^{(i)}$ and the corresponding attacked flow $V^{(i)}$. Thus $V^{(0)} = V$ is the original flow, and $I_1^{(0)} = I_1$ is the unperturbed image. Since our first attack step is when $i = 1$ we have $\delta I_1^{(0)} = 0$.

Consider the second mask, $M_{perturb}$, with L non-zero entries, of the pixels we allow the attack to perturb. Given an attack strength coefficient ϵ, our i-th attack step follows the IFGSM [30] and given by

$$I_1^{(i)} = I_1^{(i-1)} + \delta I_1^{(i)}$$
$$\delta I_1^{(i)} = \epsilon \cdot M_{perturb} \cdot Sign\left(\nabla l_{total}(V^{(i-1)}, V)\right) \tag{4}$$
$$V^{(i)} = f_{flow}(I_1^{(i)}, I_2),$$

where $Sign$ returns the negative or positive sign of its input.

In each attack step i we create a small perturbation to the first image $\delta I_1^{(i)}$. As shown in Fig. 2 this perturbation is only applied in pixels (x, y) where $M_{perturb}(x, y) = 1$. It is equal to the sign of the loss function's gradient, weighted by the attack coefficient ϵ. After computing the i'th perturbation we add it to the image from the previous step $I_1^{(i-1)}$ to get the current perturbed input $I_1^{(i)}$. Inferring on this input with the optical flow network f_{flow} results in the i'th attacked optical flow map $V^{(i)}$. The loss between this flow $V^{(i)}$ and the original flow V will then be used to compute the perturbation for the following step. It is worth to note that for the first iteration ($i = 1$) we add a small amount of white noise to the original flow V so we would have non-zero gradients.

Let us define the target L1 norm of the perturbation as $||\Delta I||$. Given the number of perturbed pixels L and an estimated number of steps n for the attack, we set ϵ according to:

$$\epsilon = \frac{||\Delta I||}{n \cdot L} \tag{5}$$

We then iteratively update our input using Eq. 4 until $||I_1^{(i)} - I_1||_1 \approx ||\Delta I||$ (up to 5%). We use $n = 2$ and $||\Delta I|| = 4 \cdot 10^{-3} (\approx 1/255)$ for most of our experiments, and will specifically state experiments with other values.

2.1 Implementation Details

Throughout our experiments we use five optical flow models to evaluate the impact of adding our consistency term on targeted category-specific adversarial attacks - HD3 [16], PPAC [8], VCN [15], RAFT [14], LFN [13]. These models are some of the top performing methods on the KITTI'15 [23] dataset. We use the published, pre-trained models, given by the authors of each of the five chosen models. Since some models published multiple checkpoints, we always use the one fine-tuned on KITTI for our attack.

All of our experiments are performed and evaluated on the KITTI 12' [47] and KITTI 15' [23] datasets. These datasets contain a semantic segmentation labeling that we employ in our attacks. We could have used any semantic segmentation method [48–50] to label each scene. This would simulate a more realistic scenario where ground truth labeling is unavailable. However, it would also introduce another source of errors which we wish to avoid in order to focus our attention on consistent attacks.

We evaluate our attack using the average *end-point-error* (EPE) metric [11], which is the average $L2$ norm of the difference between attacked and original flows. The averaging is usually done over all image pixels, but since we are particularly interested in the effect of our attack on semantic classes, we compute the EPE averaged on pixels of specific classes. Using this metric the average shift in OF prediction due to the attack can be estimated for each class of interest.

In the subsequent section we will elaborate on the results from our main experiments. These experiments will encapsulate three different attack settings. These settings differ in the perturbed pixels mask ($M_{perturb}$, defined in Eq. 4) and the pixels we aim to attack (M_{target}, defined in Eq. 1). In the first setting,

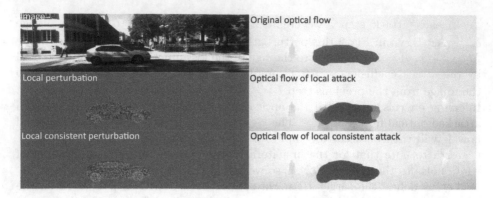

Fig. 3. A visualization of a local attack baseline and the impact of the consistency term on a vehicle instance using LFN [13] and $\|\Delta I\| = 2 \cdot 10^{-2}$. Left: the original image and the perturbation optimized by each attack. Right: the corresponding optical flows. Adding a consistency term reduces the effect on non-vehicle pixels (as can be seen below the vehicle), while still significantly changing the vehicle's optical flow.

a local attack, we perturb vehicle pixels and aim to attack the same subset of pixels. The second setting, a global attack, is where we perturb the entire image, but aim to attack vehicle pixels only. The third setting, a cross-category attack, is where we perturb the pixels of the "nature" category, and aim to attack vehicle category pixels (presented in the Supplementary material).

3   Experiments and Results

In this section, we present the experimental results obtained for the "vehicle" target category. The results for "human" target category, as well as the results obtained using the KITTI 12', are given in the supplementary.

3.1   Local Attacks

Figure 3 visualizes an example local attack ($\|\Delta I\| = 2 \cdot 10^{-2}$) experiment using the LFN model [13]. In this experiment a vehicle instance was attacked by only perturbing its pixels. Two attacks were conducted: a baseline, non -consistent, method with $\alpha = 0$ and a consistent attack with $\alpha = 10$.

Both attacks are successful in changing the car's optical flow and cause the previous right (red) moving vehicle to turn left (blue). However, the consistent attack preserves the non-targeted flow better, as can be seen by comparing the flow under the vehicle.

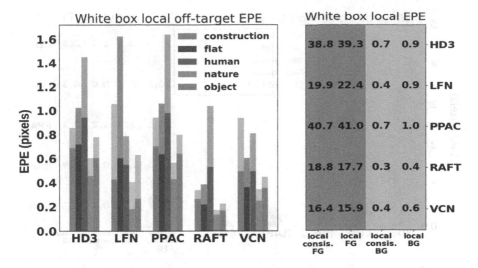

Fig. 4. Comparison between local attacks and local-consistent attacks on the KITTI dataset with $||\Delta I|| = 4 \cdot 10^{-3}$. Left - mean error caused by a local attack (transparent colors) and a consistent local attack (solid colors) over the corresponding category. Right - mean EPE for each model for the target (left) and off-target (right).

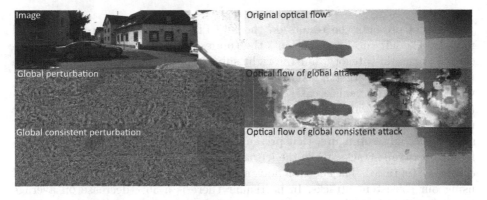

Fig. 5. A visualization of a global vehicle-targeting attack and the effect of adding a consistency term on HD3's [16] flow with $||\Delta I|| = 2 \cdot 10^{-2}$. Left: original image and perturbations. Right: the corresponding optical flows. Adding a consistency term reduces the effect on non-vehicle pixels while still significantly changing the vehicle optical flow

To quantify this effect, this experiment was expended to the entirety of the KITTI dataset. Here, for each image in the dataset we have attacked all of the vehicles in that image (by perturbing vehicle pixels). We then evaluated the mean EPE between original and attacked flow on selected categories: construction, flat, human, nature, object and vehicle.

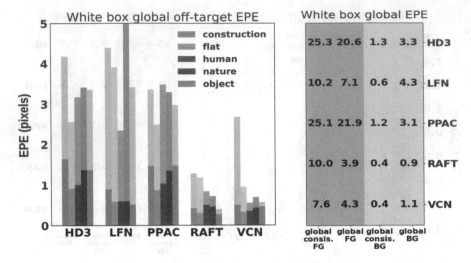

Fig. 6. Comparison between global attacks and global-consistent attacks on vehicles in the KITTI dataset with $||\Delta I|| = 4 \cdot 10^{-3}$. Left - mean error caused by a global attack (transparent colors) and a consistent global attack (solid colors) over the corresponding category. Right - mean EPE for each model for the target (left) and off-target (right).

Figure 4 presents the results on the KITTI dataset using the five selected models. The left sub-figure presents the (undesired) effect on non-targeted pixels per semantic category and the right sub-figure presents a summary for the (desired) effect on targeted pixels (left) and non-target pixels (right). We see that while the targeted vehicle category error does not vary much between attacks (right table) the non-targeted categories (left figure) suffer much less damage using a local consistent attack than our baseline non-consistent attack. The left side of the right sub-figure, that presents the targeted EPE, shows a small difference between attacks, while the left side of the right sub-figure shows a larger difference (in ratio). The effect on non-targeted categories is significantly reduced using our consistent attacks. In particular there is a 35% decrease on average (across methods) on the error induced on these categories.

3.2   Global Attacks

One of the concerns with using a local attack is that since we perturb only a subset of the image pixels, we employ a high L_∞ norm to achieve the same L_1 norm as a *global* attack that perturbs the entire image. This, in turn, causes the local attack to be more perceptible compared to a global attack. Figure 5 demonstrates this global attack in which we perturbed the entire image. The figure visually compares the results of the consistent and non-consistent attacks.

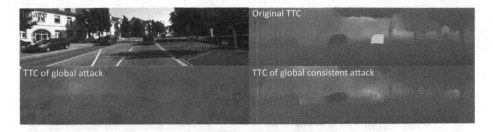

Fig. 7. Visualization of the TTC results for an AA on a vehicle instance using HD3 with $||\Delta I|| = 4 \cdot 10^{-3}$. Hot colors corresponds to shorter TTC than cold colors.

The left column presents the original image and its perturbations. Here, unlike the local attack, the entire image is perturbed. The right column presents the effect both attacks have on HD3's optical flow. For the non-consistent attack we can notice multiple non-vehicle flow segments that changed drastically, turning the naturally smooth flow of the background into a rapidly varying flow. Repeating the methodology we used for the local settings, we expand this experiment by attacking all of the vehicle category in the KITTI dataset [23], and averaging the error over the selected classes.

Figure 6 presents the result of attacking all vehicles in a global setting over the KITTI dataset, for our five OF models with $||\Delta I|| = 4 \cdot 10^{-3}$. Similarly to the local case, the left and right side of the right sub-figures demonstrate the effect on the non-targeted pixels. The left side of the right sub-figure presents the effect of the targeted pixels. The resulting targeted vehicle category error is higher when using a consistent attack. Moreover, the non-targeted categories suffer significantly less damage using a global consistent attack than the baseline non-consistent attack. Thus, for example, using the consistent attack results in a 60% stronger effect on the targeted category (averaged across models), while removing 60% of the unwanted optical flow change on the remaining categories (averaged across models).

3.3 Time-To-Collision (TTC)

As discussed in Sect. 1, we emphasize the significance of adversarial attacks on OF models by their possible impact on TTC algorithms [1–4]. We chose the state-of-the-art TTC algorithm presented in [1], which uses OF to compute a per-pixel TTC. The model is supplied with the attacked OF instead of the original OF computed by the pre-trained VCN (without fine-tuning).

An attack on a vehicle instance, which is visualized in Fig. 7, demonstrates the impact of the original flow, the global consistent and global non-consistent attacked flows, on the TTC. The TTC values are log-scaled and color-coded, where hot colors (redish) encode lower TTC than colder colors (yellowish-whitish). The attacked vehicle, which is yellow (high TTC) in the original flow,

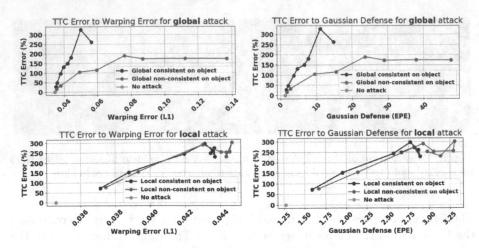

Fig. 8. TTC error to AA detection score. The top and bottom rows correspond to the global and local attacks, respectively. The Y-axis in all graphs corresponds to the TTC error, while the X-axis corresponds to the AA detection score using warping error and Gaussian defense for the left and right graphs, respectively. The graph is created using attacks with magnitude $||m \cdot 10^{-3}||$ for $m \in \{0.2, 0.4, 1.2, 2, 3.2, 4, 6, 8\}$.

is significantly darker (low TTC) in both the consistent and non-consistent attacks. Importantly, the backgrounds of the original and consistent attack are quite similar, while the background of the not-consistent attack is very different.

As argued in Sect. 1, the effect of the off-target consistency loss term allows a better tradeoff between the impact on the TTC and a AA detection score, where by "detection score" we mean the output of an AA detection method. An example for this tradeoff would be that an attacked input with the same detection score will result in a higher average TTC impact. To quantify this tradeoff we've used three AA detection methods.

Warping Error: The difference between I_1 and I_2 warped using the (attacked) OF V. That is, $||W_V(I_2) - I_1||_1$, where $W_V(I_2)$ warps I_2 using the OF V. The warping error is often used as an OF confidence measure [51]. Naturally, such confidence measure may be used to estimate an AA detection score.

The Gaussian/Median Defenses [52]: The OF error (EPE) between the predicted flow and the flow from Gaussian/Median smoothed versions of the same images. That is, let $V'(I_1', I_2)$ be an attacked flow, and $V_K' = V'(K(I_1'), I_2)$ be an attacked flow (using the same attack) with I_1' smoothed using a 3×3 Gaussian/Median kernel K before the OF computation. $||V' - V_K'||_1$ is used to estimate the detection score. Such defense methods were used as AA detection methods in [52].

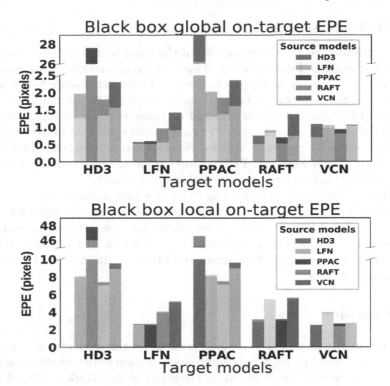

Fig. 9. "Black-box" setting results for the global attacks (top) and local attacks (bottom) on vehicles with $||\Delta I|| = 4 \cdot 10^{-3}$. Bar height signify the mean EPE over the vehicle category caused by an attack created using the source models (color-coded) on the target model (x-axis). Consistent and non-consistent attacks are marked by solid colors and transparent colors, respectively. (Color figure online)

The graphs of the error in TTC as a function of the AA detection score are presented in Fig. 8. (The median defense is presented in the supplementary material.) We measure the error in TTC as an average percentage of difference relative to the original TTC; that is, $|T_A - T_O|/T_O$, where T_A and T_O are the TTCs of the attacked and original flows, respectively. The graphs are created from 8 AA with different magnitudes, where all 5 OF models (in a white-box settings) are averaged per attack magnitude. In all 3 cases, the global consistent attack is superior to the global non-consistent attack in both the detection score (lower in X axis), and in impact on TTC (higher Y axis). In the local attack the trend is similar, however, the gap is much smaller. To conclude, the off-target consistency loss term is effective in terms of the TTC - detection score tradeoff.

3.4   Black-box Attacks and Transferability

Finally, we evaluated the transferability of the consistent attacks for the global and local attacks. To this end, we used each of the chosen models to attack the vehicle category in every image pair in the KITTI dataset. Each model was then evaluated on the adversarial datasets generated using the remaining models. The mean EPE over the vehicle targeted category for each attack is presented in Fig. 9. We use transparent colors to visualize non-consistent attacks and solid colors to visualize for consistent attacks. We note that attacks created using HD3 seem to have a high impact on PPAC and vice versa, which could be related to HD3 and PPAC having most of their architecture shared.

Similar to the results presented in the white box settings, the local attack impact does not vary much with the addition of the consistency term. However, for the global case we observe a significant increase in the targeted impact transferred to other models. If we examine the results on RAFT, adding the consistency term resulted in a 44% increase in black-box attack strength, averaged across targeted models.

4   Discussion

To summarize, we presented a new methodology for targeted adversarial attacks against optical flow models. We introduced a new term to the attack, called 'consistency term', which is used to reduce the effect of the attack on the off-target pixels. In three different settings: local, global and cross category (supplementary), adding the consistency term to the loss reduces the impact on non-targeted object. Adding the term either preserves or increases the effect on the targeted category. Moreover, we have demonstrated that for some of the settings using a consistent attack results in a more transferable attack. Finally, we have showed that for a TTC downstream task these attacks have a better detection - impact tradeoff, with an impact as high as 3x higher for the same detection score.

In our experiments we observe an obvious difference between the local and the global setting, where the effect on the non-targeted object is much more apparent in the global setting. In this setting, the danger of negatively impacting the rest of the scene is much greater since we directly change the non-target pixels. Adding the consistency term allows us to introduce global perturbations with a smaller effect on the resulting non-targeted optical flow. We also note that, in slight contrast to [38], our attacked models, which all have pyramid-like feature encoders are attacked successfully. A further analysis is left for future work.

An interesting follow-up for our work would be utilizing adversarial targeted attacks as a data augmentation technique for model training, which was demonstrated effective [53]. Other works [39] have demonstrated that some semantic classes are easier to attack than others. By leveraging consistent adversarial targeted attacks in its augmentation procedure, models might be able to learn a more robust representation of each semantic class. This, in turn, might decrease the probability of a successful attack against them [32], and increase the ability of a model to generalize its predictions for those classes [54].

References

1. Yang, G., Ramanan, D.: Upgrading optical flow to 3D scene flow through optical expansion. In: Proceedings of the IEEE/CVF Conference on Computer Vision and Pattern Recognition, pp. 1334–1343 (2020)
2. Pedro, D., Matos-Carvalho, J.P., Fonseca, J.M., Mora, A.: Collision avoidance on unmanned aerial vehicles using neural network pipelines and flow clustering techniques. Remote Sens. **13**, 2643 (2021)
3. Blumenkamp, J.: End to end collision avoidance based on optical flow and neural networks. arXiv preprint arXiv:1911.08582 (2019)
4. Badki, A., Gallo, O., Kautz, J., Sen, P.: Binary TTC: a temporal geofence for autonomous navigation. In: Proceedings of the IEEE/CVF Conference on Computer Vision and Pattern Recognition, pp. 12946–12955 (2021)
5. Kurakin, A., Goodfellow, I.J., Bengio, S.: Adversarial machine learning at scale. CoRR abs/1611.01236 (2016)
6. Goodfellow, I.J., Shlens, J., Szegedy, C.: Explaining and harnessing adversarial examples. In: ICLR (2015)
7. Szegedy, C., et al.: Intriguing properties of neural networks. In: Bengio, Y., LeCun, Y. (eds.) ICLR (Poster) (2014)
8. Wannenwetsch, A.S., Roth, S.: Probabilistic pixel-adaptive refinement networks. In: CVPR, pp. 11639–11648. IEEE (2020)
9. Grosse, K., Manoharan, P., Papernot, N., Backes, M., McDaniel, P.: On the (statistical) detection of adversarial examples. arXiv preprint arXiv:1702.06280 (2017)
10. Tian, S., Yang, G., Cai, Y.: Detecting adversarial examples through image transformation. In: Thirty-Second AAAI Conference on Artificial Intelligence (2018)
11. Baker, S., Scharstein, D., Lewis, J.P., Roth, S., Black, M.J., Szeliski, R.: A database and evaluation methodology for optical flow. Int. J. Comput. Vis. **92**, 1–31 (2011)
12. Papernot, N., McDaniel, P., Goodfellow, I., Jha, S., Celik, Z.B., Swami, A.: Practical black-box attacks against machine learning. In: Proceedings of the 2017 ACM on Asia Conference on Computer and Communications Security, pp. 506–519 (2017)
13. Hui, T., Tang, X., Loy, C.C.: LiteFlowNet: a lightweight convolutional neural network for optical flow estimation. In: 2018 IEEE/CVF Conference on Computer Vision and Pattern Recognition, pp. 8981–8989 (2018)
14. Teed, Z., Deng, J.: RAFT: recurrent all-pairs field transforms for optical flow. In: Vedaldi, A., Bischof, H., Brox, T., Frahm, J.-M. (eds.) ECCV 2020. LNCS, vol. 12347, pp. 402–419. Springer, Cham (2020). https://doi.org/10.1007/978-3-030-58536-5_24
15. Yang, G., Ramanan, D.: Volumetric correspondence networks for optical flow. In: NeurIPS, 793–803 (2019)
16. Yin, Z., Darrell, T., Yu, F.: Hierarchical discrete distribution decomposition for match density estimation. In: CVPR, pp. 6044–6053. Computer Vision Foundation/IEEE (2019)
17. Lucas, B.D., Kanade, T.: An iterative image registration technique with an application to stereo vision. In: Proceedings of the 7th international joint conference on Artificial intelligence - vol. 2, pp. 674–679. IJCAI1981, San Francisco, CA, USA, Morgan Kaufmann Publishers Inc. (1981)
18. Horn, B.K.P., Schunck, B.G.: Determining optical flow. Artif. Intell. **17**, 185–203 (1981)
19. Sun, D., Roth, S., Black, M.: A quantitative analysis of current practices in optical flow estimation and the principles behind them. Int. J. Comput. Vision **106**, 115–137 (2014)

20. Dosovitskiy, A., et al.: FlowNet: learning optical flow with convolutional networks. In: ICCV, pp. 2758–2766. IEEE Computer Society (2015)
21. Ilg, E., Mayer, N., Saikia, T., Keuper, M., Dosovitskiy, A., Brox, T.: FlowNet 2.0: evolution of optical flow estimation with deep networks. In: CVPR, pp. 1647–1655. IEEE Computer Society (2017)
22. Sun, D., Yang, X., Liu, M.Y., Kautz, J.: PWC-NET: CNNs for optical flow using pyramid, warping, and cost volume. In: CVPR, pp. 8934–8943. IEEE Computer Society (2018)
23. Menze, M., Geiger, A.: Object scene flow for autonomous vehicles. In: CVPR, pp. 3061–3070. IEEE Computer Society (2015)
24. Hu, Y., Li, Y., Song, R.: Robust interpolation of correspondences for large displacement optical flow. In: Proceedings of the IEEE Conference on Computer Vision and Pattern Recognition (CVPR) (2017)
25. Ranjan, A., Black, M.J.: Optical flow estimation using a spatial pyramid network. In: CVPR, pp. 2720–2729. IEEE Computer Society (2017)
26. Hur, J., Roth, S.: Iterative residual refinement for joint optical flow and occlusion estimation. In: CVPR, pp. 5754–5763. Computer Vision Foundation/IEEE (2019)
27. Cho, K., van Merrienboer, B., Bahdanau, D., Bengio, Y.: On the properties of neural machine translation: encoder-decoder approaches. CoRR (2014)
28. Nguyen, A., Yosinski, J., Clune, J.: Deep neural networks are easily fooled: High confidence predictions for unrecognizable images. In: Proceedings of the IEEE Conference on Computer Vision and Pattern Recognition, pp. 427–436 (2015)
29. Moosavi-Dezfooli, S.M., Fawzi, A., Frossard, P.: DeepFool: a simple and accurate method to fool deep neural networks. In: Proceedings of the IEEE Conference on Computer Vision and Pattern Recognition, pp. 2574–2582 (2016)
30. Kurakin, A., Goodfellow, I.J., Bengio, S.: Adversarial examples in the physical world. In: ICLR (Workshop), OpenReview.net (2017)
31. Tramèr, F., Kurakin, A., Papernot, N., Goodfellow, I.J., Boneh, D., McDaniel, P.D.: Ensemble adversarial training: attacks and defenses. In: ICLR (Poster), OpenReview.net (2018)
32. Madry, A., Makelov, A., Schmidt, L., Tsipras, D., Vladu, A.: Towards deep learning models resistant to adversarial attacks. In: ICLR (Poster), OpenReview.net (2018)
33. Nezami, O.M., Chaturvedi, A., Dras, M., Garain, U.: Pick-object-attack: type-specific adversarial attack for object detection. CoRR abs/2006.03184 (2020)
34. Zhang, H., Wang, J.: Towards adversarially robust object detection. In: Proceedings of the IEEE/CVF International Conference on Computer Vision (2019)
35. Liao, Q., et al.: Fast local attack: generating local adversarial examples for object detectors. In: 2020 International Joint Conference on Neural Networks (IJCNN), pp. 1–8. IEEE (2020)
36. Fischer, V., Kumar, M.C., Metzen, J.H., Brox, T.: Adversarial examples for semantic image segmentation. In: ICLR (Workshop), OpenReview.net (2017)
37. Arnab, A., Miksik, O., Torr, P.H.S.: On the robustness of semantic segmentation models to adversarial attacks. In: CVPR, pp. 888–897. IEEE Computer Society (2018)
38. Ranjan, A., Janai, J., Geiger, A., Black, M.J.: Attacking optical flow. In: Proceedings of the IEEE/CVF International Conference on Computer Vision, pp. 2404–2413 (2019)
39. Wong, A., Cicek, S., Soatto, S.: Targeted adversarial perturbations for monocular depth prediction. In: Advances in Neural Information Processing Systems (2020)

40. Schrodi, S., Saikia, T., Brox, T.: Towards understanding adversarial robustness of optical flow networks. In: Proceedings of the IEEE/CVF Conference on Computer Vision and Pattern Recognition, pp. 8916–8924 (2022)

41. Inkawhich, N., Inkawhich, M., Chen, Y., Li, H.: Adversarial attacks for optical flow-based action recognition classifiers. arXiv preprint arXiv:1811.11875 (2018)

42. Anand, A.P., Gokul, H., Srinivasan, H., Vijay, P., Vijayaraghavan, V.: Adversarial patch defense for optical flow networks in video action recognition. In: 2020 19th IEEE International Conference on Machine Learning and Applications (ICMLA), pp. 1289–1296. IEEE (2020)

43. Yamanaka, K., Takahashi, K., Fujii, T., Matsumoto, R.: Simultaneous attack on CNN-based monocular depth estimation and optical flow estimation. IEICE Trans. Inf. Syst. **104**, 785–788 (2021)

44. Schmalfuss, J., Scholze, P., Bruhn, A.: A perturbation constrained adversarial attack for evaluating the robustness of optical flow. arXiv preprint arXiv:2203.13214 (2022)

45. Manglik, A., Weng, X., Ohn-Bar, E., Kitani, K.M.: Future near-collision prediction from monocular video: feasibility, dataset, and challenges. arXiv preprint arXiv:1903.09102 1 (2019)

46. Mori, T., Scherer, S.: First results in detecting and avoiding frontal obstacles from a monocular camera for micro unmanned aerial vehicles. In: 2013 IEEE International Conference on Robotics and Automation, pp. 1750–1757. IEEE (2013)

47. Geiger, A., Lenz, P., Urtasun, R.: Are we ready for autonomous driving? the kitti vision benchmark suite. In: Conference on Computer Vision and Pattern Recognition (CVPR) (2012)

48. Chen, L.C., Papandreou, G., Schroff, F., Adam, H.: Rethinking atrous convolution for semantic image segmentation. CoRR abs/1706.05587 (2017)

49. Chen, L.C., Papandreou, G., Kokkinos, I., Murphy, K., Yuille, A.L.: DeepLab: Semantic image segmentation with deep convolutional nets, atrous convolution, and fully connected CRFs. IEEE Trans. Pattern Anal. Mach. Intell. **40**, 834–848 (2017)

50. Chen, L.-C., Zhu, Y., Papandreou, G., Schroff, F., Adam, H.: Encoder-decoder with atrous separable convolution for semantic image segmentation. In: Ferrari, V., Hebert, M., Sminchisescu, C., Weiss, Y. (eds.) ECCV 2018. LNCS, vol. 11211, pp. 833–851. Springer, Cham (2018). https://doi.org/10.1007/978-3-030-01234-2_49

51. Weinzaepfel, P., Revaud, J., Harchaoui, Z., Schmid, C.: Learning to detect motion boundaries. In: Proceedings of the IEEE Conference on Computer Vision and Pattern Recognition, pp. 2578–2586 (2015)

52. Xu, W., Evans, D., Qi, Y.: Feature squeezing: detecting adversarial examples in deep neural networks. arXiv preprint arXiv:1704.01155 (2017)

53. Liu, L., et al.: Learning by analogy: reliable supervision from transformations for unsupervised optical flow estimation. In: Proceedings of the IEEE/CVF Conference on Computer Vision and Pattern Recognition, pp. 6489–6498 (2020)

54. Stutz, D., Hein, M., Schiele, B.: Disentangling adversarial robustness and generalization. In: Proceedings of the IEEE/CVF Conference on Computer Vision and Pattern Recognition, pp. 6976–6987 (2019)

Physical Passive Patch Adversarial Attacks on Visual Odometry Systems

Yaniv Nemcovsky, Matan Jacoby, Alex M. Bronstein, and Chaim Baskin[✉]

Department of Computer Science, Technion, Haifa, Israel
chaimbaskin@cs.technion.ac.il

Abstract. Deep neural networks are known to be susceptible to adversarial perturbations – small perturbations that alter the output of the network and exist under strict norm limitations. While such perturbations are usually discussed as tailored to a specific input, a universal perturbation can be constructed to alter the model's output on a set of inputs. Universal perturbations present a more realistic case of adversarial attacks, as awareness of the model's exact input is not required. In addition, the universal attack setting raises the subject of generalization to unseen data, where given a set of inputs, the universal perturbations aim to alter the model's output on out-of-sample data. In this work, we study physical passive patch adversarial attacks on visual odometry-based autonomous navigation systems. A visual odometry system aims to infer the relative camera motion between two corresponding viewpoints, and is frequently used by vision-based autonomous navigation systems to estimate their state. For such navigation systems, a patch adversarial perturbation poses a severe security issue, as it can be used to mislead a system onto some collision course. To the best of our knowledge, we show for the first time that the error margin of a visual odometry model can be significantly increased by deploying patch adversarial attacks in the scene. We provide evaluation on synthetic closed-loop drone navigation data and demonstrate that a comparable vulnerability exists in real data. A reference implementation of the proposed method and the reported experiments is provided at https://github.com/patchadversarialattacks/patchadversarialattacks.

Keywords: Real-world adversarial attacks · Adversarial robustness · Navigation · Robot vision

1 Introduction

Adversarial Attacks. Deep neural networks (DNNs) were the first family of models discovered to be susceptible to adversarial perturbations – small bounded-norm perturbations of the input that significantly alter the output of the model

---

Y. Nemcovsky and M. Jacoby–Equal contribution.

Supplementary Information The online version contains supplementary material available at https://doi.org/10.1007/978-3-031-26293-7_31.

L. Wang et al. (Eds.): ACCV 2022, LNCS 13847, pp. 518–534, 2023.
https://doi.org/10.1007/978-3-031-26293-7_31

[15, 30] (methods for producing such perturbations are referred to as adversarial attacks). Such perturbations are usually discussed as tailored to a specific model and input, and in such settings were shown to undermine the impressive performance of DNNs across multiple fields, e.g., object detection [32], real-life object recognition [3, 5, 35], reinforcement learning [14], speech-to-text [6], point cloud classification [34], natural language processing [7, 12, 18], video recognition [28], Siamese Visual Tracking [19], and on several regression tasks [22, 25, 36] as well as autonomous driving [10]. Moreover, adversarial attacks were shown to be transferable between models; i.e., an adversarial perturbation that is effective on one model will likely be effective for other models as well [30]. Recent studies also suggest that the vulnerability is a property of high-dimensional input spaces rather than of specific model classes [2, 11, 13].

Universal adversarial attacks are another setting where the aim is to produce an adversarial perturbation for a set of inputs [16, 23, 38]. Universal perturbations present a more realistic case of adversarial attacks, as awareness of the model's exact input is not required. In addition, the universal attack setting raises the subject of generalization to unseen data, where given a set of inputs, the universal perturbations aim to alter the model's output on out-of-sample data. In this setting, universal perturbations can also be used to improve the performance of DNNs on out-of-sample data [29].

Adversarial Attacks on Visual Odometers. Monocular visual odometry (VO) models aim to infer the relative camera motion (position and orientation) between two corresponding viewpoints. Recently, DNN-based VO models have outperformed traditional monocular VO methods [1, 4, 33, 37]. Specifically, the model suggested by Wang et al. (2020b) [33] shows a promising ability to generalize from simulated training data to real scenes. Such models usually make use of either feature matching or photometric error minimization to compute the camera motion.

Vision-based autonomous navigation systems frequently use VO models as a method of estimating their state. Such navigation systems would use the trajectory estimated by the VO to compute their heading, closing the loop with the navigation control system that directs the vehicle to a target position in the scene. Visual simultaneous localization and mapping (visual SLAM, or vSLAM for short) techniques also make use of VO models to estimate the vehicle trajectory, additionally estimating the environment map and thereby adding global consistency to the estimations [20, 24, 27, 31]. Adversarial attacks on VO models, consequently, pose a severe security issue for visual SLAM, as they could corrupt the estimated map and mislead the navigation. A recent work [8] had discussed adversarial attacks on the monocular VO model over single image pairs, and shows the susceptibility of the estimated position and orientation to adversarial perturbations.

In the present work, we investigate the susceptibility of VO models to universal adversarial perturbations over trajectories with multiple viewpoints, aiming to mislead a corresponding navigation system by disrupting its ability to spatially position itself in the scene. Previous works that discuss adversarial attacks on regression models mostly discuss standard adversarial attacks where the perturbation is inserted directly into a single image [8, 10, 19, 22, 25, 36]. In contrast, we take into consideration a time evolving process where a physical passive patch adversarial attack is inserted into the scene and is perceived differently from multiple viewpoints. This is a highly realistic settings, as we test the effect of a moving camera in a perturbed scene, and do not require direct access to the model's input. Below, we outline our main contributions.

Firstly, we produce physical patch adversarial perturbations for VO systems on both synthetic and real data. Our experiments show that while VO systems are robust to random perturbations, they are susceptible to such adversarial perturbations. For a given trajectory containing multiple frames, our attacks are aimed to maximize the generated deviation in the physical translation between the accumulated trajectory motion estimated by the VO and the ground truth. We show that inserting a physical passive adversarial patch into the scene substantially increases the generated deviation.

Secondly, we continue to produce universal physical patch adversarial attacks, which are aimed at perturbing unseen data. We optimize a single adversarial patch on multiple trajectories and test the attack on out-of-sample unseen data. Our experiments show that when used on out-of-sample data, our universal attacks generalize and again cause significant deviations in trajectory estimates produced by the VO system.

Lastly, we further test the robustness of VO systems to our previously produced universal adversarial attacks in a closed-loop scheme with a simple navigation scheme, on synthetic data. Our experiments show that in this case as well, the universal attacks force the VO system to deviate from the ground truth trajectory. To the best of our knowledge, ours is the first time the vulnerability of visual navigation systems to adversarial attacks is demonstrated, and, possibly, the first instance of adversarial attacks on closed-loop control systems.

The rest of the paper is organized as follows: Sect. 2 describes our proposed method, Sect. 3 provides our experimental results, and Sect. 4 concludes the paper.

2   Method

Below, we start with a definition of the adversarial attack setting, for both the universal and standard cases. We then describe the adversarial optimization scheme used for producing the perturbations and discuss the optimization of the universal attacks aiming to perturb unseen data.

2.1   Patch Adversarial Attack Setting

Patch Adversarial Perturbation. Let $\mathcal{I} = [0,1]^{3 \times w \times h}$ be a normalized RGB image space, for some width w and height h. For an image $I \in \mathcal{I}$, inserting a patch image $P \in \mathcal{I}$ onto a given plane in I would then be a perturbation $A : (\mathcal{I} \times \mathcal{I}) \to \mathcal{I}$. We denote $I^P = A(I, P)$. To compute I^P, we first denote the black and white albedo images of the patch P as viewed from viewpoint I, namely $I^0, I^1 \in \mathcal{I}$. The albedo images are identical except for pixels corresponding to P, which contains the minimal and maximal RGB intensity values of the patch from viewpoint I. As such, I^0 and I^1 essentially describe the dependency of I^P on the lighting conditions and the material comprising P. In addition, we denote the linear homography transformation of P to viewpoint I as $H : \mathcal{I} \to \mathcal{I}$, such that $H(P)$ is the mapping of pixels from P to I, without taking into consideration changes in the pixels' intensity. Note that H is only dependent on the relative camera motion between I and P. We now define I^P as:

$$I^P = A(I, P) = H(P) * (I^1 - I^0) + I^0 \tag{1}$$

where $*$ denotes element-wise multiplication. For a set of images $\{I_t\}$, we similarly define the perturbed set as inserting a single patch P onto the same plane in each image:

$$\{I_t^P\} = A(\{I_t\}, P) = \{H_t(P) * (I_t^1 - I_t^0) + I_t^0\} \tag{2}$$

Attacking Visual Odometry. Let $VO : (\mathcal{I} \times \mathcal{I}) \to (\mathbb{R}^3 \times so(3))$ be a monocular VO model, i.e., for a given pair of consecutive images $\{I_t, I_{t+1}\}$, it estimates the relative camera motion $\delta_t^{t+1} = (q_t^{t+1}, R_t^{t+1})$, where $q_t^{t+1} \in \mathbb{R}^3$ is the $3D$ translation and $R_t^{t+1} \in so(3)$ is the $3D$ rotation. We define a trajectory as a set of consecutive images $\{I_t\}_{t=0}^L$, for some length L, and extend the definition of the monocular visual odometry to trajectories:

$$VO(\{I_t\}_{t=0}^L) = \{\hat{\delta}_t^{t+1}\}_{t=0}^{L-1} \tag{3}$$

where $\hat{\delta}_t^{t+1}$ denotes the estimation of δ_t^{t+1} by the VO model. Given a trajectory $\{I_t\}_{t=0}^L$, with ground truth motions $\{\delta_t^{t+1}\}_{t=0}^{L-1}$ and a criterion over the trajectory motions ℓ, an adversarial patch perturbation $P_a \in \mathcal{I}$ aims to maximize the criterion over the trajectory. Similarly, for a set of trajectories $\{\{I_{i,t}\}_{t=0}^{L_i}\}_{i=0}^{N-1}$, with corresponding ground truth motions $\{\{\delta_{i,t}^{t+1}\}_{t-0}^{L_i-1}\}_{i=0}^{N-1}$, a universal adversarial attack aims to maximize the sum of the criterion over the trajectories. Formally:

$$P_a = \arg\max_{P \in \mathcal{I}} \ell(VO(A(\{I_t\}_{t=0}^L, P)), \{\delta_t^{t+1}\}_{t=0}^{L-1}) \tag{4}$$

$$P_{ua} = \arg\max_{P \in \mathcal{I}} \sum_{i=0}^{N-1} \ell(\{VO(A(I_{i,t}, P))\}_{t=0}^{L_i}, \{\delta_{i,t}^{t+1}\}_{t=0}^{L_i-1}) \tag{5}$$

where A is defined as in Eq. (2), and the scope of the adversarial attacks' universality is according to the domain and data distribution of the given trajectories set. In this formulation, the limitation of the adversarial perturbation is expressed in the albedo images I^0 and I^1 and can be described as dependent on the patch material. In contrast to standard adversarial perturbations that are tailored to a specific input, we consider the generalization properties of universal adversarial perturbation to unseen data. In such cases the provided trajectories used for perturbation optimization would differ from the test trajectories.

Task Criterion. For the scope of this paper, the target criterion used for adversarial attacks is the RMS (root mean square) deviation in the $3D$ physical translation between the accumulated trajectory motion, as estimated by the VO, and the ground truth. We denote the accumulated motion as $\delta_0^L = \delta_0^1 \cdot \delta_1^2 \cdots \delta_{L-1}^L = \prod_{t=0}^{L-1} \delta_t^{t+1}$, where the multiplication of motions is defined as the matrix multiplication of the corresponding 4×4 matrix representation: $\delta_t^{t+1} = \begin{pmatrix} R_t^{t+1} & q_t^{t+1} \\ \mathbf{0} & 1 \end{pmatrix}$.

The target criterion is then formulated as:

$$\ell_{VO}(VO(A(\{I_t\}_{t=0}^L, P)), \{\delta_t^{t+1}\}_{t=0}^{L-1}) = \|q(\prod_{t=0}^{L-1} VO(I_t^P, I_{t+1}^P)) - q(\prod_{t=0}^{L-1} \delta_t^{t+1})\|_2 \tag{6}$$

where we denote $q(\delta_0^L) = q((q_0^L, R_0^L)) = q_0^L$.

2.2   Optimization of Adversarial Patches

We optimize the adversarial patch P via a PGD adversarial attack [21] with ℓ_{inf} norm limitation. We limit the values in P to be in $[0, 1]$; however, we do not enforce any additional ϵ limitation, as such would be expressed in the albedo images. We allow for different training and evaluation criteria in both attack types, and to enable evaluation on unseen data for universal attacks, we allow for different training and evaluation datasets. In the supplementary material, we provide algorithms for both our PGD (Algorithm 1) and universal (Algorithm 2) attacks. We note that the PGD attack is a specific case of the universal in which both training and evaluation datasets comprise the same single trajectory.

Optimization and Evaluation Criteria. For both optimization and evaluation of attacks we consider one of two criteria. The first criterion, which we denote as ℓ_{RMS}, is a smoother version of the target criterion ℓ_{VO}, in which we sum over partial trajectories with the same origin as the full trajectory. Similarly, the second criterion, which we denote as ℓ_{MPRMS}, i.e., mean partial RMS, differ from ℓ_{RMS} by taking into account all the partial trajectories. Nevertheless, we take the mean for each length of partial trajectories in order to keep the factoring between different lengths as in ℓ_{RMS}. ℓ_{MPRMS} may be more suited to generalization of universal attacks to unseen data than in-sample optimization, as it takes into consideration partial trajectories that may not be relevant to the full trajectory. Formally:

$$\ell_{RMS}(VO(A(\{I_t\}_{t=0}^L, P)), \{\delta_t^{t+1}\}_{t=0}^{L-1})$$
$$= \sum_{l=1}^L \ell_{VO}(VO(A(\{I_t\}_{t=0}^l, P)), \{\delta_t^{t+1}\}_{t=0}^{l-1}) \tag{7}$$

$$\ell_{MPRMS}(VO(A(\{I_t\}_{t=0}^L, P)), \{\delta_t^{t+1}\}_{t=0}^{L-1})$$
$$= \sum_{l=1}^L \frac{1}{L-l+1} \sum_{i=0}^{L-l} \ell_{VO}(VO(A(\{I_t\}_{t=i}^{i+l}, P)), \{\delta_t^{t+1}\}_{t=i}^{i+l-1}) \tag{8}$$

3   Experiments

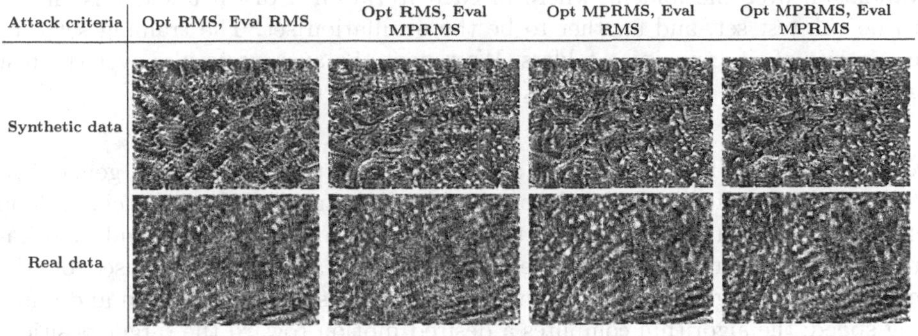

Attack criteria	Opt RMS, Eval RMS	Opt RMS, Eval MPRMS	Opt MPRMS, Eval RMS	Opt MPRMS, Eval MPRMS
Synthetic data				
Real data				

Fig. 1. Visualization of universal adversarial patches. For each dataset, and optimization and evaluation criteria, we present the universal adversarial image produced via the in-sample attack scheme.

We now present an empirical evaluation of the proposed method. We first describe the various experimental settings used for estimating the effect of the adversarial perturbations. We continue and describe the attacked VO model and the generation of both the synthetic and real datasets used in our experiments. Finally, we present our experimental results, first on the synthetic dataset and afterwards on the real dataset (Fig. 1).

3.1   Experimental Setting

In this section we describe the experimental settings used for comparing the effectiveness of our method and various baselines. We differentiate between three distinct settings: in-sample, out-of-sample and closed-loop. For each experiment we report the average value of ℓ_{VO} between the estimated and ground truth motions over the test trajectories, compared to the length of the trajectory. We compare four methods of optimizing the attack to the clean results, by taking the training and evaluation criteria to be either ℓ_{RMS} or ℓ_{MPRMS}. In all cases, we optimize the attacks for $k = 100$ iterations.

In-Sample. The in-sample setting is used to estimate the effect of universal and PGD adversarial perturbations on known data. We train and test our attack on the entire dataset. We then compare the best performing attack to the random and clean baselines, for both our universal and PGD adversarial attacks.

Out-of-Sample. The out-of-sample setting is used to estimate generalization properties of universal perturbations to unseen data. Our methodology in this setting is to first split the trajectories into several folders, each with distinct initial positions of the contained trajectories. Thereafter, we perform cross-validation over the folders, where in each iteration a distinct folder is chosen to be the test set, and another to be the evaluation set. The training set thus comprises of the remaining folders. We report the average results over the test sets. Throughout our experiments, we use 10-fold cross validation.

Closed-Loop. The closed-loop setting is used to estimate the generalization properties of the previously produced adversarial patches to a closed-loop scheme, in which the outputs of the VO model are used in a simple navigation scheme. Our navigation scheme is an aerial path follower based on the carrot chasing algorithm [26]. Given the current pose, target position and cruising speed, the algorithm computes a desired motion toward the target position. We then produce trajectories, each with a distinct initial and target position, with the motions computed iteratively by the navigation scheme based on the provided current position. The ground truth trajectories are computed by providing the current position in each step as the aggregation of motions computed by the navigation scheme. The estimated trajectories for a given patch, clean or adversarial, however, are computed by providing the current position in each step as the aggregation of motions estimated by the VO, where the viewpoint in each step corresponds to the aggregation of motions computed by the navigation scheme. We chose this navigation scheme to further assess the incremental effect of our adversarial attacks, as any deviation in the VO estimations directly affects the produced trajectory.

3.2 VO Model

The VO model used in our experiments is the TartanVO [33], a recent differentiable VO model that achieved state-of-the-art performance in visual odometry benchmarks. Moreover, to better generalize to real-world scenarios, the model was trained over scale-normalized trajectories in diverse synthetic datasets. As the robustness of the model improves on scale-normalized trajectories, we supply it with the scale of the ground truth motions. The assumption of being aware of the motions' scale is a reasonable one, as the scale can be estimated to a reasonable degree in typical autonomous systems from the velocity. In our experiments, we found that the model yielded plausible trajectory estimates over the clean trajectories, for both synthetic and real data.

3.3 Data Generation

Synthetic Data. To accurately estimate the motions using the VO model, we require a photo-realistic renderer. In addition, the whole scene is altered for each camera motion, mandating re-rendering for each frame. An online renderer is impractical for optimization, in terms of computational overhead. An offline renderer is sufficient for our optimization schemes, and only our closed-loop

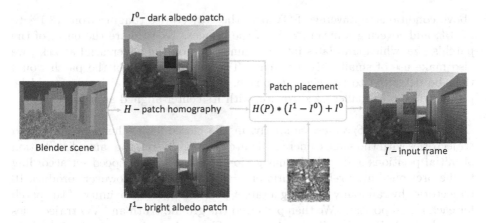

Fig. 2. Synthetic frame generation. The attack patch P is projected via the homography transformation H and is incorporated into the scene according to the albedo images I_0 and I_1.

test requires an online renderer. We, therefore, produce the data for optimizing the adversarial patches offline, and make use of an online renderer only for the closed-loop test. The renderer framework used is Blender [9], a $3D$ modeling and rendering package. Blender enables photo-realistic rendered images to be produced from a given $3D$ scene along with the ground truth motions of the cameras. In addition, we produce high quality, occlusion-aware masks, which are then used to compute the homography transformation H of the patch to the camera viewpoints. In the offline data generation of each trajectory $\{I_t\}_{t=0}^{L}$, we produce $\{I_t^0\}_{t=0}^{L}$, $\{I_t^1\}_{t=0}^{L}$, $\{H_t\}_{t=0}^{L}$, as in Eq. (2), as well as the ground truth camera motions δ_t^{t+1}. For the closed-loop test, for each initial position, target position and pre-computed patch P, we compute the ground truth motions δ_t^{t+1} and their estimation by the VO model $\hat{\delta}_t^{t+1}$, as described in Sect. 3.1. The frame generation process is depicted in Fig. 2. We produced the trajectories in an urban $3D$ scene, as in such a scenario, GPS reception and accuracy is poor, and autonomous systems rely more heavily on visual odometry for navigation purposes. The patch is then positioned on a square plane at the side of one of the buildings, in a manner that resembles a large advertising board.

Offline Rendered Data Specifics. We produced 100 trajectories with a constant linear velocity norm of $v = 5[\frac{m}{s}]$, and a constant $2D$ angular velocity sampled from $v_\theta = \mathcal{N}(0,3)[\frac{deg}{s}]$. Each trajectory is nearly $10[m]$ long and contains 60 frames at 30 fps. The trajectories are evenly divided between 10 initial positions, with the initial positions being distributed evenly on the ring of a right circular cone with a semi-vertical angle of $10°$ and a $50[m]$-long axis aligned with the patch's normal. We used a camera with a horizontal field-of-view (FOV) of $80°$ and 640×448 resolution. The patch is a $30[m]$ square, occupying, under the above conditions, an average FOV over the trajectories ranging from 18.1% to 27.3%, and covering a mean 22.2% of the images. To estimate the effect of the patch's size, which translates into a ℓ_0 limitation on the adversarial attacks, we also make use of smaller sized patches. The outer margins of the patch would then be defaulted to the clean I_0 image, and the adversarial image would be projected onto a smaller sized square, with its center aligned as before.

Closed-Loop Data Specifics. Similarly, in the closed-loop scheme we produced trajectories with the same camera and patch configuration, the same distribution of initial positions and with the navigation scheme cruising speed set according to the previous linear velocity norm of $v = 5[\frac{m}{s}]$. Here we, however, produce 10 trajectories by randomly selecting a target position at the proximity of the patch for each initial position. We then produced the ground truth and VO trajectories for each patch P as described in Sect. 3.1. The trajectories are each $45[m]$ long and contain 270 frames at 30 fps.

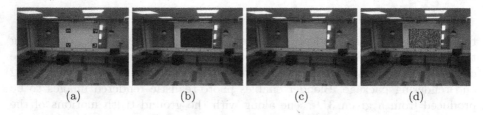

      (a)   (b)   (c)   (d)

Fig. 3. Real dataset frame generation. (a) Original image. (b+c) Black and white albedo approximations. (d) Adversarial patch projected onto the scene.

Real Data. In the real data scenario, we situated a DJI Tello drone inside an indoor arena, surrounded by an Optitrack motion capture system for recording the ground truth motions. The patch was positioned on a planar screen at the arena boundary. To compute the homography transformation H of the patch to the camera viewpoints, we designated the patch location in the scene by four Aruco markers. Similarly to the offline synthetic data generation, for each trajectory $\{I_t\}_{t=0}^{L}$ we produced $\{I_t^0\}_{t=0}^{L}$, $\{I_t^1\}_{t=0}^{L}$, $\{H_t\}_{t=0}^{L}$ as well as the ground truth camera motions δ_t^{t+1}. An example data frame is depicted in Fig. 3.

We produced 48 trajectories with a constant velocity norm of approximately $v \simeq 1[\frac{m}{s}]$. Each trajectory contained 45 frames at 30 fps with total length $l \sim \mathcal{N}(1.56, 0.15^2)[m]$. The trajectories' initial positions were distributed evenly on a plane parallel to the patch at a distance of $7.2[m]$. Not including the drone movement model, the trajectories comprised linear translation toward evenly distributed target positions at the patch's plane. We used a camera with a horizontal FOV of 82.6°, and 640×448 resolution. The patch was a $1.92 \times 1.24[m]$ rectangle, occupying, under the above conditions, an average FOV over the trajectories ranging from 6.8% to 11.2%, and covering a mean 8.8% of the images.

3.4   Experimental Results

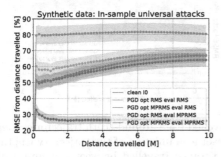 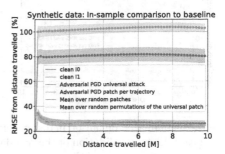

Fig. 4. Accumulated deviation in distance travelled from the ground-truth trajectories on the real dataset as a function of the trajectory length. We show a comparison of our universal attacks trained on the entire dataset (left), and a comparison of our best performing universal and PGD attacks to the clean and random perturbation baselines (right). We present mean and standard deviation over the trajectories for each trajectory length.

Synthetic Data Experiments. In Fig. 4 we show the in-sample results on the synthetic dataset. Both our universal and PGD attacks showed a substantial increase in the generated deviation over the clean and random baselines. The best PGD attack generated, after $10[m]$, a deviation of 103% in distance travelled, which is a factor of 399% from the clean I^0 baseline. For the same configuration, the best universal attack generated a deviation of 80% in distance travelled, which is a factor of 311% from the clean I^0 baseline. Moreover, the clean I^1 and random baselines show a slight decrease in the generated deviation over the clean I^0 results, including the random permutations of the best universal patch. This suggests that the VO model is affected by the structure of the adversarial patch rather than simply by the color scheme. In addition, for both the universal and PGD attacks, the best performance was achieved for $\ell_{train} = \ell_{RMS}$, where the PGD attacks showed negligible change in the choice of evaluation criterion, and $\ell_{eval} = \ell_{RMS}$ is clearly preferred for universal attacks. This supports our assumption that ℓ_{MPRMS} may be less suited for in-sample optimization.

In Fig. 5 we show the out-of-sample results on the synthetic dataset. Our universal attacks again showed a substantial increase in the generated deviation over the clean baseline, with the best universal attack generating, after $10[m]$, a deviation of 61% in distance travelled, which is a factor of 237% from the clean I^0 baseline. As for the choice of criteria, the best performance is achieved for the $\ell_{train} = \ell_{MPRMS}$ optimization criterion with a slight improvement of $\ell_{eval} = \ell_{RMS}$ as the evaluation criterion. This supports our assumption that ℓ_{MPRMS} is better suited for generalization to unseen data, and may indicate that ℓ_{RMS} is better suited for evaluation.

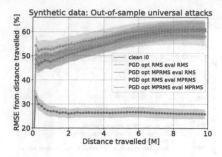 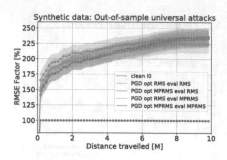

Fig. 5. Accumulated deviation in distance travelled from ground-truth trajectories over out-of-sample cross-validation of the real dataset as a function of the trajectory length. We show a comparison of the deviation in distance travelled between our universal attacks and the clean baseline (left) as well as the ratio of the deviation compared to the clean results (right). We present mean and standard deviation over the trajectories for each trajectory length.

In Fig. 6 we show the patch size comparison of the out-of-sample results on the synthetic dataset. The best performance for all patch sizes is achieved for the same choice of optimization and evaluation criteria, which supports our previous indications. Nevertheless, as the patch size is reduced, the increase in the generated deviation becomes less significant. For the $22.5[m]$ square patch, the best performing universal attack generated, after $10[m]$, a deviation of 48% in distance travelled, which is a factor of 184% from the clean baseline. Regarding the $18.75[m]$ square patch, the generated deviation decays significantly with the best performing universal attack generating, after $10[m]$, a deviation of 31% in distance travelled, which is a factor of 120% from the clean baseline.

In Fig. 7 we show the closed-loop results on the synthetic dataset. Our universal attacks showed an increase in the generated deviation over the clean baseline, which, however, was not as substantial as before as the baseline's generated deviation is already quite significant. The best performing universal attack generated, after $45[m]$, a deviation of 71%, in distance travelled, which is a factor of 112% from the clean baseline. Note that the adversarial patches that were optimized on relatively short trajectories are effective on longer trajectories in the closed-loop scheme, without any fine-tuning. We again see that the best performance is achieved for $\ell_{train} = \ell_{MPRMS}, \ell_{eval} = \ell_{RMS}$.

Real Data Experiments. In Fig. 8 we show the in-sample results on the real dataset. Similarly to the synthetic dataset, we see a substantial improvement for both our universal and PGD attacks over the clean I^0 baseline, while the clean I^1 and random baselines show a slight decrease. The best PGD attack generated, after $1.56[m]$, a deviation of 34% in distance travelled, which is a factor of 231%

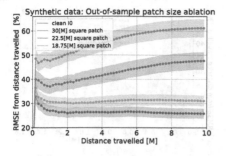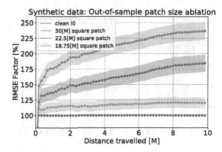

Fig. 6. A comparison of different patch sizes of the accumulated deviation in distance travelled from the ground-truth trajectories over out-of-sample cross-validation of the synthetic dataset as a function of the trajectory length. The patches are a $30[m]$, a $22.5[m]$, and $18.75[m]$ squares and occupy a FOV over the trajectories ranging from $18.1\% - 27.4\%, 8.3\% - 12.6\%, 6.1\% - 9.3\%$ respectively, and covering a mean $22.2\%, 10.2\%, 7.5\%$ of the images. We show a comparison of the deviation in distance travelled between our best performing universal attacks for each patch and the clean baseline (left) as well as the ratio of the deviation compared to the clean results (right). We present mean and standard deviation over the trajectories for each trajectory length.

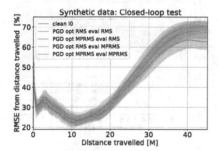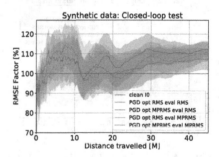

Fig. 7. Accumulated deviation in distance travelled from the ground-truth over closed-loop trajectories of the synthetic dataset as a function of the trajectory length. We show a comparison of the deviation in distance travelled between our universal attacks and the clean baseline (left) as well as the ratio of the deviation compared to the clean results (right). We present mean and standard deviation over the trajectories for each trajectory length.

from the clean I^0 baseline. For the same configuration, the best universal attack generated a deviation of 22% in distance travelled, which is a factor of 150% from the clean I^0 baseline. The increase in the generated deviation is less significant compared to the synthetic dataset, partially due to the smaller patch size as in Fig. 6.

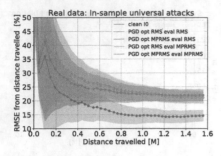
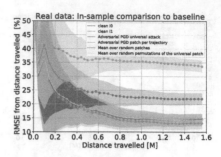

Fig. 8. Accumulated deviation in distance travelled from the ground-truth trajectories on the real dataset as a function of the trajectory length. We show a comparison of our universal attacks trained on the entire dataset (left), and a comparison of our best performing universal and PGD attacks to the clean and random perturbation baselines (right). We present mean and standard deviation over the trajectories for each trajectory length.

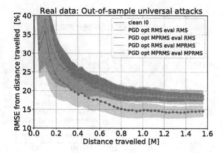
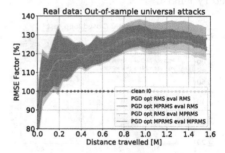

Fig. 9. Accumulated deviation in distance travelled from ground-truth trajectories over out-of-sample cross-validation of the real dataset as a function of the trajectory length. We show a comparison of the deviation in distance travelled between our universal attacks and the clean baseline (left) as well as the ratio of the deviation compared to the clean results (right). We present mean and standard deviation over the trajectories for each trajectory length.

In Fig. 9 we show the out-of-sample results on the real dataset. Our universal attacks again showed an increase in the generated deviation over the clean baseline, with the best universal attack generating, after $1.56[m]$, a deviation of 19% in distance travelled, which is a factor of 128% from the clean I^0 baseline. The best performance is again achieved for the $\ell_{train} = \ell_{MPRMS}$ optimization criterion with negligible difference in the choice of ℓ_{eval}.

4   Conclusions

This paper proposed a novel method for passive patch adversarial attacks on visual odometry-based navigation systems. We used homography of the adversarial patch to different viewpoints to understand how each perceives it and optimize the patch for entire trajectories. Furthermore, we limited the adversarial patch in the ℓ_{inf} and ℓ_0 norms by taking into account the black and white albedo images of the patch and the FOV of the patch.

On the synthetic dataset, we showed that the proposed method could effectively force a given trajectory or set of trajectories to deviate from their original path. For a patch FOV of 22.2%, our PGD attack generated, on a given trajectory, an average deviation, after $10[m]$, of 103% in distance travelled, and given the entire trajectory dataset, our universal attack produced a single adversarial patch that generated an average deviation, after $10[m]$, of 80% in distance travelled. Moreover, our universal attack generated, on out-of-sample data, a deviation, after $10[m]$, of 61% in distance travelled and in a closed-loop setting generated an average deviation, after $45[m]$, of 71% in distance travelled.

In addition, while less substantial, our results were replicated using the real dataset and a significantly smaller patch FOV of 8.8%. Nevertheless, when considering the effect with a larger patch FOV, we can expect a corresponding increase in the generated deviation as we saw in the synthetic dataset. For a given trajectory, our PGD attack generated an average deviation, after $1.56[m]$, of 34% in distance travelled, and our universal attack generated an average deviation, after $1.56[m]$, of 22% in distance travelled given the entire dataset, and on out-of-sample data generated an average deviation, after $1.56[m]$, of 19% in distance travelled.

We conclude that physical passive patch adversarial attacks on vision-based navigation systems could be used to harm systems in both simulated and real-world scenes. Furthermore, such attacks represents a severe security risk as they could potentially push an autonomous system onto a collision course with some object by simply inserting a pre-optimized patch into a scene.

Our results were achieved using a predefined location for the adversarial patch. Optimizing the location of the adversarial patches may produce even more substantial results. For example, Ikram et al. (2022) [17] showed that inserting a simple high-textured patch into specific locations in a scene produces false loop closures and thus degenerates state-of-the-art SLAM algorithms.

Acknowledgements. This project was funded by GRAND/HOLDSTEIN drone technology competition, European Research Council (ERC) under the European Union's Horizon 2020 research and innovation programme (grant agreement No. 863839), The Technion Hiroshi Fujiwara Cyber Security Research Center, and the Israel Cyber Directorate.

References

1. Almalioglu, Y., Turan, M., Saputra, M.R.U., de Gusmão, P.P., Markham, A., Trigoni, N.: SelfVIO: self-supervised deep monocular visual-inertial odometry and depth estimation. Neural Netw. **150**, 119–136 (2022)
2. Amsaleg, L.: High intrinsic dimensionality facilitates adversarial attack: theoretical evidence. IEEE Trans. Inf. Forensics Secur. **16**, 854–865 (2020)
3. Athalye, A., Engstrom, L., Ilyas, A., Kwok, K.: Synthesizing robust adversarial examples. In: Dy, J., Krause, A. (eds.) Proceedings of the 35th International Conference on Machine Learning. Proceedings of Machine Learning Research, PMLR, Stockholmsmässan, Stockholm Sweden, 10–15 July 2018, vol. 80, pp. 284–293 (2018). http://proceedings.mlr.press/v80/athalye18b.html
4. Bian, J., et al.: Unsupervised scale-consistent depth and ego-motion learning from monocular video. In: Advances in Neural Information Processing Systems, vol. 32 (2019)
5. Brown, T.B., Mané, D., Roy, A., Abadi, M., Gilmer, J.: Adversarial patch. arXiv preprint arXiv:1712.09665 (2017). http://arxiv.org/abs/1712.09665
6. Carlini, N., Wagner, D.: Audio adversarial examples: targeted attacks on speech-to-text. In: 2018 IEEE Security and Privacy Workshops (SPW), pp. 1–7. IEEE (2018)
7. Chaturvedi, A., KP, A., Garain, U.: Exploring the robustness of NMT systems to nonsensical inputs. arXiv preprint arXiv:1908.01165 (2019). http://arxiv.org/abs/1908.01165
8. Chawla, H., Varma, A., Arani, E., Zonooz, B.: Adversarial attacks on monocular pose estimation. arXiv preprint arXiv:2207.07032 (2022)
9. Blender Online Community: Blender - a 3D modelling and rendering package. Blender Foundation, Stichting Blender Foundation, Amsterdam (2018). http://www.blender.org
10. Deng, Y., et al.: An analysis of adversarial attacks and defenses on autonomous driving models. In: 2020 IEEE International Conference on Pervasive Computing and Communications (PerCom), pp. 1–10. IEEE (2020)
11. Dube, S.: High dimensional spaces, deep learning and adversarial examples. arXiv preprint arXiv:1801.00634 (2018)
12. Gao, J., Lanchantin, J., Soffa, M.L., Qi, Y.: Black-box generation of adversarial text sequences to evade deep learning classifiers. In: 2018 IEEE Security and Privacy Workshops (SPW), pp. 50–56. IEEE (2018)
13. Gilmer, J., et al.: Adversarial spheres. arXiv preprint arXiv:1801.02774 (2018)
14. Gleave, A., Dennis, M., Wild, C., Kant, N., Levine, S., Russell, S.: Adversarial policies: attacking deep reinforcement learning. In: International Conference on Learning Representations (2020). https://openreview.net/forum?id=HJgEMpVFwB
15. Goodfellow, I.J., Shlens, J., Szegedy, C.: Explaining and harnessing adversarial examples. arXiv preprint arXiv:1412.6572 (2014). http://arxiv.org/abs/1412.6572
16. Hendrik Metzen, J., Chaithanya Kumar, M., Brox, T., Fischer, V.: Universal adversarial perturbations against semantic image segmentation. In: Proceedings of the IEEE International Conference on Computer Vision, pp. 2755–2764 (2017)
17. Ikram, M.H., Khaliq, S., Anjum, M.L., Hussain, W.: Perceptual Aliasing++: adversarial attack for visual slam front-end and back-end. IEEE Robot. Autom. Lett. **7**(2), 4670–4677 (2022)
18. Jin, D., Jin, Z., Zhou, J.T., Szolovits, P.: Is BERT really robust? A strong baseline for natural language attack on text classification and entailment. In: Proceedings of the AAAI Conference on Artificial Intelligence, vol. 34, pp. 8018–8025 (2020)

19. Li, Z., et al.: A simple and strong baseline for universal targeted attacks on Siamese visual tracking. IEEE Trans. Circuit. Syst. Video Technol. **32**(6), 3880–3894 (2021)
20. Macario Barros, A., Michel, M., Moline, Y., Corre, G., Carrel, F.: A comprehensive survey of visual slam algorithms. Robotics **11**(1), 24 (2022)
21. Madry, A., Makelov, A., Schmidt, L., Tsipras, D., Vladu, A.: Towards deep learning models resistant to adversarial attacks. arXiv preprint arXiv:1706.06083 (2017)
22. Mode, G.R., Hoque, K.A.: Adversarial examples in deep learning for multivariate time series regression. In: 2020 IEEE Applied Imagery Pattern Recognition Workshop (AIPR), pp. 1–10. IEEE (2020)
23. Moosavi-Dezfooli, S.M., Fawzi, A., Fawzi, O., Frossard, P.: Universal adversarial perturbations. In: Proceedings of the IEEE Conference on Computer Vision and Pattern Recognition, pp. 1765–1773 (2017)
24. Mur-Artal, R., Montiel, J.M.M., Tardos, J.D.: ORB-SLAM: a versatile and accurate monocular SLAM system. IEEE Trans. Robot. **31**(5), 1147–1163 (2015)
25. Nguyen, A.T., Raff, E.: Adversarial attacks, regression, and numerical stability regularization. arXiv preprint arXiv:1812.02885 (2018)
26. Perez-Leon, H., Acevedo, J.J., Millan-Romera, J.A., Castillejo-Calle, A., Maza, I., Ollero, A.: An aerial robot path follower based on the 'carrot chasing' algorithm. In: Silva, M.F., Luís Lima, J., Reis, L.P., Sanfeliu, A., Tardioli, D. (eds.) ROBOT 2019. AISC, vol. 1093, pp. 37–47. Springer, Cham (2020). https://doi.org/10.1007/978-3-030-36150-1_4
27. Pinkovich, B., Rivlin, E., Rotstein, H.: Predictive driving in an unstructured scenario using the bundle adjustment algorithm. IEEE Trans. Control Syst. Technol. **29**(1), 342–352 (2020)
28. Pony, R., Naeh, I., Mannor, S.: Over-the-air adversarial flickering attacks against video recognition networks. In: Proceedings of the IEEE/CVF Conference on Computer Vision and Pattern Recognition, pp. 515–524 (2021)
29. Salman, H., Ilyas, A., Engstrom, L., Vemprala, S., Madry, A., Kapoor, A.: Unadversarial examples: designing objects for robust vision. In: Advances in Neural Information Processing Systems, vol. 34 (2021)
30. Szegedy, C., et al.: Intriguing properties of neural networks. arXiv preprint arXiv:1312.6199 (2013). http://arxiv.org/abs/1312.6199
31. Triggs, B., McLauchlan, P.F., Hartley, R.I., Fitzgibbon, A.W.: Bundle adjustment — a modern synthesis. In: Triggs, B., Zisserman, A., Szeliski, R. (eds.) IWVA 1999. LNCS, vol. 1883, pp. 298–372. Springer, Heidelberg (2000). https://doi.org/10.1007/3-540-44480-7_21
32. Wang, D., Li, C., Wen, S., Nepal, S., Xiang, Y.: Daedalus: breaking non-maximum suppression in object detection via adversarial examples. arXiv preprint arXiv:1902.02067 (2019). http://arxiv.org/abs/1902.02067
33. Wang, W., Hu, Y., Scherer, S.: TartanVO: a generalizable learning-based VO. arXiv preprint arXiv:2011.00359 (2020)
34. Xiang, C., Qi, C.R., Li, B.: Generating 3D adversarial point clouds. In: The IEEE Conference on Computer Vision and Pattern Recognition (CVPR) (2019)
35. Xu, K., et al.: Evading real-time person detectors by adversarial T-shirt. arXiv preprint arXiv:1910.11099 (2019). http://arxiv.org/abs/1910.11099
36. Yamanaka, K., Matsumoto, R., Takahashi, K., Fujii, T.: Adversarial patch attacks on monocular depth estimation networks. IEEE Access **8**, 179094–179104 (2020)

37. Yang, N., Stumberg, L.V., Wang, R., Cremers, D.: D3VO: deep depth, deep pose and deep uncertainty for monocular visual odometry. In: Proceedings of the IEEE/CVF Conference on Computer Vision and Pattern Recognition, pp. 1281–1292 (2020)
38. Zhang, C., Benz, P., Lin, C., Karjauv, A., Wu, J., Kweon, I.S.: A survey on universal adversarial attack. arXiv preprint arXiv:2103.01498 (2021)

PatchFlow: A Two-Stage Patch-Based Approach for Lightweight Optical Flow Estimation

Ahmed Alhawwary[✉], Janne Mustaniemi, and Janne Heikkilä

Center for Machine Vision and Signal Analysis, University of Oulu, Oulu, Finland
{ahmed.alhawwary,janne.mustaniemi,janne.heikkila}@oulu.fi

Abstract. The deep learning-based optical flow methods have shown noticeable advancements in flow estimation. The dense optical flow map offers high flexibility and quality for aligning neighbouring video frames. However, they are computationally expensive, and the memory requirements for processing high-resolution images such as 2K, 4K and 8K on resources-limited devices such as mobile phones can be prohibitive.

We propose a patch-based approach for optical flow estimation. We redistribute the regular CNN-based optical flow regression into a two-stage pipeline, where the first stage estimates an optical flow for a low-resolution image version. The pre-flow is input to the second stage, where the high-resolution image is partitioned into small patches for optical flow refinement. With such a strategy, it becomes possible to process high-resolution images when the memory requirements are not sufficient. On the other hand, this solution also offers the ability to parallelize the optical flow estimation when possible. Furthermore, we show that such a pipeline can additionally allow for utilizing a lighter and shallower model in the two stages. It can perform on par with FastFlowNet (FFN) while being 1.7x faster computationally and with almost a half of the parameters. Against the state-of-the-art optical flow methods, the proposed solution can show a reasonable accuracy trade-off for running time and memory requirements. Code is available at: https://github.com/ahmad-hammad/PatchFlow.

Keywords: Optic flow · Lightweight CNN · High-resolution video

1 Introduction

Optical flow is a long-standing problem in the computer vision field with ongoing research to date. Given two consecutive frames I1 and I2, the optical flow is per-pixel 2D motion map estimation from the first frame I1 to the next frame I2. It has a vital role in several applications such as video stabilization [36], image stitching [23,24], crowd-counting [10], super resolution [4,31], video interpolation [25], depth estimation [22], SLAM [40] and many robotics applications [19,40].

Supplementary Information The online version contains supplementary material available at https://doi.org/10.1007/978-3-031-26293-7_32.

Conventional optical flow methods [5,9,37] are often formulated as a hand-crafted optimization problem. With the advances in hardware and deep learning research, the deep learning-based optical flow methods have shown noticeable improvements to the previous conventional methods. In addition, deep neural networks harness the parallel processing of modern GPUs. Thus, they surpass the traditional methods in terms of speed and accuracy. Since the first convolutional neural network-based (CNN) method FlowNet [16], many methods have improved the architecture and the training protocols. RAFT [30] is a milestone in optical flow estimation achieving state-of-the-art results. They compute a huge 4D cost volume and update a single high-resolution (1/8) flow iteratively without warping rather than the coarse-to-fine scheme followed by previous works. On the other hand, pyramidal coarse-to-fine models, such as PWCNet [29], have lower computational and memory requirements. However, they perform worse due to the warping process included in each pyramid level. Some subsequent papers [18,33] proposed solution that builds on RAFT's achievement to lower the memory requirements but they are still relatively more complex than pyramidal models. Despite the improvements introduced by the CNN-based solutions for the optical flow task, it remains computationally expensive and memory-hungry.

The prevalence of smartphones equipped with high-end cameras capable of recording high-resolution videos makes it tempting to use them for computational photography applications that fuse information over multiple frames. Because of sudden camera movements those applications including, for example, deblurring, denoising, super-resolution, video stabilization, and 3D reconstruction, often require estimation of dense optical flow. However, regardless of mobile GPUs, smartphones are resource-limited devices, both in computational power and memory resources. Thus, the optical flow computational and memory burden make it less attractive to import such applications on those devices. Moreover, for high-resolution images such as 2K, 4K and 8K images, the computation and memory requirements can be prohibitive.

To this end, we propose a patch-based framework for optical flow estimation. We redistribute flow estimation into two stages, where the first stage estimates a flow for a low-resolution image which serves as an initial flow for the next stage. The second stage estimates flow on small equally-sized patches of the images with the guidance of the low-resolution flow. Such a strategy brings double-edged benefits. On one hand, it removes the high memory requirements for high-resolution images. On the other hand, the patch-based strategy can be leverged to speed up the optical flow estimation by parallel processing for applications that have running time priority. Our contributions can be summarized as follows:

- We propose a two-stage pipeline for patch-based optical flow regression eliminating high memory requirements for high-resolution images, especially when it comes to resource-limited devices such as mobile phones. From another aspect, it enables parallel processing when the memory permits.
- We show reasonable speed-accuracy trade-offs against variety of heavier *full-resolution* optical flow methods.
- We show that the two-stage approach enables us to incorporate a lighter and shallower network in each stage without hurting the accuracy.

2   Related Work

With advances in deep learning research, CNN-based optical flow methods have become superior to the conventional methods in terms of accuracy and inference speed. However, they draw inspiration from the steps followed in the classical pipeline. Here we review the deep learning-based optical flow estimation. FlowNet [7] was the start of the end-to-end, CNN-based optical flow methods that paved the way to improve the optical flow estimation. FlowNet proposed two Unet-shaped architectures: FlownetSimple and FlownetCorr. The main difference is that the first one does not include cost (correlation) volume computation. FlowNet2 [16] builds a heavier and more complex solution by cascading blocks of FlownetCorr and FlownetSimple for optical flow refinement. They also showed that the accuracy results improve by carefully scheduling the dataset training. Most of the CNN-based methods contain three main ingredients: feature extraction, cost volume computation, and feature decoder for optical flow prediction. The supervised CNN-based solutions can be roughly divided into two classes: the warping-based coarse-to-fine (pyramidal refinement) optical flow and the recurrent, non-warping solutions. First, we present the warping-based method.

PWCNet [29] first extracts pyramidal features, then in each pyramid level, it computes a cost volume and predicts a flow which is then refined in the finer level. In SPynet [28], they use image pyramids instead of features. In IRR [15], they turn PWCNet [29] and FlownetSimple [7] to iterative, instead of stacking blocks of large networks similar to FlowNet2 [16]. For IRR-PWCNet, the pyramidal coarse-to-fine architecture is preserved, but it is iterative in the sense that they, unlike PWCNet [29], share the weights of the decoder in each pyramid level. Thus, the number of iterations is limited by the number of levels. IRR improves the accuracy by bidirectional estimation of the flow, jointly estimating the occlusions and the bilateral filtering for more refinement for the optical field [16]. Instead of reshaping cost volume to multi-channel 2D arrays as in PWCnet [29] for instance, VCN [35] processes 4D cost volume with separable 2D kernels for efficient computation and memory consumption showing a significant improvement over previous methods. LiteFlowNet [13] was a concurrent work to PWCNet [29] and shared similar techniques with theirs. It has a smaller number of parameters than PWCNet [29] but requires a higher number of FLOPs (Floating Point Operations) while maintaining comparable accuracy. LiteFlowNet2 [14] further improves the accuracy and running time by optimizing the architecture and training protocols of LiteFlowNet [13]. LiteFlowNet3 [12] improves the accuracy, but incurs more computational complexity, by introducing cost volume modulation and flow field deformation (correction) to better handle the occlusion and homogeneous regions. For cost modulation, they estimate affine parameters that are used to transform each pixel's cost. To refine the flow, they replace the flow vector with a more accurate flow vector from the neighbouring pixels based on a confidence map and self-correlation cost volume (correlating the first image feature map with itself). FastFlownet [19] is based on PWCNet [29] architecture and modifies the three main ingredients to produce a lightweight flow network. They build a lighter feature extractor by replacing the convolution in low-scale levels by pooling layers. For the decoder, they use shuffling layers inspired by Shufflenet [41]. For the cost volume, they compute the correlation in a limited search radius range (local search grid around each pixel) and then resample the correlation grids non-uniformly such that it is dense around the centre and dilated otherwise.

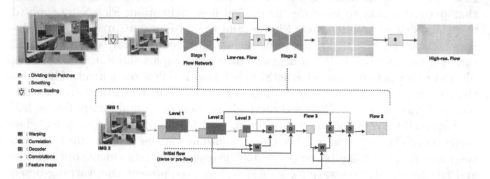

Fig. 1. Architecture of the two-stage optical flow pipeline. In the top of the image, a high-level overview of the full pipeline. The bottom image illustrates the details of the Flow Network incorporated in each stage.

In IOFPL [8], they present ideas to improve the flow estimation in the pyramidal architectures. Among those ideas, they shift from warping to sampling-based cost volume computation in pyramid levels, ameliorating the decline of the performance resulting from warping artefacts. They also propose blocking some gradient components during back-propagation for better convergence and inference performance. Before [8], Devon [26] also proposed sampling-based cost volume. Unlike the coarse-to-fine strategy in [8], they iteratively refine a fixed quarter resolution flow with a shared decoder. To handle the large motion in the fixed resolution, they used dilated search grids for the cost volume computation. Although they perform well on the Sintel [3] clean pass, they perform worse than [8] on the Sintel final pass and KITTI 2015 dataset [27]. RAFT [30] is a remarkable milestone which proposed sampling-based (no-warping), recurrent solution achieving the state-of-the-art (SOTA) results on the optical flow benchmarks when first appeared. They build 4D cost volume by computing the correlation between all pixel pairs on 1/8 resolution feature maps and then pool it recursively to produce multiscale 4d cost volume. The optical flow is iteratively refined at a fixed 1/8 resolution using a GRU-based recurrent module [6]. In each iteration, they look up (sample) the multiscale cost volume with a limited search radius based on the current estimated flow. The RAFT model has high computational and memory requirements. This is mainly due to the huge 4D multiscale volume and many refinement iterations performed on the fixed resolution flow. Assuming an image with equal width and height N, the cost volume has a computation and memory complexity of $O(N^4)$. This limits its ability to scale up to higher resolutions. For instance, a 4K image can cause an out-of-memory error on a GPU with a memory of 32GB [33]. Several subsequent methods [17,18,33,39,42] either optimized the RAFT solution while keeping a similar performance [18,33] or surpassed the performance of RAFT [17,39]. However, those methods still have high computation and memory complexity. In [33], they decrease the memory requirement by separating the 4d cost volume into 2 3D cost volumes, leading to complexity $O(N^3)$ while achieving comparable accuracy. DIP [42] is inspired by the conventional patchmatch approaches [1,2] to deal with the high memory consumption of [30]. There is also a direction that adopts the transformer models for optical flow estimation [11,34].

In the pyramidal approaches, the warping causes occlusion artefacts (copies of the occluding pixels), which may hinder the subsequent levels from picking the correct correspondences, leading to ghosting artefacts [18]. However, the coarse-to-fine often enjoys lower computational and memory requirements than RAFT and its successors. The cost volume in such methods requires memory of complexity $O(N^2R^2)$ where R is the search radius and is much smaller than N. Our work follows the pyramidal coarse-to-fine architecture. We redistribute the optical flow regression into two stages where the first stage estimate low-resolution optical flow and the second stage apply patch-based refinement. The output flow is then smoothed to produce the final result. In each stage, we incorporate a pyramidal network that estimates flow on two pyramid levels only. The resulting pipeline has lower computational complexity and memory consumption while achieving a comparable result with FastFlowNet [19] on a modified FlyingChairs dataset. Our model shows reasonable complexity-accuracy trade-offs against the SOTA methods.

3   Proposed Method

3.1   Overview

We first estimate the optical flow on a low-resolution version which can be computationally cheap compared to the full resolution. For instance, if we have 4K resolution images, we can downsample them to 1/8 and estimate the optical flow. The image pair and the pre-flow are divided into equally-sized patches where each patch is going to be refined independently. The pre-flow serves as an initial flow for the second stage. This pre-flow is utilized to warp the second image features of the coarsest level in the second stage as detailed in Sect. 3.2. After estimating the optical flow for each patch in the second stage, the patches of optical flow are concatenated and then smoothed to produce the final optical flow field. A high-level illustration of the whole pipeline is shown in the upper part of Fig. 1.

3.2   Network Architecture

The network consists of two subnetworks or stages as shown in Fig. 1. The two stages have the same architecture. We present the details of the subnetwork in this section. A subnetwork has three main components: the pyramidal feature extraction encoder, correlation (or cost) volume and optical flow decoder. The network follows the warping-based pyramidal optical flow architectures [13,29].

Pyramidal Feature Extraction. We extract features from the two images separately using a lightweight 3-level pyramidal feature encoder with tied weights as depicted in Fig. 1 in the bottom part. The feature maps in each level are $1/2$ resolution of the previous level. This results in feature maps of $1/2$, $1/4$ and $1/8$ resolutions for the first, second and third levels, respectively. Each level, except the first level, has three convolution blocks and each of them consists of a 2D convolution layer plus an activation layer. The first level has 2 convolution blocks only. The dimension of a pixel's feature vector (*i.e.* , the number of channels in the output feature maps) starting from the first level are 8, 16, and 32 channels in the last level. All the convolution layers have a kernel of size 3×3.

Computing Correlation Volume. Given the feature maps $F1$ and $F2$ of the first image and the second image, respectively, the visual similarity between two feature vectors $F1(x)$ and $F2(x)$ for a pixel located in a 2D spatial position x in the feature maps is computed by the dot product of the two vectors as the following:

$$C(x, r) = F1(x) \cdot F2(x + r)/M \tag{1}$$

where $r \in [R, -R]^2$ is a 2D offset, and R is a positive integer represents the search radius and M is the length of the feature vector. In other words, each pixel (feature vector) from the first image is correlated with a local square centered at position x in the second image and has an area of $D = (2R + 1)^2$. A cost (correlation) volume is built by computing (1) for all pixels. This is then arranged in a volume of size $H_l \times W_l \times D$ [16], where H_l and W_l are the height and width of the input feature maps in level l [16].

Feature Decoder. In level l, the decoder takes as input the concatenation of the cost volume, feature map of the first image and the upsampled flow $f_{l-1}^{\uparrow}$ from the coarser level $l - 1$. It predicts a residual flow Δf_l which is then added to the flow estimate from the coarser level $f_{l-1}^{\uparrow}$ to get the refined estimate as follows:

$$f_l = f_{l-1}^{\uparrow} + \Delta f_l \tag{2}$$

The same decoder architecture is shared across all levels, but each decoder has its own set of learning parameters (*i.e.* no weight sharing) [29]. The decoder consists of seven consecutive convolution layers with a kernel of size 3×3. All convolution layers are followed by an activation layer except the final one which outputs a 2-channel flow field. The first 4 layers output 64-channel feature maps while the subsequent two layers output 32 channels.

Flow Regression. The flow estimation from the subnetworks in our pipeline follows warping-based coarse-to-fine architectures [13, 19, 29]. We describe the process of predicting flow in one scale as the same process applies to all scales. First, we warp the feature map of the second image using the upsampled flow $f_{l-1}^{\uparrow}$ estimated from the coarser scale $l - 1$. The flow from the coarser level is upsampled using a deconvolution layer (transposed convolution) [38]. The warping module is based on a differentiable bilinear interpolation [16,28]. Then, the correlation volume is built by computing the visual similarity between the first image feature map $F_{1,l}$ at level l and the warped second image feature map $F_{2\rightarrow1,l}$. The features $F_{1,l}$ is convolved one more time and then concatenated with the correlation volume and the upsampled flow $f_{l-1}^\uparrow$ and provided to the decoder. The decoder outputs the current estimated flow f_l at level l according to (2) which is then provided to the next finer level $l+1$, and the process repeats again.

In our network, we predict the optical flow for two levels. This means that the final resulted flow is $1/4$ of the original resolution. We upsample the flow to the original resolution by bilinear interpolation.

In the first subnetwork (stage), the initial flow is zeros while in the second subnetwork the initial flow is the pre-flow from the first stage as depicted in Fig. 1.

3.3 Smoothing Filter

The concatenated optical flow may contain discontinuities near the borders of the small patches due to the independent regression of flow for each patch. To deal with these discontinuities, we perform simple smoothing operation by minimizing an energy function that has two terms:

$$E = \sum_i c(i)\|u(i) - f(i)\|_2^2 + \lambda \sum_i |\nabla^2 f(i)|^2 \tag{3}$$

where f and u are the target and the smoothed flow to be estimated, respectively, i is the pixel index, c is the confidence map and λ is the smoothing strength. The second term represents the Laplacian of the flow field. The Laplacian term encourages the flow field to be smooth while the data term encourages the flow to keep its values unchanged based on their confidence. Since all terms are quadratic, this minimization problem can be solved using any linear solver. However, it is a time-consuming process due to the high number of unknowns. Instead, we approximate the task with iterative filtering. First, we determine a Laplacian smoothing kernel θ based on λ. The kernel size gets smaller as λ decreases, meaning a weaker smoothing effect and vice versa. After that, we initially set u to be equal to the target flow field f. Then, in each iteration k, we set $u_{k+1} = g_\theta(cf - (c - o)u_k)$ where g_θ is 2D Convolution operator with kernel θ, and o is the average between the max and min values of c. This process is guaranteed to converge to the minimum of (3) as presented in [32].

3.4 Loss Function

Following [16,29], the optical flow training is supervised using the multiscale $L2$ norm between the ground truth f_j^{GT} and predicted optical flow at the jth scale for all optical flow estimation scales (levels) as follows:

$$\mathcal{L} = \sum_{j=1}^{L} \alpha_j \|f_j^{GT} - f_j\|_2 \tag{4}$$

where L is the number of optical flow prediction scales, and α_j is the loss weight for the jth scale.

4 Experiments

4.1 Training Details

The first stage in our pipeline estimates an optical flow on a downscaled version of the high-resolution image pair. However, the resolution of FlyingChairs (FC) is small (384×512) to be downscaled further. Thus, to train our pipeline, we concatenate (vertically and horizontally) multiple small FlyingChairs images to create a higher resolution image. For instance, if we want to synthesize an image of resolution 2304×2048, we concatenate 24 small FC images. We further apply several augmentation techniques similar to previous works [29,30]. Particularly, we apply geometric transformations such as random rotation, translation, scaling, shear, flipping and cropping. Such transformations effectively diversify the dataset and make the edges or the sharp discontinuities between the concatenated images appear spatially random in the high resolution image. This enables the network to be less influenced by the regular discontinuities during learning. This new modified FlyinChairs dataset (ModFC) is used to train the proposed pipeline. The size of the high-resolution images is 2304×2048 and it is randomly cropped to resolution 1024×1024. The first stage works on $1/8$ resolution (*i.e.* 128×128) while the refinement stage works on patches of size 128×128. We first train the low-resolution stage alone. Then, the second stage is initialized with the learned weights from the first stage, and the two stages are then trained in an end-to-end manner. The initial learning rate is $1e - 4$ and decreased by 0.5 at 108, 144 and 180 epochs. We train for 216 epochs and use batch size 1. In the end-to-end training, the optical flow is predicted over two scales in each stage. Therefore, the training loss weights in (4) are set to be $\alpha_1 = 0.2$, $\alpha_2 = 0.8$, $\alpha_3 = 0.008$, and $\alpha_4 = 0.03125$ where α_2 and α_4 correspond to the coarsest level in the first and the second stage, respectively.

As described in Sect. 3.1, our PatchFlow (PF) consists of two subnetworks with the same architecture, where each subnetwork has its own set of learning parameters, and the pre-flow from the first network is used to warp the second image's feature map in the coarsest level as shown at the bottom of Fig. 1. In the experiments, we compare our proposed pipeline with other options based on two design aspects: weight sharing and warping level. To make it easier for the reader to quickly grasp the training setting type, the 'S', 'H', 'T' and 'I' letters refer to weight sharing, non-shared weights, feature-level warping and image-level warping, respectively. In image-level warping, the pre-flow is used to warp the second image directly rather than its features. To this end, besides our main model, which is referred to by PF-HT, we train two other models: PF-SI and PF-ST.

Table 1. Ablation experiments. All models trained with ModFC. The patch size used in the experiments is 256×256, while the values in the brackets are based on 128×128 patches.

Variation	Sintel - train	
	Clean	Final
Ours- Full	**3.74 (3.93)**	**4.94**
Ours- w/o 1st stage	10.22	10.60
Ours- w/o smoothing	3.74 (3.94)	4.94

Smoothed   No smoothing   Smoothed   No smoothing

Fig. 2. Visualizing the effect of smoothing the optical flow field. The second frame is warped using both smoothed and non-smoothed optical flow fields and overlayed on the first frame by replacing the red channel of the second frame with its counterpart in the first frame. The small coloured crops zoom in the marked regions in the images.

4.2 Ablation Studies

We perform ablation studies on our architecture to validate the importance of some components in our proposed pipeline. We remove the first stage to check the performance when we apply the patching directly without the pre-flow estimation. In Table 1, it is shown that the performance drops significantly without the pre-flow. Also, we show the results when disabling the smoothing component. While the smoothing seems to have no significant improvements quantitatively in terms of the end-point error (EPE) [7], it has a visual effect as shown in Fig. 2. It alleviates the artefacts resulting from the discontinuities between the patches of a non-smoothed flow field used for warping the second image.

4.3 Results on ModFC

As shown in Table 2, the performance of the lighter model LF is comparable with the FFN model when trained on downscaled ModFC images (small resolution). This effectively shows that the model reduction is possible while keeping a similar performance. We noticed that just inferring the low-resolution optical flow using the original FFN model pretrained with FC which is provided by [19], gives a significantly higher EPE error of 9.44. This indicates that the network is influenced by the scale of the images used for training.

Considering the 2-stage pipeline performance, it is shown that the performance of the PF-ST is a bit better than PF-SI. This indicates that the feature-level warping is better than the image-level warping. The warping of the second image to pre-align the image pair before dividing them into patching can result in occlusion artefacts in the warped image. Consequently, this can limit the ability of the refinement network to pick the correct correspondence and create ghosting artefacts. Furthermore, the performance becomes even better when each subnetwork has its own learning parameters as shown with our model PF-HT. This helps each network to focus on learning the scale-related features.

To show that the strategy of the two-stage pipeline offers the ability to utilize a lighter architecture in the refinement stage of image patches, we train the original FFN model with a similar approach to the PF-HT model. Since the first stage estimates 1/32 resolution optical flow ($1/8 \times 1/4$), it makes sense to prune the 1/64 and 1/32 flow estimation scales from the second stage. Moreover, using the first stage flow to warp such low-resolution features (1/64) wastes the information in the pre-flow and makes it almost useless for the refinement stage. In our experiments, if we keep those scales, it produces an EPE of 5.50. Thus, we remove those scales from the second stage. We refer to this as the FFN2 model. Additionally, we report the inference performance of the original FFN model on the full resolution of ModFC (no patching and one stage). From Table 2, our PF-HT has comparable performance with the FFN2 and FFN which demonstrates the possibility of decreasing the computational complexity when adopting the two-stage strategy.

4.4   Sintel and KITTI Datasets

Next, we compare the performance of our two-stage model trained on ModFC with FFN2, FFN and RAFT using the Sintel dataset [3]. The EPE values are presented in Table 2 together with the runtime and the number of FLOPs.

The PF-HT has the best performance among the other versions (PF-ST and PF-SI) similar to the results on the ModFC dataset. On the other hand, the FFN2 model becomes worse than PF-HT. This shows that our model generalizes better. The lower performance might be an indication of overfitting because of the increased model size. Further, we test the original FFN model on the Sintel dataset by stacking two instances of the model. The first instance estimates the optical flow of 1/4 resolution, and the second estimates the residual motion on patches of size 256×256. It is slightly inferior to PF-ST which estimates 1/8 resolution for the pre-flow while being faster and smaller. This shows that reducing the pre-flow and refinement models does not hurt the accuracy.

Additionally, we report the accuracy on the full-resolution of Sintel (without patching) using the FFN and RAFT methods. Our model PF-HT has about 1.5 pixles higher EPE error than the RAFT model (with 12 iterations) while being 49x computationally less expensive. In comparison to FFN, it is about 0.5 pixels inferior in terms of accuracy, but on the other hand, 1.7 times faster. Also, note that both FFN and RAFT operate on the full resolution while our model works

Table 2. Performance comparison on the ModFC test set and Sintel-clean train dataset. LF refers to the model trained with the low-resolution images of the ModFC dataset. FFN2 is the stacking of two original FFN networks similar to the PF-HT model. For more details about FFN2 training, please refer to the text. FFN2* is the stacking of 2 blocks of the original FFN [19] model without training the 2 stages in an end-to-end fashion. The RAFT computation is done with 12 iterations. In those experiments, the Sintel dataset is cropped to a size of 384×1024. For the patch-based variants, we use patches of size 256×256 and 128×128 for Sintel and ModFC, respectively. The running time is measured on Nvidia RTX2070 GPU. The FLOPs and timing are based on resolution 512×1024.

Method	Training data	ModFC test	Sintel clean-train	Time (ms)	FLOPs (G)	Params (M)
Low Resolution						
LF	ModFC	5.86	–	–	–	0.39
FFN	ModFC	5.66	–	–	–	1.37
Patches Only						
FFN	FC	–	3.98	17	29.1	1.37
Two-stage: Low Resolution + Patches						
PF-ST	ModFC	3.49	3.77	10	17.0	0.39
PF-SI	ModFC	3.77	3.78	10	17.0	0.39
PF-HT	ModFC	3.13	3.62	10	17.0	0.78
FFN2	ModFC	3.08	3.87	19	29.2	2.23
FFN2*	FC	–	3.67	22	30.8	1.37
Full Resolution						
FFN	FC	3.09	3.11	17	29.1	1.37
RAFT	FC	–	2.14	227	827.2	5.30

on patches of size 256×256 which causes some loss of contextual information, and makes the flow estimation problem more difficult.

We further finetune our Network (PF-HT) with the FlyingThing3D (FT) dataset and evaluate on the Sintel [3] and KITTI [27] datasets to compare with the SOTA optical flow methods that operate on the full resolution images. We noticed that the FlyingThings3D is more challenging than FlyingChairs and has more motion with many objects occluding each other. We do not get a performance improvement when training the network with a similar strategy as with ModFC. Therefore, we train the network with the original dataset without concatenation and with a cropping size of 512×768. The first stage estimates optical flow on $1/8$ resolution, and the second stage operates on patches of size 256×256. The evaluation on the Sintel and KITTI datasets is shown in Table 3. We also show the parameter count, timing and FLOPs for the different optical flow methods.

As shown in Table 3, finetuning with FT leads to a slight improvement to our model on Sintel (compared to Table 1). However, it keeps similar performance

Table 3. Performance comparison on Sintel and KITTI-2015 datasets after finetuning with FlyingThings3D (FT) dataset. MC refers to ModFC. The RAFT computation is done with 12 iterations. The FLOPs is based on a resolution of 512 × 1024. The timing is done using Nvidia RTX2070, while the timing with an asterisk (after the slash) is estimated on Nvidia 1080Ti from [19]. There are methods that have timing presented on both GPUs, such that the reader can better grasp the idea of the relative performance.

Training data	Method	Sintel (train)		KITTI (train)		Params (M)	Time (ms)	FLOPs (G)
		clean	final	F1-epe	F1-all			
Patches								
MC+FT	Ours	3.64	4.88	15.60	34.57	0.78	10	17.0
Full								
FC+FT	FFN [19]	2.89	4.14	12.24	33.10	1.37	17 /11*	29.1
	LFlownetX [13]	3.58	4.79	15.81	34.90	0.90	–/35*	–
	LFlownet [13]	2.48	4.04	10.39	28.50	5.37	–/55*	327.0
	VCN-small [35]	2.45	3.63	9.43	33.40	5.20	71	73.8
	VCN [35]	2.21	3.62	8.36	25.10	6.20	206	193.0
	Flow1D [33]	1.98	3.27	6.69	22.95	5.73	181	746,2
	SPyNet [28]	4.12	5.57	–	–	1.20	–/50*	299,6
	PWCNet [29]	2.55	3.93	10.35	33.67	8.75	51/34*	187.1
	RAFT-small [30]	2.21	3.35	7.51	26.9	1.0	71	182.2
	RAFT [30]	1.43	2.71	5.40	18.12	5.3	227	827.2

gaps with FFN [19]. The Sintel final pass is more difficult than the clean pass as it includes image degradations such as motion blur, defocus blur, and atmospheric effects [3]. The KITTI dataset is more challenging real-world dataset with large displacements. It becomes even more challenging when estimating the optical flow in patches. With those difficulties, our model shows 1.47% lower accuracy compared to FFN in terms of F1-all metric. Ours is better than SPyNet [28] and comparable with LiteFlownetX [13] while being computationally more efficient. In comparison to LiteFlownet [13], PWCNet [29], RAFT-Small [30], VCN-small, VCN [35], Flow1D [33] and RAFT [30], ours shows reasonable speed-accuracy compromise.

4.5   Real Dataset with Ground-Truth Optical Flow

We also test our model on GyroFlow real dataset (GOF) [20] that has ground-truth (GT) annotations for the optical flow. The resolution of the images is 800 × 600 and there are four types of scenes: regular, fog, rain, and dark. In Table 4, we report the inference results of PF-HT, RAFT and FFN on regular scenes only. We use patches of size 256 × 256 for PF-HT. As shown in Table 4, PF-HT has a similar performance as the FFN model, while not being far from the SOTA method RAFT with 12 iterations. Figure 3 shows qualitative comparisons between our 2-stage pipeline and the FFN and RAFT methods.

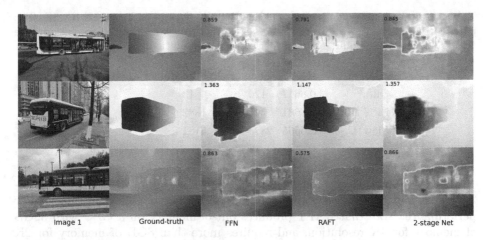

Fig. 3. Qualitative results on GOF dataset [20] for FFN, RAFT, and our 2-stage pipeline. The values indicate the EPE for each example.

Table 4. Performance comparison on GOF clean dataset in terms of EPE. For the 2-stage pipeline, the EPE value is based on patches of size 256×256, while the value in the brackets is for patches of size 128×128. Both FFN and RAFT (12 iters.) models have the full resolution as input.

Method	EPE
Ours	1.09 (1.09)
FFN	1.08
RAFT	0.78

4.6 Real Dataset Without Ground-Truth

In addition to the GOF dataset, we collected a dataset of high resolution images with 2K (1080×1920) and 4K (2048×3840) resolutions using Huawei P40 Pro phone. The dataset consists of scenes where almost all the motion is due to the camera movements. Overall, the dataset contains 8 video sequences (six 2K videos and two 4K videos). We call this dataset RealP40. Since the dataset does not contain GT optical flow annotations, we use Root Mean Squared Error (RMSE) for evaluation [21]. It is defined as one minus the normalized cross-correlation (NCC) of two pixels in neighbourhood π of size 5×5:

$$RMSE(I_1, I_2) = \sqrt{\frac{1}{N_v} \sum_{\pi} (1 - NCC(p1, p2))} \qquad (5)$$

where N_v is the total number of valid pixels in $I1$ and $I2$ (*i.e.* ignoring the black area resulted from the out-of-boundary pixels when warping $I2$), and $p1$ and $p2$ are the corresponding pixels in I1 and I2, respectively.

Table 5. Performance comparison on RealP40-2K dataset in terms of RMSE. The patch size used is 128×128, while FFN and Flow1D process the full resolution. The timing and memory requirements are based on 4K resolution.

Method	RMSE	Time (ms)	Min Memory (MB)
Ours	0.6642	120	13
FFN [19]	0.6681	213	1882
Flow1D [33]	0.6665	2885	5181

The evaluation of our model against FFN [19] and Flow1D [33] on RealP40-2K is shown in Table 5. For a 4K resolution, ours is faster while achieving similar performance. Note that RAFT produces out-of-memory on a GPU with 32GB of memory for 4K resolution, and requires more than 8GB of memory for 2K images (larger than the 8GB memory on our RTX2070) due to the huge cost volume it computes. RAFT has an alternative implementation where they compute the cost volume on-demand basis. However, it is slow in turn. Qualitative comparisons between our PF and FFN [19] on 4K videos are provided in the supplemental material. An advantage of the patch-based approach is that it requires memory as small as the patch size. For instance, suppose that we have only 1 GB in a GPU memory (such as in mobile phones or micro-controllers for example), and assuming that the memory can take up to 10 patches at a time of size 256×256 pixels from the 120 patches of a 4K image, the device can process the whole image in 12 sequential iterations. At the same time, the models processing the full resolution image can not work because of the limited memory resources. In Table 5, we show the minimum memory required (*i.e.* the peak memory consumed during inference) to process an image by a corresponding method. Ours has lower memory requirements in comparison to [19,33].

5   Conclusion

We proposed a two-stage pipeline for optical flow estimation of high-resolution images. The first stage takes as input a low-resolution version of the image pair. The pre-flow from the first stage is then fed to the second stage and utilized for warping the target frame feature patches. The second stage estimates the residual motion on the small patches of the input images. We showed that such pre-flow and refinement helps building a shallower model while achieving a comparable result with the FFN model. Thanks to the patch-based approach, the high-resolution image can be processed in limited memory and computational resources like mobile phones. The comparison with SOTA solution RAFT on Sintel and GOF datasets showed that the 2-stage solution can offer a reasonable accuracy while being significantly computationally lighter and more feasible for high-resolution images.

References

1. Barnes, C., Shechtman, E., Finkelstein, A., Goldman, D.B.: PatchMatch: a randomized correspondence algorithm for structural image editing. ACM Trans. Graph. **28**(3), 24 (2009)
2. Bleyer, M., Rhemann, C., Rother, C.: PatchMatch stereo-stereo matching with slanted support windows. In: BMVC, vol. 11, pp. 1–11 (2011)
3. Butler, D.J., Wulff, J., Stanley, G.B., Black, M.J.: A naturalistic open source movie for optical flow evaluation. In: Fitzgibbon, A., Lazebnik, S., Perona, P., Sato, Y., Schmid, C. (eds.) ECCV 2012. LNCS, vol. 7577, pp. 611–625. Springer, Heidelberg (2012). https://doi.org/10.1007/978-3-642-33783-3_44
4. Caballero, J., et al.: Real-time video super-resolution with spatio-temporal networks and motion compensation. In: Proceedings of the IEEE Conference on Computer Vision and Pattern Recognition, pp. 4778–4787 (2017)
5. Chen, Q., Koltun, V.: Full flow: optical flow estimation by global optimization over regular grids. In: Proceedings of the IEEE Conference on Computer Vision and Pattern Recognition, pp. 4706–4714 (2016)
6. Cho, K., Van Merriënboer, B., Bahdanau, D., Bengio, Y.: On the properties of neural machine translation: encoder-decoder approaches. arXiv preprint arXiv:1409.1259 (2014)
7. Dosovitskiy, A., et al.: FlowNet: learning optical flow with convolutional networks. In: Proceedings of the IEEE International Conference on Computer Vision, pp. 2758–2766 (2015)
8. Hofinger, M., Bulò, S.R., Porzi, L., Knapitsch, A., Pock, T., Kontschieder, P.: Improving optical flow on a pyramid level. In: Vedaldi, A., Bischof, H., Brox, T., Frahm, J.-M. (eds.) ECCV 2020. LNCS, vol. 12373, pp. 770–786. Springer, Cham (2020). https://doi.org/10.1007/978-3-030-58604-1_46
9. Horn, B.K., Schunck, B.G.: Determining optical flow. Artifi. Intell. **17**(1–3), 185–203 (1981)
10. Hossain, M.A., Cannons, K., Jang, D., Cuzzolin, F., Xu, Z.: Video-based crowd counting using a multi-scale optical flow pyramid network. In: Proceedings of the Asian Conference on Computer Vision (2020)
11. Huang, Z., et al.: FlowFormer: a transformer architecture for optical flow. arXiv preprint arXiv:2203.16194 (2022)
12. Hui, T.-W., Loy, C.C.: LiteFlowNet3: resolving correspondence ambiguity for more accurate optical flow estimation. In: Vedaldi, A., Bischof, H., Brox, T., Frahm, J.-M. (eds.) ECCV 2020. LNCS, vol. 12365, pp. 169–184. Springer, Cham (2020). https://doi.org/10.1007/978-3-030-58565-5_11
13. Hui, T.W., Tang, X., Loy, C.C.: LiteFlownet: a lightweight convolutional neural network for optical flow estimation. In: Proceedings of the IEEE Conference on Computer Vision and Pattern Recognition, pp. 8981–8989 (2018)
14. Hui, T.W., Tang, X., Loy, C.C.: A lightweight optical flow CNN-revisiting data fidelity and regularization. IEEE Trans. Pattern Anal. Mach. Intell. **43**(8), 2555–2569 (2020)
15. Hur, J., Roth, S.: Iterative residual refinement for joint optical flow and occlusion estimation. In: Proceedings of the IEEE/CVF Conference on Computer Vision and Pattern Recognition, pp. 5754–5763 (2019)
16. Ilg, E., Mayer, N., Saikia, T., Keuper, M., Dosovitskiy, A., Brox, T.: FlowNet 2.0: evolution of optical flow estimation with deep networks. In: Proceedings of the IEEE Conference on Computer Vision and Pattern Recognition, pp. 2462–2470 (2017)

17. Jiang, S., Campbell, D., Lu, Y., Li, H., Hartley, R.: Learning to estimate hidden motions with global motion aggregation. In: Proceedings of the IEEE/CVF International Conference on Computer Vision, pp. 9772–9781 (2021)
18. Jiang, S., Lu, Y., Li, H., Hartley, R.: Learning optical flow from a few matches. In: Proceedings of the IEEE/CVF Conference on Computer Vision and Pattern Recognition, pp. 16592–16600 (2021)
19. Kong, L., Shen, C., Yang, J.: FastFlownet: a lightweight network for fast optical flow estimation. In: 2021 IEEE International Conference on Robotics and Automation (ICRA), pp. 10310–10316. IEEE (2021)
20. Li, H., Luo, K., Liu, S.: GyroFlow: gyroscope-guided unsupervised optical flow learning. In: Proceedings of the IEEE/CVF International Conference on Computer Vision, pp. 12869–12878 (2021)
21. Li, S., Yuan, L., Sun, J., Quan, L.: Dual-feature warping-based motion model estimation. In: Proceedings of the IEEE International Conference on Computer Vision, pp. 4283–4291 (2015)
22. Li, Z., et al.: Learning the depths of moving people by watching frozen people. In: Proceedings of the IEEE/CVF Conference on Computer Vision and Pattern Recognition, pp. 4521–4530 (2019)
23. Lin, K., Jiang, N., Liu, S., Cheong, L.F., Do, M., Lu, J.: Direct photometric alignment by mesh deformation. In: Proceedings of the IEEE Conference on Computer Vision and Pattern Recognition, pp. 2405–2413 (2017)
24. Liu, S., Yuan, L., Tan, P., Sun, J.: SteadyFlow: spatially smooth optical flow for video stabilization. In: Proceedings of the IEEE Conference on Computer Vision and Pattern Recognition, pp. 4209–4216 (2014)
25. Liu, Y.L., Lai, W.S., Yang, M.H., Chuang, Y.Y., Huang, J.B.: Hybrid neural fusion for full-frame video stabilization. In: Proceedings of the IEEE/CVF International Conference on Computer Vision, pp. 2299–2308 (2021)
26. Lu, Y., Valmadre, J., Wang, H., Kannala, J., Harandi, M., Torr, P.: Devon: Deformable volume network for learning optical flow. In: Proceedings of the IEEE/CVF Winter Conference on Applications of Computer Vision, pp. 2705–2713 (2020)
27. Menze, M., Geiger, A.: Object scene flow for autonomous vehicles. In: Proceedings of the IEEE Conference on Computer Vision and Pattern Recognition, pp. 3061–3070 (2015)
28. Ranjan, A., Black, M.J.: Optical flow estimation using a spatial pyramid network. In: Proceedings of the IEEE Conference on Computer Vision and Pattern Recognition, pp. 4161–4170 (2017)
29. Sun, D., Yang, X., Liu, M.Y., Kautz, J.: PWC-NET: CNNs for optical flow using pyramid, warping, and cost volume. In: Proceedings of the IEEE Conference on Computer Vision and Pattern Recognition, pp. 8934–8943 (2018)
30. Teed, Z., Deng, J.: RAFT: recurrent all-pairs field transforms for optical flow. In: Vedaldi, A., Bischof, H., Brox, T., Frahm, J.-M. (eds.) ECCV 2020. LNCS, vol. 12347, pp. 402–419. Springer, Cham (2020). https://doi.org/10.1007/978-3-030-58536-5_24
31. Wang, L., Guo, Y., Liu, L., Lin, Z., Deng, X., An, W.: Deep video super-resolution using HR optical flow estimation. IEEE Trans. Image Process. **29**, 4323–4336 (2020)
32. Wen, R., Zhao, P.: A medium-shifted splitting iteration method for a diagonal-plus-Toeplitz linear system from spatial fractional schrödinger equations. Bound. Value Prob. **2018**(1), 1–17 (2018)

33. Xu, H., Yang, J., Cai, J., Zhang, J., Tong, X.: High-resolution optical flow from 1D attention and correlation. In: Proceedings of the IEEE/CVF International Conference on Computer Vision, pp. 10498–10507 (2021)

34. Xu, H., Zhang, J., Cai, J., Rezatofighi, H., Tao, D.: GMFlow: learning optical flow via global matching. In: Proceedings of the IEEE/CVF Conference on Computer Vision and Pattern Recognition, pp. 8121–8130 (2022)

35. Yang, G., Ramanan, D.: Volumetric correspondence networks for optical flow. In: Advances in Neural Information Processing Systems 32 (2019)

36. Yu, J., Ramamoorthi, R.: Learning video stabilization using optical flow. In: Proceedings of the IEEE/CVF Conference on Computer Vision and Pattern Recognition, pp. 8159–8167 (2020)

37. Zach, C., Pock, T., Bischof, H.: A duality based approach for realtime TV-L^1 optical flow. In: Hamprecht, F.A., Schnörr, C., Jähne, B. (eds.) DAGM 2007. LNCS, vol. 4713, pp. 214–223. Springer, Heidelberg (2007). https://doi.org/10.1007/978-3-540-74936-3_22

38. Zeiler, M.D., Krishnan, D., Taylor, G.W., Fergus, R.: Deconvolutional networks. In: 2010 IEEE Computer Society Conference on Computer Vision and Pattern Recognition, pp. 2528–2535. IEEE (2010)

39. Zhang, F., Woodford, O.J., Prisacariu, V.A., Torr, P.H.: Separable flow: learning motion cost volumes for optical flow estimation. In: Proceedings of the IEEE/CVF International Conference on Computer Vision, pp. 10807–10817 (2021)

40. Zhang, T., Zhang, H., Li, Y., Nakamura, Y., Zhang, L.: FlowFusion: dynamic dense RGB-D SLAM based on optical flow. In: 2020 IEEE International Conference on Robotics and Automation (ICRA), pp. 7322–7328. IEEE (2020)

41. Zhang, X., Zhou, X., Lin, M., Sun, J.: ShuffleNet: an extremely efficient convolutional neural network for mobile devices. In: Proceedings of the IEEE Conference on Computer Vision and Pattern Recognition, pp. 6848–6856 (2018)

42. Zheng, Z., et al.: DIP: deep inverse patchmatch for high-resolution optical flow. In: Proceedings of the IEEE/CVF Conference on Computer Vision and Pattern Recognition, pp. 8925–8934 (2022)

Tracking Small and Fast Moving Objects: A Benchmark

Zhewen Zhang, Fuliang Wu, Yuming Qiu, Jingdong Liang, and Shuiwang Li[✉]

Guilin University of Technology, Guilin 541006, China
{wufuliang,liangjingdong}@glut.edu.cn, lishuiwang0721@163.com

Abstract. With more and more large-scale datasets available for training, visual tracking has made great progress in recent years. However, current research in the field mainly focuses on tracking generic objects. In this paper, we present TSFMO, a benchmark for **T**racking **S**mall and **F**ast **M**oving **O**bjects. This benchmark aims to encourage research in developing novel and accurate methods for this challenging task particularly. TSFMO consists of 250 sequences with about 50k frames in total. Each frame in these sequences is carefully and manually annotated with a bounding box. To the best of our knowledge, TSFMO is the first benchmark dedicated to tracking small and fast moving objects, especially connected to sports. To understand how existing methods perform and to provide comparison for future research on TSFMO, we extensively evaluate 20 state-of-the-art trackers on the benchmark. The evaluation results exhibit that more effort are required to improve tracking small and fast moving objects. Moreover, to encourage future research, we proposed a novel tracker S-KeepTrack which surpasses all 20 evaluated approaches. By releasing TSFMO, we expect to facilitate future researches and applications of tracking small and fast moving objects. The TSFMO and evaluation results as well as S-KeepTrack are available at https://github.com/CodeOfGithub/S-KeepTrack.

Keywords: Visual tracking · Small and fast moving objets · Benchmark

1 Introduction

Object tracking is one of the most fundamental problems in computer vision with a variety of applications, including video surveillance, robotics, human-machine interaction, motion analysis and so forth [40–42]. Great progress has been witnessed in object tracking thanks to the successful application of deep learning to the field in recent years [29]. Despite considerable progress in the field, current researches mainly focus on tracking generic objects, while very little attention is paid to tracking small and fast moving objects. Nevertheless, small and fast moving objects are common to see in the real world. Many of them are close connection to sports. Tracking of them is crucial to meet the practical

---

Z. Zhang and F. Wu—These authors contributed equally.

L. Wang et al. (Eds.): ACCV 2022, LNCS 13847, pp. 552–569, 2023.
https://doi.org/10.1007/978-3-031-26293-7_33

and accuracy requirements of motion analysis for sports [10,36], and to provide a fairer measure of performance than that provided by human judges, and to spare referees, umpires or judges from immense pressure in making accurate split-second decisions [31,57]. It is also the key to develop automatic sports video recording and recommendation systems [28,65]. However, tracking small and fast moving objects are more challenging. There are several reasons account for this. First, small objects in sports are usually balls or objects of regular geometric shapes. Discriminative information provided by their shapes are fairly limited. Without sufficient discriminative cue of the target, existing tracking algorithms are prone to failure. Second, these objects are covered with plain or regular patterns, making the identifying of them in complex scenes very difficult, as they may be treated as parts of the background. Third, they are frequently moving at a high speed, which may cause severe motion blur in images, and they may change moving direction abruptly when at collision or being hit.

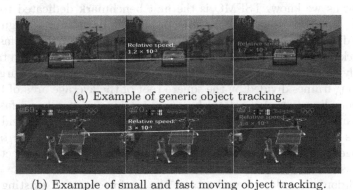

(a) Example of generic object tracking.

(b) Example of small and fast moving object tracking.

Fig. 1. Generic object tracking (a) and small and fast moving object tracking (b). In comparison, tracking of small and fast moving objects is more challenging as the objects are visually much smaller and the relative speeds are much higher.

Table 1. Comparison of average target size (in $pixel^2$) and average relative target speed (in $1/pixel$) of object tracking benchmarks.

	OTB [63]	Got-10K [26]	LasoT [18]	TrackingNet [53]	UAV123 [52]	TSFMO (ours)
Avg. target size ($\times 10^3$)	6.73	228.97	56.17	45.65	2.48	0.51
Avg. relative speed ($\times 10^{-1}$)	2.02	2.64	30.21	1.97	7.74	58.28

In addition to the above technical difficulties, another important reason that tracking small and fast moving objects is hardly touched is the lack of public available benchmarks, which are undoubtedly crucial to attack the problem and to advance the field as without which researchers are unable to effectively design and evaluate novel algorithms for improvement. Despite that there exist plenty of benchmarks for generic object tracking [18,21,26,35,44,53,59,63], there is no benchmark dedicated to tracking small and fast moving objects, especially small objects in sports. Although many of existing benchmarks consist of small and fast moving objects, the numbers of both sequences and object classes are very limited. Tracking algorithms developed to target these benchmarks are generic but not effective in dealing challenges posed by small and fast moving objects. To facilitate research on tracking small and fast moving objects, in this paper we present a dedicated dataset to serve as the testbed for fair evaluation.

1.1   Contribution

In this work, we make the first attempt to explore tracking small and fast moving objects by introducing the TSFMO benchmark for Tracking Small and Fast Moving Objects. TSFMO is made up of a diverse selection of 26 classes of sports with each containing multiple sequences. TSFMO consists of a total of 250 sequences and about 50k frames. Each sequence is manually annotated with axis-aligned bounding boxes with different attributes for performance evaluation and analysis. As far as we know, TSFMO is the first benchmark dedicated to the task of tracking small and fast moving objects, especially in sports. Figure 1 illustrates the differences between generic object tracking and small and fast moving object tracking. Compared with generic object tracking in which the target size is relatively larger and the relative speed (the displacement of the target in two neighbouring frames divided by the square root of the average area of the neighbouring bounding boxes) is relatively lower, tracking of small and fast moving objects is more challenging as the objects are visually very small and the relative speeds are much higher. A quantitative comparison of average target size and average relative target speed between four public generic object tracking benchmarks and five TSFMO is shown in Table 1.

In addition, in order to understand the performance of existing tracking algorithms and provide comparisons for future research on TSFMO, we extensively evaluated 20 state-of-the-art tracking algorithms on TSFMO. Meanwhile we conducted an in-depth analysis of the evaluation results and observed several surprising findings. First, the tracking performances of existing trackers are much lower in TSFMO than in most public benchmarks for generic object tracking, which suggests that most existing tracking methods may overlook very important factors so that they are not effective in tracking small and fast moving objects. Second, not all latest trackers whose performances rank high on OTB [63], Got-10K [26], LasoT [18], and TrackingNet [53] are highly-ranked on TSFMO, which suggests that the generalization ability of some existing trackers is questionable. So, as an unexplored or less studied problem, improving the generalization ability of tracking algorithms is valuable and interesting. These above observations imply the need to develop tracking algorithms devoted to tracking small and fast moving objects, which may also stimulate more generalizable tracking algorithms in the future.

Last but not the least, we introduce a baseline tracker in order to facilitate the development of tracking algorithms on TSFMO. The proposed tracker is based on KeepTrack [49] given that it shows the best performance among state-of-the-art trackers to be evaluated here. In view of that small objects may not have representation in deeper layers' features because of the larger receptive field in deeper layers and that combining low and high level features to boost performance has been extensively studied in tiny object detection [22,25,32], we modify the architecture of KeepTrack [49] so that low-level features are exploited to improve tracking performance. This results in the proposed tracker, which is called S-KeepTrack. The proposed S-KeepTrack outperforms all 20 state-of-the-art trackers on TSFMO. In summary, we have made the following contributions:

- We propose TSFMO, which is, as far as we know, the first benchmark dedicated to track small and fast moving objects, especially in sports.
- We evaluate 20 state-of-the-art tracking algorithms with in-depth analysis to assess their performance and provide comparisons on TSFMO.
- We develop a baseline tracker S-KeepTrack base on the KeepTrack [49] to encourage further research on TSFMO.

2   Related Works

2.1   Visual Tracking Algorithms

Visual tracking has been studied for decades with a huge literature, the comprehensive review of which is out of scope of this paper. In this section, we review two popular trends including discriminative correlation filter (DCF)-based tracking and deep learning (DL)-based tracking in the field and refer readers to [46,50] for comprehensive surveys.

Roughly stated, DCF-based trackers treat visual tracking as an online regression problem. Thanks to the Parseval theorem and the Fast Fourier Transform (FFT), DCF-based tracker can be effectively evaluated in the frequency domain and demonstrate impressive CPU speeds [41]. They started with the minimum output sum of squared error (MOSSE) filter [5]. After that great advance has been witnessed in DCF-based trackers [41]. For instance, an additional scale filter is exploited in [13,43] to deal with target scale variations. The trackers in [12,39] leverage regularization techniques to improve robustness. The approach in [40] generalize the DCF to achieve translation equivariance. The methods in [15,48] utilize deep features instead of handcrafted ones in correlation filter tracking and achieve significant improvements.

The great success of deep learning in other vision tasks motivated the mushrooming development of DL-based trackers in visual tracking in recent years. As one of the pioneering works, SiamFC [2] considered visual tracking as a general similarity-learning problem and took advantage of the Siamese network [9] to measure the similarity between target and search image. Since then, many DL-based trackers base on Siamese architectures have been proposed [24,38,64] and the tracking performances have been significantly improved. Along another line, visual tracking is divided into two sub-tasks, i.e., localization and scale estimation, which are solved, respectively, by an online classifier and an offline intersection-over-union (IoU) network [3,11,14,49].

2.2   Visual Tracking Benchmarks

As standards by which the performances of tracking methods are measured or judged, tracking benchmarks, undoubtedly, are crucial for the development of visual tracking. Existing benchmarks can be roughly divided into two types: generic benchmarks and specific benchmarks [19].

Generic Benchmarks. A generic tracking benchmark is usually designed for tracking objects in general scenes. OTB-2013 [63] is the first generic benchmark with 50 sequences and later extended to OTB-2015 with 100 sequences. TC-128 [44] consists of 128 colorful sequences, and is used to study the impact of color information on tracking performance. VOT [35] organizes a series of tracking competitions with up to 60 sequences. NfS [21] concerns about videos of high frame rate. NUS-PRO [37] collects 365 videos and primarily addresses tracking rigid objects. TracKlinic [20] includes 2,390 videos to evaluate tracking algorithms under various challenges. Recently, many large-scale benchmarks have been proposed to provide training data for developing DL-based trackers. OxUvA [59] provides 366 videos aiming for long-term tracking in the wild. TrackingNet [53] collects a large-scale dataset consisting of more than 30K sequences for deep tracking. GOT-10k [26] provides 10K sequences with rich motion trajectories. LaSOT offers 1,400 long-term videos in [18] and later introduces additional 150 sequences and a new evaluation protocol for unseen objects in [17].

Specific Benchmarks. In addition to generic visual tracking benchmarks, there exist specific benchmarks for particular goals. UAV123 [52] consists of 123 sequences captured by unmanned aerial vehicle (UAV) for low altitude UAV target tracking. VOT-TIR [33] is from VOT and focuses on object tacking in RGB-T sequences, aiming at taking advantage of RGB and thermal infrared images simultaneously. CDTB [47] and PTB [56] are designed to assess tracking performance on RGB-D videos, D indicating depth images. TOTB [19] collects 225 videos from 15 transparent object categories and focus on transparent objects.

Despite of the availability of the above benchmarks, they mainly focus on tracking objects of common sizes and relatively slow speeds. Tracking of small and fast moving objects, especially in sports, has received very little attention. The most important reason, we think, is the lack of public available benchmarks, which motivates our proposal of TSFMO.

2.3   Dealing with Small and Fast Moving Objects in Vision

Small objects here refer to objects with smaller physical sizes in the real world and occupying areas less than and equal to 32×32 pixels [58], while fast moving objects refer to the ones that may move over a distance exceeding its size within the exposure time [54]. Small and fast moving objects are common to see in the real-world, and a significant amount of researches have been devoted to deal with them. For example, the methods of [30,51] studied the problem of small object detection utilizing hand-engineered features and shallow classifiers in aerial images. The approach in [27] combined detection and tracking and integrated them into an adaptive particle filter to handle small object localization, but it was evaluated on mere two testing videos for case study. To the best of our knowledge, Chen et al. [7] are perhaps the first to introduce a small object detection (SOD) dataset, an evaluation metric, and provide a baseline score in order to explore small object detection. The work of [1] presented an algorithm for detecting and tracking small dim targets in Infrared (IR) image sequences base on the frequency and spatial domain information. The work of [54] presented a method for detecting and tracking fast moving objects and provided a new dataset consisting of 16 sports videos for evaluation. In [45], an aggregation signature was proposed for small object tracking and 112 sequences were collected for evaluation. The method in [72] implemented a segmentation network that performs near real-time detection and tracking of fast moving objects and introduced a synthetic physically plausible fast moving object sequence generator for training purpose. The method of [67] investigated the problem of small and fast moving object detection and tracking in sports video sequences, using only motion as a cue for detection and multiple filter banks for tracking.

Our work is related to [1, 27, 45, 54, 67, 72] but different in: (1) TSFMO focuses on tracking small and fast moving objects, while other works concentrate on either tracking small objects [1, 27, 45], or tracking fast moving objects [54, 72]. (2) Although both focus on tracking small and fast moving objects, TSFMO provides a diverse benchmark of hundreds challenging sequences for evaluation, while in [67] only a small number of sequences is provided, which are captured with a stationary camera under indoor conditions where the background remains stationary. Such data is way far from real applications.

3 Tracking Small and Fast Moving Objects

3.1 Video Collection

We select 26 small and fast-moving object categories to construct TSFMO, including badminton, baseball, basketball, beach volleyball, bowling, boxing, curling, discus, football, gateball, golf, hammer, handball, ice hockey, indoor football, kick volleyball, pingpong, polo, ruby, shot, shuttlecock ball, snooker, squash, tennis, volleyball and water polo. Figure 3 demonstrates some sample sequences from these categories.

After determining the object categories, we search for raw sequences of each class from the Internet, as it is the source of many tracking benchmarks (e.g., LaSOT [17], GOT-10k [26], TrackingNet [53], etc.). Initially, we collected at least 10 raw videos for each class and gathered more than 280 sequences in total. We then carefully inspect each sequence for its availability for tracking and drop the undesirable sequences. Afterwards, we verify the content of each raw sequence and remove the irrelevant parts to obtain a video clip that is suitable for tracking. We intentionally limit the number of frames in each video to 900 frames, which is enough for accessing the tracker's performance on tracking small and fast-moving objects, meanwhile manageable for annotation. Eventually, TSFMO is made up of 250 sequences from 26 object classes with about 50K frames. Table 2 summarizes TSFMO, and Fig. 2 shows the average sequence length for each object category in TSFMO.

3.2 Annotation

For sequence annotation, we follow the principle used in [14]: given an initial object, for each frame in the sequence, the annotator draws/edits an axis-aligned bounding box as the tightest bound box to fit any visible part of the object if the object appears; otherwise, either an absence label or out of view (OV) or full occlusion (FOC) is assigned to the frame.

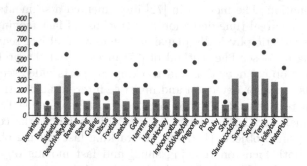

Fig. 2. Average video length for each object class in TSFMO. The red and green dots indicate the minimum and maximum frame numbers of each category. (Color figure online)

Table 2. Summary of statistics of the proposed TSFMO.

Number of videos	250	Min frames	16	Frame rate	≤30 fps	Max frames	887
Total frames	49k	Avg frames	196	Object categories	26	Avg duration	7.4 s

Adhering to the above principle, we finish the annotation in three steps, i.e., manual annotation, visual inspection, and box refinement. In the first step, each video was labelled by an expert, i.e., a student working on tracking. As annotation errors or inconsistencies is hardly avoidable in the first stage, a visual inspection is performed to verify the annotations in the second stage, which is conducted by a validation team. If the validation team do not agree on the annotation unanimously, it will be sent back to the original annotator for refinement in the third step. This three-step strategy ensures high-quality annotation for objects in TSFMO. See Fig. 3 for some examples of box annotations for TSFMO.

3.3 Attributes

In view of that in-depth analysis of tracking methods is crucial to grasp their strengths and limitations, we select twelve attributes that widely exist in video tasks and annotate each sequence with these attributes, including (1) Illumination Variation (IV), (2) Deformation (DEF), (3) Motion Blur (MB), (4) Rotation (ROT), (5) Background Clutter (BC), (6) Scale Variation (SV), which is assigned when the ratio of bounding box is outside the range [0.5, 2], (7) Out-of-view (OV), (8) Low Resolution (LR), which is assigned when the target area

Fig. 3. Example sequences of small and fast moving object tracking in our TSFMO. Each sequence is annotated with axis-aligned bounding boxes and attributes.

Table 3. Distribution of twelve attributes on the TSFMO. The diagonal (shown in **bold**) corresponds to the distribution over the entire benchmark, and each row or column presents the joint distribution for the attribute subset.

	IV	SV	DEF	MB	FM	OV	BC	LR	POC	ROT	FOC	ARC
IV	**31**	29	0	26	31	0	14	30	7	12	1	23
SV	29	**193**	33	170	191	17	49	188	53	96	8	147
DEF	0	33	**45**	39	44	12	5	38	7	26	0	41
MB	26	170	39	**221**	132	18	58	203	56	102	9	158
FM	31	191	44	132	**248**	18	58	231	56	102	9	75
OV	0	17	12	18	18	**21**	0	20	4	17	0	17
BC	14	49	5	58	58	0	**63**	59	18	21	1	40
LR	30	188	38	203	231	20	59	**235**	62	108	8	170
POC	7	53	7	56	56	4	18	62	**71**	31	1	46
ROT	12	96	26	102	102	17	21	108	31	**111**	2	80
FOC	1	8	0	9	9	0	1	8	1	2	**9**	7
ARC	23	147	41	158	75	17	40	170	46	80	7	**176**

is smaller than 900 pixels, (9) Aspect Ratio Change (ARC), which is assigned when the ratio of the bounding box aspect ratio is outside the range [0.5, 2], (10) Partial Occlusion (POC), (11) Full Occlusion (FOC), and (12) Fast Motion (FM), which is assigned when the target center moves by at least 50% of its size in last frame. Table 3 shows the distribution of these attributes on TSFMO. As can be seen, the most common challenge in TSFMO is Fast Motion. In addition, the Motion Blur and Low Resolution also present frequently in TSFMO.

4   A New Baseline: S-KeepTrack

We found that among the state-of-the-art trackers to be evaluated here Keep-Track shows the best performance, despite that it is still far from satisfactory. To facilitate the development of tracking algorithms for tracking small and fast moving objects, we present a new baseline tracker based on the KeepTrack. The proposed tracker, dubbed S-KeepTrack, combines low-level and high-level features to improve tracking performance, considering that small objects may not have representation in deeper layers' features because of their larger receptive field. In contrast to one target candidate association network of KeepTrack, S-KeepTrack has two parallel target candidate association networks that separately process the feature encodings of lower and higher level features, respectively. And the result of candidate association is a weighted combination of the result of two parallel networks. Hopefully, this combination of low-level and high-level features will build a stronger tracker than KeepTrack for tracking small and fast moving objects.

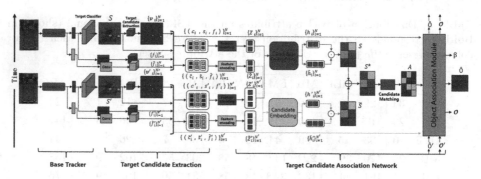

Fig. 4. Overview of the tracking pipeline of the proposed S-KeepTrack. Note that the parallel architectures plotted in the same color share parameters.

Like KeepTrack, S-KeepTrack also consists of three components: i) a base tracker that predicts the target score map s for the current frame and extracts the target candidates V and V' ($V = \{v_i\}_{i=1}^{N}$ and $V' = \{v'_i\}_{i=1}^{N'}$ denote candidate set of the current and the previous frame, respectively) by finding locations in s with high target score, ii) a target candidate extraction module that extracts for each candidate a set of features (i.e., target classifier score s_i, location c_i in the image, and two appearance cues $\tilde{f}_i$ and f_i from lower and higher level, respectively, where i indexes the ith candidate), and iii) a target candidate association network that estimates the candidate assignment probabilities between two consecutive frames. The difference between S-KeepTrack and KeepTrack lies in the parallel network architectures designed for the low level features from the backbone. An overview of our tracking pipeline is shown in Fig. 4.

Base Tracker: The base tracker is inherited from KeepTrack with the difference: the backbone of our S-KeepTrack outputs both low and high level features for target candidate extraction instead of only high level one as in KeepTrack.

Target Candidate Extraction: This module aims to build a feature representation for each target candidate. In KeepTrack, the position c_i and the target classifier score s_i are taken as two discriminative cues of the target candidate v_i, based on two observations: i) the motion of the same object from frame to frame is typically small and thus the same object has similar locations in two neighbouring frames, ii) only small changes in appearance for each object [49]. In addition, it also processes the backbone features with a single learnable convolution layer to add a more discriminative appearance-based feature f_i. Finally, each feature tuple (c_i, s_i, f_i) representing the target candidate v_i is fed into a feature encoding module to get the code [49]

$$z_i = f_i + \psi(s_i, c_i), \tag{1}$$

where ψ denotes a Multi-Layer Perceptron (MLP), mapping s_i and c_i to the same dimensional space as f_i. z_i then passes through a candidate embedding network before candidate matching is conducted. In order to build a better feature representation for each target candidate, we additionally process a low-level backbone feature with an extra learnable convolution layer in our S-KeepTrack, resulting in an extra discriminative appearance-based feature $\bar{f}_i$. In view of that the feature encoding module and the candidate embedding network have been carefully tailored to the tuple (c_i, s_i, f_i), we avoid fusing f_i and $\bar{f}_i$ to get an entangled feature, as it may not fit well with the original network and will demand modifying the feature encoding module and the candidate embedding network. Instead, from the perspective of ensemble method, we build a new tuple $(c_i, s_i, \bar{f}_i)$ to accompany (c_i, s_i, f_i). And $(c_i, s_i, \bar{f}_i)$ is fed into a new feature encoding module to get another code

$$\bar{z}_i = \bar{f}_i + \psi'(s_i, c_i), \tag{2}$$

where ψ' denotes a MLP that map s_i and c_i to the same dimensional space as $\bar{f}_i$. The feature encoding modules are followed by two parallel candidate embedding networks, which will be detailed in the following.

Candidate Embedding Network: On an abstract level, candidate association bares similarities with the task of sparse feature matching [49,55], for which KeepTrack adopted the SuperGlue [55] architecture that establishes state-of-the-art sparse feature matching performance to do candidate embedding and matching. With this method, the feature encodings $\{z_i\}_{i=1}^N$ and $\{z_i'\}_{i=1}^N$ of two neighbouring frames translate to nodes of a single complete graph with two types of directed edges: 1) self edges within the same frame and 2) cross edges connecting only nodes between the frames. In SuperGlue [55], a Graph Neural Network (GNN) is utilized to send messages in an alternating fashion across self or cross edges to produce a new feature representation for each node after every layer, in which self and cross attention are used to compute the messages for self and cross edges [49,55]. After the last message passing layer a linear projection layer extracts the final feature representation h_i for each candidate v_i [49]. In our S-KeepTrack, a new candidate embedding network is adopted to deal with the feature encoding related to the low-level features, resulting in an extra feature representation $\bar{h}_i$ for each candidate v_i.

Candidate Matching: In KeepTrack, the candidate embeddings h'_i and h_j (corresponding to two candidates $v'_i \in V'$ and $v_j \in V$, respectively.) are used to compute the similarity between v'_i and v_j by the scalar product: $S_{i,j} = \langle h'_i, h_j \rangle$. Given a match may not exist for every candidate, KeepTrack makes use of the dustbin concept [16,55] to actively match candidates that miss their counterparts to the so-called dustbin, ending up with an augmented assignment matrix A with an additional row and column representing dustbins. Note that a dustbin is a virtual candidate without any feature representation, to which a candidate corresponds only if its similarity scores to all other candidates are sufficiently low. To obtain the assignment matrix A between V and V' given the similarity matrix $S = \{S_{i,j}\}$, KeepTrack follow Sarlin et al. [55] and designed a learnable module to predict A. However, we have two parallel similarity matrice S and $\bar{S}$ in S-KeepTrack because of the parallel feature representations. Therefore, we aggregate the two similarity matrice S and $\bar{S}$ to produce a fused one S^* by the following weighted sum,

$$S^* = \omega S + (1 - \omega)\bar{S}, \tag{3}$$

where $\omega \in [0, 1]$ is the weight coefficient to balance the contributions of S and $\bar{S}$. S^* is then fed into the candidate matching module to predict the assignment matrix A as in KeepTrack.

Object Association: The object association module uses the estimated assignments to determine the object correspondences during online tracking, which follows that of KeepTrack. The idea is to keep track of every object present in each scene over time using a database $\mathcal{O}$ with each entry being an object visible in the current frame. When online tracking, the estimated assignment matrix A is used to determine which objects disappeared, newly appeared, or stayed visible and can be associated unambiguously, and to help reason the target object $\hat{o}$. Last but not least, the target detection confidence β is computed to manage the memory and control the sample weight for updating the target classifier online. This finishes the description of our S-KeepTrack. It is worth noting that the losses and training pipeline of S-KeepTrack is the same as KeepTrack. Please refer to [49] for details.

5   Evaluation

5.1   Evaluation Metrics

We use one-pass evaluation (OPE) and measure each tracker using precision, and success rate as in [18,53]. The precision measures the distance between the centers of the estimated target bounding box and the groundtruth box in pixels. Success rate is based on the intersection over union (IoU) of the estimated target bounding box and the groundtruth box, specifically, it measures the percentage of estimated target bounding boxes with IoU larger than 0.5 [18,53,63]. The precision at 20 pixels (PRC) and the area under curve (AUC) of success plot is usually used for ranking.

5.2   Trackers for Comparison

We evaluate 20 state-of-the-art trackers to understand their performance on TSFMO, including KeepTrack [49], AutoMatch [68], TransT [8], SAOT [71], SiamGAT [23], LightTrack [66], TrDiMP [60], TrSiam [60], KYS [4], HIFT [6], SuperDiMP [34], PrDiMP50 [14], PrDiMP18 [14], Ocean [69], SiamMask [61], SiamRPN++ [38], ATOM [11], DiMP50 [3], DiMP18 [3], and SiamDW [70].

5.3   Evaluation Results

Overall Performance. 20 state-of-the-art trackers and our S-KeepTrack are extensively evaluated on TSFMO. Note that existing trackers are used without any modification. The evaluation results are reported in precision and success plot, as shown in Fig. 5. As can be seen, our S-KeepTrack achieved the best results with a PRC of 0.437, AUC of 0.255. KeepTrack got the second best PRC of 0.425, and likewise KeepTrack got the second best AUC of 0.247. In comparison with the second best tracker KeepTrack, S-KeepTrack achieves improvements of 1.2% and 0.8% in terms of PRC and AUC, respectively, which evidences the effectiveness and advantage of our method of combining low level and high level features for small and fast moving object tracking.

Attribute-Based Performance. We conduct performance evaluation under twelve attributes to further analyze and understand the performances of dif-

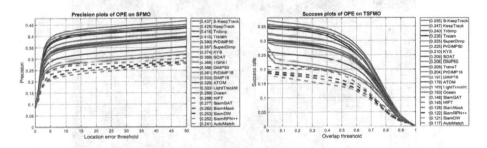

Fig. 5. Overall performance on TSFMO. Precision and success rate for one-pass evaluation (OPE) [62] are used for evaluation.

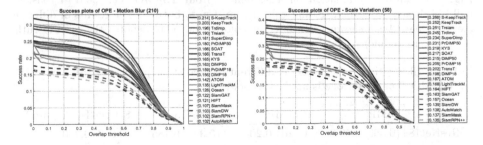

Fig. 6. Attribute-based comparison on motion blur and scale variation.

ferent trackers. Our S-KeepTrack achieves the best PRC and AUC on most
attributes. Due to space limitation, we demonstrate in Fig. 6 the success plots
for the two most frequent challenges, including motion blur and scale variation.
We observe that S-KeepTrack performs the best on both attributes. Specifically,
S-KeepTrack achieves a AUC of 0.214 on motion blur, surpassing the second best
tracker KeepTrack with AUC of 0.203 by 1.1%; on scale variation, S-KeepTrack'
AUC is 0.266, outperforming the second best tracker KeepTrack with AUC of
0.252 by 1.4%. This also supports the importance of combining low level and
high level features for small and fast moving object tracking.

Qualitative Evaluation. In Fig. 7, we show some qualitative tracking results
of our method in comparison with eight top trackers, including KeepTrack [49],
TrSiam [60], TrDiMP [60], SAOT [71], KYS [4], PrDiMP50 [14], DiMP50 [3],
DiMP18 [3]. As can be seen, only our S-KeepTrack succeeds to maintain robustness in these four examples that subject to challenges including rotation, background clutter, illumination, partial occlusion, motion blur, low resolution, and
scale variation. Specifically, all other trackers fail to keep track of the target in
polo_4, only S-KeepTrack and Trsiam succeed to track the target in basket-
ball_14, and only S-KeepTrack and KeepTrack succeed to track the target in

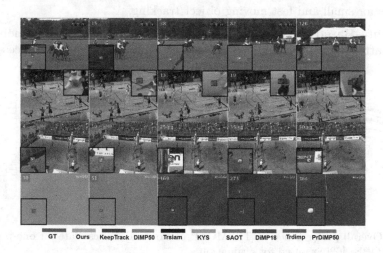

Fig. 7. Qualitative evaluation on 4 sequences from TSFMO, i.e., polo_4, basket-
ball_14, beach_volleyball_8, and golf_4 from top to bottom. The results of different
methods have been shown with different colors, and 'GT' denotes the groundtruth.
(Color figure online)

Table 4. Illustration of the impact of different backbones on precision (PRC) and AUC
on TSFMO. The better one is marked in bold.

	DiMP	PrDiMP	KeepTrack	S-KeepTrack (ours)
ResNet18	(0.355, 0.197)	(0.361, 0.204)	(0.346, 0.185)	(0.353, 0.190)
ResNet50	**(0.366, 0.206)**	**(0.390, 0.225)**	**(0.425, 0.247)**	**(0.437, 0.255)**

golf_4. We own the advantage of S-KeepTrack over other trackers, especially over KeepTrack, to the proposed method of combining low level and high level features for representation.

5.4 Ablation Study

Impact of Backbone. To study the impact of the backbone on performance of tracking small and fast moving object. We evaluate several state-of-the-art trackers with ResNet-18 and ResNet-50 as backbone separately, including DiMP [3], PrDiMP [14], KeepTrack [49], and our S-KeepTrack. Note that KeepTrack was implemented with only ResNet-50 as backbone originally. We adapt it and S-KeepTrack to support ResNet-18 as backbone for this ablation study. Table 4 shows the PRC and AUC of these trackers on TSFMO in the form of (PRC, AUC). As can be seen, in each tracker the PRC and AUC are higher with ResNet-50 than with ResNet-18. Specifically, the (PRC, AUC) of DiMP, PrDiMP, Keep-Track, and S-KeepTrack increases by (1.1%, 0.9%), (2.9%, 2.1%), (7.9%, 6.2%), (8.4%, 6.5%) when the backbone is replaced from ResNet-18 to ResNet-50. This suggests that, although low-level features is helpful for tracking small and fast moving object as demonstrated by our method, deeper backbones are crucial to learn representation that extract abstract and essential information for this tracking task.

Table 5. Illustration of the impact of the importance coefficient of the low-level and high-level features on precision (PRC) and AUC on TSFMO. Red, blue and green indicate the first, second and third place.

ω	1.0	0.9	0.8	0.7	0.6	0.5	0.4	0.3	0.2	0.1	0.0
PRC	0.424	0.417	0.415	0.424	0.421	0.427	0.428	0.428	0.437	**0.433**	**0.432**
AUC	0.247	0.241	0.240	0.246	0.243	0.247	**0.248**	**0.248**	0.255	**0.251**	**0.251**

Impact of the Importance Coefficient of the Low-Level and High-Level Features. To study the impact of the weight coefficient that balances the contributions of the similarity matrice S and $\bar{S}$ estimated with low-level and high-level features, respectively, as formulated in Eq. (3), we evaluate the proposed S-KeepTrack trained with different setting of the weight coefficient ω on TSFMO. The tried ω ranges from 0.0 to 1.0 with step size 0.1. Note that $\omega = 0.0$ and $\omega = 1.0$ mean using exclusively low-level and high-level features, respectively, where the latter is just the KeepTrack. Therefore, the larger the ω the more contributions of high-level features and the less of low-level one. The PRC and AUC of S-KeepTrack on TSFMO with respect to ω are shown on Table 5. As shown, the highest PRC and AUC occur when ω is less than or equal to 0.4, suggesting, in a sense, that the more contributions of low-level features the higher the tracking performance. Specifically, when $\omega = 0.2$, S-KeepTrack achieves the best

PRC and AUC, i.e., 0.437 and 0.255, which is used as the default setting of ω for S-KeepTrack. When $\omega = 1.0$ S-KeepTrack reduces to KeepTrack, having PRC and AUC of 0.424 and 0.247, respectively. S-KeepTrack surpasses KeepTrack on PRC and AUC by 1.3% and 0.8%, respectively, owing to introduced low-level features. Remarkably, we can observe that using low-level features is more effective than high-level features, as when $\omega = 0.0$, i.e., using low-level features exclusively, S-KeepTrack achieves PRC and AUC of 0.432 and 0.251, surpassing KeepTrack with gaps of 0.8% and 0.4% on PRC and AUC, respectively.

6   Conclusion

In this work, we explore a new tracking task, i.e., tracking small and fast moving objects. In particular, we propose the TSFMO, which is the first benchmark for small and fast moving object tracking, to our best knowledge. In addition, in order to understand the performance of existing trackers and to provide baseline for future comparison, we extensively evaluate 20 state-of-the-art tracking algorithms with in-depth analysis. Moreover, we propose a novel tracker, named S-KeepTrack, by combining low-level and high-level features to obtain a stronger tracker, which outperforms existing state-of-the-art tracking algorithms by a clear margin. Our experiments suggest that there is a big room for us to improve the performance of tracking small and fast moving objects. We believe that, the benchmark, evaluation and the baseline tracker will inspire and facilitate more future research and application on tracking small and fast moving objects.

References

1. Ahmadi, K., Salari, E.: Small dim object tracking using frequency and spatial domain information. Pattern Recogn. **58**, 227–234 (2016)
2. Bertinetto, L., Valmadre, J., Henriques, J.F., Vedaldi, A., Torr, P.H.S.: Fully-convolutional siamese networks for object tracking. In: Hua, G., Jégou, H. (eds.) ECCV 2016. LNCS, vol. 9914, pp. 850–865. Springer, Cham (2016). https://doi.org/10.1007/978-3-319-48881-3_56
3. Bhat, G., Danelljan, M., et al.: Learning discriminative model prediction for tracking. In: ICCV, pp. 6182–6191 (2019)
4. Bhat, G., Danelljan, M., Van Gool, L., Timofte, R.: Know your surroundings: exploiting scene information for object tracking. In: Vedaldi, A., Bischof, H., Brox, T., Frahm, J.-M. (eds.) ECCV 2020. LNCS, vol. 12368, pp. 205–221. Springer, Cham (2020). https://doi.org/10.1007/978-3-030-58592-1_13
5. Bolme, D.S., Beveridge, J.R., et al.: Visual object tracking using adaptive correlation filters. In: CVPR, pp. 2544–2550 (2010)
6. Cao, Z., Fu, C., et al.: HiFT: hierarchical feature transformer for aerial tracking. In: ICCV, pp. 15437–15446 (2021)
7. Chen, C., Liu, M.-Y., Tuzel, O., Xiao, J.: R-CNN for small object detection. In: Lai, S.-H., Lepetit, V., Nishino, K., Sato, Y. (eds.) ACCV 2016. LNCS, vol. 10115, pp. 214–230. Springer, Cham (2017). https://doi.org/10.1007/978-3-319-54193-8_14

8. Chen, X., Yan, B., et al.: Transformer tracking. In: CVPR, pp. 8122–8131 (2021)
9. Chicco, D.: Siamese neural networks: an overview. In: Cartwright, H. (ed.) Artificial Neural Networks. MMB, vol. 2190, pp. 73–94. Springer, New York (2021). https://doi.org/10.1007/978-1-0716-0826-5_3
10. Colyer, S.L., Evans, M., et al.: A review of the evolution of vision-based motion analysis and the integration of advanced computer vision methods towards developing a markerless system. Sports Med. Open **4**, 1–15 (2018)
11. Danelljan, M., Bhat, G., et al.: ATOM: accurate tracking by overlap maximization. In: CVPR, pp. 4655–4664 (2019)
12. Danelljan, M., Hager, G., et al.: Learning spatially regularized correlation filters for visual tracking. In: ICCV, pp. 4310–4318 (2015)
13. Danelljan, M., Hager, G., et al.: Discriminative scale space tracking. IEEE Trans. Pattern Anal. Mach. Intell. **39**(8), 1561–1575 (2017)
14. Danelljan, M., Van Gool, L. Timofte, R.: Probabilistic regression for visual tracking. In: CVPR, pp. 7183–7192 (2020)
15. Danelljan, M., Robinson, A., Shahbaz Khan, F., Felsberg, M.: Beyond correlation filters: learning continuous convolution operators for visual tracking. In: Leibe, B., Matas, J., Sebe, N., Welling, M. (eds.) ECCV 2016. LNCS, vol. 9909, pp. 472–488. Springer, Cham (2016). https://doi.org/10.1007/978-3-319-46454-1_29
16. DeTone, D., Malisiewicz, T., Rabinovich, A.: SuperPoint: self-supervised interest point detection and description. In: CVPRW, pp. 337-1–337-12 (2018)
17. Fan, H., Bai, H., et al.: LaSOT: a high-quality large-scale single object tracking benchmark. Int. J. Comput. Vis. **129**, 439–461 (2021). https://doi.org/10.1007/s11263-020-01387-y
18. Fan, H., Lin, L., et al.: LaSOT: a high-quality benchmark for large-scale single object tracking. In: CVPR, pp. 5369–5378 (2019)
19. Fan, H., Miththanthaya, H.A., et al.: Transparent object tracking benchmark. arXiv (2020)
20. Fan, H., Yang, F., et al.: TracKlinic: diagnosis of challenge factors in visual tracking. In: WACV, pp. 969–978 (2021)
21. Galoogahi, H.K., Fagg, A., et al.: Need for speed: a benchmark for higher frame rate object tracking. In: ICCV, pp. 1134–1143 (2017)
22. Gong, Y., Yu, X., et al.: Effective fusion factor in FPN for tiny object detection. In: WACV, pp. 1159–1167 (2021)
23. Guo, D., Shao, Y., et al.: Graph attention tracking. In: CVPR, pp. 9538–9547 (2021)
24. Guo, D., Wang, J., et al.: Siamese fully convolutional classification and regression for visual tracking. In: CVPR, pp. 6269–6277 (2020)
25. Hong, M., Li, S., et al.: SSPNet: scale selection pyramid network for tiny person detection from UAV images. IEEE Geosci. Remote Sens. Lett. **19**, 1–5 (2022)
26. Huang, L., Zhao, X., Huang, K.: GOT-10k: a large high-diversity benchmark for generic object tracking in the wild. IEEE Trans. Pattern Anal. Mach. Intell. **43**, 1562–1577 (2021)
27. Huang, Y., Llach, J., Zhang, C.: A method of small object detection and tracking based on particle filters. In: ICPR, pp. 1–4 (2008)
28. Jiang, J., Zhang, X.: Research on moving object tracking technology of sports video based on deep learning algorithm. In: ICISCAE (2021)
29. Jiao, L., Wang, D., et al.: Deep learning in visual tracking: a review. IEEE Trans. Neural Netw. Learn. Syst. (2021)
30. Kembhavi, A., Harwood, D., Davis, L.S.: Vehicle detection using partial least squares. IEEE Trans. Pattern Anal. Mach. Intell. **33**, 1250–1265 (2011)

31. Kerr, R.: Technologies for judging, umpiring and refereeing. In: Sport and Technology, pp. 114–134. Manchester University Press (2016)
32. Kong, T., Sun, F., et al.: FoveaBox: beyound anchor-based object detection. IEEE Trans. Image Process. **29**, 7389–7398 (2020)
33. Kristan, M., Leonardis, A., et al.: The visual object tracking VOT2017 challenge results. In: ICCVW, pp. 1949–1972 (2017)
34. Kristan, M., et al.: The eighth visual object tracking VOT2020 challenge results. In: Bartoli, A., Fusiello, A. (eds.) ECCV 2020. LNCS, vol. 12539, pp. 547–601. Springer, Cham (2020). https://doi.org/10.1007/978-3-030-68238-5_39
35. Kristan, M., Matas, J., et al.: A novel performance evaluation methodology for single-target trackers. IEEE Trans. Pattern Anal. Mach. Intell. **38**, 2137–2155 (2016)
36. Lapinski, M., Brum Medeiros, C., et al.: A wide-range, wireless wearable inertial motion sensing system for capturing fast athletic biomechanics in overhead pitching. Sensors **19**(17), 3637 (2019)
37. Li, A., Lin, M., et al.: NUS-PRO: a new visual tracking challenge. IEEE Trans. Pattern Anal. Mach. Intell. **38**, 335–349 (2016)
38. Li, B., Wu, W., et al.: SiamRPN++: evolution of siamese visual tracking with very deep networks. In: CVPR, pp. 4282–4291 (2019)
39. Li, F., Tian, C., et al.: Learning spatial-temporal regularized correlation filters for visual tracking. In: CVPR, pp. 4904–4913 (2018)
40. Li, S., Jiang, Q., Zhao, Q., Lu, L., Feng, Z.: Asymmetric discriminative correlation filters for visual tracking. Front. Inf. Technol. Electron. Eng. **21**(10), 1467–1484 (2020). https://doi.org/10.1631/FITEE.1900507
41. Li, S., Liu, Y., et al.: Learning residue-aware correlation filters and refining scale estimates with the GrabCut for real-time UAV tracking. In: 3DV, pp. 1238–1248 (2021)
42. Li, S., Liu, Y., et al.: Learning residue-aware correlation filters and refining scale for real-time UAV tracking. Pattern Recogn. **127**, 108614 (2022)
43. Li, Y., Zhu, J.: A scale adaptive kernel correlation filter tracker with feature integration. In: Agapito, L., Bronstein, M.M., Rother, C. (eds.) ECCV 2014. LNCS, vol. 8926, pp. 254–265. Springer, Cham (2015). https://doi.org/10.1007/978-3-319-16181-5_18
44. Liang, P., Blasch, E., Ling, H.: Encoding color information for visual tracking: algorithms and benchmark. IEEE Trans. Image Process. **24**, 5630–5644 (2015)
45. Liu, C., Ding, W., et al.: Aggregation signature for small object tracking. IEEE Trans. Image Process. **29**, 1738–1747 (2020)
46. Lu, H., Wang, D.D.: Online Visual Tracking. Springer, Singapore (2019). https://doi.org/10.1007/978-981-13-0469-9
47. Lukežič, A., Kart, U., et al.: CDTB: a color and depth visual object tracking dataset and benchmark. In: ICCV, pp. 10012–10021 (2019)
48. Ma, C., Huang, J.B., et al.: Hierarchical convolutional features for visual tracking. In: ICCV, pp. 3074–3082 (2015)
49. Mayer, C., Danelljan, M., et al.: Learning target candidate association to keep track of what not to track. In: ICCV, pp. 13424–13434 (2021)
50. Mazzeo, P.L., Ramakrishnan, S., Spagnolo, P.: Visual object tracking with deep neural networks (2019)
51. Morariu, V.I., Ahmed, E., et al.: Composite discriminant factor analysis. In: WCACV, pp. 564–571 (2014)

52. Mueller, M., Smith, N., Ghanem, B.: A benchmark and simulator for UAV tracking. In: Leibe, B., Matas, J., Sebe, N., Welling, M. (eds.) ECCV 2016. LNCS, vol. 9905, pp. 445–461. Springer, Cham (2016). https://doi.org/10.1007/978-3-319-46448-0_27

53. Müller, M., Bibi, A., Giancola, S., Alsubaihi, S., Ghanem, B.: TrackingNet: a large-scale dataset and benchmark for object tracking in the wild. In: Ferrari, V., Hebert, M., Sminchisescu, C., Weiss, Y. (eds.) ECCV 2018. LNCS, vol. 11205, pp. 310–327. Springer, Cham (2018). https://doi.org/10.1007/978-3-030-01246-5_19

54. Rozumnyi, D., Matas, J., et al.: The world of fast moving objects. In: CVPR, pp. 4838–4846 (2017)

55. Sarlin, P.E., DeTone, D., et al.: SuperGlue: learning feature matching with graph neural networks. In: CVPR, pp. 4937–4946 (2020)

56. Song, S., Xiao, J.: Tracking revisited using RGBD camera: unified benchmark and baselines. In: ICCV, pp. 233–240 (2013)

57. Tamir, I., Bar-eli, M.: The moral gatekeeper: soccer and technology, the case of Video Assistant Referee (VAR). Front. Psychol. **11**, 613469 (2020)

58. Tong, K., Wu, Y., Zhou, F.: Recent advances in small object detection based on deep learning: a review. Image Vis. Comput. **97**, 103910 (2020)

59. Valmadre, J., Bertinetto, L., et al.: Long-term tracking in the wild: a benchmark. ArXiv abs/1803.09502 (2018)

60. Wang, N., Zhou, W., et al.: Transformer meets tracker: exploiting temporal context for robust visual tracking. In: CVPR, pp. 1571–1580 (2021)

61. Wang, Q., Zhang, L., et al.: Fast online object tracking and segmentation: a unifying approach. In: CVPR, pp. 1328–1338 (2019)

62. Wu, Y., Lim, J., Yang, M.H.: Online object tracking: a benchmark. In: CVPR (2013)

63. Wu, Y., Lim, J., Yang, M.H.: Object tracking benchmark. IEEE Trans. Pattern Anal. Mach. Intell. **37**, 1834–1848 (2015)

64. Xu, Y., Wang, Z., et al.: SiamFC++: towards robust and accurate visual tracking with target estimation guidelines. In: AAAI, vol. 34, pp. 12549–12556 (2020)

65. Xue, Y., Song, Y., et al.: Automatic video annotation system for archival sports video. In: WACVW, pp. 23–28 (2017)

66. Yan, B., Peng, H., et al.: LightTrack: finding lightweight neural networks for object tracking via one-shot architecture search. In: CVPR, pp. 15175–15184 (2021)

67. Zaveri, M.A., Merchant, S.N., Desai, U.B.: Small and fast moving object detection and tracking in sports video sequences. In: ICME, vol. 3, pp. 1539–1542 (2004)

68. Zhang, Z., Liu, Y., et al.: Learn to match: automatic matching network design for visual tracking. In: ICCV, pp. 13319–13328 (2021)

69. Zhang, Z., Peng, H.: Ocean: object-aware anchor-free tracking. ArXiv abs/2006.10721 (2020)

70. Zhang, Z., Peng, H., Wang, Q.: Deeper and wider siamese networks for real-time visual tracking. In: CVPR, pp. 4586–4595 (2019)

71. Zhou, Z., Pei, W., et al.: Saliency-associated object tracking. In: ICCV, pp. 9846–9855 (2021)

72. Zita, A., Šroubek, F.: Tracking fast moving objects by segmentation network. In: ICPR, pp. 10312–10319 (2021)

PhyLoNet: Physically-Constrained Long-Term Video Prediction

Nir Ben Zikri(ID) and Andrei Sharf(✉)(ID)

Ben-Gurion University of the Negev, P.O.B. 653, 8410501 Be'er Sheva, Israel
asharf@cs.bgu.ac.il

Abstract. Motions in videos are often governed by physical and biological laws such as gravity, collisions, flocking, etc. Accounting for such natural properties is an appealing way to improve realism in future frame video prediction. Nevertheless, the definition and computation of intricate physical and biological properties in motion videos are challenging. In this work, we introduce PhyLoNet, a PhyDNet extension that learns long-term future frame prediction and manipulation. Similar to PhyD-Net, our network consists of a two-branch deep architecture that explicitly disentangles physical dynamics from complementary information. It uses a recurrent physical cell (PhyCell) for performing physically-constrained prediction in latent space. In contrast to PhyDNet, PhyLoNet introduces a modified encoder-decoder architecture together with a novel relative flow loss. This enables a longer-term future frame prediction from a small input sequence with higher accuracy and quality. We have carried out extensive experiments, showing the ability of PhyLoNet to outperform PhyDNet on various challenging natural motion datasets such as ball collisions, flocking, and pool games. Ablation studies highlight the importance of our new components. Finally, we show an application of PhyLoNet for video manipulation and editing by a novel class label modification architecture.

Keywords: Deep learning · Physical motion · Long-term video prediction

1 Introduction

Real-world videos often depict the natural motions and dynamics of objects and their interactions. These motions are typically governed by the physical and biological laws of nature. Accounting for such natural properties is an appealing way to improve realism in future frame video prediction. Nevertheless, the definition and computation of intricate physical and biological properties in motion videos are challenging. Thus, a key problem is to design video prediction methods able to represent the complex dynamics underlying raw data.

This research was partially supported by the Lynn and William Frankel Center for Computer Science at BGU.

Video forecasting consists of predicting the future content of a video conditioned on previous frames. In this context, a key problem is to design video prediction methods able to represent the complex dynamics underlying raw data. In this work, we focus on unsupervised video prediction of complex motions and interactions, typically governed by physical or biological laws. Our datasets lack of semantic labeling forces the network towards predicting motions only from domain knowledge in an unsupervised manner.

We introduce PhyLoNet, a PhyDNet [21] extension that learns long-term future frame prediction of natural motions and their manipulation. In our experiments, we explore natural motions and interactions such as bird flocking motions, multiple ball collisions, and different movement behaviors. We model the motion dynamics by incorporating our deep learning framework with PDE modules and a motion-aware loss. Thus, we account for physical, biological, and in general high-order priors in the motion video. Essentially, our method builds upon the two-branch PhyDNet architecture which allows disentangling the physical properties of a motion video from other factors in a latent space. We introduce a novel relative flow loss (denoted RF-loss) which together with a customized deep network architecture yields a significant improvement in future frame prediction of complex motions and interactions. We also extend our neural network design to accomplish future frame manipulation besides the classical prediction task. Thus, we incorporate motion targets and paths as additional class labels and channels in the network architecture. We show that at test time, it is possible to adjust class labels on the fly, by doing so we allow editing and controlling of the video prediction sequence. In our experiments, we show that our model is able to predict future frames with significantly higher accuracy than PhyDNet and produce state-of-the-art results. We also show an application of our framework for video completion and editing. To summarize our work makes the following contributions:

- We introduce a novel relative flow loss (RF-loss) that accounts for the complex motion dynamics in the video and significantly improves prediction accuracy and quality.
- We introduce PhyLoNet architecture which modifies PhyDNet architecture for improved long-term predictions and domain generalization.
- We extend PhyLoNet to allow video editing through on-the-fly prediction manipulation and control.

2 Related Work

Predicting the next frame of a video has received growing interest in the computer vision community over the past few years. Recent works have managed to achieve state-of-the-art performances using deep neural networks for next-frame video prediction tasks. Sequence to sequence LSTM and Convolutional variants [42,44] are the core of many similar studies [9,59]. Further works explore various Recurrent Neural Network (RNNs) [31,52–55] and 2D/3D ConvNets [4,28,39,51] architectures.

This task becomes even more challenging when facing high-dimensional images, therefore predicting the geometric transformations between frames [3,9,60] or using their optical flow [22–24,27,34] reduce substantially the complexity of image generation. This approach is usually effective for single frame prediction [11], where it aims to predict the future frame according to the whole information within a frame as one representation, but it fails when predicting frames for the long-term. Another approach for handling high-dimensional inputs is by disentangling independent factors of variations in order to apply the prediction to lower-dimensional representations [8,18,55].

Relational reasoning which is often implemented with graphs [2,17,32,40,45] accounts for basic physical laws, e.g. drift, gravity, spring [30,56,57]. Still, these methods fail for general real-world video forecasting.

A promising line of work focuses on disentangling approach which factorizes the video into independent components [6,10,13,48,50]. Some works disentangles content and motion [13,50,58] while others disentangle deterministic and stochastic factors [6]. [51] propose a generative adversarial network for the future frame prediction based on foreground-background mask disentanglement. Another foreground-background disentanglement approach is to use segmentation masks [29] extracted from a semantic segmentation network and use it to increase the attention over each instance individually [58].

Physics and PDEs. Exploiting prior physical knowledge is another appealing way to improve prediction models. Solving PDEs with DNNs [35,36,41] has grown a lot of attention in recent years, more specifically, a connection between PDEs and CNN's [25,26] shows that it is possible to learn filters that resemble set of differential orders, combining them with temporal based DNN's like LSTM [43] significantly improves physical dynamics prediction in latent space.

Some works [21,33,61] find physical models insufficient and propose to combine physical models with data-driven models in order to achieve all the data within a frame. [1] approach uses multiple physical models in order to extract different physical properties, those models are controlled by a Transformers [7,49] network which implements the Mixture of Experts concept for selecting the best physical models that represent the video dynamics the most.

Leveraging physical knowledge is not enough for general video forecasting. Despite the fact that it can learn a broad class of PDEs, there are a lot of domains where this physical prior might not fit.

Dedicated Loss. Dedicated loss functions [5,12] and the ability of Generative Adversarial Networks to generate high-quality frames [20,28,51] have been investigated. However, combining GANs with prior information, such as physical models, remains an open topic for research.

In this work, we design a novel motion loss that is based on optical flow and denoted as *Relative Flow Loss* (RF-loss). The main advantage of deep optical flow is its differentiability trait, it allows utilizing it in deep neural networks as a loss that relates directly to the object's speed and direction. In our context, we use our RF-Loss to track and differentiate an object's motion distortion in an unsupervised manner.

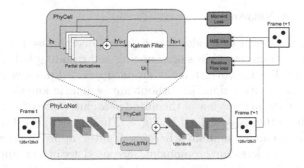

Fig. 1. The PhyLoNet architecture. An image frame is first encoded into a latent vector h_t, transferred into the PhyCell and ConvLSTM cells whose outputs are summed and decoded into the next frame. (Color figure online)

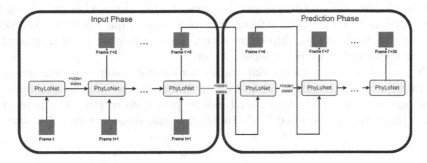

Fig. 2. The unrolled PhyLoNet architecture.

Recent works show that it is possible to compute the optical flow using various DNN architectures [14,15,37,46]. These are mostly based on pyramid structures that learn different order filters which enable to capture of different receptive fields. Another approach is based on recurrent units and per-pixel features [47].

3 Technical Details

In this work, we focus on learning long-term motion video prediction and manipulation of future frames. Our PhyLoNet network is based on PhyDNet [21]. The network consists of a two-branch deep architecture that explicitly disentangles physical dynamics from unknown complementary information (see Fig. 1). One branch is built from ConvLSTM cell in order to retain the residual information within a video. The second branch is built from a Physical cell (PhyCell) that aims to learn the physical dynamics for long-term predictions using the deep Kalman mechanism. In contrast to [21] we customize the encoder-decoder architecture and remove skip connections in order to allow more accurate and longer predictions. Similar to [21] we use an image moment loss and MSE loss between the ground-truth and the model's prediction. The third loss is our novel relative flow loss which also allows more accurate and longer predictions.

3.1 PhyLoNet Network

PhyLoNet is trained over raw video frames in an unsupervised way. In our case, we experiment with motion videos of bouncing and colliding balls, bird flocking, and a pool game. The network trains in an unsupervised manner on raw motion videos of a specific motion domain without any semantic knowledge or labels.

Figure 2 presents an overview of the unrolled architecture of our network. The PhyLoNet architecture is described in Fig. 1. The physical cell learns the PDEs underlying our dynamics domain where each filter approximates PDE coefficients of different orders. Image moments are used to better approximate the PDE coefficients. Both prediction h'_{t+1} and observation u_t (the encoded ground truth frame) are inserted into a deep Kalman filter [19] for both estimation improvement and long-term predictions.

The encoder-decoder architecture in PhyLoNet is different than PhyDNet. Instead of using a U-Net based encoder-decoder [38] which relies heavily on the skip-connections for generating future frames, we use simple CNN based encoder-decoder [62] without any skip connections. This approach allows a generation with more flexibility from the latent vector.

Nevertheless, removing the skip connections might result in poor frame quality. In order to preserve the perceptual information in the video and increase the frame quality, we use the perceptual loss as presented in [16]. The perceptual loss is based on a pre-trained VGG-16 network and constructed as follows:

$$L_{perceptual} = \alpha \cdot L_{content} + \beta \cdot L_{style} \qquad (1)$$

$L_{perceptual}$ is the perceptual loss, $L_{content}$ and L_{style} are the content and style loss controlled by the coefficients α and β respectively.

$$L_{content} = \sum_{l \in F_c} ||P_l(G) - P_l(P)||_2^2 \qquad (2)$$

$$L_{style} = \sum_{l \in F_s} ||\psi_l(G) - \psi_l(P)||_F^2 \qquad (3)$$

P_l defines the feature vector of layer l extracted from VGG-16 network, G is the ground-truth frame and P is the prediction frame. ψ_l is the Gram matrix of feature vector l and F stands for Frobenius distance. F_s and F_c are the style and content layers from the VGG-16 network. The intuition behind the perceptual loss is to extract the style and content features from a pre-trained network instead of using hand-crafted labels, by doing so, the training process remains unsupervised.

3.2 Relative Flow Loss

Let $[g_1, g_2, ...g_N]$ be the ground truth frame sequence and Let $[p_1, p_2, ...p_N]$ be the predicted frames sequence (see Fig. 3). OF is the optical flow, which is the

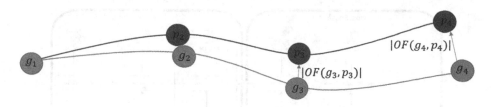

Fig. 3. The relative flow loss (RF-Loss) illustration. The loss measures the flow between the ground truth motion frames (g2–g4) and the predicted motion frames (p2–p4). (Color figure online)

pattern of apparent motion. The optical flow is a 2D vector field where each vector is a displacement vector showing the movement of points from the first frame to the second. The relative flow loss is defined as the optical flow between corresponding ground truth and predicted frames along the video sequence. This is an effective loss in the unsupervised learning process since frames lack any semantic annotations and labels.

In the traditional optical flow loss (OF-loss), the optical flow is calculated between two consecutive frames:

$$L_{of} = \sum_{t=1}^{N-1} ||OF(g_t, g_{t+1}) - OF(p_t, p_{t+1})||^2 \qquad (4)$$

In contrast, RF-loss is calculated as follows:

$$L_{rf} = \sum_{t=1}^{N} |OF(g_t, p_t)| \qquad (5)$$

where the optical flow is calculated by the RAFT network [47] and N is the number of predicted frames.

The motivation behind the relative flow loss is to increase the penalty for frames with objects that deviate from their original route at further time steps. Standard optical flow loss is focused locally on two consecutive images and measures the directional differences and deviations. Instead, the relative flow loss measures the prediction error between pairs of ground truth and the corresponding predicted frame. Thus, instead of measuring pixel-level differences locally, our loss accounts for global deviations and motions between ground truth and predicted frames. Figure 3 illustrates the RF-loss calculation. Blue/orange balls refer to corresponding (w.r.t. time step) GT/predicted frames respectively. We calculate at each time step the optical flow between ground truth and prediction, then we take the absolute value as the RF-loss, which resembles the global motion structural error between the frames.

3.3   Video Manipulation

Video manipulation refers to the ability to influence the predicted motion inside the video. To achieve this, we allow our network to obtain class labels along

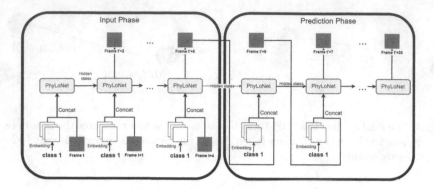

Fig. 4. The label constrained PhyLoNet. The **class 1** label is fed for video prediction editing.

the prediction process that are concatenated to the frames. See Fig. 4 for the network architecture.

Manipulating the video toward the desired class requires assimilating a label into the frame representation. LSTMs do not accept all frames at once, instead, it performs an iterative feeding of the previous states while keeping a "memory" of the important parts of each hidden state. Therefore, in order to create substantial assimilation of the label, we embed the class label in the same shape as the image and concatenate it to the image. The encoder then accounts for both the label and the image when encoding them together into the latent space.

In Fig. 4 we illustrate the PhyLoNet architecture for video manipulation. First, we embed our desired class label and reshape it into our image dimension, then we perform the concatenation and forward it to the PhyLoNet for the next frame prediction. The embeddings are for both the input and prediction phases.

3.4 Loss Functions

Figure 1 presents in red blocks the loss functions at time step t. The total loss function of our network is defined as

$$L_{total} = \gamma \cdot L_{moment} + \delta \cdot L_{perceptual} + \epsilon \cdot L_{rf} \tag{6}$$

where L_{moment} is the image moment loss, $L_{perceptual}$ is the perceptual loss, and L_{rf} is our proposed relative flow loss. We set γ, δ, ϵ to be 1, 1 and 0.00005 respectively. The coefficients α and β in $L_{Perceptual}$ are set to 0.1 and 0.05.

4   Results

We have run several video prediction and manipulation experiments, comparisons, and ablation studies to evaluate our network. To generate a ground truth, we have generated motion video datasets using various simulations. Then, we trained our network for each motion video dataset on an RTX-2080 GPU and predicted future frames. In the video manipulation experiments, we trained on a GTX-1080 GPU. All models in experiments are trained with a batch size of 4 and image size of 128×128 except the experiment of **Sin-Moving-Ball-Labeled** dataset which has a batch size of 8 and image size of 64×64. We used an Adam optimizer with a learning rate of $1e - 4$. Our code is available at https://github.com/nirey10/PhyLoNet.

4.1 Datasets

We have generated the following synthetic datasets for our experiments.

Ball-Collisions - The data is generated by the Unity physics engine. Each video consists of 3 balls at random positions within a solid bounding frame. Each ball is initialized with a random speed vector. Balls are moving upon a fraction-less surface, which means no energy loss. This Dataset consists of 1800 training videos and 200 test videos. Each video is composed of 30 frames.

Moving-Ball-Labeled - This dataset consists of labeled videos of 3 categories, go-left, go-middle, and go-right. Each category consists of 300 videos of 30 frames each. Videos in each category depict random routes towards the desired target from different camera angles and for 3 different balls.

Sin-Moving-Ball-Labeled - This dataset is similar to the **Moving-Ball-Labeled** dataset with one main difference regarding the ball motion. Specifically, we introduce specific motion characteristics for each label. In the go-left, balls have a straight movement behavior, in the go-middle label, balls have a small sinus amplitude in their movement, and in the go-right label balls have a big sinus amplitude movement behavior.

Flocking - The flocking dataset consists of a simulation of 30 birds moving around as a flock based on swarm optimization. Flocking is simulated using the Pygame library. This dataset contains 1800 videos for training and 200 for test. Each video is composed of 100 frames.

Pool - The Pool dataset consists of a simulated pool game. The data is generated by the Unity engine. It consists of 2000 videos of 30 frames each separated into 2 classes. Class 0 represents "miss" and class 1 represent "score". The black ball is the cue ball and the blue ball (target) is placed randomly in front of the cue ball. The pot is an orange rectangle surrounded by walls on each side. The videos consist of various ball initializations in term of position, speed, and heading degree.

4.2 Ball Collisions

The ball collision dataset demonstrates physical dynamics and collision interactions between balls, surrounded by solid frames, and other balls. In this experiment, the system obtains 5 frames as input and then predicts the next 100 frames.

Figure 5 shows prediction comparisons between PhyDNet (top) and Phy-LoNet (bottom). Figures show several predicted frames overlayed on top of each

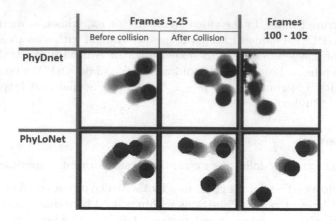

Fig. 5. Three balls collision prediction comparison. Frames 5–25 demonstrate short-term ball motion before and after the collision, while frames 100–105 show long-term prediction.

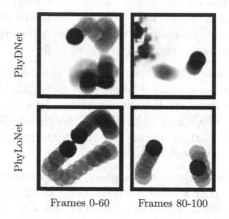

Fig. 6. Demonstration of the PhyDNet and PhyLoNet model's generalization over 2 balls collision.

other with opacity for demonstration of motion prediction. Given the initial 5 input frames, both models are able to predict ball collisions at frames 5–25. A closer look will show that the physical dynamics are much more realistic in our method, see the motion angle of 2 topmost balls after the collision, as well as the ball shape conservation. Frames 100–105 show the quality differences of our long-term prediction compared to PhyDNet, we preserve both physical dynamics and residual information intact.

4.3   Ball Collisions Generalization

We test the generalization capability of our network. For this purpose, we create a dataset consisting of 2 ball collision scenes and evaluate its prediction quality

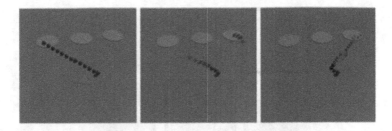

Fig. 7. Ball movement manipulation from ground truth route (left image) towards the go-right target using PhyDNet (middle image) and our method (right image).

over our pre-trained network that was trained on the 3 balls dataset (Ball-Collisions dataset).

Figure 6 compares generalization of both PhyDNet and PhyLoNet on 2 balls motion over 100 frames. As can be seen, PhyDNet generates frames with 3 balls at the very first predictions and is unable to generalize for 2 balls while our network is able to produce high-quality and realistic results in terms of both collision dynamics and the structural preservation. Frames 80–100 show that our generalization holds for long-term prediction as PhyLoNet predicts realistic and high-quality ball motions while PhyDNet contains severe artifacts and inadequate dynamics.

4.4 Video Manipulation

In order to demonstrate video manipulation, we experiment with the **Moving-Ball-Labeled** and **Sin-Moving-Ball-Labeled** datasets. Specifically, ball motions consist of additional class labels according to the video content. Thus, given an initial input motion sequence that relates to one of the classes, the network aims at predicting future frames according to a different class label chosen by the user while maintaining the consecutive path of the moving objects.

In Fig. 7 the ground truth ball motion was go-left target. Taking only the first 5 frames of the motion, PhyDNet and PhyLoNet were given also a go-right label, in order to manipulate the network prediction.

PhyDNet was unable to manipulate the ball movement towards the right target, instead, the ball disappears in-between and comes back near the right circle in the last frames. In contrast, our network predicted a continuous ball movement towards the right target. Artifacts are due to a lack of residual connections which reduce output quality but allow freedom in the frame generation. In addition, the sharp trajectory changes after 5 input frames indicate the powerful control of our model.

Similarly, we demonstrate the ability to manipulate ball movement also w.r.t. their style and not only their target. We use our **Sinus-ball-Movement-Labeled** dataset in order to train on three different movement class labels: straight, small sinus amplitude, and big sinus amplitude movements.

Fig. 8. PhyLoNet video manipulation of motion target and style. Each row shows a different ground-truth label (black framed images) and its manipulation of the other class labels.

In Fig. 8 the bold images in the diagonal are the ground-truth videos before manipulation. Thus, we show three different samples where each sample has a different target and style.

We then take each ground truth sample and manipulate it to a different target (rows). Our model is able to augment the original route towards the new destination label together with the corresponding movement style w.r.t. sinus amplitude getting bigger as we move to the "right" label.

4.5   Flocking

In this experiment, we test video prediction, in pixel space, of flocking motions generated by a swarm optimization. We train our network on the **Flocking** dataset. In the prediction step, input for our models is solely 5 frames and we predict the next 100 frames.

Figure 9 shows the flocking ground truth motion (left col) and a comparison between PhyDNet (mid-col) and our network (right-col) flocking motion prediction. The top row is the merged frames and the bottom row is the predicted flocking motions emphasized by colored vectors.

PhyDnet was unable to predict accurate flocking behavior such as cohesion. The green and red motions are performing cohesion in GT but they do not continue with the average direction in the prediction. Similarly, the blue motion is interfered by other birds and does not continue toward the dedicated target.

In contrast, our model predicts accurate cohesions for all 3 cases in this scene while keeping the speed and average heading intact.

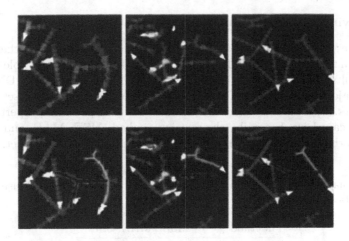

Fig. 9. Flocking motion prediction. Left-column is the GT motion, PhyDNet prediction is middle-column and our PhyLoNet prediction is right-column. (Color figure online)

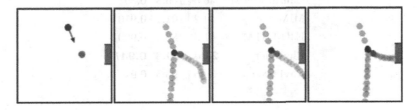

Fig. 10. Demonstration of a ball's route change towards the class label "score". The original class label of the sample is "miss". (Color figure online)

4.6 Pool

We present another video manipulation experiment on our **Pool** dataset. The data consists of pool ball interactions labeled with "score" and "miss" categories. We then feed the model with 5 frames as input and predict the next 25 frames according to the desired label category as given by the user.

Figure 10 demonstrates manipulation results for a pool ball after its collision with the cue ball. The orange rectangle is the "pot" surrounded by 2 solid walls on each side. Left-to-right, starting from an initial setup the blue ball does not move until the collision occurs, the initial cue ball's movement is demonstrated using an arrow. The original GT ball "miss" movement is shown (mid-left) followed by PhyDNet (mid-right) and PhyLoNet (right) "score" manipulation.

Our model is capable of manipulating the original scene towards our desired label, it augments the ball's route directly to the orange "pot" and produces a "score" labeled sample. The PhyDNet results are very similar to the ground truth and the model is unable to manipulate the video toward our desired class.

4.7   Moving-MNIST Comparison

Table 1 presents quantitative results of PhyLoNet compared to other baseline methods on the Moving-MNIST dataset. Although PhyLoNet was able to achieve better results from most of the baselines it just arrived second to PhyDNet. There are two major reasons for that. First, Moving-MNIST is not a dataset suited for our network. I.e, our method focuses on motions with physical properties, which is not the case here. Secondly, the metrics applied, MSE, MAE, and SSIM, do not reflect the performances on future physical interactions (see Sect. 4.8).

Table 1. Quantitative results of PhyLoNet compared to baseline models using Moving-MNIST dataset.

Method	MSE	MAE	SSIM
ConvLSTM	103.3	182.9	0.707
PredRNN	56.8	126.1	0.867
Causal LSTM	46.5	106.8	0.898
MIM	44.2	101.1	0.910
E3D-LSTM	41.3	86.4	0.920
PhyDNet	**24.4**	**70.3**	**0.947**
PhyLoNet	34.5	93.5	0.921

4.8   Ablation Study

We show the effect of our network architecture and our proposed loss in the ablation study experiments. We refer to PhyDNet as the original architecture [21] and PhyLoNet as our proposed architecture. The loss functions we incorporate are the Relative Flow loss (RF-Loss) and the original Optical Flow loss as described in Sect. 4.2.

Relative Flow Loss for Complex Dynamics Prediction. We evaluate the effect of our relative flow loss compared to the standard optical flow approach and the original PhyDNet model. We perform an ablation study on the 3-ball collision dataset. Since small deviations in collision returning angles cause large deviations over time, we compare only the first 30 frames after the collision.

In Fig. 11, we compare the performances of PhyDNet, PhyLoNet with RF-Loss, and PhyLoNet with Optical Flow Loss models. The left graph shows MSE error with GT. We observe that the MSE method is quite general metric and does not reflect the performances on future physical interactions, especially for multi-object collisions. In order to perform a better, customized evaluation of our method performance we define a new metric that tracks each ball position over time. The object tracking is implemented using OpenCV's CSRT object tracker. We compare the cumulative Euclidean distances of every ball centroid between the predicted and ground-truth frames. In Fig. 11 (right) we can see the model's comparison using our multi-object tracking metric.

In all metrics PhyLoNet outperforms PhyDNet, indicating the strong effect of the encoder-decoder over long-term prediction. The RF-Loss also shows a significant contribution to the model performance. Last, we see that RF loss outperforms the optical flow loss for the prediction of complex physic dynamics.

Relative Flow Loss for Video Manipulation. In this study, we evaluate the effect of our relative flow loss on the **Pool** dataset which combines both physical interactions, collisions, and video manipulation. In this context, we evaluate the balance between accurate physical dynamics prediction and the flexibility to perform manipulations.

Figure 12 presents two different manipulation examples. Each example is originally labeled as a "miss" and the red arrow marks the ground-truth motion vector of the blue ball after the collision. The left image in each example shows the "score" manipulations using PhyLoNet without RF-Loss, while the right image shows PhyLoNet with RF-Loss. Both models are able to manipulate the video towards the "score" label. Nevertheless, PhyLoNet with RF-Loss is more faithful to the GT data and therefore the ball motion is more restricted by the original dynamics.

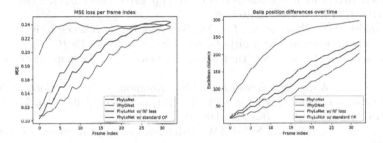

Fig. 11. Ablation study results for the **Ball-Collision-3** dataset using two different metrics

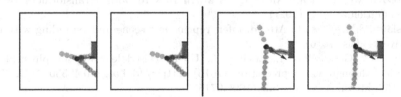

Fig. 12. Two examples (separated by a line) of manipulating the "miss" label into a "score" label using two different models: PhyLoNet without RF-Loss (left image) vs. PhyLoNet with RF-Loss (right image). (Color figure online)

5   Conclusions and Limitations

We introduce an enhanced PhyDNet design for unsupervised long-term video dynamics prediction, the PhyLoNet. We introduce a novel loss that accounts for complex dynamics. We also introduce a model to control and manipulate videos by changing their class label. The results demonstrate that our model is capable of handling complicated physical dynamics for long-term prediction. We believe this work is the basis of further physically constrained video prediction tasks and its contributions can be applied to more complicated motion domains.

In terms of limitations, we have tested our models only on synthetic data with a smooth static background. This is because we aimed at focusing on the physical and biological dynamics instead of dealing with computer vision tasks such as FG/BG separation, background stabilization and etc. More complex video settings and in general in-the-wild datasets are currently left for future work.

References

1. Aoyagi, Y., Murata, N., Sakaino, H.: Spatio-temporal predictive network for videos with physical properties. In: 2021 IEEE/CVF Conference on Computer Vision and Pattern Recognition Workshops (CVPRW), pp. 2268–2278 (2021). https://doi.org/10.1109/CVPRW53098.2021.00256
2. Battaglia, P.W., Pascanu, R., Lai, M., Rezende, D., Kavukcuoglu, K.: Interaction networks for learning about objects, relations and physics (2016)
3. Brabandere, B.D., Jia, X., Tuytelaars, T., Gool, L.V.: Dynamic filter networks (2016)
4. Byeon, W., Wang, Q., Srivastava, R.K., Koumoutsakos, P.: ContextVP: fully context-aware video prediction (2017). https://doi.org/10.48550/ARXIV.1710.08518. https://arxiv.org/abs/1710.08518
5. Cuturi, M., Blondel, M.: Soft-DTW: a differentiable loss function for time-series (2017)
6. Denton, E., Birodkar, V.: Unsupervised learning of disentangled representations from video (2017)
7. Dosovitskiy, A., et al.: An image is worth 16×16 words: transformers for image recognition at scale (2021)
8. Eslami, S.M.A., et al.: Attend, infer, repeat: fast scene understanding with generative models (2016)
9. Finn, C., Goodfellow, I., Levine, S.: Unsupervised learning for physical interaction through video prediction (2016). https://doi.org/10.48550/ARXIV.1605.07157. https://arxiv.org/abs/1605.07157
10. Gal, Y., Ghahramani, Z.: Dropout as a Bayesian approximation: representing model uncertainty in deep learning (2015)
11. Gao, H., Xu, H., Cai, Q.Z., Wang, R., Yu, F., Darrell, T.: Disentangling propagation and generation for video prediction (2019)
12. Guen, V.L., Thome, N.: Shape and time distortion loss for training deep time series forecasting models (2019)
13. Hsieh, J.T., Liu, B., Huang, D.A., Fei-Fei, L., Niebles, J.C.: Learning to decompose and disentangle representations for video prediction (2018)

14. Hui, T.W., Tang, X., Loy, C.C.: LiteFlowNet: a lightweight convolutional neural network for optical flow estimation (2018)
15. Ilg, E., Mayer, N., Saikia, T., Keuper, M., Dosovitskiy, A., Brox, T.: FlowNet 2.0: evolution of optical flow estimation with deep networks (2016)
16. Johnson, J., Alahi, A., Fei-Fei, L.: Perceptual losses for real-time style transfer and super-resolution. In: Leibe, B., Matas, J., Sebe, N., Welling, M. (eds.) ECCV 2016. LNCS, vol. 9906, pp. 694–711. Springer, Cham (2016). https://doi.org/10.1007/978-3-319-46475-6_43
17. Kipf, T., Fetaya, E., Wang, K.C., Welling, M., Zemel, R.: Neural relational inference for interacting systems (2018)
18. Kosiorek, A.R., Kim, H., Posner, I., Teh, Y.W.: Sequential attend, infer, repeat: generative modelling of moving objects (2018)
19. Krishnan, R.G., Shalit, U., Sontag, D.: Deep Kalman filters (2015)
20. Kwon, Y.H., Park, M.G.: Predicting future frames using retrospective cycle GAN. In: 2019 IEEE/CVF Conference on Computer Vision and Pattern Recognition (CVPR), pp. 1811–1820 (2019). https://doi.org/10.1109/CVPR.2019.00191
21. Le Guen, V., Thome, N.: Disentangling physical dynamics from unknown factors for unsupervised video prediction. In: Computer Vision and Pattern Recognition (CVPR) (2020)
22. Li, Y., Fang, C., Yang, J., Wang, Z., Lu, X., Yang, M.-H.: Flow-grounded spatial-temporal video prediction from still images. In: Ferrari, V., Hebert, M., Sminchisescu, C., Weiss, Y. (eds.) ECCV 2018. LNCS, vol. 11213, pp. 609–625. Springer, Cham (2018). https://doi.org/10.1007/978-3-030-01240-3_37
23. Liang, X., Lee, L., Dai, W., Xing, E.P.: Dual motion GAN for future-flow embedded video prediction (2017)
24. Liu, Z., Yeh, R.A., Tang, X., Liu, Y., Agarwala, A.: Video frame synthesis using deep voxel flow (2017)
25. Long, Z., Lu, Y., Dong, B.: PDE-Net 2.0: learning PDEs from data with a numeric-symbolic hybrid deep network. J. Comput. Phys. **399**, 108925 (2019). https://doi.org/10.1016/j.jcp.2019.108925
26. Long, Z., Lu, Y., Ma, X., Dong, B.: PDE-Net: learning PDEs from data (2018)
27. Luo, Z., Peng, B., Huang, D.A., Alahi, A., Fei-Fei, L.: Unsupervised learning of long-term motion dynamics for videos (2017)
28. Mathieu, M., Couprie, C., LeCun, Y.: Deep multi-scale video prediction beyond mean square error (2015)
29. Mo, S., Cho, M., Shin, J.: InstaGAN: instance-aware image-to-image translation (2019)
30. Mrowca, D., et al.: Flexible neural representation for physics prediction (2018)
31. Oliu, M., Selva, J., Escalera, S.: Folded recurrent neural networks for future video prediction. In: Ferrari, V., Hebert, M., Sminchisescu, C., Weiss, Y. (eds.) Computer Vision – ECCV 2018. LNCS, vol. 11218, pp. 745–761. Springer, Cham (2018). https://doi.org/10.1007/978-3-030-01264-9_44
32. Palm, R.B., Paquet, U., Winther, O.: Recurrent relational networks (2017)
33. Pan, T., Jiang, Z., Han, J., Wen, S., Men, A., Wang, H.: Taylor saves for later: disentanglement for video prediction using Taylor representation. Neurocomputing **472**, 166–174 (2022)
34. Patraucean, V., Handa, A., Cipolla, R.: Spatio-temporal video autoencoder with differentiable memory (2015)
35. Raissi, M.: Deep hidden physics models: deep learning of nonlinear partial differential equations. J. Mach. Learn. Res. **19**(1), 932–955 (2018)

36. Raissi, M., Perdikaris, P., Karniadakis, G.E.: Physics informed deep learning (part II): data-driven discovery of nonlinear partial differential equations (2017)
37. Ranjan, A., Black, M.J.: Optical flow estimation using a spatial pyramid network (2016)
38. Ronneberger, O., Fischer, P., Brox, T.: U-Net: convolutional networks for biomedical image segmentation. In: Navab, N., Hornegger, J., Wells, W.M., Frangi, A.F. (eds.) MICCAI 2015. LNCS, vol. 9351, pp. 234–241. Springer, Cham (2015). https://doi.org/10.1007/978-3-319-24574-4_28
39. Rudy, S.H., Brunton, S.L., Proctor, J.L., Kutz, J.N.: Data-driven discovery of partial differential equations. Sci. Adv. **3**(4), e1602614 (2016)
40. Sanchez-Gonzalez, A., et al.: Graph networks as learnable physics engines for inference and control (2018)
41. Seo, S., Liu, Y.: Differentiable physics-informed graph networks (2019)
42. Shi, X., Chen, Z., Wang, H., Yeung, D.Y., Wong, W.K., Woo, W.C.: Convolutional LSTM network: a machine learning approach for precipitation nowcasting (2015). https://doi.org/10.48550/ARXIV.1506.04214. https://arxiv.org/abs/1506.04214
43. Shi, X., Chen, Z., Wang, H., Yeung, D.Y., Wong, W.K., Woo, W.C.: Convolutional LSTM network: a machine learning approach for precipitation nowcasting. In: Proceedings of the 28th International Conference on Neural Information Processing Systems, NIPS 2015, vol. 1, pp. 802–810. MIT Press, Cambridge (2015)
44. Srivastava, N., Mansimov, E., Salakhutdinov, R.: Unsupervised learning of video representations using LSTMs (2015). https://doi.org/10.48550/ARXIV.1502.04681. https://arxiv.org/abs/1502.04681
45. van Steenkiste, S., Chang, M., Greff, K., Schmidhuber, J.: Relational neural expectation maximization: unsupervised discovery of objects and their interactions (2018)
46. Sun, D., Yang, X., Liu, M.Y., Kautz, J.: PWC-Net: CNNs for optical flow using pyramid, warping, and cost volume (2018)
47. Teed, Z., Deng, J.: RAFT: recurrent all-pairs field transforms for optical flow. In: Vedaldi, A., Bischof, H., Brox, T., Frahm, J.-M. (eds.) ECCV 2020. LNCS, vol. 12347, pp. 402–419. Springer, Cham (2020). https://doi.org/10.1007/978-3-030-58536-5_24
48. Tulyakov, S., Liu, M.Y., Yang, X., Kautz, J.: MoCoGAN: decomposing motion and content for video generation (2017)
49. Vaswani, A., et al.: Attention is all you need (2017)
50. Villegas, R., Yang, J., Hong, S., Lin, X., Lee, H.: Decomposing motion and content for natural video sequence prediction (2018)
51. Vondrick, C., Pirsiavash, H., Torralba, A.: Generating videos with scene dynamics (2016)
52. Wang, Y., Gao, Z., Long, M., Wang, J., Yu, P.S.: PredRNN++: towards a resolution of the deep-in-time dilemma in spatiotemporal predictive learning (2018). https://doi.org/10.48550/ARXIV.1804.06300. https://arxiv.org/abs/1804.06300
53. Wang, Y., Jiang, L., Yang, M.H., Li, L.J., Long, M., Fei-Fei, L.: Eidetic 3D LSTM: a model for video prediction and beyond. In: ICLR (2019)
54. Wang, Y., et al.: PredRNN: a recurrent neural network for spatiotemporal predictive learning (2021). https://doi.org/10.48550/ARXIV.2103.09504. https://arxiv.org/abs/2103.09504
55. Wang, Y., Zhang, J., Zhu, H., Long, M., Wang, J., Yu, P.S.: Memory in memory: a predictive neural network for learning higher-order non-stationarity from spatiotemporal dynamics (2018). https://doi.org/10.48550/ARXIV.1811.07490. https://arxiv.org/abs/1811.07490

56. Watters, N., Tacchetti, A., Weber, T., Pascanu, R., Battaglia, P., Zoran, D.: Visual interaction networks (2017)
57. Wu, J., Lu, E., Kohli, P., Freeman, W.T., Tenenbaum, J.B.: Learning to see physics via visual de-animation. In: Proceedings of the 31st International Conference on Neural Information Processing Systems, NIPS 2017, Red Hook, NY, USA, pp. 152–163. Curran Associates Inc. (2017)
58. Wu, Y., Gao, R., Park, J., Chen, Q.: Future video synthesis with object motion prediction (2020)
59. Xu, J., Ni, B., Li, Z., Cheng, S., Yang, X.: Structure preserving video prediction. In: 2018 IEEE/CVF Conference on Computer Vision and Pattern Recognition, pp. 1460–1469 (2018). https://doi.org/10.1109/CVPR.2018.00158
60. Xue, T., Wu, J., Bouman, K.L., Freeman, W.T.: Visual dynamics: probabilistic future frame synthesis via cross convolutional networks (2016)
61. Yin, Y., et al.: Augmenting physical models with deep networks for complex dynamics forecasting. J. Stat. Mech. Theory Exp. **2021**(12), 124012 (2021). https://doi.org/10.1088/1742-5468/ac3ae5
62. Zhu, J.Y., Park, T., Isola, P., Efros, A.A.: Unpaired image-to-image translation using cycle-consistent adversarial networks (2020)

Video Object Segmentation via Structural Feature Reconfiguration

Zhenyu Chen[1], Ping Hu[2], Lu Zhang[1], Huchuan Lu[1], You He[3(✉)], Shuo Wang[4], Xiaoxing Zhang[4], Maodi Hu[4], and Tao Li[4]

[1] Dalian University of Technology, Dalian, China
dlutczy@mail.dlut.edu.cn, lhchuan@dlut.edu.cn
[2] Boston University, Boston, USA
pinghu@bu.edu
[3] Naval Aeronautical University, Yantai, China
youhe_nau@163.com
[4] Meituan, Beijing, China
{wangshuo28,zhangxiaoxing,humaodi,litao19}@meituan.com

Abstract. Recent memory-based methods have made significant progress for semi-supervised video object segmentation, by explicitly modeling the semantic correspondences between the target frame and the historical ones. However, the indiscriminate acceptance of historical frames into the memory bank and the lack of fine-grained extraction for target objects may incur high latency and information redundancy in these approaches. In this paper, we circumvent the challenges by developing a Structural Feature Reconfiguration Network (SFRNet). The proposed SFRNet consists of two core sub-modules, which are Global-temporal Attention Module (GAM) and Local-spatial Attention Module (LAM). In GAM, we exploit self-attention-based encoders to capture the target objects' temporal context from historical frames. The LAM then reconfigures features with the current frame's spatial structural prior, which reinforces the objectness of foreground objects and suppresses the interference from background regions. By doing so, our model reduces the reliance on the large memory bank containing redundant historical frames, while instead effectively segmenting video objects with spatio-temporal context aggregated from a small set of key frames. We conduct extensive experiments with benchmark datasets, and the results demonstrate our method's favorable performance against the state-of-the-art approaches. The code will be available at https://github.com/zy5037/SFRNet.

Keywords: Video object segmentation · Structural feature reconfiguration · Global-temporal attention · Local-spatial attention

1 Introduction

Video Object Segmentation (VOS) aims to segment out the interested objects along the video sequence. It has received great attention recently because of its

L. Wang et al. (Eds.): ACCV 2022, LNCS 13847, pp. 588–605, 2023.
https://doi.org/10.1007/978-3-031-26293-7_35

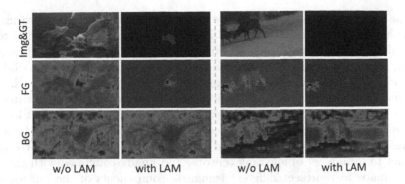

Fig. 1. Visualization of key feature channels. We observe that a target frame is structured as foreground (FG) and background (BG) regions by different feature channels. Our Local-spatial Attention Module (LAM) exploits the current frame's spatial structure to better extract discriminative structural feature representations, hence alleviating the demand for heavy historical memory.

benefits for applications like video surveillance, video editing, and multimedia analysis. In this paper, we focus on addressing semi-supervised video object segmentation, where the target objects are manually annotated in the first frame.

Thanks to the recent advances in deep learning techniques, state-of-the-art methods in VOS have achieved significant progress. Early methods typically propagate object masks over time via motion cues like optical flow [12,21–23,42, 58] or adopting an online learning strategy [5,20,24,27,45,49] to finetune on the first frame with annotations. However, the motion-based mask propagation may accumulate errors and online finetuning suffers from very low efficiency. Recently, matching-based methods have emerged as a promising solution for this task [8, 30,44,57,59]. Among these approaches, Space-Time Memory (STM) network [33] achieves great success, by extracting the spatio-temporal context from a memory bank, which is typically large and redundant to ensure effectiveness. In order to optimize the memory efficiency, several follow-ups of STM [11,37,38,46,50] have been proposed with improved encoding [38,46] and matching [11,37,50] process. Though significant progress toward this direction has been made, maintaining memory with both efficiency and effectiveness is still very challenging, due to the difficulties in balancing between the memory capacity and quality.

In this work, we circumvent the challenge by proposing a Structural Feature Reconfiguration Network (SFRNet), which alleviates the reliance on large memory banks by exploiting the spatial structural composition of testing frame. Given a video frame, we aim to discriminate foreground objects and background regions as different pixel sets that are spatially structured by their underlying semantic coherence. Therefore, the segmentation of video objects should not only benefit from pixel-level space-time correlations for referred object, but also a video frame's own spatial structural compositions as illustrated in Fig. 1. Based on this, we design SFRNet with two core components including a Global-temporal Attention Module (GAM) and a Local-spatial Attention Module (LAM).

The GAM extracts pixel-level spatio-temporal context from historical key frames with a Transformer-based architecture. And the LAM is adopted to further enhance the feature representations with spatial composition priors of the testing frame. To explicitly extract and represent a target frame's spatial composition, we draw inspiration from image subspace composition [9,40], where the deep model is trained to encode visual components in images as discriminative low-rank tensors [4,41,61]. Specifically, in LAM we explicitly construct low-rank feature maps by first collecting the contextual feature basis along the spatial dimensions of the feature maps. These basis are then aggregated via Kronecker Product to form a set of low-rank tensors, and finally combined with the input feature maps to represent different semantic components of the images. With end-to-end optimization, LAM is able to separate and encode visual information at object/region-level as illustrated in Fig. 1, hence achieving robustness for the quality of the temporal context aggregation in GAM. By combining GAM and LAM, our framework avoids heavy overheads caused by maintaining a large amount of memory, while achieving effectiveness in encoding and extracting visual objects in videos. Extensive experiments are performed to analyze the proposed method, and the results on multiple datasets [34,35,52] show that our proposed SFRNet can effectively segment video objects.

In summary, we have the following contributions:

- We propose a Local-spatial Attention Module that characterizes deep features with the structural composition prior to better extract video objects from the background.
- We develop a Structural Feature Reconfiguration Network (SFRNet), which utilizes space-time context aggregation and spatial structural composition of target objects, to relieve the dependency on heavily accumulated historical frames, and achieves effective video object segmentation.
- We conduct extensive experiments to demonstrate the effectiveness of the proposed method. Our SFRNet achieves the favorable performance against the state-of-the-art approaches on multiple datasets including *DAVIS16*, *DAVIS17*, and *YouTube-VOS*.

2   Related Work

Video Object segmentation. Learning video object segmentation via deep neural networks receives growing attention recently. To improve the model's generalization, early methods [5,24,27,45,49] usually rely on an online learning scheme that finetunes the deep model with the annotated first frame during testing. Despite the improvement in accuracy, the online finetuning process is quite time-consuming and hard to be applied in real-world applications. In recent years, with the advances in dense prediction tasks, STM [33] and CFBI [57] propose to build robust space-time correspondence modules and show great breakthroughs in performance against previous online approaches. They thus become the new baselines in VOS for further promotion on the accuracy [15,16,50] or efficiency [11,46]. The matching-based methods [15,44] like CFBI [57] usually build a multi-context feature matching mechanism between the query frame and key frames (typically the first frame and the recent frame) to encode the long-range similarity in semantics and the short-range similarity in appearance.

On the contrary, memory-based methods [11,19,28,46,50] like STM [33] aim to learn the pixel-wise space-time correspondence between the current frame and the historical ones. STM [33] introduces a memory mechanism, which resorts the non-local module to model the correspondence between query frame and memory frames. Through non-local module, the long-range dependencies among different frames can be established. However, simply including all the previous frames into the memory bank without selection would lead to memory explosion and a heavy computational burden. This motivates some recent attempts to improve the memory encoding strategy in STM. For example, SwiftNet [46] and AFB-URR [28] propose to further filter the redundant pixels and merge the similar ones in memory storage to alleviate the memory growth issue. RDE-VOS [25] instead limits the memory to a constant size to improve the model efficiency. RPCMVOS [53] suppress error propagation through a correction mechanism to avoid error accumulation. STCN [11] proposes to simplify the non-local calculation by replacing the original cosine distance with the L2 distance.

HMMN [38] builds a pyramid memory network where the multi-level features are incorporated to capture robust spatio-temporal correspondence. In this paper, we propose an effective memory encoding framework, in which the robust space-time coherence can be built on the key frames.

Transformer in Videos. Transformer [43] was originally proposed as a sequence-to-sequence model for machine translation and has become the mainstream baseline for natural language processing. Recently, it has been successfully applied to many computer vision tasks [6,14,29,48,60] and shown convincing performance w.r.t convolutional neural networks. Inspired by this, many attempts are made to explore the effectiveness of Transformer in video tasks. For example, TimeSformer [2] and STARK [54] extends the original Transformer to establish the spatio-temporal self-attention in video sequences. TransT [7] introduces Transformer to enhance the intra-correlation and inter-correlation, respectively. Recently, some Transformer-based models are proposed to tackle the VOS task. JOINT [32] incorporates the Transformer with an update mechanism for integrating transductive and inductive information into a unified framework. In this paper, we propose to build the Global-temporal Attention Module and Local-spatial Attention Module based on Transformer, for enhancing the structural representation of the target objects to achieve robust segmentation.

3   Method

We propose a Structural Feature Reconfiguration Network (SFRNet) for effective and efficient video object segmentation by enhancing the spatial structural representation of the referred objects. Given the ground-truth mask at the first frame, the proposed network aims to predict the masks in the following video sequence. The framework is shown in Fig. 2. We use two separate encoders to capture the

592   Z. Chen et al.

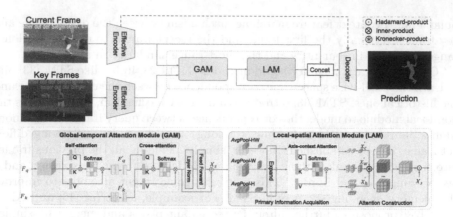

Fig. 2. An overview of our framework. Our network consists of an Effective Encoder and an Efficient Encoder for the feature extraction of query frame and key frames, respectively. The Global-temporal Attention Module (GAM) is used to construct the space-time correspondence at pixel level. The Local-spatial Attention Module (LAM) is used to enhance the structural composition of the target instances. Finally, the output of the two modules together with the skip-connections from effective encoder are fed to the decoder for mask generation.

embeddings for the current input I_t and the key frames I_k, respectively. To fix the memory storage issue, we use the first frame and a recent frame to form the key frames, *i.e.*, $I_k = \{I_1, I_r\}$. Following [11,46], we implement an Effective Encoder based on ResNet50 [17] to extract the features with rich semantics and spatial details for the current frame. Meanwhile, an Efficient Encoder is built on ResNet18 [17] to swiftly incorporate the memory embeddings for key frames. To capture historical information of the target objects, the concatenation of RGB images and predicted masks are fed to the efficient encoder as in [33].

With the extracted features of the current frame $F_q \in \mathbb{R}^{H \times W \times C}$ and key frames $F_k \in \mathbb{R}^{2 \times H \times W \times C}$, SFRNet achieves memory encoding and video object segmentation via a Global-temporal Attention Module (GAM) and a Local-spatial Attention Module (LAM). The GAM is proposed to extract information for target object from the key-frame set, and a Transformer based attention formulation is designed to capture the globally long-term correlation for pixels across frames. To improve the robustness of captured features, we build LAM to further enhance with the referred objects' spatial structural composition, which strengthens the objectness of foreground objects in the current frame and suppresses the interference from background noise. The structural enhanced features, together with GAM output and skip-connections, are fed to the decoder model for mask generation.

We adopt a commonly used decoder architecture as [11,33,38], which stacks several refinement modules to incorporate the skip-connections from Effective Encoder.

3.1   Global-temporal Attention Module

To implement GAM, we leverage the advanced Transformer [43], which shows superior capability to model the global and long-range context in dense prediction tasks [6,60]. We start by briefly reviewing the multi-head attention module in Transformer, which is the core unit of the proposed GAM. Given the spatially flattened d_m-channel feature map $X \in \mathbb{R}^{N \times d_m}$, the attention formulation of each head is as follows:

$$h_i = Softmax(\frac{Q_i K_i^\top}{\sqrt{d_k}})V_i \ . \tag{1}$$

where $Q_i \in \mathbb{R}^{N \times d_n}$, $K_i \in \mathbb{R}^{N \times d_n}$, and $V_i \in \mathbb{R}^{N \times d_n}$ are the *Query*, *Key*, and *Value* vectors respectively converted from the input vector X via several linear layers, d_k is a scaling factor equal to the output channel number d_n. The attention formulation in Eq. 1 can be performed in the form of multiple heads to produce richer representations.

$$MultiHead(Q, K, V) = Concat(h_1, ..., h_L) \cdot W^o. \tag{2}$$

where $Concat(\cdot)$ denotes the concatenation along the channel dimension, L is the number of heads, and $W^o \in \mathbb{R}^{(L \cdot d_n) \times d_m}$ is a linear layer that fuses the output vectors of the multi-head attention.

With the above-mentioned multi-head attention module, we build the Global-temporal Attention Module as in the left part of Fig. 2. Given the feature maps F_q of the current frame (also known as query frame), we first calculate the self-attention to exploit the spatial correlation. Specifically, we flatten the feature map $F_q \in \mathbb{R}^{H \times W \times C}$ over spatial dimensions and get a set of vectors $F_q' \in \mathbb{R}^{N \times C}$, with $N = H \times W$. The flattened vectors are then transferred to *Query*, *Key*, and *Value* vectors to formulate the multi-head attention as described in Eq. 1 and 2.

With the aggregated feature from the current frame, we then construct cross-context attention to capture target objects' temporal context from key frames. The cross-context attention is also based on multi-head attention, yet implemented on the enhanced current feature F_q' and the key frames' features F_k.

To better extract spatio-temporal context from key frames, we construct multiple stages of attention to aggregate information in an aggressive way. In detail, the flattened F_q' as *Query* and flattened F_k' are taken as *Key* and *Value* in cross-context attention of the first layer. In subsequent stages, the outputs of previous layer are sent to the self-attention for feature aggregation. The cross-context attention of subsequent stages takes the aggregated feature as *Query* and transform F_k' as *Key* and *Value* and finally generates the outputs of GAM, which are denoted as X_s.

3.2   Local-Spatial Attention Module

Many efforts have been made to improve the efficiency of memory encoding and matching [11,28,46]. Yet it is still very challenging to balance between the efficiency and capacity, *i.e.*, a large memory contains redundant information incurring heavy computational cost, and a small memory bank benefits computational efficiency yet may lack sufficient information for extracting details of video objects. We argue that this challenge can be circumvented by exploiting the target objects structural composition in spatial domain of the current testing frame, which helps to reduce model's reliance on the quality of historical information, and thus avoiding large-size memory while keeping effectiveness in representations. We propose to learn to model and exploit the spatial structure of visual components in the current frame. Inspired by the tensor low-rank decomposition of deep features [9,40], we recompose the feature maps via a set of low-rank components, which condense the key structural components for the objects and background regions. To this end, we build a Local-spatial Attention Module to characterize the output of cross-context features by GAM with region-level structure prior. The framework of LAM is shown in the right part of Fig. 2, which consists of two sub-modules: Primary Information Acquisition and Attention Construct Module.

Primary Information Acquisition (PIA). The goal of PIA is to learn to extract the discriminative basis vectors for the structural subspaces of the current frame. The architecture of PIA is shown in Fig. 3. Taking the cross-context feature $X_s \in \mathbb{R}^{H \times W \times C}$ from GAM as input, the PIA module first generates a set of spatial basis, which are later utilized to generate low-rank discriminative components. Specifically, we implement the average pooling on X_s along the *Height*, *Width*, and the full spatial dimensions to extract the contextual information of the target object and the current frame,

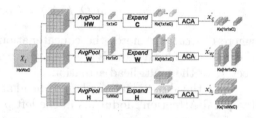

Fig. 3. Illustration of the Primary Information Acquisition (PIA) module. After pooling and expanding, we obtain the structural features in different spatial dimensions. "ACA" denotes the Axis-context Attention as described in Sect. 3.2.

$$x_h = AvgPool_H(X_s),$$
$$x_w = AvgPool_W(X_s), \tag{3}$$
$$x_c = AvgPool_{HW}(X_s).$$

where $AvgPool_H(\cdot)$, $AvgPool_W(\cdot)$ and $AvgPool_{HW}(\cdot)$ indicate the average pooling operation along height, width and spatial dimension, respectively. With this formulation, the cross-context feature can be compressed into three groups of basis feature vectors, which are expressed as $x_c \in \mathbb{R}^{1 \times 1 \times C}$, $x_h \in \mathbb{R}^{H \times 1 \times C}$ and $x_w \in \mathbb{R}^{1 \times W \times C}$. To further enhance the representation ability of the basis vectors, we feed them into a 1×1 convolutional layer and expand it by K times to obtain K groups of basis for each spatial dimension, $x'_c \in \mathbb{R}^{K \times 1 \times 1 \times C}$, $x'_h \in \mathbb{R}^{K \times H \times 1 \times C}$ and $x'_w \in \mathbb{R}^{K \times 1 \times W \times C}$.

Axis-Context Attention (ACA). After obtaining the semantic basis from PIA, we further build an ACA to enhance intra-correlation among the extracted basis. We implement this Axis-context Attention based on the multi-head self-attention. Taken $x'_h \in \mathbb{R}^{K \times H \times 1 \times C}$ as example, we flatten it to shape $KH \times C$ to produce *Query*, *Key* and *Value* features. Then, self-attention as defined in Eq. 1 and 2 is applied. Following the multi-head attention layer, we further add a norm layer and a feed forward network to enhance the fitting ability of the subspace characteristics. Note that the ACA is also applied to x'_w and x'_c in a similar manner.

Attention Construct Module (ACM). With PIA and ACA, we can obtain the feature basis for different spatial dimensions, which can be utilized to represent the structural composition of the target objects in the testing frame. To transform the structural prior to the original input (*i.e.*, X_s from Global-temporal Attention Module), we further propose an Attention Construct Module. The architecture of ACM

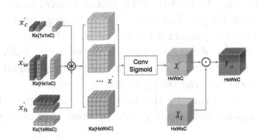

Fig. 4. Illustration of Attention Construct Module (ACM). X' indicates the reconstructed attention multiplied by X_s for generating final output F_o.

is shown in Fig. 4. Taking the feature basis of each group as input, the ACM first performs the low-rank structural component reconstruction by

$$x'_i = x'_{c_i} \odot x'_{h_i} \odot x'_{w_i}. \tag{4}$$

where $x'_i \in \mathbb{R}^{H \times W \times C}$ indicates the combined feature of each group and $\odot$ is Kronecker Production. By this means, the obtained feature x'_i is reconstructed to the original shape. Then, the reconstructed structural components are combined via weighted summation,

$$x' = \sum_{k=1}^{K} \alpha_k \cdot x'_k. \tag{5}$$

where $\alpha = \{\alpha\}_{k=1}^{K}$ are learnable weights. A Sigmoid function is then applied to convert the x' into a 3D attention matrix X'. Finally, the attention matrix X' is applied to the cross-context feature X_s to construct the output feature F_o of the local-spatial attention module, which will be fed to the decoder for mask generation. As shown in Fig. 1, compared with the baseline model, the generated structural feature shows great capability in capturing the discriminative embeddings of the target object and suppressing the inferential instances from the background.

4 Experiments

4.1 Dataset and Evaluation Metric

We evaluate the proposed SFRNet on three benchmark datasets for video object segmentation including DAVIS2016 [34], DAVIS2017 [35], and YouTube-VOS [52]. DAVIS2016 contains 50 high-quality videos with per-frame fine-grained annotations. In this dataset, the multiple instances of the video sequence are grouped as one object for segmentation. DAVIS2017 is an extension version of DAVIS2016, and consists of 60 sequences for training and 30 sequences for testing. In DAVIS-2017, instance-level video object segmentation is evaluated in each frame. YouTube-VOS is a large scale VOS dataset, which contains 3471 video sequences for training and 474/507 videos for validation in the 2018/2019 version of dataset. Compared with the DAVIS benchmark, videos in YouTube-VOS are more challenging with large variations in object motion, deformation and cluttered background.

To evaluate the proposed model, we adopt three metrics including mean region similarity $(\mathcal{J})$, mean contour accuracy $(\mathcal{F})$ and their average $(\mathcal{J}\&\mathcal{F})$.

4.2 Implementation Details

Parameter Settings. In our model, the number of low-rank structural component K is set to 16 in LAM, the head numbers L in Transformer is set to 4, and the number of attention layers in GAM is set to 3. We apply fixed *Sine* spatial positional embedding in the self-attention and the cross-attention.

Training Details. Following previous methods [11,28,33,37,38,46,50], we conduct a two-stage training process. At first, we pretrain the model using static image datasets [13,26,39,47,55], by constructing synthetic video data through affinity transformation and image augmentation operations. The learning rate is

Table 1. Comparison results on DAVIS2017 validation set.

Method	FRTM [36]	PReMVOS [31]	LWL [3]	STM [33]	CFBI [57]	CoVOS [51]	GraphMem [30]	KMN [37]	JOINT [32]	RDE [25]	HMMN [38]	STCN [11]	SFRNet
$\mathcal{J}\&\mathcal{F}$ ↑	76.7	77.8	81.6	81.8	81.9	82.4	82.8	82.8	83.5	84.2	84.7	85.4	**85.9**
$\mathcal{J}$ ↑	73.9	73.9	79.1	79.2	79.1	79.7	80.2	80.0	80.8	80.8	81.9	82.2	**82.7**
$\mathcal{F}$ ↑	79.6	81.7	84.1	84.3	84.6	85.1	85.2	85.6	86.2	87.5	87.5	88.6	**89.1**

Table 2. Comparison results on DAVIS2016 validation set.

Method	OSMN [56]	FEELVOS [44]	FRTM [36]	CINN [1]	CoVOS [51]	STM [33]	CFBI [57]	KMN [37]	HMMN [38]	RDE [25]	SFRNet (ours)
$\mathcal{J}\&\mathcal{F}$ ↑	73.5	81.7	83.5	84.2	89.1	89.3	89.4	90.5	90.8	91.1	**91.3**
$\mathcal{J}$ ↑	74.0	81.1	83.6	83.4	88.5	88.7	88.3	89.5	89.6	89.7	**90.5**
$\mathcal{F}$ ↑	72.9	82.2	83.4	85.0	89.6	89.9	90.5	91.5	92.0	**92.5**	92.1
FPS	9.3	2.9	17.7	-	39.6	8.4	7.2	-	11.6	40.0	20.4

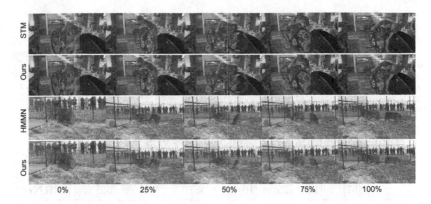

Fig. 5. Qualitative examples on DAVIS 2017 valid, and YouTube-VOS 2019 valid sets. The two examples show the comparisons of our method with STM and HMMN. Our method enables robust mask prediction in different scenarios.

set to 1e-5 in the first 150k iterations and decreases by 1/10 for the next 150k iterations. At the second stage, the pretrained model is further finetuned on YouTube-VOS and DAVIS2017. The initial learning rate is 1e-5 and decays by 1/10 at 125k iterations. The model converges after 150k iterations. The whole training process is conducted with 4 NVIDIA RTX 2080Ti GPUs. We use Adam optimizer with a batch size of 16 in pretraining and 8 at the finetuning stage. All the images and video frames are resized to 384×384 during training. The bootstrapped cross-entropy loss is applied for model optimization.

Testing Details. During testing, we use the first frame annotation and a recent prediction to form the key frames. The input images are kept at their original resolution for prediction. To avoid frequently running the efficient encoder on key frames, which decreases models' efficiency, we update the key frame features at fixed frequencies.

We compare the results on DAVIS2017 under different update frequencies in Table 7 and found that setting updating frequency to be 3 works best for our method.

4.3 Comparison to State-of-the-Art

DAVIS Datasets. We first compare the performance on DAVIS2017 for multi-instance video object segmentation. The results on the validation split of DAVIS-2017 are shown in Table 1. As we can see, our SFRNet achieves superior performance against previous online learning based methods and memory-based methods. To further verify the generalization of our model, we conduct comparison experiments on the test-dev of DAVIS2017 in supplementary material, our model again achieves better performance against previous state-of-the-art approaches. We also evaluate the performance for single-object on DAVIS2016

Table 3. Comparison results on YouTube-VOS 2018 validation set.

Method	PReMVOS [31]	FRTM [36]	CoVOS [51]	STM [33]	AFB-URR [28]	CFBI [57]	KMN [37]	LWL [3]	LCM [18]	HMMN [38]	STCN [11]	JOINT [32]	SFRNet
$\mathcal{G}$ ↑	66.9	72.1	79.0	79.4	79.6	81.4	81.4	81.5	82.0	82.6	83.0	83.1	**83.6**
$\mathcal{J}_S$ ↑	71.4	72.3	79.4	79.7	78.8	81.1	81.4	80.4	82.2	82.1	81.9	81.5	**82.4**
$\mathcal{F}_S$ ↑	75.9	76.2	83.6	84.2	83.1	85.8	85.6	84.9	86.7	87.0	86.5	85.9	**87.2**
$\mathcal{J}_U$ ↑	56.5	65.9	72.6	72.8	74.1	75.3	75.3	76.4	75.7	76.8	77.9	**78.7**	78.1
$\mathcal{F}_U$ ↑	63.7	74.1	80.4	80.9	82.6	83.4	83.3	84.4	83.4	84.6	85.7	86.5	**86.7**

Table 4. Comparison results on YouTube-VOS 2019 validation set.

Method	CFBI [57]	SST [15]	RDE [25]	MiVOS [10]	HMMN [38]	STCN [11]	JOINT [32]	SFRNet
$\mathcal{G}$ ↑	81.0	81.8	81.9	82.4	82.5	82.7	82.8	**83.3**
$\mathcal{J}_S$ ↑	80.6	80.9	81.1	80.6	81.7	81.1	80.8	**82.0**
$\mathcal{F}_S$ ↑	85.1	-	85.5	84.7	86.1	85.4	84.8	**86.5**
$\mathcal{J}_U$ ↑	75.2	76.6	76.2	78.2	77.3	78.2	**79.0**	78.2
$\mathcal{F}_U$ ↑	83.0	-	84.8	85.9	85.0	85.9	**86.6**	**86.6**

in Table 2. As we can see, our method achieves favorable performance in terms of both accuracy and speed.

YouTube-VOS. Tables 3 and 4 compare the proposed SFRNet with other state-of-the-art methods on the 2018/2019 validation sets of YouTube-VOS. Our method surpasses these existing top competitors with better overall accuracy on this benchmark. Among the existing approaches, JOINT [32] achieves the best $\mathcal{J}_U$ accuracy for unseen objects. This is because JOINT adopts an online learning strategy, which achieve better performance for unseen categories of objects but sacrifices the efficiency. Without testing-time finetuning on the first frame, SFR-Net achieves competitive accuracy for unseen objects compared to JOINT [32], and outperforms all the other methods, which demonstrates the superior generalization ability of our model.

Qualitative Results. Figure 5 lists the visual comparison between our SFR-Net with STM [33] and HMMN [38]. As we can see, the proposed SFRNet is able to effectively capture the structural representation of the target objects for high-quality mask generation. Besides, the proposed Global-temporal Attention Module and Local-spatial Attention Module cooperate to propose discriminative space-time correlation, which shows great effect in distinguishing the target object from interfered instances.

4.4 Method Analysis

In the following, we provide detailed analysis to demonstrate the effectiveness of the designs and modules in our method. For experiments in this part, we use the training splits of DAVIS2017 and YouTube-VOS for model training and report results on DAVIS2017 validation set.

Table 5. The effectiveness analysis of Global-temporal Attention and Local-spatial Attention Module on DAVIS2017 validation set.

	STM	GAM	LAM	TREnc	$\mathcal{J}\&\mathcal{F}$	$\mathcal{J}$	$\mathcal{F}$	FPS
M1	✓				81.2	77.8	84.6	16.4
M2	✓		✓		82.5	79.4	85.7	17.8
M3		✓			82.9	79.5	86.3	9.1
M4		✓	✓		83.5	80.3	86.7	13.1
M5	✓			✓	81.7	78.7	84.8	15.4

Table 6. Performance comparison between key-frame-based model and memory-based model.

	Architecture	MEM	KF	$\mathcal{J}\&\mathcal{F}$	$\mathcal{J}$	$\mathcal{F}$	FPS
M1	STM	✓		81.2	77.8	84.6	16.4
M2	STM		✓	80.8	77.3	84.3	21.5
M3	STM+LAM		✓	82.5	79.4	85.7	17.8
M4	GAM	✓		82.9	79.5	86.3	9.1
M5	GAM		✓	82.0	78.5	85.5	13.8
M6	GAM+LAM		✓	83.5	80.3	86.7	13.1

Table 7. Impact of key-frames update frequency on final accuracy.

DAVIS2017	Every 1	Every 2	Every 3
$\mathcal{J}\&\mathcal{F}$	84.0	85.3	85.9
DAVIS2017	Every 4	Every 7	Every 10
$\mathcal{J}\&\mathcal{F}$	85.7	83.2	83.0

Table 8. Impact of the number for structural components.

	$K = 4$	$K = 8$	$K = 16$	$K = 20$
$\mathcal{J}\&\mathcal{F}$	83.3	82.9	83.5	83.2
$\mathcal{J}$	80.1	79.8	80.3	80.3
$\mathcal{F}$	86.5	86.1	86.7	86.1

Effectiveness of Global-Temporal Attention Module. We verify the effectiveness of GAM by comparing its performance with a baseline model based on STM [33]. Compared with GAM, the STM baseline only contains a single head self-attention architecture. Based on the results of M1 and M3 in Table 5, we can observe that GAM achieves 1.7% improvement on $\mathcal{J}\&\mathcal{F}$ against the STM baseline. Besides, when combined with LAM, the GAM still outperforms non-local module by 1.0% on $\mathcal{J}\&\mathcal{F}$ (see M2 and M4 in Table 5). These results demonstrate that the multi-layer based GAM achieves more robust results than the STM baseline on the VOS task.

Effectiveness of Local-Spatial Attention Module. At first, we show the impact of LAM in Table 5. As shown, the application LAM (M2 and M4) helps to improve $\mathcal{J}\&\mathcal{F}$ by 1.3% and 0.6% for both the non-local baseline (M1) and GAM baseline (M3), respectively. To demonstrate the design of LAM, we compare it with a plain Transformer Encoder. As indicated by M1 and M5 in Table 5, adding a Transformer Encoder upon non-local module improves $\mathcal{J}\&\mathcal{F}$ by 0.5%. Yet, replacing the Transformer Encoder with LAM (M2 in Table 5) can further improve $\mathcal{J}\&\mathcal{F}$ by 0.8% and the speed from 15.4 FPS to 17.8 FPS. This demonstrates the effectiveness and efficiency of the LAM module.

Efficiency with Key-Frames. In this part, we analyze the efficiency of SFR-Net. As shown in Table 2, SFRNet achieves state-of-the-art accuracy at the speed of 20.4 FPS, which is faster than most of the previous methods. As introduced in the previous section, in SFRNet we adopt only two key frames for recording the historical information, which is in contrast to traditional memory-based methods that always add new frames into the memory bank. As shown in Table 6,

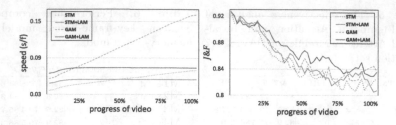

Fig. 6. Accuracy and time analysis for different methods on DAVIS2017 validation set.

Table 9. Impact of key-frames update frequency on final accuracy.

Memory frames	2	4	6
$\mathcal{J}\&\mathcal{F}$	85.9	86.1	86.3
GFLOPS	91.6	138.3	185.0

Table 10. Impact of different LAM attention on the output of GAM.

	Sigmoid	Add	Multi
$\mathcal{J}\&\mathcal{F}$	83.5	82.3	83.2

for both STM-only and GAM-only baselines, adopting two-key-frame strategy (M2 and M5) achieves faster speeds yet lower accuracy. This shows that a large memory bank is critical for effectiveness while bringing in more computational costs, which poses the challenge of balancing between capacity and efficiency in previous memory-based designs. In contrast, as indicated by M3 and M6, our proposed LAM with key frame based strategy keeps fast speeds while achieving the best accuracy. In Fig. 6, we compare efficiency and accuracy over time under different model designs. As we can see, for both STM-based and GAM-based model, the proposed LAM leads to better accuracy as well as faster and more stable speed over time. This again demonstrates the LAM's effectiveness to leverage local spatial composition of testing frames, which improves models' ability to effectively extract information from a small set of historical frames.

Update of Key-Frames. In Table 7, we investigate the impact of updating key-frames frequency. We notice that setting frequency every 3 frames performs best for both datasets, and either a lower or higher frequency doesn't perform better.

Larger Memory Bank and LAM Attention. As shown in Table 9, increasing the memory frames can still improve accuracy, yet decrease computational efficiency. With only 2 memory frames, the proposed SFRNet can already achieve state-of-the-art accuracy in the experiments showed above. This demonstrates that GAM and LAM work complementarily to boost the effectiveness and alleviate the dependence on large memory banks. In Table 10, we compare the different LAM attention operation on the output feature of GAM and the sigmoid operation performs the best.

Analysis of Feature Basis in LAM. We analyze the feature basis extracted in LAM for constructing the low-rank structural component. As discussed in

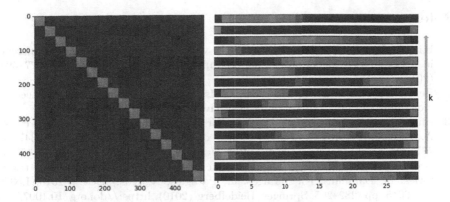

Fig. 7. Visualization of feature basis in LAM. The left side shows the affinity between the different feature basis. The right side visualizes a group of feature basis. Examples here are based on the features basis for "Height" dimension.

Sect. 3.2, we have K discriminative structural components produced by K groups of feature basis collected from different spatial dimensions. In Fig. 7, we visualize correlations between all $K \cdot H$ feature basis for the "Height" dimension. Note that these basis are organized by each of K groups. As we can see, the basis of the same group has a high correlation, and the basis of different groups presents a low correlation. This shows the orthogonality between different basis groups, which demonstrates that the constructed structural components are representing different aspects of structural composition for the target object as well as the scene. One group of the feature basis are also visualized in Fig. 7. In Table 8, we also vary the number of structural components in LAM and find that our model achieves the highest accuracy when $K = 16$.

5   Conclusion

We present SFRNet as a novel and effective framework for semi-supervised video object segmentation. In SFRNet, we first introduce a Global-temporal Attention Module (GAM) based on self-attention modules to capture the target objects' temporal context across frames. Then, the Local-spatial Attention Module (LAM) is proposed to further reconfigure features with a testing frame's spatial structural prior, so as to reinforce the objectness of foreground objects and suppress the interference from background regions. GAM and LAM work complementarily to extract target objects from video frames. Extensive experiments are conducted to analyze the effectiveness of SFRNet. The results demonstrate that our method achieves state-of-the-art results on multiple VOS benchmarks.

References

1. Bao, L., Wu, B., Liu, W.: CNN in MRF: video object segmentation via inference in a CNN-based higher-order Spatio-temporal MRF. In: CVPR, pp. 5977–5986 (2018)
2. Bertasius, G., Wang, H., Torresani, L.: Is space-time attention all you need for video understanding? arXiv preprint arXiv:2102.05095 (2021)
3. Bhat, G., et al.: Learning what to learn for video object segmentation. In: Vedaldi, A., Bischof, H., Brox, T., Frahm, J.-M. (eds.) ECCV 2020. LNCS, vol. 12347, pp. 777–794. Springer, Cham (2020). https://doi.org/10.1007/978-3-030-58536-5_46
4. Brox, T., Malik, J.: Object segmentation by long term analysis of point trajectories. In: Daniilidis, K., Maragos, P., Paragios, N. (eds.) ECCV 2010. LNCS, vol. 6315, pp. 282–295. Springer, Heidelberg (2010). https://doi.org/10.1007/978-3-642-15555-0_21
5. Caelles, S., Maninis, K.K., Pont-Tuset, J., Leal-Taixé, L., Cremers, D., Van Gool, L.: One-shot video object segmentation. In: CVPR (2017)
6. Carion, N., Massa, F., Synnaeve, G., Usunier, N., Kirillov, A., Zagoruyko, S.: End-to-end object detection with transformers. In: Vedaldi, A., Bischof, H., Brox, T., Frahm, J.-M. (eds.) ECCV 2020. LNCS, vol. 12346, pp. 213–229. Springer, Cham (2020). https://doi.org/10.1007/978-3-030-58452-8_13
7. Chen, X., Yan, B., Zhu, J., Wang, D., Yang, X., Lu, H.: Transformer tracking. In: CVPR (2021)
8. Chen, Y., Pont-Tuset, J., Montes, A., Van Gool, L.: Blazingly fast video object segmentation with pixel-wise metric learning. In: CVPR (2018)
9. Cheng, B., Liu, G., Wang, J., Huang, Z., Yan, S.: Multi-task low-rank affinity pursuit for image segmentation. In: ICCV (2011)
10. Cheng, H.K., Tai, Y.W., Tang, C.K.: Modular interactive video object segmentation: interaction-to-mask, propagation and difference-aware fusion. In: Proceedings of the IEEE/CVF Conference on Computer Vision and Pattern Recognition (2021)
11. Cheng, H.K., Tai, Y.W., Tang, C.K.: Rethinking space-time networks with improved memory coverage for efficient video object segmentation. arXiv preprint arXiv:2106.05210 (2021)
12. Cheng, J., Tsai, Y.H., Wang, S., Yang, M.H.: SegFlow: joint learning for video object segmentation and optical flow. In: ICCV (2017)
13. Cheng, M.M., Mitra, N.J., Huang, X., Torr, P.H., Hu, S.M.: Global contrast based salient region detection. IEEE Trans. Pattern Anal. Mach. Intell. **37**(3), 569–582 (2014)
14. Dosovitskiy, A., et al.: An image is worth 16 × 16 words: transformers for image recognition at scale. arXiv preprint arXiv:2010.11929 (2020)
15. Duke, B., Ahmed, A., Wolf, C., Aarabi, P., Taylor, G.W.: SSTVOS: sparse spatiotemporal transformers for video object segmentation. In: CVPR (2021)
16. Ge, W., Lu, X., Shen, J.: Video object segmentation using global and instance embedding learning. In: CVPR (2021)
17. He, K., Zhang, X., Ren, S., Sun, J.: Deep residual learning for image recognition. In: CVPR (2016)
18. Hu, L., Zhang, P., Zhang, B., Pan, P., Xu, Y., Jin, R.: Learning position and target consistency for memory-based video object segmentation. In: CVPR (2021)
19. Hu, P., Caba, F., Wang, O., Lin, Z., Sclaroff, S., Perazzi, F.: Temporally distributed networks for fast video semantic segmentation. In: CVPR (2020)

20. Hu, P., Liu, J., Wang, G., Ablavsky, V., Saenko, K., Sclaroff, S.: Dipnet: dynamic identity propagation network for video object segmentation. In: WACV (2020)
21. Hu, P., Wang, G., Kong, X., Kuen, J., Tan, Y.P.: Motion-guided cascaded refinement network for video object segmentation. In: CVPR (2018)
22. Hu, P., Wang, G., Kong, X., Kuen, J., Tan, Y.P.: Motion-guided cascaded refinement network for video object segmentation. In: IEEE Transactions on PAMI (2019)
23. Hu, Y.-T., Huang, J.-B., Schwing, A.G.: Unsupervised video object segmentation using motion saliency-guided spatio-temporal propagation. In: Ferrari, V., Hebert, M., Sminchisescu, C., Weiss, Y. (eds.) ECCV 2018. LNCS, vol. 11205, pp. 813–830. Springer, Cham (2018). https://doi.org/10.1007/978-3-030-01246-5_48
24. Khoreva, A., Benenson, R., Ilg, E., Brox, T., Schiele, B.: Lucid data dreaming for object tracking. CoRR abs/1703.09554 (2017). http://arxiv.org/abs/1703.09554
25. Li, M., Hu, L., Xiong, Z., Zhang, B., Pan, P., Liu, D.: Recurrent dynamic embedding for video object segmentation. In: CVPR (2022)
26. Li, X., Wei, T., Chen, Y.P., Tai, Y.W., Tang, C.K.: FSS-1000: a 1000-class dataset for few-shot segmentation. In: CVPR (2020)
27. Li, X., Loy, C.C.: Video object segmentation with joint re-identification and attention-aware mask propagation. In: Ferrari, V., Hebert, M., Sminchisescu, C., Weiss, Y. (eds.) ECCV 2018. LNCS, vol. 11207, pp. 93–110. Springer, Cham (2018). https://doi.org/10.1007/978-3-030-01219-9_6
28. Liang, Y., Li, X., Jafari, N., Chen, Q.: Video object segmentation with adaptive feature bank and uncertain-region refinement. arXiv preprint arXiv:2010.07958 (2020)
29. Liu, Z., et al.: Swin transformer: Hierarchical vision transformer using shifted windows. arXiv preprint arXiv:2103.14030 (2021)
30. Lu, X., Wang, W., Danelljan, M., Zhou, T., Shen, J., Van Gool, L.: Video object segmentation with episodic graph memory networks. In: Vedaldi, A., Bischof, H., Brox, T., Frahm, J.-M. (eds.) ECCV 2020. LNCS, vol. 12348, pp. 661–679. Springer, Cham (2020). https://doi.org/10.1007/978-3-030-58580-8_39
31. Luiten, J., Voigtlaender, P., Leibe, B.: PReMVOS: proposal-generation, refinement and merging for video object segmentation. In: Jawahar, C.V., Li, H., Mori, G., Schindler, K. (eds.) ACCV 2018. LNCS, vol. 11364, pp. 565–580. Springer, Cham (2019). https://doi.org/10.1007/978-3-030-20870-7_35
32. Mao, Y., Wang, N., Zhou, W., Li, H.: Joint inductive and transductive learning for video object segmentation. In: ICCV (2021)
33. Oh, S.W., Lee, J.Y., Xu, N., Kim, S.J.: Video object segmentation using space-time memory networks. In: ICCV (2019)
34. Perazzi, F., Pont-Tuset, J., McWilliams, B., Van Gool, L., Gross, M., Sorkine-Hornung, A.: A benchmark dataset and evaluation methodology for video object segmentation. In: CVPR (2016)
35. Pont-Tuset, J., Perazzi, F., Caelles, S., Arbeláez, P., Sorkine-Hornung, A., Van Gool, L.: The 2017 davis challenge on video object segmentation. arXiv:1704.00675 (2017)
36. Robinson, A., Lawin, F.J., Danelljan, M., Khan, F.S., Felsberg, M.: Learning fast and robust target models for video object segmentation. In: CVPR (2020)
37. Seong, H., Hyun, J., Kim, E.: Kernelized memory network for video object segmentation. In: Vedaldi, A., Bischof, H., Brox, T., Frahm, J.-M. (eds.) ECCV 2020. LNCS, vol. 12367, pp. 629–645. Springer, Cham (2020). https://doi.org/10.1007/978-3-030-58542-6_38

38. Seong, H., Oh, S.W., Lee, J.Y., Lee, S., Lee, S., Kim, E.: Hierarchical memory matching network for video object segmentation. In: ICCV (2021)
39. Shi, J., Yan, Q., Xu, L., Jia, J.: Hierarchical image saliency detection on extended CSSD. IEEE Trans. Pattern Anal. Mach. Intell. **38**(4), 717–729 (2015)
40. Tang, C., Yuan, L., Tan, P.: LSM: learning subspace minimization for low-level vision. In: CVPR (2020)
41. Tao, L., Porikli, F., Vidal, R.: Sparse dictionaries for semantic segmentation. In: Fleet, D., Pajdla, T., Schiele, B., Tuytelaars, T. (eds.) ECCV 2014. LNCS, vol. 8693, pp. 549–564. Springer, Cham (2014). https://doi.org/10.1007/978-3-319-10602-1_36
42. Tsai, Y.H., Yang, M.H., Black, M.J.: Video segmentation via object flow. In: CVPR (2016)
43. Vaswani, A., et al.: Attention is all you need. In: NeurIPS, pp. 5998–6008 (2017)
44. Voigtlaender, P., Chai, Y., Schroff, F., Adam, H., Leibe, B., Chen, L.: FEELVOS: fast end-to-end embedding learning for video object segmentation. In: CVPR (2019)
45. Voigtlaender, P., Leibe, B.: Online adaptation of convolutional neural networks for video object segmentation. In: BMVC (2017)
46. Wang, H., Jiang, X., Ren, H., Hu, Y., Bai, S.: SwiftNet: real-time video object segmentation. In: CVPR (2021)
47. Wang, L., Lu, H., Wang, Y., Feng, M., Ruan, X.: Learning to detect salient objects with image-level supervision. In: CVPR (2017)
48. Wang, W., et al.: Pyramid vision transformer: a versatile backbone for dense prediction without convolutions. arXiv preprint arXiv:2102.12122 (2021)
49. Xiao, H., Feng, J., Lin, G., Liu, Y., Zhang, M.: Monet: deep motion exploitation for video object segmentation. In: CVPR (2018)
50. Xie, H., Yao, H., Zhou, S., Zhang, S., Sun, W.: Efficient regional memory network for video object segmentation. In: CVPR (2021)
51. Xu, K., Yao, A.: Accelerating video object segmentation with compressed video. In: Proceedings of the IEEE/CVF Conference on Computer Vision and Pattern Recognition (2022)
52. Xu, N., et al..: YouTube-VOS: a large-scale video object segmentation benchmark. arXiv preprint arXiv:1809.03327 (2018)
53. Xu, X., Wang, J., Li, X., Lu, Y.: Reliable propagation-correction modulation for video object segmentation. In: Proceedings of the AAAI Conference on Artificial Intelligence (2022)
54. Yan, B., Peng, H., Fu, J., Wang, D., Lu, H.: Learning spatio-temporal transformer for visual tracking (2021)
55. Yang, C., Zhang, L., Lu, H., Ruan, X., Yang, M.H.: Saliency detection via graph-based manifold ranking. In: CVPR (2013)
56. Yang, L., Wang, Y., Xiong, X., Yang, J., Katsaggelos, A.K.: Efficient video object segmentation via network modulation. In: CVPR (2018)
57. Yang, Z., Wei, Y., Yang, Y.: Collaborative video object segmentation by foreground-background integration. In: Vedaldi, A., Bischof, H., Brox, T., Frahm, J.-M. (eds.) ECCV 2020. LNCS, vol. 12350, pp. 332–348. Springer, Cham (2020). https://doi.org/10.1007/978-3-030-58558-7_20
58. Zhang, L., Lin, Z., Zhang, J., Lu, H., He, Y.: Fast video object segmentation via dynamic targeting network. In: ICCV (2019)

59. Zhang, L., Zhang, J., Lin, Z., Měch, R., Lu, H., He, Y.: Unsupervised video object segmentation with joint hotspot tracking. In: Vedaldi, A., Bischof, H., Brox, T., Frahm, J.-M. (eds.) ECCV 2020. LNCS, vol. 12359, pp. 490–506. Springer, Cham (2020). https://doi.org/10.1007/978-3-030-58568-6_29
60. Zheng, S., et al.: Rethinking semantic segmentation from a sequence-to-sequence perspective with transformers. In: CVPR (2021)
61. Zohrizadeh, F., Kheirandishfard, M., Kamangar, F.: Image segmentation using sparse subset selection. In: WACV (2018)

Document Image Analysis

BorderNet: An Efficient Border-Attention Text Detector

Juntao Cheng$^{(\boxtimes)}$, Liangru Xie, and Cheng Du

AI R&D Department, Kingsoft Office, Beijing, China
{chengjuntao1,xieliangru,ducheng}@wps.cn

Abstract. Recently, segmentation-based text detection methods are quite popular in the scene text detection field, because of their superiority for text instances with arbitrary shapes and extreme aspect ratios. However, the vast majority of the existing segmentation-based methods are difficult to detect curved and dense text instances due to principle of these methods. In this paper, we propose a novel text detection method named BorderNet. The key idea of BorderNet is making full use of border-center information to detect the curve and dense text. Furthermore, a efficient Multi-Scale Feature Enhancement Module is proposed to improve the scale and shape robustness by enhancing features of different scales adaptively. Our method outperforms SOTA on multiple datasets, achieving 89% accuracy on ICDAR2015 and 87.1% accuracy on Total-Text. What's more, we can maintain 84.5% accuracy on DAST1500.

Keywords: First keyword · Second keyword · Another keyword

1 Introduction

Scene texts often appear in a variety of application, and provide a wealth of important information, such as intelligent office, visual search, scene understanding, automatic driving and other application directions. Therefore, reading scene text images is extremely important. Text detection that locates the text position is a very important part of reading text. Scene text detection faces greater challenges than general object detection owing to the extreme aspect ratio of the text, irregular shapes, different scales and other factors.

Owing to the development of object detection and segmentation based on deep learning in recent years, scene text detection has made great progress. Scene text detection can be roughly divided into three categories: Regression-based methods, Segmentation-based methods and Hybrid methods. Hybrid methods merge the advantages of segmentation and regression so that they complement each other. Regression-based methods and some hybrid methods can achieve excellent performance on benchmark testsets. However, they have a huge bottleneck which assume text instances have a linear shape. Hence, horizontal or multi-oriented quadrilaterals are used to represent text boxes. In addition, their performance in detecting text with irregular shapes such as curved and dense

---

J. Cheng and L. Xie—Authors contribute equally.

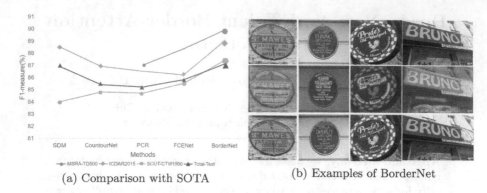

(a) Comparison with SOTA   (b) Examples of BorderNet

Fig. 1. (a) The figure shows the comparison with recent SOTA methods such as SDM [28], CountourNet [26], PCR [3] and FCENet [33] on various benchmark datasets, in terms of accuracy (F-measure). Our proposed BorderNet can achieve higher accuracy. (b) The figure visualizes the entire inference process of BorderNet. The first row is the input image, the second row is the heatmap during network inference, and the third row is the final text detection result.

text drops significantly. In contrast, segmentation-based scene text detectors generally have an advantage in detecting text instances with irregular shapes and extreme aspect ratios due to their pixel-level representation and local prediction. Although segmentation-based methods can accurately predict text regions, it is difficult to separate close text instances. Many recent segmentation-based methods focus on how to separate the segmented text region into multiple text instances, such as SOTD [27], TextField [29], TextMountain [34]. Furthermore, most segmentation-based methods require complex post-processing to aggregate text regions, resulting in considerable time consumption. For example, PSENet [23] proposes a progressive scale expansion algorithm for post-processing, which integrates feature maps of multiple scales, resulting in a large time cost. In addition, although most segmentation-based methods using feature pyramids [10] or UNet [20] structure fuse multi-scale features to get higher accuracy, there is a semantic gap between different layers. Therefore, forcibly merging features of different scales will reduce the ability of multi-scale feature expression and cause feature redundancy.

The detection of curved and dense text quickly and accurately still faces severe challenges. The border of curved and dense text are easy to stick, which is difficult to distinguish by common methods. Therefore, this study proposes a text detector based on border learning (BorderNet) for curved and dense text. Our approach proposes to directly use image segmentation methods to learn the center and border regions of text, and use border regions to separate different text instances. In order to improve the efficiency and accuracy of the algorithm, based on DBNet [7] and DBNet++ [9], we integrate Differentiable Binarization (DB) into boundary learning, which greatly reduces the number of calculations used for post-processing. To fuse multi-scale features more fully, we propose a multi-scale feature enhancement module (MSFEM), which enhances the features of each scale to make it more fully fused. MSFEM contains two feature fusions, one occurs when deep layers are fused with shallow layers gradually, and the other occurs when shallow layers are fused with deep layers gradually. In the fusion process, the multi-scale features with stronger semantics can be obtained by introducing the multi-head channel attention module (MHCA) which enhances the features of each layer. By directly fusing the enhanced multi-scale features, the center and border regions of the text can be better segmented. As shown in Fig. 2 below, our method can detect curved and dense text better than other methods. Our method surpasses SOTA methods on various benchmark datasets such as ICDAR2015 and Total-Text. Our method reaches 89% on the ICDAR2015 and 87.1% on the Total-Text. It is evident from the experiments that by introducing border learning and fully fusing multi-scale features, the model has higher accuracy, stronger robustness, and better efficiency.

Fig. 2. Comparison with other methods. Fig a is our method, fig b is result from FCENet [33], fig c is result from FAST [2], fig d is result from PSENet [23].

The main contributions of this study are as follows: 1) We propose a text detector based on border learning, which directly uses the image segmentation method to learn the center and border regions of the text, and uses the border regions to separate different text instances which can solve the challenge of curved and dense text. 2) In order to detect scene text more accurately and faster, we proposed MSFEM, which can enhance multiple features of different scales to make feature fusion more fully and obtain higher semantic information.

2 Related Work

Scene text detection has received extensive attention recently, and many new methods have emerged. These methods can be roughly divided into three categories: regression-based methods, segmentation-based methods, and hybrid methods.

Regression-Based Methods. Most of regression-based methods directly regress the text instance boxes. CTPN [21] divides the text instance into many small vertical text regions, regresses each text box directly, and then connects text boxes using a recurrent neural network. However, the regression anchor of CTPN [21] is vertical, and this method can only detect horizontal or slightly multi-oriented text. To solve this problem, RRPN [17] basing on the framework of Faster RCNN [19] integrates the rotated anchor to achieve the effect of detecting multi-oriented text boxes. However, this method uses a rectangular text box, which makes it difficult for text box to get close to the border of multi-oriented text. Then DMPNet [11] is proposed, which regresses the quadrilateral directly and the text box can be positioned close to the border of multi-oriented text. Although the development of regression-based methods has matured, and has good effects on horizontal and multi-oriented texts, there are still great challenges in the case of curved and dense text.

Segmentation-Based Method. The segmentation-based method regards text detection as a text region segmentation problem, and usually requires pixel-level prediction and post-processing algorithms to obtain the text box. The border position of the text can be found, and distinguishing between text regions and non-text regions is very important in this type of method. SOTD [27] introduces the text border learning method, which uses the text border region as the third category for semantic segmentation. But there has a strong dependence on feature extraction ability of model. TextField [29] and TextMountain [34] achieve the purpose of detecting dense text by modeling the text center and text border. However, this requires complex modeling and post-processing, which is a time-consuming process. Although segmentation-based methods can detect horizontal, multi-oriented and curved texts well, dealing with dense texts quickly and efficiently still faces considerable challenges.

Hybrid Methods. It is also worth mentioning that some other methods utilize segmentation to classify text/non-text pixels and then localize the text via bounding box regression. For example, East [32] and Deep Regression [6] predict a rotated rectangle or quadrilateral for each pixel, but can not solve the detection problem on curved text. Mask TextSpotter [16] uses instance segmentation methods for detection, which can detect curved text. But it does not perform well on both curved and dense text, which have time consuming and low accuracy. Other hybrid methods [31] require complex and time-consuming post-processing [18]to remove duplicate prediction boxes.

In text detection, how to learn text borders quickly, efficiently and precisely is very important for detecting curved and dense text. Most of the above methods have relatively complex post-processing, which will bring a large time cost. Therefore, we propose a text detection method based on border learning and design a more concise framework. Meanwhile, we introduce a feature Fusion module, which can simplify post-processing and enable the network to learn where the text region is quickly and precisely.

3   Methodology

3.1   Overall Architecture

Based on the fully convolutional network [14], we design the overall architecture in the form of an Encoder-Decoder. The specific structure includes a feature extraction module, MSFEM and Decoder module. In the feature extraction module, we use the ConvNeXt [12]as the backbone to learn the features of text images. Because the size distribution of text may vary greatly, it is difficult for single-layer features to adapt to texts of different scales. Therefore, MSFEM is introduced to fuse multi-scale features by means of feature pyramids. To improving the fusion quality of MSFEM, we design a module named MHCA. The features of different scales of stage 2, stage 3, stage 4 and stage 5 are fed into MSFEM to obtain multiple enhanced scale features through MHCA, and then concat them together and fully fuse them through a layer of convolution to obtain multi-scale fusion features with stronger semantics. Finally, the obtained multi-scale fusion features are fed into Decoder module. Decoder module first predicts the shrinkage probability map for the center region of the text and applies border learning to obtain the probability map for the text border region, and gets the threshold map corresponding to the predicted probability map. Afterwards, the DB module is introduced to optimize the learning process, and obtains the final result by fusing the probability map and the threshold map. Figure 3 shows the overall network of BorderNet.

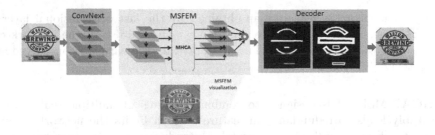

Fig. 3. BorderNet network structure.

3.2   MSFEM

In the process of extracting text image features, we found that features of different scales have different receptive fields and can pay attention to texts of different scales. Shallow features have smaller receptive fields and can perceive the details of small texts, but can not capture global information. Deep features have a large receptive field and can capture global information, but can not perceive detailed information. Therefore, for dense text detection, we need to capture the feature information of each scale to the greatest extent, and fully integrate the features of different scales. To achieve this purpose, we introduce MSFEM.

We first build MSFEM based on the feature pyramid. In this module, the feature pyramid gradually upsample deep features and fuse with shallow features. To fully fuse and amplify the feature information extracted from each scale to speed up the learning efficiency of the network, the fused features obtained from each scale are passed through MHCA to learn the attention weight, and multiplied by the original scale feature to obtain the filtered fusion feature. Therefore, to extract useful multi-scale feature information to a greater extent, the filtered fusion features from each scale are gradually down-sampled and fused with the deep features. The operation of fused refers to concat&conv. The particular network structure of MSFEM is shown in Fig. 4.

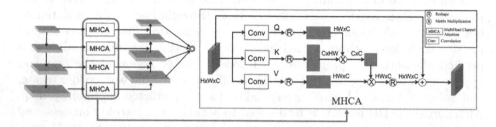

Fig. 4. MSFEM network structure.

By visualizing the feature map obtained by the last convolutional layer of MSFEM (as shown in Fig. 4), we can see that after MSFEM, the network can focus better on the text region. The network can also learn well on curved and dense text.

MHCA. MSFEM is designed to combine features of multiple scales, which inevitably leads to redundancy in features, which limits the network's ability to locate object rapidly and accurately. Introducing an attention mechanism can solve this problem. As some methods proved that transformer is a successful module based on attention mechanism, where multi-head self-attention (MHSA) can adaptively learn more types of features and filter key features in the spatial dimension. However, the computational time of MHSA is positively correlated with the required memory and input resolution, which is very resource-intensive. Inspired by Restormer [30], we propose the MHCA module, which is a multi-head channel attention mechanism that can expand the feature learning space, learn more feature types, and enhance the expressive ability of the model. Moreover, the MHCA module is an attention module with linear complexity. It builds the attention map not in the spatial dimension but in the feature channel dimension. Hence, MHCA is not affected by the input resolution, and the computational cost is greatly reduced. We use gradcam to visualize the feature map obtained by the last convolution layer of MSFEM, as shown in Fig. 5.

Fig. 5. Visualize the feature map obtained by the last convolutional layer of MSFEM. Fig a, b, c and d are samples.

The MHCA module first uses a 1×1 convolution on the input to communicate the information between each channel, and then a 3×3 depthwise separable convolution to obtain spatial local context information on each channel. Then, query (Q), key (K) and value (V) are generated from the obtained features. Multi-head of MHCA is achieved by dividing it into multiple heads based on the number of channels. MHCA changes the dimensions of Q and K, and then dot-multiply each other to form a channel-based attention map. After that, the attention map performs dot multiplication with V to determine the attention feature for the current head. Finally, attention feature maps of multiple heads are fused to obtain the module output. The calculation process of attention can be expressed as:

$$Attention(Q, K, V) = V \cdot Softmax(\frac{K \cdot Q}{\alpha}) \tag{1}$$

where α is a learnable scale parameter to control the weights of the dot product of K and Q.

3.3 Decoder

In this section, we design two key modules, including Convolution Map(CM) module and Fusion module. The main function of the CM module is to map the feature map obtained in the previous part to a single-channel grayscale image with a value range of [0,1]. We designed the CM module consist of a convolution and two deconvolutions. The convolution part is responsible for feature mapping, the deconvolution part is responsible for restoring the image size without losing effective information, and finally obtaining the corresponding grayscale image. The Fusion module is used to fuse the multiple grayscale images through the DB module to get the final result. The Fusion module is used for training to improve the accuracy of the network and can be unused during inference phase to reduce inference time and memory consumption.

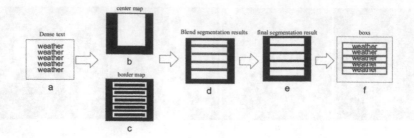

Fig. 6. The fusion process of border map and center map. Fig a is the dense text to be detected, fig b and fig c are the center map corresponding to the text center region and the border map corresponding to the text edge region. Then, the two maps are mixed to obtain fig e as shown in fig d, and finally the boxes of dense text in fig f are obtained based on fig e.

The decoder uses the feature map from the MSFEM to predict the approximate position of the text. The feature maps are respectively created by two CM modules to create center map and border map, which are probability maps for predicting the text center and border regions, and then the probability maps are fed into the Fusion module. In the Fusion module, we first map the feature to the threshold map through the CM module. The threshold map obtained at this time corresponds to the probability map obtained earlier. The center map corresponds to the threshold map of the text border region. The border map corresponds to the threshold map of the text center region. The corresponding probability map and threshold map are learned by DB optimization. Finally, the results of the optimized learning are integrated to get the text detection results.

In this section, we not only retain the commonly used center map for the center region of the text, but also design border map for learning the text border to enhance the accuracy of text border learning. Because center map is fused with border map, the detection results of dense text can be well separated, improving the accuracy. The whole process is shown in Fig. 6.

In addition to learning the center and border regions of the text image, the model also learns the center point of the text as well as the region of the text border, to estimate the size and location of the text more accurately. The design is shown in Fig. 7.

Label Generation. A total of 4 kinds of labels are required in our network design, namely the label corresponding to the border map, the label corresponding to the center map, the threshold map label corresponding to the border map, and the threshold map label corresponding to the center map.

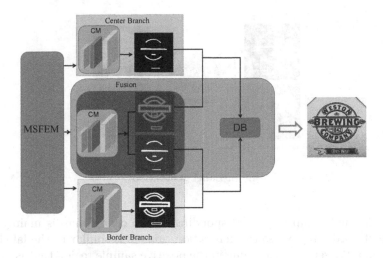

Fig. 7. Decoder network structure.

The label of the border map mainly describes the border of the text. Given an image of text, each polygonal region of its text is described by a set of line segments, as shown in the following formula:

$$G = \{S_k\}_{k=1}^n \tag{2}$$

where n is the number of vertices. Next, the polygon G is reduced to G1 by using the Vatti clipping algorithm [22]. The positive sample region obtained by subtracting G1 from G is the border region of the text. The shrinking offset D is usually calculated from the perimeter and region of the original polygon, which can be expressed as:

$$D = \frac{A(1 - r^2)}{L} \tag{3}$$

where r is the shrinkage ratio, which is empirically set to 0.4.

The label of the center map is used to describe the center region of the text in the image. The positive sample region generated by reducing the above-mentioned original polygon G to polygon G1 is the center region of the text. This process draws on the generation method of the probability map of PSENet [23] and DBNet [7].

The threshold map label corresponding to the border map is mainly aimed at the center region of the text. Hence, the generation process is similar to the label of the center map. After subtracting the reduced polygon G1 from the original polygon G to obtain the positive sample region, the region is dilated to obtain the edge of center region in text. Based on the edge, it assigns values to pixels from far to near, and the farther away from the edge, the smaller the value of the pixel.

618   J. Cheng et al.

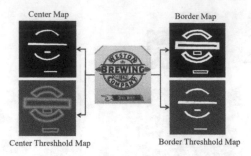

Fig. 8. Label generation.

The threshold map label corresponding to the center map is mainly aimed at the text border region, so the generation process is similar to the label of the border map. We set the pixel value in the positive sample region that is obtained after reducing the polygon to 1, and other regions are set to 0.3. The schematic diagram of the various labels generated is shown in Fig. 8.

4 Experiments

4.1 Datasets

In our experiments, we first pre-train on a large multi-language dataset and then finetune on a relatively small dataset. We chose the MLT-2017 as the pre-training dataset, which is a large real-world dataset with 9 languages. In the MLT-2017, there are a total of 7,200 training images, 1,800 validation images, and 9,000 testing images. We used a total of 9,000 images in the training and validation datasets for pre-training.

In downstream tasks for finetuning, we used public datasets such as ICDAR2015, MSRA-TD500, SCUT-CTW1500, Total-Text, and DAST1500 datasets to test the effectiveness of our ideas.

The characteristics of these types of datasets are different. We test on text datasets with different characteristics to comprehensively evaluate the effectiveness and verify the superiority of our method. Furthermore, it shows that our method has significant advantages over other popular methods in detecting curved and dense text.

4.2 Implementation Details

Training. Our training strategy is to first train 100 epochs on the MLT-2017 dataset, use F1 as the evaluation criterion, and select the optimal epoch training result as the pre-trained model. Then, based on the pre-trained model obtained above, fine-tuning is performed on different real datasets for 1,200 epochs. Among them, we follow the poly learning rate decay strategy, and the initial learning rate is set to 0.0002. In addition, we set the batch-size at training time to 16 and use the Adam optimizer with parameter decay of 0.05 to speed up the training network convergence. In terms of training data augmentation, we follow common augmentation methods: (1) random rotation with an angle variation range of (–10, 10); (2) random cropping of images; (3) random flipping. To improve the efficiency of training, all images are scaled to 640 × 640 for training.

Table 1. Comparison table of ablation experiment results.

Backbone	Border Branch	MSFEM	ICDAR2015				DAST1500			
			P	R	F	FPS	P	R	F	FPS
ConvNeXt-Tiny	✗	✗	89.7	84.6	87.1	10	86.4	79.1	82.6	39
ConvNeXt-Tiny	✓	✗	89.5	85.9	87.6	8	87.8	79.1	83.2	38
ConvNeXt-Tiny	✗	✓	90.4	87.1	88.7	6	83.5	83.3	83.4	25
ConvNeXt-Tiny	✓	✓	92.4	86.0	89.0	5	86.9	82.1	84.5	23

Inference. During the inference phase, we maintain the aspect ratio of the test image and scale the image by the appropriate shortest side. The center map and border map obtained from the two CM modules and the corresponding probability map results obtained in the Fusion module are used for inference. Originally, the DB module will use Eq. 4 to generate center binary image and border binary image to get the final result. To speed up the inference speed, the Fusion module can be discarded and only the results of the center map and border map are combined for inference. The specific fusion method is as follows:

$$output = centermap \cdot (1 - bordermap > t) \tag{4}$$

where t is the pixel classification threshold of the border map.

4.3 Ablation Study

To verify the effect of border learning and the multi-scale feature augmentation module on curved and dense text, we construct an ablation experiment on the ICDAR2015 with horizontal and multi-orientation text and DAST1500 with curved and dense text. The experimental results are shown in Table 1.

Border Branch. It can be seen from Table 1, border branch can strengthen border learning and improve the overall text detection effect. Border branch improves the accuracy of the network on ICDAR2015 and DAST1500 by 0.5% and 0.6%, respectively. Moreover, border branch does not bring about a large time consumption, and the speed in the inference phase does not change much. The comparison result with and without border branch is shown in Fig. 9.

Fig. 9. Comparison results with and without border branch. Fig a and fig c are the results without border branch, and fig b and fig d are the corresponding results with border branch.

MSFEM. MSFEM can fully fuse and amplify the feature information extracted at each scale, and strengthen feature semantic. As shown in Table 1, when joining MSFEM to the network improve ICDAR2015 and DAST1500 by 1.6% and 0.8%, respectively. When joining MSFEM and border branch to the network can improve both ICDAR2015 and DAST1500 by 1.9%.

Backbone. In this study, we use ConvNeXt [12] as backbone, which is an enhanced version of ResNet [5], and its effect of use in various visual fields has been affirmed in recent years. To be fair with other methods, we use the tiny version of ConvNeXt [12] as backbone. The computation of the tiny version is similar to ResNet-50 [5] network, and the basic feature extraction ability is stronger.

4.4   Comparisons with Previous Methods

Comparisons with previous methods is conducted on four datasets (including curved text and multi-oriented text).

Curved Text Detection. We demonstrate the robustness of our method on two datasets with curved text (Total-Text and SCUT-CTW1500). As shown in Table 2, our method achieves accuracy that exceeds state-of-the-art methods.

Total-Text. The dataset is word-level annotated. In the inference phase, the test image is scaled according to the shortest side of 800. As can be seen from a and b in the Fig. 10, our method can accurately detect irregular text at the word-level. Furthermore, our method achieves F-measure accuracy of 87.1% shown in Table 2.

SCUT-CTW1500. The dataset is sentence-level annotated. In the inference phase, the test image is scaled according to the shortest side of 800. The visualization results are shown in c and d of Fig. 10, our method can detect the border of text instances more precisely. As can be seen from Table 2, compared with other SOTA methods, the accuracy of our method exceeds by 2.1%.

Fig. 10. Visualization of Text results.Fig a and fig b are the results of Total-Text, fig c and fig d are the results of SCUT-CTW1500. Fig e and fig f are the results of ICDAR2015, fig g and fig h are the results of MSRA-TD500.

Multi-Orientation Text Detection. As is evident from the above, BorderNet can significantly outperform other methods in curved text detection. To further verify the ability of BorderNet to detect text of arbitrary shapes, we perform validation on the ICDAR2015 and MSRA-TD500 datasets, proving that BorderNet can still achieve competitive results on multi-oriented text detection tasks.

ICDAR2015. The dataset is word-level annotated. In the inference phase, the test image is scaled according to the shortest side of 1152. The visualization results are shown in a and b of Fig. 10, some difficult samples of different scales can still be detected accurately. As can be seen from Table 2, BorderNet achieves 89.0% accuracy, outperforming other SOTA methods by 1.7%.

MSRA-TD500. The dataset is sentence-level annotated. In the inference phase, to improve the inference speed, the test image is scaled according to the shortest side of 736. The visualization results are shown in c and d of Fig. 10. BorderNet can accurately detect long text lines with multi-Oriented characteristic. As is evident from the Table 2, BorderNet achieves accuracy of 89.9% in F-measure, surpassing other SOTA methods by 2.7%.

From the above, it can be seen that our method can accurately detect both line-annotated and word-annotated multi-oriented text. This further proves the stability and versatility of BorderNet.

Table 2. Comparison table of experiments results with SOTA.

Methods	Ext	Total-Text				SCUT-CTW1500				ICDAR2015				MSRA-TD500			
		P	R	F	FPS	P	R	F	FPS	P	R	F	FPS	P	R	F	FPS
ATRR [25]	Syn	80.9	76.2	78.5	–	80.1	80.2	80.1	–	89.2	86.0	87.6	–	85.2	82.1	83.6	10.0
CTPN [21]	–	–	–	–	–	60.4	53.8	56.9	7.14	74.2	51.6	60.9	7.1	–	–	–	–
EAST [32]	–	50.0	36.2	42.0	–	78.7	49.1	60.4	21.2	83.6	73.5	78.2	13.2	87.3	67.4	76.1	–
RRD [8]	–	–	–	–	–	–	–	–	–	88.0	80.0	83.8	6.5	87.0	73.0	79.0	10.0
MCN [13]	–	–	–	–	–	–	–	–	–	72.0	80.0	76.0	–	88.0	79.0	83.0	–
PixelLink [4]	–	–	–	–	–	–	–	–	–	85.5	82.0	83.7	–	83.0	73.2	77.8	3.0
TextSnake [15]	Syn	82.7	74.5	78.4	–	67.9	85.3	75.6	1.1	84.9	80.4	82.6	1.1	83.2	73.9	78.3	1.1
TextField [29]	Syn	81.2	79.9	80.6	–	83.0	79.8	81.4	–	84.3	83.9	84.1	1.8	87.4	75.9	81.3	–
PSENet [23]	MLT	84.0	78.0	80.0	3.9	84.8	79.7	82.2	3.9	86.9	84.5	85.7	1.6	–	–	–	–
CRAFT [1]	Syn	87.6	79.9	83.6	–	86.0	81.1	83.5	–	89.8	84.3	86.9	–	88.2	78.2	82.9	8.6
PAN [24]	Syn	89.3	81.0	85.0	39.6	86.4	81.2	83.7	39.8	84.0	81.9	82.9	26.1	84.4	83.8	84.1	30.2
FAST [2]	MLT	90.5	82.5	86.3	46.0	87.2	80.4	83.7	66.5	89.9	84.4	87.0	15.7	90.9	83.0	86.7	56.8
DBNet [7]	Syn	87.1	82.5	84.7	32.0	86.9	80.2	83.4	22.0	91.8	83.2	87.3	12.0	91.5	79.2	84.9	32.0
DBNet++ [9]	Syn	88.9	83.2	86.0	28.0	87.9	82.8	85.3	26.0	90.9	83.9	87.3	10.0	91.5	83.3	87.2	29.0
BorderNet(Ours)	MLT	89.4	84.9	**87.1**	19.0	87.8	87.0	**87.4**	19.0	92.4	86.0	**89.0**	5.0	90.9	88.8	**89.9**	21.0

5   Limitation

Because BorderNet is a segmentation-based method, it can not handle cases where a text instance is centered inside another text instance. Although our method improves the accuracy of the border region and helps to distinguish text instances that are close together, it can not handle the case when a text instance locate in the center of another text, which is also a common problem with segmentation-based methods. In the future, we will explore the idea of instance segmentation to solve this problem by fusing the advantages of detection and segmentation.

6   Conclusion

In this study, we propose an novel and efficient framework for detecting scene text of curved and dense. The framework can detect scene text by learning text border regions. For detecting scene text more accurately and faster, we propose MSFEM, which makes the feature fusion more reasonable and improves the detection efficiency. This method has surpassed the accuracy of other methods on four benchmark datasets of text detection with different text instances, particularly for distinguishing text instances with close distances and arbitrary shapes. In the future, we will simplify our network while ensuring accuracy, making the network more lightweight. Moreover, we will focus on distinguishing dense or even sticky text instances to improve the detection accuracy on this type of text.

References

1. Baek, Y., Lee, B., Han, D., Yun, S., Lee, H.: Character region awareness for text detection. In: Proceedings of the IEEE/CVF Conference on Computer Vision and Pattern Recognition, pp. 9365–9374 (2019)
2. Chen, Z., Wang, W., Xie, E., Yang, Z., Lu, T., Luo, P.: Fast: searching for a faster arbitrarily-shaped text detector with minimalist kernel representation. arXiv preprint arXiv:2111.02394 (2021)
3. Dai, P., Zhang, S., Zhang, H., Cao, X.: Progressive contour regression for arbitrary-shape scene text detection. In: Proceedings of the IEEE/CVF Conference on Computer Vision and Pattern Recognition, pp. 7393–7402 (2021)
4. Deng, D., Liu, H., Li, X., Cai, D.: PixelLink: detecting scene text via instance segmentation. In: Proceedings of the AAAI Conference on Artificial Intelligence, vol. 32 (2018)
5. He, K., Zhang, X., Ren, S., Sun, J.: Deep residual learning for image recognition. In: Proceedings of the IEEE Conference on Computer Vision and Pattern Recognition, pp. 770–778 (2016)
6. He, W., Zhang, X.Y., Yin, F., Liu, C.L.: Deep direct regression for multi-oriented scene text detection. In: Proceedings of the IEEE International Conference on Computer Vision, pp. 745–753 (2017)
7. Liao, M., Wan, Z., Yao, C., Chen, K., Bai, X.: Real-time scene text detection with differentiable binarization. In: Proceedings of the AAAI Conference on Artificial Intelligence, vol. 34, pp. 11474–11481 (2020)
8. Liao, M., Zhu, Z., Shi, B., Xia, G.S., Bai, X.: Rotation-sensitive regression for oriented scene text detection. In: Proceedings of the IEEE Conference on Computer Vision and Pattern Recognition, pp. 5909–5918 (2018)
9. Liao, M., Zou, Z., Wan, Z., Yao, C., Bai, X.: Real-time scene text detection with differentiable binarization and adaptive scale fusion. IEEE Transactions on Pattern Analysis and Machine Intelligence (2022)
10. Lin, T.Y., Dollár, P., Girshick, R., He, K., Hariharan, B., Belongie, S.: Feature pyramid networks for object detection. In: Proceedings of the IEEE Conference on Computer Vision and Pattern Recognition, pp. 2117–2125 (2017)
11. Liu, Y., Jin, L.: Deep matching prior network: Toward tighter multi-oriented text detection. In: Proceedings of the IEEE Conference on Computer Vision and Pattern Recognition, pp. 1962–1969 (2017)
12. Liu, Z., Mao, H., Wu, C.Y., Feichtenhofer, C., Darrell, T., Xie, S.: A convnet for the 2020s. In: Proceedings of the IEEE/CVF Conference on Computer Vision and Pattern Recognition, pp. 11976–11986 (2022)
13. Liu, Z., Lin, G., Yang, S., Feng, J., Lin, W., Goh, W.L.: Learning Markov clustering networks for scene text detection. arXiv preprint arXiv:1805.08365 (2018)
14. Long, J., Shelhamer, E., Darrell, T.: Fully convolutional networks for semantic segmentation. In: Proceedings of the IEEE Conference on Computer Vision and Pattern Recognition, pp. 3431–3440 (2015)
15. Long, S., Ruan, J., Zhang, W., He, X., Wu, W., Yao, C.: TextSnake: a flexible representation for detecting text of arbitrary shapes. In: Ferrari, V., Hebert, M., Sminchisescu, C., Weiss, Y. (eds.) ECCV 2018. LNCS, vol. 11206, pp. 19–35. Springer, Cham (2018). https://doi.org/10.1007/978-3-030-01216-8_2

16. Lyu, P., Liao, M., Yao, C., Wu, W., Bai, X.: Mask TextSpotter: an end-to-end trainable neural network for spotting text with arbitrary shapes. In: Ferrari, V., Hebert, M., Sminchisescu, C., Weiss, Y. (eds.) Computer Vision – ECCV 2018. LNCS, vol. 11218, pp. 71–88. Springer, Cham (2018). https://doi.org/10.1007/978-3-030-01264-9_5

17. Ma, J., et al.: Arbitrary-oriented scene text detection via rotation proposals. IEEE Trans. Multimedia **20**(11), 3111–3122 (2018)

18. Neubeck, A., Van Gool, L.: Efficient non-maximum suppression. In: 18th International Conference on Pattern Recognition (ICPR2006), vol. 3, pp. 850–855. IEEE (2006)

19. Ren, S., He, K., Girshick, R., Sun, J.: Faster R-CNN: towards real-time object detection with region proposal networks. In: Advances in Neural Information Processing Systems 28 (2015)

20. Ronneberger, O., Fischer, P., Brox, T.: U-Net: convolutional networks for biomedical image segmentation. In: Navab, N., Hornegger, J., Wells, W.M., Frangi, A.F. (eds.) MICCAI 2015. LNCS, vol. 9351, pp. 234–241. Springer, Cham (2015). https://doi.org/10.1007/978-3-319-24574-4_28

21. Tian, Z., Huang, W., He, T., He, P., Qiao, Yu.: Detecting text in natural image with connectionist text proposal network. In: Leibe, B., Matas, J., Sebe, N., Welling, M. (eds.) ECCV 2016. LNCS, vol. 9912, pp. 56–72. Springer, Cham (2016). https://doi.org/10.1007/978-3-319-46484-8_4

22. Vatti, B.R.: A generic solution to polygon clipping. Commun. ACM **35**(7), 56–63 (1992)

23. Wang, W., et al.: Shape robust text detection with progressive scale expansion network. In: Proceedings of the IEEE/CVF Conference on Computer Vision and Pattern Recognition, pp. 9336–9345 (2019)

24. Wang, W., et al.: Efficient and accurate arbitrary-shaped text detection with pixel aggregation network. In: Proceedings of the IEEE/CVF International Conference on Computer Vision, pp. 8440–8449 (2019)

25. Wang, X., Jiang, Y., Luo, Z., Liu, C.L., Choi, H., Kim, S.: Arbitrary shape scene text detection with adaptive text region representation. In: Proceedings of the IEEE/CVF Conference on Computer Vision and Pattern Recognition, pp. 6449–6458 (2019)

26. Wang, Y., Xie, H., Zha, Z.J., Xing, M., Fu, Z., Zhang, Y.: ContourNet: taking a further step toward accurate arbitrary-shaped scene text detection. In: proceedings of the IEEE/CVF Conference on Computer Vision and Pattern Recognition, pp. 11753–11762 (2020)

27. Wu, Y., Natarajan, P.: Self-organized text detection with minimal post-processing via border learning. In: proceedings of the IEEE International Conference on Computer Vision, pp. 5000–5009 (2017)

28. Xiao, S., Peng, L., Yan, R., An, K., Yao, G., Min, J.: Sequential deformation for accurate scene text detection. In: Vedaldi, A., Bischof, H., Brox, T., Frahm, J.-M. (eds.) ECCV 2020. LNCS, vol. 12374, pp. 108–124. Springer, Cham (2020). https://doi.org/10.1007/978-3-030-58526-6_7

29. Xu, Y., Wang, Y., Zhou, W., Wang, Y., Yang, Z., Bai, X.: TextField: learning a deep direction field for irregular scene text detection. IEEE Trans. Image Process. **28**(11), 5566–5579 (2019)

30. Zamir, S.W., Arora, A., Khan, S., Hayat, M., Khan, F.S., Yang, M.H.: Restormer: efficient transformer for high-resolution image restoration. In: Proceedings of the IEEE/CVF Conference on Computer Vision and Pattern Recognition, pp. 5728–5739 (2022)

31. Zhang, C., et al.: Look more than once: an accurate detector for text of arbitrary shapes. In: Proceedings of the IEEE/CVF Conference on Computer Vision and Pattern Recognition, pp. 10552–10561 (2019)
32. Zhou, X., et al.: East: an efficient and accurate scene text detector. In: Proceedings of the IEEE Conference on Computer Vision and Pattern Recognition, pp. 5551–5560 (2017)
33. Zhu, Y., Chen, J., Liang, L., Kuang, Z., Jin, L., Zhang, W.: Fourier contour embedding for arbitrary-shaped text detection. In: Proceedings of the IEEE/CVF Conference on Computer Vision and Pattern Recognition, pp. 3123–3131 (2021)
34. Zhu, Y., Du, J.: TextMountain: accurate scene text detection via instance segmentation. Pattern Recogn. **110**, 107336 (2021)

CMT-Co: Contrastive Learning with Character Movement Task for Handwritten Text Recognition

Xiaoyi Zhang[1], Jiapeng Wang[1], Lianwen Jin[1,2]($\boxtimes$), Yujin Ren[1], and Yang Xue[1]

[1] South China University of Technology, Guangzhou, China
lianwen.jin@gmail.com, yxue@scut.edu.cn
[2] SCUT-Zhuhai Institute of Modern Industrial Innovation, Zhuhai, China

Abstract. Mainstream handwritten text recognition (HTR) approaches require large-scale labeled data for training to achieve satisfactory performance. Recently, contrastive learning has been introduced to perform self-supervised training on unlabeled data to improve representational capacity. It minimizes the distance between the positive pairs while maximizing their distance to the negative ones. Previous studies typically consider each frame or a fixed window of frames in a sequential feature map as a separate instance for contrastive learning. However, owing to the arbitrariness of handwriting and the diversity of word length, such modeling may contain the information of multiple consecutive characters or an over-segmented sub-character, which may confuse the model to perceive semantic clues information. To address this issue, in this paper, we design a character-level pretext task termed *Character Movement Task*, to assist word-level *contrastive learning, namely CMT-Co*. It moves the characters in a word to generate artifacts and guides the model to perceive the text content by using the moving direction and distance as supervision. In addition, we customize a data augmentation strategy specifically for handwritten text, which significantly contributes to the construction of training pairs for contrastive learning. Experiments have shown that the proposed CMT-Co achieves competitive or even superior performance compared to previous methods on public handwritten benchmarks.

Keywords: Self-supervised learning · Pretext task · Handwritten text recognition · Contrastive learning

1 Introduction

Handwritten text recognition (HTR) is a vital field in computer vision [2, 5, 16, 29, 32, 35–39]. Most current methods for handwritten text recognition

Supplementary Information The online version contains supplementary material available at https://doi.org/10.1007/978-3-031-26293-7_37.

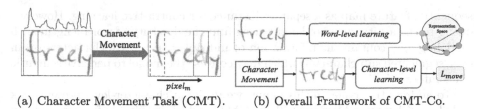

(a) Character Movement Task (CMT).   (b) Overall Framework of CMT-Co.

Fig. 1. Illustration of CMT and our method CMT-Co. CMT-Co uses the CMT as an auxiliary task to assist contrastive learning, which contains word-level learning and character-level learning. The red arrow in the representation space represents minimizing the distance of positive pairs, while the black dashed arrows in the representation space represent maximizing the distance between negative samples. (Color figure online)

[8,25,26,41,42] require full supervision, which not only takes much annotation time but also needs expensive costs. Moreover, with the advent of the era of big data and the development of network information technology, data acquisition is becoming easier, resulting in an exponential increase in the amount of unlabeled data. Therefore, it is necessary to explore how to effectively utilize them without manual annotation.

Self-supervised learning provides a solution to this problem and has been widely studied [3,6,7,11,15,20,21,31,33,46]. It aims to learn representations from the data itself, without manual annotation. Features learned through the self-supervised process are then fine-tuned in specific downstream tasks to speed up convergence or achieve better performance, while greatly reducing the amount of data annotation.

Early self-supervised methods often designed pretext tasks [9,11,18,22,31, 33,44,46]. They focused on discovering tasks in which labels can be derived from prior knowledge or manual modification of data. For example, Gidaris et al. [11] rotated the original image by specific angles, and then let the network deal with the rotation prediction task. It is worth noting that almost no pretext task has been specifically designed for handwritten text recognition.

In recent years, self-supervised methods based on contrastive learning have received considerable attention [3,4,6,7,11,15,17,24,30]. Contrastive learning aims to minimize the distance between the positive pairs while maximizing their distance to the negative ones to achieve feature learning. A few related studies have introduced this concept into the field of handwritten text recognition [1, 23]. Aberdam et al. [1] proposed SeqCLR, which creates positive and negative samples by sliding a window over sequential feature maps and then performs contrastive learning on them. This method must guarantee sequence alignment between different augmented views of the same image, and consequently, the data augmentation method of SeqCLR is limited. Liu et al. [23] proposed PerSec, which learns both low- and high-level features through contrastive learning from shallow and deep feature maps. It aims to enable each element of sequential features to distinguish itself from the context. These methods for handwritten text recognition typically consider each frame or a fixed window of frames in a sequential feature map as a separate instance for contrastive learning. However, owing to the arbitrariness of handwriting and the diversity of word length, such modeling may contain the information of multiple consecutive characters or an over-segmented sub-character, which may confuse the model to perceive semantic clues information.

To address this issue, in this paper, we design a character-level pretext task called the Character Movement Task (CMT), which is suitable for handwritten text. The handwritten text has unique prior knowledge, such as its vertical and horizontal projection distributions, which were applied to the traditional document text line and word segmentation methods [14,34]. For handwritten word images, we first use vertical projection distribution, denoted by the red line in Fig. 1(a), to estimate the approximate position of each character. Then, some characters are selected to move, and the artifact is generated. Note that the artifact does not change the meaning of the word. Finally, the network is required to predict the moving direction and distance. To solve the Character Movement Task, the network needs to recognize the moving characters and achieve the purpose of character-level feature learning.

To utilize both word-level and character-level semantic information, we adopt CMT as an auxiliary task to assist the contrastive learning of the entire word image, termed *CMT-Co*. In addition, to enhance the effectiveness of contrastive learning, we customize a data augmentation strategy for handwritten text, named Text-Aug. It includes four aspects: affine transformation, stroke jitter, stroke overlap, and stroke thickness. Text-Aug can sufficiently provide variety for contrastive learning frameworks.

Experimental results have shown that our method achieves competitive or even superior performance on public handwritten benchmarks compared to the previous self-supervised method for text recognition.

The main contributions of the paper are as follows:

(1) We propose a character-level pretext task for handwritten text, called the Character Movement Task (CMT), which is combined with the word-level contrastive learning process. To the best of our knowledge, this is the first pretext task in the field of handwritten text recognition.
(2) We propose a data augmentation strategy named Text-Aug, including affine transformation, stroke jitter, stroke overlap, and stroke thickness, to further unlock the power of our framework.
(3) The overall frameworks CMT-Co achieve competitive or even superior performance compared to the previous self-supervised method in terms of representation quality and downstream fine-tuning of handwritten text.

2   Related Works

2.1   Pretext Task

Early self-supervised methods often designed self-supervised tasks, also called pretext tasks [9,11,31,33,46]. Self-supervised tasks have been studied extensively [10,44]. They focus on discovering tasks in which labels can be derived from prior knowledge or manual modification of data. For example, in image classification, Gidaris *et al.* [11] rotated the original image by specific angles and then let the network deal with the rotation prediction task. Doersch *et al.* [9] extracted random pairs of patches from each image and drove the network to predict the position of the second patch relative to that of the first.

However, to the best of our knowledge, currently in the field of handwritten text recognition, there is no method for designing a pretext task based on unique prior knowledge of the handwritten text.

2.2   Contrastive Learning

In addition to designing specific pretext tasks, recent self-supervised methods based on contrastive learning have shown significant potential [1,3,6,12,15,23]. SimCLR [3] generates negative samples through a large batch size. Furthermore, because of rich data augmentation, it learns sufficient representations after contrastive learning and shows pleasing performance in downstream tasks. MoCo [15] designed a symmetric structure, and then one party generated negative samples through momentum update, which not only did not need to store all the data in advance but also did not require a large batch size to generate negative samples. BYOL [12] designed an asymmetric structure, and a pleasing performance could be achieved without negative samples.

In the field of handwritten text recognition, SeqCLR [1] improves upon SimCLR [3], which uses the instance-mapping function on the sequential feature map to generate positive and negative samples. Multiple instances are generated from a single image through a sliding window. The same position in the same image with different data augmentation is a positive pair. Different positions and other images are negative samples. However, this method must ensure sequence alignment; otherwise, the same position of the same image with different data augmentation may not contain the same characters, which greatly limits the data augmentation method of SeqCLR. PerSec [23] learned both low- and high-level features through contrastive learning from shallow and deep feature maps within one image. It aims to enable each element of the sequential features to distinguish itself from the context. In general, the instance of the aforementioned contrastive learning for handwritten text recognition may contain the information of multiple consecutive characters or an over-segmented sub-character, thus confusing the model in perceiving semantic information.

Therefore, this paper will use prior knowledge of handwritten word text to design a character-level Character Movement Task, and then assist the word-level learning of the entire word image in contrastive learning to achieve better results on handwritten text.

3   Method

3.1   Text-Aug

It is known from recent self-supervised studies on contrastive learning [3,6,12, 15,40] that data augmentation plays an important role in feature representation

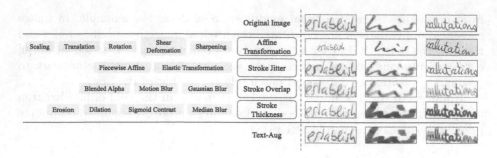

Fig. 2. Composition of Text-Aug and examples of different parts.

learning. Thus, we designed a data augmentation strategy suitable for handwritten text.

It is well known that text is composed of strokes. For handwritten text, the same word or character may have different sizes due to different writers' styles. A stroke of the same character may also have different degrees of bending or jitting distortions. Moreover, when people write, they may not erase the wrongly written characters and choose to overwrite the correct characters directly on them, which will result in many overlapping strokes. Due to different writing equipment and writing strength, strokes of the same character may be inconsistent in thickness, and handwriting may be separated or stuck together.

Therefore, for the above situation, we design *Text-Aug* for handwritten text. It includes four types of text augmentation: affine transformation, stroke jitter, stroke overlap, and stroke thickness. Figure 2 gives some examples of these four types of text augmentation in Text-Aug. The affine transformations include scaling, translation, rotation, shear deformation, and sharpening. This data augmentation provided text images of different scales, positions, and brightness. The stroke jitter includes piecewise affine and elastic transformation. They can simulate the bending and jittering of strokes in a text. Data augmentation for stroke overlap simulates the overlapping and blurring of strokes in text images. It includes blended alpha, motion blur, and Gaussian blur. The stroke thickness contains erosion, dilation, sigmoid contrast, and median blur. These augmentations can change the thickness of the strokes and may also produce modifications in character sticking and separation.

It can be seen that Text-Aug greatly improves the diversity of the text, which can enable the network to distinguish more variable texts. Pseudo-code and more examples of Text-Aug are shown in the Appendix.

3.2   Character Movement Task

It can intuitively know some prior knowledge from the handwritten word image, such as the vertical projection distribution, which can give the approximate location of the characters in the image. The Character Movement Task (CMT) first roughly locates the characters through the vertical projection distribution,

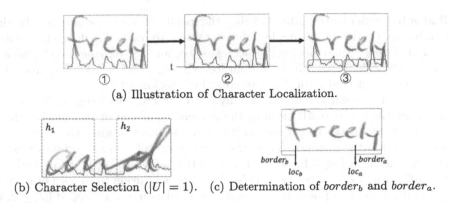

(a) Illustration of Character Localization.

(b) Character Selection ($|U| = 1$).   (c) Determination of $border_b$ and $border_a$.

Fig. 3. Process illustration of Character Movement Task (CMT).

then moves the characters, and finally, drives the network to predict the movement direction and the moving distance of the characters. To solve the Character Movement Task, the network needs to recognize the moving characters and achieve the purpose of character-level feature learning. The process of the CMT is shown in Algorithm 1 and explained in detail below. We assume that each word image I has characters.

Character Localization. Because this task requires moving characters, we first need to locate the position of each character in the word image. For the handwritten grayscale word image, it is resized to $H \times W$ and then adaptively binarized and normalized to $[0, 1]$. Note that the value of the area where the characters are located is one. Finally, row summation is performed to obtain the vertical projection distribution Sta of the text image. As shown in the first image of Fig. 3(a), the red line represents the vertical projection distribution, indicating the projected cumulative value of the character pixel at the corresponding column position. According to Sta, we can locate the approximate position of the characters in the word image.

However, writers usually have continuous strokes between different characters in the word. Therefore, to approximately eliminating the interference of stroke adhesion, we set the number less than t in Sta to zero, where t takes the second smallest value in Sta. As shown in the second image of Fig. 3(a), the blue line represents the value of t, and the position below the blue line is set to zero, resulting in the third image of Fig. 3(a). We define a continuous region with non-zero projected values as the character block region u. Taking the third image in Fig. 3(a) as an example, there are three continuous regions with non-zero projection values, implying that there are three character block regions, $U = \{u_1, u_2, u_3\}$. It can be inferred from the process of character block region generation that each character block region contains characters.

Character Selection. After locating the positions of the characters in the word image, we then need to select the characters to move. We define the center position of the selected moving characters as loc_b and the moving target position as loc_a. The characters will move from loc_b to loc_a. The position loc_b and loc_a are randomly selected from the character block region set U.

If $|U| = 1$, it means that there is only one character block region, $U = u_1$. If the movement is too small (such as the movement of one or two pixels), there is no difference to the naked eye, so it is unreasonable to force the network to predict the moving distance. Thus, the characters will move a certain distance here. As shown in Fig. 3(b), we first denote the front 40% of u_1 as h_1 and the back 40% of u_1 as h_2. Then, we randomly select a position from each of h_1 and h_2. Finally, these two positions are randomly used as loc_a and loc_b. If $|U| \geq 2$, it means there are two or more character block regions. In this case, two character block regions u_b and u_a are randomly selected from U as the character block region before the movement and the character block region after the movement. Then, we randomly select a location from u_b as loc_b and randomly select a location from u_a as loc_a.

After determining the center position of the selected moving characters, we need to specify the width of the characters. Firstly, we randomly sample a value from $[\frac{0.15}{2}W, \frac{0.25}{2}W]$ and regard it as half of the initial moving character area width, denoted as w_{ini}, where W is the image width. Then the minimum distance between loc_b and the image border is $border_b$, and the minimum distance between loc_a and the image border is $border_a$, as shown in Fig. 3(c). Half of the final moving character area width w_{move} is the minimum value among w_{ini}, $border_b$, and $border_a$.

The selected character area is denoted as

$$img_b = I[0 : H, loc_b - w_{move} : loc_b + w_{move}] \tag{1}$$

The original image of the area to which the characters are moved is

$$img_a = I[0 : H, loc_a - w_{move} : loc_a + w_{move}] \tag{2}$$

Finally, the selected character area is superimposed on the image at a scale of $1 - \lambda$, which means that

$$img_a = \lambda img_a + (1 - \lambda)img_b \tag{3}$$

Algorithm 1. Process of the Character Movement Task

Input: Grayscale image I

Output: Image MI after character movement, Label y

1. Obtain the vertical projection distribution Sta of image I.
2. Obtain the character block region set U.
3. Randomly select the position loc_b and loc_a from the character block region set U. The characters will move from loc_b to loc_a.
4. Determine the width w_{move} of the character movement, and then superimposes the selected character area on the target position to get image MI.
5. Determine the label y of the character movement.

The rest of I remains unchanged. We set image I after character movement as MI, which does not change the meaning of the initial word image I.

Loss Function. Character Movement Task is defined as a classification task. The label y is given by the formula (4).

$$y = pixel_m + W \qquad (4)$$

where W denotes the image width and $pixel_m = loc_a - loc_b$. When $pixel_m = 0$, there is no movement in the image, while $pixel_m < 0$, means that the character moves to the left; and $pixel_m > 0$, means that the character moves to the right.

Because the image has been resized before the character movement, the value range of $pixel_m$ is $[-W, W]$. Thus, the number of categories for classification is $2W + 1$. The loss in the Character Movement Task is given by Eq. (5).

$$\mathcal{L}_{move} = - \sum_{i=1}^{N} y_i \log p_i \qquad (5)$$

where

$$p_i = \frac{e^{F(MI_i)}}{\sum_{j=1}^{2W+1} e^{F(MI_j)}} \qquad (6)$$

where F represents the encoder and the multilayer perceptron and N represents the mini-batch size.

Figure 4 shows the $pixel_m$ distribution generated by the Character Movement Task on the IAM training set. It can be seen that the range of movement to the left and right is roughly equal and the distribution is roughly balanced. Figure 5 shows some examples of the Character Movement Task.

3.3 Overall Framework

Our CMT-Co method uses CMT as a character-level auxiliary task to assist in the learning of the entire word-level image in contrastive learning. MoCo v2 [6] is

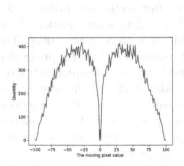

Fig. 4. Distribution of the $pixel_m$ generated by the CMT.

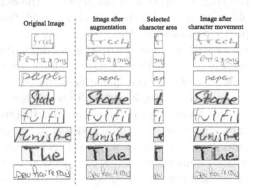

Fig. 5. Examples of character movement.

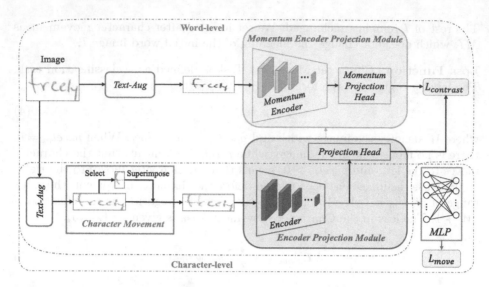

Fig. 6. Overall framework of our method CMT-Co. CMT-Co uses the CMT as a character-level auxiliary task to assist the learning of the entire word-level image in contrastive learning. Noted that the labels used to calculate the loss L_{move} are generated by the Character Movement module.

selected as the basic architecture. The overall architecture of CMT-Co is shown in Fig. 6. In CMT-Co, the high-level semantic information of the word can be learned through the contrastive learning of the whole-word image, and the low-level single-character information can be learned through the CMT.

We use Text-Aug to augment the text images. However, due to the overlapping phenomenon of the Character Movement Task, we remove the data augmentation method that produces overlapping characters to prevent ambiguity and interference with the Character Movement Task learning. Similar to MoCo [15], we predefine a queue with a large length and initialize the queue randomly. In the training process of the network, the current mini-batch is enqueued and the oldest mini-batch is dequeued. For each feature vector generated by the encoder projection module, the positive sample is the feature vector generated by the momentum encoder projection module after different data augmentation of the same image and the negative samples are the feature vectors in the queue. The feature vector in the queue is extracted by the momentum encoder projection module; therefore, it should be relatively consistent with the features extracted by the encoder projection module. Inspired by MoCo v2 [6], the momentum encoder projection module adopts the same structure as the encoder projection module, but it does not share parameters. Formally, we denote the parameters of the momentum encoder projection module as θ_v and those of the encoder projection module as θ_q, θ_v is initialized by θ_q and updated by

$$m\theta_v + (1 - m)\theta_q \rightarrow \theta_v \tag{7}$$

where $m \in [0, 1)$ is a momentum coefficient.

The encoder and the momentum encoder are CNN-based networks. The projection and momentum projection heads are multilayer perceptrons with a hidden layer, and they are used to map the visual representation to the contrastive space. The multilayer perceptron (MLP) for the Character Movement Task consists of two fully connected layers, which mainly convert the visual feature map into a vector for classification.

The total loss of CMT-Co can be formulated as (8).

$$\mathcal{L} = \mathcal{L}_{contrast} + \alpha \mathcal{L}_{move} \tag{8}$$

where

$$\mathcal{L}_{contrast} = -\log \frac{\exp(MI_q \cdot k_+/\tau)}{\sum_{i=1}^{C} \exp(MI_q \cdot k_i/\tau)} \tag{9}$$

Here, C is the queue size, τ is a temperature hyper-parameter [43], MI_q is the feature vector output by the encoder projection module, k_+ is the feature vector output by the momentum encoder projection module and k_i is the feature vector in the queue. α is a hyper-parameter that is adopted for weighting loss items.

4   Experiments

4.1   Implementation Details

Datasets and Metrics. We conduct our experiments on public handwritten benchmarks, which are IAM [28], RIMES [13], and CVL [19]. We use word-level accuracy as the evaluation metric for all experiments. We follow the dataset setting of SeqCLR [1] for pre-training and downstream fine-tuning, which only uses the training set when training. All images are resized to 32×100 as input size.

Pre-training and Downstream Fine-Tuning Settings. We conduct all experiments on one A40 GPU with a mini-batch size of 256. For *pre-training*, we adopt the same settings as MoCo v2 [6]. We use SGD as the optimizer with an initial learning rate of 0.03. For the CMT, the superposition scale λ in formula (3) is 0.7. The hyper-parameter α in formula (8) is 0.2. The momentum coefficient m in formula (7) is 0.999. The momentum encoder and the encoder is ResNet29 [8], which is the same as the backbone of SeqCLR [1]. The MLP for

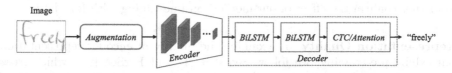

Fig. 7. Framework for downstream task training. We follow the framework of Seq-CLR [1].

Table 1. Representation quality. We freeze the encoder and only train the decoder. **Bold** represents the best result, and underlined represents the second-best result. "Ft Aug." represents the augmentation during downstream fine-tuning. "Seq. Aug." represents the augmentation in SeqCLR [1]. "MoCo v2†" indicates that our data augmentation method Text-Aug is used in pre-training.

Method	Ft Aug.	Decoder	IAM	RIMES	CVL
Baseline	Seq. Aug. [1]	CTC	29.9	22.7	24.3
SimCLR [3]			4.0	10.0	1.8
SeqCLR [1]			39.7	63.8	66.7
MoCo v2† [6]			<u>51.7</u>	**66.9**	<u>71.1</u>
CMT-Co(Ours)			**53.1**	66.0	**72.1**
Baseline	Seq. Aug. [1]	Attention	33.9	28.7	35.0
SimCLR [3]			16.0	22.0	26.7
SeqCLR [1]			51.9	**79.5**	<u>74.5</u>
MoCo v2† [6]			<u>56.3</u>	<u>75.1</u>	74.1
CMT-Co(Ours)			**58.2**	74.7	**74.7**

the Character Movement Task is two fully connected layers. The projection head and the momentum projection head are multilayer perceptrons with a hidden layer. We train the model for 68K iterations for each dataset, which takes about 20 h.

For *downstream fine-tuning*, we follow the settings of SeqCLR [1] for a fair comparison. Specifically, we use the "encoder-decoder" paradigm in text recognition, in which there are two types of decoders: CTC-based decoder and attention-based decoder, as shown in Fig. 7. We use Adadelta [45] optimizer with an initial learning rate of 2. To further demonstrate the effectiveness of Text-Aug, we also present results using our data augmentation approach for downstream fine-tuning.

4.2 Experimental Results

In this section, we quantitatively verify the effectiveness of the proposed method. We compare our method with previous self-supervised methods based on contrastive learning (i.e. SeqCLR [1], PerSec [23]) on the handwritten dataset. We first conduct experiments on the representation quality to validate the representational ability of the encoder parameters learned from the pre-training stage. We then conduct experiments on a certain proportion of labeled training data to further confirm the effect of our method on fine-tuning with few data.

Representation Quality. To examine the quality of encoder representations learned by pre-training, we follow SeqCLR [1] and PerSec [23], which freeze the parameters of the encoder and only train the decoder. The decoder is all randomly initialized. For Baseline, we initialize the encoder randomly, but for other methods, we use pre-trained parameters to initialize the encoder.

Table 2. Downstream Task Fine-tuning. We train the entire model including the encoder and the decoder. **Bold** represents the best result, and <u>underlined</u> represents the second-best result. "Ft Aug." represents the augmentation during downstream fine-tuning. "Seq. Aug." represents the augmentation in SeqCLR [1]. "MoCo v2†" indicates that our data augmentation method Text-Aug is used in pre-training.

Method	Ft Aug.	Decoder	IAM			RIMES			CVL		
			5%	10%	100%	5%	10%	100%	5%	10%	100%
PerSec-CNN [23]	Luo et al. [27]	CTC	–	–	77.9	–	–	–	–	–	<u>78.1</u>
SimCLR [3]	Seq. Aug. [1]		15.4	21.8	65.0	36.5	52.9	84.5	52.1	62.0	74.1
SeqCLR [1]			31.2	44.9	76.7	61.8	71.9	<u>90.1</u>	66.0	71.0	77.0
Baseline			35.8	44.2	78.6	56.7	67.7	87.0	63.0	70.6	77.7
MoCo v2† [6]			<u>42.2</u>	51.7	79.6	63.2	72.6	89.4	67.3	72.7	77.8
CMT-Co (Ours)			41.7	<u>52.0</u>	<u>79.9</u>	<u>63.8</u>	<u>72.7</u>	89.3	<u>68.1</u>	<u>72.8</u>	78.0
CMT-Co (Ours)	Text-Aug		**47.7**	**56.0**	**80.1**	**67.1**	**75.4**	**90.1**	**71.9**	**74.5**	**78.2**
PerSec-CNN [23]	Luo et al. [27]	Attention	–	–	80.8	–	–	–	–	–	**80.2**
SimCLR [3]	Seq. Aug. [1]		22.7	32.2	70.7	49.9	60.9	87.8	59.0	65.6	75.7
SeqCLR [1]			40.3	52.3	79.9	**70.9**	<u>77.0</u>	**92.5**	<u>73.1</u>	74.8	77.8
Baseline			40.5	49.8	79.5	62.1	72.7	90.0	68.4	73.8	77.3
MoCo v2† [6]			44.2	53.3	80.7	66.8	75.7	90.8	70.2	74.9	78.6
CMT-Co (Ours)			<u>45.5</u>	<u>53.4</u>	<u>81.3</u>	66.9	76.0	90.4	71.5	<u>75.1</u>	78.5
CMT-Co (Ours)	Text-Aug		**50.4**	**55.8**	**81.9**	<u>68.8</u>	**77.6**	<u>91.2</u>	**73.6**	**76.2**	<u>78.7</u>

Table 1 shows the representation quality comparison with our method, SimCLR [3], and SeqCLR [1] on IAM, RIMES, and CVL datasets. Combined with our data augmentation Text-Aug in pre-training, "MoCo v2†" can achieve even better performance than SeqCLR [1], which is specifically designed for text recognition in most datasets. "MoCo v2†" is improved by 12% on the IAM dataset on the CTC-based decoder method compared to SeqCLR [1]. Then, assisted by CMT, CMT-Co achieves better performance on IAM and CVL datasets compared to "MoCo v2†". This proves that our Character Movement Task designed with prior knowledge of the handwritten text is indeed beneficial for self-supervised representation learning. Finally, our method CMT-Co outperforms SeqCLR [1] on most datasets, with the most significant improvement on the IAM dataset with a gain of 13.4%.

Downstream Task Fine-Tuning. To further demonstrate the pre-trained model representation ability, we explore the performance of the model on a small amount of labeled data. In Table 2, we present the results for 5%, 10%, and 100% of the labeled training data, respectively. In fine-tuning, we train the entire model, including the encoder and decoder. The encoder and decoder parameters of the baseline are initialized randomly. The encoders of other methods load the pre-trained parameters, and the decoder parameters are initialized randomly.

Table 2 shows the downstream task fine-tuning comparison with our method, SimCLR [3], SeqCLR [1], and PerSec [23]. It can be seen that in most datasets, combined with our data-augmented Text-Aug in pre-training, "MoCo v2†" can achieve comparable performance to SeqCLR [1] and PerSec [23], which are specially designed for text recognition. Assisted by CMT, CMT-Co has further

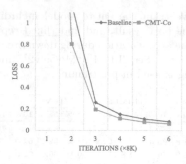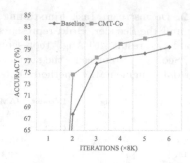

Fig. 8. Training loss and word-level accuracy between baseline and CMT-Co during downstream fine-tuning with an attention-based decoder on the IAM dataset.

improvements on most scales of datasets. Eventually, when augmenting data using Text-Aug during fine-tuning, our method improves again, achieving best or second-best results on all scales of datasets. With the dual power of Text-Aug and CMT, our method achieves promising performance on the IAM dataset.

Figure 8 shows training loss and word-level accuracy between baseline and CMT-Co during downstream fine-tuning with an attention-based decoder on the IAM dataset. It can be seen that the encoder parameters pre-trained by CMT-Co indeed speed up the convergence and achieve better results.

4.3 Ablation Study

In this section, we first explore the effect of important parts in CMT-Co, including CMT and data augmentation Text-Aug. Then, we perform ablation of the important parameters λ and loss function in CMT, where λ determines the superposition scale. More ablations to verify the effectiveness of the proposed method are shown in the Appendix.

We conduct ablation experiments on the IAM dataset. We use attention for the decoder and the data augmentation in SeqCLR [1] during fine-tuning. Table 3 explores the ablation of two parts of CMT-Co. It can be seen that Text-Aug performs better than "MoCo v2-Aug." and "Seq. Aug.", verifying the effectiveness

Table 3. The ablation of data augmentation and CMT. "Pre-training Aug." represents the augmentation during pre-training. "MoCo v2-Aug." represents the augmentation in MoCo v2 [6]. "Seq. Aug." represents the augmentation in SeqCLR [1].

Method	Pre-training Aug.	Pretext Task	Accuracy
MoCo v2 [6]	MoCo v2-Aug. [6]	✗	80.2
MoCo v2* [6]	Seq. Aug. [1]	✗	79.8
MoCo v2† [6]	Text-Aug	✗	80.7
CMT-Co (Ours)	Text-Aug	CMT	**81.3**

Table 4. The ablation of λ.

λ	0.1	0.3	0.5	0.7	0.9
Accuracy	80.7	80.0	80.5	**81.3**	80.9

Table 5. The ablation of different loss functions for the CMT.

Loss	CE	MSE
Accuracy	**81.3**	72.5

of our Text-Aug strategy. Assisted by the pretext task CMT, the performance of the CMT-Co method is further improved, showing the effectiveness of CMT. The influence of λ that determines the superposition scale is shown in Table 4. The performance of CMT-Co reaches its best when λ is 0.7.

Although in this paper, we define the Character Movement Task as a classification task, it is also reasonable for CMT to be defined as a regression task. Therefore, we also explored its effect as a regression task. The ablation of different loss functions of the Character Movement Task is studied in Table 5. "CE" refers to the cross-entropy loss, which means we regard CMT as a classification task, and the equation is shown in (5). "MSE" refers to the mean-squared loss, which means we regard CMT as a regression task, and the equation is shown in (10).

$$\mathcal{L}_{move} = \|y_{pred} - y\|^2 \tag{10}$$

where y_{pred} is the prediction of the network and y is generated by CMT.

It can be seen that when CMT is used as a classification task, the performance is better. We believe that CMT is different from reconstruction and generation tasks, which reduce the penalty when the predicted distance gap becomes smaller. To better learn the representation of the character, the position of the character should be accurately predicted.

5 Conclusion

In this paper, we propose a Character Movement Task (CMT) to learn character-level features based on prior knowledge of the handwritten text. Our method CMT-Co uses the CMT as a character-level auxiliary task to assist the learning of the entire word-level image in contrastive learning. Furthermore, to better improve the performance of contrastive learning, we also propose a data augmentation strategy named Text-Aug which is suitable for handwritten text. Experiments show that our method can achieve comparable performance or even better performance than existing self-supervised methods for text recognition.

In the future, it is worth exploring more effective text-related pretext tasks and multi-task self-supervised learning methods. In addition, we hope that this work can inspire more research on self-supervised learning for text recognition, which has not been well investigated in the literature.

Acknowledgment. This research is supported in part by NSFC (Grant No.: 61936003), GD-NSF (no. 2017A030312006, No. 2021A1515011870), Zhuhai Industry

Core and Key Technology Research Project (no. ZH22044702200058PJL), and the Science and Technology Foundation of Guangzhou Huangpu Development District (Grant 2020GH17).

References

1. Aberdam, A., et al.: Sequence-to-sequence contrastive learning for text recognition. In: CVPR, pp. 15302–15312 (2021)
2. Bhunia, A.K., Ghose, S., Kumar, A., Chowdhury, P.N., Sain, A., Song, Y.Z.: MetaHTR: towards writer-adaptive handwritten text recognition. In: CVPR, pp. 15830–15839 (2021)
3. Chen, T., Kornblith, S., Norouzi, M., Hinton, G.: A simple framework for contrastive learning of visual representations. In: ICML, pp. 1597–1607 (2020)
4. Chen, T., Kornblith, S., Swersky, K., Norouzi, M., Hinton, G.E.: Big self-supervised models are strong semi-supervised learners. In: NeurIPS, pp. 22243–22255 (2020)
5. Chen, X., Jin, L., Zhu, Y., Luo, C., Wang, T.: Text recognition in the wild: a survey. ACM Comput. Surv. **54**(2), 1–35 (2021)
6. Chen, X., Fan, H., Girshick, R.B., He, K.: Improved baselines with momentum contrastive learning. CoRR abs/2003.04297 (2020)
7. Chen, X., He, K.: Exploring simple Siamese representation learning. In: CVPR, pp. 15750–15758 (2021)
8. Cheng, Z., Bai, F., Xu, Y., Zheng, G., Pu, S., Zhou, S.: Focusing attention: towards accurate text recognition in natural images. In: ICCV, pp. 5076–5084 (2017)
9. Doersch, C., Gupta, A., Efros, A.A.: Unsupervised visual representation learning by context prediction. In: ICCV, pp. 1422–1430 (2015)
10. Gan, Y., Han, R., Yin, L., Feng, W., Wang, S.: Self-supervised multi-view multi-human association and tracking. In: ACM International Conference on Multimedia, pp. 282–290 (2021)
11. Gidaris, S., Singh, P., Komodakis, N.: Unsupervised representation learning by predicting image rotations. In: ICLR (2018)
12. Grill, J.B., et al.: Bootstrap your own latent-a new approach to self-supervised learning. In: NeurIPS, vol. 33, pp. 21271–21284 (2020)
13. Grosicki, E., El Abed, H.: ICDAR 2009 handwriting recognition competition. In: ICDAR, pp. 1398–1402 (2009)
14. Ha, J., Haralick, R.M., Phillips, I.T.: Document page decomposition by the bounding-box project. In: ICDAR, pp. 1119–1122 (1995)
15. He, K., Fan, H., Wu, Y., Xie, S., Girshick, R.: Momentum contrast for unsupervised visual representation learning. In: CVPR, pp. 9729–9738 (2020)
16. Ingle, R.R., Fujii, Y., Deselaers, T., Baccash, J., Popat, A.C.: A scalable handwritten text recognition system. In: ICDAR, pp. 17–24 (2019)
17. Jaiswal, A., Babu, A.R., Zadeh, M.Z., Banerjee, D., Makedon, F.: A survey on contrastive self-supervised learning. Technologies **9**(1), 2 (2020)
18. Kinakh, V., Taran, O., Voloshynovskiy, S.: ScatSimCLR: self-supervised contrastive learning with pretext task regularization for small-scale datasets. In: ICCV Workshops, pp. 1098–1106 (2021)
19. Kleber, F., Fiel, S., Diem, M., Sablatnig, R.: CVL-DataBase: an off-line database for writer retrieval, writer identification and word spotting. In: ICDAR, pp. 560–564 (2013)
20. Kolesnikov, A., Zhai, X., Beyer, L.: Revisiting self-supervised visual representation learning. In: CVPR, pp. 1920–1929 (2019)

21. Kolesnikov, A., Zhai, X., Beyer, L.: Revisiting self-supervised visual representation learning. In: Proceedings of the IEEE/CVF Conference on Computer Vision and Pattern Recognition (CVPR), pp. 1920–1929 (2019)
22. Li, W., Wang, G., Fidon, L., Ourselin, S., Cardoso, M.J., Vercauteren, T.: On the compactness, efficiency, and representation of 3D convolutional networks: brain parcellation as a pretext task. In: Niethammer, M., et al. (eds.) IPMI 2017. LNCS, vol. 10265, pp. 348–360. Springer, Cham (2017). https://doi.org/10.1007/978-3-319-59050-9_28
23. Liu, H., et al.: Perceiving stroke-semantic context: hierarchical contrastive learning for robust scene text recognition. In: AAAI, pp. 1702–1710 (2022)
24. Liu, X., et al.: Self-supervised learning: generative or contrastive. IEEE Trans. Knowl. Data Eng. **35**, 1 (2021)
25. Liu, Y., Chen, H., Shen, C., He, T., Jin, L., Wang, L.: ABCNet: real-time scene text spotting with adaptive Bezier-curve network. In: CVPR, pp. 9806–9815 (2020)
26. Luo, C., Jin, L., Sun, Z.: MORAN: a multi-object rectified attention network for scene text recognition. Pattern Recogn. **90**, 109–118 (2019)
27. Luo, C., Zhu, Y., Jin, L., Wang, Y.: Learn to augment: joint data augmentation and network optimization for text recognition. In: CVPR, pp. 13746–13755 (2020)
28. Marti, U.V., Bunke, H.: The IAM-database: an English sentence database for offline handwriting recognition. Int. J. Doc. Anal. Recogn. **5**(1), 39–46 (2002)
29. Michael, J., Labahn, R., Grüning, T., Zöllner, J.: Evaluating sequence-to-sequence models for handwritten text recognition. In: ICDAR, pp. 1286–1293 (2019)
30. Misra, I., Maaten, L.V.D.: Self-supervised learning of pretext-invariant representations. In: CVPR, pp. 6706–6716 (2020)
31. Noroozi, M., Favaro, P.: Unsupervised learning of visual representations by solving Jigsaw puzzles. In: Leibe, B., Matas, J., Sebe, N., Welling, M. (eds.) ECCV 2016. LNCS, vol. 9910, pp. 69–84. Springer, Cham (2016). https://doi.org/10.1007/978-3-319-46466-4_5
32. Parvez, M.T., Mahmoud, S.A.: Offline Arabic handwritten text recognition: a survey. ACM Comput. Surv. **45**(2), 1–35 (2013)
33. Pathak, D., Krahenbuhl, P., Donahue, J., Darrell, T., Efros, A.A.: Context encoders: feature learning by inpainting. In: CVPR, pp. 2536–2544 (2016)
34. Ptak, R., Żygadło, B., Unold, O.: Projection-based text line segmentation with a variable threshold. Int. J. Appl. Math. Comput. Sci. **27**(1), 195–206 (2017)
35. Puigcerver, J.: Are multidimensional recurrent layers really necessary for handwritten text recognition? In: ICDAR, pp. 67–72 (2017)
36. Sánchez, J.A., Bosch, V., Romero, V., Depuydt, K., De Does, J.: Handwritten text recognition for historical documents in the tranScriptorium project. In: DATeCH, pp. 111–117. ACM (2014)
37. Sánchez, J.A., Romero, V., Toselli, A.H., Vidal, E.: ICFHR2014 competition on handwritten text recognition on tranScriptorium datasets (HTRtS). In: ICFHR, pp. 785–790 (2014)
38. Sánchez, J.A., Romero, V., Toselli, A.H., Vidal, E.: ICFHR2016 competition on handwritten text recognition on the read dataset. In: ICFHR, pp. 630–635 (2016)
39. Sánchez, J.A., Romero, V., Toselli, A.H., Villegas, M., Vidal, E.: ICDAR2017 competition on handwritten text recognition on the read dataset. In: ICDAR, pp. 1383–1388 (2017)
40. Thoker, F.M., Doughty, H., Snoek, C.G.: Skeleton-contrastive 3D action representation learning. In: ACM International Conference on Multimedia, pp. 1655–1663 (2021)

41. Wang, K., Babenko, B., Belongie, S.: End-to-end scene text recognition. In: ICCV, pp. 1457–1464 (2011)
42. Wang, T., et al.: Implicit feature alignment: learn to convert text recognizer to text spotter. In: CVPR, pp. 5973–5982 (2021)
43. Wu, Z., Xiong, Y., Yu, S.X., Lin, D.: Unsupervised feature learning via non-parametric instance discrimination. In: CVPR, pp. 3733–3742 (2018)
44. Yan, J., Wang, J., Li, Q., Wang, C., Pu, S.: Self-supervised regional and temporal auxiliary tasks for facial action unit recognition. In: ACM International Conference on Multimedia, pp. 1038–1046 (2021)
45. Zeiler, M.D.: ADADELTA: an adaptive learning rate method. CoRR abs/1212.5701 (2012)
46. Zhang, R., Isola, P., Efros, A.A.: Colorful image colorization. In: Leibe, B., Matas, J., Sebe, N., Welling, M. (eds.) ECCV 2016. LNCS, vol. 9907, pp. 649–666. Springer, Cham (2016). https://doi.org/10.1007/978-3-319-46487-9_40

Looking from a Higher-Level Perspective: Attention and Recognition Enhanced Multi-scale Scene Text Segmentation

Yujin Ren[1], Jiaxin Zhang[1], Bangdong Chen[1], Xiaoyi Zhang[1], and Lianwen Jin[1,2]

[1] South China University of Technology, Guangzhou, China
{msjxzhang,eebdchen}@mail.scut.edu.cn, lianwen.jin@gmail.com
[2] SCUT-Zhuhai Institute of Modern Industrial Innovation, Zhuhai, China

Abstract. Scene text segmentation, which aims to generate pixel-level text masks, is an integral part of many fine-grained text tasks, such as text editing and text removal. Multi-scale irregular scene texts are often trapped in complex background noise around the image, and their textures are diverse and sometimes even similar to those of the background. These specific problems bring challenges that make general segmentation methods ineffective in the context of scene text. To tackle the aforementioned issues, we propose a new scene text segmentation pipeline called Attention and Recognition enhanced Multi-scale segmentation Network (ARM-Net), which consists of three main components: Text Segmentation Module (TSM) generates rectangular receptive fields of various sizes to fit scene text and integrate global information adequately; Dual Perceptual Decoder (DPD) strengthens the connection between pixels that belong to the same category from the spatial and channel perspective simultaneously during upsampling, and Recognition Enhanced Module (REM) provides text attention maps as a prior for the segmentation network, which can inherently distinguish text from background noise. Via extensive experiments, we demonstrate the effectiveness of each module of ARM-Net, and its performance surpasses that of existing state-of-the-art scene text segmentation methods. We also show that the pixel-level mask produced by our method can further improve the performance of text removal and scene text recognition.

Keywords: Scene text segmentation · Deep neural network

1 Introduction

As an important constituent of image pre-processing, text segmentation was once the foundation of text detection and recognition. With mature applications

L. Wang et al. (Eds.): ACCV 2022, LNCS 13847, pp. 643–659, 2023.
https://doi.org/10.1007/978-3-031-26293-7_38

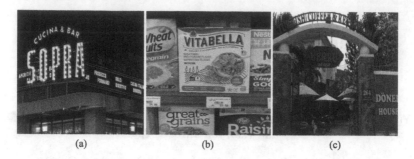

(a)   (b)   (c)

Fig. 1. Three specific issues in scene text segmentation [4]: (a) various text scales; (b) scattered text distribution; (c) background distraction.

of deep neural networks (DNNs) in optical character recognition (OCR), pixel-level (stroke) text segmentation is rarely used in traditional text-related vision tasks. However, some more fine-grained scenarios have emerged recently, such as text editing [14,25,39] and text removal [17,44]. They require segmentation to obtain precise pixel-level text masks in advance, which can be used to separate texts from complex backgrounds. In response to these new demands, scene text segmentation has gradually regained researchers' attention [1,23,32,40,41].

It is difficult to obtain satisfactory results by directly transferring general segmentation methods to scene text, as there are specific issues that must be addressed in scene text segmentation: (1) Scene text is non-convex and prone to exhibiting drastic differences in scale, making it challenging to segment structural details of texts with various styles. (2) The uneven distribution of scene text in the image makes it easy for text that appears in inconspicuous locations, especially text whose texture appears less frequently, to be ignored by the segmentation network. (3) Scene text is trapped in complex background noise and sometimes has similar textures with them, which may lead to ambiguity in segmentation results (Fig. 1).

Although existing scene text segmentation approaches partially solve the aforementioned problems to some extent, they still have major shortcomings. SMANet [1] used a pooling operation to obtain multi-scale text features, but it cannot preserve the resolution of the feature map, making it unsuitable for retaining the spatial location information of scene text. MGNet [32] adopted a semi-supervised training strategy and used polygon-level mask annotations to provide a prior for pixel-level text segmentation, which is helpful to confirm text location, but lacks a subtle network design for scene text. TexRNet [40] well-designed a refinement network after the segmentation backbone, exploiting cosine similarity to correct those infrequent pixels that are misclassified. It can indeed produce a better segmentation effect by considering the characteristics of scene text, but requires complex character-level annotations for its discriminator.

In this study, we propose a text-tailored segmentation pipeline called ARM-Net, which jointly focuses on both low-level appearance information and higher-level text semantic information. It is worth noting that low-level and high-level features in segmentation network are collectively referred to as low-level text appearance information to differentiate them from text semantic information. We optimize low-level text appearance information by rethinking the classical encoder-decoder structure of the segmentation network. In the feature encoding stage, the proposed Text Segmentation Module (TSM) is used for modeling sophisticated text segmentation features by accommodating global and local perspectives. It assigns equal attention weight to global texts to reactivate those with rare textures because of their strong semantic association with the dominant text. Moreover, it also adapts irregular multi-scale scene text to eliminate the interference of background noise and thus capture more effective local features. In the decoding recovery stage, pixels are progressively aggregated into specific classes during upsampling. Slight deviations in deep feature maps may result in inaccurate and distorted segmentation, especially on scene text with an arbitrary shape. Accordingly, we propose a Dual Perceptual Decoder (DPD), whose parameters can be dynamically adjusted to spatial and channel contents. Aiming to take full advantage of text characteristics, we explore the essential differences between text and generic scenes (background noise in scene text segmentation), explaining why human beings rarely struggle with how to distinguish between them. The key, we believe, is that text is no longer treated as simple graphic symbols after people endow them with specific meanings. To imitate the human behavioral patterns, we design an innovative Recognition Enhanced Module (REM) to introduce higher-level text semantic information that provides text attention maps as prior knowledge to promote text discrimination.

To summarize, our main contributions are three-fold:

1. We propose an end-to-end trainable model, ARM-Net, which exploits a combination of low-level text appearance information and higher-level text semantic information to facilitate segmentation.
2. Extensive experiments demonstrate the effectiveness of ARM-Net, which achieves superior performance on three mainstream scene text segmentation benchmarks, and each module plays significant role.
3. Experiments on downstream tasks illustrate that the addition of pixel-level masks generated by ARM-Net can improve the effectiveness of text removal and the accuracy of scene text recognition.

2 Related Work

2.1 Semantic Segmentation

Semantic segmentation, one of the traditional tasks in computer vision, aims to predict a correct category for each pixel. With the development of deep learning, many methods based on the DNN are distinguishable from traditional graph algorithms such as MRF [31] and CRF [15]. Since the FCN [19] firstly adopted fully convolutional network in semantic segmentation, numerous works based on the encoder-decoder structure [8,24,46] have emerged.

In order to overcome the dilemma of limited receptive fields, the importance of multi-scale features has been continuously emphasized. PSPNet [45] fused features of different scales through pooling operations to aggregate contextual information in different regions. DeepLab [2,3] introduced the atrous convolution operation to obtain multi-scale features without changing the resolution of feature maps. HRNet [35] performed repetitive fusion by exchanging information on parallel multi-resolution sub-networks so that the network can maintain high-resolution representation.

As the self-attention mechanism [29] has shown extraordinary value in natural language processing (NLP), researchers have applied it to semantic segmentation to obtain the long-range dependency. DANet [5] associated spatial attention and channel attention to acquire broader contextual information. Employing a similar strategy, CCNet [9] and Axial-DeepLab [33] proposed Criss-Cross Attention and Axial-Attention, which both use fewer pixels to participate in the attention calculation so as to reduce computing costs. EMANet [16] abandoned the process of computing attention maps on a full graph and utilized the Expectation Maximization (EM) algorithm instead to iterate over a set of bases and then performed the attention mechanism on it.

2.2 Scene Text Segmentation

Previously developed scene text segmentation methods mostly use thresholds or low-level features to binarize scene text images, making it difficult to produce satisfactory results due to the complexity of scenes and textures. Recently, several approaches based on deep learning have been explored. Bonechi et al. [1] labeled COCO [30] and MLT [22] by machine for pre-training, and proposed SMANet, which combines the pooling pyramid with the attention mechanism to form a multi-scale attention module. Wang et al. [32] proposed a mutually guided dual-task network that uses polygon-level masks (bounding boxes) for semi-supervision which can be easily obtained from scene text detection datasets. They pointed out that pixels outside a polygon-level mask do not belong to the pixel-level mask. So, polygon-level masks can serve as a filter to guide the generation of pixel-level masks, and vice versa. Xu et al. [40] made a new text segmentation dataset, TextSeg, which contains 4,024 images with comprehensive annotations. They made some special designs in their TexRNet to refine the output from the aforementioned segmentation networks such as DeepLabV3+ [3] and HRNet [35] to improve their performance on scene text segmentation. TexRNet firstly guarantees that high-confidence regions are reliable by calculating the modified cosine-similarity between the text class and background class. It uses the key features pooling and attention-based similarity checking to activate text regions that may be ignored owing to low-confidence in the initial prediction.

3   Methodology

3.1   Pipeline

As shown in Fig. 2, our proposed ARM-Net consists of three main components: Text Segmentation Module (TSM), Dual Perceptual Decoder (DPD) and Recognition Enhanced Module (REM). We adopt ResNet-50 [6] as backbone for feature extraction. Extracted features are fed into the TSM and transformed to dense multi-scale text segmentation features with a global view. Then, the DPD dynamically aggregates text and background pixels according to content information from the spatial and channel perspective during upsampling. We also blend low-level features from the backbone with each stage of the DPD to acquire more visual details. In addition, the REM is applied as an auxiliary cue and brings in higher-level text semantic information to enhance segmentation features. Text attention maps generated by the REM indicate where the segmentation network should focus.

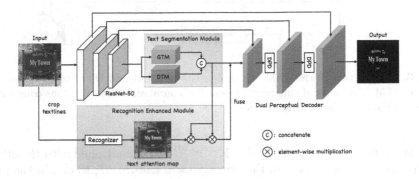

Fig. 2. Pipeline of our proposed ARM-Net.

3.2   Text Segmentation Module

For scene text segmentation, local and global information are like both sides of a scale, and a reasonable balance needs to be achieved between the two to succeed better performance. Accordingly, our proposed TSM integrates global correlations and local details adequately to obtain more effective segmentation representations, as illustrated in Fig. 3.

Scene texts are frequently scattered in images, and they are sometimes submerged in the complex background noise; so, those texts with small size or rare textures are easily ignored by the segmentation network. We propose a Global Text Module (GTM), which draws on the core idea of CCNet [9] to model dependencies of the entire image, while avoiding a surge of computation and parameters. Unlike the non-local network [38] that calculates the correlation matrix between each pixel in the feature map spatially, we only perform the self-attention for each pixel in the horizontal and vertical directions that the pixel belongs to, as shown in the red dashed box of Fig. 3.

In the concrete, after 1×1 convolution layers, we obtain three feature maps Q, K and V. The affinity matrix $A = \varphi\left(Q_u \cdot K_u^T\right)$ between spatial locations is obtained by element-wise multiplying each position of Q (i.e. Q_u) with the vector set $K_u \in \mathbb{R}^{C' \times (H+W-1)}$ and then applying a softmax function, $\varphi(\cdot)$. Here K_u denotes the set of positions in K that are in the same row or column as u.

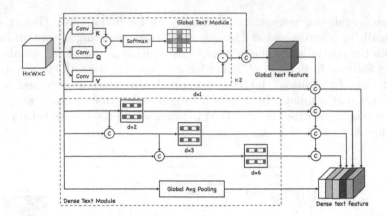

Fig. 3. Architecture of the TSM, which consists of GTM and DTM two sub-modules. In the DTM, blue and red squares represent horizontal and vertical atrous convolution, respectively, and d is the dilation rate.

Similar to the above operation, $V_u \in \mathbb{R}^{C \times (H+W-1)}$ is multiplied with the affinity matrix A. Then, we add it to the primary local feature, H, to obtain the pixel-wise contextual augmented feature representation, H'. Here we implement Criss-Cross Attention twice to deliver the information of one pixel into all paths. The above operation can be expressed as follows:

$$H'_u = \varphi \left(Q_u \cdot K_u^T \right) \cdot V_u + H_u .$$ (1)

As scene texts are mostly in rectangular or curved shape and their scale is extremely different, we propose a Dense Text Module (DTM) on the basis of atrous convolution [2]. The operation of atrous convolution is equivalent to inserting $d - 1$ zeros between two adjacent weights of the filter, where d is the dilation rate. This approach can expand the range of the receptive field while maintaining the resolution of the feature map.

As illustrated in the blue dashed box of Fig. 3, the DTM heuristically performs atrous convolution in two directions, which separately creates horizontal or vertical zero padding. Afterwards, the DTM cascades these atrous convolution layers with the dilation rate from low to high. In this way, we obtain denser feature representations and rectangular receptive fields with various aspect ratio, which are more appropriate for irregular scene text. When the dilation rates of the two directions are equal, the receptive field is of a regular shape. Stacking n atrous convolution layers can obtain a larger equivalent kernel size K:

$$K = \sum_{i=1}^{n} K_i - (n - 1) .$$ (2)

Moreover, the receptive field size, R, increases linearly with dilation rates:

$$R = (d - 1) \times (K - 1) + K .$$ (3)

Following Eqs. (2) and (3), the number of R in the DTM is at most 26, which sufficiently alleviates the problem wherein the fixed shape receptive field is not appropriate for multi-scale scene text.

The DTM outputs dense multi-scale text features with abundant sizes of the receptive field. Each feature is concatenated with the augmented feature representation H' produced by the GTM to combine global information adequately.

3.3 Dual Perceptual Decoder

In segmentation methods that utilize the encoder-decoder structure, decoder plays an important role in rebuilding image from features. However, traditional upsampling methods have their limitations. Nearest and bilinear interpolation only consider adjacent positions, and deconvolution is restrained by fixed kernel size and weights. To better cope with the problem that are neglected by previous scene text segmentation methods, we propose the DPD, as shown in Fig. 4.

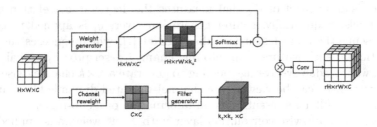

Fig. 4. Structure of the DPD. The upper branch is Spatial Context-Aware Module, and the branch below is Channel Semantic-Aware Module.

The upper branch in Fig. 4 is a Spatial Context-Aware Module, which is inspired by CARAFE [34] and the dynamic filter network [11]. Given a $C \times H \times W$ feature map, F, the Spatial Context-Aware Module can dynamically expand the feature map to $C \times rH \times rW$ according to the context information of different objects in the spatial dimension, where r is the up-sampling rate. We first employ a convolution layer with kernel size k_e on segmentation features as a weight generator through which the feature map becomes $H \times W \times C'$. Here, $C' = r^2 k_u^2$, and k_u is the kernel size during up-sampling. The weight generator aggregates the spatial context information within the $k_e \times k_e$ receptive field. Then, we reshape the channel dimension and spatial dimension to obtain a $k_u \times k_u$ weight matrix, $W_u \in \mathbb{R}^{rH \times rW \times k_u^2}$, which can satisfy each position on $C \times rH \times rW$ as an individual weight. Note that each $k_u \times k_u$ kernel is normalized by a spatial softmax function. This makes the kernel values sum to one and has no effect on the feature distribution. Finally, we calculate the upsampled output, F_s, as follows:

$$F_s = \sum_{i=-n}^{n} W_i \cdot F_i . \tag{4}$$

From another perspective, each channel of high-level features can be regarded as a class-specific response, and semantic responses belonging to the same category should be associated with each other during upsampling. By assigning channel weights according to their interdependencies, associated channels are emphasized, and interfering channels are suppressed simultaneously, thereby improving the discriminative capacity of the model.

To fulfill the above purpose, we design the Channel Semantic-Aware Module as illustrated in the lower branch of Fig. 4. Taking segmentation feature $F \in \mathbb{R}^{C \times H \times W}$ as the input, we first calculate its channel-wise relationship on it. Concretely, we reshape F to $\mathbb{R}^{C \times N}$, and then multiply it with its transpose matrix. We normalize the output through a softmax layer to obtain the weight matrix $W_c \in \mathbb{R}^{C \times C}$. W_c represents the interrelationship between channels and can also be seen as a channel attention map.

$$w_c = \frac{exp(F_i \cdot F_i^T)}{\sum_{i=1}^{C} exp(F_i \cdot F_i^T)} , \tag{5}$$

where w_c is an element of W_c that measures the degree of correlation between two channels. Sequentially, a filter generation layer, g, is applied to generate channel semantic-aware features. Unlike SENet [7], which squeezes the spatial information of each feature map into one channel descriptor by global average pooling, we use adaptive average pooling to generate a $k \times k$ channel-wise feature F_c. Afterward, F_c can be viewed as a channel-weighted dynamic filter, and each position in the filter represents general information of a sub-region.

Finally, a depth-wise convolution layer with F_c as weights is applied to F_s. Through a 1×1 convolution layer to fuse the channel information, upsampled output is obtained. Replacing traditional upsampling methods with the DPD allows us to rearrange activation responses of scene text elaborately according to the spatial and channel information.

3.4 Recognition Enhanced Module

When the texture is similar to the background, scene text is difficult to be identified accurately by the segmentation network. We think it is because the segmentation network only utilizes low-level appearance information such as text structure and color. To alleviate this problem, the network needs to be taught that text has higher-level meanings that go beyond general objects and symbols, just like human beings do. Consequently, we propose an REM, which can highlight text regions in the whole image and deliver higher-level text semantic information to segmentation features, as shown in Fig. 5.

In the training phase, we first crop textlines from the image and feed them into a pre-trained DAN [37] recognizer whose parameters are frozen. The reason why we choose DAN is that it mitigates attention drift problem through the Convolutional Alignment Module (CAM). Other attention-based recognizers [21, 27] are also appropriate here. The cropped textline images are unified into $M \times 128 \times 32$, where M is the total number of textlines in one batch. Attention maps produced by DAN are used to align character positions at each time

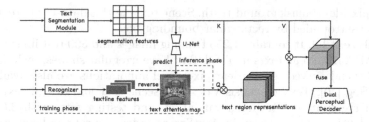

Fig. 5. Architecture of the REM. Portions in the red dotted box only participate in the training phase, and during inference, the text attention map is predicted by U-Net.

step, which are implicitly supervised by word annotations. Summing all steps' attention maps along the channel can obtain a textline attention map. Furthermore, we put textline attention maps into an all-black background according to their initial positions to acquire the text attention map of the whole image, which can indicate text distribution. Pixels with a higher probability in the text attention map are more likely to be candidates of text regions. For a better relation estimation, we follow the operation in the self-attention mechanism [29]. Formally, we treat the text attention map as Q and segmentation features as K, V. After computing the relation matrix between the text attention map, F_Q, and segmentation features, F_K, we assign it as text region representations to the corresponding position of segmentation features, F_V. As a result, the segmentation feature with high similarity to the corresponding position on the text attention map is highlighted while that with low similarity is restrained. By this means, we obtain the enhanced text features, T_e, as follows:

$$T_e = \frac{exp(F_K \cdot F_Q^T)}{\sum_{i=1}^{n} exp(F_K \cdot F_Q^T)} \cdot F_V . \tag{6}$$

Since the coordinate information of texts is not available in the inference phase, we employ a lightweight U-Net [24] to predict text attention maps by using segmentation features under the supervision of the l_1 loss. In this way, we can not only impose a more direct constraint on segmentation features but also replace text attention maps' output by recognizer with that prediction, enabling an image-to-image inference process.

4 Experiment

4.1 Datasets and Implementation Details

We conduct experiments on the following three benchmarks, all of which have high-quality pixel-level annotations that can be used to supervise the training phase of segmentation network.

1) **ICDAR-2013** [12]: This is a dataset for the ICDAR 2013 'Robust Text Reading' competition, which contains 229 training images and 233 test images

with pixel-level mask ground-truth. Scene texts in this dataset are regular and can be surrounded by rectangular bounding box.

2) **Total-Text** [4]: It contains 1,255 training images and 300 test images. Unlike ICDAR-2013, scene texts in Total-Text have irregular shapes, including rectangular and curved texts. Clear pixel-level annotations are also available.

3) **TextSeg** [40]: TextSeg consists of 4,024 scene text and design text images, which are split into training, validation, and testing sets, with 2,646, 340, and 1,038 images, respectively. TextSeg provides accurate and comprehensive annotations including word-wise and character-wise polygon-level masks as well as pixel-level masks and transcriptions.

We train each model on a NVIDA V100 GPU for 100 epochs and use the Adam [13] optimizer. The first five epochs are warm-ups, and the remaining epochs employ poly decayed learning rates, where the initial learning rate 1e–4 is multiplied by $\left(1 - \frac{iter}{max_{iter}}\right)^{power}$ with power = 0.9.

4.2   Comparison with Existing Methods

We compare our proposed ARM-Net with some state-of-the-art methods, including three semantic segmentation methods [3, 24, 35] and three text segmentation methods [1, 32, 40]. We use the Precision(P), Recall(R), and F-score(F) of the foreground text to quantitatively evaluate the performance of the network, where $F = \frac{2P \cdot R}{P+R}$ denotes the harmonic mean of P and R. The results of three evaluation metrics are presented in Table 1. ARM-Net outperforms the state-of-the-art methods in pixel-level scene text segmentation on all three benchmarks. Furthermore, the inference speed of ARM-Net (4.35fps) is faster than TexRNet (1.22 fps), which shows that our method has a better efficiency.

Table 1. Quantitative results between ARM-Net and other segmentation methods.

Methods	Total-Text			ICDAR 2013			TextSeg		
	P	R	F	P	R	F	P	R	F
U-Net [24] (MICCAI'15)	79.6	69.7	74.3	74.6	53.9	62.6	89.0	77.4	82.8
DeepLabV3+ [3] (ECCV'18)	80.2	76.5	78.3	77.4	63.2	69.6	91.4	90.9	91.2
HRNetV2 [35] (TPAMI'20)	81.4	78.0	79.7	72.8	69.5	71.1	91.9	90.5	91.2
SMANet [1] (PRL'20)	86.6	73.9	77.5	74.4	73.8	71.3	–	–	–
MGNet [32] (TIP'20)	83.3	81.6	80.5	79.0	77.0	74.5	–	–	–
TexRNet+DeepLabV3+ [40] (CVPR'21)	–	–	84.4	–	–	83.5	–	–	92.1
TexRNet+HRNetV2-W48 [40] (CVPR'21)	–	–	84.8	–	–	85.0	–	–	92.4
ARM-Net (Ours)	**87.1**	**83.8**	**85.4**	**88.9**	**81.6**	**85.1**	**92.8**	**92.6**	**92.7**

Several typical qualitative examples are presented in Fig. 6, where images contain a variety of styles/shapes/arrangements of text, as well as other complex background distractions such as illumination. It is obvious that our approach overcomes these difficulties and produces more accurate results.

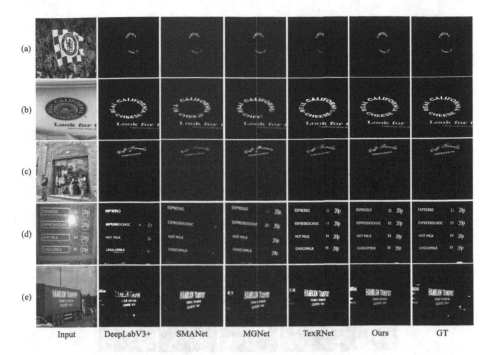

| Input | DeepLabV3+ | SMANet | MGNet | TexRNet | Ours | GT |

Fig. 6. Text segmentation visualization results on Total-Text(a~c) and ICDAR-2013(d,e). From left to right, each column is input, masks predicted by DeepLabV3+, SMANet, MGNet, TexRNet, our ARM-Net, and ground truth, respectively.

4.3 Ablation Studies

In this section, we conduct ablation experiments to verify the effectiveness of each module in our method. All ablation experiments are performed on Total-Text. Note that when conducting ablation experiments within one module, we ensure that other modules are in the best settings.

We first investigate three main components of ARM-Net. The baseline is a simple FCN [19], beginning with which, we add the TSM after the encoder, replace the decoder with the DPD, and introduce the REM progressively. The results in Table 2 exhibit a continuous upward trend with the introduction of each module. When including TSM and DPD, the F-score increases 2.7%, while this gain is 2.5% when we use REM only. Moreover, the ARM-Net with all three modules achieves the best performance, with an increase in F-score of more than 3% compared to the baseline. This suggest that higher-level semantic information is an effective supplement and is as critical as low-level appearance information.

In addition, the samples in Fig. 7 provide more intuitive evidence that ARM-Net is highly adaptable to the scene text segmentation task. Because of the comprehensive consideration of low-level text appearance information and higher-level text semantic information through TSM, DPD, and REM, our approach

Table 2. Effectiveness experiment on three main modules of ARM-Net.

TSM	DPD	REM	P	R	F
×	×	×	85.4	79.4	82.3
✓	×	×	85.7	83.4	84.6
✓	✓	×	85.9	**84.2**	85.0
×	×	✓	85.8	83.7	84.8
✓	✓	✓	**87.1**	83.8	**85.4**

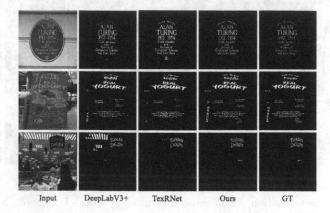

Input   DeepLabV3+   TexRNet   Ours   GT

Fig. 7. Segmentation results for typical difficult samples. From top to bottom, each row is large variations in the scale and shape of text, scattered distribution of text throughout the image, and small text hidden in complex background noise, respectively.

achieves satisfactory segmentation results, without many serious cases of misclassification such as incomplete, blurred text structure and unfiltered background.

Better Low-Level Text Appearance Information. We conduct experiments in Table 3 to investigate the effectiveness of GTM and DTM. It can be seen that in scene text segmentation, both global attention information and local multi-scale information are beneficial for scene text segmentation. The shape and size of scene text are different from those of general objects; so, the dilation rate of DCM needs to be carefully designed to prevent degradation problem, that is, as the dilation rate increases, fewer weights are applied to valid feature regions. Consequently, we empirically set both the horizontal and vertical dilation rate to 1,2,3,6 based on the experimental results in Table 4.

Higher-level Text Semantic Information. REM aims to deliver higher-level semantic information from the recognizer to the segmentation network. As demonstrated in Table 2: the addition of REM increases the F-score by a further 0.4% thus achieving optimal performance. As shown in Fig. 8, the role of REM is

Table 3. Comparison on the impact of GTM and DTM to text segmentation.

GTM	DTM	P	R	F
×	×	85.9	80.1	82.9
✓	×	87.0	80.5	83.6
×	✓	**87.2**	82.3	84.7
✓	✓	87.1	**83.8**	**85.4**

Table 4. Ablation experiment on dilation rate combinations of DTM.

Dilation rate	P	R	F
1,12,18,24	86.3	81.6	83.9
1,6,12,18	85.6	82.7	84.1
1,3,6,12	86.0	83.3	84.6
1,2,3,6	**87.1**	**83.8**	**85.4**

manifested in three aspects: (1) filtering out the misclassified background noise; (2) reactivating text regions that have been ignored by segmentation network; (3) distinguishing negative samples whose texture is close to that of the target text, e.g. dot, symbols (e.g. '&'), and Chinese characters outside the ground truth. Characters that are not of concern to the recognizer are suppressed by REM along with the background, which is a similar pattern as that of humans, who treat unrecognized words as graphic symbols.

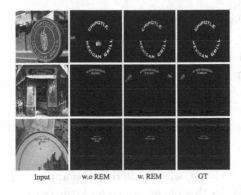

Fig. 8. Visualized results of using the REM (w. REM) and not (w.o REM).

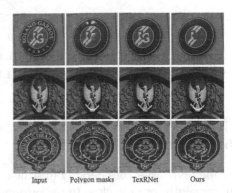

Fig. 9. Downstream application on text removal task.

4.4 Downstream Tasks

Downstream tasks such as text editing and text removal can benefit from fine-grained pixel-level text masks. Here we take text removal task as an example to demonstrate the application value of our ARM-Net. We feed the segmentation result as mask into DeepfillV2 [42], one of the state-of-the-art inpainting networks, to generate a text-free image. As shown in Fig. 9, the inpainting image using mask predicted by our method avoid suffering from serious smudging and erasing mistakes, which appear in inpainting results using polygon masks from ground truth, and TexRNet (column 2 and 3). As DeepfillV2 utilizes the segmentation mask as a guidance, any misclassified pixels of text will be omitted and those of background will be erased incorrectly. Such text removal results also reflects the superior segmentation performance of ARM-Net.

Apart from low-level image tasks, we observe that segmentation results also boost the performance of text recognition. Here we choose CRNN [26], a widely used method for text recognition, as the baseline. To utilize segmentation results, we simply concatenate pixel-level text masks and original images along the channel dimension, then feed it into the CRNN, which is initialized with an official pre-trained model, and fine-tune the first layer further with other parameters fixed. For a fair comparison, we train the model on the synthetic text dataset Synth90k [10] and validate it on ICDAR-2003 [20], ICDAR-2013 [12] and SVT [36], follow the setting of [26,32]. The experimental results in Table 5 illustrate that the inclusion of pixel-level text masks generated by ARM-Net leads to an improvement in recognition accuracy of over 2% on both ICDAR-2003 and SVT, and of 1.2% on ICDAR-2013. The performance not only exceeds that of the CRNN baseline but is also better than the MGNet [32] method.

Table 5. Downstream experiment on scene text recognition.

Method	ICDAR-2003	ICDAR-2013	SVT
CRNN [26]	89.4	86.7	80.8
CRNN+MGNet [32]	91.4	87.7	82.8
CRNN+ARM-Net (Ours)	**91.7**	**87.9**	**82.9**

5 Conclusion

Scene text varies considerably in scale and shape, with some textures appearing infrequently, or even close to backgrounds. In this paper, we rethink the essence of scene text segmentation task and propose an effective end-to-end neural network, ARM-Net. The proposed TSM and DPD capture better low-level text appearance information, while the REM incorporates higher-level text semantic information as a complement. And by jointly exploiting both, we implement an optimization for the segmentation network. Quantitative and qualitative experiments demonstrate that our model outperforms state-of-the-art segmentation networks. We also show promising results that using text segmentation masks from ARM-Net on text removal and text recognition downstream tasks. In the future, we will investigate end-to-end networks for segmentation and recognition to further improve the performance of text segmentation in extreme scenarios.

Acknowledgement. This research is supported in part by NSFC (Grant No.: 61936003), GD-NSF (no.2017A030312006, No.2021A1515011870), Zhuhai Industry Core and Key Technology Research Project (no. ZH22044702200058PJL), and the Science and Technology Foundation of Guangzhou Huangpu Development District (Grant 2020GH17)

References

1. Bonechi, S., Bianchini, M., Scarselli, F., Andreini, P.: Weak supervision for generating pixel-level annotations in scene text segmentation. Pattern Recogn. Lett. **138**, 1–7 (2020)
2. Chen, L.C., Papandreou, G., Kokkinos, I., Murphy, K., Yuille, A.L.: DeepLab: Semantic image segmentation with deep convolutional nets, atrous convolution, and fully connected CRFs. IEEE Trans. Pattern Anal. Mach. Intell. **40**(4), 834–848 (2017)
3. Chen, L.-C., Zhu, Y., Papandreou, G., Schroff, F., Adam, H.: Encoder-decoder with atrous separable convolution for semantic image segmentation. In: Ferrari, V., Hebert, M., Sminchisescu, C., Weiss, Y. (eds.) ECCV 2018. LNCS, vol. 11211, pp. 833–851. Springer, Cham (2018). https://doi.org/10.1007/978-3-030-01234-2_49
4. Ch'ng, C.K., Chan, C.S.: Total-text: a comprehensive dataset for scene text detection and recognition. In: 2017 14th IAPR International Conference on Document Analysis and Recognition, vol. 1, pp. 935–942. IEEE (2017)
5. Fu, J., et al.: Dual attention network for scene segmentation. In: Proceedings of the IEEE/CVF Conference on Computer Vision and Pattern Recognition, pp. 3146–3154 (2019)
6. He, K., Zhang, X., Ren, S., Sun, J.: Deep residual learning for image recognition. In: Proceedings of the IEEE Conference on Computer Vision and Pattern Recognition, pp. 770–778 (2016)
7. Hu, J., Shen, L., Sun, G.: Squeeze-and-excitation networks. In: Proceedings of the IEEE Conference on Computer Vision and Pattern Recognition, pp. 7132–7141 (2018)
8. Huang, H., et al.: UNet 3+: a full-scale connected UNet for medical image segmentation. In: ICASSP 2020–2020 IEEE International Conference on Acoustics, Speech and Signal Processing (ICASSP), pp. 1055–1059. IEEE (2020)
9. Huang, Z., Wang, X., Huang, L., Huang, C., Wei, Y., Liu, W.: CCNet: criss-cross attention for semantic segmentation. In: Proceedings of the IEEE/CVF International Conference on Computer Vision, pp. 603–612 (2019)
10. Jaderberg, M., Simonyan, K., Vedaldi, A., Zisserman, A.: Synthetic data and artificial neural networks for natural scene text recognition. arXiv preprint arXiv:1406.2227 (2014)
11. Jia, X., De Brabandere, B., Tuytelaars, T., Gool, L.V.: Dynamic filter networks. In: Advances in Neural Information Processing Systems 29 (2016)
12. Karatzas, D., et al.: ICDAR 2013 robust reading competition. In: 2013 12th International Conference on Document Analysis and Recognition, pp. 1484–1493. IEEE (2013)
13. Kingma, D.P., Ba, J.: Adam: a method for stochastic optimization. arXiv preprint arXiv:1412.6980 (2014)
14. Krishnan, P., Kovvuri, R., Pang, G., Vassilev, B., Hassner, T.: TextStyleBrush: transfer of text aesthetics from a single example. arXiv preprint arXiv:2106.08385 (2021)
15. Lafferty, J., Mccallum, A., Pereira, F.: Conditional random fields: Probabilistic models for segmenting and labeling sequence data. In: Proceedings of ICML (2002)
16. Li, X., Zhong, Z., Wu, J., Yang, Y., Lin, Z., Liu, H.: Expectation-maximization attention networks for semantic segmentation. In: Proceedings of the IEEE/CVF International Conference on Computer Vision, pp. 9167–9176 (2019)

17. Liu, C., Liu, Y., Jin, L., Zhang, S., Luo, C., Wang, Y.: EraseNet: end-to-end text removal in the wild. IEEE Trans. Image Process. **29**, 8760–8775 (2020)
18. Liu, R., et al.: An intriguing failing of convolutional neural networks and the coord-conv solution. In: Advances in Neural Information Processing Systems 31 (2018)
19. Long, J., Shelhamer, E., Darrell, T.: Fully convolutional networks for semantic segmentation. In: Proceedings of the IEEE Conference on Computer Vision and Pattern Recognition, pp. 3431–3440 (2015)
20. Lucas, S.M., et al.: ICDAR 2003 robust reading competitions: entries, results, and future directions. Int. J. Doc. Anal. Recogn. **7**(2), 105–122 (2005)
21. Luo, C., Jin, L., Sun, Z.: MORAN: a multi-object rectified attention network for scene text recognition. Pattern Recogn. **90**, 109–118 (2019)
22. Nayef, N., et al.: ICDAR 2017 robust reading challenge on multi-lingual scene text detection and script identification-RRC-MLT. In: 2017 14th IAPR International Conference on Document Analysis and Recognition, vol. 1, pp. 1454–1459. IEEE (2017)
23. Rong, X., Yi, C., Tian, Y.: Unambiguous scene text segmentation with referring expression comprehension. IEEE Trans. Image Process. **29**, 591–601 (2019)
24. Ronneberger, O., Fischer, P., Brox, T.: U-Net: convolutional networks for biomedical image segmentation. In: Navab, N., Hornegger, J., Wells, W.M., Frangi, A.F. (eds.) MICCAI 2015. LNCS, vol. 9351, pp. 234–241. Springer, Cham (2015). https://doi.org/10.1007/978-3-319-24574-4_28
25. Roy, P., Bhattacharya, S., Ghosh, S., Pal, U.: STEFANN: scene text editor using font adaptive neural network. In: Proceedings of the IEEE/CVF Conference on Computer Vision and Pattern Recognition, pp. 13228–13237 (2020)
26. Shi, B., Bai, X., Yao, C.: An end-to-end trainable neural network for image-based sequence recognition and its application to scene text recognition. IEEE Trans. Pattern Anal. Mach. Intell. **39**(11), 2298–2304 (2016)
27. Shi, B., Yang, M., Wang, X., Lyu, P., Yao, C., Bai, X.: Aster: an attentional scene text recognizer with flexible rectification. IEEE Trans. Pattern Anal. Mach. Intell. **41**(9), 2035–2048 (2018)
28. Sun, K., et al.: High-resolution representations for labeling pixels and regions. arXiv preprint arXiv:1904.04514 (2019)
29. Vaswani, A., et al.: Attention is all you need. In: Advances in Neural Information Processing Systems 30 (2017)
30. Veit, A., Matera, T., Neumann, L., Matas, J., Belongie, S.: COCO-text: dataset and benchmark for text detection and recognition in natural images. arXiv preprint arXiv:1601.07140 (2016)
31. Wang, C., Komodakis, N., Paragios, N.: Markov random field modeling, inference & learning in computer vision & image understanding: a survey. Comput. Vis. Image Underst. **117**(11), 1610–1627 (2013)
32. Wang, C., et al.: Semi-supervised pixel-level scene text segmentation by mutually guided network. IEEE Trans. Image Process. **30**, 8212–8221 (2021)
33. Wang, H., Zhu, Y., Green, B., Adam, H., Yuille, A., Chen, L.-C.: Axial-DeepLab: stand-alone axial-attention for panoptic segmentation. In: Vedaldi, A., Bischof, H., Brox, T., Frahm, J.-M. (eds.) ECCV 2020. LNCS, vol. 12349, pp. 108–126. Springer, Cham (2020). https://doi.org/10.1007/978-3-030-58548-8_7
34. Wang, J., Chen, K., Xu, R., Liu, Z., Loy, C.C., Lin, D.: CARAFE: content-aware reassembly of features. In: Proceedings of the IEEE/CVF International Conference on Computer Vision, pp. 3007–3016 (2019)
35. Wang, J., et al.: Deep high-resolution representation learning for visual recognition. IEEE Trans. Pattern Anal. Mach. Intell. **43**(10), 3349–3364 (2020)

36. Wang, K., Babenko, B., Belongie, S.: End-to-end scene text recognition. In: 2011 International Conference on Computer Cision, pp. 1457–1464. IEEE (2011)
37. Wang, T., et al.: Decoupled attention network for text recognition. In: Proceedings of the AAAI conference on artificial intelligence, vol. 34, pp. 12216–12224 (2020)
38. Wang, X., Girshick, R., Gupta, A., He, K.: Non-local neural networks. In: Proceedings of the IEEE Conference on Computer Vision and Pattern Recognition, pp. 7794–7803 (2018)
39. Wu, L., et al.: Editing text in the wild. In: Proceedings of the 27th ACM International Conference on Multimedia, pp. 1500–1508 (2019)
40. Xu, X., Zhang, Z., Wang, Z., Price, B., Wang, Z., Shi, H.: Rethinking text segmentation: a novel dataset and a text-specific refinement approach. In: Proceedings of the IEEE/CVF Conference on Computer Vision and Pattern Recognition, pp. 12045–12055 (2021)
41. Xu, X., Qi, Z., Ma, J., Zhang, H., Shan, Y., Qie, X.: BTS: a bi-lingual benchmark for text segmentation in the wild. In: Proceedings of the IEEE/CVF Conference on Computer Vision and Pattern Recognition, pp. 19152–19162 (2022)
42. Yu, J., Lin, Z., Yang, J., Shen, X., Lu, X., Huang, T.S.: Free-form image inpainting with gated convolution. In: Proceedings of the IEEE/CVF International Conference on Computer Vision, pp. 4471–4480 (2019)
43. Yuan, Y., Chen, X., Wang, J.: Object-contextual representations for semantic segmentation. In: Vedaldi, A., Bischof, H., Brox, T., Frahm, J.-M. (eds.) ECCV 2020. LNCS, vol. 12351, pp. 173–190. Springer, Cham (2020). https://doi.org/10.1007/978-3-030-58539-6_11
44. Zhang, S., Liu, Y., Jin, L., Huang, Y., Lai, S.: EnsNet: ensconce text in the wild. In: Proceedings of the AAAI Conference on Artificial Intelligence, vol. 33, pp. 801–808 (2019)
45. Zhao, H., Shi, J., Qi, X., Wang, X., Jia, J.: Pyramid scene parsing network. In: Proceedings of the IEEE Conference on Computer Vision and Pattern Recognition, pp. 2881–2890 (2017)
46. Zhou, Z., Siddiquee, M.M.R., Tajbakhsh, N., Liang, J.: Unet++: redesigning skip connections to exploit multiscale features in image segmentation. IEEE Trans. Med. Imaging 39(6), 1856–1867 (2019)

Big Data, Large Scale Methods

Unsupervised Online Hashing
with Multi-Bit Quantization

Zhenyu Weng$^{(\boxtimes)}$ (iD) and Yuesheng Zhu (iD)

Communication and Information Security Lab, Shenzhen Graduate School, Peking
University, Shenzhen, China
{wzytumbler,zhuys}@pku.edu.cn

Abstract. Online hashing methods aim to update hash functions with
newly arriving data streams, which can process large-scale data online.
To this end, most existing methods update projection functions online
and adopt a single-bit quantization strategy that quantizes each projected component with one bit. However, single-bit quantization results
in large information loss in the quantization process and thus cannot
preserve the similarity information of original data well. In this paper,
we propose a novel unsupervised online hashing method with multi-bit
quantization towards solving this problem, which consists of online data
sketching and online quantizer learning. By maintaining a small-size data
sketch to preserve the streaming data information, an orthogonal transformation is learned from the data sketch to make the components of
the streaming data independent. Then, an optimal quantizer is learned
to adaptively quantize each component with multiple bits by modeling
the data distribution. Therefore, our method can quantize each component with multiple bits rather than one bit to better preserve the data
similarity online. The experiments show that our method can achieve
better search accuracy than the relevant online methods for approximate
nearest neighbor search.

Keywords: Online hashing · Unsupervised hashing · Multi-bit
quantization

1 Introduction

With the development of feature representation methods, especially deep learning methods [12,24], images and videos are represented by high-dimensional
features that can obtain high-level semantic information. Conventional nearest neighbor search methods [1,23] are inefficient for high-dimensional features
as a consequence of the curse of dimensionality. The difficulty of finding the
exact nearest neighbors in the high-dimensional space leads to the emergence of
Approximate Nearest Neighbor (ANN) search methods [19,25] using a compact
data representation for high-dimensional data. Hashing methods [26,27] are one
type of widely-used ANN search methods. The goal of hashing methods is to
learn a binary-code representation which can preserve the similarity structure

L. Wang et al. (Eds.): ACCV 2022, LNCS 13847, pp. 663–678, 2023.
https://doi.org/10.1007/978-3-031-26293-7_39

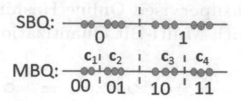

Fig. 1. A comparison example between Single-Bit Quantization (SBQ) and Multi-Bit Quantization (MBQ) for the projected component. Here MBQ adopts two bits. c_i is a group centroid to represent the data points in the group.

of data in the original feature space. Using a binary representation for data can not only reduce the database storage but also improve the search efficiency.

Traditional hashing methods [3,6,21,22,25] are mostly designed to learn hash functions offline from a fixed collection of training data. However, they cannot deal with the streaming data. In many real-world applications, data is available continuously in streaming fashion. When new data arrives, the data distribution changes and these methods have to accumulate all the data to re-learn hash functions, which is time-consuming. Therefore, online hashing methods [4,8,16, 16,17,28,31,33] have attracted much research attention recently, which can learn the hash functions online from the streaming data.

According to whether there is label information provided in the learning process, online hashing methods can be categorized into two groups, supervised online hashing methods [16,17,28,31,32,36] and unsupervised online hashing methods [13,33]. In this paper, we focus on unsupervised online hashing methods as it is expensive to collect a large number of labeled samples in real-world applications.

Unsupervised online hashing methods [13,33] usually adopt a learning strategy containing two stages: projection stage and quantization stage [11]. They maintain a small-size data matrix to preserve the characters of streaming data online and learn the projection functions from the data matrix. Then, a Single-Bit Quantization (SBQ) strategy is adopted to quantize each projected component with one bit. However, the recent research [5,30] shows that SBQ results in large information loss in the quantization process and cannot preserve the similarity structure of original data well, as shown in Fig. 1.

In this paper, we propose an unsupervised online hashing method with Multi-Bit Quantization (MBQ) to address the above problem. In our method, a data sketch is maintained from the streaming data and an orthogonal transformation is learned from the data sketch to reduce statistical dependence among the data components. By modeling the distribution of each component, a bit allocation algorithm is designed to adaptively allocate bits to each component, and an optimal quantizer is learned by independently quantizing each component. As shown in Fig. 1, by learning to quantize each component with multiple bits rather than one bit online, our method have a much bigger distance space than online hashing methods with SBQ and can better preserve the data similarity. The experiments on two widely-used datasets validate the effectiveness of our method.

2 Related Work

Since directly learning best binary codes for a given database is proven as a NP-hard problem [29], most hashing methods adopt a learning strategy containing two stages: projection stage and quantization stage [11]. In the projection stage, several projected components of real values are generated. In the quantization stage, the components generated from the projection stage are quantized into binary codes. Many hashing methods focus on the projection stage and adopt the widely-used SBQ strategy to quantize each projected component [3,21,22,25]. However, the recent research [5,10,11,30,34] shows that replacing the SBQ strategy by the MBQ strategy can result in a better search accuracy. For example, the literature [11] quantizes each projected component with multiple bits and calculates Manhattan distance between the binary codes for nearest neighbor search. In addition to quantize each component with the same number of bits, some methods [5,30] develop quantizers to adaptively quantize each projected component with a certain number of bits according to the data distribution. Although the above hashing methods with MBQ can obtain the satisfactory search performance for the static database, they cannot deal with the streaming data. When new data arrives, the data distribution changes and these MBQ methods have to accumulate all the data to re-learn the quantizer, which is time-consuming. Therefore, we propose an novel unsupervised online hashing method to learn the multi-bit quantizer online. To our best knowledge, this is the first work to adopt the MBQ strategy for online hashing.

In parallel to hashing methods, Product Quantization (PQ) methods [19] are another type of ANN search methods. PQ methods [2,7,9,19] represent each high-dimensional feature vector by a Cartesian product of several quantized values. Recently, some online PQ methods [14,18,35] are also developed for streaming data. As PQ methods are related to hashing methods, we will compare our method with online unsupervised PQ methods to validate the effectiveness of our method.

3 Online Hashing with Multi-Bit Quantization

This section presents the proposed method, Online Hashing with Multi-Bit Quantization (OHMBQ), to quantize streaming data with multiple bits. The proposed OHMBQ consists of two key designs. The first is to reduce the dependence among data components by learning an orthogonal transformation from streaming data. The second is to learn an optimal quantizer by modeling the data distribution of the transformed components. A theoretical proof for learning the optimal quantizer is provided. More details are described in the following subsections.

3.1   Preliminary About Multi-Bit Quantization

Assume there is a set of data points $\mathbf{X} = [\mathbf{x}_1, ..., \mathbf{x}_N], \mathbf{x} \in \mathbb{R}^D$, where N is the number of data points and D is the data dimensionality. Quantization is the process of dividing a large set (or a continuous set) of data points into groups and learning the group's centroid to represent the group. Given the quantization level m, the quantizer $q : \mathbb{R}^D \to Z$ where $Z = \{0, 1, ..., m-1\}$ is characterized by the input space partition $Q(z) = \{\mathbf{x} : q(\mathbf{x}) = z\}$ and the group centroid $\mathbf{c}(z) \in \mathbb{R}^D$ for $z \in Z$. To obtain a binary representation for the data point, the index z is transformed to be a binary expression $\mathbf{t}(z) \in \{0, 1\}^b$ where b is the number of bits, and $m = 2^b$.

The quality of the quantization is measured in terms of its average distortion,

$$A = E[d(\mathbf{x}, \mathbf{c}(q(\mathbf{x})))], \tag{1}$$

where $d : \mathbb{R}^D \times \mathbb{R}^D \to \mathbb{R}$ is the distortion function which takes on the metric of Euclidean distance in the nearest neighbor search.

As [3] denotes, to minimize the average distortion A, the optimal quantizer is characterized by the following properties:

$$\begin{cases} Q(z) = \{\mathbf{x} : d(\mathbf{x}, \mathbf{c}(z)) \leq d(\mathbf{x}, \mathbf{c}(z')), \forall z' \in Z\} \\ \mathbf{c}(z) = \arg\min_{\mathbf{x}'} E_{\mathbf{x}}[d(\mathbf{x}, \mathbf{x}')|x \in Q(z)] \end{cases} \tag{2}$$

It is a challenging problem to quantize the high-dimensional data into hundreds of groups. Especially, the quantization level of the quantizer $m = 2^b$ grows exponentially on the bit number b. It is impossible to collect a sufficient number of training examples to span the quantized space which comprises hundreds of bits.

To address the challenge, multi-bit quantization methods assume the data distribution $p(\mathbf{x})$ is independent in its components after projecting the data onto a new space with the projection matrix learned in the projection stage. Then, the metric is of the form $d(\mathbf{x}, \mathbf{x}') = \sum_i d_i(x_i, x_i')$, where x_i are the components of $\mathbf{x}$. Therefore, a minimum distortion quantizer can be obtained by forming the Cartesian product of the independently quantized components. That is, let

$$\mathbf{q}(\mathbf{x}) = (q_1(x_1), q_2(x_2), ..., q_D(x_D)) = \mathbf{z}, \tag{3}$$

where $\mathbf{z} = (z_1, ..., z_D), z_i = q_i(x_i)$

Given a query $\tilde{\mathbf{x}}$, the distance between the query and the data point is calculated as

$$d(\tilde{\mathbf{x}}, \mathbf{q}(\mathbf{x})) = \sum_i d_i(\tilde{x}_i, c_i(q_i(x_i))). \tag{4}$$

According to Eq. (4) and Fig. 1, hashing methods with MBQ obviously have a much bigger distance space than hashing methods with SBQ and thus can achieve better search accuracy.

Although multi-bit quantization methods have achieved promising search performance on the static database, they cannot be directly applied on the

---

Algorithm 1. Zero-Mean Sketching

Input: data chunk $\mathbf{D}$, sketch matrix $\mathbf{P}$, mean value μ, cumulative number n
Output: sketch matrix $\mathbf{P}$
1: Sketch $[\mathbf{D} - \bar{\mathbf{D}}, \sqrt{\frac{n n_D}{n + n_D}}(\bar{\mathbf{D}} - \mu)]$ into $\mathbf{P}$ with FD where $\bar{\mathbf{D}}$ is the mean value of $\mathbf{D}$
    and n_D is the size of the sketch $\mathbf{D}$
2: $\mu = \frac{n\mu}{n + n_D} + \frac{n_D \bar{\mathbf{D}}}{n + n_D}$
3: $n = n + n_D$

---

streaming environment where the data distribution varies when data keeps growing. Specifically, the existing techniques for reducing the dependence between data components and learning an optimal quantizer require knowing the data distribution in advance. In the following, we describe how to reduce the dependence between data components and learn an optimal quantizer online in our method.

3.2 Online Transform Coding

For the static database, the assumption that the components of $\mathbf{x}$ are independent can be addressed by transform coding [3], which seeks a transformation to reduce statistical dependence among the components. This is typically done through Principal Component Analysis (PCA) where an orthogonal transformation matrix $\mathbf{U}$ is obtained by concatenating the top eigenvectors of covariance matrix $\mathbf{XX}^T$. Here, $\mathbf{X}$ is zero-mean data. To learn an orthogonal transformation from the streaming data that can approximate the orthogonal transformation in PCA, we adopt Online Sketching Hashing (OSH) [13] to maintain a small-size sketch online to preserve the character of the streaming data and learn the transformation matrix from the data sketch.

Assume data comes in chunks. New data chunk $\mathbf{D}_t$ arrives at round t. $\mathbf{X}_t = [\mathbf{D}_1, ..., \mathbf{D}_t]$ denotes the data matrix accumulated from round 1 to round t, μ_t is the mean of $\mathbf{X}_t$. Since the data should be zero-mean in PCA, Online Sketching Hashing (OSH) [13] develops a zero-mean data sketching method based on Frequent Directions(FD) [15] to learn a small-size sketch $\mathbf{P} \in \mathbb{R}^{D \times k}$ where k is the size of the sketch such that $\mathbf{PP}^T \approx (\mathbf{X}_t - \mu_t)(\mathbf{X}_t - \mu_t)^T$, which is summarized in Algorithm 1. The input to Algorithm 1 is the data chunk $\mathbf{D}$, the data sketch $\mathbf{P}$, the mean value of the cumulative data μ, and the cumulative number of the data n. More details can be found in [13].

By applying Algorithm 1 we can obtain an orthogonal transformation matrix $\mathbf{U}$ by taking the top eigenvectors from $\mathbf{PP}^T$ for learning the quantizer.

3.3 Online Quantizer Learning

Assume the data has been transformed by the transformation matrix $\mathbf{U}$ learnt from the data sketch and the dependence among the components has been reduced. To obtain a b-bit binary code for representing the data point, we need to determine the bit number for each component, *i.e.*, the quantization level for each component, at first. Then, for each component, we need to separate the input space into groups and learn the centroid of each group.

Bit allocation is a process of determining the number of data components and determining the quantization level of each component. The number of the bits is fixed. When more bits are allocated to the component to preserve the information of the single component, the number of the selected components become smaller, which results in the information loss, and vice versa. The literature [30] determines the bit number for each component according to the data distribution of each component. In the streaming environment, since the data distribution keeps changing and accumulating all the data is time-consuming, this bit allocation algorithm cannot be used for the streaming data. The literature [3] allocates bits to each component proportionally according to the standard deviation of the component. It can be used in the streaming environment. However, this method treats each component independently. In our method, we design a new bit allocation algorithm from the perspective of the accumulative component energy.

Since b bits can be allocated to at most b components, *i.e.*, one bit for one component, b components that have the maximum information are chosen by $\mathbf{Y} = \mathbf{U}^T(\mathbf{X}_t - \mu_t)$ where $\mathbf{U} \in \mathbb{R}^{D \times b}$ is obtained by taking the top b eigenvector from $\mathbf{P}\mathbf{P}^T$. The standard deviation δ_i of the ith component corresponds to the ith eigenvalue. Each column $\mathbf{y} \in \mathbb{R}^b$ of $\mathbf{Y}$ has b components.

As it is time-consuming to obtain the data distribution by accumulating all the data when new data arrives, we hope to model the data distribution of each component. Figure 2 shows the data distribution of each component after transformation on the CIFAR-10 dataset [20] for each round. From the figure, we can see that the data distribution for the component tends to be a Gaussian distribution after a few rounds. Hence, we assume each component is subject to the Gaussian distribution $N(0, \delta_i^2)$. Inspired by PCA which selects eigenvectors according to the accumulative component energy, we want to choose a small number of components L while achieving a reasonably high percentage $\alpha \in [0, 1]$ of accumulative component energy (standard deviation). The total energy of the components is $G = \sum_{i=1}^{b} \delta_i$. And the smallest L is chosen so that the cumulative energy g for L components is above the certain threshold, which is

$$g = \sum_{i=1}^{L} \delta_i \geq \alpha G. \tag{5}$$

Since each component has the identical distribution after normalizing the variance, the remaining $b - L$ bits are allocated proportionally on the remaining standard deviations of the components, which is summarized on Algorithm 2. After choosing L components and allocating one bit to each component, the remaining $b - L$ bits are allocated for $b - L$ rounds. In each round, we can find the component index i that has the largest remaining standard deviation h_i by enumeration. Then, the allocated bit number for the ith component is

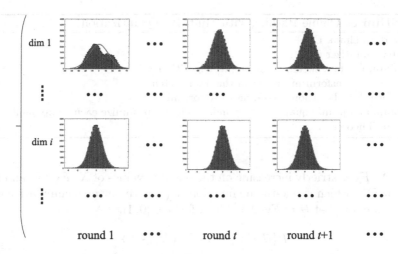

Fig. 2. The data distribution of each component after transformation on CIFAR-10 with the rounds increasing.

Algorithm 2. Bit Allocation

Input: standard deviations of components $\{\delta_i\}_{i=1}^b$, α
Output: bit allocation $\{l_i\}_{i=1}^L$
 1: Find the smallest L such that $\sum_{i=1}^L \delta_i \geq \alpha \sum_{i=1}^b \delta_i$
 2: Initialize $h_i \leftarrow \frac{\delta_i}{2}$ and $l_i \leftarrow 1$
 3: **for** each round in $b - L$ rounds **do**
 4:     $i \leftarrow \arg\max_{i'} h_{i'}$
 5:     $l_i \leftarrow l_i + 1$
 6:     $h_i \leftarrow h_i/2$
 7: **end for**

---

incremented by 1 and the remaining standard deviation h_i is divided by 2 as the quantization level is doubled.

After bit allocation, we need to find the optimal quantizer for each component. [3] and [30] find the space division and the centroids by using all the data points, which is difficult in the streaming environment. As each component is subject to the Gaussian distribution, we have the following theorem to minimize the quantization error for each component.

Theorem 1. Assume X be a random variable subject to a Gaussian distribution $N(0, \delta^2)$ and F_X is the cumulative distribution function. Then, the quantizer that minimizes the quantization error is obtained by

$$\begin{cases} Q(z) = \{X : F_X^{-1}(\frac{z}{2^l}) \leq X \leq F_X^{-1}(\frac{z+1}{2^l})\} \\ c(z) = F_X^{-1}(\frac{z \times 2+1}{2^{l+1}}) \end{cases}, \tag{6}$$

where l is the number of bits allocated to the component, $z \in Z = \{0, 1, ..., 2^l - 1\}$, and F_X^{-1} is the inverse cumulative distribution function.

Algorithm 3. Online Hashing with Multi-Bit Quantization

Input: data chunk **D**
Output: quantizer **q**
1: Maintain a data sketch according to Algorithm 1
2: Learn an transform matrix from the data sketch
3: Perform bit allocation according to Algorithm 2
4: Obtain an optimal quantizer **q** by independently minimize each component according to Theorem 1

Proof. As F_X is strictly increasing on the possible values of X, F_X has an inverse function F_X^{-1} which is one-to-one function. F_X^{-1} is the inverse cumulative distribution function. Let $U = F_X(X)$. Then, for $u \in [0, 1]$,

$$\begin{aligned} P\{U \le u\} &= P\{F_X(X) \le u\} \\ &= P\{U \le F_X^{-1}(u)\} \\ &= F_X(F_X^{-1}(u)) = u. \end{aligned} \tag{7}$$

Obviously, U is uniform random variable on [0,1]. For U, according to Eq. (2), the optimal quantizer is obviously characterized by

$$\begin{cases} Q(z) = \{U : \frac{z}{m} \le U \le \frac{z+1}{m}\} \\ c(z) = \frac{z \times 2 + 1}{m \times 2} \end{cases}, \tag{8}$$

where $z \in Z = \{0, 1, ..., m-1\}$ and m is the quantization level.

Therefore, as F_X has an one-to-one inverse function F_X^{-1} and $U = F_X(X)$, the optimal quantizer for X is characterized by

$$\begin{cases} Q(z) = \{X : F_X^{-1}(\frac{z}{m}) \le X \le F_X^{-1}(\frac{z+1}{m})\} \\ c(z) = F_X^{-1}(\frac{z \times 2 + 1}{m \times 2}) \end{cases}. \tag{9}$$

3.4 Algorithm Analysis

Our method, Online Hashing with Multi-Bit Quantization (OHMBQ), is summarized in Algorithm 3. OHMBQ includes four steps, data sketching, learning a transform matrix, bit allocation, and learning optimal quantizers. Specifically, assume there is a stream of data chunks, $\mathbf{D}_1, \mathbf{D}_2, ..., \mathbf{D}_t$. For each chunk, our method learns a data sketch from the streaming data according to Algorithm 1. With the data sketch, a transform matrix is learned to reduce the dependence among streaming data components. Then, the bit allocation for each component is learned to make a good tradeoff between the information loss arising from reducing data components and the quantization error arising from quantization level per data component according to Algorithm 2. With the quantization level (*i.e.*, bit number) for each component, the optimal quantizer for each component to minimize the quantization error is learned according to Theorem 1.

Time Complexity. As shown in Algorithm 3, OHMBQ includes four steps. As indicated in [13], the time complexity of data sketching is $O(Dkn_t)$, where n_t is the size of accumulated time at round t, D is the data dimensionality, and k is the size of data sketch. For each round, the time complexity of learning a transform matrix is $O(Dk^2 + k^3)$. The time complexity of bit allocation at one round is $O(bL)$ where L is the number of components and b is the number of bits. For the step of quantizing each component, the inverse cumulative distribution function is implemented by the icdf function in MATLAB, and the time complexity of quantizing L components at one round is $O(2^l L)$ where l is the maximum bit number for the component. In the experiments, the largest value of l is 5. Therefore, the cumulative time complexity of learning the quantizer at round t is $O(Dkn_t + tDk^2 + tk^3 + tbL + 2^l tL)$ where l is usually smaller or equal to 5. The time complexity is close to that of OSH [13].

The process of encoding the database in our method is composed of projecting the data into a low-dimensional space and mapping each component value to the corresponding quantized value. Assume the size of the database is N. The time complexity of projection is $O(NDL)$ and the time complexity of mapping is $O(2^l NL)$. Therefore, the time complexity of encoding the database is $O(NDL + 2^l NL)$ where l is usually smaller or equal to 5 in the experiments.

4 Experiments

4.1 Experimental Setting

We adopt two widely-used datasets to evaluate the proposed method: CIFAR-10 [20] and GIST1M [9]. CIFAR-10 dataset includes 60,000 32×32 images. Each image is represented by a 4096-dimensional feature extracted from the last convolutional layer in the VGG-net [24]. In CIFAR-10, 10,000 images are randomly selected as the queries, and the rest is used for training and search. GIST1M includes 1,000,000 960-dimensional GIST features and 1,000 queries. For each query, the top K Euclidean nearest neighbors is set as the ground-truth. For both datasets, K is set to 1,000. Following [33,35], mean Average Precision (mAP) and pre@100 are used as the measurements to evaluate the performance of ANN search. pre@100 is the precision of the top 100 retrieved data points. The experiments are repeated five times and the average results are reported.

4.2 Comparison to Other Online Methods

We compare our method, OHMBQ, with Online Product Quantization (OPQ) [35], Online Optimized Product Quantization (OOPQ) [18], Online Sketching Hashing (OSH) [13] and Online Spherical Hashing (OSpH) [33]. OPQ and OOPQ are online unsupervised product quantization method for Approximate Nearest Neighbor (ANN) search. We carefully implement OPQ and OOPQ according to the descriptions from [35] and [18], respectively. OSH and OSpH are online unsupervised hashing methods for ANN search. The source codes of

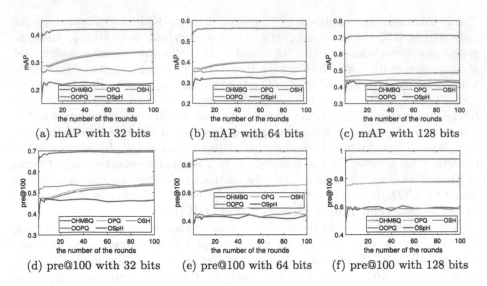

Fig. 3. The search results of different methods for the early 100 chunks on CIFAR-10.

OSH and OSpH are publicly available. The experiments in [33,35] show that Mutual Information Hashing (MIH) [4] and Online Kernel Hashing (OKH) [8], two state-of-the-art online supervised hashing methods that take the pairwise pseudo-labels as the input, are inferior to OPQ and OSpH in the unsupervised environment. Hence, we do not compare our method with these online supervised hashing methods in the unsupervised environment.

Following the setting in [13,33], we divide the training data evenly into chunks of 100 data points for both datasets to simulate the streaming environment. Figure 3 shows the search results of the online methods in the early 100 chunks from 32 bits to 128 bits on CIFAR-10. We can see that OHMBQ has achieved better search accuracy than other methods from the beginning for mAP and pre@100. For pre@100, OOPQ and OPQ are inferior to OSH at first and better than OSH later for 32bits. For mAP, OOPQ and OPQ are better than OSH from the beginning. The gap between OHMBQ and other methods is huge on CIFAR-10. Figure 4 shows the search results of the online methods in the early 100 chunks from 32 bits to 128 bits on GIST1M. According to Fig. 3 and Fig. 4, we can see that OHMBQ can achieve a better performance since the beginning compared to other methods as OHMBQ combines the advantages of online sketching from OSH and large distance space from online product quantization methods.

Table 1 shows the search results of online methods on CIFAR-10 after all the chunks are received. According to the results in the table, OHMBQ, OOPQ, and OPQ are better than OSH and OSpH for both mAP and pre@100. Compared with OOPQ and OPQ, OHMBQ further improves the performance by more than 10 percent for both mAP and pre@100. Table 2 shows the search results of

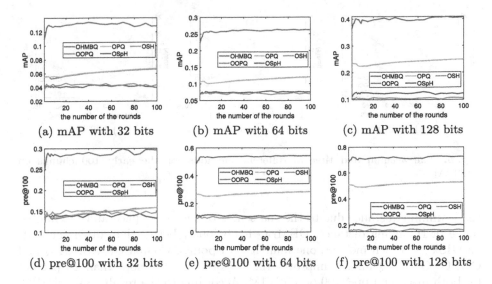

Fig. 4. The search results of different methods for the early 100 chunks on GIST1M.

Table 1. The search results on CIFAR-10.

	mAP			pre@100		
	32 bits	64 bits	128 bits	32 bits	64 bits	128 bits
OHMBQ	**0.423**	**0.562**	**0.711**	**0.687**	**0.843**	**0.941**
OOPQ	0.361	0.425	0.498	0.574	0.683	0.784
OPQ	0.348	0.406	0.474	0.554	0.656	0.757
OSpH	0.228	0.324	0.434	0.468	0.427	0.596
OSH	0.273	0.348	0.437	0.537	0.428	0.570

Table 2. The search results on GIST1M.

	mAP			pre@100		
	32 bits	64 bits	128 bits	32 bits	64 bits	128 bits
OHMBQ	**0.129**	**0.263**	**0.403**	**0.296**	**0.542**	**0.721**
OOPQ	0.091	0.147	0.265	0.208	0.336	0.547
OPQ	0.084	0.140	0.260	0.195	0.323	0.542
OSpH	0.041	0.078	0.120	0.136	0.120	0.206
OSH	0.043	0.072	0.105	0.147	0.099	0.159

online methods on GIST1M after all the chunks are received. According to the results in the table, OHMBQ, OOPQ and OPQ are also better than OSH and OSpH for both mAP and pre@100. As denoted by [3], the distance between the original data can be better estimated by using the quantized values than using

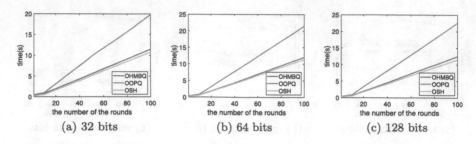

Fig. 5. The accumulated time of different methods for the early 100 chunks on GIST1M.

the Hamming distance due to the sparseness of the code space and the limited range of Hamming distance. OHMBQ is obviously better than other methods on GIST1M which includes one million data points. Compared with OPQ and OOPQ, OHMBQ further improves the performance by more than 20 percent for both mAP and pre@100 on GIST1M. According to the results, as OHMBQ, OPQ, and OOPQ have a larger distance space than OSH and OSpH, OHMBQ, OPQ, and OOPQ achieve better search accuracy. Compared with OPQ and OOPQ, OHMBQ can achieve better search accuracy, which demonstrates the effectiveness of OHMBQ.

Apart from search accuracy, we compare OHMBQ with OOPQ and OSH in terms of the accumulated time of online learning from streaming data. According to the above results, OOPQ is the second best method in terms of search accuracy. OHMBQ and OSH both maintain the data sketch from the streaming and learn the functions from the data sketch. Figure 5 shows the accumulated time of the online methods learning from streaming data in the early 100 chunks from 32 bits to 128 bits on GIST1M. The experiments are run on a computer with a CPU of Intel(R) Xeon(R)W-2223 at 3.60 GHz with 32 GB RAM. According to the results in the figure, we can find that although three methods have similar time costs for initialization at the beginning, OOPQ is obviously slower than OHMBQ and OSH when the chunks of data increase. OHMBQ and OSH have close time costs from 32 bits and 128 bits, which justifies the time complexity analysis from Sect. 3.4.

4.3   Bit Allocation Analysis

As described in Sect. 3.3, bit allocation is a tradeoff between dimension reduction and quantization levels. The number of bits is fixed. When more components are selected, the number of bits allocated to each component is smaller, resulting in more information loss in quantizing the component. When more bits are allocated to each component, the number of the selected components is smaller, resulting in more information loss in dimension reduction. Figures 6 and 7 shows the search performance comparison with different α values on CIFAR-10 and GIST1M, respectively. The number of the selected components is increasing

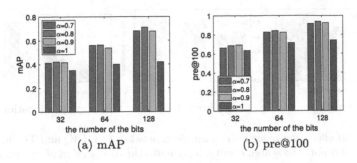

(a) mAP

(b) pre@100

Fig. 6. The search results of OHMBQ with different α values on CIFAR-10.

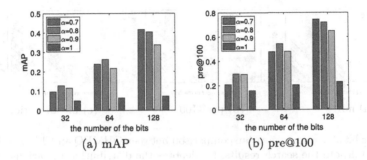

(a) mAP

(b) pre@100

Fig. 7. The search results of OHMBQ with different α values on GIST1M.

from $\alpha = 0.7$ to $\alpha = 1$. Meanwhile, the search accuracy increases at first and then decreases. From the results, we can see that taking $\alpha = 0.8$ can achieve good performance. For simplicity, we take $\alpha = 0.8$ for all the experiments in this paper.

Meanwhile, we compare our method with the bit allocation algorithm in Transform Coding (TC) [3], a bit allocation algorithm for the static database. Figures 8 and 9 show the comparison between OHMBQ and TC on CIFAR-10 and GIST1M, respectively. From Fig. 8(a)(b) and Fig. 9(a)(b), we can see that the bit allocation algorithm in our method can achieve higher search accuracy than that in TC in terms of mAP and pre@100. Figure 8(c) and Fig. 9(c) shows the distribution of various numbers of allocated bits over data components. The number in the legend denotes the number of the allocated bits. From Fig. 8(c) and Fig. 9(c), we can see that the number of the selected components in our method is smaller than that in TC, and each component can have more bits in our method. Among the bit distribution for the components, 2 bits per component takes a majority in our method while 1 bit per component takes a majority in TC.

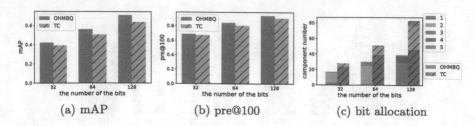

(a) mAP   (b) pre@100   (c) bit allocation

Fig. 8. The bit allocation algorithm comparison between OHMBQ and TC on CIFAR-10. (a) and (b) denote the search results. (c) denotes the distribution of various numbers of allocated bits over data components. The number in the legend of (c) denotes the number of allocated bits.

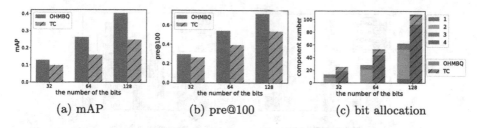

(a) mAP   (b) pre@100   (c) bit allocation

Fig. 9. The bit allocation algorithm comparison between OHMBQ and TC on GIST1M. (a) and (b) denote the search results. (c) denotes the distribution of various numbers of allocated bits over data components. The number in the legend of (c) denotes the number of allocated bits.

5   Conclusions

In this paper, we propose a simple but effective online hashing method with multi-bit quantization for approximate nearest neighbor search. By learning an orthogonal transformation to make the components of the streaming data independent and modelling the statistical properties of the components, our method can learn the optimal quantizer online from the streaming data. The complexity analysis shows that the time complexity of learning the quantizer and encoding the database in our method is close to that of Online Sketching Hashing [13]. The experiments show that our method can achieve a huge search accuracy improvement compared with other online hashing methods based on SBQ as our method has a larger distance space. At the same time, compared with the online product quantization methods, our method can improve the search accuracy on the high-dimensional feature vectors by more than 10 percent in most of the cases, especially on the dataset that includes one million feature vectors.

Acknowledgements. This work was supported in part by the Nature Science Foundation of China under Grant 62006007, and in part by the National Innovation 2030 Major S&T Project of China under Grant 2020AAA0104203.

References

1. Andoni, A., Indyk, P.: Near-optimal hashing algorithms for approximate nearest neighbor in high dimensions. In: FOCS, pp. 459–468 (2006)
2. Babenko, A., Lempitsky, V.: Additive quantization for extreme vector compression. In: CVPR, pp. 931–938 (2014)
3. Brandt, J.: Transform coding for fast approximate nearest neighbor search in high dimensions. In: CVPR, pp. 1815–1822 (2010)
4. Cakir, F., He, K., Bargal, S.A., Sclaroff, S.: MIHash: online hashing with mutual information. In: ICCV, pp. 437–445 (2017)
5. Cao, Y., Qi, H., Gui, J., Li, K., Tang, Y.Y., Kwok, J.T.Y.: Learning to hash with dimension analysis based quantizer for image retrieval. TMM **23**, 3907–3918 (2021)
6. Chen, Y., Wang, S., Lu, J., Chen, Z., Zhang, Z., Huang, Z.: Local graph convolutional networks for cross-modal hashing. In: Proceedings of the 29th ACM International Conference on Multimedia, pp. 1921–1928 (2021)
7. Douze, M., Sablayrolles, A., Jégou, H.: Link and code: fast indexing with graphs and compact regression codes. In: CVPR, pp. 3646–3654 (2018). https://doi.org/10.1109/CVPR.2018.00384
8. Huang, L., Yang, Q., Zheng, W.: Online hashing. TNNLS **29**(6), 2309–2322 (2018)
9. Jegou, H., Douze, M., Schmid, C.: Product quantization for nearest neighbor search. TPAMI **33**(1), 117–128 (2011)
10. Kong, W., Li, W.J.: Double-bit quantization for hashing. In: AAAI (2012)
11. Kong, W., Li, W.J., Guo, M.: Manhattan hashing for large-scale image retrieval. In: SIGIR, pp. 45–54 (2012)
12. Krizhevsky, A., Sutskever, I., Hinton, G.E.: ImageNet classification with deep convolutional neural networks. In: NeurIPS, pp. 1097–1105 (2012)
13. Leng, C., Wu, J., Cheng, J., Bai, X., Lu, H.: Online sketching hashing. In: CVPR, pp. 2503–2511 (2015)
14. Li, P., Xie, H., Min, S., Zha, Z.J., Zhang, Y.: Online residual quantization via streaming data correlation preserving. IEEE Trans. Multimedia **24**, 981–994 (2022). https://doi.org/10.1109/TMM.2021.3062480
15. Liberty, E.: Simple and deterministic matrix sketching. In: KDD, pp. 581–588. ACM (2013)
16. Lin, M., Ji, R., Liu, H., Sun, X., Chen, S., Tian, Q.: Hadamard matrix guided online hashing. IJCV **128**(8), 2279–2306 (2020)
17. Lin, M., et al.: Fast class-wise updating for online hashing. In: TPAMI (2020)
18. Liu, C., Lian, D., Nie, M., Xia, H.: Online optimized product quantization. In: 2020 IEEE International Conference on Data Mining (ICDM), pp. 362–371. IEEE (2020)
19. Matsui, Y., Uchida, Y., Jégou, H., Satoh, S.: A survey of product quantization. ITE Trans. Media Technol. Appl. **6**(1), 2–10 (2018)
20. Norouzi, M., Blei, D.M.: Minimal loss hashing for compact binary codes. In: ICML, pp. 353–360 (2011)
21. Shen, F., Xu, Y., Liu, L., Yang, Y., Huang, Z., Shen, H.T.: Unsupervised deep hashing with similarity-adaptive and discrete optimization. TPAMI **40**(12), 3034–3044 (2018)
22. Shen, Y., et al.: Auto-encoding twin-bottleneck hashing. In: CVPR, pp. 2818–2827 (2020)
23. Silpa-Anan, C., Hartley, R.: Optimised kD-trees for fast image descriptor matching. In: CVPR, pp. 1–8 (2008)

24. Simonyan, K., Zisserman, A.: Very deep convolutional networks for large-scale image recognition. CoRR abs/1409.1556 (2014)
25. Wang, J., Liu, W., Kumar, S., Chang, S.: Learning to hash for indexing big data – a survey. IEEE **104**(1), 34–57 (2016)
26. Wang, J., Zhang, T., Song, J., Sebe, N., Shen, H.T.: A survey on learning to hash. TPAMI **40**(4), 769–790 (2018)
27. Wang, J., Kumar, S., Chang, S.F.: Semi-supervised hashing for large-scale search. TPAMI **34**(12), 2393–2406 (2012)
28. Wang, Y., Luo, X., Xu, X.S.: Label embedding online hashing for cross-modal retrieval. In: Proceedings of the 28th ACM International Conference on Multimedia, pp. 871–879 (2020)
29. Weiss, Y., Torralba, A., Fergus, R.: Spectral hashing. In: NeurIPS, pp. 1753–1760 (2009)
30. Weng, Z., Sun, Z., Zhu, Y.: Asymmetric hashing with multi-bit quantization for image retrieval. Neurocomputing **207**, 71–77 (2016)
31. Weng, Z., Zhu, Y.: Online hashing with efficient updating of binary codes. In: AAAI, pp. 12354–12361 (2020)
32. Weng, Z., Zhu, Y.: Online hashing with bit selection for image retrieval. IEEE Trans. Multimedia **23**, 1868–1881 (2021)
33. Weng, Z., Zhu, Y., Lan, Y., Huang, L.K.: A fast online spherical hashing method based on data sampling for large scale image retrieval. Neurocomputing **364**, 209–218 (2019)
34. Xie, H., Mao, Z., Zhang, Y., Deng, H., Yan, C., Chen, Z.: Double-bit quantization and index hashing for nearest neighbor search. TMM **21**(5), 1248–1260 (2019). https://doi.org/10.1109/TMM.2018.2872898
35. Xu, D., Tsang, I.W., Zhang, Y.: Online product quantization. TKDE **30**(11), 2185–2198 (2018)
36. Zhan, Y.W., Wang, Y., Sun, Y., Wu, X.M., Luo, X., Xu, X.S.: Discrete online cross-modal hashing. Pattern Recogn. **122**, 108262 (2022)

Compressed Vision for Efficient Video Understanding

Olivia Wiles[(⊠)], João Carreira, Iain Barr, Andrew Zisserman,
and Mateusz Malinowski

DeepMind, London, UK
{oawiles,mateuszm}@deepmind.com

Abstract. Experience and reasoning occur across multiple temporal scales: milliseconds, seconds, hours or days. The vast majority of computer vision research, however, still focuses on individual images or short videos lasting only a few seconds. This is because handling longer videos require more scalable approaches even to process them. In this work, we propose a framework enabling research on hour-long videos with the same hardware that can now process second-long videos. We replace standard video compression, e.g. JPEG, with neural compression and show that we can directly feed compressed videos as inputs to regular video networks. Operating on compressed videos improves efficiency at all pipeline levels – data transfer, speed and memory – making it possible to train models faster and on much longer videos. Processing compressed signals has, however, the downside of precluding standard augmentation techniques if done naively. We address that by introducing a small network that can apply transformations to latent codes corresponding to commonly used augmentations in the original video space. We demonstrate that with our compressed vision pipeline, we can train video models more efficiently on popular benchmarks such as Kinetics600 and COIN. We also perform proof-of-concept experiments with new tasks defined over hour-long videos at standard frame rates. Processing such long videos is impossible without using compressed representation.

Keywords: Video · Long-video · Compression · Representation

1 Introduction

Most computer vision research focuses on short time scales of two to ten seconds at 25 fps (frames-per-second) because vision pipelines do not scale well beyond that point. Raw videos are enormous and must be stored compressed on a disk; after loading them from a disk, they are decompressed and placed in a device memory before using them as inputs to neural networks. In this setting, and with current hardware, training models on minute-long raw videos can take prohibitively long or take too much physical memory. Even loading such

Supplementary Information The online version contains supplementary material available at https://doi.org/10.1007/978-3-031-26293-7_40.

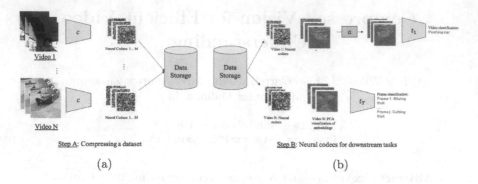

Fig. 1. The compressed vision pipeline. Videos are first compressed using a *neural compressor* c to produce codes. These are stored on a disk and the original videos can be discarded. The neural codes are directly used to train video tasks $t_1 \ldots t_T$. We can optionally augment these codes with augmented versions using an *augmentation network* a (here we show a flipping augmentation). Note that within our framework, all the computations are done in a more efficient compressed space as decompression is not required at any stage of the pipeline.

videos onto GPU or TPU might become infeasible, as it requires decompressing and transferring, often over the bandwidth-limited network infrastructure. While previous work has considered using classic video or image compressed codes (such as JPEG or MPEG) directly as input to their models [1–4], this generally requires specialised neural network architectures.

In this work, we propose and investigate a new, efficient and more scalable video pipeline – *compressed vision* – which preserves the ability to use most of the state-of-the-art data processing and machine learning techniques developed for videos. The pipeline, described in more detail in Sect. 3, has three components. First, we train a *neural compressor* to compress videos. Second, we optionally use *augmentation network* to transform the compressed space for doing augmentations. Third, we *directly* apply standard video backbone architectures on these neural codes to train and evaluate on standard video understanding tasks (thereby avoiding costly decompression of the videos). As our framework is modular, each component could be replaced with a more efficient variant.

Performing augmentations (e.g. *spatial cropping* or *flipping*) is an important component of many pipelines used to train video models but are impossible to perform directly in the compressed space. Therefore, we face the following dilemma. We either give up on augmentations or we decompress the codes and do the transformations in the pixel space. However, if we decide on the latter, we loose some benefits of the compressed space. Decompressed signals expands the space and if they are long enough they cannot fit to a GPU or TPU memory anymore. Moreover, even though a *neural compressor* yields superior quality at higher compression rates to JPEG or MPEG, it has large decoders that take even more time and space; i.e., neural decompression is slow.

To overcome the last challenge, we propose an *augmentation network* – a small neural network that acts directly on latent codes by transforming them according to some operation. In brief, the *augmentation network* for spatial cropping takes crop coordinates and a tensor of latent codes as inputs. Next, it outputs the modified latents that approximate the ones obtained by spatially cropping the video frames. Note that, unlike [5], we *learn* how to augment in the compressed space as opposed to cropping the compressed tensor. As a result, we can train an *augmentation network* for a wider variety of augmentations, such as changing the *brightness* or *saturation* or even performing *rotations*; all these are difficult or impossible by directly manipulating the tensor.

Our approach has the following benefits. First, it allows for standard video architectures to be directly applied on these neural codes, as opposed to devising new architectures, as in [3] which trains networks directly on MPEG representations. Second, as demonstrated in Sect. 3.2, we can apply augmentations directly on the latent codes *without* the need to decompressing them. This significantly impacts training time and saves memory. With these two properties, we can use standard video pipelines with minimal modifications and achieve competitive performance to operating on raw videos (RGB values).

In summary, we show that neural codes generalise to a variety of datasets and tasks (whole-video and frame-wise classification), and are better at compression than JPEG or MPEG. To enable augmentations in the latent space, we train and evaluate a separate network, which conditioned on transformation arguments, outputs transformed latents. We also demonstrate our framework on much longer videos. Here, we collected a large set of ego-centric videos recorded by tourists walking in different cities[1]. These videos are long and continuous, and last between 30 min to ten hours. One can see a few samples of our results on the website[2]. We plan to update it in the future, e.g., to include the source code.

2   Related Work

Our work is built upon the following prior work.

Operating on a Compressed Space. Directly using compressed representations for downstream tasks for video or image data has primarily been studied by considering standard image and video codecs such as JPEG or MPEG [1–3], DCT [4,6] or scattering transforms [7]. However, in general these approaches require devising novel architectures, data pipelines, or training strategies in order to handle these representations. Another strategy is to learn representations that are *invariant* to a range of transformations, as in [8]. However, this requires a-priori knowledge of the downstream tasks to be invariant to. In our case, we do not have this knowledge, as we want to have the same representation to be used for a variety of potentially unknown tasks.

Discrete Representations for Compression. Some modalities such as language are inherently discrete and naturally benefit from neural discrete representations [9,10]. However, the same representation has become increasingly more common in generative modelling of images [11,12] and videos [13,14]. In our work, we

---

[1] The reader can get a feel for the dataset here https://youtu.be/MIzp8Wrj44s?t=131.

[2] Project website: https://sites.google.com/view/compressed-vision.

demonstrate a different use of vector quantization: to compress videos. Thus, we can directly train classifiers on the resulting compressed space, without the need for decompression, making one-hour long training and inference feasible.

Augmentations Using Latents. In the image domain, it has been demonstrated that augmentations in an embedded space learned via a GAN [15–17] or an encoder-decoder model [18] can be used to improve classification performance. In the video setting, [5] have applied augmentations directly in a latent space, but only considered cropping of the tensor – an operation that can readily be designed. Instead, we propose a learnable approach to approximate augmentations in a latent space, so that we can generalize the approach to arbitrary augmentations.

Video Understanding. Understanding long videos encompasses many important sub-problems including action or event understanding and reasoning over longer spatio-temporal blocks. It is also computationally demanding. On these tasks, transformer-like architectures [19–24] could potentially compete with spatio-temporal convolutional networks (3D CNNs) [25–29] that reason about time and space locally. However, videos in current datasets are often too short [30–34] and thus these data hungry architectures are not so useful. This work demonstrates how neural compression can be used to scale architectures to operate on much longer sequences.

Long Sequence Processing. Long-Range Arena [35] was designed to challenge the ability of transformer-like architectures to model long-term dependencies and their efficiency. These problems are well handled directly by efficient transformers [12,36–38]. Walking Tours also sets up a benchmark for efficient and long-term video understanding by probing networks on their ability to handle long sequences where input signals are high-dimensional.

3   Compressed Vision

Here, we introduce our *compressed vision* pipeline. Next, we discuss each component of the pipeline.

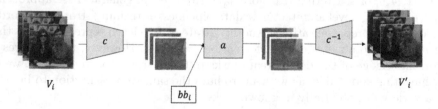

Fig. 2. Augmentation Network. We train a network a that, conditioned on the bounding box coordinates of the desired spatial crops, performs that spatial crop directly on the latent codes (these embeddings are visualised using PCA). As shown here, after decompression, the resulted video corresponds to spatially-cropped original video. Even though cropping is our running example, our methodology extends to potentially arbitrary augmentations.

3.1   Our Pipeline

The typical setup used for training video networks is motivated by that videos are stored in a compressed form. To train neural networks, these videos need to be loaded into a memory and decompressed; taking up space and time.

Instead, it would be preferable to operate directly on the neural codes. Other methods that have investigated this setup used the MPEG encoding directly [3]. However, this requires devising specific architectures and pipelines to handle compressions in that form. As a result, a significant amount of work and progress in the last ten years to apply neural networks to video data cannot be directly applied to these representations. Instead, our pipeline creates compressed tensors that are amenable to being loaded efficiently and passed directly as input to standard video pipelines with minimal, if any, changes in hyper-parameters.

An overview of our pipeline is given in Fig. 1. It operates in three stages. First, we have a *neural compressor* c that is used to compress videos a-priori and store them on a disk or other data carriers (Fig. 1a). After learning this network, there is no further need to touch the original videos and in fact they can be removed to free-up space. Second, for any downstream task, we have a downstream network t_i that is applied to these neural codes to solve that task (Fig. 1b). These tasks may be varied (e.g. frame prediction, reconstruction, classification, etc.) but they all use the same neural codec. Additionally, as we use spatial tensors, many standard architectures can be used while only changing hyper-parameters of the model. Finally, we also learn an *augmentation network* a that (Fig. 2), conditioned on a neural code R, can transform it into a new representation $R' = c(X)$. When decompressed using a decoder c^{-1}, R' approximates an augmentation transformation A, e.g. a spatial crop, acting on the corresponding rgb-valued video X. That is, we are aiming at $c^{-1}(a(c(X))) \approx A(X)$.

3.2   Learning Neural Codecs

To learn c, we build on a standard VQ-VAE encoder-decoder model [10]. The VQ-VAE model uses an encoder to map images into a spatial tensor. Next, each vector in the spatial tensor is compared with a sequence of embeddings using nearest neighbours. Indices that correspond to the chosen embeddings are stored as codes. These codes represent the input images, which can now be discarded: we need only store codes for each image and the fixed-length sequence of embeddings, which we call the codebook. To decode, the codes are mapped to the corresponding embeddings from the codebook, and passed to the decoder. To extend such a representation to videos simply requires modifying the encoding and decoding networks. Instead of using 2D spatial convolutions, we use 3D convolutions to obtain a 3D tensor of codes. We can additionally use additional codebooks for improved performance at the cost of using more memory.

Discussion. There are three factors controlling compression rate. First, the size $T_T \times T_H \times T_W$ of the tensor learned by the encoder controls its spatio-temporal size. Second, the number of codebooks T_C controls the number of codes stored at each location in the tensor. Finally, the number of codes K in the codebook controls the number bits required to store a code. An RGB-video, $I_T \times I_H \times I_W \times 3$, gives a compression rate: $c_r = \frac{I_T I_H I_W * 3 * \log_2 256}{T_T T_H T_W T_C \log_2 K}$. Although we focus on compressing space, we provide promising time-compression results in the supplementary.

Training General Representations. We train the discrete codes using an auto-encoding task and l_1 reconstruction loss. The codes should encode a neural codec that approximates the original video while removing redundancy. As a result, we will be able to use it for a variety of (potentially unknown) downstream tasks (e.g. classification). Moreover this representation should be useful for finding augmentations in that space as different videos (including augmented ones) will map to a separate representation. Note that the intuition here is *not* that we are learning an invariant representation as in [8], which would impede the ability to learn augmentations in the representation space.

Encoder/Decoder Architecture. First, we create spatio-temporal patches of the videos. Next, we use strided 3D CNNs with inverted ResNet blocks [39] to obtain a spatial tensor of vectors. We independently quantize the resulting vectors. This gives intermediate representations that are reversed using strided transposed convolutions for decoding. Further details about the architecture and hyperparameters are given in the supplementary material.

3.3   Video Tasks with Neural Codes

Our final aim is to use the neural codes as the input for arbitrary video tasks. As our neural codes have a spatio-temporal structure, we can train standard modern architectures, such as S3D [29], directly on this representation for the downstream task. During training, we sample codes that correspond to videos, and obtain the corresponding embeddings using the codes as indices. We input these tensors to our downstream model, e.g. S3D [29]. To maintain the same spatial resolution as the original S3D, we modified the strides of the convolutions. For example, for a compressed tensor of size 28×28 we use a stride of 1 (versus 2 for a 224×224 image).

3.4   Augmentations in the Compressed Space

In a standard vision pipeline, being able to augment the input at train and evaluation time often leads to large boosts in performance. Here we describe how a similar procedure can be applied to our neural codes for similar gains.

The central idea is to learn to approximate augmentations but in the compressed space. We use a relatively small neural network, which we call an *augmentation network*, to learn such an approximation. Let A be an augmentation transformation on the input video X. For instance, $A(X)$ can spatially crop the input signal X. Given X and the bounding box bb describing the coordinates of the crop, we train a neural network $a(\cdot)$ to perform the equivalent transformation but in the compressed space. If c and c^{-1} denote encoder and decoder, we want to maintain the following relationship: $A_{bb}(X) = c^{-1}(a(c(X), bb))$.

Downstream Tasks. As in a traditional pipeline, we can apply augmentations at both train and test time to boost performance. In our pipeline, we proceed as follows and use an *augmentation network* trained, for example, to predict spatial crops. At train time, for each compressed video clip X in a batch, we randomly select a bounding box bb and then apply the transformation $a(X, bb)$ to obtain a transformed version of the video clip X' which is the input to the downstream network. At test time, for a given compressed video clip X, we linearly interpolate N bounding box coordinates. Each of these are used to transform the video clip to create N augmented versions of each video clip. The predictions for all N clips are averaged to obtain the final prediction for that clip.

Training. We train the *augmentation network* after learning the neural codec (so there is no requirement that these steps are done on the same datasets) and keep c and c^{-1} fixed. To train, we select a given augmentation class (such as spatial cropping) which is parameterised by some set of values (such as the bounding box for spatial cropping). We then create training pairs by randomly selecting pairs: a video X and a bounding box bb. We apply the augmentation to the video to get its augmented version X'. Finally, we minimize the following reconstruction loss (we use an l_1 loss) to train the corresponding transformation network:

$$||A(X) - c^{-1}(a(c(X), bb))||_{l_1}. \tag{1}$$

We find that using an l_1 loss is sufficient for good results on downstream tasks; we do not require more complicated training pipelines that use adversarial losses. Note that we only train a and all other transformations or networks are frozen. $a(\cdot)$ can then be applied to any video to simulate the given augmentation.

Implementation. We use a multi-layer perceptron (MLP) with three hidden layers and a two-layer transformer [40] to represent $a(\cdot)$. The MLP is applied to a tensor representing an external transformation, e.g., bounding box coordinates or binary values describing when flipping. This creates a representation that we condition on. Obtained features have the same number of channels as the neural codes. We then broadcast them over the spatial and temporal dimensions of the compressed tensor and concatenate along the channel dimension. The result is passed through the transformer to obtain a transformed representation of the same dimension as the neural codes but conditioned on the external transformation. Please see the supplementary material for the precise details.

4   Experiments

Here, we evaluate our generic pipeline using compressed video representations with latent augmentations. We test three things: (1) the utility of our neural codes for achieving high quality performance on a variety of downstream tasks; (2) the memory and speed improvements using our neural codes over using the original video frames; and (3) the ability to use standard tools, such as augmentations, within our framework to recreate the generic vision pipeline.

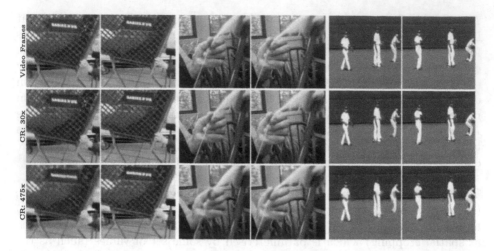

Fig. 3. Reconstruction results at different compression levels. At 30x compression (second row), the results are virtually indistinguishable from the original videos (first row). At 475x compression (third row), the codes lose higher frequency detail but still capture the overall structure of the scene.

We investigate this by first determining how well the compressed tensors can be used to reconstruct the original videos in Sect. 4.2 before discussing their utility on a variety of downstream tasks in Sects. 4.3–4.6. We also investigate transfer of neural codecs if they are trained on another dataset. Next, we investigate how well our trainable transformations model augmentations in Sect. 4.5. Finally, we discuss the speed of a classification model when used with neural codes in Sect. 4.7. We show more details in the supplementary material.

4.1   Datasets

We pre-train our models on one of two datasets: Kinetics600 [30,41] and our internal dataset of recordings of tourists visiting different places. We named this dataset WalkingTours. We evaluate our models on three datasets: Kinetics600 (video classification), WalkingTours (hour long understanding), and COIN (framewise video classification) [42]. Further details are in the supplementary.

4.2   Reconstructions

We first investigate the quality of our neural codes for the task of reconstruction. Note, however, that in our work the final performance on downstream tasks is more important than the reconstruction benchmark; in the spirit of [8]. Nonetheless, we would expect these numbers to be correlated with downstream performance. We can also use reconstruction for introspection and debugging.

Baselines. We compare our results to using a JPEG encoding. This is a commonly used compression scheme in video pipelines, and even though it ignores

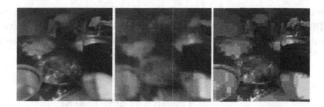

Fig. 4. Neural vs MPEG vs JPEG codecs. Our neural codecs (left) are better than MPEG (middle) and JPEG (right) at higher compression rates ($\approx$180 CR).

the time dimension, it offers quite good quality at low compression rates. However, unlike the frame-based JPEG, our compressors have access to neighbouring information in order to learn better representations at potentially larger compression rates. We use the OpenCV library [43] to obtain the JPEG encodings for various compression levels. We also compare the quality of our compression mechanism to MPEG, which is better than JPEG at higher compression rates.

Results. We report the performance of our compression model when reconstructing the original videos in Table 1 and visualise reconstructions for varying compression levels in Fig. 3. We use standard reconstruction metrics for various levels of compression to demonstrate the reconstruction and compression trade-offs. We can see that we can get low reconstruction error for large compression rates and that these reconstructions capture the higher level structure of the scene and are good quality. We also observe much higher degradation in quality for higher compression rates in JPEG and MPEG than our neural codec. Our visualizations in Fig. 4 confirm the quantitative results in Table 1 that our codec is better than MPEG at high compression rates. This demonstrates that our representations should be informative enough to use for downstream tasks.

4.3   Video-Level Classification on Kinetics600

Next, we investigate directly using our neural codes for downstream tasks.

Baselines. We compare to the baseline of just using the original RGB frames. This is an upper bound of what we would expect to achieve. In our case, instead of inputting a tensor of frames into the downstream network for classification, we input the compressed tensors directly. To apply the S3D architecture to these compressed tensors we simply modify the stride and kernel shapes of the network (hyperparameters of the model). Note that such small modifications are common in the research on videos, e.g. when using 'space-to-depth' tricks.

Results. Results are reported in Table 2. As we can see, using 30x compression leads to a small ($\sim$1% drop in performance) and we can even use on the order of 256x or 475x compression with only a 5% drop in performance. In this research, we have simplified the pipeline, e.g. we use fewer iterations than commonly used, hence, our results are below SOTA using the same model.

Table 1. Reconstruction error for codecs. We compare our approach to using JPEG and MPEG encodings. We report three standard reconstruction metrics (PSNR, SSIM, the mean absolute error (MAE)) on Kinetics600 at different compression rates (CRs). For MAE, lower is better, for others, higher is better.

	Kinetics600		
	PSNR ↑	SSIM ↑	MAE ↓
JPEG CR~30	36.4	94.1	0.013
JPEG CR~90	25.1	70.2	0.045
JPEG CR~180	22.5	63.1	0.057
MPEG CR~30	33.2	89.6	0.034
MPEG CR~90	38.7	82.4	0.026
MPEG CR~180	23.7	67.3	0.054
CR~30	38.6	97.6	0.008
CR~236	30.8	89.8	0.019
CR~384	30.0	88.4	0.019
CR~768	29.0	85.4	0.022

We also investigate how well the neural codes transfer: if we use neural codes trained on a different dataset, say WalkingTours, how well do they transfer to Kinetics600. We find in Table 2 that representations transfer between datasets; i.e. using a representation trained on WalkingTours leads to a similar result.

Table 2. Classification accuracy on Kinetics600. We report Top-1 accuracy on K600 when using neural codes trained one either K600 or WalkingTours. We experiment with different levels of compression (different compression rates (CRs)). CR~1 denotes original RGB frames (without compression).

Evaluated on K600			
Trained on K600		Trained on WalkingTours	
CR	Top-1 ↑	CR	Top-1 ↑
CR~1	73.1	CR~1	73.1
CR~30	72.2	CR~30	71.3
CR~475	68.2	CR~256	68.4

4.4   Frame-Level Classification on COIN

We next investigate the performance of our neural codes on a different task: framewise prediction. Unlike in the video classification task, this requires being able to predict localised information. A representation that throws away too much information would struggle on this task. Here we investigate the utility of our neural codes on this task.

Setup. We use the same RGB baseline and setup as in the Kinetics600 case (Sect. 4.3). As we found in Sect. 4.3 that training the neural codes on a different dataset led to no loss of performance, we use our neural codes trained on either Kinetics600 or WalkingTours when training the downstream model.

Results. We report per frame accuracy in Table 3 for both datasets and different compression rates. We compute means and standard deviations over the test set. We find that, surprisingly there is virtually no loss of performance at a 30x compression rate and only minimal loss of performance at higher compression rates (e.g. 256x, 475x). Indeed, at 30x compression our model trained on Kinetics600 is even performing better than using the original RGB frames. The choice of training dataset for a neural codec has minimal impact on the downstream performance.

Table 3. Downstream classification accuracy on COIN. We report Top-1 and Top-5 accuracy on COIN when using neural compression trained on either K600 or WalkingTours. We experiment with different levels of compression (different compression rates (CRs)). CR~1 denotes original RGB frames.

Evaluated on COIN					
Trained on K600			Trained on WalkingTours		
CR	Top-1 ↑	Top-5 ↑	CR	Top-1 ↑	Top-5 ↑
CR~1	44.3 ± 0.3	71.7 ± 0.1	CR~1	44.3 ± 0.3	71.7 ± 0.1
CR~30	45.5 ± 0.4	73.2 ± 3.7	CR~30	44.6 ± 0.3	71.9 ± 0.3
CR~475	41.7 ± 0.5	65.6 ± 0.5	CR~256	42.7 ± 0.3	67.1 ± 0.3

4.5 Augmentations in the Compressed Space

We also investigate whether we can harness standard techniques in a vision pipeline to push performance further. A standard technique when achieving high quality results is to use spatial and temporal cropping at both training and evaluation time. While our previous results assume that we can simply save various augmented versions of the dataset, that is not feasible for larger datasets.

Setup. Here we investigate whether we can use an *augmentation network* to augment neural codes, as described in Sect. 3.4. We train the augmentation network on Kinetics600, and use it to augment neural codes when training a downstream model. In our experiments, we use the compression model trained on Walking-Tours with 30x compression rate. We focus on spatial cropping and flipping – standard augmentations used in video tasks to improve performance.

For spatial cropping, we proceed as follows. We take a video of size 256×256 and randomly select a 224×224 crop. This defines our bounding box which we use to learn the *augmentation network*. At train time, we randomly select a crop.

Table 4. Using our learnt network for augmentation. We report Top-1 accuracy on K600. We experiment with different numbers of spatial crops at resolution 224×224 and with flipping at resolution 256×256. When using spatial crops, all models are trained using augmented spatial crops.

	Crop size	Num of temporal clips			
		1	2	4	8
224 central crop	224	60.6	62.1	67.8	69.6
224 NN Crops (4 spatial crops) [5]	224	60.7	63.0	68.1	69.0
Ours (2 spatial crops)	224	61.6	64.1	69.1	**70.1**
Ours (3 spatial crops)	224	61.3	63.9	68.9	69.6
Ours (4 spatial crops)	224	**61.9**	**64.4**	**69.3**	69.6
256 central crop	256	60.8	62.4	68.2	68.9
Ours (with flipping at train)	256	61.7	64.4	68.5	70.0
Ours (with flipping at train and eval)	256	**62.9**	**65.2**	**69.0**	**70.2**

At evaluation time, we pool over linearly spaced crops to obtain the final logits. To investigate the utility of our approach, we investigate how we can improve performance at test time with additional, learned spatial crops (after training the downstream model with augmented crops).

For flipping, we take a video of size 256×256. At train time we either flip or not. At evaluation time, we pool over the original clip and its flipped version. We evaluate whether flipping (at train, eval or both) improves performance.

Baselines. For spatial cropping, we compare to two baselines. First, we consider simply using 224×224 central crops. Second, we crop directly the compressed tensor by using nearest neighbours sampling according to the bounding box resized to that spatial resolution; this baseline is similar to the approach in [5]. For flipping, we compare performance to using a 256×256 central crop.

Results. We visualise our learned augmentations in Fig. 5 for spatial cropping and flipping. As can be seen, these augmentations closely match the ground truth transformation. This observation is quantitatively confirmed by the SSIM scores (0.96) between the results obtained by directly cropping in the RGB-space and using our learned *augmentation network*. As a reference the same score between unchanged videos and doing crops in the RGB-space yields only 0.44 SSIM score, showing that the ground-truth and learned crops are indeed highly correlated. We draw similar conclusions about the other two augmentations. Regarding downstream tasks, we report results in Table 4 for Kinetics600. As we can see, learning to approximate augmentations improves over all baselines, demonstrating that our pipeline can leverage standard data augmentation techniques to push performance on downstream tasks. An additional benefit of our learned approach is that we can model transformations that cannot be defined by a simple transformation over embeddings, e.g. flipping or changing the brightness. The appendix has more such examples (e.g. saturation and rotation).

Fig. 5. Learned transformations for augmentations. The top row shows the original frames for three videos; the next two rows show these frames after applying our *augmentation network* for spatial cropping with two different bounding box inputs; and the bottom row shows them after applying our *augmentation network* for flipping. These results are obtained by applying our learned *augmentation network* to the corresponding compressed codes and decompressing.

Note that the setup for these results differ from those in Table 2. In Table 2, we augment the videos before creating codes by using random spatial cropping. Here we *only* use the learned augmentation at train and test time to augment the neural codes with transformed versions. Using learned augmentations leads to a 2% drop in performance while improving over the baselines. It is because we investigate augmentations after applying central crops; effectively reducing the space of possible combinations. This is mostly an engineering limitation as the existing pipelines do cropping before assembling data into batches.

Finally, we also show that we can train our *augmentation network* to do other transformations that are unnecessary for the classification tasks, but could potentially be useful for other problems. These results also show how universal our methodology is. Figures 6 and 8 in the supplementary material show the brightness transformation. Figure 9 shows two other challenging transformations: rotations and changes in saturation. As we can see all these transformations are successfully learned. Note that all these transformations are conducted in the latent space for convenience and we decode them afterwards for visualisation.

4.6   Long-Term Video Predictions

Here, we discuss our results on long-term video prediction using WalkingTours as the dataset. Towards this goal we have created a task *past-future* that does not require supervision. *Past-future* provides a query to the network and asks the question if the short video clip (5 s) has been observed by the network at some point in the past. Sampled clips always come from the same video but have different time stamps. We do the inference, at train and test times, in the causal setting, where the network only sees inputs seen in the past and has no access to the future frames. All the past frames form a compressed memory and are directly accessible to the network at the query time.

Our downstream model differs from the models used above as it requires access to the memory. For that we use a transformer architecture that can effectively query the memory. We also experiment with different memories. Non-parametric *slot* refers to directly keeping all the past frames in the memory (but in the compressed form). We also use *LSTM* as the memory; it transforms a variable-length inputs into a fixed-length vector representation. Finally, *none* denotes no memory. We provide a more detailed explanation of the architecture and the task in the supplementary material. We have conducted experiments on 30 min long videos (training and inference). Our results give 99.6% for *slot*, 78.2% for *LSTM* and 52.9% for *none* (which is equivalent to random chance). Using one-hour long videos, with our best configuration, leads to a drop in performance of 66%, demonstrating there are still gains to be made for long video understanding.

4.7   Speed Requirements

Here, we investigate whether we can train architectures faster; as our representations are smaller spatially and in total memory size. To perform this comparison we run the forward pass 100 times and report the mean and standard deviation of the time it takes on a Tesla V100. We compare the speed when using a compression rate of 256 versus no compression rate. Using compression rate of 256 leads to a forward pass that is on average 0.052 ± 0.002 seconds versus 0.089 ± 0.001 when using no compression. Thus, we can achieve about 2x speed-up with minimal loss of accuracy. The supplementary material has more comprehensive study.

5   Conclusions

This work recreates a standard video pipeline using a neural codec. After learning our neural codes, we can (1) use competitive, modern architectures and (2) apply data augmentation directly on the neural codes. The benefits of using our compressed setup is that we can (1) reduce memory requirements; (2) improve training speed; and (3) generalise to minute or even hour long videos. In order to investigate the feasibility of our approach on hour long videos, we introduced a dataset and task to explore this setting.

Our work is the first, to our knowledge, to demonstrate how to use neural codes for a set of, potentially unknown, downstream tasks by leveraging standard video pipelines, including data augmentation in the form of learned transformations on the compressed space. There are many avenues of future work to improve video understanding at scale. Some directions include improving the compression, scaling to multi-day videos, and collating tasks and datasets for such long horizon timescales.

Acknowledgements. We thank Sander Dieleman for feedback on the paper. We thank Meghana Thotakuri for her help with open sourcing this work.

References

1. Gueguen, L., Sergeev, A., Kadlec, B., Liu, R., Yosinski, J.: Faster neural networks straight from JPEG. In: Advances in Neural Information Processing Systems (NeurIPS) (2018)
2. Ehrlich, M., Davis, L.S.: Deep residual learning in the jpeg transform domain. In: Proceedings of International Conference on Computer Vision (ICCV) (2019)
3. Wu, C.Y., Zaheer, M., Hu, H., Manmatha, R., Smola, A.J., Krähenbühl, P.: Compressed video action recognition. In: Proceedings of the IEEE Conference on Computer Vision and Pattern Recognition (CVPR), pp. 6026–6035 (2018)
4. Xu, K., Qin, M., Sun, F., Wang, Y., Chen, Y.K., Ren, F.: Learning in the frequency domain. In: Proceedings of the Conference on Computer Vision and Pattern Recognition (CVPR) (2020)
5. Patrick, M., et al.: Space-time crop & attend: improving cross-modal video representation learning. In: International Conference on Computer Vision (ICCV) (2021)
6. Nash, C., et al.: Transframer: arbitrary frame prediction with generative models. arXiv preprint arXiv:2203.09494 (2022)
7. Oyallon, E., Belilovsky, E., Zagoruyko, S., Valko, M.: Compressing the input for CNNs with the first-order scattering transform. In: Proceedings of the European Conference on Computer Vision (ECCV) (2018)
8. Dubois, Y., Bloem-Reddy, B., Ullrich, K., Maddison, C.J.: Lossy compression for lossless prediction. In: Advances in Neural Information Processing Systems (NeurIPS) (2021)
9. Mnih, A., Gregor, K.: Neural variational inference and learning in belief networks. In: International Conference on Machine Learning, pp. 1791–1799. PMLR (2014)
10. Oord, A.V.D., Vinyals, O., Kavukcuoglu, K.: Neural discrete representation learning. arXiv preprint arXiv:1711.00937 (2017)
11. Esser, P., Rombach, R., Ommer, B.: Taming transformers for high-resolution image synthesis. arXiv preprint arXiv:2012.09841 (2020)
12. Ramesh, A., et al.: Zero-shot text-to-image generation. arXiv preprint arXiv:2102.12092 (2021)
13. Walker, J., Razavi, A., Oord, A.V.D.: Predicting video with VQVAE. arXiv preprint arXiv:2103.01950 (2021)
14. Yan, W., Zhang, Y., Abbeel, P., Srinivas, A.: VideoGPT: video generation using VQ-VAE and transformers. arXiv preprint arXiv:2104.10157 (2021)
15. Chai, L., Zhu, J.Y., Shechtman, E., Isola, P., Zhang, R.: Ensembling with deep generative views. In: Proceedings of the Conference on Computer Vision and Pattern Recognition (CVPR) (2021)
16. Jahanian, A., Chai, L., Isola, P.: On the "steerability" of generative adversarial networks. In: International Conference on Learning Representations (ICLR) (2020)
17. Härkönen, E., Hertzmann, A., Lehtinen, J., Paris, S.: GANspace: discovering interpretable GAN controls. In: Advances in Neural Information Processing Systems (NeurIPS) (2020)
18. DeVries, T., Taylor, G.W.: Dataset augmentation in feature space. arXiv preprint arXiv:1702.05538 (2017)
19. Bello, I., Zoph, B., Vaswani, A., Shlens, J., Le, Q.V.: Attention augmented convolutional networks. In: Proceedings of the IEEE/CVF International Conference on Computer Vision, pp. 3286–3295 (2019)

20. Bertasius, G., Wang, H., Torresani, L.: Is space-time attention all you need for video understanding? arXiv preprint arXiv:2102.05095 (2021)
21. Fan, H., et al.: Multiscale vision transformers. arXiv preprint arXiv:2104.11227 (2021)
22. Huang, Z., Wang, X., Huang, L., Huang, C., Wei, Y., Liu, W.: CCNet: criss-cross attention for semantic segmentation. In: Proceedings of the IEEE/CVF International Conference on Computer Vision, pp. 603–612 (2019)
23. Vaswani, A., et al.: Attention is all you need. In: Advances in Neural Information Processing Systems (NeurIPS) (2017)
24. Wang, X., Girshick, R., Gupta, A., He, K.: Non-local neural networks. In: Proceedings of the IEEE Conference on Computer Vision and Pattern Recognition, pp. 7794–7803 (2018)
25. Carreira, J., Zisserman, A.: Quo Vadis, action recognition? A new model and the kinetics dataset. In: Conference on Computer Vision and Pattern Recognition (CVPR) (2017)
26. Feichtenhofer, C., Fan, H., Malik, J., He, K.: SlowFast networks for video recognition. In: Proceedings of the IEEE International Conference on Computer Vision, pp. 6202–6211 (2019)
27. Stroud, J., Ross, D., Sun, C., Deng, J., Sukthankar, R.: D3D: distilled 3D networks for video action recognition. In: The IEEE Winter Conference on Applications of Computer Vision, pp. 625–634 (2020)
28. Tran, D., Bourdev, L., Fergus, R., Torresani, L., Paluri, M.: Learning spatiotemporal features with 3D convolutional networks. In: Proceedings of the IEEE International Conference on Computer Vision, pp. 4489–4497 (2015)
29. Xie, S., Sun, C., Huang, J., Tu, Z., Murphy, K.: Rethinking spatiotemporal feature learning for video understanding. arXiv preprint arXiv:1712.04851 (2017)
30. Kay, W., et al.: The kinetics human action video dataset. arXiv preprint arXiv:1705.06950 (2017)
31. Gu, C., et al.: AVA: a video dataset of Spatio-temporally localized atomic visual actions. In: Proceedings of the IEEE Conference on Computer Vision and Pattern Recognition, pp. 6047–6056 (2018)
32. Kuehne, H., Jhuang, H., Garrote, E., Poggio, T., Serre, T.: HMDB: a large video database for human motion recognition. In: International Conference on Computer Vision (ICCV) (2011)
33. Sigurdsson, G.A., Gupta, A., Schmid, C., Farhadi, A., Alahari, K.: Charades-ego: a large-scale dataset of paired third and first person videos. arXiv preprint arXiv:1804.09626 (2018)
34. Soomro, K., Zamir, A.R., Shah, M.: UCF101: a dataset of 101 human actions classes from videos in the wild. arXiv preprint arXiv:1212.0402 (2012)
35. Tay, Y., et al.: Long range arena: a benchmark for efficient transformers. arXiv preprint arXiv:2011.04006 (2020)
36. Kitaev, N., Kaiser, Ł., Levskaya, A.: Reformer: the efficient transformer. arXiv preprint arXiv:2001.04451 (2020)
37. Wang, S., Li, B., Khabsa, M., Fang, H., Ma, H.: Linformer: self-attention with linear complexity. arXiv preprint arXiv:2006.04768 (2020)
38. Zaheer, M., et al.: Big bird: transformers for longer sequences. arXiv preprint arXiv:2007.14062 (2020)
39. Sandler, M., Howard, A., Zhu, M., Zhmoginov, A., Chen, L.C.: MobileNetV2: inverted residuals and linear bottlenecks. In: Proceedings of the IEEE Conference on Computer Vision and Pattern Recognition, pp. 4510–4520 (2018)

40. Dosovitskiy, A., et al.: An image is worth 16x16 words: transformers for image recognition at scale. In: International Conference on Learning Representations (ICLR) (2020)
41. Carreira, J., Noland, E., Banki-Horvath, A., Hillier, C., Zisserman, A.: A short note about kinetics-600. arXiv preprint arXiv:1808.01340 (2018)
42. Tang, Y., et al.: COIN: a large-scale dataset for comprehensive instructional video analysis. In: Conference on Computer Vision and Pattern Recognition (CVPR) (2019)
43. Bradski, G.: The openCV library. Dr. Dobb's J. Softw. Tools **25**(11), 120–123 (2000)

A²: Adaptive Augmentation for Effectively Mitigating Dataset Bias

Jaeju An[1], Taejune Kim[1], Donggeun Ko[4], Sangyup Lee[1],
and Simon S. Woo[1,2,3(✉)]

[1] Department of Computer Science and Engineering, Sungkyunkwan University,
Seoul, South Korea
{anjaeju,taemo,sangyup.lee,swoo}@g.skku.edu
[2] Department of Artificial Intelligence, Sungkyunkwan University, Seoul,
South Korea
[3] Department of Applied Artificial Intelligence, Sungkyunkwan University,
Seoul, South Korea
[4] Department of Applied Data Science, Sungkyunkwan University, Seoul,
South Korea
seanko@g.skku.edu

Abstract. Recently, deep neural networks (DNNs) have become the de facto standard to achieve outstanding performances and demonstrate significant impact on various computer vision tasks for real-world scenarios. However, the trained networks can often suffer from overfitting issues due to the unintended bias in a dataset causing inaccurate, unreliable, and untrustworthy results. Thus, recent studies have attempted to remove bias by augmenting the bias-conflict samples to address this challenge. Yet, it still remains a challenge since generating bias-conflict samples without human supervision is generally difficult. To tackle this problem, we propose a novel augmentation framework, Adaptive Augmentation (A²), based on a generative model that help classifiers learn debiased representations. Our framework consists of three steps: 1) extracting bias-conflict samples from a biased dataset in an unsupervised manner, 2) training a generative model with the biased dataset and adapting the learned biased distribution to the extracted bias-conflict samples' distribution, and 3) augmenting bias-conflict samples by translating bias-align samples. Therefore, our classifier can effectively learn the debiased representation without human supervision. Our extensive experimental results demonstrate that A² effectively augments bias-conflict samples, mitigating widespread bias issues. The code is available in here (https://github.com/anjaeju/A2-Adaptive-Augmentation-for-Effectively-Mitigating-Dataset-Bias).

Keywords: Computer vision · Debiasing · Image translation

1 Introduction

Recently, deep neural networks (DNNs) have achieved great success across various research fields, including image classification [21], object detection [34],

---

Supplementary Information The online version contains supplementary material available at https://doi.org/10.1007/978-3-031-26293-7_41.

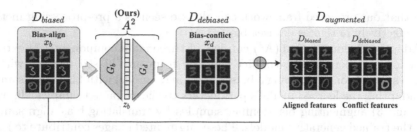

Fig. 1. Illustration of our A^2 framework. A^2 projects bias-align samples x_b, and outputs translated bias-conflict samples x_d. The classifier trained with D_{biased} produces 33.36%, while the classifier trained with $D_{augmented}$ produces 67.47% accuracy on bias-conflict samples in Colored MNIST dataset.

semantic segmentation [22], and even image generation [8]. Nonetheless, overfitting, a well-known problem in DNNs, causes the models to produce inaccurate and unreliable results leading to failure in making proper decisions [31]. This phenomenon is closely related to "dataset bias," where unintended bias exists in the training dataset. In particular, unintended bias indicates that a large number of samples appear with similar task-irrelevant features in a visual context.

Consider the case where we need to classify frog images with different backgrounds. As such, frogs and the background is assigned as *task-relevant feature* and *task-irrelevant feature*, respectively. Most of the frogs could be located in a green pond background (*bias-align*), while a few could be positioned in a swamp or an asphalt road against the green pond (*bias-conflict*). Since the models tend to learn task-irrelevant features (easy-to-learn) as a cue for the labels [19], the models fail to properly learn the task-relevant features, raising questions on the actual performance in the presence of different biases.

To handle such aforementioned bias issue, numerous research has been proposed to 'debias' the dataset bias by defining specific bias types [16,20,27] or learning debiased representation without explicitly defining bias types [2,3,5,6, 30]. Recent advancements in debiasing have demonstrated that augmentation is one of the most promising approaches for mitigating dataset bias [19]. However, augmenting bias-conflict samples without human supervision remains challenging due to the complex properties of different biases, such as texture or shape [19]. Therefore, given the difficulties, we aim to build a practical debiasing method that generates bias-conflict samples without any prior knowledge.

In this paper, we first conduct a set of preliminary experiments to illustrate the importance of augmenting bias-conflict samples, when applying augmentation in biased settings. We find that augmenting only bias-conflict samples significantly improves the classification performance. Based on this finding, we propose a novel augmentation framework, Adaptive Augmentation (A^2), for augmenting bias-conflict samples in an unsupervised manner. Figure 1 illustrates our approach, an end-to-end debiasing pipeline to prevent dataset bias by translating bias-align samples into bias-conflict samples which effectively increases the number of bias-conflict samples to prevent overfitting to task-irrelevant features.

Note that our proposed framework can also be seen as a pre-processing method that effectively mitigates dataset bias.

Adaptive Augmentation (A^2) consists of three main components: 1) extracting a few numbers of bias-conflict samples from any biased dataset without human supervision, 2) training a biased generative model with bias-align samples and adapting the learned model's parameters to the extracted samples' distribution, and 3) augmenting bias-conflict samples by translating bias-align samples with the trained generative models. These augmented images contribute to learning task-relevant features for a biased classifier. We demonstrate that our A^2 outperforms the baselines in comprehensive debiasing benchmark datasets through extensive experiments. Moreover, we confirm that our method performs effectively in an extremely biased setting, where we have very few bias-conflict samples in each dataset. The contributions of our work are summarized as follows:

- We propose a novel debiasing augmentation framework, A^2, which leverages an unsupervised algorithm for extracting bias-conflict samples and exploits few-shot adaptation by adjusting the distribution of the biased generative model to bias-conflict distribution.
- We evaluate the performance of our A^2 through quantitative and qualitative analysis for both synthetic and real-world datasets. We demonstrate that our method achieves the state-of-the-art performance in biased settings.
- We investigate the reason for performance improvement through comprehensive and carefully constructed ablation studies. We believe that our approach has a broader impact by presenting a new application of generative models to solve challenging bias issues for a variety of computer vision applications.

2   Related Work

2.1   Benchmark Datasets for Debiasing

Recently, synthetic or real-world datasets have been created and released publicly to foster the debiasing research field, as shown in Fig. 2.

First, Colored MNIST (CMNIST) and Corrupted CIFAR10 (CCIFAR10) are synthetic datasets built by manually injecting distinct biases into existing MNIST and CIFAR10 datasets. CMNIST dataset is created by adding a color bias to the MNIST [18,19] dataset, as shown in Fig. 2a. Instead of adding color bias like CMNIST, CCIFAR10 is constructed by applying ten distinct noise corruption to each of the labels in CIFAR10 [12,17], as depicted in Fig. 2b. While these two synthetic datasets have been extensively used in previous studies, the challenge for mitigating real-world bias remains, as synthetic datasets are relatively simple to cover real-world bias, such as age or gender.

Second, Biased FFHQ (BFFHQ) and Biased Action Recognition (BAR) datasets are released to mitigate biases in real-world data. BFFHQ [19] is curated from the FFHQ [14] dataset, which contains high-quality images of human faces. From FFHQ, the BFFHQ dataset selects age as a task-relevant feature and gender as a task-irrelevant feature. Accordingly, the majority of the young are

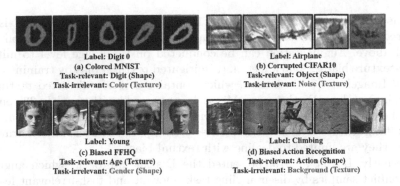

Label: Digit 0
(a) Colored MNIST
Task-relevant: Digit (Shape)
Task-irrelevant: Color (Texture)

Label: Airplane
(b) Corrupted CIFAR10
Task-relevant: Object (Shape)
Task-irrelevant: Noise (Texture)

Label: Young
(c) Biased FFHQ
Task-relevant: Age (Texture)
Task-irrelevant: Gender (Shape)

Label: Climbing
(d) Biased Action Recognition
Task-relevant: Action (Shape)
Task-irrelevant: Background (Texture)

Fig. 2. Sample images from benchmark datasets. We provide task-relevant and task-irrelevant features in each dataset, where each class is represented by a group of images comprising four bias-align samples and a bias-conflict sample.

women as shown in Fig. 2c. On the other hand, the BAR [23] consists of real-world images of six action classes in distinct places, assuming that the classifier is likely to be biased in background (texture) bias. For instance, as shown in Fig. 2d, most bias-align samples from the *climbing* label contain a climber climbing a rock cliff. In contrast, the bias-conflict sample exhibits a climber with a glacier cliff. Unlike synthetic datasets containing texture biases, real-world datasets have both texture and shape biases which are relatively difficult to handle.

2.2 Existing Methods for Debiasing

Numerous previous studies have been performed to mitigate dataset biases. We mainly investigate existing debiasing methods in the context of augmentation: debiasing without data augmentation and with augmentation.

Debiasing without Augmentation. Debasing can be performed through an explicit definition of the bias type [7,16,20,23,27,30]. Li and Vasconcelos et al. [16] and Kim et al. [20] demonstrated that a specific color bias could be relieved by normalizing the biased distribution or utilizing biased RGB values as a clue for classification. Sagawa et al. [27] proposed the groupDRO, a debiasing method which clusters subgroups in the dataset with explicit supervision. GroupDRO is then further improved with various modifications [7,26,32]. Moreover, Kim et al. [16] employed a regularization loss to inhibit the model from learning unwanted bias in the dataset. However, none of the above are applicable in real-world datasets since defining whole bias types is unfeasible when dealing with large-scale real-world datasets [19]. In contrast, recent approaches widely adopt the debiasing method without defining prior knowledge of the bias. Nam et al. [23] proposed LfF, a state-of-the-art method for simultaneously training biased and debiased models to amplify the influence of bias-align samples on biased models and bias-conflict samples on debiased models, respectively.

Debiasing with Augmentation. Debiasing with augmentation is less explored compared to the debiasing approaches without augmentation mentioned above. Augmentation can be conducted on the image level to mitigate shape-texture biases. Agarwal et al. [1] mitigated textural bias by training on the 'Styled ImageNet' [6,13] dataset, which contains severely distorted textures of the ImageNet data. Furthermore, MixStyle, proposed by Zhou et al. [33], encouraged the classifier to extract more generalized features against the texture bias by shuffling feature-level statistics. These techniques, however, have a drawback in that they are limited to dealing with textual bias.

Recently, Lee et al. [19] presented the DisEnt approach, which augments bias-conflict samples by disentangling task-relevant and task-irrelevant features and permuting each bias feature vector to the other task-relevant feature vectors. Motivated by this observation, we design an augmentation framework that adapts task-irrelevant features regardless of the bias type to address both texture and shape biases without supervision, enhancing the performance of classification methods.

3 Importance of Augmenting Bias-conflict Samples

3.1 Overview

Data augmentation is a crucial way to boost the performance of DNNs, which applies various transformations to the original data, and compensates for the lack of datasets [28,29]. Various augmentation techniques are available and have been shown to improve the model performance; however, when applied to biased datasets, augmentation might amplify the bias in the dataset by increasing bias-align samples. Therefore, we conduct the following experiments assuming that indiscriminate augmentation can degrade the classification performance:

- Case 1: Augmenting only bias-align samples.
- Case 2: Augmenting only bias-conflict samples.
- Case 3: Augmenting both bias-align and conflict samples.

We can find that Case 1 and 2 augment bias-align and bias-conflict samples, respectively, whereas Case 3 augments both bias-align and bias-conflict samples. These case studies can validate the degree of bias amplification according to the augmented sample type. The above experiments show the effect of augmentation methods for each case while using only the simple augmentations (i.e., the random crop and rotation) that do not affect the image's texture.

Dataset and Classifier. We demonstrate our method's performance on two synthetic and one real-world datasets, CMNIST, CCIFAR10, and BFFHQ. For CMNIST and CCIFAR10, we use bias ratios 99.5% and 95%, respectively. The model's performance is evaluated using a test set composed solely of bias-conflict samples. See Appendix A.1 for more details.

Table 1. Performance comparison on the bias-conflict test sets. Each case indicates different augmentation scenarios. We observe performance degradation in Cases 1 and 3 but performance improvement in Case 2. It denotes that applying data augmentation in a biased setting can cause the bias to exacerbate. We report the average accuracy over three runs. Bold indicates the best accuracy.

Dataset	Bias Ratio	Baseline	Case 1	Case 2	Case 3
Colored MNIST	99.5	33.36	24.89	**37.07**	33.76
	95	83.88	82.70	**84.10**	79.96
Corrupted CIFAR10	99.5	13.23	12.77	**13.31**	12.46
	95	27.37	27.27	**28.08**	27.60
Biased FFHQ	99.5	43.93	42.20	**45.67**	45.13

Results. Table 1 summarizes the model's performance in each case mentioned above. The *baseline* column indicates how well the vanilla model performs on each test set. The model's performance in Case 1 (augmenting bias-align samples) declines by up to 8%, indicating that the dataset's bias has been amplified due to the augmented bias-align samples. In contrast, Case 2 (augmenting bias-conflict samples) illustrates that the model's performance increases by up to 4%, demonstrating that the dataset's bias has diminished. However, in Case 3 (augmenting both bias-align and bias-conflict samples), the model's performance remains close to the vanilla model without making any meaningful improvements. Therefore, it is desirable to augment only bias-conflict samples in biased settings. These findings established the validity of our hypothesis. With this motivation, we propose A^2 to augment bias-conflict samples effectively.

4   Design of A^2 Framework

This section describes a novel augmentation framework, Adaptive Augmentation (A^2). First, we introduce our extraction method that selects a few bias-conflict samples from a biased dataset in an unsupervised manner. Second, we describe the detailed design of A^2, which utilizes a generative model and few-shot adaptation. Finally, we explain the training scheme to build a debiased classifier with an augmented dataset. We present the pipeline of our approach in Fig. 3.

4.1   Extracting Bias-conflict Samples

Suppose we can train a classifier to be biased regardless of the bias type (e.g., different bias types presented in Sect. 2.1). Then, the biased classifier predicts a target label with high confidence for unseen bias-align samples because the representation of the biased sample is similar to the trained biased samples [2, 19]. On the contrary, the classifier predicts with a low confidence for unseen bias-conflict sample, because its representation is different from the bias-align samples. This result leads to a higher cross-entropy loss for bias-conflict samples compared to the bias-align samples. In that case, we can easily identify bias-conflict samples with the higher cross-entropy loss [2,19]. Therefore, we need

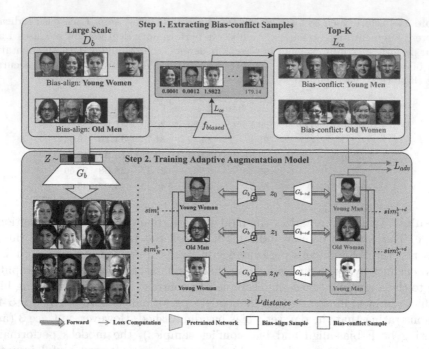

Fig. 3. Overview of our proposed debiasing method, Adaptive Augmentation (A^2) model. Our A^2 consists of two steps: 1) extracting bias-conflict samples and 2) training the Adaptive Augmentation model. In Step 1, we extract bias-conflict samples by sorting top-k L_{ce} values calculated by the biased classifier f_{biased}. Next, we adapt a biased generative model to the extracted bias-conflict distribution by minimizing L_{adv} and $L_{distance}$.

an unsupervised method to train a classifier to become biased without human supervision or pre-defined bias types.

To enable a classifier to be biased in an unsupervised manner, we need to keep emphasizing the impact of bias-align samples during training. Therefore, we employ generalized cross-entropy (GCE) loss [23] that amplifies the bias of the neural network without human-supervision. The equation of GCE loss is defined as follows:

$$GCE(p(x;\theta),y) = \frac{1 - p_y(x;\theta)^q}{q}, \tag{1}$$

$$\frac{\partial GCE(p(x;\theta),y)}{\partial \theta} = p_y(x;\theta)^q \cdot \frac{\partial CE(p_y(x;\theta),y)}{\partial \theta}, \tag{2}$$

where $p(x;\theta)$ indicates the softmax output of a classifier θ, $p_y(x;\theta)$ represents the probability assigned to the target variable y with $q \in (0,1]$ that is a hyper-parameter controlling the degree of amplification. The gradient of GCE loss emphasizes the gradient of CE loss by multiplying the probability p_y (i.e., confidence score). Since the task-irrelevant features are easy-to-learn [23] in the early stages of training, the classifier first learns biased representations from the bias-align samples, producing higher confidence in the bias-align samples than that of bias-conflict samples. Thus, classifiers trained with GCE loss learn task-irrelevant features from biased images regardless of the type of bias each dataset has, ensuring that the classifiers become biased toward the aligned task-irrelevant features. With this assumption, we propose simple, yet powerful bias-conflict sample extraction algorithm, depicted in Step 1 of Fig. 3. Our algorithm is described as follows: we train our classifier with GCE loss and calculate CE loss for all samples in the training dataset to sort the loss values. After training, we extract the top-k samples as bias-conflict samples for utilizing them when adapting a generative model.

4.2 Training Generative Models

We elaborate on the training procedure of the bias adaptive generator that translates bias-align samples into bias-conflict samples, as described in Step 2 of Fig. 3. First, we need a pretrained generative model G_b trained on a biased dataset D_b. For G_b, we adopt a powerful generative model, StyleGAN2 [15], as our backbone for generating images (Note that any generative model can be used in this procedure, where we used StyleGAN2 since it is one of the high-performing GAN methods). G_b learns the distribution of D_b, by mapping 512 low-dimensional noise vector $z \sim p_z(z)$ to biased image $x_b \sim D_b$. The biased generative model can be obtained by using a GAN training procedure, with a discriminator D_m as follows:

$$L_{adv}(G_b, D_m) = D_m(G_b(z)) - D_m(x_b), \quad (3)$$

$$G_b^* = \arg \min_{G_b} \max_{D_m} \mathbb{E}_{z \sim p_z(z),\ x_b \sim D_b} L_{adv}(G_b, D_m), \quad (4)$$

where G_b^* denotes optimal weights of the model and the detailed training procedure is explained by Karras et al. [15]. After training the biased distribution, our G_b can generate samples that have bias-aligned representation, depicted in the left side of Step 2 in Fig. 3. Second, we convert biased generator (G_b) to debiased generator (G_d) by adapting the learned distribution to the bias-conflicting distribution with the extracted bias-conflict samples and all of the bias-align samples. In this case, we need to preserve the learned relationship between the generated samples from G_b and G_d during adaptation, as maintaining relative pairwise distances prevents mode collapse [4,10,24,25]. To this end, we leverage the state-of-the-art few-shot adaptation method [24]. We initialize G_b and G_d with the pretrained weight of G_b, and sample N noise vectors $\{z_n\}_0^N$. Then, we forward z_i into the networks for obtaining the ith generated samples $G_b(z_i)$ and $G_d(z_i)$. To match the relationship of generated samples from each network, we need to represent the relationship as probability distribution and minimize the distance of the two distributions. Therefore, we convert the generated samples into probability distributions using cosine similarity as follows:

$$y_i^{b,l} = Softmax(\{sim(G_b^l(z_i), G_b^l(z_j))\}_{\forall j \neq i}), \tag{5}$$

$$y_i^{d,l} = Softmax(\{sim(G_d^l(z_i), G_d^l(z_j))\}_{\forall j \neq i}), \tag{6}$$

where l and sim indicates the lth activation layers and cosine similarity, respectively. Now, we can preserve the relationship of the generated samples by minimizing $y_i^{b,l}$ and $y_i^{d,l}$ through KL-divergence:

$$L_{distance}(G_d, G_b) = \sum_{l,i} D_{KL}(y_i^{d,l} \| y_i^{b,l}), \tag{7}$$

where the generated bias-align and bias-conflict samples have similar distribution. Thus, our final objective consists of two terms as follows:

$$G_d^* = \arg \min_{G_d} \max_{D_m} \mathbb{E}_{z \sim p_z(z), \ x_d \sim D_d} L_{adv}(G_d, D_m) + \lambda L_{distance}(G_d, G_b), \tag{8}$$

where D_d denotes the extracted bias-conflict samples, $L_{distance}$ for preserving the learned relationship, and L_{adv} for converting the learned distribution into another distribution.

4.3   Learning Debiased Representation with Adaptive Augmentation

After G_b and G_d converges, we generate debiased representation of bias-align sample x_d through an image-to-image translation approach as shown in the Fig. 1. Our A^2 takes the bias-align sample as an input, and projects it into the latent space (i.e., sample to latent), forming the biased latent vector z_b using G_b. The z_b contains label and bias information. Then, we forward z_b into G_d to generate bias-conflict representation of the input image (i.e., latent to sample). By keeping the label information for each x_b from the projected latent vector z_b, we do not need to label the generated samples manually (see Appendix A.2 for more details). After translating all x_b to debiased sample x_d, we obtain the debiased dataset D_d. By adding D_d to the original dataset, we have $D_{augmented} = D_b \cup D_d$ that supports to learn task-relevant features more effectively than training solely D_b, as task-irrelevant features are not aligned. Finally, we train the debiased classifier $f_{debiased}$ with $D_{augmented}$ using CE loss as follows:

$$f_{debiased}^* = \arg \min_f \frac{1}{N} \sum_{i=1}^N CE(f(x), y), \tag{9}$$

where we use $f_{debiased}^*$ for predicting bias-align and bias-conflict samples.

5   Experiments

5.1   Experimental Setup

Datasets. We use both synthetic (Colored MNIST and Corrupted CIFAR10) and real-world (Biased FFHQ and Biased Action Recognition) datasets.

We report bias-align and bias-conflict classification accuracy (%) on each test set, qualitative analysis on augmented samples, and ablation studies for our A². We further assess our A² with one-shot scenario, where only one bias-conflict sample per class exists, for demonstrating the validity of the proposed method in extremely biased settings; in this case, we adopt Colored MNIST, Corrupted CIFAR10, and Biased FFHQ datasets. See Appendix A.1 for more details (e.g., the number of samples, bias ratios, or description of each dataset).

Baseline. We compare the performance of our A² with state-of-the-art debiasing methods, including empirical risk minimization (ERM). The unsupervised debiasing and augmentation techniques are taken into consideration when choosing the baselines. Detailed descriptions of each baseline method are provided below:

1) ERM. Empirical risk minimization (ERM) indicates a classifier trained with only original CE loss [16], not exploiting any debiasing scheme. ERM performance is treated as a reference to other debiasing methods.

2) LfF. LfF is a state-of-the-art method proposed by Nam et al. [23], utilizing weighted cross-entropy loss for bias-conflict samples in an end-to-end manner. Specifically, LfF trains two networks, a biased classifier for calculating relative difficulty and a debiased classifier for learning debiased representation. After training, the only debiased classifier is used to predict.

3) DisEnt. DisEnt proposed by Lee et al. [19] is the first to apply augmentation approach to the debiasing method; specifically, DisEnt is an extended version of LfF in terms of augmentation. They follow the training mechanism of LfF and introduce the additional swapping function of the feature vectors from the biased classifier.

Implementation Details. We adopt common implementations for the debiasing process: we use a 3-layer MLP network with 100 hidden units for the CMNIST dataset, and Resnet18 [11] network for the other datasets. For training the biased classifier, we set the controlling degree q as 0.7 in Eq. 1. We set $k = 10$ for our entire experiments for extracting bias-conflict except one-shot testing scenario. See Appendix A.3 for more details.

6 Results

6.1 Performance on Benchmark Datasets

Real-World Datasets. The results are presented in Table 2, where we provide both bias-align and bias-conflict performances.

For the BFFHQ dataset, the ERM model correctly identified bias-align samples, but incorrectly predicted bias-conflict samples, with 39.87% and 50.80% in the one-shot and 99.5% bias setting, respectively. On the other hand, our classifier, trained with the augmented dataset, outperforms ERM and other state-of-the-art debiasing methods by a large margin, producing 47.87% and 56.73%

Table 2. Performance comparison on both real-world and synthetic datasets. We evaluate both bias-align and bias-conflict test sets to show that each method can learn debiased representation without performance degradation for bias-align samples. We report the average accuracy over three runs. Bold and underlined numbers indicate the best performance and the second best performance, respectively.

Dataset Type	Dataset	Bias Ratio	Bias-align				Bias-conflict			
			ERM	LfF	DisEnt	(Ours)	ERM	LfF	DisEnt	(Ours)
Real-world	Biased FFHQ	One-Shot	99.33	98.67	**99.33**	87.73	39.87	39.47	40.40	**47.87**
		99.5%	99.40	96.67	97.13	**98.93**	50.80	55.73	52.67	**56.73**
	BAR	100%	N/A				63.10	67.84	66.46	**71.15**
Synthetic	Colored MNIST	One-Shot	98.40	**97.80**	96.50	97.21	14.22	22.06	18.33	**23.70**
		99.5%	99.90	83.70	68.33	**93.03**	33.36	53.43	60.96	**67.47**
		99%	99.90	86.65	78.35	**97.14**	57.28	61.75	75.99	70.68
		98%	99.83	90.21	90.64	**99.57**	73.24	68.65	79.69	76.93
		95%	99.67	85.99	98.21	**99.20**	83.88	85.13	87.29	86.09
	Corrupted CIFAR10	One-Shot	97.60	94.97	96.93	82.40	9.03	9.44	13.39	**15.45**
		99.5%	96.87	82.47	89.37	**90.03**	13.23	14.43	14.58	**15.96**
		99%	97.27	85.67	95.63	82.37	13.46	19.01	19.97	**21.45**
		98%	96.57	78.93	95.17	88.53	17.62	**26.50**	23.11	21.79
		95%	95.10	72.93	92.37	**93.70**	27.37	**35.20**	30.39	28.23

accuracy in the one-shot and 99.5% bias settings, respectively. We can clearly observe that our method is much more effective when the bias setting is severe (one-shot and 99.5%). Furthermore, our A^2 outperforms other baselines in the most severe case, where the bias ratio is set to 100% in the BAR dataset. We believe this is because our method easily adapts bias-conflict features in the severe bias setting. Overall, our augmentation process demonstrated its effectiveness in a severely biased environment by converting bias-align samples into bias-conflict samples, effectively assisting the model to learn the debiased representations in real-world datasets.

Synthetic Datasets. To further demonstrate the effectiveness of augmentation methods, we evaluate the classification performance in controlled environments with synthetic datasets. For the CMNIST dataset, our classifier outperformed two baselines (ERM and LfF) with bias-conflict samples across all bias ratios. Furthermore, our classifier successfully predicted bias-align samples even after learning the debiased representation, producing on par performance compared to the ERM model. DisEnt performed better than our method in three cases (99%, 98%, and 95%); however, our classifier still outperforms DisEnt with a large margin in extremely biased environments (one-shot and 99.5%). For the CCIFAR10 dataset, our classifier also achieved the state-of-the-art performance in severely biased settings (one-shot, 99.5%, and 99%), and produced on par performance compared to baselines in other cases. In some cases, DisEnt has

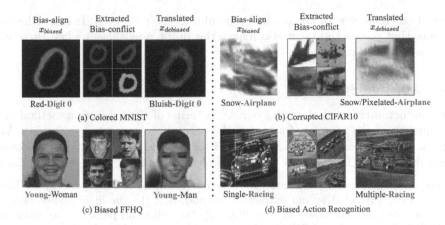

Red-Digit 0   Bluish-Digit 0   Snow-Airplane   Snow/Pixelated-Airplane

(a) Colored MNIST   (b) Corrupted CIFAR10

Young-Woman   Young-Man   Single-Racing   Multiple-Racing

(c) Biased FFHQ   (d) Biased Action Recognition

Fig. 4. Qualitative evaluation for the translated images via A^2 with 99.5% bias ratio. The 2nd column (middle images) from each dataset indicates the samples extracted by our extraction algorithm. The blue text indicates label information for each image. We observe that the translated images successfully include bias-conflict features while retaining their respective labels.

better accuracy on bias-align samples. We believe that the cases where DisEnt performed better are due to the difficulty of learning debiased representations of bias-conflict samples. In fact, the performance difference between the bias-align and bias-conflict test set in these cases is lower than other baselines across all datasets. Overall, our approach performed better than other approaches across different datasets with various bias ratios, achieving the best performance among one-shot and 99.5% settings and the second best performance among most cases.

6.2 Qualitative Analysis on Augmented Samples

We examine translated images used in our main experiments with 99.5% of bias-align samples. Figure 4 shows results on the bias-align sample for each dataset and the corresponding translated image using the extracted samples. We observed that the translated images have bias-conflict representations against bias-align samples. For example, red-colored digit 0 in CMNIST is translated to a bluish color digit 0.

Interestingly, the translated image has a mixture of several colors from the extracted bias-conflict samples, which implies our A^2 reflects the features from bias-conflict samples. We also observed this successful translation in BFFHQ. Nevertheless, a failure case was also observed in CCIFAR10 when enforcing all noise biases from the conflicting samples to the given image, challenging to recognize the original object. For the BAR dataset, where training samples are all aligned with a bias, we observed that there is an unintended bias in the aligned samples (e.g., most of the images have single racing cars, but extracted images have several cars). Therefore, our A^2 converted a single car image to several car images with the original red-colored car preserved. It is worthwhile to note that our method can also relieve the unintended biases when there are no bias-conflict samples.

6.3 Ablation Study

We conduct ablation studies with our A^2 in terms of the extraction method and the quality of the translated samples, the results are presented in Table 3. We report all valid cases for CMNIST, CCIFAR10, and BFFHQ datasets except BAR dataset, as BAR does not have bias-conflict samples in the training data.

Table 3. A^2 ablation experiments with benchmark datasets. We report (a) the bias-conflict ratio extracted by our extraction method, and (b) the high quality ratio of the translated samples based on the GIQA [9]. We utilize all possible bias ratios for CMNIST, CCIFAR10 and BFFHQ datasets. We cannot include BAR dataset, as it does not have bias-conflict samples in training set.

Bias Ratio	CMNIST	CCIFAR10	BFFHQ	Bias Ratio	CMNIST	CCIFAR10	BFFHQ
99.5	100	80	60	99.5	98.37	97.03	98.52
99.0	100	100	N/A	99.0	98.95	96.11	N/A
98.0	100	100	N/A	98.0	98.61	96.67	N/A
95.0	91	100	N/A	95.0	98.66	97.95	N/A

| (a) Bias-Conflict Ratio (%) | (b) High-Quality Ratio (%) |

Bias-Conflict Ratio of Extracted Samples. To verify the effectiveness of extracting bias-conflict samples, we measure the proportion of bias-conflict samples among the extracted samples across all datasets. The results are reported in Table 3a. The bias-conflict ratio is surprisingly high, showing mostly from 80% to 100% for CMNIST and CCIFAR10 datasets; however, the ratio was relatively low in the BFFHQ dataset.

We noticed that extracting whole bias-conflict samples in real-world datasets with our proposed method remains challenging. It is difficult because of the highly entangled representations, where correlated complex attributes, such as age and gender, cannot be fully disentangled by the proposed biased classifier. However, we believe that our unsupervised extraction method offers a simple yet practical approach for extracting bias-conflict samples, by efficiently extracting bias-conflict samples without any supervision. Moreover, we successfully enhanced the classification performance in real-world datasets, achieving state-of-the-art results.

High-Quality Ratio of Translated Samples. We measure the percentage of high-quality samples in the translated samples for all datasets in order to assess the quality of the translated bias-conflict samples generated by our A^2. We use generated image quality assessment (GIQA) proposed by Gu et al. [9] for this analysis. Specifically, we use the GMM-GIQA among the GIQA family (refer to the Sect. 3.2 for more details [9]).

The GMM-GIQA value is in the range of $[0, 1]$, and higher value indicates higher quality. We define a generated image as *high-quality* when it gets the GMM-GIQA value over 0.5. As reported in Table 3b, the majority of samples that generated by our method are evaluated as *high-quality* samples in all bias ratios, producing over 96% results.

We noticed that our framework could enhance the classification performance, providing lower bounds and valid results without using the quality assessment method. Furthermore, our framework can be improved by reducing errors with the GIQA method (e.g., the GIQA method can be used as an 'image picker,' selecting only high-quality images).

7 Discussion

Limitations. We observed possible limitations of our model during experiments. Since our method requires a biased generative model, dependence on the generative model may arise. However, developing generative models is one of the most fast-growing research fields in machine learning; there is a great possibility for future improvement when combined with our novel approach.

Future Work and Broader Impacts. We plan to extend our framework to prevent and filter out badly generated samples by using GIQA metric as the 'image picker' to select the high-quality samples. As machine learning models become deeply embedded in diverse aspects of our daily lives, it is crucial to ensure they produce accurate, reliable, and trustworthy results. In this context, we believe that our approach can contribute to various computer vision applications, such as media or database platforms, by debiasing the massive data for protecting machine learning models from bias.

8 Conclusion

In this work, we proposed a novel augmentation framework, A^2, which effectively learns biased representations through image-level augmentation to address dataset biases. Our framework is derived from our findings that augmenting bias-conflict samples is crucial in biased contexts. Thus, A^2 effectively augments bias-conflict samples through image-to-image translation methods integrated with an unsupervised extraction algorithm. We demonstrated the performance and effectiveness of our augmentation framework through extensive experiments. We believe our work can contribute to building more accurate and trustworthy computer vision applications by effectively preventing bias, predominately occurring in real-world datasets.

Acknowledgements. We thank members of DASH Lab. for the helpful feedback. This work was partially supported by the Basic Science Research Program through National Research Foundation of Korea (NRF) grant funded by the Korean Ministry of Science and ICT (MSIT) under No. 2020R1C1C1006004 and Institute for Information & communication Technology Planning & evaluation (IITP) grants funded by the Korean MSIT: (No. 2022-0-01199, Graduate School of Convergence Security at Sungkyunkwan University), (No. 2022-0-01045, Self-directed Multi-Modal Intelligence for solving unknown, open domain problems), (No. 2022-0-00688, AI Platform to Fully Adapt and Reflect Privacy-Policy Changes), (No. 2021-0-02068, Artificial Intelligence Innovation Hub), (No. 2019-0-00421, AI Graduate School Support Program at Sungkyunkwan University), and (No. 2021-0-02309, Object Detection Research under Low Quality Video Condition).

References

1. Agarwal, V., Shetty, R., Fritz, M.: Towards causal VQA: revealing and reducing spurious correlations by invariant and covariant semantic editing. In: Proceedings of the IEEE/CVF Conference on Computer Vision and Pattern Recognition, pp. 9690–9698 (2020)
2. Bahng, H., Chun, S., Yun, S., Choo, J., Oh, S.J.: Learning de-biased representations with biased representations. In: International Conference on Machine Learning, pp. 528–539. PMLR (2020)
3. Cadene, R., Dancette, C., Cord, M., Parikh, D., et al.: RUBi: reducing unimodal biases for visual question answering. In: Advances in Neural Information Processing Systems 32 (2019)
4. Chen, T., Kornblith, S., Norouzi, M., Hinton, G.: A simple framework for contrastive learning of visual representations. In: International Conference on Machine Learning, pp. 1597–1607. PMLR (2020)
5. Clark, C., Yatskar, M., Zettlemoyer, L.: Don't take the easy way out: ensemble based methods for avoiding known dataset biases. arXiv preprint arXiv:1909.03683 (2019)
6. Geirhos, R., Rubisch, P., Michaelis, C., Bethge, M., Wichmann, F.A., Brendel, W.: Imagenet-trained CNNs are biased towards texture; increasing shape bias improves accuracy and robustness. arXiv preprint arXiv:1811.12231 (2018)
7. Goel, K., Gu, A., Li, Y., Ré, C.: Model patching: closing the subgroup performance gap with data augmentation. arXiv preprint arXiv:2008.06775 (2020)
8. Goodfellow, I., et al.: Generative adversarial nets. In: Advances in Neural Information Processing Systems 27 (2014)
9. Gu, S., Bao, J., Chen, D., Wen, F.: GIQA: generated image quality assessment. In: Vedaldi, A., Bischof, H., Brox, T., Frahm, J.-M. (eds.) ECCV 2020. LNCS, vol. 12356, pp. 369–385. Springer, Cham (2020). https://doi.org/10.1007/978-3-030-58621-8_22
10. He, K., Fan, H., Wu, Y., Xie, S., Girshick, R.: Momentum contrast for unsupervised visual representation learning. In: Proceedings of the IEEE/CVF Conference on Computer Vision and Pattern Recognition, pp. 9729–9738 (2020)
11. He, K., Zhang, X., Ren, S., Sun, J.: Deep residual learning for image recognition. In: Proceedings of the IEEE Conference on Computer Vision and Pattern Recognition, pp. 770–778 (2016)
12. Hendrycks, D., Dietterich, T.: Benchmarking neural network robustness to common corruptions and perturbations. arXiv preprint arXiv:1903.12261 (2019)
13. Huang, X., Belongie, S.: Arbitrary style transfer in real-time with adaptive instance normalization. In: Proceedings of the IEEE International Conference on Computer Vision, pp. 1501–1510 (2017)

14. Karras, T., Laine, S., Aila, T.: A style-based generator architecture for generative adversarial networks. In: Proceedings of the IEEE/CVF Conference on Computer Vision and Pattern Recognition, pp. 4401–4410 (2019)
15. Karras, T., Laine, S., Aittala, M., Hellsten, J., Lehtinen, J., Aila, T.: Analyzing and improving the image quality of styleGAN. In: Proceedings of the IEEE/CVF Conference on Computer Vision and Pattern Recognition, pp. 8110–8119 (2020)
16. Kim, B., Kim, H., Kim, K., Kim, S., Kim, J.: Learning not to learn: training deep neural networks with biased data. In: Proceedings of the IEEE/CVF Conference on Computer Vision and Pattern Recognition, pp. 9012–9020 (2019)
17. Krizhevsky, A., Hinton, G., et al.: Learning multiple layers of features from tiny images (2009)
18. LeCun, Y., Cortes, C.: MNIST handwritten digit database (2010). http://yann.lecun.com/exdb/mnist/
19. Lee, J., Kim, E., Lee, J., Lee, J., Choo, J.: Learning debiased representation via disentangled feature augmentation. In: Advances in Neural Information Processing Systems 34 (2021)
20. Li, Y., Vasconcelos, N.: Repair: removing representation bias by dataset resampling. In: Proceedings of the IEEE/CVF Conference on Computer Vision and Pattern Recognition, pp. 9572–9581 (2019)
21. Lu, D., Weng, Q.: A survey of image classification methods and techniques for improving classification performance. Int. J. Remote Sens. **28**(5), 823–870 (2007)
22. Minaee, S., Boykov, Y.Y., Porikli, F., Plaza, A.J., Kehtarnavaz, N., Terzopoulos, D.: Image segmentation using deep learning: a survey. IEEE Transactions on Pattern Analysis and Machine Intelligence (2021)
23. Nam, J., Cha, H., Ahn, S., Lee, J., Shin, J.: Learning from failure: de-biasing classifier from biased classifier. Adv. Neural. Inf. Process. Syst. **33**, 20673–20684 (2020)
24. Ojha, U., et al.: Few-shot image generation via cross-domain correspondence. In: Proceedings of the IEEE/CVF Conference on Computer Vision and Pattern Recognition, pp. 10743–10752 (2021)
25. Van den Oord, A., Li, Y., Vinyals, O.: Representation learning with contrastive predictive coding. arXiv e-prints pp. arXiv-1807 (2018)
26. Puli, A.M., Zhang, L.H., Oermann, E.K., Ranganath, R.: Out-of-distribution generalization in the presence of nuisance-induced spurious correlations. In: International Conference on Learning Representations (2021)
27. Sagawa, S., Koh, P.W., Hashimoto, T.B., Liang, P.: Distributionally robust neural networks for group shifts: on the importance of regularization for worst-case generalization. arXiv preprint arXiv:1911.08731 (2019)
28. Shorten, C., Khoshgoftaar, T.M.: A survey on image data augmentation for deep learning. J. big data **6**(1), 1–48 (2019)
29. Van Dyk, D.A., Meng, X.L.: The art of data augmentation. J. Comput. Graph. Stat. **10**(1), 1–50 (2001)
30. Wang, H., He, Z., Lipton, Z.C., Xing, E.P.: Learning robust representations by projecting superficial statistics out. arXiv preprint arXiv:1903.06256 (2019)
31. Ying, X.: An overview of overfitting and its solutions. J. Phys. Conf. Ser. **1168**, 022022 (2019). IOP Publishing (2019)
32. Zhou, C., Ma, X., Michel, P., Neubig, G.: Examining and combating spurious features under distribution shift. In: International Conference on Machine Learning, pp. 12857–12867. PMLR (2021)

33. Zhou, K., Yang, Y., Qiao, Y., Xiang, T.: Domain generalization with mixStyle. In: International Conference on Learning Representations (2021). https://openreview. net/forum?id=6xHJ37MVxxp
34. Zou, Z., Shi, Z., Guo, Y., Ye, J.: Object detection in 20 years: a survey. arXiv preprint arXiv:1905.05055 (2019)

Deep Active Ensemble Sampling
for Image Classification

Salman Mohamadi[✉], Gianfranco Doretto, and Donald A. Adjeroh

West Virginia University, Morgantown, WV, USA
sm0244@mix.wvu.edu

Abstract. Conventional active learning (AL) frameworks aim to reduce the cost of data annotation by actively requesting the labeling for the most informative data points. However, introducing AL to data hungry deep learning algorithms has been a challenge. Some proposed approaches include uncertainty-based techniques, geometric methods, implicit combination of uncertainty-based and geometric approaches, and more recently, frameworks based on semi/self supervised techniques. In this paper, we address two specific problems in this area. The first is the need for efficient exploitation/exploration trade-off in sample selection in AL. For this, we present an innovative integration of recent progress in both uncertainty-based and geometric frameworks to enable an efficient exploration/exploitation trade-off in sample selection strategy. To this end, we build on a computationally efficient approximate of Thompson sampling with key changes as a posterior estimator for uncertainty representation. Our framework provides two advantages: (1) accurate posterior estimation, and (2) tune-able trade-off between computational overhead and higher accuracy. The second problem is the need for improved training protocols in deep AL. For this, we use ideas from semi/self supervised learning to propose a general approach that is independent of the specific AL technique being used. Taken these together, our framework shows a significant improvement over the state-of-the-art, with results that are comparable to the performance of supervised-learning under the same setting. We show empirical results of our framework, and comparative performance with the state-of-the-art on four datasets, namely, MNIST, CIFAR10, CIFAR100 and ImageNet to establish a new baseline in two different settings.

1 Introduction

Active learning (AL) has consistently played a central role in domains where labeling cost is of great concern. The core idea of AL frameworks revolves around learning from small amounts of annotated data and sequentially choosing the most informative data sample or batch of data samples to label. To this end, after initial training using available labeled data, an acquisition function is utilized to leverage the model's uncertainty in order to explore the pool of unlabeled data for most informative data points. In parallel with advancements in AL, in the recent years, deep learning has gained tremendous attention due to its emergence as a high-performing approach, primarily conditioned on the availability

© The Author(s), under exclusive license to Springer Nature Switzerland AG 2023
L. Wang et al. (Eds.): ACCV 2022, LNCS 13847, pp. 713–729, 2023.
https://doi.org/10.1007/978-3-031-26293-7_42

714 S. Mohamadi et al.

of large amounts of training data. An interesting challenge is how to efficiently incorporate data-hungry deep learning tools into supposedly data-efficient AL frameworks.

Adjusting AL algorithms for deep neural networks has been very challenging, where extending the model complexity/capacity to that of CNNs ultimately ended up with either a poor performance, or some minor improvements at the cost of querying almost all samples. On the other hand, sequential training of such expressive models as well as extending the framework to high dimensional data injects even more complexity [1–3]. This challenge was relatively under-explored, until a breakthrough work by Gal et al. [4], which essentially considered the problem of incorporating deep learning into AL for high dimensional data as highly connected with that of uncertainty representation. They thus approached the problem from the perspective of uncertainty representation in deep learning for AL, and developed a Bayesian AL framework for image data. Later work (such as [5]), however, argued that the approach exhibits poor scalability to big datasets due to its limited model capacity .

Another approach that also relied on uncertainty representation, is ensemble-based AL [5]. Here, an ensemble of classifiers is used, where the classifiers independently learn from the data in parallel. The major drawback is the poor diversity (lack of exploration) even with larger ensembles. Our approach, while enjoying the power of ensembles, solves this problem by offering an inherent exploration/exploitation trade-off as classifiers maintain some dependency in the form of a shared prior. Apart from uncertainty representation, another set of emerging methods that primarily rely on geometrical data representation [6] showed improved performance in deep AL. However, similar to [7], we empirically observed that these geometric approaches typically suffer from performance degradation as the class diversity (number of classes) increases. Another recent approach is the work reported in [7] where they take advantage of adversarial training to provide improved performance over previous methods. We empirically find that their work provided a balanced performance on datasets at different scales and diversity. As we will show later, our proposed model outperforms this approach in multiple settings with significant margins, with results approaching that of supervised learning models in some cases.

In the first part of the paper, primarily motivated to efficiently integrate the advantages of uncertainty and geometrical representations, we propose an approach built upon approximate Thompson sampling. On one hand, this provides an improved representation of uncertainty over unlabeled data, and on the other hand, supports an inherent tune-able exploration/exploitation trade-off for diverse sampling [8,9]. Unlike conventional ensemble-based methods whose performance tend to saturate quickly, under our tuneable model, adding a few more classifiers tends to improve the uncertainty and geometric representation. To mitigate the general sample diversity problem of ensemble models (see [5,10]), we use an inclusive sample selection strategy. Our framework showed a noticeable improvement over the state-of-the-art, with performance approaching those of supervised learning methods. Further, we explore the scope and scale of model efficiency improvements brought about by our proposed techniques.

Briefly, due to the exploration/exploitation trade-off, Thompson sampling is expected to improve both predictive uncertainty and sample diversity by computing, sampling, and updating a posterior distribution. A serious consideration, however, is that, for more expressive models such as deep convolutional neural networks (CNNs) designed for high dimensional data, Thompson sampling makes the process computationally difficult. This is primarily because computation of the posterior distribution over CNNs is complex by nature. Inferences based on Laplace approximations or Markov chain Monte Carlo approaches would be two possible alternatives. However, both approaches are still very expensive in terms of computational cost [11–13]. Lu et al. [13] argue that due to the compatibility of Thompson sampling with sequential decision and updating, an approximate version of Thompson sampling could be a promising solution. Accordingly, we build an ensemble model relying on an efficient approximate of Thompson sampling, which improves the state-of-the-art. Interestingly, this model possesses both the advantage of uncertainty based deep AL approaches (exploiting most uncertain samples), and of geometric solutions (exploring for more diverse though not necessarily highly uncertain samples).

In the second part of the paper, we investigate a new line of efforts/arguments revolving around the idea of boosting AL frameworks using self/semi supervised learning techniques. We substantiate and unify these arguments and also design and perform extensive experiments on multiple baselines to assess this approach as a new general training protocol for AL frameworks. This enables our approach to be compared against recent boosted AL frameworks.

Briefly, our key contributions in this paper are as follows:

- A new framework for deep AL which enables an exploration/exploitation trade-off for sample selection and hence offers the advantages of both uncertainty-based and geometry-based methods.
- A new general training protocol for visual AL approaches, developed by substantiating and unifying recent arguments on boosting AL using self/semi supervised learning, and experimentally evaluating this approach on multiple recent baselines. We compare our framework against two sets of baselines to show its performance.

2 Background and Preliminaries

Background: Early efforts on AL with image data considered mainly kernel-based approaches [14–16]. Later, AL methods with image data using CNN included uncertainty-based approaches [4,5,17–19], geometry-based approaches [6], or their combination [7], e.g., based on adversarial training. Generally speaking, uncertainty-based approaches focus on finding most uncertain samples to label, with the potential downside of less diversity in sample selection, while geometric approaches tend to weigh on diversity of samples, resulting in performance degradation in cases of very diverse datasets (with large number of classes). Most recently, in a relatively different setting, Gao et al. [20] leveraged semi-supervised learning while Bengar et al. [21] applied self-supervised learning (SSL) techniques to deliver a significant performance improvement. We will compare our proposed approach against these related work, on the same problem settings. Some other recent work in this general area of modern AL with high dimensional data can be found in [22–27]. Though these are relevant, they are not as closely related to our approach.

SSL: As the second contribution of this work relates to SSL we briefly review the literature. Briefly, SSL is one of the closest modern problem domains to AL with zero labeling effort policy. Here, the goal is to leverage all unlabeled data to train a network for a pretext task so as to prepare the network for a downstream task, usually with small amounts of data [28]. Until recently, a major set of SSL baselines were contrastive baselines relying on contrasting augmented views of a sample with each other (positive contrastive pairs) and with views of other samples (negative contrastive pairs) [29,30]. Newer baselines such as [31,32], a.k.a non-contrastive approaches, rely on contrasting positive pairs, needless of contrasting negative pairs. Recently, Ermolov et al. [33] reported a non-contrastive method based on whitening the embedding space, which was effective, yet conceptually simple. We adopt this approach in this work.

Preliminary: We describe these two major paradigms below.

1. Uncertainty-Based Techniques: Two categories of well-known deep learning techniques for uncertainty representation and estimation include ensemble-based techniques (non-Bayesian) [18,19] and Monte-Carlo (MC) dropout (Bayesian) [4,17]. In ensemble-based methods, an ensemble of N identically structured neural networks are trained using identical training data D_{tr}, where the different random values are applied for weight initialization w_i. For a given class c out of multiple classes and input X, we then have:

$$p(y = c|x, D_{tr}) = \frac{1}{N} \sum_{i=1}^{i=N} p(y = c|x, w_i) \tag{1}$$

However, MC-dropout trains a network with dropout, and during test, implements T forward passes, each individually with a new dropout mask, resulting in T sets of weights w_t. Given input x, the average of all T softmax vectors represents the output for a desired class c.

$$p(y = c|x, D_{tr}) = \frac{1}{T} \sum_{t=1}^{t=T} p(y = c|x, w_t) \tag{2}$$

Here we briefly describe some popular effective uncertainty-based acquisition functions [4,5] or their approximation for ensemble-based approaches, MC dropout, and our proposed framework, all based on uncertainty sampling.

A. Selecting samples with highest predictive entropy [34].

$$H[y|x, D_{tr}] := - \sum_c (\frac{1}{N} \sum_n p(y = c|x, w_n)) . \log(\frac{1}{N} \sum_n p(y = c|x, w_n)) \tag{3}$$

B. Selecting samples with highest mutual information between their predicted labels and the weights, BALD [4,35], which was initially applied in [4] with T forward passes in MC-dropout. It can be analogously rewritten for an ensemble with N members by replacing T with N.

$$I[y; w|x, D_{tr}] := H[y|x, D_{tr}] - \frac{1}{T} \sum_t \sum_c -p(y = c|x, w_t) . \log p(y = c|x.w_t) \quad (4)$$

C. Highest Variation Ratio [36] as a measurement of non-modal predicted class labels, where f_m is the number of modal class predictions [5].

$$VR := 1 - f_m/N \quad (5)$$

We used this acquisition function in our proposed DAES framework.

2. Geometry-Based Techniques: Geometric or representation-based methods primarily rely on density-based acquisition functions. Typical examples include REPR [37], and Core-Set [38]. With a total of n samples, at each iteration Core-Set selects a fixed number of samples, that minimize the upper bound on the distance between point x_i in n samples, and x_j, its closest neighbour in selected subset o. The acquisition function of Core-Set is given as follows: $s = argmax_{i \in [n]\emptyset} \min_{j \in o} dist(x_i, x_j)$. See [37] for that of REPR.

3. Other Techniques: Other methods include implicit combination of uncertainty and geometry approaches, such as in [7], which designs a minimax game in the context of adversarial training. There are also methods that have used the power of pre-trained models such as [20], and to a less extent [39].

3 Deep Active Ensemble Sampling

Our work is primarily inspired by the reports in [4,13,19] towards finding an uncertainty-diversity trade-off. In particular, we propose a tuneable trade-off between uncertainty-wise exploitation of samples vs exploration of less uncertain, but more diverse samples.

3.1 Thompson Sampling for AL

Contextual Bandit: Thompson sampling was primarily developed as a heuristic to address the Multi-armed bandit (MAB) problem, aiming for a trade-off between exploration and exploitation in sequential decision making. The core idea of Thompson sampling has a Bayesian essence (See Algorithm 1). Unlike greedy algorithms that mostly lean toward exploitation, Thompson sampling draws random samples from a posterior distribution to fine-tune between exploration and exploitation. See [8,9] for related work on low dimensional data. New attempts towards using Thompson sampling for efficient estimation of posterior distribution for more complex models such as CNNs revealed an immediate need to find a computationally tractable approximation [13].

Deep AL: Assuming a pool-based AL setting, we initially have a set of unannotated data $U_0 = \{x_1, x_2, ...x_n\}$ and a small set of annotated data A_0, where at each iteration, an algorithm, known as *acquisition function*, looks into the whole set of unlabeled data to select a number of samples and pass them to an Oracle for labeling. In deep learning backed AL with high dimensional data such as images, the goal is to adjust the model to enable learning from a relatively small initial training set, and accordingly select a subset of most informative unlabeled data samples (in terms of uncertainty and diversity) to be labeled.

3.2   Ensemble Sampling

From a geometry perspective, one ideal estimation of the desired posterior space in AL framework could be represented by a direct sum over the space. Along this line, some methods such as [24] propose splitting the input space to improve uncertainty sampling associated with the posterior distribution. Hinton et al. [40] noted the fact that data points are generated by natural sources that actually inject limited complexity, rather than random sources with unlimited complexity. Therefore, unlike a random source that practically enables sampling from an infinite space, the natural source can be represented with a direct sum over the posterior space $\mathbf{S}$ with any *finite* number of summands Q: $\mathbf{S} = \mathbf{S_1} \oplus \mathbf{S_2} \oplus ... \oplus \mathbf{S_Q}$, where $\mathbf{S_i}$ represents the i-th subspace. Later we will see that compared with regular ensembles, ensemble sampling is closer to this direct sum as it allows a better exploration of whole representation space.

In the case of AL on a neural network with weights θ, let's say the network represents the mapping $g_\theta : \mathcal{R}^W \mapsto \mathcal{R}^K$ (W is the dimensionality of input) and the goal is to sequentially choose a fixed number of samples d_t from a pool $\mathcal{D}$ of K samples as input at each time $t = 0, 1, ...T$, where $\mathcal{D} \subseteq \mathcal{R}^W$, such that it leads to desirable output. Accordingly, with each set of samples d_t selected from $\mathcal{D}$ at time $t = 0, 1, ..., T$, an output $g_\theta(d_t)$ and random variable $w_t \sim N(0, \sigma_w^2 I)$ form the observation $y_t = g_\theta(d_t) + w_t$ which allows to update a reward $r_t = r(w_t)$ sequentially. Supposing that we have a prior on θ, $\theta \sim N(\mu_0, \Sigma_0)$, the model will become much more prone to uncertainty. Therefore, at each time t, the neural network will be fitted by d_t, y_t, and the samples are selected with the goal of converging to a trade off between immediate desirable outputs (minimizing the loss) and reducing uncertainty in θ.

With the problem presented in as above, an algorithm is required to incorporate Thompson sampling in this new context. In the case of linear bandit problem, since the conventional Thompson sampling yields an efficient solution, no approximation to Thompson sampling is needed. However, in case of neural networks, the conventional form of Thomson sampling could be computationally expensive. This calls for a more efficient implementation in terms of approximate Thompson sampling. Accordingly, Lu et al. [13] introduce an ensemble of N networks with a shared prior on their weights, as an approximate Thompson sampling. This allows efficient posterior estimation on complex models such as neural network.

3.3   Algorithms for CNNs

Here we represent ensemble sampling as an efficient approximation of Thompson sampling for neural networks. In fact, unlike in simpler cases such as linear bandit, exact Bayesian inference can not easily be performed effectively for neural networks, which necessitates an efficient approximation. First, we present the algorithm for Thompson sampling (Algorithm 1 (taken from [41])). Then, we discuss ensemble sampling as its efficient approximation, and present the algorithm for Deep Active Ensemble Sampling (Algorithm 2).

More precisely on Thompson sampling, let's assume $\mathcal{X}$ is a finite set of data points $x_1, ..x_n$, where selecting a data point x_t (or a number of data points) at time t yields a randomly generated output y_t based on a conditional probability distribution $q(.|x_t)$. Accordingly, a known function $r_t = r(y_t)$ is defined to capture the reward for the selected data point. This reward can be interpreted as a negative loss. At the beginning, the decision maker gets initialized with a prior p on θ, and as it starts to explore, updates its uncertainty representation. While greedy algorithms generally use expected value of θ with respect to p to produce model parameters $\hat{\theta}$, Thompson sampling relies on random sampling from p. Next, the algorithm will choose data points maximizing the expected reward presented as follows:

$$\mathbb{E}_{q_{\hat{\theta}}}[r(y_t)|x_t = x] = \sum_o q_{\hat{\theta}}(o|x)r(o) \tag{6}$$

Subsequently p is updated by conditioning on $\hat{y}_t$, and for θ coming from a finite set, relying on Bayes rule we will have:

$$\mathbb{P}_{p,q}(\theta = u|x_t, y_t) = \frac{p(u)q_u(y_t|x_t)}{\sum_v p(v)q_v(y_t|x_t)} \tag{7}$$

Algorithm 1 (taken from [41]) captures the above steps. As noted, this will be very time consuming, especially for neural networks.

Algorithm 1 Thompson($\mathcal{X}, p, q, r$)

1: **for** $t = 1, 2, ..., T$ **do**
2:     Sample $\hat{\theta} \sim p$
3:     $x_t \leftarrow \arg\max_{x \in \mathcal{X}} \mathbb{E}_{q_{\hat{\theta}}}[r(y_t)|x_t = x]$
4:     Input chosen x_t and observe y_t
5:     $p \leftarrow \mathbb{P}_{p,q}(\theta \in .|x_t, y_t)$
6: **end for**

As an efficient approximate Thompson sampling for neural networks, we use ensemble sampling, where we employ an ensemble of M networks and set priors on the weights, as presented in Algorithm 2. All networks will be trained on identical data samples while the initial shared priors on the weights makes a connection between them. Algorithm 2 is inspired by [13], with key adjustments to make the approximate Thompson sampling adaptable to the AL framework. These changes include (1) the optimization process of ensembles; (2) selecting a set of samples rather than one sample; (3) we replace the original concept of maximizing reward in the algorithm with minimizing the loss, namely, $\bar{L}(\theta)$ in our deep active learning framework. Accordingly, it is important to mention that the optimization of the method need not to be combinatorial as in the case with combinatorial contextual bandits. Moreover, sample selection is sequential in which, each iteration of sample selection provides a batch of samples ranked by the acquisition function. Unlike classical ensemble-based approaches, the proposed deep active ensemble sampling (DAES) not only puts a joint prior on the weights of the networks (all sampled from one prior distribution rather than individual priors), but also jointly optimizes the members of an ensemble.

Algorithm 2 Deep Active Ensemble Sampling (M)

1: Ensemble $En_M\left(g, \mathcal{N}(\mu, \sigma^2)\right)$: $g(\theta_1), ..., g(\theta_M)$; Labeled Set: S_l^t; Unlabeled Set: S_u^t
2: **for** $t = 1, 2, ... , T$ **do**
3:     Train over S_l^t: En_M: $g(\theta_{1,t}), ..., g(\theta_{M,t})$
4:     Optimize: $arg\,min_{\theta_{i,t}}(L^t) = arg\,min_{\theta_{i,t}}(L(\theta_{1,t}) + ... + L(\theta_{M,t}))$
5:     Batch b^t selection by fixed En_M: $En_M(S_u^t)$, $VR = (1 - \frac{f_m}{M})$ via Eqn (5)
6:     Update Training Set: $S_l^{t+1} = S_l^t + b^t$
7: **end for**

3.4 DAES with Self-trained Knowledge Distillation

Consistent with primary focus of AL on less annotation effort and with the goal of establishing a new standard AL training protocol, we empirically evaluate a simple training technique which inherently empowers any active learner, regardless of the underlying approach. While this is inspired by the recent trend in [20, 21], we also argue that using pre-training, here SSL pre-training, enables any AL framework to better model uncertainty over the data, or to capture the geometry of the data, due to the prior knowledge attained by SSL. To ensure fairness of our comparisons, we apply the new training protocol to both the previous baseline AL models, and to our proposed DAES framework. The proposed training protocol could help to eventually unify this line of work with some form of knowledge distillation [42, 43].

Training Protocol: The protocol is a two step process: SSL pre-training, and then active learning using the pre-training output. (See Fig. 1). Due to huge success of SSL in learning representation from unlabeled data, we adopt a most recent SSL model suitable for our setting. Thus, our proposal for training AL models is to consider a training protocol, first a pre-training is performed on the deep network (encoder in Fig. 1) as a building block for the active learning

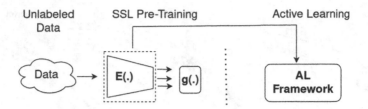

Fig. 1. SSL pre-training for deep active learning. Here, $E(.)$ is the encoder and $g(.)$ is the projection head. After pre-training, the weights of $E(.)$ will be fixed and will then be used in our AL setting, training a classifier head on top of that.

models. In this work, we tested this idea by adopting the conceptually simple, yet effective SSL model in [33] to initially train ResNet18 as the building block for the AL methods, namely, Random baseline, VAAL, Core-Set and DAES.

We explore a new setting in which a given baseline is equipped with a conceptually simple self-training as discussed above. As shown in Fig. 1, we adopt the SSL framework from [33], to leverage knowledge distilled from unlabeled data for empowering the active learner. The idea is to use whitening in SSL in order to train the encoder (ResNet18) and then freeze all layers except for head-layers which are replaced with fully connected layers to be trained.

4   Experiments and Results

We conduct two sets of experiments on images classification task to evaluate our proposed DAES framework as well as compare it against state-of-the-art models. Specifically, we mainly perform the experiments on MNIST [44], CIFAR10 and CIFAR100 [45], and ImageNet [46]. To ensure the fairness of compassion scenarios, we compare the framework against **two sets** of baselines, namely, trained from scratch, and self-trained enabled by self/semi supervised learning (SSL).

Evaluation: On CIFAR10/100 and ImageNet, starting with an initial budget of 10% labeled samples, we measure the performance on sequential training using T training iterations, where in each iteration of training we add 5% labeled data from unlabeled pool to the training set (labeled data ratio of 0.1, 0.15, 0.20, ... up to 0.35 or 0.50). We assume each training iteration is from scratch unless otherwise stated. On MNIST the initial training set is 200 samples and the evaluation is performed on acquisition budget of 100 samples. The results of all our experiments on all datasets including ImageNet are averaged over three trials.

Baselines: We compare the performance of DAES against two sets of baselines. First set of approaches, specifically trained from scratch, includes Random sampling from unlabeled pool (Random), Monte-Carlo dropout (M-C Dropout) [17], deep Bayesian active learning (DBAL) [4], Core-Set [6], Ensemble with Variation Ratio (Ens-VarR) [5], and VAAL [7]. We also design and implement another set of extensive experiments on our framework as well as some of previous baselines empowered by self-training including Random, Core-Set, and VAAL to contrast against a very recent baseline taking advantage of SSL, CSSAL [20], and also later compare with a semi-supervised baseline, REVIVA [39].

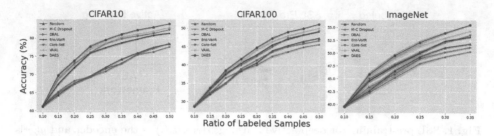

Fig. 2. Accuracy vs ratio of labeled samples from CIFAR10, CIFAR100 and ImageNet datasets.

4.1 Experimental Settings

We implemented our network architectures in Pytorch. Besides our experiments, experiments of all other competitive baselines including Random baseline, on CIFAR10, CIFAR100 and ImageNet are performed with ResNet18, with similar setting of VAAL except they used VGG16 [47]. However for MNIST, we used a three-layer (two convolutional and one fully connected) network described in [4]. Specifically an ensemble includes $N = 5$ identical classifiers unless otherwise specified. We used Xavier initialization when applicable, and we utilized Adam optimizer [48] for all experiments. All experiments start with an initial balanced budget of 10% of unlabeled training pool (6000 for MNIST, 5000 for CIFAR10/100, and 128120 for ImageNet), which is then iteratively updated by adding 5% of whole training pool. Both initial training and other sequential iterations of training continue for 100 epochs. After every update, the network is trained from scratch unless otherwise specified (i.e., incremental training). Further, unlike classical ensemble-based methods, the optimization process of all classifiers in DAES is performed jointly as one loss function. Practical considerations in case of DAES with very deep networks are discussed in ablation studies.

4.2 DAES Performance Comparison

In this section we explain the immediate results of experiments on MNIST, CIFAR10/100 and ImageNet in two comparing scenarios, namely, AL model trained from scratch, and AL on self-trained model.

1. Trained from Scratch: The conventional protocol is training from scratch.

Our results on MNIST is on par with VAAL and Core-Set where all three approaches attained $99 + \%$ accuracy with 1000 samples (1.67%) of the data. Ens-VarR, DBAL, M-C Dropout and Random baselines achieved 97.81 ± 0.12, 97.55 ± 0.18, 97.26 ± 0.14, and 95.2 ± 0.23.

On CIFAR10 as shown in Fig. 2, our framework tends to outperform other baselines including VAAL upon using more than 15% of the data, while the difference grows by adding more labeled samples. Our approach attains mean accuracy of 82.98$ and 83.93% upon using 40% and 50% of the data respectively, whereas Top-1 accuracy using 100% of data is 93.27%. Second and third highly performant methods using half of the data are VAAL and Core-Set with 82.89% and 82.31% respectively. While Ens-VarR remains fairly competitive, M-C dropout as well as DBAL are evidently underperforming.

On CIFAR100 also our method starts to outperform competitive VAAL and Core-Set approaches upon using 20 + % of the data. The accuracy difference swiftly grows by adding more samples to the point that upon using 50% of data, our method outperforms VAAL and Core-Set by 51.33% to 50.01% and 49.03%. Note that the Top-1 accuracy using full data is 75.43%. As it is clear, due to larger number of classes, Core-Set experienced performance degradation down to performing on par with Ens-VarR .

Performance on Dataset at Scale: On ImageNet as a large and more challenging dataset of 1.2+ million samples of 1000 classes, our method patently outperforms former baselines upon using 15% or more of data. Compared to Top-1 mean accuracy of 71.8% using whole data, we achieve mean accuracy of 55.57% upon using only 35% of data, which is a 1.2% improvement over VAAl, (while VAAL offers only less than 1% improvement over its former baseline, Core-Set using 35% of data). Our method improves over Random baseline by mean accuracy of 3.67%. Similar to their performance on CIFAR10 and CIFAR100, Bayesian techniques, i.e., DBAL and M-C dropout, slightly underperform Random baseline.

2. Self-training: We also evaluated the proposed use of self-supervised knowledge distillation [42] from unlabeled data as a general technique to further improve the model training process for AL methods. There are two objectives here. First, to provide a fair comparison of this SSL+AL approach when applied on our proposed DAES, and three other AL baselines (namely, VAAL, Randon and Core-Set), against two approaches [20, 39] that take advantage of knowledge distillation of unlabeled data. Second, to show that the SSL+AL protocol establishes a new standard training protocol for deep AL regardless of the underlying principle. As shown in Fig. 3, extensive experiments on Random baseline, VAAL, Core-Set and DAES on CIFAR10, CIFAR100 and ImageNet consistently confirm the performance jump due to SSL-wise leveraging of unlabeled data while still using a small percentage of labeled data. Aside from bringing some accuracy jump to VAAL, Core-Set and Random baseline, this allows our framework to outperform CSSAL [20] on CIFAR10, CIFAR100 and ImageNet by using 18 + %, 20 + % and 17% of data. As can be observed, the performance of our method on all three datasets rivals Top-1 mean accuracy attained by supervised learning (having the whole data labeled, denoted by the red line in the figure). On CIFAR10 and only using 40% of data (labeled), all approaches except for Random acquisition perform above Top-1 mean accuracy of 93.27%. On CIFAR100 (50% labeled) and ImageNet (35% labeled) all methods are competitive to supervised Top-1 mean accuracy, with our method (DAES) achieving a mean accuracy of 73.55% (compared to 75.81%) and 69.92% (compared to 71.80%). Finally,

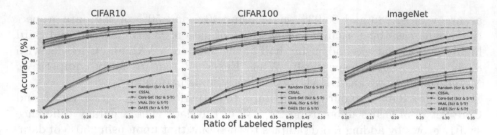

Fig. 3. Accuracy vs ratio of labeled samples from CIFAR100 and ImageNet datasets on SSL boosted networks. Top groups: results with proposed training protocol using SSL; Lower group: results with training without SSL. Red line denotes results using supervised learning with the full labeled data. (Color figure online)

compared with a recent baseline on semi-supervised learning, REVIVAL proposed in [39], on CIFAR10 and using 40% of the data, our framework performs on par with REVIVAL. On CIFAR100 our approach (using 35% of the data) performs on par with REVIVAL (using 25% of the data).

5   Ablation Study and Investigative Scenarios

In this section we discuss our ablation studies to assess the effect of model size on tuning the trade-off between performance and model capacity/complexity, DAES behaviour with deeper networks, and finally incremental training. For all methods, we used Variation Ratio as acquisition function, as it is empirically proved to be the most effective query strategy in the literature [4,5].

5.1   DAES Model Size

One main advantage of DAES is that it can provide higher accuracy by enlarging the ensemble. Ens-VarR enjoys a performance boost only when changing the 1-member ensemble to ensemble with more than one member. Unlike Ens-VarR which lacks a malleable trade-off between computational over-head and performance, meaning that adding reasonably more classifiers to the ensemble does not lead to a proportional increase in performance, we empirically assess how larger ensembles provide desirable improvement in accuracy for DAES. As shown in Fig. 4, DAES-10 with 5 additional classifiers (total of 10), approximately doubles the former accuracy improvement on ImageNet dataset (3 times the accuracy improvement that VAAL adds to Core-Set under the same experimental setting). This is while DAES-20 with 20 classifiers brings 180 + % improvement over 5-member DAES. Similar experiments on CIFAR100 also confirm the proportional improvement. CIFAR10 however enjoys relatively smaller accuracy enhancement compared with the other datasets. We suspect that the underlying cause of the source of the improvement could be due to two separate reasons. First, adding more classifiers positively impacts the model's capacity on efficient sample selection. Second, training the model on full training budget, allows classifiers to individually specialize in diverse feature representation and accordingly yields to a

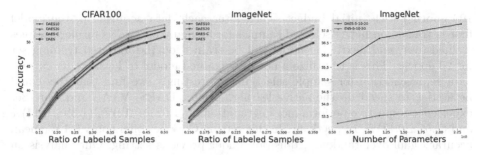

Fig. 4. Accuracy with different size of DAES and cumulative training (left and middle); and the slope of the curve representing a tunable trade-off for DAES (right).

better generalization at test time, compared to a model with fewer classifiers. The former explanation could be intuitively conceived as the performance/behaviour spectrum of ensembles with $1, 2, ..., N$ member(s) over test time.

In all experiments, cumulative training using the union of chosen samples by DAES, VAAL and Core-Set performs slightly better than DAES-20 except for CIFAR10. We see this as indicative of the superior effect of training budget size over model capacity on overall performance in this setting. Also as shown in Fig 4, compared with DAES, training VAAL with the same cumulative training budget led to lower accuracy – a clear contrast of the models' capacities.

5.2 DAES with Deeper Networks

We closely watch the training behavior of DAES with deeper networks such as ResNet50 and ResNet101. As an occasionally observed drawback, DAES built on very deep networks such as ResNet101, tends to take much longer convergence time than expected, or could even fail to converge. As a remedy, we found it helpful to initially pre-train the networks (or blocks) separately using initial training budget, and then train the ultimate ensemble built using pre-trained blocks. Applying this simple trick ensures the convergence of DAES.

On ImageNet and using 35% of data, DAES-5 with ResNet101 brings only approximately 1% mean accuracy improvement over DAES-5 with ResNet18 which is less than the improvement provided by DAES-10 with ResNet18. The same behavior was observed with CIFAR100. We suspect that adding more members (ensembles) to DAES leads to more improvement than replacing the blocks with deeper CNNs.

5.3 Incremental Training and Tunable Accuracy/Cost Trade-Off

A. Incremental Training: In a standard AL experimental setting, after updating the training set, the next training iteration starts from **scratch** (here for some 100 epochs). However, we investigate incremental training of models (VAAL, Core-Set and DAES) in which models are trained under much fewer number of epochs at each iteration while in next iteration rather than restarting the training, training continues. Specifically, we train the model for 20 epochs (formerly

100 epochs) with initial budget. Then after each data acquisition, the model first is trained on newly selected samples for as many epochs as former samples trained over, and next, the model will be trained on the updated training set for 20 epochs. This is to utilize a not fully trained model to leverage its current data representation for sample selection. Interestingly, we find that this could be a trick to speed up the active learner. Briefly, DAES, VAAL and Core-Set experience respective performance degradation of $(1.07 \pm 0.12)\%$, $(1.39 \pm 0.14)\%$ and $(1.51 \pm 0.11)\%$, respectively. In this setting Core-Set offers only 0.14% mean accuracy gain over random acquisition under previous setting. Our analysis on time complexity briefly shows that the ratio (to DAES) of average consumed time for one iteration of sample selection for DAES, VAAL, Core-Set and DAES with incremental training were 1, 0.57, 3.78, and 0.24 respectively.

B. Tunable Trade-Off: Consistent with the results in [5], we could not see much accuracy improvement with increasing the ensemble size in classical ensemble-based methods as shown in Fig. 4 (right figure). In other words, such classical methods do provide a tunable trade-off between accuracy and computational overhead. A satisfactory accuracy would be attained using 5 members, and increasing the number of members does not seem to proportionally improve the performance. However, active ensemble sampling showed a much robust performance in terms of exploiting more model capacity by adding more members to the ensemble. In fact, the Bayesian nature of active ensemble sampling in conjunction with its ensemble-designed structure allows achieving much higher accuracy by enlarging the ensemble at the cost of a proportional increase in computational overhead.

6   Conclusion and Future Work

In this paper, we introduced deep active ensemble sampling (DAES) inspired by an efficient approximation of Thompson sampling in order to combine the advantages of uncertainty-based and geometric-based approaches into one unified framework. We also examine a new training protocol formed on self-supervised knowledge distillation from unlabeled data on four baselines in order to confirm its effectiveness. Our framework is assessed on four benchmark datasets in two experimental settings to establish a new baseline. Finally we pose a few scenarios aiming for analysing DAES. We leave further theoretical and empirical analyses on DAES with asymmetric architectures for future research.

Acknowledgement. This work is supported in part by grants from the US National Science Foundation (Award #1920920, #2125872).

References

1. Cohn, D., Atlas, L., Ladner, R.: Improving generalization with active learning. Mach. Learn. **15**, 201–221 (1994)

2. Balcan, M.F., Beygelzimer, A., Langford, J.: Agnostic active learning. J. Comput. Syst. Sci. **75**, 78–89 (2009)
3. Cesa-Bianchi, N., Gentile, C., Orabona, F.: Robust bounds for classification via selective sampling. In: Proceedings of the 26th Annual International Conference on Machine Learning, pp.121–128 (2009)
4. Gal, Y., Islam, R., Ghahramani, Z.: Deep Bayesian active learning with image data. In: International Conference on Machine Learning, pp. 1183–1192. PMLR (2017)
5. Beluch, W.H., Genewein, T., Nürnberger, A., Köhler, J.M.: The power of ensembles for active learning in image classification. In: Proceedings of the IEEE Conference on Computer Vision and Pattern Recognition, pp. 9368–9377 (2018)
6. Sener, O., Savarese, S.: Active learning for convolutional neural networks: a core-set approach. In: International Conference on Learning Representations (2018)
7. Sinha, S., Ebrahimi, S., Darrell, T.: Variational adversarial active learning. In: Proceedings of the IEEE/CVF International Conference on Computer Vision, pp. 5972–5981 (2019)
8. Bouneffouf, D., Laroche, R., Urvoy, T., Feraud, R., Allesiardo, R.: Contextual bandit for active learning: active Thompson sampling. In: Loo, C.K., Yap, K.S., Wong, K.W., Teoh, A., Huang, K. (eds.) ICONIP 2014. LNCS, vol. 8834, pp. 405–412. Springer, Cham (2014). https://doi.org/10.1007/978-3-319-12637-1_51
9. Ganti, R., Gray, A.G.: Building bridges: Viewing active learning from the multi-armed bandit lens. arXiv preprint arXiv:1309.6830 (2013)
10. Melville, P., Mooney, R.J.: Diverse ensembles for active learning. In: Proceedings of the Twenty-First International Conference on Machine Learning, p. 74 (2004)
11. Chapelle, O., Li, L.: An empirical evaluation of Thompson sampling. Adv. Neural. Inf. Process. Syst. **24**, 2249–2257 (2011)
12. Brooks, S., Gelman, A., Jones, G., Meng, X.L.: Handbook of Markov Chain Monte Carlo. CRC Press (2011)
13. Lu, X., Van Roy, B.: Ensemble sampling. arXiv preprint arXiv:1705.07347 (2017)
14. Zhu, X., Lafferty, J., Ghahramani, Z.: Combining active learning and semi-supervised learning using Gaussian fields and harmonic functions. In: ICML 2003 Workshop on the Continuum from Labeled to Unlabeled Data in Machine Learning and Data Mining, vol. 3 (2003)
15. Li, X., Guo, Y.: Adaptive active learning for image classification. In: Proceedings of the IEEE Conference on Computer Vision and Pattern Recognition, pp. 859–866 (2013)
16. Joshi, A.J., Porikli, F., Papanikolopoulos, N.: Multi-class active learning for image classification. In: 2009 IEEE Conference on Computer Vision and Pattern Recognition, pp. 2372–2379. IEEE (2009)
17. Gal, Y., Ghahramani, Z.: Dropout as a Bayesian approximation: representing model uncertainty in deep learning. In: International Conference on Machine Learning, pp. 1050–1059. PMLR (2016)
18. Lakshminarayanan, B., Pritzel, A., Blundell, C.: Simple and scalable predictive uncertainty estimation using deep ensembles. arXiv preprint arXiv:1612.01474 (2016)
19. Osband, I., Blundell, C., Pritzel, A., Van Roy, B.: Deep exploration via boot-strapped DQN. Adv. Neural. Inf. Process. Syst. **29**, 4026–4034 (2016)
20. Gao, M., Zhang, Z., Yu, G., Arık, S.Ö., Davis, L.S., Pfister, T.: Consistency-based semi-supervised active learning: towards minimizing labeling cost. In: Vedaldi, A., Bischof, H., Brox, T., Frahm, J.-M. (eds.) ECCV 2020. LNCS, vol. 12355, pp. 510–526. Springer, Cham (2020). https://doi.org/10.1007/978-3-030-58607-2_30

21. Bengar, J.Z., van de Weijer, J., Twardowski, B., Raducanu, B.: Reducing label effort: self-supervised meets active learning. In: Proceedings of the IEEE/CVF International Conference on Computer Vision, pp. 1631–1639 (2021)
22. Yoo, D., Kweon, I.S.: Learning loss for active learning. In: Proceedings of the IEEE/CVF Conference on Computer Vision and Pattern Recognition, pp. 93–102 (2019)
23. Agarwal, S., Arora, H., Anand, S., Arora, C.: Contextual diversity for active learning. In: Vedaldi, A., Bischof, H., Brox, T., Frahm, J.-M. (eds.) ECCV 2020. LNCS, vol. 12361, pp. 137–153. Springer, Cham (2020). https://doi.org/10.1007/978-3-030-58517-4_9
24. Cortes, C., DeSalvo, G., Gentile, C., Mohri, M., Zhang, N.: Adaptive region-based active learning. In: International Conference on Machine Learning, pp. 2144–2153. PMLR (2020)
25. Zhang, Y., et al.: DatasetGAN: efficient labeled data factory with minimal human effort. In: Proceedings of the IEEE/CVF Conference on Computer Vision and Pattern Recognition, pp. 10145–10155 (2021)
26. Zhang, B., Li, L., Yang, S., Wang, S., Zha, Z.J., Huang, Q.: State-relabeling adversarial active learning. In: Proceedings of the IEEE/CVF Conference on Computer Vision and Pattern Recognition, pp. 8756–8765 (2020)
27. Ebrahimi, S., et al.: Minimax active learning. arXiv preprint arXiv:2012.10467 (2020)
28. Jing, L., Tian, Y.: Self-supervised visual feature learning with deep neural networks: a survey. IEEE Trans. Pattern Anal. Mach. Intell. **43**, 4037–4058 (2021)
29. Hadsell, R., Chopra, S., LeCun, Y.: Dimensionality reduction by learning an invariant mapping. In: 2006 IEEE Computer Society Conference on Computer Vision and Pattern Recognition (CVPR'06), vol. 2, pp. 1735–1742. IEEE (2006)
30. Wu, Z., Xiong, Y., Yu, S.X., Lin, D.: Unsupervised feature learning via nonparametric instance discrimination. In: Proceedings of the IEEE Conference on Computer Vision and Pattern Recognition, pp. 3733–3742 (2018)
31. Grill, J.B., et al.: Bootstrap your own latent-a new approach to self-supervised learning. Adv. Neural. Inf. Process. Syst. **33**, 21271–21284 (2020)
32. Chen, X., He, K.: Exploring simple Siamese representation learning. In: Proceedings of the IEEE/CVF Conference on Computer Vision and Pattern Recognition, pp. 15750–15758 (2021)
33. Ermolov, A., Siarohin, A., Sangineto, E., Sebe, N.: Whitening for self-supervised representation learning. In: International Conference on Machine Learning, pp. 3015–3024. PMLR (2021)
34. Shannon, C.E.: A mathematical theory of communication. Bell Syst. Tech. J. **27**, 379–423 (1948)
35. Houlsby, N., Huszár, F., Ghahramani, Z., Lengyel, M.: Bayesian active learning for classification and preference learning. arXiv preprint arXiv:1112.5745 (2011)
36. Freeman, L.C., Freeman, L.C.: Elementary applied statistics: for students in behavioral science. Wiley, New York (1965)
37. Yang, L., Zhang, Y., Chen, J., Zhang, S., Chen, D.Z.: Suggestive annotation: a deep active learning framework for biomedical image segmentation. In: Descoteaux, M., Maier-Hein, L., Franz, A., Jannin, P., Collins, D.L., Duchesne, S. (eds.) MICCAI 2017. LNCS, vol. 10435, pp. 399–407. Springer, Cham (2017). https://doi.org/10.1007/978-3-319-66179-7_46
38. Sener, O., Savarese, S.: A geometric approach to active learning for convolutional neural networks. arXiv preprint arXiv:1708.00489 7 (2017)

39. Guo, J., et al.: Semi-supervised active learning for semi-supervised models: exploit adversarial examples with graph-based virtual labels. In: Proceedings of the IEEE/CVF International Conference on Computer Vision, pp. 2896–2905 (2021)

40. Hinton, G.E., Sejnowski, T.J., et al.: Unsupervised learning: foundations of neural computation. MIT press (1999)

41. Russo, D.J., Van Roy, B., Kazerouni, A., Osband, I., Wen, Z.: A tutorial on thompson sampling. Found. Trends® Mach. Learn. **11**, 1–96 (2018)

42. Xu, G., Liu, Z., Li, X., Loy, C.C.: Knowledge distillation meets self-supervision. In: Vedaldi, A., Bischof, H., Brox, T., Frahm, J.-M. (eds.) ECCV 2020. LNCS, vol. 12354, pp. 588–604. Springer, Cham (2020). https://doi.org/10.1007/978-3-030-58545-7_34

43. Bhat, P., Arani, E., Zonooz, B.: Distill on the go: online knowledge distillation in self-supervised learning. In: Proceedings of the IEEE/CVF Conference on Computer Vision and Pattern Recognition, pp. 2678–2687 (2021)

44. LeCun, Y.: The MNIST database of handwritten digits. http://yann.lecun.com/exdb/mnist/ (1998)

45. Krizhevsky, A., Hinton, G., et al.: Learning multiple layers of features from tiny images (2009)

46. Deng, J., Dong, W., Socher, R., Li, L.J., Li, K., Fei-Fei, L.: ImageNet: a large-scale hierarchical image database. In: 2009 IEEE Conference on Computer Vision and Pattern Recognition, pp. 248–255. IEEE (2009)

47. Simonyan, K., Zisserman, A.: Very deep convolutional networks for large-scale image recognition. arXiv preprint arXiv:1409.1556 (2014)

48. Kingma, D.P., Ba, J.: Adam: A method for stochastic optimization. arXiv preprint arXiv:1412.6980 (2014)

Author Index

© The Editor(s) (if applicable) and The Author(s), under exclusive license
to Springer Nature Switzerland AG 2023
L. Wang et al. (Eds.): ACCV 2022, LNCS 13847, pp. 731–733, 2023.
https://doi.org/10.1007/978-3-031-26293-7

Printed in the United States
by Baker & Taylor Publisher Services